Conversion Factors between SI and BG Units

English	SI	SI symbol	To convert from English to SI multiply by	To convert from SI to English multiply by
		Power, Heat Rate		
horsepower	kilowatt	kW	0.7457	1.341
foot-pound/sec	watt	W	1.356	0.7376
Btu/hour	watt	W	0.2929	3.414
		Pressure		
pound/square inch	kilopascal	kPa	6.895	0.1450
pound/square foot	kilopascal	kPa	0.04788	20.89
feet of H_2O	kilopascal	kPa	2.983	0.3352
inches of Hg	kilopascal	kPa	3.374	0.2964
		Temperature		
Fahrenheit	Celsius	°C	$5/9(°F - 32)$	$9/5 \times °C + 32$
Fahrenheit	kelvin	K	$5/9(°F + 460)$	$9/5 \times K - 460$
		Velocity		
foot/second	meter/second	m/s	0.3048	3.281
mile/hour	meter/second	m/s	0.4470	2.237
mile/hour	kilometer/hour	km/h	1.609	0.6215
		Acceleration		
foot/second squared	meter/second squared	m/s^2	0.3048	3.281
		Torque		
pound-foot	newton-meter	$N \cdot m$	1.356	0.7376
pound-inch	newton-meter	$N \cdot m$	0.1130	8.85
		Viscosity, Kinematic Viscosity		
pound-sec/square foot	newton-sec/square meter	$N \cdot s/m^2$	47.88	0.02089
square foot/second	square meter/second	m^2/s	0.09290	10.76
		Flow Rate		
cubic foot/second	cubic meter/second	m^3/s	0.02832	35.32
cubic foot/second	liter/second	L/s	28.32	0.03532

Elementary Hydraulics

James F. Cruise
Department of Civil and Environmental Engineering
University of Alabama in Huntsville, Huntsville, AL

Mohsen M. Sherif
Department of Civil and Environmental Engineering
United Arab Emirates University, Al-Ain, UAE

Vijay P. Singh
Department of Biological and Agricultural Engineering
Texas A&M University, College Station, TX

THOMSON
NELSON

Australia Canada Mexico Singapore Spain United Kingdom United States

Elementary Hydraulics

by James F. Cruise, Mohsen M. Sherif, and Vijay P. Singh

Associate Vice President and Editorial Director:
Evelyn Veitch

Publisher:
Chris Carson

Developmental Editors:
Kamilah Reid Burrell/
Hilda Gowans

Permissions Coordinator:
Vicki Gould

Production Services:
RPK Editorial Services

Copy Editor:
Patricia Daly

Proofreader:
Erin Wagner

Indexer:
RPK Editorial Services

Production Manager:
Renate McCloy

Creative Director:
Angela Cluer

Interior Design:
Carmela Pereira

Cover Design:
Andrew Adams

Compositor:
Interactive Composition Corporation

Printer:
R. R. Donnelley

Printed and bound in the United States
1 2 3 4 07 06

For more information contact Nelson, 1120 Birchmount Road, Toronto, Ontario, Canada, M1K 5G4. Or you can visit our Internet site at http://www.nelson.com

Library of Congress Control Number: 2005938474

ISBN: 0-534-49483-8

North America
Nelson
1120 Birchmount Road
Toronto, Ontario M1K 5G4
Canada

Asia
Thomson Learning
5 Shenton Way #01-01
UIC Building
Singapore 068808

Australia/New Zealand
Thomson Learning
102 Dodds Street
Southbank, Victoria
Australia 3006

Europe/Middle East/Africa
Thomson Learning
High Holborn House
50/51 Bedford Row
London WC1R 4LR
United Kingdom

Latin America
Thomson Learning
Seneca, 53
Colonia Polanco
11560 Mexico D.F.
Mexico

Spain
Paraninfo
Calle/Magallanes, 25
28015 Madrid, Spain

Dedicated to our families

James F. Cruise: To my mother, Lula Grace Cruise
Mohsen M. Sherif: To Hala, Muhammad, and Mai
Vijay P. Singh: To Anita, Vinay, and Arti

Preface

Recent years have witnessed an enormous increase in public awareness of the interaction between man and environment. One indication of this awareness is seen in public debates being frequently held on virtually every aspect of the environment before any civil works project is proposed. The result has been the establishment of higher standards for environmental protection and increased emphasis on environmental quality and sustainability. Environmental consequences of human activities or civil works projects are multifaceted and require an interdisciplinary approach for their assessment.

Environmental protection and management involve a gamut of engineering and scientific disciplines. Hydraulics occupies a central position in this gamut. The recognition of the fundamental role of hydraulics in protection and management of the environment has resulted in introduction of hydraulics in undergraduate curricula of such diverse disciplines as agricultural engineering, atmospheric science, biological engineering, civil engineering, environmental sciences, earth sciences, forest sciences, military science, water resources engineering, and watershed sciences. In these curricula, hydraulics is usually taught to students who are either at the junior or senior level. In some curricula, students are exposed to hydraulics for the first time through this course and the students taking the course have not had the opportunity to be exposed to the hydraulic jargon before and have not taken a course in fluid mechanics which most civil engineering students have had in their sophomore year. In many cases, this may be the only course the students will have in hydraulics in their undergraduate schooling. Therefore, this hydraulics course must, of practical expediency, be at an elementary level, present basic concepts of fluid mechanics and their extension to hydraulics, and develop a flavor for application of hydraulics to the solution of a range of environmental problems. It is these considerations that motivated the writing of this book.

Elementary Hydraulics is written at the elementary level, requiring no previous background in hydraulics. Indeed, the student can understand the book for the most part by himself. To further help the student, at the end of each chapter a "reading aid" and a number of numerical problems are provided. Accompanying the book is a solution manual which contains solutions to these numerical problems. The book, divided into three parts, blends fluid mechanics, hydraulic science, and hydraulic engineering, which means application and design. The first part draws upon fluid mechanics and summarizes the concepts deemed essential to teaching of hydraulics. The coverage of this material is essential for students who have not taken a course in fluid mechanics. This part can be skipped by those who have taken such a course. The second part builds on the first part and discusses the science of hydraulics. This is where hydraulics normally starts for civil engineering students. The third part dwells upon the engineering practice of hydraulics and illustrates practical application of the material covered in the second part.

In an undergraduate hydraulics course, students are drawn from a range of disciplines and have obviously widely differing backgrounds and interests. The textbook used in teaching such a course must have sufficient material to appeal to a diversified class of students. This consideration also influenced the material covered in the book. Experience indicates that about 70 to 80% of the book may occupy a typical undergraduate course. Inclusion of about 20 to 30% extra material is deemed desirable to provide flexibility to the instructor in the choice of depth and breadth of coverage. Depending on the particular requirements of a course and the pace of coverage, one can choose material from different chapters of the book. It is hoped that the material covered would stimulate interest in this fascinating branch of water science and be of interest to those beginning their careers in disciplines requiring some background in hydraulics.

Introducing the definition and scope of hydraulics as well as its association with other sciences in Chapter 1, the subject matter of *Elementary Hydraulics* is divided into three parts. Part One, spanning four chapters, is comprised of fluid mechanic preliminaries. This part shows the connection between fluid mechanics and hydraulics. Chapter 2 discusses the fundamental properties and behavior of fluids. Different types of forces, motions, and energy encountered in hydraulics of flow in channels and pipes and the flow types and regimes are briefly presented in Chapter 3. In solving a wide range of hydraulic problems, a fundamental assumption of hydrostatic pressure distribution is invoked. Chapter 4 presents the principles of hydrostatics, computation of hydrostatic force and buoyancy of immersed bodies. It also covers the principles of pressure measurement using manometers.

Part Two, comprising ten chapters, constitutes the core of the book. This part can be called hydraulic science. Essential for solving hydraulic problems are the fundamental equations of continuity, momentum, and energy. Therefore, an understanding of these equations and their limitations is deemed basic. Sacrificing the mathematical rigor, these equations are derived in a simple manner in Chapter 5. Also discussed are the simplified forms of these equations and their applications, the Bernoulli equation, and the choice between the momentum equation and the energy equation. Many hydraulic problems can be solved using simple techniques. One of such techniques is dimensional analysis which also constitutes the basis of hydraulic scale models. Chapter 6 discusses methods of dimensional analysis, hydraulic scale models, types and conditions of similarity, and dominating forces.

A typical hydraulic course emphasizes pipe flow and open channel flow. Therefore, the next three chapters deal with some aspects of pipe hydraulics. One of the most fundamental concepts in hydraulics is that of flow resistance. Discussing the concept of flow resistance and the factors affecting it, Chapter 7 presents resistance equations and velocity distributions for steady uniform flow. Building on this background, Chapter 8 discusses resistance in the context of closed conduit flow either in a single pipeline or pipe networks, including laminar and turbulent flow, empirical resistance equations, and minor losses as well as water hammer. Pumps are used everywhere and everyday; therefore, an elementary knowledge of pumps is deemed important for those wanting to understand and apply hydraulics. Introducing the concept of overall efficiency, Chapter 9 goes on to discussing different types of pumps and pump selection, setting, and operation.

The next five chapters emphasize open channel under steady flow conditions. Chapter 10 introduces different types of open channels and their geometric characteristics which serve as basic information for the design stage. The sources of geometric data such as remote sensing images and digital elevation models and associated errors are discussed. Chapter 11 treats resistance in open channels. Normal depth, channel efficiency, steady uniform flow, and resistance in steady nonuniform flow are discussed in the chapter.

Energy principle in open channels constitutes the subject of Chapter 12. Specific energy, the specific energy diagram, critical flow, discharge-depth relation, and application of the energy principle are covered in this chapter. Chapter 13 presents the momentum principle in open channels. The momentum function, the momentum function diagram, hydraulic jump and its geometry, energy loss in hydraulic jump, and different types of hydraulic jumps constitute the subject matter of this chapter. Part Two is concluded with a discussion on gradually varied flow which is the subject of Chapter 14. It discusses gradually varied flow equation, water surface flow profiles, control sections, jump location, and water surface profiles in long channels of several reaches of different slopes.

The concluding part of the book, Part Three, consists of two chapters on hydraulic applications and design. This part is a sample of hydraulic engineering. Chapter 15 presents methods of computation of water surface profiles using numerical integration, direct step, and the standard step method. The well-known HEC-RAS model is introduced and the relevant applications of geographic information systems are presented. Beginning with a discussion of basic principles, Chapter 16 goes on to present design of hydraulic drainage and control structures. The discussion on design of hydraulic control structures includes different types of culverts, impoundments facilities, and channels. This chapter illustrates the application of the principles enumerated in the preceding chapters. In a sense, it is an illustrative summation of the previous chapters.

There exists a voluminous literature on the topics covered in this elementary book. A comprehensive list of references has been advertently omitted. Only those references that are deemed most pertinent have been cited. This is because the intended audience is the beginner who, for want of time, background, and experience, would have little appetite for the advanced literature. This, however, in no way reflects a lack of appreciation for either the importance or the quality of literature. Indeed this book would not have been completed without reviewing the pertinent literature.

ACKNOWLEDGMENTS

Several people reviewed portions of the book. Mr. Maqsood Ahamd of Louisiana State University in Baton Rouge, Mr. Z. Q. Deng of Shihezi University in China, Dr. L. Yilmaz of Istanbul Technical University in Turkey, and Mr. T. Drake, formerly of the U.S. Army Corps of Engineers, reviewed earlier drafts of the book or portions thereof and provided helpful comments and suggestions. Dr. Anvar Kacimov, Sultan Qaboos University, Muscat, reviewed the bulk of the book manuscript and made a number of corrections and suggestions. Inclusion of these suggestions resulted in an improvement of the book. Dr. Khaled Hamza, Cairo University and Eng. Roohe Zafar, UAE University provided several problems that have been included in different chapters of the book. The authors would also like to thank the following people for their contributions: Rajagopalan Balaji, University of Colorado, Boulder; Rolando Bravo, Southern Illinois University, Carbondale; Davyda Hammond, University of Alabama, Birmingham; M. H. Nachabe, University of South Florida; and Fred. L. Ogden, University of Connecticut. The authors would like to gratefully acknowledge their contribution. Last but not least, the authors would like gratefully acknowledge the support of their families, without which this book would not have been completed. As a small token of appreciation, the book is dedicated to them: Cruise: (Mother Lula); Sherif: (wife Hala, son Muhammad, and daughter Mai); and Singh: (wife Anita, son Vinay, and daughter Arti).

J. F. CRUISE, HUNTSVILLE, ALABAMA
M. M. SHERIF, AL-AIN, UNITED ARAB EMIRATES
V. P. SINGH, COLLEGE STATION, TEXAS

Contents

5 Governing Equations 113

6 Dimensional Analysis and Hydraulic Similarity 153

7 Flow Resistance and Velocity Distributions 192

List of Notations

A = area

a = acceleration

α = velocity correction coefficient

b = bottom width

B = top width

β = momentum correction factor, also compressibility

C = Chezy coefficient, Francis weir coefficient

C_{HW} = Hazen-Williams coefficient

D = hydraulic depth

d = pipe diameter, depth measured normal to channel bed

γ = specific weight

ξ = modulus of elasticity

E_v = bulk modulus of elasticity

e = eccentricity

e_h = hydraulic efficiency of pump

e_m = mechanical efficiency of pump

e_p = overall efficiency of pump

e_v = volumetric efficiency

ϵ = roughness element

F = force

F_n = Froude number

f = Darcy resistance coefficient

g = acceleration of gravity

h = vertical distance from water surface to a point

$\bar{h}$ = depth to centroid in vertical plane

h_{cp} = depth to center of pressure in vertical plane

I_o = moment of inertia about fluid surface

I_c = moment of inertia around the horizontal axis passing through the centroid

H = total energy head

$k_{()}$ = energy loss coefficient
L = length
M = momentum function
m = mass
n = Manning resistance coefficient
η_s = specific speed
ω = angular velocity
P = wetted perimeter
p = pressure
p_v = vapor pressure
Q = discharge
q = specific discharge
R = hydraulic radius
R_n = Reynolds number
ρ = fluid density
r = pipe radius
σ = surface tension
S_f = slope of energy grade line
S_o = bed slope
S_w = water surface slope
S.G. = specific gravity
T = time, also torque
t = horizontal component of a channel side slope
τ = shear stress
τ_o = boundary shear stress
ν = kinematic viscosity
μ = dynamic viscosity
v = point velocity
v^* = shear velocity
V = average velocity
V_{ol} = volume
V_s = specific volume
y = depth measured vertically, distance from water surface in plane
$\bar{y}$ = distance from water surface to centroid in plane
y_c = critical depth
y_{cp} = distance from the water surface to the center of pressure in the plane
z = elevation above datum

CHAPTER 1

INTRODUCTION

Niagra Falls and Niagra River, NY. (Courtesy of Dr. Mohammad Al-Hamdan.)

In the holy books of most religions, water is described as a source of life. All life on earth is dependent on water in one way or another. Water is essential for survival of individuals, populations, and nations. It sustains animal and plant life, and acts as a carrier of pollutants and purifier of waste. Water aids in transmission and storage of energy received from the sun and provides a means for transportation of goods, navigation of rivers and channels, and recreational activities. Water plays an important role in national and economic development. The study of the science of water is, therefore, vitally important.

Water is a renewable resource and follows the hydrologic cycle, as shown in Figure 1.1. The hydrologic cycle is an endless circulation and has neither beginning nor end. It is a natural machine, a constantly running distillation and pumping system. The sun supplies heat energy which, together with the force of gravity, keeps the water moving: from the atmosphere to the earth as condensation and precipitation; from the earth to the atmosphere as evaporation and transpiration, to the oceans as stream flow, and within the earth as stream flow and groundwater movement; and from the oceans to the atmosphere as evaporation (Singh, 1992).

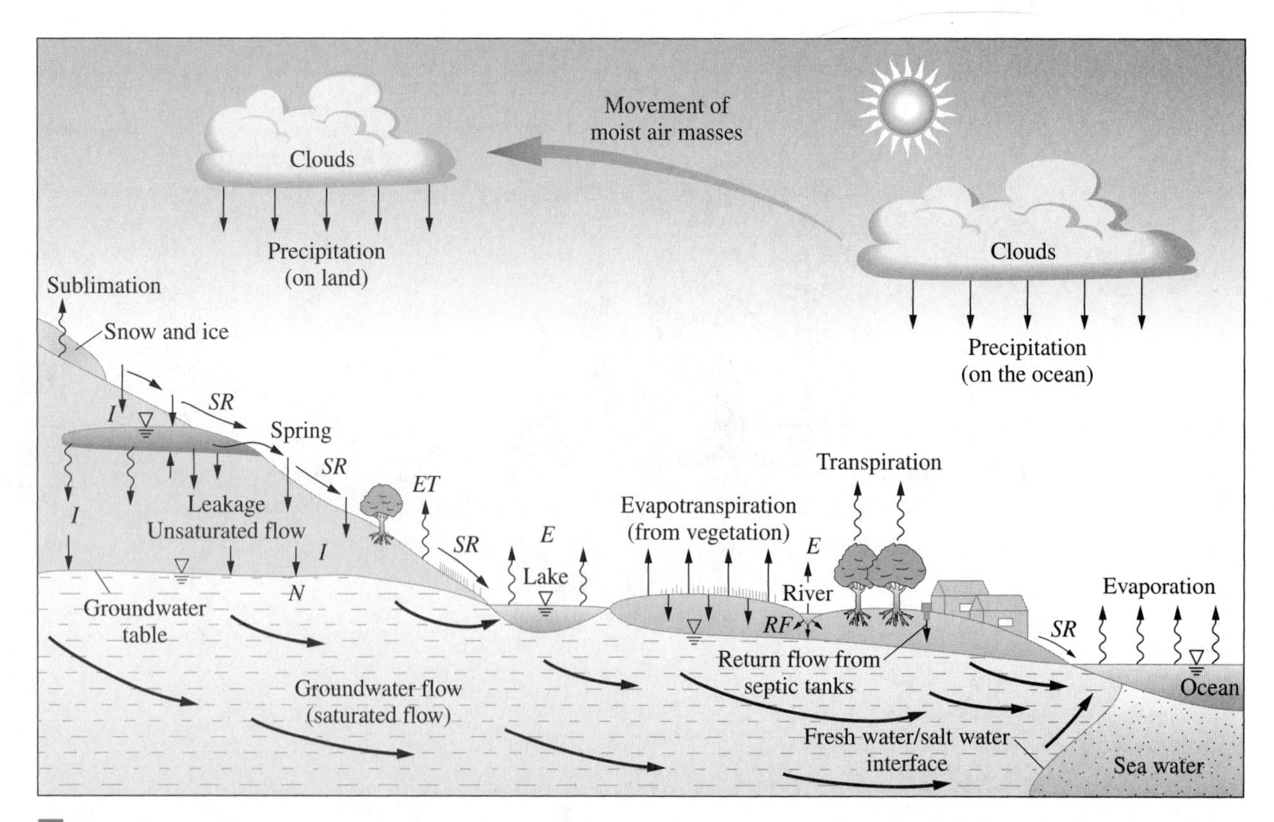

Figure 1.1 The hydrological cycle (adapted from Bear and Verruijt, 1987); *SR* = surface runoff, *E* = evaporation, *I* = infiltration, *RF* = return flow, and *N* = recharge to groundwater.

Waters of long geologic history are the waters of today; little has been added or lost through the ages since the first clouds formed and the first rains fell. Most of the global water is ocean water. Only a small portion of it is fresh. Of this fresh water supply found in streams, lakes, swamps, glaciers and icecaps, and in the geologic formations only a smaller portion is available for human consumption. Water is found over the land (surface water) as well as below it (ground water). At some places it is stationary and at others it is in motion.

1.1 Definition of Hydraulics

The word *hydraulics* has been derived from the Greek word *Hudour,* which means "water." Hydraulics may be defined as the science that deals with the mechanical behavior of water at rest or in motion. In some contexts it also includes the study of other fluids. The mechanical behavior may entail computing forces and energy associated with fluids at rest or momentum and energy of fluids in motion, or computation of water surface elevation in channels and flood plains, or calculation of discharge, velocity, and fluid potential, and sediment and pollutant transport in hydraulic conduits. The science of fluids at rest is called hydrostatics, and the science of moving fluids is called hydrodynamics; they both are embedded in hydraulics. Hydraulics is an applied science of fluid mechanics and studies, for the most part, the flow of incompressible fluids, which do not exhibit any changes in density as the pressure exerted on them is varied.

1.2 Distinction Among Hydraulics, Hydrology, and Fluid Mechanics

The science of water encompasses fluid mechanics, hydraulics and hydrology. Fluid mechanics is the study of fluids under all conditions of rest and motion. The fluids may be gaseous or liquids, including water. The fluid mechanics approach to the study of water employs basic principles of science and is therefore science based rather than empirical. It provides solutions to a wide variety of problems encountered in many fields of engineering, regardless of the physical properties of the fluids involved.

Contrasting hydraulics with fluid mechanics, the following distinctions are noted. Hydraulics deals primarily with water, whereas fluid mechanics deals with fluids of all types. A hydraulic approach is more empirical and the emphasis is on finding a solution of practical value, whereas in fluid mechanics the approach is more mathematical and the emphasis is on understanding the behavior of hydraulic systems by employing the basic principles of science (physics). Hydraulics can be construed as applied fluid mechanics; thus, fluid mechanics is the mother of hydraulics. Hydraulics has also inspired many developments in fluid mechanics, and from this prospective, fluid mechanics and hydraulics complement each other.

Hydrology is the study of space, time, and frequency characteristics of the quantity and quality of waters of the earth encompassing their occurrence, movement, distribution, circulation, storage, exploration, and development (Singh, 1992). These characteristics are determined by the relation of water to the earth.

Contrasting hydraulics with hydrology, the following distinctions are noted. Hydrology deals with water in all forms: liquid, solid, ice and snow, and water vapor, whereas hydraulics deals primarily with liquid water. Hydrology deals with occurrence of water whereas hydraulics does not. A hydrologic approach is empirical, perhaps even more than a hydraulic one.

A sharper contrast among hydraulics, fluid mechanics, and hydrology can be drawn by looking at the spatial and temporal scales commonly employed by them. In fluid mechanics the spatial scale can be very small, say, a small segment of a channel or even a point. At that scale, a detailed inquiry into temporal and frequency characteristics of the fluid is undertaken.

In hydraulics the spatial scale is much larger—the channel segment or overland flow plane or estuary. At that scale, spatial averaging of hydraulic phenomena naturally takes place. The laws of science implicitly entail empirical parameters. At that scale, investigations are made of the temporal and frequency characteristics of the behavior of flow of water.

In hydrology the spatial scale is often larger than that normally employed in hydraulics. The spatial scale is a watershed which can be as small as a parking lot or as big as the Mississippi River basin or even bigger. Indeed even the entire globe can be taken as a spatial scale of a study in hydrology. At the spatial scale of a watershed, detailed investigations of temporal and frequency characteristics of the quantity and quality of the waters of the earth are made in all aspects.

Hydrology deals with the water cycle and it connects hydrosphere, atmosphere, pedosphere and lithosphere whereas the water cycle is outside the domain of hydraulics and fluid mechanics. Because of the much larger spatial scales involved, hydrology employs far greater averaging and therefore empiricism than does hydraulics.

1.3 Classification of Hydraulics

Hydraulics can be classified based on a multitude of criteria, including (1) availability and source of water, (2) type of conduits, (3) scientific content, (4) quality of water, (5) type of environment, (6) land use, (7) solution technique, (8) focus of study, and (9) scale of study. The classification of hydraulics, based on these criteria, is shown in Table 1.1.

(1) Source of Water

Water is available on the land surface, in the vadose zone, in the saturated geologic formations, and in snowpacks and ice glaciers. On the land surface it is available in lakes, reservoirs, streams, brooks, bayous, channels, canals, and so on. Thus, hydraulics can be classified as surface-water hydraulics, vadose-zone hydraulics, groundwater hydraulics, and snow and glacial hydraulics. The surface-water hydraulics can be further classified as river hydraulics, lake or reservoir hydraulics, and canal hydraulics. Groundwater hydraulics deals with water flowing in saturated porous media under phreatic, confined or semiconfined conditions. It studies the motion of all interstitial water below the water table, however deep down it may occur.

(2) Type of Conduits

Depending on the type of conduits, hydraulics can be classified as (1) open-channel hydraulics, (2) pipe flow hydraulics, and (3) watershed hydraulics. Each branch includes physical, chemical, and biological characteristics of water. Open channel hydraulics deals with water flowing in a channel or a conduit under the force of gravity, where the water surface is subject to

TABLE 1.1 Classification of Hydraulics

Distinguishing Criteria	Source of Information	Classification
Source of water	Land surface, vadose zone, saturated geologic formations, snowpacks and glaciers	Surface-water hydraulics, vadose-zone hydraulics, groundwater hydraulics, snow and glacial hydraulics
Type of conduits	Open-channel, pipe, watershed	Open-channel hydraulics, closed conduit hydraulics, watershed hydraulics
Land surface	Plane, river, lake (or reservoir), canal, sewers	Plane-flow hydraulics, river hydraulics, canal hydraulics, sewer hydraulics, lake hydraulics
Subsurface	Unsaturated zone and saturated zone	Unsaturated zone or soil-water hydraulics, saturated zone or groundwater hydraulics
Property of water	Quantity and quality	Water-quantity hydraulics, water-quality hydraulics
Scientific content	Chemical, biological, physical	Chemical hydraulics, biological hydraulics, physical hydraulics
Type of environment	Agricultural, coastal, wetland, forest, urban, lakes, ecosystems	Agricultural hydraulics, coastal hydraulics, wetland hydraulics, forest hydraulics, urban hydraulics, lake hydraulics, ecosystem hydraulics
Land use	Agricultural, urban, rural, desert, forest, wetland, mountains, coastal	Agricultural hydraulics, urban hydraulics, rural hydraulics, coastal hydraulics, forest hydraulics, wetland hydraulics, mountain hydraulics, transportation hydraulics
Solution technique	Mathematical, Statistical, and laboratory	Mathematical hydraulics, statistical hydraulics, and laboratory hydraulics
Mathematics	Analytical, parametric, numerical	Analytical hydraulics, system hydraulics, computational hydraulics
Statistics	Empirical, probability theory, stochastic processes	Statistical hydraulics, probabilistic hydraulics, stochastic hydraulics
Focus of study	River, lakes, estuaries, coastal environments, wetlands, ecology	River hydraulics, lake hydraulics, costal hydraulics, wetland hydraulics, ecological hydraulics
Open channels	Rivers, canals, storm drains, sewers	River hydraulics, fluvial hydraulics, canal hydraulics, sewer hydraulics, storm drain hydraulics
Plane	Flood plains, parking lots, airport runways, coastal areas	Flood plain hydraulics, airport hydraulics, coastal hydraulics, urban hydraulics
Structures	Dams, reservoirs, highways, railways, water supply tanks and machines	Dam hydraulics, reservoir hydraulics, transportation hydraulics, water supply hydraulics, and hydraulic machinery
Scale	Small, medium, large	Small-scale hydraulics, medium-scale hydraulics, large scale-hydraulics

atmospheric pressure. Closed conduit hydraulics deals with water flowing under pressure in pipes or closed sections. Watershed hydraulics is related to the surface water flow on planes and in channels, subject to various land uses, such as urban, agricultural, forest, mountainous, coastal, and marsh.

(3) Properties of Water

Depending on the properties of water to be emphasized, hydraulics can be classified as (1) water-quantity hydraulics and (2) water-quality hydraulics. The water-quantity hydraulics can be called physical hydraulics. The water-quality hydraulics encompasses the study of the constituents that are carried by water, including sediments, chemicals, and microorganisms and can therefore be classified as chemical hydraulics, biological hydraulics, and physical-quality hydraulics. Fluvial hydraulics, chemical hydraulics, and biological hydraulics may constitute the newly emerging areas of environmental hydraulics.

(4) Scientific Content

Depending on the scientific direction to be emphasized, hydraulics can be classified as physical hydraulics, chemical hydraulics, and biological hydraulics. Traditionally, physical hydraulics has been the area of emphasis, largely because of its ubiquitous application to civil and water resources engineering design. However, chemical hydraulics and biological hydraulics are receiving increasing attention these days due to heightened environmental awareness.

(5) Type of Environment

Depending on the particular application of hydraulics to be emphasized, hydraulics can be classified as agricultural hydraulics, coastal hydraulics, desert hydraulics, marsh and wetland hydraulics, forest hydraulics, mountainous hydraulics, ecosystem hydraulics, lake hydraulics, and estuary hydraulics. Much of the emphasis in this classification is on surface-water hydraulics, although subsurface hydraulics is equally important in forest, agricultural, coastal, and desert environments.

(6) Land Use

The soil, vegetation, and land use play a fundamental role in shaping the hydraulic behavior of a watershed. Therefore, hydraulics can be classified, based on land use, as agricultural hydraulics, urban hydraulics, forest hydraulics, transportation hydraulics, rural hydraulics, mountainous hydraulics, wetland hydraulics, and desert hydraulics.

(7) Solution Technique

The techniques for solving hydraulic problems can be distinguished as either mathematical or statistical. The mathematical techniques can be analytical, numerical, digital, or systemic. Similarly, statistical techniques can be empirical, probabilistic or stochastic. Thus, based on solution techniques, hydraulics can be classified as mathematical hydraulics and statistical hydraulics. Mathematical hydraulics can be further classified as parametric hydraulics, numerical or computational hydraulics, digital hydraulics, and systems hydraulics. Similarly, statistical hydraulics can be classified as empirical hydraulics, probabilistic hydraulics, and stochastic hydraulics.

Mathematical hydraulics deals with problems that can be described mathematically by differential, difference, or integral equations. These equations either have closed form solutions or can be solved numerically. Probabilistic hydraulics deals with hydraulic phenomena that have uncertain parameters or variables called random variables. It uses the axioms or laws of probabilities to deal with such variables. For example, the flooding of a river town and overtopping of a dam can be considered as random phenomena. Stochastic hydraulics deals with temporal and/or spatial variability of hydraulic phenomena that are random in nature. It uses the laws of probabilities to deal with such phenomena. The pollutant concentration along a river is an example of a stochastic process.

Hydraulic modeling encompasses physical simulation of open-channel flow, closed-conduit flow, flow in fluvial channels, hydraulic structures, hydraulic machines, groundwater flow, or whatever else having a relation with water.

(8) Focus of Study

Depending on the focus of study, hydraulics can be classified as river hydraulics, coastal hydraulics, lake hydraulics, wetland hydraulics, ecological hydraulics, and hydraulic structures and machinery. Fluvial hydraulics studies the entrainment of solid material into the water body of a natural stream or a channel. It encompasses erosion, deposition, sediment transport and bed forms. Fluvial hydraulics deals with two-phase flow and interface flow in open channels and is a semiempirical subject.

Hydraulic structures is a branch of hydraulics which is concerned with structures installed on waterways to store, regulate, distribute, elevate and/or measure water flow. Some hydraulic structures are also built to enhance navigation. The objective of this branch of hydraulics is to produce a structure that will accomplish the desired result with minimum cost. Hydraulic machinery is a branch of hydraulics that deals with machines run by water under certain head or raising water to a certain level.

(9) Scale of Study

Depending on the scale of study, hydraulics can be classified into small-scale hydraulics, medium-scale hydraulics, and large-scale hydraulics. Examples of small-scale hydraulics are hydraulic studies in flumes; flow over spillways, weirs, and so on; flow through orifices and culverts and the like; flow in pipes; and so on. Examples of medium-scale hydraulics are hydraulics of parking lots, drainage ditches, pipe networks, flow routing in canals or channels segments, and so on. Large-scale hydraulics is exemplified by hydraulics of rivers, dams, estuary, flood control works, large irrigation canals, diversions, barrages, levees, and so on.

1.4 Subject Matter of Hydraulics

The scope of hydraulics is vast and is, perhaps, best summed up by the classification of hydraulics given in Table 1.1. The study of hydraulics is of vital importance in sustainable development of a nation. It covers a wide field of applications in many aspects of our life. It not only elevates the standard of our living through various hydraulic projects, such as the Salt River project, which supplies water and power to much of the Valley of the Sun near Phoenix, Arizona, but also mitigates disasters of nature, such as droughts and floods.

Hydraulics deals with water which is either at rest or in motion above the land or below it, flows naturally or artificially, or carries physical or chemical constituents. Hydraulics encompasses two of the three major systems of the hydrologic cycle: the surface water system and the subsurface water system. Hydraulics does not deal with the atmospheric system in which water is encountered in the form of precipitation, evapotransiparation, rainfall, snowfall, hail, sleet, dew, drizzle, fog, and so on. The subject of hydraulics covers the study of water in open channels, closed conduits, watersheds, oceans and open seas, lakes and reservoirs, estuaries, wetlands, snowpacks, soils, groundwater, and so on. Hydraulics also encompasses the study of constituents that are carried by water, including sediments, chemicals, and microorganisms. The size and dimensions of hydraulic systems may vary from a small current meter, a few centimeters in size, to a channel, several hundreds of kilometers long and tens of square meters in cross section. Generally, hydraulic systems and structures are relatively massive as compared with the projects of other engineering branches Hwang and Haughtalen (1996).

1.5 Environmental and Water Resources Problems Involving Hydraulic Applications

Water is essential to life. Without it, human survival could not be possible. The extent to which modern society depends on water is apparent in the variety of ways in which water is used. Hydraulics touches every human life in some manner. To some, it is simply a need for drinking water, and to others, the need for water might be economic or just for convenience (Singh, 1992). Development and utilization of water resources require conception, planning, construction, and operation of facilities to store, distribute, regulate, control, and utilize water. Hydraulics is concerned with utilization of water resources for various economic needs. It deals with a wide range of practical problems in the field of environmental and water resources, especially the behavior of water in waterways. Generally, hydraulics is introduced to alter the natural behavior of a water source (i.e., river, lake, underground reservoir) and adapt it to the benefit of national economy and to protect the environment. Two aspects of water usage are of prime concern in hydraulics (Grishin, 1987): (1) consumptive (i.e., the use of water for various purposes, such as water supply, irrigation, water delivery, land reclamation, and so on, where water is drawn from various sources); and (2) exploitative, such as electric power generation, navigation, fishing, and so on.

Environmental and water resources problems involving application of hydraulics include flood control, drought mitigation, water supply, pollution control, urban development, industrial development, agricultural production, design of hydraulic works, energy resources development, land conservation, environmental impact assessment, land use change, forest and wildlife management, military operations, rural development, navigation, recreation, and fisheries. These problems are addressed through an integrated approach requiring application of the fundamental knowledge of many disciplines, in addition to hydraulics.

Hydraulics is needed for flood control. A flood occurs when a lake, reservoir, or stream is unable to convey the amount of water it receives. The result is an inundation of what is usually a dryland. Floods are sometimes caused by the failure of hydraulic structures, such as dams, levees, and dykes. Natural floods are, however, more common. The problem of flooding is defined by its areal extent, duration, intensity, and damage. Hydraulic structures are thus designed to mitigate flooding and flood damage.

A drought occurs when the demands for water exceed the stored water considerably in a certain region over an extended period (several years) of time. Construction of water impoundments, groundwater pumping, and interbasin transfer of water are some of the ways involving hydraulic applications to mitigate droughts. A water supply scheme must provide sufficient water of acceptable quality to serve its intended purpose, be in urban, agricultural or industrial setting. The disruption in water supply should be minimum. Hydraulics aids the design and construction of water supply schemes as well as sewage systems.

Water is an efficient and economical carrier of pollutants and undesirable materials. It dilutes the waste and disposes that waste by natural processes, to a certain extent. Of course, there is a limit to the amount of waste that can be absorbed by any water course, including rivers, lakes, reservoirs, and seas. Environmental hydraulics deals with such issues.

Water is essential for urban and industrial development. Urban planning and development involves construction of houses, schools, sports and recreational facilities, roads, culverts, drainage systems, water supply schemes, and so on. Industry has an enormous thirst for water and, therefore, hydraulic applications are numerous in industrial development.

The largest user of water is agriculture. Vast irrigation projects that carry water long distances are common. Open channel hydraulics is used to design systems of canals and drains to ensure water availability and proper soil drainage. Groundwater hydraulics is used to design the network of wells for a farm located away from a surface-water source. Dams, culverts, spillways, bridge crossings, dykes, levees, diversions, channel-improvement works, drainage works, and so on are typical hydraulic works required for water-resources development and management.

Development of groundwater resources requires an understanding of groundwater hydraulics. Aquifer's safe yield and drawdown are estimated through the knowledge of hydraulics of flow through porous media. In coastal areas, groundwater resources are threatened by seawater intrusion. This problem is further exacerbated by excessive pumping of groundwater. Hydraulics of two-phase flow is thus used to determine the safe yield of coastal aquifers without tangible encroachment by seawater.

Hydraulic machines are used extensively in modern life. Turbines convert the water power into mechanical power, which is then converted into electrical power via generators. Hydroelectric power plants are considered to be a lot safer than other types of electrical power plants and, most importantly, do not have significant detrimental environmental consequences. In the United States, hydroelectric plants generate about 15% of the nation's electricity.

Pumps constitute a major element in water delivery systems. Unlike turbines, pumps convert mechanical power into water power. Pumping stations are constructed in open channels (canals and drains), sewage systems, and so on to control flow. Pumps are used in most of the buildings to elevate water to upper stories. In groundwater, pumps are used to deliver water from wells and to pump it through water supply systems. Pumps are also used to prevent low lying areas from flooding. For example, some parts of New Orleans are below sea level and water has to be pumped to keep these parts from getting flooded as happened during Hurricane Katrina in September 2005.

In the United States, Canada, Europe, and various other countries, river navigation is widespread, for water provides the most economical means for transportation of bulky raw materials, such as coal, pulped paper, lumber, and minerals. There are nearly 26,000 miles of improved inland waterways for navigation in the United States (Singh, 1992). In order to maintain navigability, a minimum depth of flow has to be maintained in the river. A system of locks and dams is built on the river which monitors river traffic. Hydraulics is thus employed to enhance conditions for navigation and to design the required hydraulic structures.

Recreation is turning into a major commerce industry and recreational requirements these days are an important consideration in the development of multipurpose water resources projects. Hydraulics assists with the design and operation of these facilities.

Hydraulics is used in many military tactical operations. A knowledge of river hydraulics is required to determine if river crossing during military operations would be safe or not. Dam breaching and the resulting damage are important in planning tactical offenses against enemy as well as adequate defense (Wickham and Jourdan, 1985; Jourdan and Sullivan, 1987). A good example illustrating the employment of hydraulics in military operations is one of Dutch flooding their valley during World War I.

There are thousands of sports fishermen who each year cast their lures into lakes and rivers in all parts of the world. The fish grow if good quality water is available sufficiently. The increasingly polluted condition of many lakes and streams has had a deleterious effect on both the quantity and type of fish available for sports and commerce. Environmental hydraulics is concerned with such problems.

1.6 Classification of Hydraulic Problems

Hydraulic problems, as shown in Table 1.2, can be classified into five groups: (1) detection (or instrumentation), (2) prediction, (3) forecasting, (4) identification, and (5) simulation. Solution of these problems requires modeling involving several steps: (1) definition of the problem, (2) scope (objective) of the problem, (3) representation of the hydraulic system, (4) model development, (5) solution technique, (6) scientific basis of the model, (7) acquisition of data (laboratory and/or field), (8) model calibration (parameter estimation), (9) model validation, and (10) error analysis and model reliability. The solution of the problem may vary with the type of the hydraulic system, which may be a river section, canal section, pipe section, estuary, drainage ditch, and so on. There is a distinction between forecasting and prediction problems. Forecasting refers to determining the output of a hydraulic system ahead of time, with a certain probability. Forecasting of river stages is an example. Prediction refers to determining the output regardless of time (i.e., present or past), given system representation and input.

From the standpoint of finding a solution or modeling, hydraulic problems can also be distinguished, depending on the scale, the domain of analysis, location of the forcing function (input), location of output, direction of input or output, and dimensionality. Domains of analysis are time, space, and frequency. Scales are temporal, spatial and statistical. Temporal scales are exemplified by second, minutes, hours, days, weeks, months, years, and so on.

TABLE 1.2 Classification of Hydraulic Problems

Input	Hydraulic System (Function)	Output	Problem
Unknown	Known	Known	Detection
Known	Known	Unknown	Prediction
Known	Known	Unknown	Forecasting
Known	Unknown	Known	Identification
Known	Unknown	Unknown	Simulation/modeling

Examples of spatial scales are channel element, pipe section, well section, river, estuary, etc. Another scale can be statistical, as for example, moments, cumulants, spectral properties, etc. A hydraulic problem involves a combination of a scale and a domain for its analysis, and the method of solution (model) to be developed depends on this combination.

In a similar vein, specification of flow, direction and dimensionality defines a hydraulic problem and its scope. Inflows and outflows can occur at a point or in a distributed manner. Similarly inflows can be one dimensional, two-dimensional or three-dimensional; and they can occur in the downstream direction, upstream direction, or transverse direction. Hydraulic problems can be surficial, subsurficial, or both.

1.7 Scientific Approach to Investigating Hydraulic Problems

A scientific approach to solving a hydraulic problem involves specification of three elements as shown in Figure 1.2 (Dooge, 1973; Singh, 1996): (1) hydraulic system representation, encompassing (i) geometry, (ii) physical laws, and (iii) initial and boundary conditions; (2) input (or sources and sinks), and (3) output. A brief discussion of each of these elements is in order.

1.7.1 Geometry for Representation of Hydraulic System

The first element in representing a hydraulic system is the system geometry. A hydraulic system refers to a segment of a conduit-pipe, channel, estuary, plane, aquifer, or soil-containing a finite mass of fluid and distinguishes this mass from all other matters and surroundings. The geometry of a conduit through which water flows has a great effect on the flow characteristics. The cross-sectional area of a water conduit may vary from a few square millimeters in groundwater flow to some tens of square meters in natural streams. Pressurized flow is usually encountered in circular sections. Open channel flow is encountered in different types of sections, such as rectangular, trapezoidal, triangular, circular, and irregular sections. Groundwater flow takes place in the pores of underground media. These pores have varying geometric shapes. The geometry of the conduits may vary with space. Therefore, different flow patterns are encountered in different reaches. In such cases, the problem is much more complicated and a different approach should be employed.

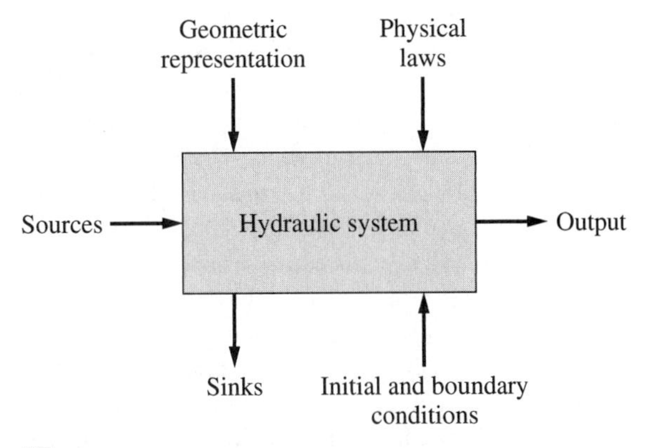

Figure 1.2 Elements of a hydraulic system (Singh, 1996).

The hydraulic system may consist of a set of interactive elements (Mays and Tung, 1992). Its boundary determines whether an element is a part of the system or of the environment and there can be interrelationships among the system elements, inputs and outputs. The boundaries of a hydraulic system may form a closed surface which may vary with time. However, mass does not. One kilogram of liquid may be contained within certain boundaries, for example, a cylinder. As the piston moves, the boundaries of the system change, while the mass of the liquid does not. Hydraulic systems may contain infinitesimal or large finite masses of fluids, depending on the problem under investigation.

1.7.2 Governing Equations (Physical Laws)

All hydraulic processes are governed by a set of basic principles, which are actually common to all physical phenomena. These basic principles constitute the constraints that hydraulic processes must conform to and are applicable to all hydraulic systems. These can be summarized as (Singh, 1996): (1) external constraints on the process encompassing principles of (a) conservation of mass, (b) conservation of momentum (linear or angular), and (c) conservation of energy; (2) internal mechanics of the process encompassing (a) law of entropy increase, and (b) law of space-time-mass dimensionality.

The conservation principles are universally valid except for the processes involving mass-energy conversion. In such cases, mass and energy principles are combined into a broader law of mass-energy conservation.

A word about the internal mechanics of the flow process may be appropriate here. The term entropy is used to describe the disorder of a system. For a closed system, the entropy always tends to increase. In thermodynamics, entropy is usually defined in terms of the irreversible component of heat. In hydraulics, for a fluid to maintain its flow it has to perform work to overcome frictional forces resisting flow. This involves conversion of flow energy (mechanical) into heat energy through friction, and this conversion of energy is irreversible. Thus the increase in entropy in a hydraulic system usually occurs due to this irreversible conversion of flow energy into heat energy through friction. The heat energy is conducted through the fluid and its bounding walls (system geometry) into the atmosphere and is eventually dissipated through space. The amount of energy so deployed is a function of the fluid, the system geometry, and the types of forces or energy affecting the system. The quantitative formulation of this function can be called the process equation, which is what leads to the resistance equation in hydraulics.

Thus, it can be concluded that every flow process is a conservation process, according to the energy conservation principle (the first law of thermodynamics) and a decay process according to the entropy-increase principle (the second law of thermodynamics). Because the flow process operates in the universe, which is a continuum of mass, time, and space, the process must be expressible in space, time, and mass dimensions.

To summarize, the science of hydraulics is based on three fundamental laws of physics (Chadwick and Morfett, 1993): (1) conservation of mass, (2) conservation of energy; and (3) conservation of momentum. The law of conservation of mass states that matter (mass) can be neither created nor destroyed, although it may undergo some transformation in its state (e.g., via chemical processes). When a hydraulic study does not include such processes, the law is reduced to a simple form of mass conservation. The law of conservation of mass is mathematically expressed as continuity equation or mass balance equation.

The law of conservation of energy states that energy can be neither created nor destroyed, however, it can be transformed from one form to the other. This is the first law of thermodynamics. Static energy can be transformed to kinetic energy and vice versa; that is, kinetic energy may transform to static energy. In open channels, reduction in the water level (static energy) is normally associated with an increase in the flow velocity (kinetic energy). Likewise, as the water level increases, the velocity of flow decreases assuming constant channel width. It should be noted that friction in hydraulic systems does not cause loss of energy; rather it converts some energy into heat. Therefore, no energy can be damnified. The law of conservation of energy is mathematically expressed as an energy equation. Frequently, this equation is expressed in terms of energy head. A special form of this equation is the Bernoulli equation which will be used extensively in this text.

The law of conservation of momentum or Newton's second law of motion stipulates that a body in motion cannot gain or lose momentum unless some external force is applied. In other words, the force acting on a control volume is equal to the rate of change of momentum. This law is mathematically expressed as momentum equation, sometimes referred to as the equation of motion or dynamic equation. A simpler linear momentum equation will be used in the text.

Hydraulic processes may entail just flow of water or flow of water as well as transport of contaminants—physical, chemical, and/or biological. The flow of water can be analyzed by using the aforementioned laws of mass, momentum, and energy. Because contaminants are carried by water, contaminant transport can be analyzed by conservation of mass of contaminants and a relation between contaminant flux and concentration (referred to as flux law), in addition to the equations needed for analysis of water flow.

1.7.3 Initial and Boundary Conditions

When solving the abovementioned governing equations of a hydraulic system, an infinite number of possible solutions for flow pattern are obtained. Each of these corresponds to a particular problem. Only one particular solution is correct with respect to a certain problem. Additional information (e.g., domain geometry, physical parameters, etc.) is thus needed to define an individual problem. To that end, the initial and boundary conditions are needed. Initial conditions describe the initial state of the hydraulic system or unknowns in the study domain. Consider, for example, the problem of computing the flow stage in a river reach during flooding. The flow stage prior to flooding is the initial condition in this case. Realistic initial conditions can be obtained from field observations (Sherif and Singh, 1996). Boundary conditions describe how the water in the study domain interacts with its surroundings. For a channel segment these may be defined as either upstream or/and downstream boundary conditions.

1.7.4 Inflow and Outflow

Inflow and outflow are defined as the mass or volume of water which enters or leaves a hydraulic system within a certain period of time. Mass can be neither created nor destroyed. The discharge flowing into a system minus the discharge flowing out must be equal to the change of the volume stored inside the system. This defines the volume balance which can be written as

$$(V_{ol})_{in} - (V_{ol})_{out} = \text{change in storage} \tag{1.1}$$

where $(V_{ol})_{in}$ and $(V_{ol})_{out}$ are the volumes of water entering and leaving the system, respectively, in a certain period of time. In some cases, where no change in storage is possible, as in many pressurized systems, the right side of the above equation (1.1) is reduced to zero. Therefore,

$$(V_{ol})_{in} - (V_{ol})_{out} = 0 \tag{1.2}$$

which means that what goes into the system must come out.

1.7.5 Considerations in Formulation of Governing Equations

The boundaries of a hydraulic system separate the system from its surroundings. Generally, no crossflow is allowed through the system boundaries, yet under certain circumstances the system may interact with its surroundings. In hydraulics, we usually study the flow of water through open channels, hydraulic machines, porous media, and so on. It is therefore difficult, if not impossible, to focus our attention on a fixed quantity of mass. Instead, we focus our attention on a volume in space through which water flows. Thus, we employ what is called as a control volume for formulation of governing equations. The control volume is an arbitrary volume in space through which water flows. The control surface is the geometric boundary of the control volume which may be real or imaginary. For example, streamlines are imaginary lines drawn in the flow field so that they are tangent to the direction of flow at every point at any given time, as shown in Figure 1.3a. A path line is defined as a line traced out by a moving fluid particle. In steady flow where variables (e.g., velocity, discharge, or flow depth) are constant in time, pathlines coincide with streamlines. Since streamlines are tangent to the velocity vector at every point, no flow can cross a streamline. Similarly, a set of streamlines may form an imaginary tube, called streamtube, as shown in Figure 1.3b. The internal surface of a pipe line may act as a streamtube, since the streamlines must be parallel to that surface. A short stream tube, shown in Figure 1.3b, may act as a control volume. The inside boundaries of the pipe form a real physical control surface. The vertical portions of the control surface are imaginary and may be located arbitrarily for purposes of calculation. The control surface should be clearly defined a priori since it has a direct effect on the method of solution.

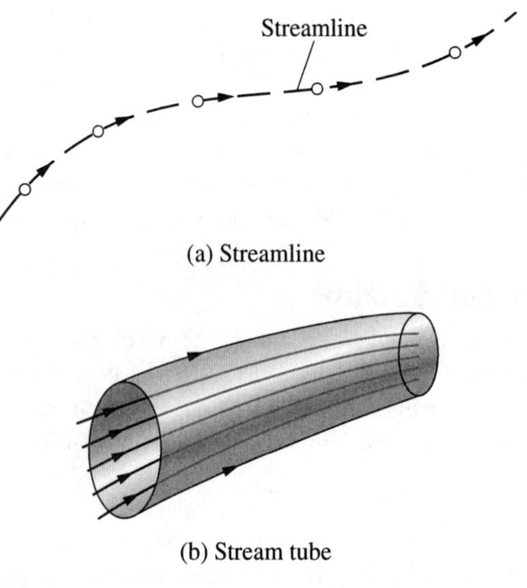

Figure 1.3 Streamlines and stream tubes.

1.7.6 Dimensionality of Flow: One-, Two-, and Three-Dimensional Flow

Natural flows are commonly three-dimensional in that flow characteristics (e.g., velocity and pressure) vary with respect to three directions: x, y, and z. At any point in the flow field, the velocities v_x, v_y, and v_z measured in the x, y, and z directions, respectively, are not equal. The flow characteristics at the same point may also vary with time. Dealing with three-dimensional flow fields is a very difficult task (if not impossible) because of the lack of proper information (e.g., data) and tedious mathematics involved when solving three-dimensional differential equations. Therefore, most of the hydraulic problems are approximated by one- or two- dimensional representations. This greatly simplifies the governing equations and (most importantly) the resulting solution is satisfactory. Moreover, minor adjustments may be introduced to one- and two-dimensional representations (or equations) to incorporate three-dimensional problems. For example, in steady uniform flow in pressurized systems, the velocity and pressure variations across the pipe may be ignored. Therefore, the flow may be characterized by a stream line along the center line of the pipe and the flow is, thus, reduced to one-dimensional flow.

1.8 Simplification of Scientific Approach

Solving a hydraulic problem requires a quantitative treatment of the hydraulic system under consideration. Because hydraulic processes are complex, the corresponding system representation is simplified in many ways, such as lumping of processes in space and/or time, using systems approach, reducing the three-dimensional nature to a one-dimensional treatment, and so on. The laws of science applied to hydraulic systems are the laws of conservation of mass, energy, and momentum. These laws are expressed, respectively, as continuity equation, energy equation, and momentum equation. For many hydraulic problems it is sufficient to utilize only two equations (i.e., continuity and energy). Frequently, approximations of the equation of motion along with the continuity equation are found to be adequate.

In general, for any hydraulic system the governing equations are the law of the conservation of mass and a flux law. In some cases these laws can be combined to produce a single governing equation, whereas in other cases the equations given by these laws constitute a system that has to be solved, often numerically. In the case of surface flow the flux law is represented by a momentum equation or a simplified variant thereof. In groundwater flow, the flux law is stated by Darcy's law, which when coupled with the continuity equation produces a diffusion equation. In solute transport, Fick's law is the flux law, which in combination with the continuity equation yields the hydrodynamic dispersion equation.

When expressed for a defined system geometry, the governing equations involve system characteristics that may vary in space and time. For example, the roughness parameter appearing in open channel flow varies in space. The hydraulic conductivity of a saturated flow varies in space, and that in an unsaturated flow varies in space and time. Although the governing equations are expressible for an arbitrary geometry, the hydraulic system has a fixed geometry. Consequently, these equations must be expressed for the geometry of the system or the geometry that optimally represents the system. For example, the cross-sectional area of a river reach is represented by a rectangular, parabolic, or trapezoidal shape. Boundaries of hydraulic systems are usually made to coincide with solid boundaries as far as possible. For other parts, they are taken normal to the flow direction to simplify the solution.

A hydraulic system may have sources and sinks that have to be specified. Flow at the upstream point and lateral inflow are examples of sources. Infiltration, lateral outflow, evaporation, and so on exemplify sinks. These have to be expressed in quantitative terms. The sources and sinks often vary in space and time. The usual simplification is to assume these sources and sinks to be uniform in space. Another simplification may be made about the temporal variability.

When a hydraulic system is subject to known sources, it provides for sinks, if any, and produces output, depending on the initial state of the system. This initial state is expressed in terms of initial conditions. The system output may have to satisfy some other conditions, called boundary conditions. The initial and boundary conditions are usually considered known a priori; however, for some systems, some of the boundaries are not known beforehand and have to be determined as part of the solution. Such boundaries are called free boundaries. The usual simplification is to assume the initial condition to be uniform throughout the hydraulic system. Similarly, the boundary conditions are assumed to be simple and one-dimensional.

1.9 Hydraulic Modeling

Hydraulic problems are usually investigated using laboratory experimentation or mathematical analyses. In other words, models of hydraulic systems are constructed and tested. As a result, there exists a multitude of models for investigation of hydraulic systems.

1.9.1 Classification of Hydraulic Models

There are two broad classes of models: (1) material models and (2) symbolic or formal models. A material model is the representation of the real hydraulic system by another system, which has similar properties but is easier to work with. Material models can be classified as iconic or "look-alike" models and analog models. Examples of iconic material models are laboratory channel or flume, laboratory watershed, etc. An analog model does not physically resemble the prototype system but depends on the correspondence between the symbolic models describing the prototype system and the analog system. Examples include the Hele-Shaw model for a coastal aquifer and electrical analog models for watershed response.

A formal model is a symbolic expression in logical terms of an idealized, relatively simple situation sharing the structural properties of the original system. Formal models are also referred to as symbolic or intellectual. Most popular formal models are mathematical models. A mathematical model expresses the system behavior by a set of equations, perhaps together with logical statements expressing relationships between parameters and variables. Three broad classes of mathematical models can, based on the amount of science involved, be distinguished: (1) empirical models, (2) conceptual models, and (3) hydrodynamic models. These models can be either deterministic, stochastic, or part deterministic and part stochastic. The models in the first category are data-based and the amount of science, as such, is reflected indirectly through data. Newly emerging models in this category are artificial neural networks and fuzzy logic. In the second category, certain simplifications regarding the hydraulic system are made, and, as a result, simpler forms of scientific laws are employed. A popular simplification is to lump the spatial variability and then express the continuity equation (law of conservation of mass) in the form of a volume balance reflecting only temporal variability. The third group of models employs the laws of conservation of mass, momentum, and energy, but when these equations are used in their complete form the models become unwieldly. Therefore, certain assumptions are made and simplifications employed to render models tractable.

1.9.2 Hydraulic Problems Formulated as Boundary-Value Problems

Conceptual and hydrodynamic models are based on formulating hydraulic problems as boundary value problems. A boundary-value problem is a mathematical formulation of the hydraulic problem under consideration. In a sense, this is a mathematical model of the problem. Four steps are involved in mathematically modeling the problem: (1) examination of the hydraulic problem, (2) replacement of the hydraulic problem by an equivalent mathematical problem, (3) solution of the mathematical problem, and (4) physical interpretation of the mathematical results.

To define a transient boundary-value problem for flow in a hydraulic system, one needs to know (1) the size and shape of the flow region or hydraulic system, (2) equations governing the flow within the region, (3) the conditions around the boundaries of the flow region, (4) initial conditions in the flow region, (5) the spatial distribution of the hydraulic parameters that control the flow, and (6) a mathematical method of solution. For steady-state systems, requirement (4) is not needed in boundary-value problems.

1.9.3 Hydraulic Data Collection and Database Management

The data needed for hydraulic analysis and modeling can be divided into geometric data and hydraulic data. Geometric data include depth, width, cross-section, length, slope, side slopes, grain size, roughness heights, bed forms, sinuosity, and meandering features for rivers, channels, and flumes. Hydraulic data include velocity, discharge, pressure, forces, and energy loss. Geometric data are collected using hydrographic surveys and remote sensing and satellite technologies. The data so collected can be huge and must be properly archived and managed so that they can be easily accessed and used. Currently a number of database management (DBM) techniques are available and indeed DBM constitutes a fascinating area of computer sciences.

1.9.4 Hydraulic Data Quality Assurance

Before undertaking a hydraulic analysis, data need to be processed carefully with respect to precision and accuracy, homogeneity and consistency, completeness and length of record. Indeed many times, it is the data that dictate the type of analysis to be performed rather than the availability of a hydraulic model itself.

Measurements should be sufficiently accurate or unbiased and precise or certain. The errors associated with measurements can be characterized with respect to their accuracy and precision. Precision refers to how closely individual measured values agree with each other. Thus, precision encompasses (1) the number of significant figures representing a quantity or (2) the spread in repeated measurements of a particular value. Imprecision, also called uncertainty, refers to the magnitude of the scatter. Inaccuracy, sometimes called bias, is defined as the systematic deviation from the truth (i.e., the true value). Thus, hydraulic data can be (1) inaccurate and imprecise, (2) accurate and imprecise, (3) inaccurate and precise, and (4) accurate and precise. The data error represents both the inaccuracy and imprecision.

Errors in measurements result from three sources: (a) instrumental defects, (b) improper siting or location of the measurement device, and (c) human errors. The resulting errors are of two types: systematic and random. Random errors occur when the data show no tendency toward either overestimation or underestimation of measured values for a number of successive time intervals. Random errors in the data will undoubtedly produce random errors in the output. Systematic errors occur when the error tends to persist over a series of time intervals without changing sign. Systematic errors will be reflected in the incorrect parameter values. The importance of one type of error relative to the other depends on the problem at hand. Nonlinearity of hydraulic processes complicates treatment of the mechanism by which errors in data are transferred to model parameters and combined with input errors in the test period to produce errors in the simulated output. All errors transmit part of their magnitudes to model parameters and then to model results, but each one does so differently.

Many hydraulic analyses require a long-term record of data. It is observed that measuring devices are moved from their original location for one reason or another, and this, in turn, affects the consistency of measurements. If the record is not consistent then it must be corrected for inconsistency. There are graphical and statistical techniques, including the double mass curve analysis, the von Newmann ratio test, cumulative deviation test, likelihood ratio test, run test, among others for testing the consistency of data (Singh, 1988).

It is not uncommon in hydraulics that measurement records are incomplete. Breaks may vary in length from one or two days to several years. It is often necessary to estimate the missing data in order to utilize partial records. This is especially important in data-sparse regions. Several methods are available for estimating missing data, including the arithmetic average method, the normal ratio method, the inverse distance method, modified inverse distance method, linear programming, isohyetal method, the Lagrange method, interpolation methods, maximum entropy spectral methods, finite element method, kriging, among others (Singh, 1988).

Hydraulic data should be reduced to a common homogeneous condition. All hydraulic time series modeling approaches assume that the data reflect a stationary process. This means that the historical data should be transformed to reflect natural conditions so the natural process can be reliably modeled. This step may encompass correction of historical data for systematic errors, filling in of missing records, extension of data, and the reduction of data to the natural condition.

1.9.5 Hydraulic Data Processing

The amount of hydraulic data collected for a river system can be large. To process such large data bases, spatially visualize them and prepare maps, a very powerful technique has been developed in recent years and has become a mainstay for spatial data processing and for investigating spatial phenomena. This technique is called the Geographic Information System (GIS). A GIS is a spatial database that can be used to visualize or make maps and analyze data. It is also an Information System. Developments in different spatial data input devices, such as Global Positioning Systems (GPS), have dramatically increased the application of GIS to environmental and water resources problems. As a result, a number of software packages are available. One of the most popular GIS software packages is ArcView. The GIS can be broken into four major parts: (1) spatial data input which gets information such as maps and field data into the GIS; (2) spatial data manipulation by which maps and spatial data can be shown; (3) spatial data analysis in which different spatial analysis techniques can be run to identify patterns; and

(4) spatial data visualization. Spatial data are meant to be viewed as maps—the GIS allows the user to change these maps interactively—and add on different spatial layers. These maps can be displayed on screen, or output to a printer, allowing for large poster-maps to be produced. The information system component of the GIS is a useful way of working with large amounts of incoming and existing data.

1.9.6 Error in Hydraulic Modeling

By definition, a model is only an approximation of the prototype hydraulic system. Therefore, all models are less than perfect. There are different sources of error: (1) modeling errors, for models are approximate idealized representations of hydraulic systems; (2) parameter errors, for model parameters are estimated from experimental or historical data and are thus subject to data and estimation errors; (3) truncation errors, for model solutions entail truncation errors because of the need to use transcendental operations and implicit definitions; and (4) roundoff errors, for computing procedures introduce roundoff errors in their performance of elementary operations. Of these the first two are of major importance. The errors can be either random or systematic. The relative significance of the error component depends on the hydraulic system to be modeled and the scale at which it is modeled. By changing the scale, a system can be entirely deterministic or stochastic or mixed. Thus, a scale can serve as a basis to choose a particular type of model for the system. Most natural hydraulic systems have both deterministic and stochastic elements. Therefore, it is only logical that a hydraulic system is modeled with a proper mix of deterministic and stochastic models.

1.9.7 Sensitivity Analysis

Once a model is developed, it is important to analyze the changes in model results due to changes in model components, the input, or the parameters. Sometimes it may also be desirable to analyze the rate of change in the system output with respect to the change in one input or system parameter. The study of these changes constitutes what is referred to as sensitivity analysis which, in general, is divided into two parts: (1) parametric sensitivity and (2) component sensitivity. Parametric sensitivity can be used to determine an appropriate algorithm for estimation of model parameters, whereas component sensitivity can be employed to examine if some components of a model can be dropped without seriously undermining its predictive value and to ascertain the relative importance of individual components within the model.

1.9.8 Choosing a Hydraulic Model

Choosing an appropriate model from a group of models entails a number of steps (Singh, 1988):

1. Define the problem.
2. Specify the objective.
3. Study the data availability.
4. Determine the computing facilities available.
5. Specify the economic and social constraints.
6. Choose a particular class of hydraulic models.
7. Select a particular type of model from the given class.
8. Calibrate the model (that is, estimate the parameters).

9. Evaluate the performance of the model.
10. If the chosen model is satisfactory, then use the model for prediction purposes. If not, then go to step 6 and repeat the process until a satisfactory model is found.
11. Embed the model in a more general engineering or water resources model.

It should be emphasized that step 9 is crucial. Only when a model has been objectively calibrated and evaluated, can it be applied to a specific problem with the assurance that the best use is being made of the data and that something is known about the order of magnitude of the accuracy of prediction.

1.10 Scientific Foundations for the Study of Hydraulics

The study of hydraulics requires knowledge of many of the basic principles of engineering mechanics, geomorphology, geology, soil science, physics, chemistry, biology, mathematics, and statistics. For example, the study of flow in a river is strongly dependent on the river morphology which shapes the geometry, meandering, braiding, and other fluvial characteristics of the river, through which the flow takes place. An understanding of the river morphology is, therefore, needed to visualize the dynamic evolution of the river. Geology in general and geomorphology in particular provide us with a qualitative understanding of the river characteristics and river flow. Physics, chemistry, mathematics, and statistics provide the tools for quantitative analysis. The body of laws that control the flow in a river is fluid mechanics—a branch of physics. The analysis of natural chemical evolution of flow of water and of the behavior of contaminants in river flow requires use of some of the principles of chemistry.

Hydraulics is a quantitative science and therefore hydraulic analysis rests on an understanding of mathematics, both analytical and numerical. Much of hydraulics deals with natural systems whose behavior entails inherent uncertainty. It is therefore no surprise that statistics is becoming one of the main dialects of hydraulics. Information-based technologies, such as artificial neural networks, fuzzy logic, and artificial intelligence, are also becoming popular these days for solving hydraulic problems.

1.11 Dimensions

In the study of hydraulics, one deals with a variety of physical characteristics. It is, therefore, important to define a system for describing these characteristics both qualitatively and quantitatively. The qualitative aspect defines the nature of the characteristics (e.g., length, volume, discharge, and force). The quantitative aspect provides a numerical measure of the characteristics. The quantitative description requires both a number and a standard by which various quantities can be compared (Munson et al., 1990).

The basic dimensions required in the study of hydraulics are the length (L), the time (T), the mass (M), and the temperature (θ). Alternatively, the force (F) may be used instead of the mass (M). Newton's second law of motion states that force is equal to mass times acceleration, which may be written as $F \equiv MLT^{-2}$, where the sign $\equiv$ is used to indicate the dimensions. Two sets of fundamental dimensions can thus be employed to express the dimensions of any physical parameter: the mass system (L, T, M, and θ) and the force system (L, T, F, and θ). All other dimensions can then be expressed in terms of these fundamental dimensions by means of definitions and laws.

The best way to determine the dimensions of any parameter is to go back to its definition. For example, the discharge (Q) is defined as volume per unit time (i.e., $Q \equiv L^3T^{-1}$). Likewise, the pressure (P) is defined as the force per unit area (i.e., $P \equiv FL^{-2}$), which can be written in the mass system as $P \equiv ML^{-1}T^{-2}$. Also, the torque (Tq) is defined as the rate of momentum (or momentum flux) multiplied by the radius of curvature. Dimensionally, torque is equal to the dimensions of force times the dimension of length. In a dimensional form it is written as $Tq \equiv FL$, which can be expressed in the mass system as $Tq \equiv ML^2T^{-2}$. Table 1.3 provides a list of dimensions of common physical parameters.

All equations must be dimensionally homogeneous, which means that the dimensions of the left side of an equation must be the same as those of the right side. Naturally, one cannot add or equate unlike physical quantities. For example, length cannot be added to volume or discharge. All summation terms in any equation must have the same dimensions. Some parameters are dimensionless (e.g., the slope (S), the strain (ε), etc.).

TABLE 1.3 Dimensions of Common Physical Parameters

Type	Quantity	Mass System	Force System
Geometric	Length	L	L
	Area	L^2	L^2
	Volume	L^3	L^3
	Moment of inertia (area)	L^4	L^4
Kinematic	Time	T	T
	Velocity	LT^{-1}	LT^{-1}
	Angular velocity	T^{-1}	T^{-1}
	Angular acceleration	T^{-2}	T^{-2}
	Frequency	T^{-1}	T^{-1}
	Discharge	L^3T^{-1}	L^3T^{-1}
	Kinematic viscosity	L^2T^{-1}	L^2T^{-1}
Dynamic	Force	MLT^{-2}	F
	Density	ML^{-3}	$FL^{-4}T^2$
	Energy	ML^2T^{-2}	FL
	Work	ML^2T^{-2}	FL
	Moment of force	ML^2T^{-2}	FL
	Torque	ML^2T^{-2}	FL
	Pressure	$ML^{-1}T^{-2}$	FL^{-2}
	Stress	$ML^{-1}T^{-2}$	FL^{-2}
	Modulus of elasticity	$ML^{-1}T^{-2}$	FL^{-2}
	Momentum	MLT^{-1}	FT
	Power	ML^2T^{-3}	FLT^{-1}
	Specific weight	$ML^{-2}T^{-2}$	FL^{-3}
	Surface tension	MT^{-2}	FL^{-1}
	Moment of inertia (mass)	ML^2	FLT^2
	Dynamic viscosity	$ML^{-1}T^{-1}$	$FL^{-2}T$

Example 1.1

A commonly used equation for calculating the velocity of uniform flow, V, through prismatic channels is

$$V = C\sqrt{RS_o}$$

where C is the Chezy coefficient; R is the hydraulic radius defined as the cross-section area, A, divided by the wetted perimeter, P; and S_o is the bed slope. Find the dimensions of the Chezy coefficient which satisfy the condition of homogeneity of the equation.

Solution: The dimensions of the various terms in the above equation are:

$$V = \text{length/time} \equiv LT^{-1}$$

$$R = \text{Area/wetted perimeter} = A/P \equiv L^2L^{-1} \equiv L$$

$$S_o = \text{bed slope} \equiv \text{dimensionless}$$

Equating the dimensions on the left-hand side and right-hand side of the equation, one gets

$$LT^{-1} \equiv CL^{1/2}$$

Therefore,

$$C \equiv L^{1/2}T^{-1}$$

This means that the dimensions of the Chezy coefficient must be $L^{1/2}T^{-1}$.

1.12 Systems of Units

Engineers need to communicate with their peers not only through carefully defined words but also through numerical descriptors of the magnitude of certain quantities. The magnitude of every dimensional quantity must be expressed by a number and an associated unit of measurement. The magnitude of a quantity, such as length, depends on the system of units used to describe the quantity (e.g., we can describe length in terms of centimeters, feet, miles, etc.). It makes a difference whether a pipe diameter is 1 meter or 1 foot. We must, therefore, not only define quantities carefully, but we must be equally careful to use a set of consistent measurement units that is universally defined and accepted. It is, therefore, important to have a quantitative measure of any given quantity. The units of measurement must be defined for all basic quantities (length, time, mass, force, and temperature). The units of certain quantities are composed of other units. For example, the units of velocity combine the units of length and time. Similarly, the units of discharge combine the units of volume and time, and can be expressed in terms of the units of length and time. There are several systems of units in use; however two systems are much more common among scientists and engineers: The International System (SI) and the British Gravitational (BG) system.

1.12.1 International System (SI)

The SI, Système International d'Unités, has been adopted world wide since the early 1980s. The United States is unique in clinging to the British Gravitational (BG) system, whereas most countries of the world have switched to the SI system. However, it is expected that the SI system will prevail in the United States as well within the next few years. Many countries have declared the SI system to be the only legally accepted system. In the SI system, the unit of length is meter (m),

TABLE 1.4 Prefixes for Forming Multiples and Fractions of SI Units

Factor by Which Unit Is Multiplied	Prefix	Symbol
10^{12}	tera	T
10^{9}	giga	G
10^{6}	mega	M
10^{3}	kilo	k
10^{2}	hecto	h
10	deka	da
10^{-1}	deci	d
10^{-2}	centi	c
10^{-3}	milli	m
10^{-6}	micro	μ
10^{-9}	nano	n
10^{-12}	pico	p
10^{-15}	femto	f
10^{-18}	atto	a

the unit of time is second (s), the unit of mass is kilogram (kg), and the unit of temperature is Kelvin (K). The Kelvin scale (K) is related to the Celsius scale (°C) through the formula

$$°K = °C + 273.15$$

In the SI system, the force is a secondary dimension, and its unit, the Newton (N), is defined from Newton's second law of motion as

$$1\ N = (1\ kg)\ (1\ m/s^2)$$

A Newton (N) is defined as the force required to accelerate one kilogram of mass at the rate of one meter per second squared. A one kg mass weighs 9.81 N under the standard gravity (9.81 m/s^2). A dyne is the force required to accelerate a mass of one gram one centimeter per second squared.

In the metric system lengths are measured in millimeters (mm), centimeters (cm), meters (m) or kilometers (km). Areas are measured in square centimeters (cm^2), square meters (m^2), or hectares (100 m $\times$ 100 m = 10^4 m^2). The hectare is equivalent to about 2.5 acres. In the SI system, centimeters are not a unit of length. Lengths should be measured in meters. The unit of work in SI is Joule (J), which is the work done when the point of application of a 1-N force is displaced through a 1-m distance in the direction of the force. Thus, 1 Joule is equal to 1 Newton-meter (i.e., J = N-m). It can also be expressed as kg(m/s)^2. The unit for power is Watt (W), which is equivalent to a Joule per second (i.e., W = J/s = N-m/s). Prefixes for forming multiples and fractions of SI units are given in Table 1.4.

1.12.2 British Gravitational (BG) System

In the United States, the (BG) system is often called the (FPS) system, where F, P, and S refer to foot, pound, and second, respectively. In the British Gravitational (BG) system, the unit of length is foot (ft), the unit of time is second (s), the unit of force is pound (lb), and the unit of temperature is degree Fahrenheit (°F), whereas the absolute temperature unit is degree Rankine (°R). The Rankine scale is related to the Fahrenheit scale through the formula

$$°R = °F + 459.67$$

In the BG system the mass is a secondary dimension. The unit of mass, slug, is defined from Newton's second law of motion as

$$\text{Force} = (\text{mass})\,(\text{acceleration})$$
$$1\text{ lb} = (1\text{ slug})\,(1\text{ ft/s}^2)$$

Therefore, a 1-lb force, when acting on a 1 slug of mass, will produce an acceleration of 1 ft/s^2. The weight, W, in the BG system is given by

$$W(\text{lb}) = m(\text{slugs})\ g(\text{ft/s}^2)$$

The inside cover provides factors for conversion from one system of units to the other. In this text, both the SI system and the BG system will be used with more emphasis on the SI system. It is important to use a consistent system of units throughout the solution of a given problem. Numerous errors may be introduced through the use of inconsistent units. One must never mix meters with feet or Newtons with slugs. To eliminate this source of error, the following simple rules should be kept in mind:

1. Check the homogeneity of each term in a given equation.
2. Never write a quantity without its dimensions.
3. Convert the dimensions of the problem to the dimensions required in the answer and use one system of dimensions throughout the solution.

Example 1.2

The total water supply of the world is about 326 million cubic miles. Express this supply in cubic feet, gallons, cubic meters, acre-feet, cubic kilometers, kilometers-meters, cubic meters, liters, and tons.

Solution: Total water supply = 326×10^6 mi^3:

$$\text{acre-feet} = 326 \times 10^6\text{ mi}^3 \times \left(\frac{5280}{1\text{ mile}}\right)^3 = 4.8 \times 10^{19}\text{ ft}^3$$
$$= 4.8 \times 10^{19}\text{ ft}^3 \times \frac{1\text{ acre-ft}}{43{,}560\text{ ft}^3} = 1.1 \times 10^{15}\text{ acre-ft}$$
$$= 1.1 \times 10^{15}\text{ acre-ft} \times \frac{3.25 \times 10^5\text{ U.S. gallon}}{1\text{ acre-ft}}$$
$$= 3.58 \times 10^{20}\text{ U.S. gallons}$$
$$= 4.8 \times 10^{19}\text{ ft}^3 \times \frac{1\text{ km}^3}{[3281\text{ ft}]^3} = 1.36 \times 10^9\text{ km}^3$$
$$= 1.36 \times 10^9\text{ km}^3 \times \frac{10^3\text{ m}}{1\text{ km}} = 1.36 \times 10^{12}\text{ km}^2\text{-m}$$
$$= 1.36 \times 10^{12}\text{ km}^2\text{-m} \times \frac{10^6\text{ m}^2}{1\text{ km}^2} = 1.36 \times 10^{18}\text{ m}^3$$
$$= 1.36 \times 10^{18}\text{ m}^3 \times \frac{1000\text{ liters}}{1\text{ m}^3} = 1.36 \times 10^{21}\text{ liters}$$
$$= 3.58 \times 10^{20}\text{ U.S. gallons} \times \frac{9.2\text{ lb}}{1\text{ U.S. gallons}} \times \frac{1\text{ ton}}{2000\text{ lb}}$$
$$= 1.65 \times 10^{18}\text{ tons}$$

In the United States, 1 ton is 2000 lb. Therefore,

Total water supply in U.S. tons = 1.646×10^{18} tons.

Example 1.3

A certain object weights 200 N at the earth's surface. Determine the mass of the object (in kilograms) and its weight (in Newtons) when located on a planet with an acceleration of gravity equal to 3.0 m/s^2.

Solution: The weight of the object at the earth's surface, where the acceleration of gravity is equal to 9.81 m/s^2, is 200 N. Applying Newton's second law,

$$200 \text{ (N)} = m \text{ (kg) } 9.81 \text{ (m/s}^2\text{)}$$

Hence, the mass of the object = (200/9.81) = 20.387 kg.

This object of mass (20.387 kg) when moved to another planet where the acceleration of gravity is equal to (3 m/s^2) will weigh

$$W = 20.387 \times 3 = 61.16 \text{ N}$$

One should note that the weight of an object may vary from one place to the other as long as there is a change in the acceleration of gravity. Mass itself does not change.

1.13 Scope and Organization of the Book

The subject of hydraulics introduced in the preceding sections is much larger than the scope of this elementary hydraulics book. The subject matter of this book is divided into three parts. The first part deals with fluid mechanic preliminaries encompassing fundamental properties of fluids and flow types; forces, motions, and energy; and hydrostatics. The second part, which constitutes the core of the book, includes a discussion of hydraulic principles or it might as well be called hydraulic science. After introducing governing equations and dimensional analysis, the pipe hydraulics is treated first. Part Two includes a discussion of pipe flow, resistance, and pumps. The remainder of the second part is devoted to the hydraulics of open channels, encompassing channel geometry, channel resistance, uniform flow, energy in open channels, the momentum principle and gradually varied flow. The last part of the book deals with hydraulic applications and design, including computation of water surface profiles, and design of some hydraulic controls and structures. It is hoped that the student will appreciate that the study of hydraulics is fundamental to our development and a career in hydraulics will be a challenging, fascinating and satisfying one.

READING AID

1.1. Define hydraulics. What are different disciplines of the science of hydraulics? Define each discipline and give a practical example of each.
1.2. What is the difference between hydrostatics and hydrodynamics?
1.3. Give five areas of hydraulic applications and five types of hydraulic structures.
1.4. What are the basic elements of a hydraulic system?
1.5. Define a streamline and a stream tube. Give an example of a stream tube.
1.6. What is meant by a path line? Give two examples in which pathlines coincide with streamlines.
1.7. What is meant by the control surface and the control volume? Give an example with an illustration.
1.8. Give at least a couple of examples of different sections through which water may flow. Does the geometry of a conduit affect the flow pattern?
1.9. What is meant by entropy? Does entropy in closed systems usually increase or decrease?

1.10. State the fundamental laws of physics on which the science of hydraulics is based.
1.11. What is meant by initial and boundary conditions? What is the best way to obtain realistic initial conditions?
1.12. What is meant by three-dimensional flow? Give an example of each of three-, two-, and one-dimensional flows.
1.13. Give five examples of environmental and water resources problems involving hydraulic applications.
1.14. Explain the role of hydraulic machines in modern life. Are they used as well in the development of groundwater?
1.15. Give examples of some rivers used for navigation and transportation in the United States. State some of the hydraulic structures commonly installed to enhance river navigation.
1.16. Name the two commonly used systems of measurement. How are they related?
1.17. Write down the dimensions of discharge, acceleration, force, work, power, surface tension, and dynamic viscosity via the two systems of measurement.
1.18. Define mass flux and momentum flux and give their units.
1.19. What is the hydrologic cycle? Discuss the driving forces of the hydrologic cycle and explain its significance on our environment.
1.20. Name the equations that express the laws of conservation of mass, momentum, and energy in hydraulic systems.
1.21. What are the main steps involved in mathematical modeling?
1.22. What are the requirements for defining a transient boundary-value problem?
1.23. Define the sources of error in hydraulic models. Which errors are more significant?
1.24. State the usefulness of parametric sensitivity and component sensitivity in hydraulic models.
1.25. Define the steps that are needed to choose the appropriate model from a group of models.
1.26. State the main and secondary dimensions in the SI and BG systems. What are the units of mass and force in each?
1.27. Define the Newton and write down the relationship between the Newton, the kilogram, and the acceleration of gravity.
1.28. Write the relationship between the Kelvin and Celsius for measuring temperatures in SI system. Write an equation to relate the Rankine scale to the Fahrenheit scale in the GB system.

Problems

1.1. A certain control volume of porous media received a discharge of 0.1 m^3/s during a period of 10.0 minutes. The volume of outflow was measured to be 58.0 m^3 through the same period of time. Determine the change in storage in the control volume using equation (1.1).

1.2. In Problem 1.1, if the change in storage in the control volume is zero, what is the percentage of error encountered in measuring the outflow?

1.3. A certain reach of an open channel 1 mile in length is excavated through permeable soil. The rate of inflow to this reach is 100.0 ft^3/s and the rate of outflow is 95 ft^3/s. How much of the water is being lost through this reach every year?

1.4. A cylindrical tank is supplied with water at a flow rate of 80 liter/min. Two outlet pipes are connected to the base of the tank extracting water at flow rates of 50 and 35 liter/min, respectively. Calculate the rate of change of the water volume inside the tank per min. If the diameter of the tank is 60 cm, determine the change in the water level after one hour.

1.5. Water is flowing from a tap 8 mm in diameter and is collected in a 10 liter bottle. Determine the exit velocity of the water from the tap if the bottle is filled in (a) 30 s, (b) 5 min.

1.6. The dimensionless parameter Reynolds number (R_n) is defined as $R_n = Vd/v$, where V is the water velocity, d is the diameter of the pipe, and v is the kinematic viscosity of water (1.005×10^{-6} m^2/s). Determine R_n for the two cases (a and b) given in Problem 1.5.

1.7. The friction losses through a 10.0 km horizontal pipe of a certain diameter is given by the relation

$$h_f = 0.005 \text{ L}$$

where h_f is the head loss due to friction and L is the length of the pipe. The total energy head at the end of the pipe is 21.0 m. Estimate the total energy head at the pipe inlet.

1.8. Prove that the following ratios are dimensionless:

$$\frac{Vy}{v}, \frac{\rho Vy}{\mu}, \frac{V}{\sqrt{gy}}, \frac{p}{\rho V^2}, \frac{gy}{V^2}, \frac{p}{\gamma y}, \frac{\sigma}{\gamma y^2}, \frac{Q}{y^2\sqrt{gy}}$$

where V is the velocity, y is the flow depth, v is the kinematic viscosity, ρ is the fluid density, μ is the dynamic viscosity, p is the pressure, γ is the specific weight, σ is the surface tension, and Q is the discharge.

1.9. A commonly used equation for calculating the velocity of uniform flow in open channels (v) is

$$V = \frac{1}{n} R^{2/3} S_o^{1/2}$$

where v is the velocity of uniform flow, n is the Manning coefficient, R is the hydraulic radius, and S_o is the bed slope. Find the dimensions of the Manning coefficient to ensure that the equation is homogeneous.

1.10. From Example 1.1 and Problem 1.9, find the relationship between the Chezy and Manning coefficients. Check the homogeneity of this relationship.

1.11. The water power developed by a pump is given by the equation $P = \gamma QH$, where P is the power, γ is the specific weight, Q is the discharge, and H is the water head raised by the pump. Check the condition of homogeneity of the equation.

1.12. The Darcy equation for groundwater flow is written as

$$q = -K\frac{\partial h}{\partial x}$$

where q is the specific discharge (LT^{-1}), K is the hydraulic conductivity, and $\partial h/\partial x$ is the gradient of the piezometric head. Find the dimensions of K.

1.13. Determine the dimensions of the coefficients A and B in the following equation:

$$\frac{\partial^2 x}{\partial t^2} + A\frac{\partial x}{\partial t} + Bx = 0$$

where x is the length and t is the time.

1.14. Determine whether the following equation is dimensionally homogeneous or not.

$$F = 16.83\, \mu vd + 29.78\, \rho Q^2 d^{-2}$$

where F is the force, μ is the dynamic viscosity, v is the velocity, d is the diameter, ρ is the density, and Q is the discharge.

1.15. The head loss for a pipe flow is expressed as

$$h_f = f\frac{L}{d}\frac{V^2}{2g}$$

where f is the friction coefficient, L and d are the length and diameter of the pipe, respectively, v is the average velocity, and g is the acceleration of gravity. Determine the units of the friction factor, f. If the system of units is changed from SI to BG, explain how this will effect the value of f.

1.16. A person weighs 150 lb at the earth's surface. Determine the person's mass in slugs and kilograms.

1.17. The discharge of water through an open channel is 10.0 m^3/s. What is the discharge in liters/min, gallons/min? What is the volume of water (in gallons) delivered by this channel per year?

1.18. A certain object weighs 150.0 Newtons at the earth's surface. Determine the mass of the object (in kilograms) and its weight (in Newtons) when located on a planet with an acceleration of gravity equal to 4.0 m/s^2.

1.19. A person having a total mass of 80.0 kg rides an elevator. Determine the force (in Newton) exerted on the floor of the elevator when it is accelerating upward at 5.0 m/s^2.

1.20. In Problem 1.19, if the elevator is accelerating downwards at 5.0 m/s^2, determine the force (in Newtons) exerted on the floor.

1.21. If the force exerted on the floor of the elevator in Problem 1.19 is zero, what will be the direction and the acceleration of the elevator?

1.22. A tank of oil has a mass of 30 slugs. Determine its weight in pounds and in Newtons at the earth's surface.

1.23. If the tank in Problem 1.22 is been placed on the surface of the moon where the gravitational attraction is one-sixth of that at the earth's surface, what would be its weight in Newtons?

1.24. How many hectares and acres in an area of 1 km^2?

1.25. Convert the following:

a. Discharge of 60 ft^3/min to lit/s
b. Force of 20 pounds to dynes
c. Pressure intensity of 30 lb/in^2 to N/m^2
d. Specific weight of 62.4 lb/ft^3 to N/m^3
e. Dynamic viscosity of 255 dyne-s/cm^2 to lb-s/ft^2
f. Dynamic viscosity of 100 gm/cm-s to slug/ft-s

1.26. Making use of the conversion factors given in Appendix A, express the following quantities in BG units: (a) 16.7 km, (b) 9.30 N/m^3, (c) 1.95 slugs/ft^2, (d) 0.045 N-m/s, (e) 6.35 mm/hr.

1.27. Express the following quantities in SI units using the conversion factors in Appendix A: (a) 10.25 in/min, (b) 4.20 slugs, (c) 4.0 lb, (d) 82.2 ft/s^2, and (e) 0.0315 lb-s/ft^2.

1.28. Verify the conversion factors for (a) acceleration, (b) density, (c) specific weight, and (d) dynamic and kinematic viscosity. Use the basic conversion relations, 1 ft = 0.3048 m, 1 lb = 4.4482 N, and 1 slug = 14.594 kg.

1.29. Water flows from a large drainage pipe at a rate of 1350 gal/min. What is the flow rate in m^3/s and lit/min?

1.30. The viscosity of a certain fluid is 3.5×10^{-4} poise. Determine its viscosity in both SI and BG units.

References

Chadwick A., and Morfett J., 1993. *Hydraulics in Civil and Environmental Engineering.* Chapman & Hall, Inc., New York, 10119.

Dooge, J. C. I., 1973. Linear theory of hydrologic systems. Technical Bulletin No. 1468, 327 pp., Agricultural Research Service, U.S. Department of Agriculture, Washington, D.C.

Gerhart, P. M., Gross, R.J., and Hochstein, J.I., 1992. *Fundamentals of Fluid Mechanics,* 2nd ed. Addison-Wesley Publishing Company, Inc.

Grishin, M. M., editor, 1987. *Hydraulic Structures,* English translation from the Russian by Dang, P. K., Mir Publishers, Moscow.

Hwang, N. H. C. and Haughtalen P. J. 1996. *Fundamentals of Hydraulic Engineering Systems.* 3rd ed. Prentice Hall, Englewood Cliffs, N. J.

Jourdan, M. R., and Sullivan, G. J., 1987. Tactical Uses of Induced Flooding. *The Military Engineer,* 79 (516): 433–444.

Mays, L. W., and Tung, Y. K., 1992. *Hydrosystems Engineering and Management.* McGraw-Hill.

Munson, B. R., Young, D. F., and Okiishi, T. H., 1990. *Fundamentals of Fluid Mechanics.* John Wiley & Sons, New York.

Sherif, M. M., and Singh, V. P., 1996. Saltwater Intrusion, Chapter 10 in *Hydrology of Disasters,* Singh V. P. editor, Water Science and Technology Library, Kluwer Academic Publisher.

Singh, V. P., 1988. *Hydrologic Systems, Vol.1: Rainfall-Runoff Modeling.* Prentice Hall, Englewood Cliffs, N. J.

Singh, V. P., 1992. *Elementary Hydrology.* Prentice Hall, Englewood Cliffs, N. J.

Singh, V. P., 1996. *Kinematic Wave Modeling in Water Resources: Surface Water Hydrology.* John Wiley & Sons, New York.

Wickham, M. P., and Jourdan, M. R., 1985. Military Hydrology Program: Tactical Environment Application. *The Military Engineer,* 77 (502): 468–469.

CHAPTER 2

FLUID PROPERTIES

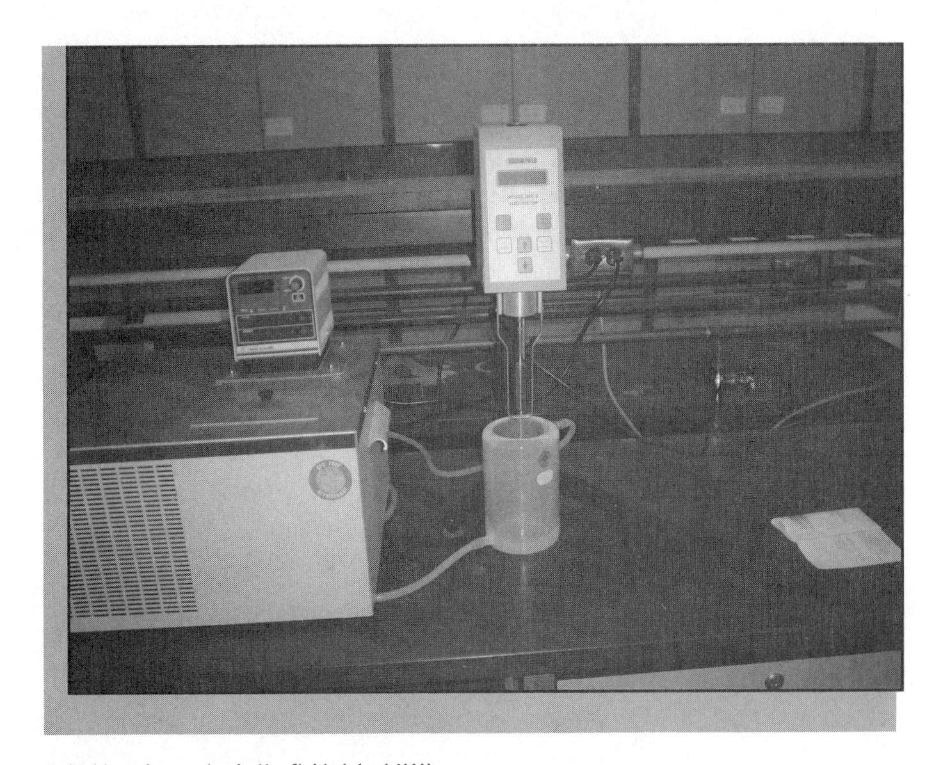

A desktop viscometer in the fluids lab at UAH.

It is useful to first understand what a fluid is. In general, matter is classified by the physical form of its occurrence known as phase. The phase can be liquid, solid, or gas or vapor. Fluids comprise matter either in the liquid phase or in the gas or vapor phase. To understand the difference among solid, liquid, and gas, it is recalled that all materials are composed of molecules which are not fixed in space but move about relative to each other. The molecules of a solid are closer together than those of a liquid and even more so than those of a gas. The molecular structure of a material shows that a solid has more densely spaced molecules with large intermolecular cohesive forces that permit the solid to maintain its shape and not be easily deformed. In case of a liquid (water, oil, etc.), the molecules are spaced farther apart and intermolecular (or attractive) forces are smaller than for solids. As a result, liquids are easily deformed (but not compressed). Gases have even larger molecular spacing, are easily deformed and intermolecular forces are negligible. At normal temperatures and pressures, the spacing between molecules is on the order of 10^{-6} *mm for gases and on the order of* 10^{-7} *mm for liquids. The number of molecules per cubic millimeter is on the order of* 10^{18} *for gases and* 10^{21} *for liquids. In gases, molecular spacing is an order of magnitude greater than the individual molecules, whereas in liquids it is of the same order as the size of molecules.*

A fluid may be either a gas or a liquid. A gas is very compressible and tends to expand indefinitely if all external pressure is removed. On the other hand, a liquid is relatively incompressible and does not expand indefinitely when all pressure, except that of its own vapor pressure, is removed. The reason is that the cohesion between molecules holds them together. A vapor is a gas and its temperature and pressure are such that it is near the liquid phase. The volume of a gas or vapor is greatly affected by changes in pressure or temperature or both.

In fluid mechanics, the distinction among solids, liquids, and gases is based on the way they deform under the action of an external load. Thus, a fluid is defined as a substance that deforms continuously under a shearing stress (force per unit area) regardless of its magnitude. It will be clear later in the text that the magnitude of the stress depends on the rate of angular deformation. On the other hand, a solid will deform by an amount (usually small) proportional to the stress applied but will not deform continuously. When the shape of a solid is altered when subjected to an external force, the tangential stresses between adjacent particles tend to restore the shape to its original configuration. For a liquid, these stresses depend on the velocity of deformation and vanish as the velocity tends to zero. When the fluid motion ceases, the tangential stresses disappear and the fluid does not regain its original shape.

Physical properties of fluids are of fundamental importance, for they influence the behavior of hydraulic systems (Henderson, 1966). A property is a characteristic of a fluid which is invariant when the fluid is in a particular state. In each state the condition of the fluid is unique and is characterized by its properties. Fluid properties can be distinguished into (1) extensive properties and (2) intensive properties. An extensive property is a property whose value depends on the amount of the fluid present. For example, total volume, total energy, and total weight are extensive properties. On the other hand, an intensive property is a property whose value is independent of the amount of the fluid present. For example, temperature, pressure, viscosity, and surface tension are intensive properties, for they are independent of the amount of the fluid. Likewise, volume per unit mass, energy per unit mass, and weight per unit volume are also intensive properties. Intensive properties apply to a particle of the fluid. Therefore, their values may change from particle to particle over any finite volume or mass in a system. In this chapter, both types of properties are discussed. When considering fluid properties, fluids are considered to have properties of a continuum, although they are composed of discrete particles. In other words, their properties are defined in terms of the behavior exhibited by the aggregate, not in terms of individual molecular properties. It is acknowledged here that the aggregate properties of a continuum depend on the molecular structure of the fluid and the intermolecular forces. Thus, the fluid behavior is characterized by considering the average, or macroscopic, value of the property of interest, where the average is evaluated over a small volume containing a large number of particles. This characterization involves neglecting the motion of individual molecules for two reasons. First, individual molecular interactions are not vital to developing an understanding of the macroscopic flow behavior of hydraulic interest. Second, describing molecular motion is not practical for lack of data and is computationally unwieldy. Furthermore, to account for the net effect of molecular interactions, fluid properties, such as density, pressure, viscosity, and others, are introduced which are described in this chapter.

Many fluid properties vary with temperature, pressure, and so on. Although these properties are slightly interdependent, of primary concern in hydraulics is how the numerical magnitudes of these properties change with temperature. For example, consider a large lake in the United States. The change of density with temperature causes lake water to stratify in summer with warmer water on the surface. During fall, the surface water temperature drops rapidly and the water sinks toward the lake bottom. The warmer water near the bottom rises to the surface resulting in overturn of the lake in fall. In winter, when the water temperature falls below 4°C, the lake surface freezes while the warm water remains at the bottom. The winter stratification is followed by overturn of the lake in spring (Hwang and Haughtalen, 1996). Similarly, variations of surface tension directly affect the evaporation loss from large water bodies, while variations of water viscosity affect fluids in motion under viscous force.

There are three fundamental properties of fluids, including density, viscosity, and compressibility (Chow, 1959). Many other fluid properties can be expressed in terms of these fundamental properties. For example, surface tension can be expressed in terms of density, and flow and conduit properties. The main properties of fluids that affect the behavior of hydraulic systems are presented in what follows.

2.1 Measure of Fluid Mass and Weight

2.1.1 Density

The fluid density is a measure of the mass or the number of individual molecules per unit volume multiplied by their respective masses. In other words, it is a measure of the concentration of matter or heaviness. Thus, the density of a fluid (ρ) is defined as the mass per unit volume of the fluid; its dimensions are ML^{-3}:

$$\rho = \frac{m}{V_{ol}} \tag{2.1}$$

where m is the mass and V_{ol} is the volume. ρ is also called the mass density. The density is a property inherent to the molecular structure of the fluid. It appears in most of the governing equations regarding fluid motion because both gravitational force and momentum are proportional to mass. In the BG system, ρ has units of slugs/ft^3 and in the SI units it has units of kg/m^3.

There can be a large variation in the value of density of different fluids. For liquids, changes in pressure and temperature exercise only a small effect on the value of density. Water is one of the few substances that expand as they freeze. Water reaches its maximum density of 1000 kg/m^3 at 4°C or 1.94 slugs/ft^3 at 40°F. The variation in water density with temperature is given in Tables 2.1 and 2.2. The density of water depends on its contents of total dissolved salts. Thus, at the same temperature the density of seawater is higher than the density of freshwater.

The specific volume, V_s, is defined as the volume per unit mass; it is, thus, the reciprocal of the mass density

$$V_s = \frac{V_{ol}}{m} = \frac{1}{\rho} \tag{2.2}$$

2.1.2 Specific Weight

The weight, W, of a fluid is defined as the mass multiplied by the gravitational acceleration, g:

$$W = mg \tag{2.3}$$

The specific weight, γ, is defined as the weight per unit volume (FL^{-3}):

$$\gamma = \frac{W}{V_{ol}} \tag{2.4}$$

This is also referred to as the weight density. The specific weight can also be expressed as

$$\gamma = \left(\frac{m}{V_{ol}}\right)g = \rho g \tag{2.5}$$

Equation (2.5) expresses the relation between mass density and weight density. The specific weight is used to characterize the weight of a fluid system. Weights in the SI system are measured in Newtons (N) and in pounds (lb) in the BG (FPS) system. In the BG units, the specific weight has units of lb/ft^3 and the in SI units it has units of N/m^3. For a water density of

TABLE 2.1 Physical Properties of Water in SI Units at Atmospheric Pressure

Temperature, T (°C)	Vapor Pressure, p_v (kN/m², abs)	Density, ρ (kg/m³)	Specific Weight, γ (kN/m³)	Dynamic Viscosity, μ (N-s/m²) × 10^{-3}	Kinematic Viscosity, ν (m²/s) × 10^{-6}	Surface Tension,* σ (N/m)	Bulk Modulus of Elasticity, E_v (kN/m²) × 10^6
0	0.61	999.8	9.805	1.781	1.785	0.0756	2.02
5	0.87	1000.0	9.807	1.518	1.519	0.0749	2.06
10	1.23	999.7	9.804	1.307	1.306	0.0742	2.10
15	1.70	999.1	9.798	1.139	1.139	0.0735	2.14
20	2.34	998.2	9.789	1.002	1.003	0.0728	2.18
25	3.17	997.0	9.777	0.890	0.893	0.0720	2.22
30	4.24	995.7	9.764	0.798	0.800	0.0712	2.25
40	7.38	992.2	9.730	0.653	0.658	0.0696	2.28
50	12.33	988.0	9.689	0.547	0.553	0.0679	2.29
60	19.92	983.2	9.642	0.466	0.474	0.0662	2.28
70	31.16	977.8	9.589	0.404	0.413	0.0644	2.25
80	47.34	971.8	9.530	0.354	0.364	0.0626	2.20
90	70.10	965.3	9.466	0.315	0.326	0.0608	2.14
100	101.33	958.4	9.399	0.282	0.294	0.0589	2.07

*Surface tension of water in contact with air.

TABLE 2.2 Physical Properties of Water in BG Units at Atmospheric Pressure

Temperature, T (°C)	Vapor Pressure, p_v (psia)	Density, ρ (slug/ft³)	Specific Weight, γ (lb/ft³)	Dynamic Viscosity, μ (lb-s/ft²) × 10^{-5}	Kinematic Viscosity, ν (ft²/s) × 10^{-5}	Surface Tension,* σ (lb/ft) × 10^{-3}	Bulk Modulus of Elasticity, E_v (psi) × 10^3
32	0.09	1.940	62.42	3.746	1.931	5.18	293
40	0.12	1.940	62.43	3.229	1.664	5.14	294
50	0.18	1.940	62.41	2.735	1.410	5.09	305
60	0.26	1.938	62.37	2.359	1.217	5.04	311
70	0.36	1.936	62.30	2.050	1.059	4.98	320
80	0.51	1.934	62.22	1.799	0.930	4.92	322
90	0.70	1.931	62.11	1.595	0.826	4.86	323
100	0.95	1.927	62.00	1.424	0.739	4.80	327
120	1.69	1.918	61.71	1.168	0.609	4.67	333
140	2.89	1.908	61.38	0.981	0.514	4.54	330
160	4.74	1.896	61.00	0.838	0.442	4.41	326
180	7.51	1.883	60.58	0.726	0.385	4.27	318
200	11.52	1.868	60.12	0.637	0.341	4.13	308
212	14.70	1.860	59.83	0.593	0.319	4.04	300

*Surface tension of water in contact with air.

1000 kg/m³ and the acceleration of gravity $g = 9.81$ m/s², the specific weight of water is 9810 N/m³. If the density of water is 1.94 slugs/ft³ and $g = 32.17$ ft/s², then the specific weight of water is 62.4 lb/ft³. The density, ρ, is absolute in the sense that it depends on the mass which is constant and does not vary from one location to the other. On the other hand, like the weight of an object, the specific weight, γ, is not absolute, as it depends on the acceleration of gravity, g, which may vary from one location to the other as well as from one elevation to the other. Because of its dependence on g, the specific weight is not a true fluid property.

2.1.3 Specific Gravity

The specific gravity, S.G., of a fluid is defined as the ratio of the density of the fluid to the density of water at some specified temperature. Normally the temperature is taken as 4°C (39.2°F). At this temperature, the density of water is 1.94 slugs/ft^3 or 1000 kg/m^3. One can also define the specific gravity as the specific weight of the liquid divided by the specific weight of water. Algebraically,

$$S.G. = \frac{\rho_{\text{fluid}}}{\rho_w} = \frac{\gamma_{\text{fluid}}}{\gamma_w} \tag{2.6}$$

The specific gravity is a dimensionless number. The specific weight of water, γ_w, at 4°C and under atmospheric pressure is 9810 N/m^3.

It is noted that density, specific weight, and specific gravity are all interrelated. This means that from a knowledge of one of the three, others can be computed.

Example 2.1

If 1 m^3 of oil has a mass of 860 kg, calculate its specific weight, density, and specific gravity.

Solution:

$$\gamma = \frac{W}{V_{\text{ol}}} = \frac{860}{1} \times 9.81 = 8436.6 \text{ N/m}^3$$

$$\rho = \frac{\gamma}{g} = \frac{8436.6}{9.81} = 860 \text{ kg/m}^3$$

$$S.G. = \frac{\gamma_{\text{oil}}}{\gamma_w} = \frac{8436.6}{9810} = 0.86$$

Example 2.2

A container weighs 5000 lb when filled with 6 ft^3 of a certain liquid. The empty weight of the container is 500 lb. What is the density of the liquid?

Solution: The net weight of the liquid = 5000 − 500 = 4500 lb.

$$\gamma = \frac{W}{V_{\text{ol}}} = \frac{4500}{6} = 750 \text{ lb/ft}^3$$

$$\rho = \frac{\gamma}{g} = \frac{750}{32.2} = 23.3 \text{ slugs/ft}^3$$

2.2 Viscosity

The fluid properties discussed in the preceding section measure the heaviness of the fluid but do not characterize the fluid behavior. For example, two fluids (such as water and oil) can approximately have the same value of density but behave quite differently when they flow.

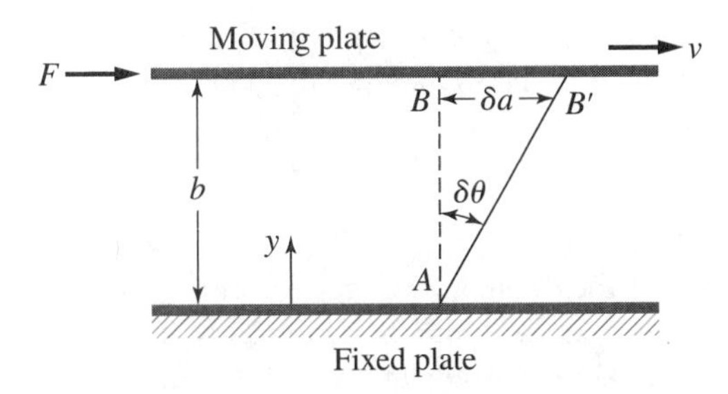

Figure 2.1 Shear stress in a fluid between two plates.

Additional properties are needed to uniquely characterize the fluid or the fluidity of the fluid. One of these properties is viscosity which is a measure of the fluid resistance to shear or angular deformation. Friction forces in a fluid result from the cohesion and momentum interchange between molecules in the fluid. An increase in temperature decreases this molecular cohesion and the consequent resistance to motion, and this effect is exhibited by a reduction in the fluid viscosity.

Consider that two parallel plates, as shown in Figure 2.1, are placed such that the lower one is fixed while the upper one is movable. If a fluid is placed between the two plates and the upper plate is subjected to a force F, such that it moves with a velocity v, then the fluid in contact with the upper plate will move with the same velocity v, while the fluid in contact with the lower plate will remain at rest. The fluid velocity between the two plates will be dependent on the distance b between the plates [i.e., the fluid velocity, $v = v(y)$ at any point y is as shown in Figure 2.1]. A velocity gradient of dv/dy which can be approximated as v/b will be developed. In other complex cases, such a linear gradient may not exist.

After a time increment of δt (since the upper plate has started to move), an imaginary line AB would rotate by a small angle $\delta\theta$, then

$$\delta\theta \approx \tan \delta\theta = \frac{\delta a}{b} \tag{2.7}$$

where $\delta a = v\delta t$. Hence,

$$\delta\theta = \frac{v\delta t}{b} \tag{2.8}$$

The rate of change of $\delta\theta$ with time is defined as the rate of shearing strain, ε:

$$\varepsilon = \lim_{\delta t \to \infty} \frac{\delta\theta}{\delta t} \tag{2.9}$$

which is equal to

$$\varepsilon = \frac{v}{b} = \frac{dv}{dt} \tag{2.10}$$

Experimental investigations have indicated that the shear stress, τ, increases with the increase of the velocity of the upper plate v (i.e., with the increase of the applied force, F). The shear stress is equal to F/A, where A is the area of the upper plate in contact with the fluid. Hence, one can write

$$\tau \propto \varepsilon \tag{2.11}$$

or

$$\tau \propto \frac{dv}{dy} \tag{2.12}$$

For common fluids, such as water and oil, experiments have indicated that the shear stress and the rate of shearing strain (velocity gradient) can be related through Newton's law of viscosity expressed as

$$\tau = \mu \frac{dv}{dy} \tag{2.13}$$

where the factor μ is the dynamic or absolute viscosity, or simply the fluid viscosity. It has the dimensions of FTL^{-2}, as seen from equation (2.13). The units of μ are slug/ft-s or lb-s/ft^2 in the FPS system, and kg/m-s or N-s/m^2 in the SI system. Plots of τ versus dv/dy would often indicate a linear relation with a slope equal to viscosity. The viscosity varies from one fluid to the other; for a particular fluid, it varies with temperature. Fluids for which the shearing stress varies linearly with the time rate of change of the shearing strain are called Newtonian fluids. Few fluids are non-Newtonian. In the study of hydraulics, we only deal with Newtonian fluids, mainly water. The study of non-Newtonian fluids is beyond the scope of this text.

The fluid viscosity varies slightly with pressure variation but considerably with temperature changes. A 40°C increase in water temperature causes a 1% decrease in its density while the same increase in temperature causes a 40% decrease in its viscosity.

Quite often, viscosity and density are combined and the kinematic viscosity, ν, is defined as the dynamic viscosity divided by the fluid density; its dimensions are L^2T^{-1}:

$$\nu = \frac{\mu}{\rho} \tag{2.14}$$

If the dynamic viscosity is expressed in the metric system in dyne-s/cm^2 (or poise), then the kinematic viscosity is expressed in cm^2/s (or stoke). As indicated in Chapter 1, a dyne is 1 g × 1 cm/s^2. Tables 2.1 and 2.2 present the values of dynamic and kinematic viscosities of water at different temperatures.

Viscosity can be used as a basis to classify fluids into (1) ideal and (2) real fluids. An ideal fluid is the one in which there is no friction, that is, its viscosity is zero. In such a fluid, the internal forces at any internal section are always normal to the section. This is also true when the fluid is in motion. This would mean that the forces are purely pressure forces. Such a fluid does not exist in reality. In a real fluid, tangential or shearing forces always exist whenever motion occurs. This gives rise to fluid friction, for these forces oppose the movement of one particle passing over the other. These friction forces are due to fluid's viscosity. Likewise, fluids are classified as Newtonian or non-Newtonian as explained earlier.

Example 2.3

The velocity distribution within a fluid moving between two parallel fixed plates is given by

$$\frac{v}{v_{max}} = 1 - \left(\frac{2y}{b}\right)^2$$

where v is the fluid velocity at a point y (units of length) from the origin, v_{max} is the maximum fluid velocity, and b is the spacing between the two plates. If the origin is placed midway between the two plates, $v_{max} = 50$ cm/s, $b = 0.2$ cm, and the dynamic viscosity μ of the fluid is 0.01 poise, (a) draw the velocity and shear stress distribution, and (b) calculate the shear stress at the upper plate.

Solution: From the equation given in the problem, one can write

$$v = v_{max}\left[1 - \left(\frac{2y}{b}\right)^2\right] = 50(1 - 100y^2)$$

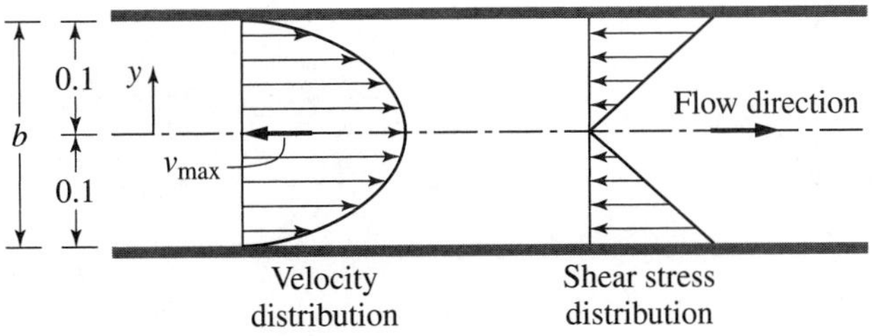

Figure 2.2 Velocity distribution and shear stress (Example 2.3).

The shear stress is given by equation (2.13), where dv/dy is obtained as

$$\frac{dv}{dy} = v_{max}[-200y] = 50 \times (-200y) = -10000y$$

Therefore,

$$\tau = 0.01(-10000y) = -100y$$

The negative sign means that the shear stress is a resistance force (i.e., is acting against the flow direction). Substituting for different values of y between 0 and 0.1 in the two equations for v and τ, the following values are obtained.

y (cm)	0	0.025	0.05	0.075	0.1
v (cm/s)	50	46.875	37.5	21.875	0
τ (dyne/cm^2)	0	−2.5	−5.0	−7.5	−10.0

The velocity distribution and shear stress are plotted in Figure 2.2. The shear stress at the upper plate where $y = 0.1$ cm is -10.0 dynes/cm^2.

Example 2.4

A crude oil with a viscosity of 8.95×10^{-4} lb-s/ft^2 is contained between two parallel plates. The bottom plate is fixed and the upper one moves when applying a force F as shown in Figure 2.3. If the distance between the two plates is 0.3 in, what is the value of F to cause the upper plate to move at a velocity of 3.6 ft/s? Take the effective area of the upper plate as 180 in^2.

Solution:

$$\tau = \frac{F}{A} = \mu \frac{dv}{dy} = \mu \frac{v}{b}$$

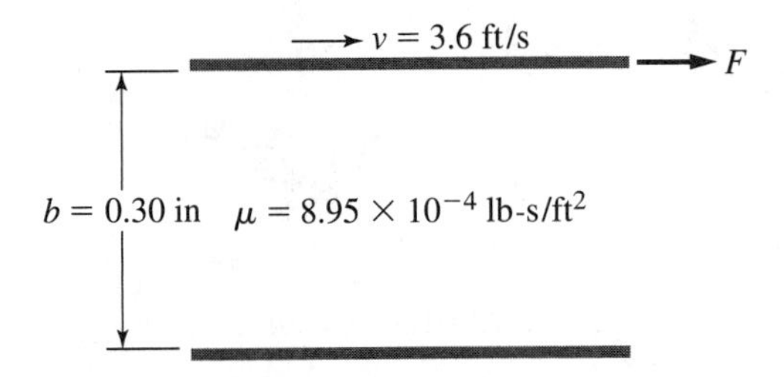

Figure 2.3 Force F applied to the upper plate (Example 2.4).

Therefore,

$$\tau = 8.95 \times 10^{-4} \frac{\text{lb-s}}{\text{ft}^2} \times \frac{3.6 \text{ ft/s}}{(0.3/12)\text{ft}} = 0.12888 \text{ lb/ft}^2$$

Because

$$\tau = \frac{F}{A}$$

$$0.128888 \frac{\text{lb}}{\text{ft}^2} = \frac{F\,(\text{lb})}{(180/144)}, \quad F = 0.1611 \text{ lb}$$

Example 2.5

A 15-kg block slides down a smooth inclined surface as shown in Figure 2.4. The thin gap (0.2 mm) between the block and the surface is filled with SAE 30 oil at a temperature of 60°F. The area of contact between the block and the oil is 0.15 m^2. Find the terminal velocity of the block, assuming a linear velocity distribution in the gap.

Solution:
For SAE 30 oil at 60°F, the dynamic viscosity, μ, is 3.8×10^{-1} N-s/m^2.

The weight of the block, W = m.g = 15 kg × 9.81 m^2/s = 147.15 N.

The weight component parallel to the surface ($W \sin 25$) is in balance with the resisting shear force, F (Figure 2.4). Therefore,

$$F = W \sin 25 = 147.15 \sin 25° = 62.19 \text{ N}$$

$$\tau = \frac{F}{A} = \frac{62.19 \text{ N}}{0.15 \text{ m}^2} = 414.6 \text{ N/m}^2$$

$$\tau = \mu \frac{dv}{dy} = \mu \frac{v}{b}$$

$$414.6 \text{ N/m}^2 = 3.8 \times 10^{-1} \left(\frac{\text{N-s}}{\text{m}^2}\right) \frac{v(\text{m/s})}{0.2 \times 10^{-3}(\text{m})}$$

Solving the preceding equation, we get $\nu = 0.218$ m/s

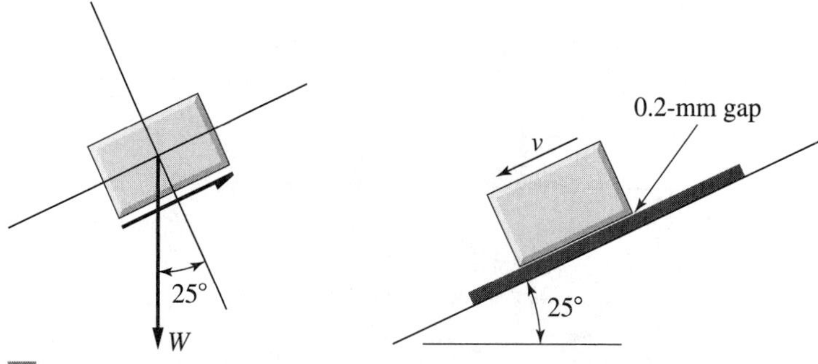

Figure 2.4 A block sliding on a smooth inclined surface (Example 2.5).

2.3 Compressibility of Fluids

All fluids, in reality, are compressible. However, most liquids exhibit very limited changes or compressibility in their volumes and hence densities as they are exposed to higher pressures. Such liquids are called incompressible. However, sound waves, which are really pressure waves, travel through them, giving an indication of the elasticity of liquids. Contrary to fluids, most gases are compressible. In many hydraulic applications, we deal with incompressible fluids, mainly water. In problems involving water hammer, the compressibility of the liquid needs to be taken into account, however.

2.3.1 Bulk Modulus

The bulk modules, E_v, is a measure of the fluid compressibility. It indicates the ease with which the volume and hence the density of a fluid changes with pressure. The bulk modulus is defined as

$$E_v = -\frac{dp}{dV_{ol}/V_{ol}} \tag{2.15}$$

where dp is the change in pressure needed to create a change in volume, dV_{ol} is the resulting change in volume, and V_{ol} is the initial volume. The negative sign indicates that as the pressure increases the volume decreases. Note that $m = \rho V_{ol}$, $V_{ol} = m/\rho$, $dV_{ol} = -m\,d\rho/\rho^2$, and $dV_{ol}/V_{ol} = -d\rho/\rho$. Therefore, the bulk modulus is also expressed as

$$E_v = \frac{dp}{d\rho/\rho} \tag{2.16}$$

where $d\rho$ is the change in fluid density due to the change in pressure (dp) and ρ is the initial density.

Integration of equation (2.16) with the initial condition that $\rho = \rho_0$ at $p = p_0$ yields

$$\rho = \rho_0 \exp[(p - p_0)/E_v] \tag{2.17a}$$

If p_0 is taken as the atmospheric pressure, then equation (2.17a) can be written as

$$\rho = \rho_0 \exp[p/E_v] \tag{2.17b}$$

The bulk modulus has the same dimensions as those for pressure (FL^{-2}) (i.e., it has the units of N/m^2 in the SI system and lb/ft^2 in the FPS units). The value of E_v at normal atmospheric pressure and temperature is 312,000 psi. The bigger the bulk modulus the more incompressible the fluid is, that is, it takes a higher pressure to create a small change in volume. For practical applications in hydraulics, water is considered incompressible. At atmospheric pressure and 60°F, it would require 3120 psi to compress a unit volume of water by 1%. This result is a representative of the compressibility of liquids. The atmospheric air is about 20,000 times more compressible than water. Water is about 100 times more compressible than steel. The bulk modulus of water is a function of both the temperature and pressure as shown in Table 2.3. The bulk modulus for fluids is quite analogous to the modulus of elasticity for solids; the former is defined on a volume basis while the latter is defined on the basis of one dimensional strain. The concept of bulk modulus is mainly applied to liquids, since for gases the compressibility is very high such that the value of E_v is not a constant. It is also applied in groundwater hydraulics, especially when dealing with land subsidence.

TABLE 2.3 Bulk Modulus of Water ($E_v/10^9$) in N/m^2 at Different Temperature and Pressure

Pressure $\times$ 10^5 N/m^2	Temperature 0°C	10°C	20°C	50°C
1 - 25	1.93	2.03	2.07	
25 - 50	1.96	2.06	2.13	
50 - 75	1.99	2.14	2.23	
75 - 100	2.02	2.16	2.24	
100 - 500	2.13	2.27	2.34	2.43
500 - 1000	2.43	2.57	2.67	2.77
1000 - 1500	2.84	2.91	3.00	3.11

Example 2.6

A liquid is compressed in a cylinder of volume of 2.0 ft^3 under a pressure of 100 psi. When compressed at 180 psi, it has a volume of 1.8 ft^3. Estimate the bulk modulus of the liquid.

Solution:

$$dp = p_2 - p_1 = 180 - 100 = 80 \text{ psi} = 11520 \text{ psf}$$

$$dV_{ol} = V_{ol2} - V_{ol1} = 1.8 - 2.0 = -0.2 \text{ ft}^3$$

$$E_v = \frac{-11520}{\frac{-0.2}{2}} = 115{,}200 \text{ psf}$$

2.3.2 Compressibility

The compressibility is a material property that describes a change in volume, or strain, induced in the material under an applied stress. Stress is imparted to a fluid through the fluid pressure. Thus, compressibility is simply the inverse of the modulus of elasticity, and is defined as strain/stress, rather than stress/strain. In hydraulics, it is necessary to define two compressibility terms, one for water and one for the solid material, say porous material in case of an earth dam, a levee, or a saturated geologic formation. An increase in pressure leads to a decrease in the volume V_{ol} of a given mass of water. Thus, the compressibility of water, β, is defined as

$$\beta = -\frac{dV_{ol}/V_{ol}}{dp} \tag{2.18a}$$

where the negative sign is necessary for the compressibility to be a positive number. Equation (2.18a) expresses a linear relation between the volumetric strain dV_{ol}/V_{ol} and the stress induced in the water by the change in pressure dp. Interpreted another way, β is the slope of the line relating strain to stress for water. This slope is usually assumed constant for the range of pressures encountered in hydraulics, although the slope is influenced a little bit by temperature. Dimensionally, the inverse of the dimensions of pressure or stress yields the dimensions of compressibility. Noting that the volume can be expressed in terms of density and the mass remains constant under the applied stress, one can write equation (2.18a) as

$$\beta = -\frac{d(m/\rho)/(m/\rho)}{dp} = \frac{d\rho/\rho}{dp} \tag{2.18b}$$

Integration of equation (2.18b) yields the equation of state for water:

$$\rho = \rho_0 \exp[\beta(p - p_0)] \tag{2.19}$$

where ρ_0 is the water density at the datum pressure p_0. Taking p_0 as the atmospheric pressure, equation (2.19) becomes

$$\rho = \rho_0 \exp[\beta p] \tag{2.20}$$

Equation (2.20) shows that for water to be considered as incompressible, $\beta = 0$ and $\rho = \rho_0$. Thus, depending on whether the value of β is zero or not, fluids can be classified as incompressible or compressible.

Example 2.7

The density of seawater is 1025 kg/m^3 at sea level. Determine the density of seawater on ocean floor 1200 m deep, where pressure is 15200.5 kN/m^2. Bulk modulus of elasticity = 2.2×10^9 N/m^2.

Solution: The difference in pressure between a point at a depth of 1200 m below the sea water level and a point at the water surface is: $dp = p - p_{atm}$, $dp = 15200.5 - 10.33 \times 9.81 = 15099.163$ kN/m^2.

$$dp = -E_v \frac{dV_{ol}}{V_o}$$

$$\frac{dV_{ol}}{V_o} = -\frac{dp}{E_v} = -\frac{15099163}{2.2 \times 10^9} = -0.00686$$

$$dV_{ol} = V_{ol} - V_o = \frac{m}{\rho_1} - \frac{m}{\rho_0}$$

$$\frac{dV_{ol}}{V_o} = \frac{\rho_o}{\rho} - 1$$

$$\rho = \frac{\rho_o}{1 + \dfrac{dV_{ol}}{V_o}} = \frac{1025}{1 + (-0.00686)} = 1032 \text{ kg/m}^3$$

2.4 Thermal Expansion

The density of a fluid may change with the change of temperature. Most fluids expand when the temperature is increased. This means that the density decreases when the temperature increases. The coefficient of thermal expansion (α_T) represents the ability of a fluid to expand or compress under the increase or decrease of the temperature. For a fluid with a initial volume of

V_{ol}, increasing the temperature by dT would cause an increase in the volume by dV_{ol}. Assuming constant pressure, the coefficient of thermal expansion is expressed as:

$$\alpha_T = \frac{dV_{ol}/V_{ol}}{dT} = \frac{1}{V_{ol}}\frac{dV_{ol}}{dT} \tag{2.21}$$

Since the mass of the fluid can be assumed constant, α_T can be expressed in terms of the fluid density, ρ, as

$$\alpha_T = \frac{1}{\rho}\frac{d\rho}{dT} \tag{2.22}$$

At atmospheric condition, α_T for water is approximately equal to $1.5 \times 10^{-4}\,°K^{-1}$ or $9 \times 10^{-5}\,°R^{-1}$.

2.5 Surface Tension

Liquids have cohesion and adhesion which are forms of molecular attraction. Cohesion permits a liquid to resist tensile stress, whereas adhesion permits it to adhere to another body. A steel needle will float on water surface if placed gently on the surface. Small droplets of mercury deform spherically when placed on a smooth surface. A fluid in a capillary tube would exhibit a rise or a drop above or below the free fluid surface. These phenomena are due to unbalanced cohesive forces acting on the fluid molecules at the fluid surface. The tension developed on the fluid surface supports the needle. The cohesive force on the surface of mercury tends to hold all molecules together in a compact shape. The adhesive force between fluids and solid surfaces causes the capillary phenomenon.

Interior molecules in fluids are in balance as they are surrounded by other molecules and all are attached to each other equally. Molecules along the fluid surface are subject to a cohesive force from one side. The consequence of these unbalanced forces on the surface is surface deformation as if a skin or membrane is created. The effect of surface tension is to reduce the surface of a free body of liquid to a minimum and to expand the surface area of molecules which have to be brought to the surface from the bulk of the liquid against the unbalanced attraction pulling the surface molecules inward (Douglas et al., 1995).

At the interface between two immiscible liquids, forces develop at the liquid surface which cause the surface to be like a membrane or skin stretched over the fluid mass. This skin is actually not present but is only imaginary. The attraction between molecules forms an imaginary film capable of resisting tension at the interface. The liquid property that creates this is known as surface tension. To explain it another way, in any liquid, every surface molecule experiences a strong attraction perpendicular to the surface which leads to a tendency to deform any but a plane surface. For example, water drops in air will spontaneously deform into spheres—the shape for minimum surface for a given volume of liquid. On the other hand, such a contracted surface extends by inclusion of more molecules but work will have to be performed. This work equals the free energy of the surface, the additional potential energy per unit surface due to the surface molecules. This free energy is measured by the surface tension.

The surface tension, σ, is defined as the intensity of the molecular attraction per unit length along any line in the surface. It has the dimensions of force per unit length, FL^{-1}, and

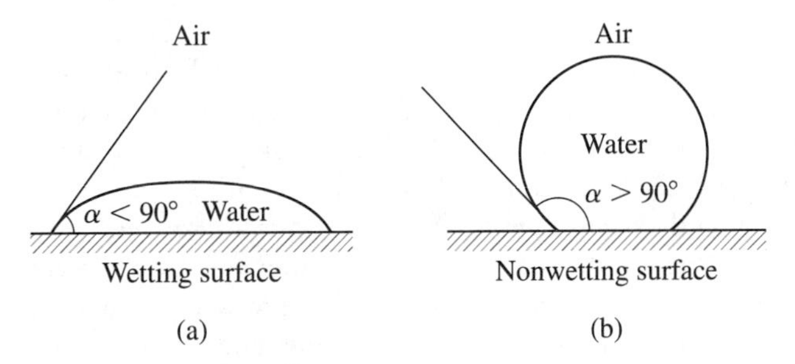

Figure 2.5 Water droplet in wetting and nonwetting surfaces.

is measured in dyne/cm or N/m. It is a physical property of the fluid and depends on the temperature. The surface tension of pure water decreases with increasing temperature as given in Tables 2.1 and 2.2.

The adhesive force between fluids and solid surfaces varies according to the nature of both the liquid and the solid surface. If the adhesive force between the liquid and the solid surface is greater than the internal cohesion between liquid molecules, the liquid tends to spread over and wet the surface as shown in Figure 2.5a. Conversely, if the internal cohesion is greater, a small drop forms as shown in Figure 2.5b (Hwang and Haughtalen, 1996).

Capillarity is caused by both cohesion and adhesion. When the effect of cohesion is less than that of adhesion, the liquid will wet a solid surface with which it is in contact and rise to the point of contact. On the other hand, if the effect of cohesion is greater than that of adhesion, the liquid surface will be depressed at the point of contact. Water wets the surface of a glass but mercury does not. If a capillary tube is placed into a free surface of water, the water inside the tube will rise as shown in Figure 2.6a. This rise is due to capillarity. The smaller the diameter of the tube the higher the rise of water inside the tube. This shows that the adhesive force between the water and the capillary tube is greater than the internal cohesion between water molecules. If the water is replaced with mercury, the surface of mercury inside the tube will fall below the free surface outside the tube as shown in Figure 2.6b. In case of mercury, the internal cohesion is greater than the adhesive force. If the liquid wets the tube, the level of the liquid will rise in the tube. If it does not, the level of the liquid in the tube will be depressed below the free surface (Douglas et al., 1995). The capillary rise or depression, h,

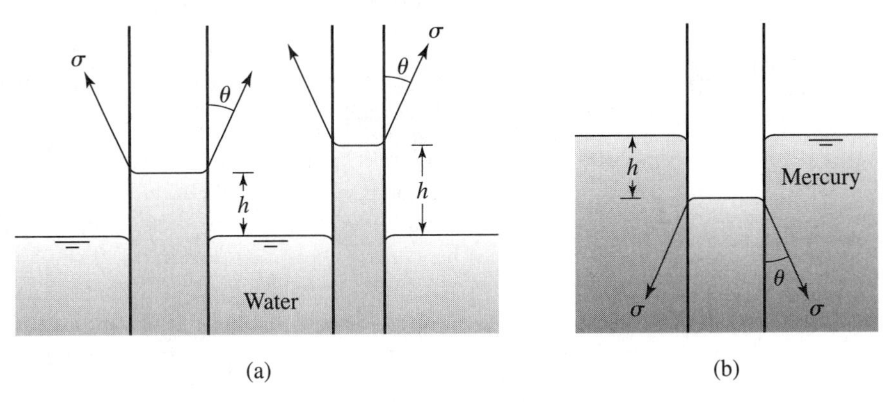

Figure 2.6 Capillary action in a capillary tube.

can be determined by balancing the adhesive force between the liquid and the solid surface with the weight of liquid column above or below the free surface.

The pressure inside a drop of liquid can be computed using the free-body diagram. If the spherical drop is cut in half, the force developed around the edge due to surface tension is $2\pi r\sigma$, where r is the inner radius of the tube. This force must be balanced by the pressure p acting on the circular area, πr^2. Thus, $2\pi r\sigma = p\pi r^2$, yielding $p = 2\sigma/r$, where p is the difference between the pressure inside the drop and the external pressure. This shows that the pressure inside the drop is greater than the pressure surrounding the drop.

The nature of both the liquid and the solid surface determines the angle θ between the liquid film and the vertical solid surface. The rise or drop in the liquid surface below the free surface is obtained by balancing the vertical component of the surface tension force around the edge of the film with the weight of the raised or lowered liquid column. It depends on the value of surface tension, the tube radius, the specific weight of the fluid, and the angle of contact between the fluid and the tube. The smaller the angle θ, the more the difference is between fluid levels inside and outside of the tube. Neglecting the small volume of liquid above or below the base of the curved meniscus (Figure 2.6a), and equating the lifting force created by the surface tension ($\pi d \sigma \cos\theta$) to the gravity force or weight ($\gamma\, \pi d^2 h/4$) for equilibrium, one can write

$$(\sigma\pi d)\cos\theta = \frac{\pi d^2}{4}(\gamma h) \tag{2.23}$$

where σ is the surface tension, γ is the specific weight, h is the capillary rise, and d is the inner diameter of the vertical tube. Hence, equation (2.23) yields

$$h = \frac{4\sigma\cos\theta}{\gamma d} \tag{2.24}$$

Equation (2.24) computes the approximate capillary rise or depression in a tube. For a clean tube, $\theta = 0°$ for water and about 140° for mercury. For tube diameters larger than 12 mm, capillary effects are negligible. The angle of contact is a function of both the liquid and the surface. The height is inversely proportional to the tube radius. When the tube radius is small, the capillary action may become pronounced.

In many hydraulic applications, surface tension effects are considered negligible. However, they may be important in hydraulics of unsaturated soils, small hydraulic model studies, breakup of liquid jets, formation of bubbles and drops, and so on.

Example 2.8

Determine the capillary rise of water between two vertical parallel glass plates 4.0 mm apart assuming that the angle θ is equal to 12°.

Solution: For the balance between the forces on water inside the two plates, it is essential that the weight of the water column is equal to the surface tension force. Assume that the length of the glass plates is L and let $\sigma = 7.28 \times 10^{-2}$ N/m for temperature = 20 degrees C, and $\gamma = 9789$ N/m^3. Then $Lbh\gamma_w = 2\sigma L\cos\theta$, where b is the distance between the two plates. Hence,

$$h = \frac{2\sigma\cos\theta}{b\gamma_w} = \frac{2 \times 7.28 \times 10^{-2} \times \cos 12°}{0.004 \times 9789} = 0.00364 \text{ m} = 0.364 \text{ cm}$$

Example 2.9

Determine the minimum diameter of a glass tube that can be used to measure liquid levels, if the capillary rise in the tube is not to exceed 0.166 inch knowing that the surface tension of the liquid under investigation is 5.03×10^{-3} lb/ft and its angle of contact, θ, is 12°. The specific weight of the liquid is 62.4 lb/ft^3.

Solution: Using equation (2.24),

$$0.166 = \frac{4 \times 5.03 \times 10^{-3} \times \cos 12°}{62.4 \times d}$$

Therefore, $d = 0.001899$ ft $= 0.0228$ in.

2.6 Vapor Pressure

All liquids, such as water, gasoline, wine, vinegar, and so on, evaporate or vaporize when open to the atmosphere. A liquid evaporates when molecules from its surface escape into the atmosphere. This is made possible when the liquid molecules have sufficient momentum to overcome the intermolecular cohesive forces. If a closed container is partially filled with a liquid and the space left is evacuated to form a vacuum, then escaping molecules will form and a pressure (partial) to be exerted by the molecules will develop in the space. This pressure increases until the number of molecules leaving the liquid surface equals the number of molecules reentering the surface. This condition is called the equilibrium condition and the pressure exerted by the vapor is the vapor pressure or saturation pressure. At this condition the liquid and vapor coexist in equilibrium in the container. The vapor pressure of a liquid depends on temperature. As temperature increases, the molecular activity increases and as a result the vapor pressure increases. Values of vapor pressure for water at various temperatures are given in Tables 2.1 and 2.2. There is a wide variation in vapor pressure among liquids as shown in Table 2.4. Mercury has the lowest vapor pressure and that may explain why it is particularly suitable for use in barometers.

At any given temperature the pressure on the liquid surface may be higher than the saturation pressure. When this liquid surface pressure reaches the vapor pressure vapor bubbles start forming within the liquid mass and boiling is initiated. This means that the boiling point of a liquid depends on the imposed pressure as well as the temperature. Thus saturation pressure may be called the boiling pressure for a given temperature. Water at standard atmospheric pressure boils when the temperature is 212°F (or 100°C). The vapor pressure of water at 212°F is 14.7 psi(abs), where abs indicates absolute pressure. At a 10,000-ft elevation above

TABLE 2.4 Vapor Pressure of 5 Liquids at 20°C (68°F)

Liquid	Pressure (psia)	Pressure (N/m^2 abs)	Pressure (mbar, abs)
Mercury	0.000025	0.17	0.0017
Water	0.339	2340	23.4
Kerosine	0.26	3200	32
Carbon tetrachloride	1.76	12,100	121
Gasoline	8	55,000	550

the mean sea level the atmospheric pressure is 10.1 psi(abs) and the corresponding boiling temperature is 193°F. Hence the saturation pressure of water is 10.1 psi(abs) and the water will boil at 193°F. This may explain why it takes longer to cook at higher elevations. The boiling can be induced by raising the temperature at a given pressure or by lowering the pressure at a given temperature. By reducing the pressure, boiling can be caused to occur at temperatures well below the boiling point at atmospheric pressure. For example, water will boil at a temperature of 60°C, if the pressure is reduced to 0.22 bar or 0.2 atm. Vapor pressure of liquids is of practical significance in pump-piping systems, barometers, cavitation, and so on.

READING AID

2.1. Explain, with the aid of an example, how the physical properties of fluids affect their behavior.
2.2. State the difference between extensive and intensive fluid properties and give examples for each.
2.3. Define density, specific weight, specific volume, and specific gravity. Write the relation between density and specific weight.
2.4. Do the density and specific weight change under different gravity?
2.5. What makes the density of water from different sources differ? At what temperature does the water have maximum density?
2.6. Define dynamic and kinematic viscosities. How are they related to each other? Write their dimensions.
2.7. Does the viscosity of liquids differ more with temperature or with pressure? What is the difference between Newtonian and non-Newtonian fluids?
2.8. What is meant by compressible and incompressible fluids? Are liquids more compressible than gases?
2.9. Define bulk modulus and write its dimensions. In hydraulic applications is water considered a compressible or incompressible fluid?
2.10. State the relation between the bulk modulus and the compressibility.
2.11. Define the coefficient of thermal expansion.
2.12. Define surface tension and give its dimensions.
2.13. What creates surface tension in liquids? How does the surface tension vary with temperature?
2.14. Give two phenomena in which the effect of surface tension is significant.
2.15. What factors affect the capillary rise in capillary tubes? Develop an equation to evaluate the capillary rise in a small tube.
2.16. Define vapor pressure. Does vapor increase or decrease when the temperature increases?
2.17. Explain why it takes longer to cook at higher elevations.

Problems

2.1. The specific weight of a certain liquid is 90.30 lb/ft^3. Determine its density and specific gravity.

2.2. The density of a jet fuel is 760 kg/m^3. Determine the specific weight and specific gravity of this fuel.

2.3. If 1.0 m^3 of oil has a mass of 890 kg, calculate its specific weight (γ), density (ρ), and specific gravity (S.G.).

2.4. A container has a 6 m^3 volume capacity and weights 1500 N when empty and 56,000 N when filled with a liquid. What is the density of the liquid?

2.5. A cylindrical tank with a diameter of 80 cm is partially filled with SAE 10W oil (ρ = 870 kg/m^3). If the mass of the oil in the tank is 340 kg, determine the height of the oil in the tank.

2.6. An overhead tank has length, width and height of 2 m, 2.5 m, and 1.5 m, respectively. If the water is filled up to three-quarters of its height, determine the mass and weight of the water inside the tank.

2.7. If a similar tank, as in Problem 2.6, is filled to the same level with oil with a mass of 4000 kg, determine the density and specific gravity of the oil.

2.8. If 6.2 m^3 of oil weights 52,980 N, calculate its density and specific gravity.

2.9. If the density of a certain liquid is 1.82 slug/ft^3, determine its specific weight (γ) in SI units and its specific gravity (S.G.).

2.10. The specific gravity of mercury at 80°C is 13.4. Determine its density and specific weight at this temperature. Express your answer in both SI and BG units.

2.11. What is the dynamic viscosity of water at 212°F? Express your answer in both SI and BG units.

2.12. The kinematic viscosity and specific gravity of liquid are 3.8×10^{-4} m^2/s and 0.88, respectively. What is the dynamic viscosity of the liquid in SI units?

2.13. A liquid has a specific weight of 55 lb/ft^3 and a dynamic viscosity of 2.90 lb-s/ft^2. Determine its kinematic viscosity.

2.14. A cylindrical water tank is suspended vertically by its sides. The tank has a 2.5 m diameter and is filled with 30°C water to 0.75 m in height. Determine the force exerted on the tank bottom.

2.15. The viscosity of a certain fluid is 3.5×10^{-4} poise. Determine its viscosity in both SI and BG units.

2.16. The dynamic viscosity (μ) for water at 20°C is 0.01 poise and its specific gravity (S.G.) is 0.98. Find its kinematic viscosity (ν).

2.17. What is kinematic viscosity (ν) of a liquid with a viscosity (μ) of 0.006 Pa-s and density of 870 kg/m^3 (Pa is pascal).

2.18. What is the kinematic viscosity of liquid with a viscosity 0.002 Pa-s and a specific gravity of 0.8?

2.19. A certain oil flows through a 2.5 in diameter pipe at 60°F with a mean velocity of 6 ft/s. Determine the Reynold's number, R_n, for this flow. The oil has a density of 1.77 slug/ft^3 and a dynamic viscosity of 8×10^{-3} lb-s/ft^2.

2.20. A solid cube with a side length of 0.4 ft and a weight of 110 lb slides down a smooth surface that is inclined at 45° from the horizontal (Figure P2.20). The block slides on a film of oil with a viscosity of 1.85×10^{-2} lb-s/ft^2. If the velocity of the block is 1.2 ft/s, determine the thickness of the oil film assuming a linear velocity distribution in the film.

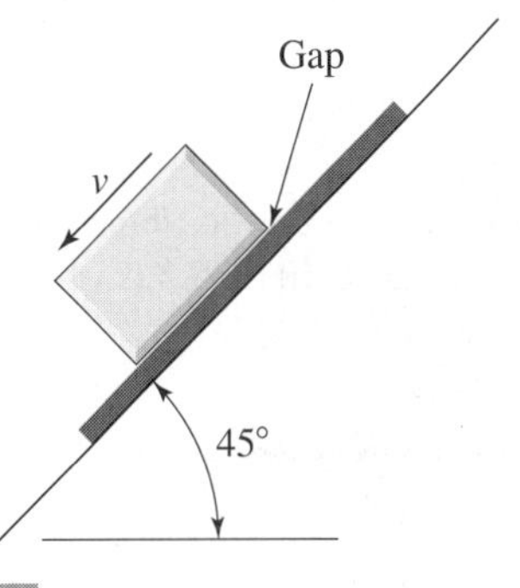

Figure P2.20

2.21. Two fixed plates are set at a distance of 12 mm apart. A large, movable thin plate is located between the two fixed plates a distance of 4 mm from the lower plate and 8 mm from the upper plate as shown in Figure P2.21. The space between the upper fixed plate and the movable plate is filled with a fluid with a dynamic viscosity of 0.03 N-s/m^2. The other space between the lower fixed plate and the movable one is filled with another fluid with a dynamic viscosity of 0.015 N-s/m^2. Determine the magnitude and direction of the shear stress that act on the fixed plates when the speed of the movable plate is 5 m/s. Assume linear velocity distribution in the two oil fields and neglect the thickness of the two plates.

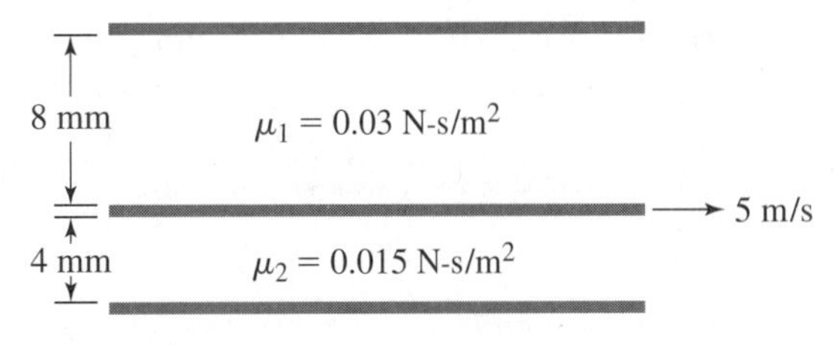

Figure P2.21

2.22. A plate measuring 20 cm × 20 cm is pulled horizontally through SAE10 oil at v = 0.15 m/s as shown in Figure P2.22. The oil temperature is 40°C. Find the force F.

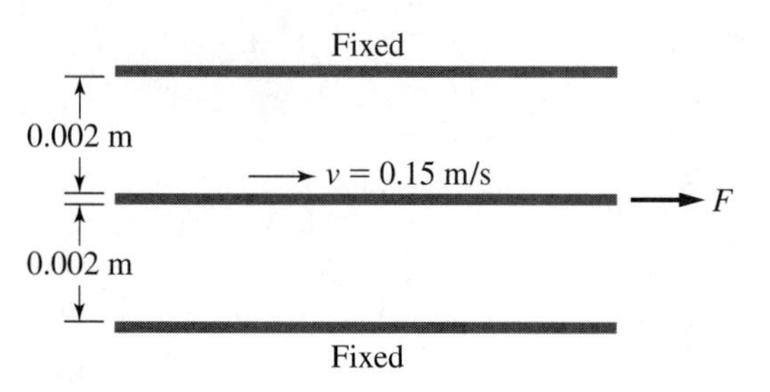

Figure P2.22

2.23. The plate in Problem 2.22 is moved with a velocity of 0.15 m/s to the right but the top and bottom plates are moved to the left at 0.10 m/s, as shown in Figure P2.23. Find the force *F*.

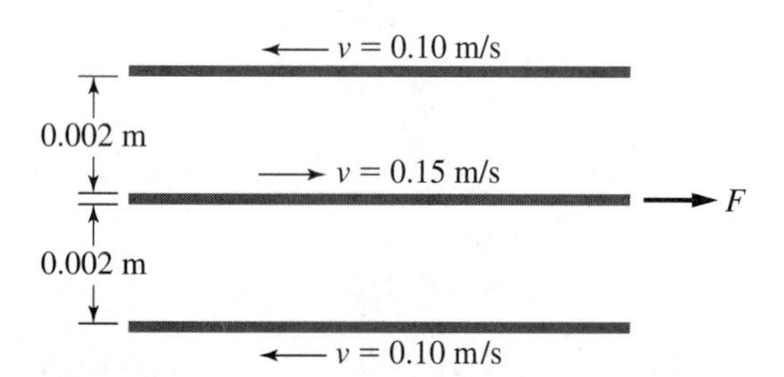

Figure P2.23

2.24. Two parallel plates, one moving at 4 m/s and the other fixed, are separated by a 5-mm-thick layer of oil with a specific gravity of 0.8 and a kinematic viscosity of 1.25×10^{-4} m^2/s. What is the average shear stress in the oil?

2.25. If the fixed plate in Problem 2.24 is moved with a speed of 2 m/s in the opposite direction of the movement of the second plate, what is the average shear stress in oil for this case?

2.26. Three large plates are separated by thin layers of ethylene glycol and water as shown in Figure P2.26. The top plate moves to right at 1.80 m/s. At what speed and in what direction must the bottom plate be moved to hold the center plate stationary?

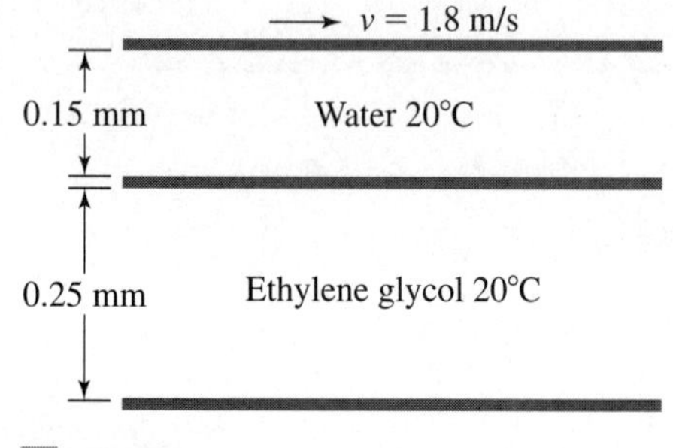

Figure P2.26

2.27. Oil ($\mu = 0.0004$ lb-s/ft^2) flows in the boundary layer as shown in Figure P2.27. Calculate the shear stress at (a) the plate surface, (b) 0.01 ft above the plate surface.

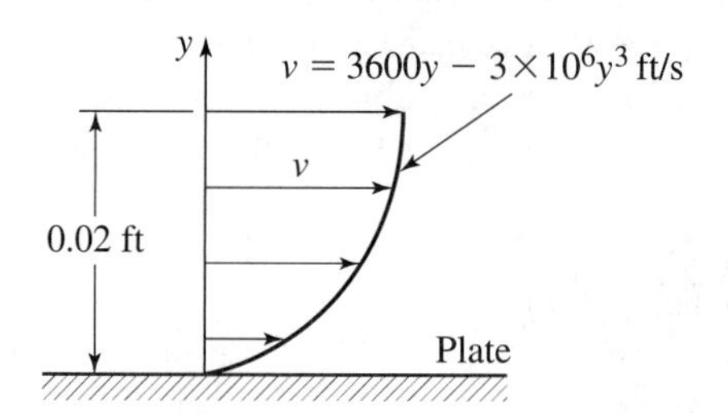

Figure P2.27

2.28. A plate is sliding down a plane inclined at an angle of 30° as shown in Figure P2.28. The plate weighs 15 lb, measures 25 in $\times$ 30 in, and has an average velocity of 0.65 ft/s. Determine the thickness of the SAE10 oil between the plate and the plane if the oil temperature is 50°F. Assume linear velocity distribution in the film.

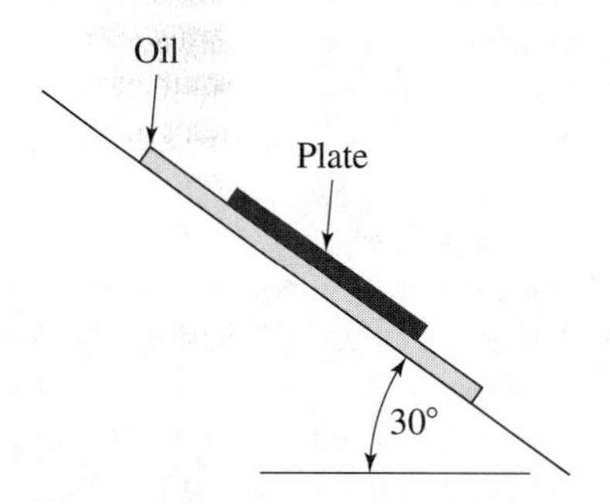

Figure P2.28

2.29. A plate, 0.6 mm distance from a fixed plate, moves at 0.3 m/s and requires a force per unit area of 4 Pa to maintain this speed. Determine the viscosity of the substance between the two plates.

2.30. If the velocity distribution of a viscous liquid ($\mu = 0.9$ N-s/m^2) over a fixed boundary is given by $v = 0.68y - y^2$ in which v is the velocity in m/s at a distance y (m) above the boundary surface. Determine the shear stress at the surface and at $y = 0.34$ m.

2.31. A thin plate is moving vertically between two parallel plates with a constant velocity of 20 m/s. The distance between the boundaries is 0.5 cm and the plate moves at a distance of 0.2 cm from one side of the boundaries. If the area of the plate is 1 m^2 and $\mu = 0.08$ poise, find the weight and mass of the plate.

2.32. A piston 11.96 cm in diameter and 14 cm long works in a cylinder of 12 cm in diameter. If the lubricating oil which fills the space between them has a viscosity of 0.65 poise, calculate the speed with which the piston will move through the cylinder when the axial load of 0.86 N is applied.

2.33. A 50-mm-diameter steel cylinder 600 mm long falls, because of its own gravity force, at a uniform rate of 0.2 m/s inside a tube of slightly larger diameter. A castor-oil film of constant thickness fills the space between the cylinder and the tube. Determine the clearance between the tube and the cylinder. The temperature is 38°C and the relative density of steel = 7.85.

2.34. A cylinder of radius 0.2 m and length 1.3 m rotates concentrically inside a fixed cylindrical sleeve of radius 0.21 m. Find the viscosity of oil that fills the space between them if a torque of 1.2 N-m is required to maintain a rotation of 80 rpm.

2.35. The lower end of a vertical shaft rests on a foot bearing. The shaft is 1.5 m diameter and is separated from the bearing by an oil film of 0.08 cm thickness and $\mu = 1.5$ poise. What will be the power absorbed by the shaft when it rotates at a speed of 400 rpm?

2.36. A piston moves in a cylinder with a clearance of 1 mm around the piston, as shown in Figure P2.36. The oil viscosity is 0.85 kg/m-s. Determine the force required to move the piston with a constant speed of 6.0 m/s.

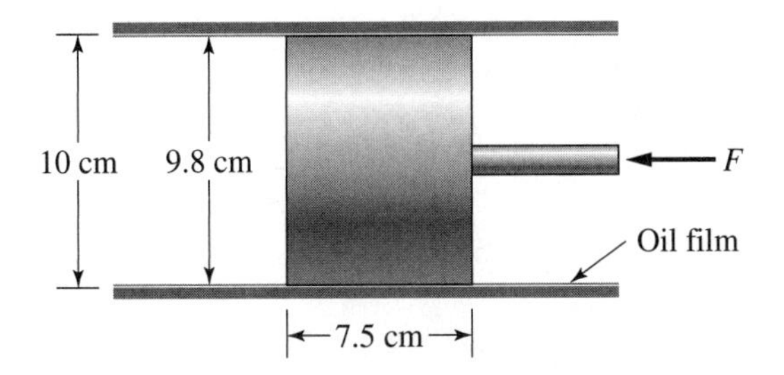

Figure P2.36 A moving piston in a cylinder.

2.37. A piston 3 kg in mass is sliding down in a cylinder due to gravity as shown in Figure P2.37. The gap between the piston and cylinder is filled with oil with a viscosity of 0.29 kg/m-s. The piston diameter is 15 cm and the clearance between the piston and the cylinder is 0.4 mm. Determine whether the piston is accelerating or decelerating when it is moving with a speed of (a) 5 m/s, and (b) 0.5 m/s.

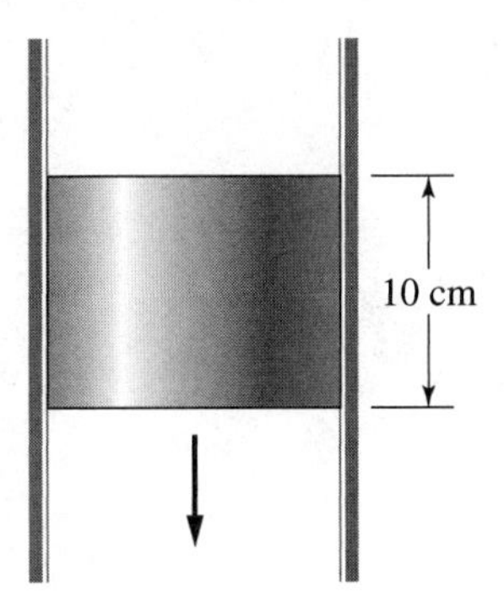

Figure P2.37 A freely falling piston inside a cylinder.

2.38. A liquid compressed in a cylinder has a volume of 1 liter at a pressure of 100 N/m^2 and a volume of 995 cm^3 at a pressure of 200 N/m^2. What is its bulk modulus of elasticity?

2.39. A closed steel tank has a volume of 5.0 ft^3. If the bulk modulus of elasticity of water is 300,000 psi, how many pounds of water can the tank hold at 2000 psi?

2.40. A liquid compressed in a cylinder has a volume of 995 cm^3 at 2000 kN/m^2 and a volume of 990 cm^3 at 3000 kN/m^2. What is its bulk modulus of elasticity?

2.41. Water has a volume of 1 m^3 at 40 bar and a volume of 0.99 m^3 at 425 bar. Find the bulk modulus of elasticity.

2.42. Assuming that the bulk modulus of elasticity of water is 1.15×10^6 kN/m^2 at standard atmospheric conditions. Determine the increase of pressure necessary to produce a 2% reduction in the volume at the same temperature.

2.43. Find the change in volume of one cubic meter of water when it is subjected to an increase in pressure of 1840 kN/m^2. Take the bulk modulus of elasticity as 2.16×10^6 kN/m^2.

2.44. An empty cylinder (filled with air) is sealed at 1 atmospheric pressure during winter when the ambient temperature is −15°C. What will be the pressure inside the cylinder during summer when the ambient temperature is 50°C?

2.45. To what height above the reservoir level will water (at 20°C) rise in a glass tube, as shown in Figure P2.45, if the inside diameter of tube is 2 mm?

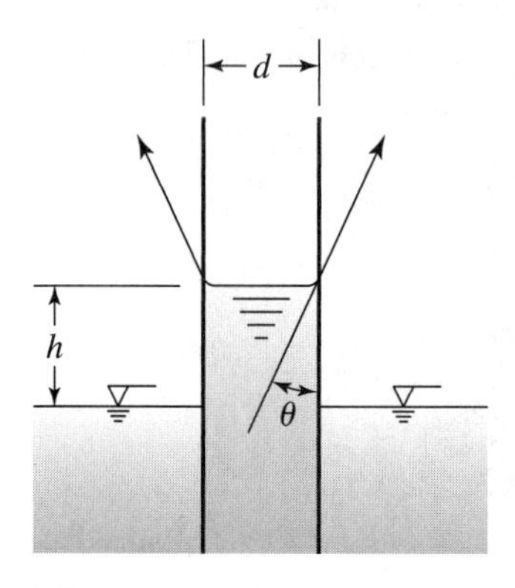

Figure P2.45

2.46. Estimate the height to which water at 26.7°C will rise in a capillary tube of diameter 2.5 mm. Take surface tension (σ) = 0.0718 N/m.

2.47. A capillary tube, having an inside diameter of 6 mm, is dipped in water at 20°C. Determine the height of water which will rise in the tube. Take the specific weight of water at 20°C = 998 kg/m^3. Take surface tension (σ) = 0.08 N/m and angle of contact (α) as 60°.

2.48. A U-tube is made up of two capillaries of diameters 1.5 mm and 2.0 mm, respectively. The U-tube is kept vertically and partially filled with water of surface tension 0.07 N/m and zero contact angle. Calculate the difference in the level caused by the capillarity.

2.49. Calculate the capillary effect in mm in a glass tube of 6 mm diameter when immersed in (i) water, and (ii) mercury, both liquids being at 20°C. Assume σ to be 73×10^{-3} N/m for water and 0.5 N/m for mercury. The contact angles for water and mercury are zero and 130°, respectively.

2.50. Show that the gauge pressure within a liquid droplet varies inversely with diameter of the droplet.

2.51. A small drop of water is in contact with air and has a diameter of 0.06 mm. If the pressure within the droplet is 466 Pa greater than the atmosphere, what is the value of surface tension?

2.52. Calculate the internal pressure of 25-mm-diameter soap bubble if the surface tension in the soap film is 0.5 N/m.

2.53. Calculate the gauge pressure and the absolute pressure within (a) a droplet of water 0.3 cm in diameter, and (b) a jet of water 0.3 cm in diameter. Assume the surface tension of water 0.064 N/m.

2.54. What is the pressure intensity within a free jet of water 0.02-inch diameter, if the surface tension of water is 0.005 lb/ft?

2.55. Calculate the maximum capillary rise in a 1-mm-diameter glass tube for (a) water at 20°C, and (b) kerosene at 20°C ($\theta = 5°$).

References

Chow, V. T., 1959. *Open-Channel Hydraulics.* McGraw-Hill, New York.

Douglas, J. F., Gasiorek, J. M., and Swaffield, J. A., 1995. *Fluid Mechanics* (3rd ed.). John Wiley & Sons, Inc., New York.

Henderson, F. M., 1966. *Open Channel Flow.* Macmillan, New York.

Hwang, N. H. C. and Haughtalen, P. J., 1996. *Fundamental of Hydraulic Engineering Systems* (3rd ed.). Prentice Hall, Englewood Cliffs, N.J.

CHAPTER 3

FORCES, MOTIONS, AND ENERGY

Hurricane Katrina storm surge in New Orleans, LA. (Courtesy of Don McClosky.)

All hydraulic systems satisfy a set of external and internal constraints. The external constraints are constituted by the equations of mass, momentum, and energy conservation. Indeed these constraints are satisfied by all processes occurring in various systems and are not limited to those in hydraulic systems. The equation of momentum conservation involves different types of forces and the equation of energy conservation involves different types of energy. There are also internal constraints that hydraulic systems must satisfy. One of these constraints leads to what is referred to as the process equation. In hydraulics, the process equation may be in the form of an energy loss function or any other equivalent form. The process equation expresses the relation between various types of energy acting on a fluid mass within the system boundaries and the conversion of some energy into nonrecoverable heat energy. In most practical problems, the interaction between different types of energy is very complex and it is difficult to analytically derive the process equation. This equation therefore is developed empirically, mainly through experimental measurements. In a few cases, such as in the case of laminar flow in pipes, the empirical relations are well established, and the resulting equations can, thus, be regarded as completely rational.

The process equation, also called the flow equation, when derived from dimensional considerations, exhibits a relation between different types of energy and different types of forces. In this relation, a particular type of force corresponds to a particular type of energy and vice versa. Although not simple, there is also a connection between forces and motion. Different forces play different roles and their relative magnitudes vary, depending on the hydraulic system under consideration. The relative magnitudes of these forces also determine the type and state of flow. This chapter summarizes different types of forces, combinations of forces through dimensionless numbers, motions, and energy encountered in hydraulics and alludes to relations existing between them.

3.1 Hydraulic Parameters

To understand the behavior of hydraulic systems and their characteristics, it is important to recognize the different types of parameters involved. Hydraulic systems involve three types of parameters: geometric, kinematic, and dynamic. Geometric parameters describe the dimensions and boundaries of the system. Examples of geometric parameters include the cross-sectional area of flow, A; the hydraulic depth, D; the hydraulic radius, R; the width of flow, B; the hydraulic roughness, ε; the grain size diameter, d; a representative dimension in the flow direction, L; or any other dimension which may affect the fluid behavior. Geometric parameters have the dimensions of length, L, or its multiplication. They do not contain either time, mass or force. Regardless of the way a hydraulic system is designed, at least one geometric parameter is always involved.

Kinematic parameters define the motion of the fluid, and hence must encompass time but not force or mass. The fluid velocity, v, is an example of a kinematic parameter. Other examples of kinematic parameters are acceleration, time of travel, celerity, etc. The velocity parameter can also be replaced by the time, T, required for flow to move a given distance L since $T = L/v$. It can also be replaced by discharge, Q, since $Q = Av$.

Dynamic parameters represent mass and force or energy that affect the flow process. Examples of dynamic parameters include force, F; pressure, p; mass density, ρ; specific weight, γ; shear stress τ; work, W; viscosity, μ; surface tension, σ; compressibility, β; modulus of elasticity, E; and so on.

A complete description of a hydraulic system in general involves dynamic parameters. In many instances, however, a simplification involving kinematic parameters may suffice.

3.2 Forces

The behavior of a hydraulic system is governed by different types of forces. Each force may create or dominate a certain type of motion. In hydraulic applications, the forces are distinguished as (*i*) inertial force, (*ii*) potential (pressure and position) force, (*iii*) viscous force, (*iv*) gravitational force, (*v*) surface force (tension), and (*vi*) elastic force (compression). In hydrodynamics, these forces in hydraulic systems are generally categorized into inertial forces and applied forces. Applied forces include internal and external forces, gravity forces, pressure forces and viscous forces.

3.2.1 Inertia Force

According to Newton's second law of motion, a constant mass M to have a motion with an acceleration of $d\mathbf{V}/dt$ would require a force $\mathbf{F}$. This statement is expressed mathematically as

$$\mathbf{F} = M\left(\frac{d\mathbf{V}}{dt}\right) \tag{3.1}$$

where $\mathbf{V}$ is the velocity vector. Equation (3.1) represents a vectorial relationship and is true for both the magnitude and the direction. The product $M(d\mathbf{V}/dt)$ is defined as the inertial force, which characterizes the natural resistance of matter to any change in its state (either in motion or at rest). When the inertial force is equated to the applied force, the momentum equation is obtained.

The mass M in equation (3.1) is the mass of a unit volume of fluid, that is,

$$M = \rho \cdot (\text{unit of volume}) = \rho \tag{3.2}$$

where ρ is the density. Hence, equation (3.1) can also be written on a unit volume basis as

$$\mathbf{F} = \rho\left(\frac{d\mathbf{V}}{dt}\right) \tag{3.3}$$

where $\mathbf{F}$ is now the force per unit volume.

3.2.2 Applied Forces: Internal and External Forces

Applied forces on an elementary mass of fluid are comprised of intenal forces and external forces. Internal forces result from the interaction of the interior points of the considered mass of the fluid. These are in balance and their sum is equal to zero, as for each action there is an equivalent reaction. However, recognition of these internal forces is important as the resulting work is not equal to zero. For example, the head loss in a pipe is the result of the work of internal viscous forces.

External forces include surface forces, body forces, capillary forces and geostrophic forces. Surface forces result from forces acting on the outer boundary of the considered volume. They are mainly caused by molecular attraction and their effect is rather confined to a very limited layer near the boundary. These surface forces can be further divided into (1) normal forces due to pressure, and (2) shearing forces due to viscosity. Both of these two forces also exist within the fluid but their sum is equal to zero.

The inertial and external forces develop due to an external field such as gravity or a magnetic field which acts on each element of the considered volume in a given direction (Le Méhauté, 1976). For this reason they are called body forces. For the majority of problems, only gravity force has to be considered which can generally be assumed to act in the vertical direction. In rare cases, such as for tidal motion and oceanic circulation, the acceleration of gravity should be considered radial. Body forces are also referred to as volume forces and always act in the same direction on its mass and are not balanced.

Capillary forces develop due to the molecular attraction between fluid particles and between the fluid particles and the media. They are of practical importance for groundwater flow in phreatic aquifers, flow in the vadose zone, and a few other applications. They diminish quickly away from the boundaries of fluid particles. The geostrophic force caused by the Coriolis acceleration due to the earth's rotation is sometimes considered as a body force.

3.2.3 Gravity or Body Force

Similar to the inertial forces, in the case of gravity action, the gravity force per unit volume (F_g) is equal to the fluid weight:

$$F_g = \rho g$$
$$\text{Hence, } F_g = Mg \text{ (for volume } = 1) \tag{3.4a}$$

where g is the acceleration due to gravity. The gravity force is independent of the fluid state and has a constant value whether the fluid is in motion or at rest. It acts downward in the vertical direction. The gravitational force per unit mass is the acceleration of gravity. In an inclined channel, g has three components g_x, g_y, and g_z in the x, y, and z directions, respectively, and can be expressed as

$$\mathbf{g} = g_x\mathbf{i} + g_y\mathbf{j} + g_z\mathbf{k} \tag{3.4b}$$

where **i**, **j**, and **k** are unit vectors. In vectorial form, the gravity force is expressed as

$$\mathbf{F_g} = -\nabla(\rho g Z) \tag{3.5}$$

where Z is the vertical. In a horizontal open channel, $g_x = 0$, $g_y = 0$, and $F_{g_z} = -\rho g = -\partial(\rho g Z)/\partial z$. In vectorial form, $F_g = -\text{grad}(\rho g Z)$, as given in equation (3.5), since $F_{g_x} = -\partial(\rho g Z)/\partial x = 0$, and $F_{g_y} = -\partial(\rho g Z)/\partial y = 0$.

The body force in the x direction $= \rho\, dV_{\text{ol}} g_x$

$$= \rho dV_{\text{ol}}\left(-g\frac{\partial Z}{\partial x}\right) = -\rho g\frac{\partial Z}{\partial x}dV_{\text{ol}} \tag{3.6}$$

where dV_{ol} is the unit volume, and $g_x = -g(\partial Z/\partial x)$ = the component of g in the x direction.

If Z is the vertical in the (x–z)-plane as we have assumed, it is seen from Figure 3.1 that

$$Z = Z_x + Z_z = -x\sin\theta i + z\cos\theta k \tag{3.7}$$

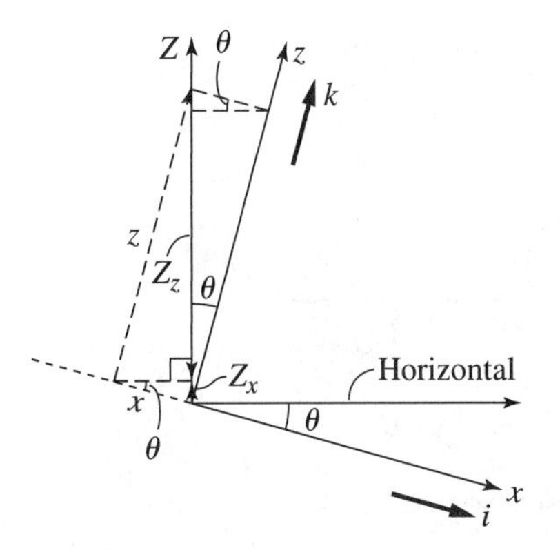

Figure 3.1 Analysis of body force.

where i and k are unit vectors in the x and z directions, respectively. Equation (3.7) can be recast as

$$Z = z\cos\theta - x\sin\theta \tag{3.8}$$

Therefore,

$$\frac{\partial Z}{\partial x} = -\sin\theta = -S_o \tag{3.9}$$

where S_o is the slope of the channel bed. Thus, the gravity force per unit volume in the x direction becomes $F_{g_x} = \rho g S_o$.

Example 3.1

A circular water tank is filled with water up to a depth of 5 meters. The radius of the tank is 6.5 meters. Compute the force acting on the base of the tank.

Solution: First, we calculate the force per unit area using $F = \rho g z$, where $\rho = 1000$ kg/m^3, $g = 9.81$ m/s^2, and $z = 5$ m. Therefore, the force per unit area, F, is

$F = 1000 \times 9.81 \times 5 = 49050$ kg-m/s^2/m^2

The base area of the tank is $\pi r^2 = 6.5 \times 6.5 \times \pi = 42.25\pi$ m^2

Total force on the base $= 49050 \times 42.25\pi = 6.51 \times 10^6$ N

3.2.4 Pressure Force

Pressure is measured as force per unit area. Consider a volume of matter isolated as a free body. The forces acting on the volume are the surface forces over every element of area bounding the volume. A surface force will, in general, have two components: perpendicular and parallel to the surface. At any point, the perpendicular component per unit area is known as the normal stress. If the stress is of compression type, it is called pressure intensity or simply pressure. Thus, pressure is used to indicate the normal force per unit area at a given point on a plane within the fluid mass of interest. Pressure forces result from the normal components of the molecular forces near the boundary of the volume under consideration. Consider a two-dimensional element (in a stagnant fluid), as shown in Figure 3.2. Because the fluid is at rest, both inertial and

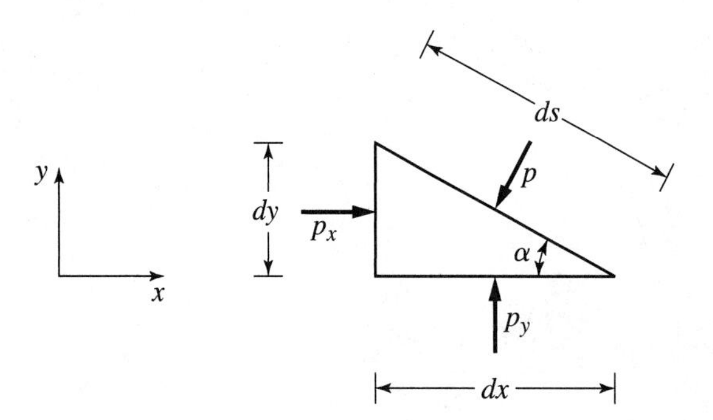

Figure 3.2 Pressure magnitude is independent of the areal orientation.

viscous forces are absent while gravity and pressure forces are present. Since this element is in balance, the summation of the components of forces in the x and y directions should be equal to zero. Therefore, in the x direction

$$p_x \, dy - p \, ds \sin \alpha = 0 \tag{3.10a}$$

and in the y direction

$$p_y \, dx - p \, ds \cos \alpha = \frac{\rho g \, dx \, dy}{2} \tag{3.10b}$$

knowing that $dy = ds \sin \alpha$ and $dx = ds \cos \alpha$. Dividing both sides of equation (3.10b) by dx, we obtain

$$p_y - p = \frac{\rho g \, dy}{2} \tag{3.10c}$$

As dy tends to zero, the right side approaches zero and we get

$$p = p_x = p_y \tag{3.11}$$

Equation (3.11) shows that the pressure is independent of the direction and hence is a scalar; however, the gradient of the pressure force is a vector. Similarly, the force caused by pressure against an area (which is a vector) changes direction as the normal to the area under consideration changes direction. The force associated with a given pressure acting on a unit of area is the product of pressure and incremental area and has the direction of the normal to the area. The magnitude of pressure at a point is computed by dividing the normal force against an infinitesimally small area by the area. At a point in the interior of a fluid mass, the direction of the pressure force depends on the orientation of the plane cut through the point. Thus, we state Pascal's law: The pressure at a point in a fluid at rest or in motion is independent of direction as long as there are no shearing stresses present. For moving fluids in which there is relative motion between particles leading to development of shearing stresses, the normal stress at a point, which corresponds to pressure in fluids at rest, is not necessarily equal in all directions. In such cases, the pressure at any point is defined as the average of any three mutually perpendicular stresses at that point.

Now we consider how the pressure in a fluid with no shearing stress varies from point to point. Consider a fluid element as shown in Figure 3.3. The pressure force acting on the side

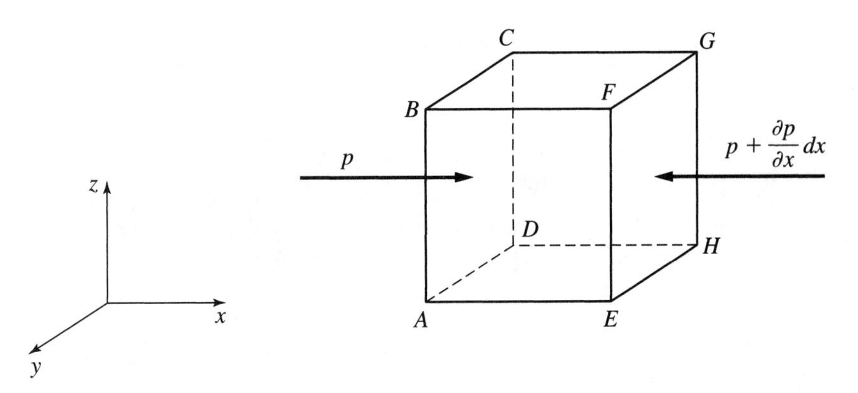

Figure 3.3 Pressure forces in a unit of volume.

ABCD is $p\,dy\,dz$. The pressure force acting on the side *EFGH* is $\left(p + \frac{\partial p}{\partial x} dx\right) dy\,dz$. The pressure force in the x direction is, therefore,

$$p\,dy\,dz - \left(p + \frac{\partial p}{\partial x} dx\right) dy\,dz = -\frac{\partial p}{\partial x} dx\,dy\,dz \tag{3.12}$$

Equation (3.12) gives the difference of pressure forces acting in opposite directions. Also, for the y and z directions the difference in pressure forces is $-\left(\frac{\partial p}{\partial y}\right) dx\,dy\,dz$ and $-\left(\frac{\partial p}{\partial z}\right) dx\,dy\,dz$, respectively. Hence, the rate of change of pressure force per unit volume is written vectorially as $-\nabla(p)$. It is worth noting that the motion of fluid particles does not depend on the absolute value of p but only on the gradient of p. This fact is of great value in hydraulic scale models. Pressure is usually measured relative to an absolute zero value called absolute pressure or relative to the atmospheric pressure at the location of measurement called gage pressure. Thus, gage pressure equals the absolute pressure minus the atmospheric pressure. The standard atmospheric pressure is defined as 14.696 (abs) lb/in^2 and 101.33 kPa (abs).

Example 3.2

Compute the total pressure force on the walls of the tank in Example 3.1.

Solution: The pressure force will vary from zero at the top of water surface to a maximum of ρgh at the bottom, where h is the depth of water. The force on a section with width w (perimeter of the cylinder) will be given as

$$F = 0.5\,\rho gh \times wh = 0.5\,w\rho gh^2, \text{ where } w = 2\pi r$$

$$\begin{aligned}\text{Total force on the wall} &= (2\pi r/2)\rho gh^2 \\ &= \pi(6.5) \times (1000) \times (9.81) \times (5)^2\ \text{m(kg/m}^3\text{)(m/s}^2\text{)m}^2 \\ &= 5 \times 10^6\ \text{N}\end{aligned}$$

3.2.5 Pressure and Gravity Forces

The pressure and gravity forces can be combined and their total per unit volume is

$$\text{grad}(p) + \text{grad}(\rho gz) = \text{grad}(p + \rho gz) \tag{3.13}$$

where grad denotes the gradient. In hydrostatics, the sum of these two forces is constant. Dividing by ρg, equation (3.13) can also be expressed in terms of head with the dimensions of length, L, as

$$\text{grad}\left(\frac{p}{\rho g}\right) + \text{grad } z = \text{grad}\left(\frac{p}{\rho g} + z\right)$$

where the term $p/(\rho g)$ is the pressure head, the term z is the gravitational head, and the term $(p + \rho g z)/(\rho g)$ is the piezometric head. Noting that $\gamma = \rho g$, the piezometric head can also be expressed as $\left(\frac{p}{\gamma} + z\right)$.

3.2.6 Viscous Forces

The forces exerted between fluid particles are pressure forces due to molecular attraction and friction forces due to viscosity. The friction force per unit area in a given direction is the shear stress, τ, and is assumed to be zero for ideal or perfect fluid. Otherwise, the shear stress, resulting from fluid viscosity, is assumed to be proportional to the coefficient of dynamic viscosity, μ, and to the rate of angular deformation:

$$\tau = \mu \frac{dv}{dy} \tag{3.14}$$

Equation (3.14) is Newton's law of viscosity. Transfer of molecular momentum and viscosity causes the shear stress; it is a scalar. The set of shear stresses at a point together with the normal stresses at the point constitute a tensor.

Consider a two-dimensional element of the fluid given in Figure 3.4. The friction force on side AB of length dx is

$$\tau dx = \mu\left(\frac{\partial v}{\partial y}\right) dx \tag{3.15}$$

The velocity at point C is $\left(v + \frac{\partial v}{\partial y} dy\right)$. Thus, the friction force on the side DC is

$$\left(\tau + \frac{\partial \tau}{\partial y} dy\right) dx = \mu \frac{\partial}{\partial y}\left(v + \frac{\partial v}{\partial y} dy\right) dx = \mu \frac{\partial v}{\partial y} dx + \mu \frac{\partial^2 v}{\partial y^2} dy\, dx \tag{3.16}$$

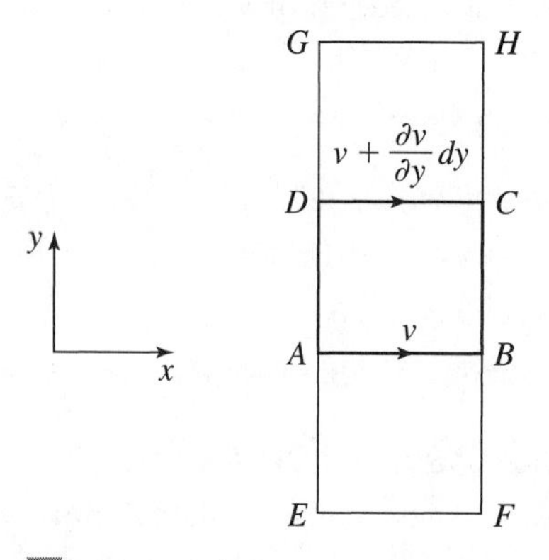

Figure 3.4 A two-dimensional fluid element.

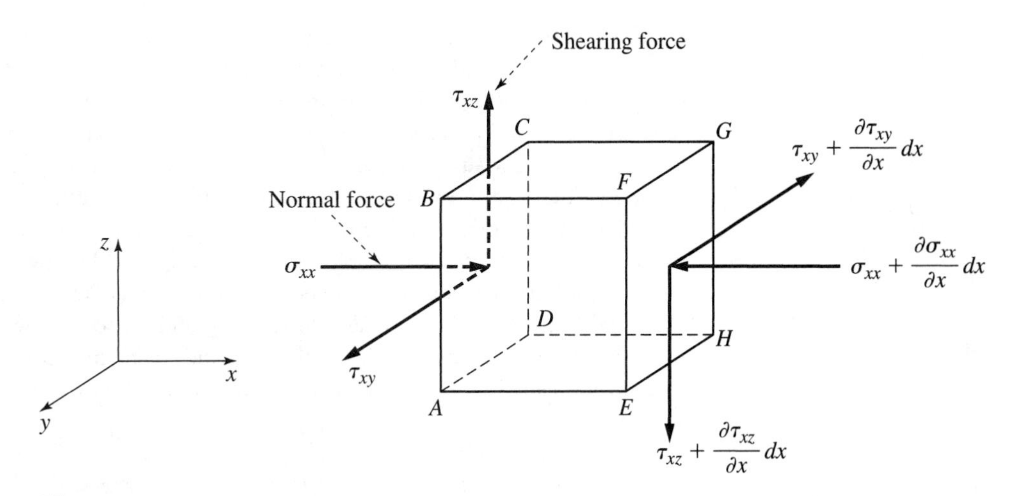

Figure 3.5 Surface forces.

The viscous forces on *AB* and *DC* act in opposite directions. Therefore, the net shearing force is

$$\frac{\partial \tau}{\partial y} dx\, dy = \mu \frac{\partial^2 v}{\partial y^2} dx\, dy \tag{3.17}$$

The friction force per unit area is, thus, obtained by dividing equation (3.17) by the area $dx\,dy$:

$$\frac{\partial \tau}{\partial y} = \mu \frac{\partial^2 v}{\partial y^2} \tag{3.18}$$

In a similar manner, the components of the frictional force per unit volume for an incompressible fluid in the *x*, *y*, and *z* directions can be obtained. In other words, equations similar to equation (3.18) can be derived for a three dimensional fluid element as shown in Figure 3.5.

Example 3.3

A wooden box with a base of 1 m by 0.5 m is being dragged on a smooth oily floor. If the box is separated by a thin oil film of 0.2 mm, what is the force needed to maintain a uniform speed of 2.5 m/s. The viscosity of the oil is 0.15 N-s/m^2.

Solution: Assuming a linear velocity distribution in the oil, the force per unit area is given by equation (3.14), where $\mu = 0.15$ N-s/m^2, $dv/dy = 2.5/0.2 \times 10^{-3}$.

$$\tau = 0.15 \times 2.5/(0.2 \times 10^{-3}) \text{ N-s/m}^2 \text{ m/s/m} = 1875 \text{ N/m}^2$$

Total force required to pull the box $= 1875 \times 1 \times 0.5 = 937.5$ N

3.3 Motions

Fluid particles are, in general, under three types of primary motions and deformations: (a) translation, with velocity components $V(u, v, w)$; (b) dilatation, the variation of velocity components in their own direction; and (c) rotation and angular deformation, the variation of velocity components with respect to a direction normal to their own direction. Mathematically, the motion of the fluid particles along their paths is obtained by superposition of these primary motions (Singh, 1996).

To each kind of motion of fluid particles, there is an inertial force. The relation between the kind of motion and the corrsponding inertial force is determined by equation (3.1). The body forces, including gravity and pressure forces, exist whether the fluid is in motion or at rest. The fluid motion does not depend on the absolute value of the pressure force but only on its gradient. The surface forces, comprising pressure and friction forces, apply to any kind of motion (i.e., viscous or inviscid, laminar or turbulent, compressible or incompressible).

Translatory motion implies no spatial dependence of the velocity components and can be along a straight line or a curved line. An example is the flow of particles along parallel and straight lines with a constant velocity (uniform flow). When a rectangular fluid element with edges, dx, dy, and dz, parallel to the x, y, and z axes is considered, under translatory motion the edges remain parallel to the axes and maintain a constant length. The amount of translation in time dt is $u\,dt$ in the x direction, $v\,dt$ in the y direction, and $w\,dt$ in the z direction.

Fluid particles may be deformed, and each particle may have a particular motion that may be markedly different from the motion of other particles. The forces exerted between fluid particles are the pressure and frictional forces. Deformation of fluid particles can be distinguished as dilatational motion (or linear) and angular. In the case of linear deformation, the velocity components change along their direction. For example, in a converging flow the velocity tends to increase along the path of particles, whereas the opposite is true in a diverging flow. The velocities of dilatational deformation per unit length are $\partial u/\partial x$, $\partial v/\partial y$, $\partial w/\partial z$. These derivatives of velocity do not depend on time. The derivatives of velocity with respect to directions other than their own are not considered. In a two-dimensional case (with $x + dx$ and $y + dy$ as initial coordinates), the sum $\partial u/\partial x + \partial v/\partial y$ gives the total rate of dilatational deformation, which is the rate of change of area per unit area. The amount of linear deformation is $(\partial u/\partial x)dx\,dt$ in the x direction, $(\partial v/\partial y)dy\,dt$ in the y direction, and $(\partial w/\partial z)\,dz\,dt$ in the z direction.

In the case of an angular deformation or shear strain, the angular velocities for a two-dimensional flow are $\partial u/\partial y$ and $\partial v/\partial x$, and these derivatives of velocity do not depend on time. The sum of angular velocities $(\partial u/\partial y) + (\partial v/\partial x)$ defines the rate of angular deformation if the two deformations exist simultaneously. When a square fluid element is considered, the bisectors of the angles made by the edges of the square fluid element tend to remain parallel to their position. In other words, $\partial u/\partial y = \partial v/\partial x$, the bisectors do not rotate, and there is angular deformation. When this is not true, the motion is rotational. An example is flow around a bend in a pipe. In the absence of friction (assumed), the velocity tends to be greater on the inside than on the outside of the bend. For deformation without rotation, $(\partial u/\partial x) + (\partial v/\partial y) \neq 0$ and $(\partial u/\partial y) - (\partial v/\partial x) = 0$ (Singh, 1996).

To understand rotation, a two-dimensional motion is considered, again. The angular velocities of deformation are $\partial u/\partial y$ and $\partial v/\partial x$. The rotation of a particle is proportional to the difference between these angular velocities, for $(\partial u/\partial y) - (\partial v/\partial x) \neq 0$. The bisectors change their direction and no longer remain parallel to their initial position. In that case, there is rotation with angular deformation or rotation only. In the case of rotation only (without deformation), $(\partial u/\partial x) + (\partial v/\partial y) = 0$ and $(\partial u/\partial y) - (\partial v/\partial x) \neq 0$. For both rotation and deformation, $(\partial u/\partial y) - (\partial v/\partial x) \neq 0$ and $(\partial u/\partial x) + (\partial v/\partial y) \neq 0$.

In a two-dimensional flow, the rate of angular deformation can be expressed as $[(\partial u/\partial y) + (\partial v/\partial x)]/2$ in the x and y directions, and the amount of angular deformation as the (rate) times $(dy\,dt)$ in the x direction, and the (rate) times $(dx\,dt)$ in the y direction.

3.4 Relation between Forces and Motion

A fluid particle can be subject to three types of motions: translation, dilatation, and rotation and angular deformation. Motions are caused by the applied forces which, when equated to inertial forces, yield the momentum equation. For two-dimensional flow, the elementary components of velocity of a fluid particle are summarized as follows:

Translation: u, v

Dilatational deformation: $\frac{\partial u}{\partial x} dx \qquad \frac{\partial v}{\partial y} dy$

Shear deformation: $\frac{1}{2}\left(\frac{\partial u}{\partial y} + \frac{\partial v}{\partial x}\right) dy \qquad \frac{1}{2}\left(\frac{\partial u}{\partial y} + \frac{\partial v}{\partial x}\right) dx$

Rotation: $-\frac{1}{2}\left(\frac{\partial v}{\partial x} - \frac{\partial u}{\partial y}\right) dy \qquad \frac{1}{2}\left(\frac{\partial v}{\partial x} - \frac{\partial u}{\partial y}\right) dx$

To each of these velocity components corresponds a component of acceleration, which when multiplied by ρ, yields a component of inertial force. Inertial forces may be distinguished into two types according to the type of acceleration or elementary motion considered. These are (Le Méhauté, 1976):

1. Local acceleration—corresponding to variation of the velocity of translation or the derivative of the velocity with respect to time. Local acceleration characterizes an unsteady motion (i.e., motion where the velocity at a given point changes with respect to time). Local acceleration results from changes in the translatory motion of a fluid particle imposed by external forces.
2. Convective acceleration—corresponding to the velocity of deformation and rotation or derivative of velocity with respect to space. Convective acceleration characterizes a nonuniform flow (i.e., when the velocity at a given time changes with respect to distance). It is sometimes called field acceleration. Convective acceleration results from any linear or angular deformation, or from a change in the rotation of fluid particles, imposed by external forces.

3.5 Energy

Energy is defined as the capacity for doing work and is, therefore, measured in the same units as work. Work is defined as the product of force times the distance. Thus, energy is measured in terms of N-m in the SI system and in terms of lb-ft in the FPS system. In hydraulic systems, energy may exist in different forms according to the applied forces. The dynamic terms encountered in any problem represent the type of energy (or force) that may have an internal effect on the process (Morris and Wiggert, 1972). Frequently, energy in hydraulics is expressed in terms of head or the equivalent depth of water, the amount of energy per unit weight of the fluid. In that case, the energy head is expressed in terms of m or ft.

3.5.1 Kinetic Energy

The kinetic energy is always present in a moving fluid, and is specified by the mass density, ρ, of the fluid. Mass can be considered as that form of energy locked up in a matter of the fluid itself, and it is characterized by the property of inertia. The inertial force by Newton's law is the force required to produce the acceleration a on the mass M and is expressed as $\mathbf{F} = M\mathbf{a}$. The internal force per unit area is, thus,

$$\frac{Ma}{A} \propto L^3 \frac{\rho}{L^2}\left(\frac{V}{2T}\right) = \frac{1}{2}\rho\left(\frac{L}{T}\right)V = \frac{1}{2}\rho V^2 \tag{3.19a}$$

where L and T denote length and time dimensions, respectively, and area and velocity being represented by these dimensions. T is the time required for the force to accelerate the flow from rest to average velocity V. The term $\frac{1}{2}\rho V^2$ is the kinetic energy per unit volume or $MV^2/2$. All of these relationships are implicit when mass density, ρ, is included in the analysis. Thus, the kinetic energy is related to the inertial force. By dividing equation (3.19a) by the specific weight γ, the kinetic energy head (KE Head) can be expressed as

$$\text{KE Head} = \frac{V^2}{2g} \tag{3.19b}$$

3.5.2 Potential Energy

The potential energy is comprised of two types: (1) the elevation energy or position energy and (2) pressure energy. The potential energy per unit volume is equal to the force per unit area resulting from fluid pressure and position. Of course, if the total potential energy is the same throughout the length of a hydraulic conduit, then the fluid is, by the hydrostatic equation, at rest. For a moving fluid, a difference in the potential energy must exist in the direction of flow. For flow in open channels, the pressure at the outer surface would be atmospheric throughout the channel length. The potential energy head (PE Head) is expressed as the sum of the elevation and the pressure head:

$$\text{PE Head} = h + \frac{p}{\gamma} \tag{3.20}$$

where h is the elevation above a specified datum, p is the pressure and γ is the specific weight. The difference in the potential energy head is expressed by the term $\Delta(p + \gamma h)$.

3.5.3 Viscous Energy (or Shear or Friction)

Flow in hydraulic systems is always resisted by internal shearing stresses between fluid particles and between fluid particles and flow boundaries. These frictional effects must be overcome by work done by the fluid's mechanical energy, and this results in the conversion of a part of this energy into heat energy which is then conducted through the boundary walls into the environment. The inability to reclaim this heat energy for reconversion to mechanical energy in the fluid results from the second law of thermodynamics and it is therefore treated as "friction loss." The rate of this energy loss depends on the molecular structure of the particular fluid and is specified in terms of the viscosity, μ. The viscous energy per unit volume is expressed as τ and the viscous energy head (VE Head) can be expressed as

$$\text{VE Head} = \frac{\tau}{\gamma} \tag{3.21}$$

3.5.4 Gravitational Energy (or Weight)

All masses respond to a gravitational field in such a way that the force of gravity or weight is equal to Mg, where g is the acceleration of gravity. This effect can be specified in any hydraulic system by introducing the fluid specific weight, γ. The gravitational energy is the product of γ and the distance the fluid is to be raised, say h. In terms of head, this will amount to h itself.

3.5.5 Surface Energy (Tension)

Surface tension results from the difference in mutual attraction between particles near a surface as compared with those farther in the liquid. The surface tension energy can be understood as the work required to form a free surface (or more appropriate, an interface between two fluids) against the tendency for mixing or dispersal. This work brings the molecules to the surface and requires an expenditure of energy. This also is a molecular phenomenon. The surface energy per unit area of the interface is equivalent to the surface tensile force per unit length along any line in the interface and is called the coefficient of surface tension, specified in terms of the unit surface tension, σ. Thus, σ has dimensions of energy per unit area or force per unit length (N/m or FL^{-1}). The value of surface tension depends on the temperature as well as other fluid the liquid is in contact with.

A common phenomenon associated with surface tension is the rise or fall of a liquid in a capillary tube. If a small open tube is inserted into water, the water level in the tube will rise above the water level outside the tube. This occurs because there is an attraction (adhesion) between water molecules and the wall of the tube, which is strong enough to overcome the mutual attraction (cohesion) between the water molecules. In this case the liquid is said to wet the wall surface. The height, h, to which the water will rise is determined by the value of σ, the tube radius, r, the specific weight of the liquid γ, and the angle of contact, θ, between the liquid and the tube. The weight of the liquid column in the tube is $\gamma\pi r^2 h$. The vertical force due to surface tension is $2\pi r\sigma \cos\theta$. Since these two forces are in equilibrium, one obtains

$$h = \frac{2\sigma \cos\theta}{\gamma r} \tag{3.22}$$

Equation (3.22) shows that when $\theta > \pi/2$, the liquid is nonwetting and when $\theta < \pi/2$, the liquid is wetting. Mercury, for example, has a contact angle of 130 to 150°. Surface tension plays a role in the movement of liquids in soils and other porous media, flow of thin films, formation of drops and bubbles, breakup of liquid jets, and so on.

3.5.6 Elastic Energy (or Compression)

Fluids do not resist tension, but they do respond to compressive stresses by a reduction in volume and a corresponding storage of elastic energy, which can, of course, accomplish work when released. This effect is measured by the bulk modulus of elasticity of the fluid, defined as the ratio of an increment of pressure to the corresponding unit change in density, and shows how compressible the fluid is. Thus, the compressibility is defined in terms of the bulk modulus of elasticity and has the dimensions of pressure or force per unit area. The elastic (or compression) energy can be expressed as the compressibility times the distance.

Example 3.4

Water is flowing in a pipe with a mean uniform velocity of 2.5 ft/s. The diameter of the pipe is 3 ft. Calculate the kinetic energy per unit reach of the pipe.

Solution: Cross-sectional area of the pipe $= \pi r^2 = \pi \times 1.5^2 = 7.069$ ft^2

Kinetic energy $= 0.5\, mV^2$

$m = 62.4 \times A \times 1 = 441.1$ slugs

Kinetic energy $= 0.5 \times 441.1 \times 2.5 \times 2.5 = 1378.4$ slugs-ft^2/s^2

Example 3.5

A constant depth of 12 ft is maintained in a storage tank through an overflow orifice. An outlet valve is provided at a level 2 ft above the base of the tank. If the friction losses are neglected, what is the velocity of water at the outlet?

Solution: In the absence of friction losses, the potential energy is equal to the kinetic energy. Thus,

$$mgh = 0.5mV^2$$

This yields $V = (2gh)^{1/2} = (2 \times 32.2 \times 10)^{1/2} = 25.38$ ft/s.

3.6 Relation between Force and Energy

It is clear from the foregoing discussion that each type of energy uniquely corresponds to a particular type of force. It is worth recalling that the energy per unit volume is equivalent to the force per unit area, or energy per unit area is equal to force per unit length. This correspondence is as follows:

Energy	Force
Kinetic energy	Inertial force
Potential energy	Pressure and position
Gravitational energy	Weight
Viscous energy	Shear force
Surface energy	Surface tension
Elastic energy	Compression

3.7 Mass, Momentum, and Energy Fluxes

Flux of a quantity is normally defined as the quantity per unit area. For example, flux of water flow is the volumetric rate of flow per unit cross-sectional area. In hydraulics, flux is also defined as a quantity per unit time. For example, the mass per unit time is the mass flux, momentum per unit time is momentum flux, and energy per unit time is energy flux. It may then be noted that the momentum flux has the units of force and the energy flux the units of power. Consider flow in an open channel with cross-sectionally averaged velocity V, discharge Q (volume of flow per unit time), and cross-section area A. Then the mass flux at a point on the channel is

$$\text{Mass flux} = \text{Mass/time} = \rho KAV = \rho KQ$$

where K is the mass distribution factor.

$$\text{Momentum flux} = \text{Momentum/time} = \rho\beta AV^2 = \rho\beta QV$$

where β is the momentum distribution factor.

$$\text{Kinetic energy flux} = \text{Kinetic energy/time} = \rho\alpha AV^3 = \rho\alpha QV^2$$

where α is the energy distribution factor.

The concept of flux is used in derivation of hydraulic equations presented in Chapter 5.

3.8 Significance of Relative Magnitudes of Forces or Energy

In open channel flow, the main forces are gravity, inertial, pressure, and viscous forces. The surface tension force may influence the behavior of the flow under certain circumstances. However, in most open channel flow problems encountered in practice, the surface tension force does not play a significant role and is normally ignored. The flow in a conduit is governed by the relative magnitudes of forces rather than actual force magnitudes. Indeed the same holds for geometric parameters, that is, the relative geometric sizes affect flow patterns, not individual sizes. This concept forms the basis of hydraulic similitude widely used in experimental hydraulics. Since the inertial force is always present in a flowing fluid, it is appropriate to express ratios of the inertial force to other forces. These ratios shed considerable light on the flow behavior and give rise to dimensionless numbers commonly used in hydraulics. The unit inertial force can be represented as ρV^2. Although each ratio involves forces, it can be modified to kinematic parameters (ratio of two velocities) by dividing the numerator and the denominator by the mass density and then taking the square root. In this manner, the denominator in each ratio represents a velocity which can be a measure of the force it is associated with. This will be clear in what follows.

3.8.1 Effect of Viscosity

The effect of viscosity on the flow behavior is represented by the ratio of inertial forces to viscous forces. This ratio is given by the Reynolds number defined as

$$R_n = \frac{V^2 D \rho}{\mu v} = \frac{DV}{\mu/\rho} = \frac{DV}{\nu} \tag{3.23}$$

where ρ is the density of water, V is the mean velocity of flow, D is a characteristic length (e.g., hydraulic radius, R, of the conduit $= \dfrac{\text{Area } (A)}{\text{wetted perimeter } (P)}$), μ is the dynamic viscosity, and ν is the kinematic viscosity, which is equal to μ/ρ. When used in conjunction with equation (3.14), equation (3.23) leads to an expression of shear velocity v^* defined as

$$v^* = \sqrt{\frac{\mu V}{\nu D}} = \sqrt{\tau_0/\rho} \tag{3.24}$$

The flow is considered laminar if the viscous forces dominate the flow behavior. In such a flow, the water particles move in streamlines as if infinitesimally thin layers of fluid slide over each other. If the water particles are moving in a random manner and do not follow certain smooth paths, then the flow is turbulent. In turbulent flows, viscous forces are relatively small when compared with inertial forces. In some cases, both the viscous force and the inertia force significantly contribute to the state of flow. The flow is then transitional or mixed. The transition from laminar flow to turbulent flow in open channels occurs for Reynolds number, R_n, of about 500, in which R_n is based on the hydraulic radius as the characteristic length. As R_n, exceeds 2000, the flow is fully turbulent.

In real practice, laminar flows in open channels are rare. The appearance of smooth and glassy surfaces and streams to an observer is by no means an indication that the flow is laminar. Most probably, the surface velocities are lower than those required to form capillary waves. Laminar conditions are encountered in flows of thin sheets of water over the ground surface or in hydraulic models. It should be noted that when modeling open channel flows in laboratories the state of flow (laminar, transition, or turbulent) should be maintained as in the prototype. Otherwise, the dynamic similarity between the prototype and its model will not be satisfied and incorrect results will be obtained.

Example 3.6

A triangular channel with 2 m of water depth and a 1:1 side slope, carries a discharge of 6 m^3/s. If the kinematic viscosity of the water is 1.003×10^{-6} m^2/s, determine whether the flow is laminar or turbulent.

Solution: $A = ty^2 = 4.0$ m^2, $P = 5.657$ m, $R = A/P = 0.707$ m, and $v = Q/A = 1.5$ m/s

$$R_n = vR/\nu = \frac{1.5 \times 0.707}{1.003 \times 10^{-6}} = 1.057 \times 10^6$$

The flow is, thus, turbulent.

3.8.2 Effect of Gravity

The effect of gravity on the flow behavior is represented by the ratio of inertial forces to gravity forces. This ratio is given by the Froude number, F_n, defined as

$$F_n^2 = \frac{\rho V^2}{\gamma D} \quad \text{or} \quad F_n = \frac{V}{\sqrt{\gamma D/\rho}} = \frac{V}{\sqrt{gD}} \tag{3.25}$$

where V is the mean velocity of flow, g is the acceleration due to gravity, and D is a characteristic length (e.g., hydraulic depth, D, of the conduit $= \frac{\text{Area } (A)}{\text{Top width } (B)}$). For rectangular sections the hydraulic depth is equal to the depth of flow. If the Froude number, F_n, is equal to unity, i.e., the flow velocity is equal to the velocity of gravity wave having a small amplitude, then equation (3.25) can be written as

$$V = \sqrt{gD} \tag{3.26}$$

and the flow is said to be in a critical state. The critical velocity, V_c, is then equal to the square root of (gD). If the flow velocity is less than the critical velocity (i.e., F_n is less than unity), then the flow is called subcritical flow (or streaming or tranquil). If the flow velocity is greater than the critical velocity, i.e., F_n is greater than unity, then the flow is called supercritical flow (or rapid or shooting). Thus, according to Froude number, F_n, the flow is classified as subcritical if $F_n < 1$; critical if $F_n = 1$; and supercritical if $F_n > 1$. Since open channel flow is mainly controlled by the gravity effect, hydraulic models for open channels should be designed under the condition of the same gravity effect. The Froude number of the model must be equal to that of the prototype.

Example 3.7

A trapezoidal channel, with a 3 ft bed width, 1 ft water depth, and 1:1 side slope, carries a discharge of 50 ft^3/s. Determine whether the flow is supercritical or subcritical.

Solution:

$A = by + ty^2 = 4.0$ ft^2

$B = b + 2ty = 5.0$ ft

$D = A/B = 0.80$ ft

$V = Q/A = 50/4 = 12.5$ ft/s

$F_n = 12.5/(32.2 \times 0.8)^{0.5} = 2.46$

The flow is, thus, supercritical.

Example 3.8

The flow rate in a 4.0 m wide and 1.0 m deep rectangular channel is 10.0 m^3/s. The kinematic viscosity of the flowing water is 1.10×10^{-6} m^2/s. Calculate the Froude and Reynolds numbers. What is the flow regime?

Solution: $b = 4.0$ m, $y = 1.0$ m, $Q = 10.0$ m^3/s, $v = 1.10 \times 10^{-6}$ m^2/s. Therefore,

$$A = by = 4.0 \text{ m}^2$$

$$P = b + 2y = 4.0 + 2.0 = 6.0 \text{ m}$$

$$R = A/P = 4.0/6.0 = 0.667 \text{ m}$$

$$V = Q/A = 10.0/4.0 = 2.5 \text{ m/s}$$

$$F_n = \frac{2.5}{\sqrt{9.81 \times 1}} = 0.798$$

which is less than 1 and the flow is, therefore, subcritical.

$$R_n = \frac{VR}{v} = \frac{2.5 \times 0.667}{1.1 \times 10^{-6}} = 1.516 \times 10^6$$

which shows that the flow is turbulent. Thus, the flow is subcritical and turbulent.

3.8.3 Effect of Pressure and Position Force

The effect of pressure force and elevation can be evaluated by taking the ratio of these forces to inertial forces. This ratio is given by the Euler number and is defined as

$$E_n = \frac{\Delta(p + \gamma h)}{\rho V^2/2} \tag{3.27}$$

When the numerator in equation (3.27) is divided by ρ and then the square root is taken, the result is the efflux velocity which one gets from a pressurized chamber where the entire potential energy in the escaping liquid is converted to the kinetic energy in the jet. The Euler number is essentially the ratio of the potential energy difference to the kinetic energy. The potential energy difference may occur due to any or all other forms of energy in the hydraulic system. That is, the fluid pressure may result from gravity, elasticity, surface tension, viscosity or a combination of these energy forms. Since the Euler number is always present in the fluid, it is taken as a dependent parameter in derivation of the flow process equation.

3.8.4 Effect of Surface Tension

The effect of surface tension on the flow can be evaluated by the ratio of the inertial force to the surface tension force which is given by the Weber number, expressed as

$$W_n = \frac{\rho v^2}{\sigma/D} = \left[\frac{v}{\sqrt{\sigma/\rho D}}\right]^2 = \frac{\rho D V^2}{\sigma} \tag{3.28}$$

The term $(\sigma/\rho D)^{0.5}$ represents the velocity of a capillary wave, in which the dimension D can be shown to be proportional to the wave length, λ. In open channel flows, this ratio is seldom used.

3.8.5 Effect of Elastic (or Compressive) Force

The effect of compression energy or force on the flow behavior can be evaluated by taking the ratio of the inertial force to the compression force, E. This ratio gives the Cauchy number expressed as

$$C_n = \left[\frac{\rho V^2}{E}\right] \tag{3.29}$$

The square root of the above ratio defines the Mach number, expressed as

$$M_n = \left[\frac{\rho V^2}{E}\right]^{0.5} = \frac{V}{\sqrt{E/\rho}} = \frac{V}{c} \tag{3.30}$$

where c is the celerity denoting the velocity of a pressure wave (e.g., sound) in the fluid. In hydraulic machinery or water hammer in pipes, this is important, but in routine open channel flows it is hardly used.

3.9 Regimes of Flow

It has been shown that according to the Reynolds number, the flow is classified into laminar, transient and turbulent flows. In other words, this classification is based on the relative effects of inertial and viscous forces. According to the Froude number, the flow is classified into subcritical, critical and supercritical flows. This classification is based on the relative effects of gravity forces. When the effects of inertial, viscous and gravity forces are combined, the flow in open channels can be classified into four regimes as shown in Figure 3.6:

1. Subcritical-laminar regime, where F_n is less than unity and R_n is less than 500
2. Subcritical-turbulent regime, where F_n is less than unity and R_n is greater than 2000
3. Supercritical-laminar regime, where F_n is greater than unity and R_n is less than 500
4. Supercritical-turbulent regime, where F_n is greater than unity and R_n is greater than 2000

The depth-velocity relationship for different flow regimes in wide open channels is presented in Figure 3.6. The bold line is for $F_n = 1$, whereas the hatched band indicates the transitional zone between laminar and turbulent flows. To the left of the bold line the flow is subcritical (or tranquil) and to the right of it the flow is supercritical (or shooting). On the other hand, below the hatched band the flow is laminar and above it the flow is turbulent. The whole area is divided, therefore, into four portions each of which represents a certain regime. Two other minor regimes may be interpreted from Figure 3.6. These are the subcritical transitional regime and supercritical transitional regime. However, these two minor regimes as well as the subcritical laminar and the supercritical laminar are not commonly encountered in real applications. These flow regimes can only be encountered where the depths of flow are very thin as perhaps in "sheet flow."

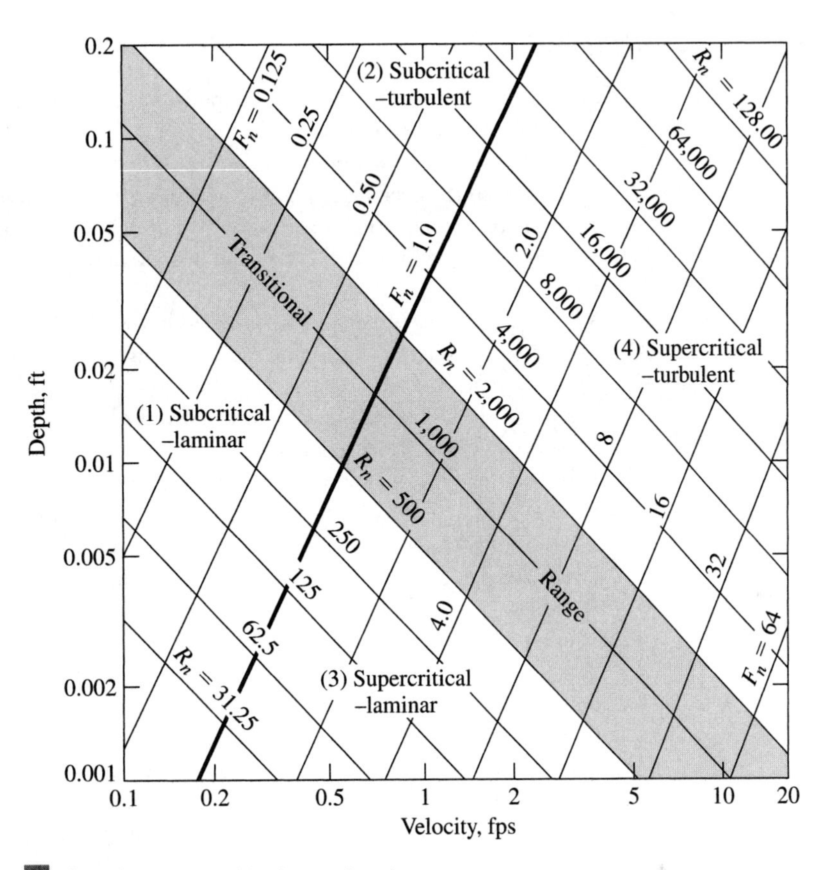

Figure 3.6 Regimes of flow in open channels.

Example 3.9

The flow rate in a 4 ft wide, 2 ft deep rectangular channel is 50 cfs. The kinematic viscosity of the water is 1.217×10^{-5} ft²/s. Calculate the Froude and Reynolds numbers. What is the flow regime?

Solution: $b = 4$ ft, $y = 2$ ft, $Q = 50$ cfs, $\nu = 1.217 \times 10^{-5}$ ft²/s. Therefore,

$$A = by = 8 \text{ ft}^2$$

$$P = b + 2y = 4 + 4 = 8 \text{ ft}$$

$$R = A/P = 8/8 = 1 \text{ ft}$$

$$V = Q/A = 50/8 = 6.25 \text{ ft/s}$$

$$F_n = \frac{V}{\sqrt{gy}} = \frac{6.25}{\sqrt{32.2(2)}} = 0.78, \text{ which is less than 1, so the flow regime is subcritical}$$

$$R_n = \frac{VR}{\nu} = \frac{6.25(1)}{1.217 \times 10^{-5}} = 5.136 \times 10^5, \text{ which shows that the flow is turbulent.}$$

Therefore, the flow regime is subcritical-turbulent.

READING AID

3.1. State the types of constraints that hydraulic systems must satisfy.
3.2. Define the process equation.
3.3. Does the process equation represent an internal or external constraint in hydraulic systems? Can the head loss function be a form of the process equation?
3.4. How the process equation is generally developed? Can it be developed analytically? Why?
3.5. Give an example for a case where the process equation can be regarded as completely rational.
3.6. What are the types of parameters involved in hydraulic systems?
3.7. Define the geometric, kinematic, and dynamic parameters in hydraulic systems. Give three examples of each.
3.8. What are the different types of forces involved in hydraulic systems? How are these forces categorized in hydrodynamics?
3.9. Write down Newton's second law of motion. Express it in a mathematical form.
3.10. Explain how the internal forces in hydraulic systems develop. Why does the sum of internal forces equal zero?
3.11. Give examples of external forces. Name the surface forces that are attributed to pressure and viscosity.
3.12. What is meant by body or volume forces? Are these forces always in balance?
3.13. What are the causes of capillary and geostrophic forces?
3.14. Does the gravity force vary with the state of a fluid whether it is in motion or at rest? Express the gravity force in a vector form.
3.15. Is the pressure force a scalar or a vector? Does the motion of fluid particles depend on the absolute value of the pressure or on its gradient?
3.16. Write an equation for the shear stress resulting from fluid viscosity (Newton's law of viscosity).
3.17. What are the three types of primary motions and deformations for fluid particles? Define each.
3.18. Under translatory motion the edges of a rectangular fluid element remain parallel to the axes and maintain a constant length. Elaborate whether this statement is true or not.
3.19. In a converging flow field, does the particle velocity tend to increase or decrease along its path?
3.20. Give an example of flow with an angular deformation or shear strain.
3.21. Define the relation between the rotation of a particle and the angular velocities of deformations.
3.22. What are the two types of acceleration that distinguish internal forces?
3.23. What are the different types of energy in hydraulic systems as classified according to the applied forces?
3.24. Define the bulk modulus of elasticity.
3.25. Define the relation between force and energy.
3.26. Is the flow expected to be subcritical or supercritical in the following situations? (1) Flow just before water control structures, and (2) flow just after sluice gates of a small opening.
3.27. Is the flow expected to be uniform or nonuniform for frictionless open channels? Why?
3.28. What criteria are employed to classify flows in open channels into laminar, turbulent, subcritical, critical, and supercritical?
3.29. What are the different types of flow regimes in open channels?

Problems

3.1. A 3 m × 4 m rectangular tank is filled with water up to a depth of 5 m. Compute the force on the base of the tank.

3.2. Compute the total pressure force on the walls of the tank in Problem 3.1.

3.3. A box with a base of 0.8 m × 0.6 m is being dragged on smooth oily floor. If the box is separated by a thin oil film of 0.25 mm, what is the force required to maintain a uniform speed of 2.75 m/s. The viscosity of the oil is 0.20 N-s/m^2.

3.4. If the box in Problem 3.3 is being dragged with a force of 1440 N on a 0.2 mm oil film of the same type, what would be its velocity?

3.5. Water is flowing in a 1.0 ft diameter pipe with a mean velocity of 8.5 ft/s. Calculate the kinetic energy per unit length of the pipe.

3.6. A constant depth of 6 ft is maintained in a storage tank through an overflow orifice. An outlet valve is provided at 1.0 ft above the base of the tank. If the friction loss is neglected, what is the velocity of water at the outlet?

3.7. A 3 m wide and 2 m deep rectangular channel carries a discharge of 9 m^3/s. If the kinematic viscosity of the water is 1.12×10^{-6} m^2/s, determine whether the flow is laminar or turbulent.

3.8. A trapezoidal channel with a 6 ft bed width, 3 ft water depth and 1:1 side slope, carries a discharge of 250 ft^3/s. Determine whether the flow is supercritical or subcritical.

3.9. In the following situations, is the flow laminar or turbulent? (1) Flow in a narrow rectangular channel with a flow velocity of 2.0 m/s and a depth of flow of 1.0 m. Assume the hydraulic radius as $y/2$. (2) Sheet flow with a flow velocity of 0.08 m/s and a depth of flow of 2.0 mm. The hydraulic radius is equal to y in this case.

3.10. The flow rate in a 10.0 ft wide and 3 ft deep rectangular channel is 225.0 ft^3/s. Is this flow subcritical or supercritical?

3.11. Is the flow laminar or turbulent in Problem 3.10? Take the kinematic viscosity, ν as 1.217×10^{-5} ft^2/s.

3.12. The flow rate in an 8.0 m wide and 0.5 m deep rectangular channel is 2.0 m^3/s. What is the regime of the flow?

3.13. The discharge in a trapezoidal channel is 200 ft^3/s. The channel has a bottom width of 5.0 ft and a side slope of 3:2. The depth of flow is 4 ft. Determine the Froude number and Reynolds number (take the kinematic viscosity, ν, as 1.059×10^{-5} ft^2/s). What is the regime of the flow?

3.14. A triangular channel with side slopes of 1:1 and a depth of flow of 2.0 m, issues a discharge of 20.0 m^3/s. What is the regime of flow?

3.15. A rectangular channel 4.0 ft wide carries 100 ft^3/s at a depth of 1.5 ft. Is the flow subcritical or supercritical? For the same discharge, what depth will give critical flow?

3.16. Water is flowing at a rate of 200 ft^3/s in a triangular channel with a 6 ft water depth and 1:1 side slope. If the kinematic viscosity of water is 1.1×10^{-5} ft^2/s, calculate the Froude and Reynolds numbers and determine the flow regime.

3.17. An overhead tank of 0.5 m height is at an elevation of 15 m from the ground. Determine the maximum velocity of the jet from an outlet of this tank 2 m above the ground.

3.18. A cubic tank of 64 m^3 volume is supported by 4 rectangular pillars, each pillar having a base area of 0.1 m^2. If the tank is completely filled with water, calculate the force exerted by each pillar on the ground. Neglect the weight of the tank materials and the pillars.

3.19. An earth dam holds river water up to a height of 120 m. Determine the maximum kinetic energy per unit volume available to the water jet for running a turbine if 5 m head is required to exit water from the turbine housing. Also find the velocity of the jet.

3.20. The pressure inside a pipeline at a certain point, 8.0 m above the ground surface, is 90 kPa when there is no flow (zero velocity). Determine the possible velocity of the water jet at ground level.

3.21. A block of mass 6 kg is lying on an inclined plane covered with a film of oil (μ = 0.29 kg/m-s) of thickness

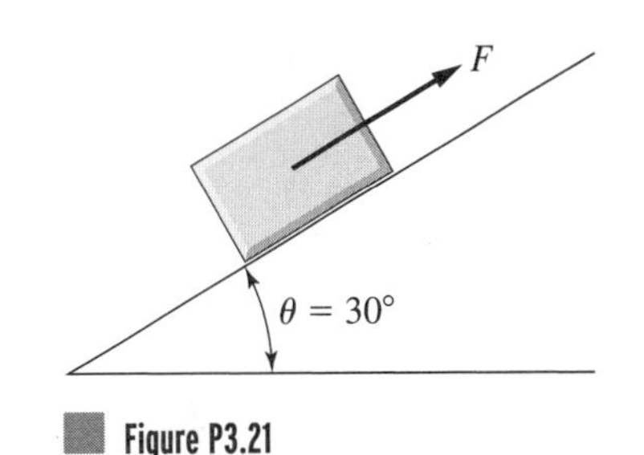

Figure P3.21

0.1 mm, as shown in Figure P3.21. Surface area of the block is 45 cm^2. Determine the force, *N*, required to move this block upward with a velocity of 2 m/s.

3.22. Determine the velocity of the block in Problem 3.21 if the force is applied downwards with a magnitude of 5 *N*.

3.23. A water jet with a velocity of 8 m/s is coming out of a nozzle facing vertically upward. Determine the height up to which this jet will rise.

3.24. A pressure gage attached to a straight horizontal pipe is showing a reading of 248.4 kPa. The flow rate in the pipe is determined using a flow meter and was found to be 35 liter/s. If the diameter of the pipe is 8 cm, determine the Euler number for this pipe flow.

3.25. Water is seeping though a porous dam at the rate of 1.5 m/day. If the average diameter of the dam pores is 0.06 mm, determine the Weber number for the flow.

3.26. A large wind tunnel fan is designed for high speed flows at the room temperature of 20°C. The fan rotating at 6000 rpm. If the blade size from the center of rotation is 1.5 m, determine the Mach number for the velocity at the tip of the fan blade. Also determine the corresponding Cauchy number. *Hint:* Speed of sound in air = 20.04 $(T)^{0.5}$ m/s, where T is temperature in Kelvin.

References

Le Méhauté, B., 1976. *An Introduction to Hydrodynamics and Water Waves.* Springer-Verlag. New York.

Morris, H. M. and Wiggert, J. M., 1972. *Applied Hydraulics in Engineering.* Wiley, New York.

Singh V.P., 1996. *Kinematic Wave Modeling in Water Resources: Surface-Water Hydrology.* Wiley, New York.

CHAPTER 4

HYDROSTATICS

Norris Dam and Reservoir, Clinch River, USA. (Courtesy of John Sohlen/Visuals Unlimited.)

Hydrostatics deals with fluids either at rest or in motion in such a way that there is no relative motion between adjacent particles. Shearing stresses between fluid particles are thus absent and the pressure forces are the only forces acting on the surfaces of the fluid particles. The term pressure *is used to indicate the normal force per unit area at a given point on a given plane within the fluid. The focus of this chapter is to investigate pressure and its variation throughout a fluid and to elaborate the effect of pressure on submerged surfaces. The buoyancy and stability of submerged and floating bodies are discussed. Manometers, as basic pressure measurement devices, are also presented.*

4.1 Pressure at a Point

Keeping the definition of pressure in mind, a question arises as to how the pressure at a point varies with the orientation of the plane passing through the point. To that point, consider an arbitrary triangular wedge-shaped element of fluid as at an arbitrary location within a fluid mass as shown in Figure 4.1. In the absence of shearing stresses, the only external forces acting on the wedge are the pressure and the weight. Although the fluid is at rest, one can also generalize the analysis by considering the fluid element to have accelerated motion as long as it moves as a rigid body, meaning that there is no relative motion between adjacent fluid elements and shearing stresses are absent. In the figure, z is the vertical direction, x is the longitudinal direction, and y is the transverse horizontal direction. The faces of the element are $\delta x\,\delta y$, $\delta y\,\delta z$, $\delta x\,\delta z$, and $\delta y\,\delta s$. The average pressures on these faces are p_z, p_x, p_y, and p_s, respectively. These pressure forces must be multiplied by appropriate areas to get the corresponding forces generated by them. It is assumed that accelerations in the x, y, and z directions, respectively, are a_x, a_y, and a_z. For simplicity, the forces in the y direction are not shown. Applying Newton's second law of motion, the forces in the x direction can be written as

$$\sum F_x = p_x\,\delta y\,\delta z - p_s\,\delta y\,\delta s \sin\theta = \rho\frac{\delta x\,\delta y\,\delta z}{2}a_x \tag{4.1a}$$

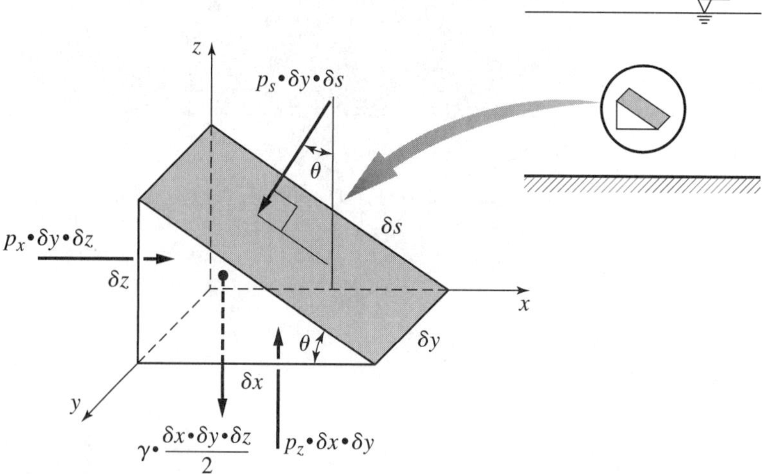

Figure 4.1 Forces on an arbitrary wedge-shaped triangular fluid element.

Similarly, the force in the z direction can be expressed as

$$\sum F_z = p_z\,\delta x\,\delta y - p_s\,\delta y\,\delta s\cos\theta - \gamma\frac{\delta x\,\delta y\,\delta z}{2} = \rho\frac{\delta x\,\delta y\,\delta z}{2}a_z \tag{4.1b}$$

where ρ is the density and γ is the specific weight. Referring to Figure 4.1, $\delta x = \delta s\cos\theta$ and $\delta z = \delta s\sin\theta$. Inserting these quantities in equations (4.1a) and (4.1b), one obtains

$$p_x - p_s = \rho a_x\frac{\delta x}{2} \tag{4.2a}$$

$$p_z - p_s - \frac{\gamma\,\delta z}{2} = \rho a_z\frac{\delta z}{2} \tag{4.2b}$$

In order to get the values of pressure at a point, the faces of the element are allowed to go to zero in the limit while keeping the angle θ unchanged. Thus, equations (4.2a) and (4.2b) yield

$$p_x = p_s \quad\text{and}\quad p_z = p_s \quad\text{or}\quad p_x = p_z = p_s \tag{4.2c}$$

Equation (4.2c) shows that the pressure at a point in fluid at rest or in motion is independent of direction, provided there are no shearing stresses present. This statement is known as Pascal's law, after the French mathematician Blaise Pascal (1623–1662). In the presence of shear stresses, Pascal's law will not be valid, and the pressure at a point is thus defined as the average of three mutually perpendicular normal stresses at that point (Munson et al., 1990). This concept will be clear in the ensuing discussion.

4.2 Pressure Field

The variation of pressure from point to point within the mass of a fluid needs to be considered under the condition that there are no shearing stresses present. To that end, consider a small rectangular element removed from an arbitrary position within the fluid mass, as shown in Figure 4.2. As before, the forces acting on this element are surface forces due to pressure and a body force equal to the weight of the element. Denoting the pressure at the center of the element by p, the average pressure on the faces of the element (a short distance away) can then be expressed using a Taylor series expression and neglecting higher order terms. This leads to the expression in terms of p and its derivatives. For simplicity, the surface forces in the y direction are not shown. Using the same notation as before, the resultant surface force in the x direction can be expressed as

$$\delta F_x = \left(p - \frac{\partial p}{\partial x}\frac{\delta x}{2}\right)\delta y\,\delta z - \left(p + \frac{\partial p}{\partial x}\frac{\delta x}{2}\right)\delta y\,\delta z = -\frac{\partial p}{\partial x}\delta x\,\delta y\,\delta z \tag{4.3a}$$

Similarly, the resultant surface forces in the y and z directions can be expressed, respectively, as

$$\delta F_y = -\frac{\partial p}{\partial y}\delta x\,\delta y\,\delta z, \quad \delta F_z = -\frac{\partial p}{\partial z}\delta x\,\delta y\,\delta z \tag{4.3b}$$

The resultant surface force acting on the element is the vectorial sum of the above three forces:

$$\delta\mathbf{F}_s = -\left(\frac{\partial p}{\partial x}\mathbf{i} + \frac{\partial p}{\partial y}\mathbf{j} + \frac{\partial p}{\partial z}\mathbf{k}\right)\delta x\,\delta y\,\delta z = -\nabla p\,\delta x\,\delta y\,\delta z \tag{4.3c}$$

where **i**, **j**, and **k** are unit vectors in the x, y, and z directions, and ∇ is the notation for the del vector operator.

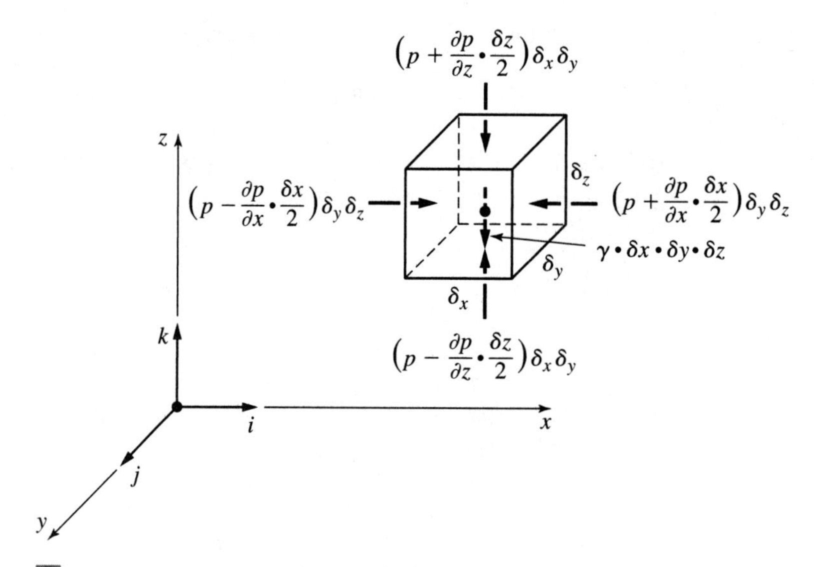

Figure 4.2 Surface and body forces acting on a small fluid element.

The resultant surface force per unit volume can be as

$$\frac{\delta \mathbf{F}_s}{\delta x\, \delta y\, \delta z} = -\nabla p \tag{4.4}$$

The weight, **W**, of the element which acts in the downward direction (i.e., negative *z*) can be written as

$$\delta \mathbf{W} = -\gamma\, \delta x\, \delta y\, \delta z\, \mathbf{k} \tag{4.5}$$

Taking *a* as the acceleration of the element, δm as the mass of the element, and applying Newton's second law of motion, one obtains

$$\sum \delta \mathbf{F} = \delta \mathbf{F}_s + \delta \mathbf{W} = \delta m\, \mathbf{a} \tag{4.6}$$

Equation (4.6) can be written as

$$-\nabla p\, \delta x\, \delta y\, \delta z - \gamma\, \delta x\, \delta y\, \delta z\, \mathbf{k} = \rho\, \delta x\, \delta y\, \delta z\, \mathbf{a} \tag{4.7}$$

Equation (4.7) reduces to

$$-\nabla p - \gamma \mathbf{k} = \rho \mathbf{a} \tag{4.8}$$

Equation (4.8) is the general equation of motion for a fluid with no shearing stresses.

4.3 Variation of Pressure in a Fluid at Rest

For fluid at rest, $\mathbf{a} = 0$. This reduces equation (4.8) to

$$\nabla p + \gamma \mathbf{k} = 0, \quad \text{or} \quad \frac{\partial p}{\partial x} = 0, \ \frac{\partial p}{\partial y} = 0, \ \frac{\partial p}{\partial z} = -\gamma \tag{4.9}$$

Equation (4.9) shows that the pressure does not vary with location in the horizontal *x-y* plane. However, it does vary with the location below the fluid surface. Thus, the fundamental equation for the variation of pressure in a fluid at rest is given by an ordinary differential equation as (Fox and McDonald, 1985)

$$\frac{dp}{dz} = -\gamma \tag{4.10}$$

where γ is the specific weight that may vary with depth, z (measured from the bottom up to a point under consideration), and p is the pressure. The pressure gradient in the vertical direction (or with elevation) is negative (i.e., as the elevation increases, the pressure decreases). There is no requirement for γ to be constant in equation (4.10). Since γ is the product of the mass density (ρ) and the acceleration due to gravity (g), these for most cases are considered constant. Thus, the pressure gradient will be linear for a homogeneous fluid with constant specific weight. This permits direct integration of equation (4.10):

$$\int_{p_1}^{p_2} dp = -\gamma \int_{z_1}^{z_2} dz \quad \text{or} \quad p_1 - p_2 = \gamma(z_2 - z_1) \quad \text{or} \quad p_1 = \gamma h + p_2 \tag{4.11}$$

where p_i is the pressure at the vertical elevation z_i, $i = 1, 2$, and h is the difference between the two elevations z_1 and z_2 or the depth of the fluid measured downward from the point z_2. Thus, the pressure difference between two points can also be written in terms of head as

$$h = \frac{p_1 - p_2}{\gamma} \tag{4.12}$$

In other words, h is the height of a column of fluid of specific weight γ to produce a pressure difference of $p_1 - p_2$. Thus, any pressure difference can be expressed in terms of the height of a water or mercury column. If $p_2 = p_0$ is the atmospheric pressure, then p at any point is

$$p = \gamma h + p_0 \tag{4.13a}$$

Pressure is measured with respect to the atmospheric pressure. Thus, if $p_0 = 0$ then

$$p = \gamma h \tag{4.13b}$$

If an irregular surface is immersed in a fluid, then the average pressure, p_{av}, on the immersed surface is the sum of the pressures, p_i, on the different strips multiplied by the areas of the corresponding strips, a_i, divided by the total area of the surface, that is,

$$P_{av} = \frac{\sum P_i a_i}{\sum a_i} \tag{4.14}$$

Example 4.1

Calculate the pressure at a point located 50 m below the sea level. The seawater has a mass density, ρ_s, of 1025 kg/m^3.

Solution: From equation (4.13b), $p = \gamma h$. In this example, $h = 50.0$ m. Therefore,

$$p = 1025 \times 9.81 \times 50 = 1.0055.25 \text{ N/m}^3 \times 50 \text{ m} = 502.76 \text{ kN/m}^2$$

TABLE 4.1 Properties of the U.S. Standard Atmosphere at the Mean Sea Level

Property	SI Units	BG Units
Temperature (T)	288.14°K (15°C)	518.67°R (59°F)
Pressure (p)	101.33 kPa (abs)	2116.2 lb/ft^2 (abs) 14.696 lb/in^2 (abs)
Density (ρ)	1.225 kg/m^3	0.002377 slug/ft^3
Specific weight (γ)	12.014 N/m^3	0.07647 lb/ft^3
Viscosity (μ)	1.789×10^{-5} N-s/m^2	3.737×10^{-7} lb-s/ft^2

4.4 Standard Atmosphere

The pressure in the earth's atmosphere varies with the altitude above mean sea level. It is desirable to have measurements of pressure versus altitude over the specific range of conditions, including temperature, reference pressure, etc. The U.S. *standard atmosphere* is an idealized representation of middle-latitude, year-round mean conditions of the earth's atmosphere. If one considers the temperature profile for the U.S. standard atmosphere, it is found that the temperature decreases with altitude in the troposphere (region nearest the earth's surface up to about 11 km), then becomes essentially constant in the stratosphere and then starts increasing with elevation in the next layer. The properties of the U.S. standard pressure at sea level are as given in Table 4.1. The acceleration of gravity at sea level is 9.807 m/s^2 (32.174 ft/s^2).

4.5 Hydrostatic Force on Immersed Surfaces

Immersed surfaces are subject to hydrostatic pressures. The deeper the surface below the fluid surface, the greater the intensity of the pressure. This can be felt easily while diving in a swimming pool. The pressure varies linearly with depth if the fluid is incompressible. The immersed surface may generally be horizontal, vertical, or inclined.

4.5.1 Horizontal Surface

Consider a horizontal surface immersed in a liquid as shown in Figure 4.3. The total weight of the fluid above this surface is equal to the volume of the liquid above the surface multiplied

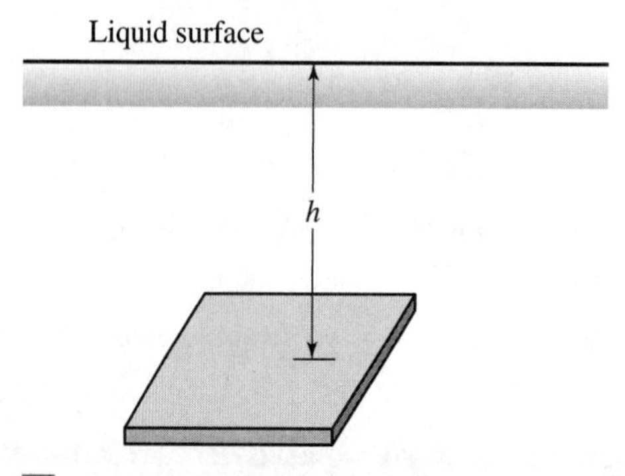

Figure 4.3 A horizontal surface immersed in a liquid.

by the specific weight of the liquid. The immersed surface is, thus, subject to a vertical force equal to the total weight of the fluid above it. Therefore,

$$F = \gamma A h \tag{4.15}$$

where F is the force acting on the immersed surface, γ is the specific weight of the liquid, A is the area of the surface, and h is the depth of the horizontal surface measured from the liquid surface.

Example 4.2

A square tank 4 ft long contains water up to a depth of 3 ft. Calculate the total force acting on the base of that tank.

Solution: Area of the base $= 4 \times 4 = 16\ \text{ft}^2$. Using equation (4.15), the total force is

$$F = \gamma A h = 62.4\ (\text{lbs/ft}^3) \times 16\ (\text{ft}^2) \times 3\ (\text{ft}) = 2995.2\ \text{lb}.$$

4.5.2 Vertical Surface

As previously stated, the pressure varies with the depth below the fluid surface. Therefore, a vertically immersed surface will be subject to different pressure heads according to the point of interest. Consider a vertical plane surface immersed in a liquid as shown in Figure 4.4. This vertical surface can be divided into a certain number of parallel strips as shown in the figure. A strip δh which has a width b and its center located at a depth h from the surface of the liquid is considered. The pressure at the center of this strip, p_o, is given as

$$p_o = \gamma h \tag{4.16}$$

The strip has an area of $\delta A = b\,\delta h$. Therefore, the force on that strip is given by

$$F_o = p_o\,\delta A = \gamma h b\,\delta h \tag{4.17}$$

The total force on that surface can be obtained by integration of equation (4.17) as

$$F = \sum \gamma b h\,\delta h = \int \gamma b h\,dh \tag{4.18}$$

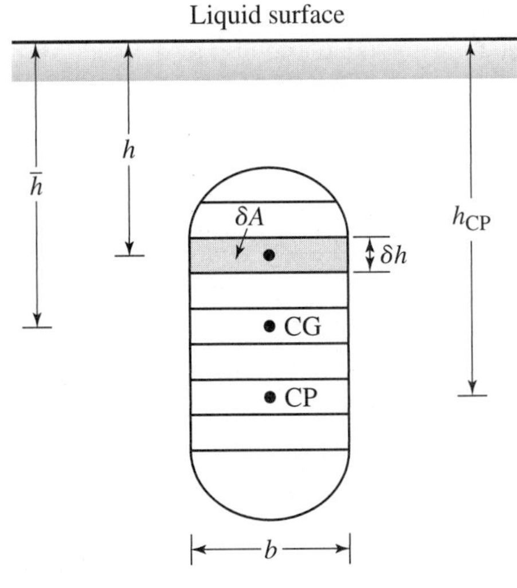

Figure 4.4 A vertical surface immersed in a liquid.

Glen Canyon Dam, Colorado River, USA. (Courtesy of Doug Sokell/Visuals Unlimited.)

The integration of $hb\,dh$ is equal to the first moment of the area around the surface of the liquid, that is,

$$\int hb\,dh = A\bar{h} \tag{4.19}$$

where A is the area of the immersed surface and $\bar{h}$ is the vertical distance between the centroid of the surface under consideration and the liquid surface. The total force is thus given as

$$F = \gamma A\bar{h} \tag{4.20}$$

Another simple way to prove equation (4.20) is by dividing the vertical area into horizontal strips of areas $a_1, a_2, \ldots$ and a_n. The corresponding distances from the liquid surface to the center of these areas are $h_1, h_2, \ldots$ and h_n. The force acting on any strip (i) can be calculated as

$$F_i = \gamma a_i h_i \tag{4.21}$$

The total force on the vertical surface is, thus, given as

$$F = \gamma a_1 h_1 + \gamma a_2 h_2 + \cdots + \gamma a_n h_n$$

or

$$F = \gamma \sum_{i=1}^{n} a_i h_i = \gamma A\bar{h} \tag{4.22}$$

Example 4.3

A rectangular door 3 ft × 4 ft closes an opening in a vertical face of a submarine. The center of the door is located 300 ft below the seawater level. The specific weight of the seawater is 64 lb/ft^3. Calculate the total force acting on that door.

Solution: The area of the door = 3 × 4 = 12 ft^2. Using equation (4.20), the total force acting on the door is

$$F = \gamma A\bar{h} = 64 \text{ lb/ft}^3 \times 12 \text{ ft}^2 \times 300 \text{ ft} = 230{,}400 \text{ lb}$$

Since the pressure is greater over the lower portion of an immersed surface, the resultant pressure force will act at some point below the center of gravity of that surface. The point at which the resultant force is located is called the center of pressure, CP, and is expressed in terms of depth, h_{CP}, below the surface of the liquid. To locate the center of pressure, CP, consider the immersed surface shown in Figure 4.4. The sum of the moments of the pressure forces around the surface of the liquid is given as

$$M = \int \gamma b h^2\,dh = \gamma I_o \tag{4.23}$$

where the integral represents the second moment of the area or the moment of inertia about the liquid surface and is denoted by I_o.

Also, the total sum of the moments of pressure forces is equal to Fh_{CP}. Therefore,

$$M = \gamma I_o = F h_{CP} \tag{4.24}$$

or

$$\gamma I_o = \gamma A\bar{h} h_{CP} \tag{4.25}$$

Hence,

$$h_{CP} = \frac{I_o}{A\bar{h}} \tag{4.26}$$

The parallel axis theorem stipulates that for any given area, A, the moment of inertia of this area, I_o, around any horizontal axis is equal to its moment of inertia around the horizontal axis passing through its centroid, I_c, plus the area multiplied by the square of the distance $\bar{h}$ between the two axes. Hence,

$$I_o = I_c + A\bar{h}^2 \tag{4.27}$$

where I_c is the moment of inertia of the surface about a horizontal axis passing through its center of gravity and $\bar{h}$ is the distance between the liquid surface and the center of gravity of the immersed surface.

Substituting equation (4.27) into equation (4.26) yields

$$h_{CP} = \frac{I_c}{A\bar{h}} + \bar{h} \tag{4.28}$$

Equation (4.28) implies that the center of pressure, CP, of an immersed surface is always below its center of gravity, CG, by a distance $I_c/A\bar{h}$.

4.5.3 Inclined Surface

Consider now the more general case in which the submerged plane surface is inclined by an angle θ with the liquid surface, as illustrated in Figure 4.5. The area can have any arbitrary shape. The objective is to determine the direction, location, and magnitude of the resultant force acting on one side of the area. For ease of computation, also consider an axis y along the inclined surface and an axis O perpendicular to the y axis. The whole area can be divided into

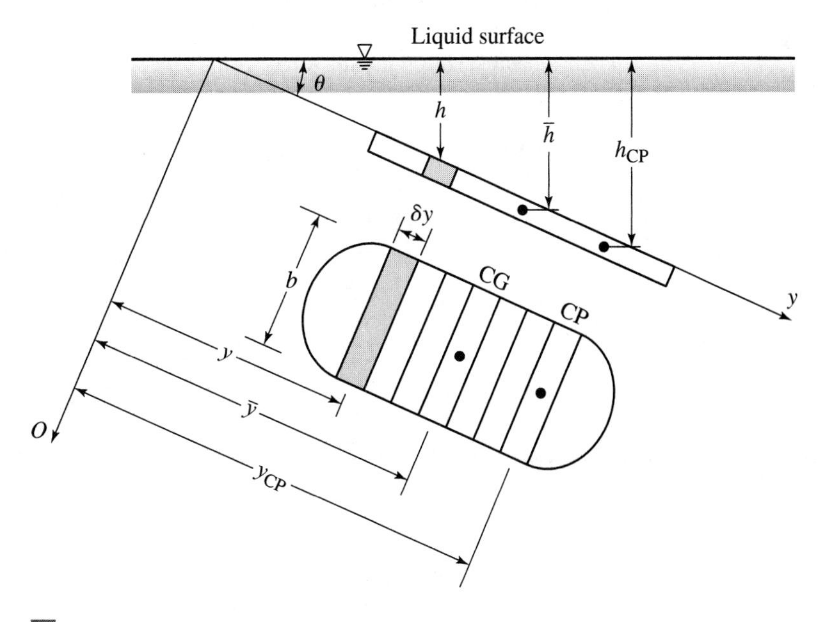

Figure 4.5 An inclined surface immersed in a liquid.

small strips, each having an area of δA. The force acting on δA is $\delta F = \gamma h\, \delta A$ and is perpendicular to the surface. The resultant force can be obtained by *summing* these differential forces over the entire area. This can be expressed as

$$F = \sum \gamma h\, \delta A = \int_A \gamma y \sin\theta\, dA \tag{4.29}$$

where y is measured perpendicular to the O axis. The liquid specific weight, γ, and the angle θ are constant. Therefore, equation (5.29) can be written as

$$F = \gamma \sin\theta \int_A y\, dA \tag{4.30}$$

where the quantity $\int_A y\, dA$ is the first moment of the area with respect to the O axis, that is,

$$\int_A y\, dA = \bar{y}A \tag{4.31}$$

where $\bar{y}$ is the y coordinate of the centroid of the inclined surface measured from the O axis. Equation (4.30) can be rewritten as

$$F = \gamma A \bar{y} \sin\theta = \gamma A \bar{h} \tag{4.32}$$

where $\bar{h}$ is the vertical distance from the liquid surface to the centroid of the immersed surface. As seen from equation (4.32), the total pressure force is dependent on the specific weight of the liquid, the area of the immersed surface, and the depth of the centroid below the liquid surface. It does not, however, depend on the angle θ. The resultant force is perpendicular to the surface.

The moment of the resultant force around the O axis must be equal to the moment of the distributed pressure forces around the same axis, or

$$F y_{CP} = \sum \delta F y = \int_A \gamma y^2 \sin\theta\, dA \tag{4.33}$$

Substituting equation (4.32) into equation (4.33), one can write

$$y_{CP} = \frac{\int_A y^2 dA}{A\bar{y}} \tag{4.34}$$

The integral, $\int_A y^2 dA$, is the second moment of area or the moment of inertia, I_o, around the O axis. Therefore,

$$y_{CP} = \frac{I_o}{A\bar{y}} \tag{4.35}$$

Using the parallel axis theorem, I_o is expressed as

$$I_o = I_c + A\bar{y}^2 \tag{4.36}$$

where I_c is the moment of inertia of the area around an axis passing through its centroid and parallel to the O axis. Hence,

$$y_{CP} = \frac{I_c}{A\bar{y}} + \bar{y} \tag{4.37}$$

Note that in equation (4.37) y_{CP} and $\bar{y}$ are in the orientation of the object and are measured in the plane of the object. The vertical distance, h_{CP}, between the liquid surface and the center of pressure, CP, is given as

$$h_{CP} = y_{CP} \sin \theta \tag{4.38}$$

As the depth of submergence increases, $\bar{y}$ in equation (4.37) increases and the point through which the resultant force acts (the center of pressure) moves closer to the centroid of the area. In some cases where plane surfaces are placed at greater depths in liquids, the first term on the right hand side of equation (4.37) can be ignored for simplicity. Geometric properties of some common areas are given in Figure 4.6.

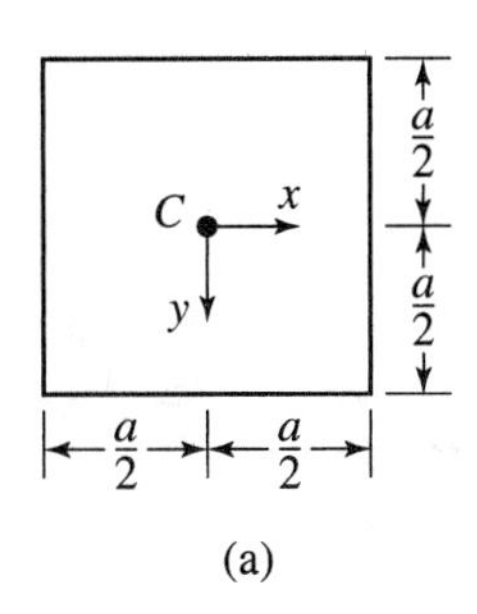

$A = a^2$

$I_{xc} = I_{yc} = \frac{1}{12}a^4$

$I_{xyc} = 0$

(a)

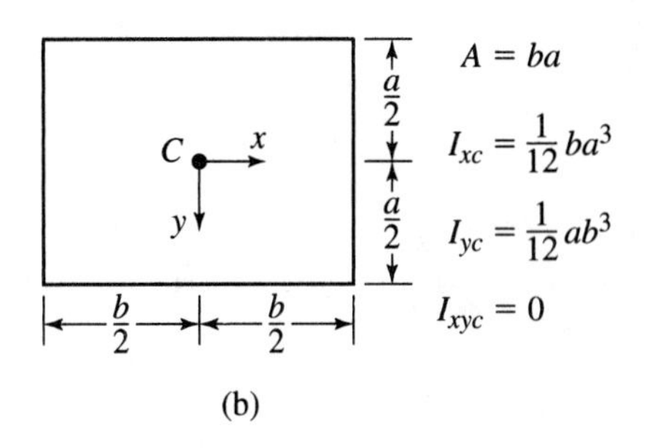

(b)

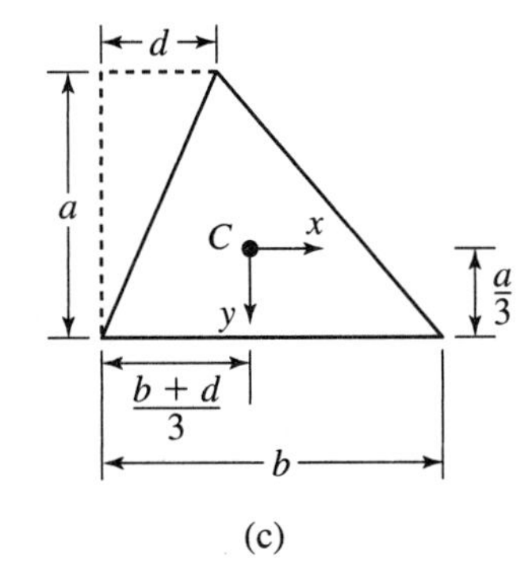

$A = \frac{ab}{2}$

$I_{xc} = \frac{ba^3}{36}$

$I_{xyc} = \frac{ba^2}{72}(b - 2d)$

(c)

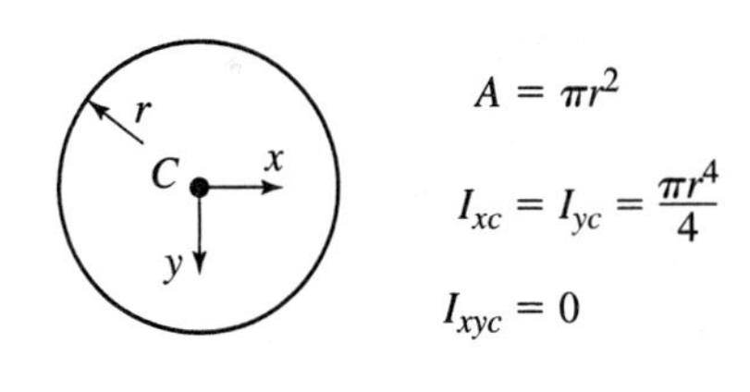

(d)

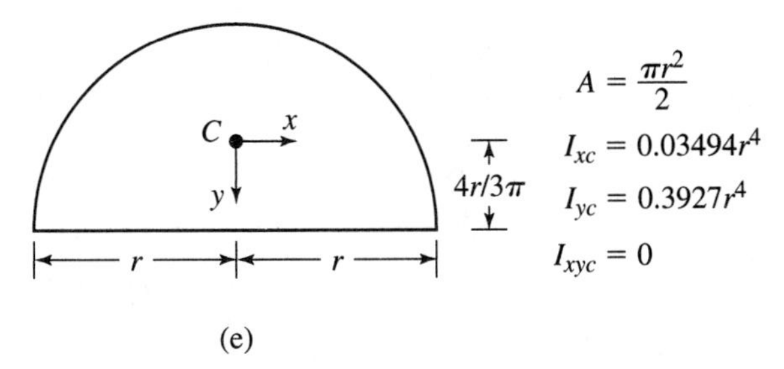

$A = \frac{\pi r^2}{2}$

$I_{xc} = 0.03494r^4$

$I_{yc} = 0.3927r^4$

$I_{xyc} = 0$

(e)

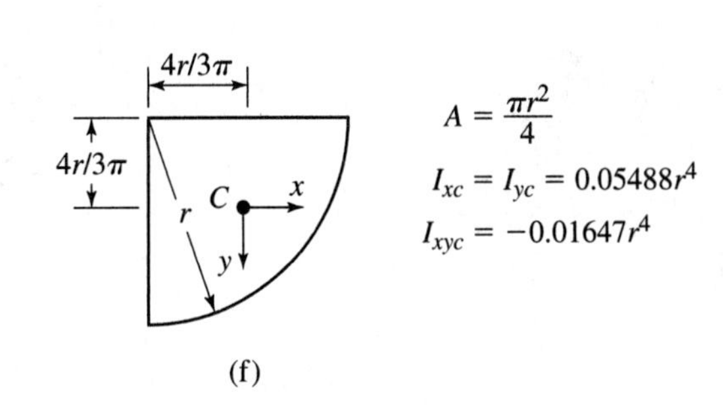

(f)

Figure 4.6 Geometric properties of some common areas.

Example 4.4

A 2 m square gate is located in the inclined wall of a large reservoir containing water, as shown in Figure 4.7. The wall is inclined at an angle of 45°. The gate operates on a shaft along its horizontal axis. For a water depth of 5 m above the shaft, calculate the resultant force exerted on the gate and its location. What is the moment required to open this gate?

Solution: The resultant force, F, can be calculated from equation (4.32) as $F = 1000\ (\text{kg/m}^3) \times 9.81\ (\text{m/s}^2) \times 4\ (\text{m}^2) \times 5\ (\text{m}) = 196{,}200\ \text{kg-m/s}^2 = 196{,}200$ N. The center of pressure, CP, at which the resultant force acts, is located below the water surface by $h_{\text{CP}} = y_{\text{CP}} \sin 45°$, where

$$y_{\text{CP}} = \frac{I_c}{A\bar{y}} + \bar{y}$$

in which

$$I_c = \frac{a^4}{12} = \frac{(2)^4}{12} = 4/3 = 1.333\ \text{m}^4, \text{ and}$$

$\bar{y} = 5.0/\sin 45° = 7.07$ m. Therefore,

$$y_{\text{CP}} = \frac{1.333\ (\text{m}^4)}{4 \times 7.07\ (\text{m}^3)} + 7.07\ (\text{m}) = 7.117\ \text{m}$$

The center of pressure, CP, is located at a distance of $y_{\text{CP}} - \bar{y} = 0.047$ m below the gate center measured along the y axis. Thus, the gate is subject to a force of 196.2 kN acting at the center of pressure. The moment required to open the gate is calculated as

$$M = Fy_{\text{CP}} - F\bar{y} = F(y_{\text{CP}} - \bar{y}) = 196{,}200 \times 0.047 = 9221.4\ \text{N-m}$$

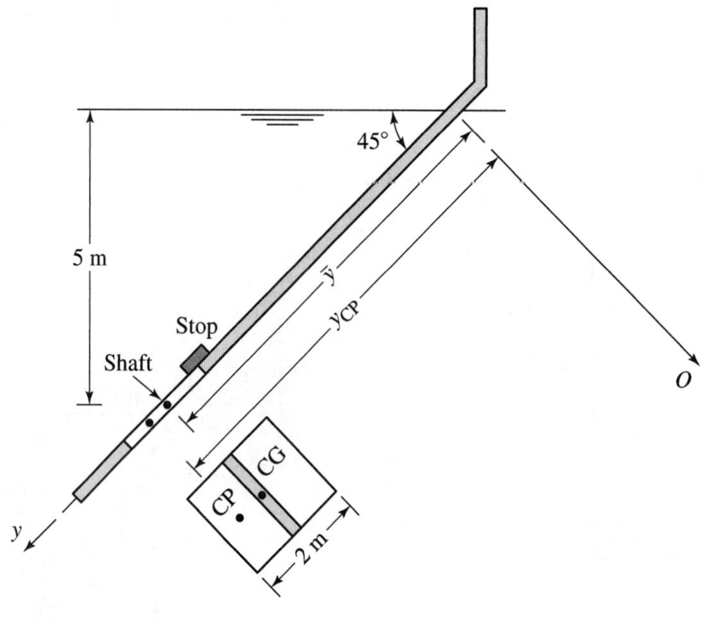

Figure 4.7 A square gate fixed in an inclined wall (Example 4.4).

4.6 Graphical Representation of Hydrostatic Forces

The graphical representation of the pressure forces on a plane surface is informative in that the magnitude and location of the resultant force can be obtained. To illustrate, consider the hydrostatic pressure distribution on the vertical wall of a tank of width b, as shown in Figure 4.8a. The liquid inside this tank has a specific weight, γ, and its surface is subject to atmospheric pressure. The pressure varies linearly with depth with a maximum value of γh at the bottom, where h is the depth of water. The average value of the pressure p_{av} is $\gamma h/2$ and is located at a depth $h/2$. The center of pressure at which the resultant force acts is located at a distance $h/3$ above the bottom. The resultant force, F, acting on the side wall of the tank is

$$F = p_{av} \cdot A = \gamma(h/2)A \tag{4.39}$$

where A is the area of the vertical wall of the tank.

A three-dimensional pressure distribution is given in Figure 4.8b. The side wall of the tank is now considered the base and altitudes at its different points represent pressures. This volume is called the pressure prism. The volume of the prism is equal in magnitude to the resultant pressure force acting on the side wall. Thus,

$$F = \text{volume of the pressure prism} = (1/2)\gamma h\, bh = \gamma(h/2)A \tag{4.40}$$

The same procedure can be used for immersed surfaces that do not extend to the liquid surface as shown in Figure 4.9a. In such cases, the cross section of the pressure prism is trapezoidal. Nevertheless, the resultant force is equal in magnitude to the volume of the pressure prism and acts on the center of pressure of the prism. That pressure prism can be divided into two parts, $ABDE$ and BCD, as shown in Figure 4.9b. The total pressure force, F_R, is equal to

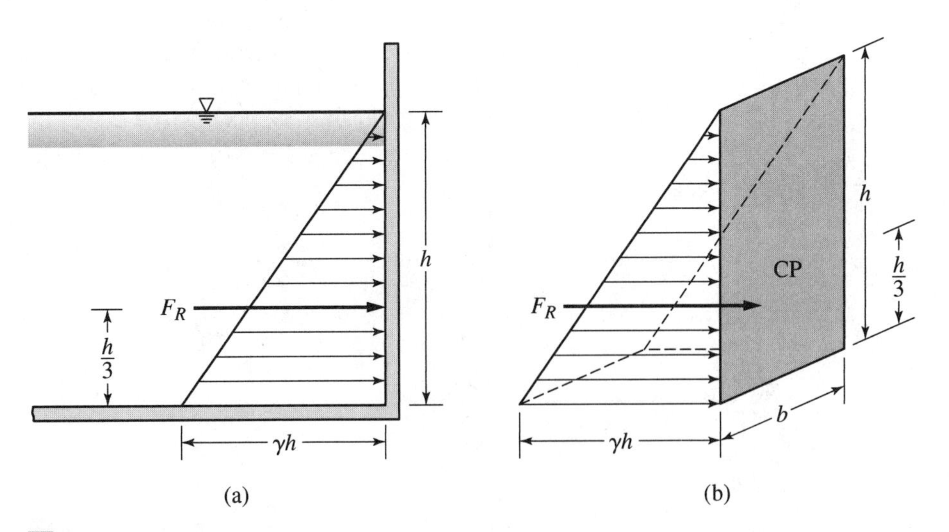

Figure 4.8 Graphical representation of hydrostatic force. (a) Hydrostatic pressure distribution on a vertical wall of a tank. (b) Three-dimensional pressure distribution.

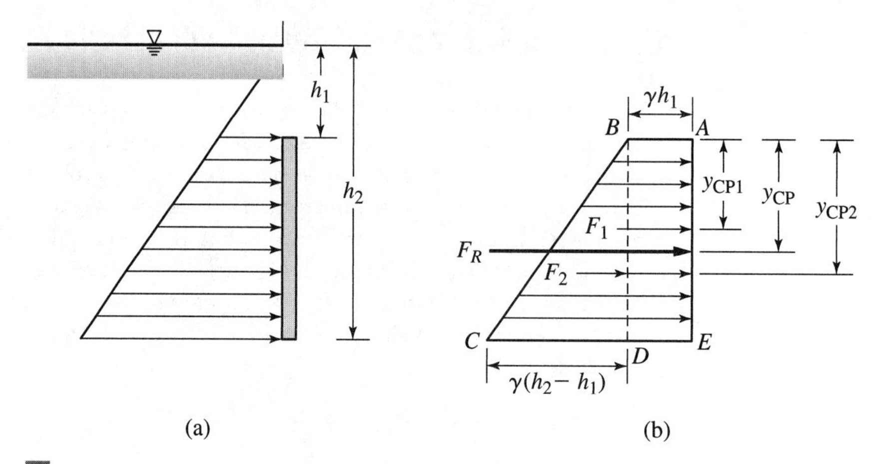

Figure 4.9 Graphical representation of hydrostatic force on immersed surfaces. (a) Cross section of the pressure prism. (b) Resultant force and center of pressure.

the resultant of the forces acting on the divided parts, that is,

$$F_R = F_1 + F_2 \tag{4.41}$$

The center of pressure (y_{CP}) can be obtained from

$$F_R\, y_{CP} = F_1\, y_{CP1} + F_2\, y_{CP2} \tag{4.42}$$

where y_{CP}, y_{CP1}, and y_{CP2} are the distances between the liquid surface and the centers of pressure of the entire prism, the *ABDE* prism, and the *BCD* prism, respectively. Note that since in this instance the object is oriented in the vertical plane, the y distances are equivalent to the vertical h distances given in Section 4.5.2. The nomenclature y is employed to keep the discussion general. For inclined surfaces pressure prisms are generally triangular or trapezoidal for surfaces not extending to the surface of the liquid, as shown in Figure 4.10. Although, distances (y) are generally measured along the inclined surface, pressures vary with the vertical distances (h).

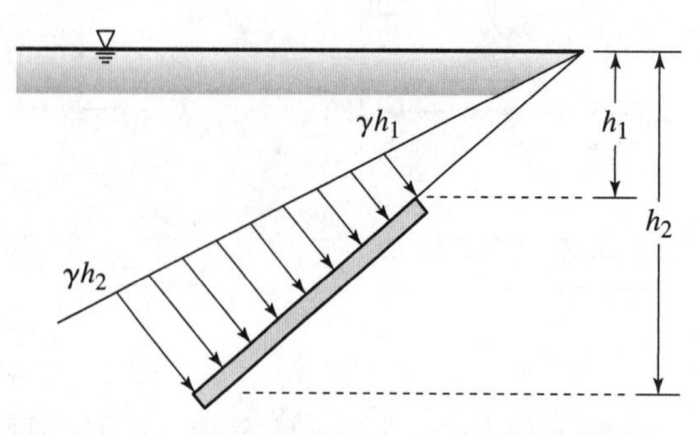

Figure 4.10 Pressure prisms in inclined surfaces.

Example 4.5

Find the magnitude and location of the resultant force acting on the side wall of a 1 m square box tank of 2 m depth. Half of the depth is filled with a liquid of a density of 1500 kg/m^3, while the other half is filled with water.

Solution: The tank is filled one meter deep with a liquid of a density 1500 kg/m^3 and the other meter is filled with water, as shown in Figure 4.11a. The pressure prism is shown in Figure 4.11b, which can be divided into three parts *ADE, DBFE,* and *EFC.* To draw the pressure prism, we need to first evaluate the pressure *DE* (which is equal to *BF*) and the pressure *FC.* Therefore,

$$DE = BF = \gamma_w h_1 = 1000\ (\text{kg/m}^3) \times 9.81\ (\text{m/s}^2) \times 1\ (\text{m}) = 9810\ \text{kg-m/s}^2\text{-m}^2\ (\text{N/m}^2)$$

and

$$FC = \gamma_2 h_2 = 1500\ (\text{kg/m}^3) \times 9.81\ (\text{m/s}^2) \times 1\ (\text{m}) = 14{,}715\ \text{kg-m/s}^2\text{-m}^2\ (\text{N/m}^2)$$

The pressure forces, F_1, F_2, and F_3, acting on the centroid of the three different prisms, can be calculated from the volumes of the corresponding pressure prisms. Therefore,

$$F_1 = 1/2 \times 1000 \times 9.81 \times 1.0 \times 1.0 = 4905\ \text{kg-m/s}^2\ (\text{N})$$

$$F_2 = 1000 \times 9.81 \times 1.0 \times 1.0 = 9{,}810\ \text{kg-m/s}^2\ (\text{N}), \text{ and}$$

$$F_3 = 1/2 \times 1500 \times 9.81 \times 1.0 \times 1.0 = 7{,}357.5\ \text{kg-m/s}^2\ (\text{N})$$

The resultant force F is equal to the sum of the above forces. Thus, $F_R = 22{,}072.5$ kg-m/s^2 (N) The location of the resultant force can be obtained from equation (4.42). Therefore,

$$F_R y_{CP} = F_1 y_{CP1} + F_2 y_{CP2} + F_3 y_{CP3}$$

where y_{CP1}, y_{CP2}, and y_{CP3} are equal to 0.67 m, 1.50 m, and 1.67 m, measured from the surface of the liquid, respectively. Thus,

$$y_{CP} = \frac{4905\ (\text{N}) \times 0.67\ (\text{m}) + 9810\ (\text{N}) \times 1.5\ (\text{m}) + 7{,}357.5\ (\text{N}) \times 1.67\ (\text{m})}{22{,}072.5\ (\text{N})}$$

$$= 1.37\ \text{m}$$

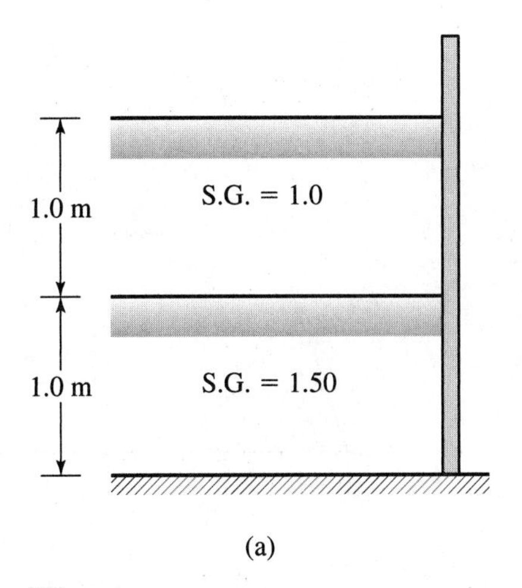

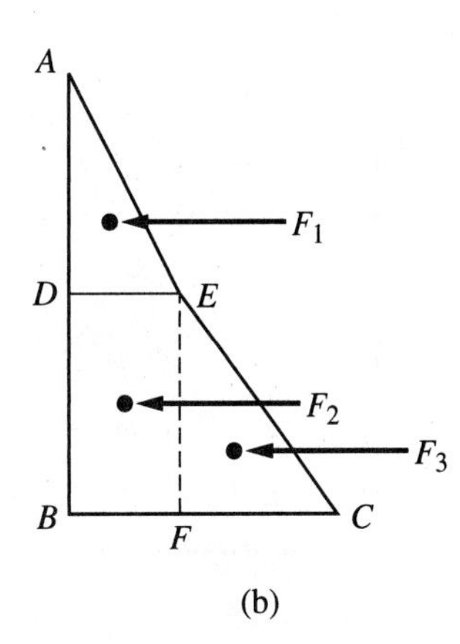

Figure 4.11 (a) The tank given in Example 4.5. (b) The pressure prism.

Example 4.6

A triangular spillway is used to control the flow of water from a reservoir as shown in Figure 4.12a. When the water level is 3 ft above the crest as shown, what is the hydrostatic force exerted on the face of the spillway per foot and where is its point of actions?

Solution: The pressure diagram is shown in Figure 4.12b. Thus, the hydrostatic force in the x direction is $F_H = \gamma \bar{h} A = 62.4 \text{ lbs/ft}^3 \times 13 \text{ ft} \times 20 \text{ ft} = 16224 \text{ lb/ft}$. The location of the force is determined by breaking the trapezoidal pressure distribution into an equivalent combination of a rectangular distribution and a triangular distribution, as shown in the figure. Thus,

$$F_1 y_{CP1} + F_2 y_{CP2} = F_R y_{CP}$$

Referring to Figure 4.12b, $a = 3\gamma = 187.2 \text{ lbs/ft}^2$, $b = 23\gamma = 1435.2 \text{ lbs/ft}^2$, and $b - a = 20\gamma = 1248 \text{ lbs/ft}^2$. Then,

$$F_1 = 20a = 3744 \text{ lb/ft}; \quad \text{and} \quad F_2 = 10(b - a) = 12480 \text{ lb/ft}.$$

On checking, it is seen that $F_1 + F_2 = 16224$ lb/ft. Then, taking moments about the plane through the water surface,

$$3744\left[\left(\frac{1}{2}20\right) + 3\right] + 12480\left[\left(\frac{2}{3}20\right) + 3\right] = 16224\, y_{CP}; \text{ or } y_{CP} = 15.56 \text{ ft.}$$

One can check this result my noting that the resultant force must act through the centroid of the trapezoidal pressure distribution, which, measured from the base, is given by $Z_{CP} = \dfrac{h(2a + b)}{3(a + b)}$; or $Z_{CP} = 7.44$ ft. Note that Z_{CP} must be equal to $23 - y_{CP}$.

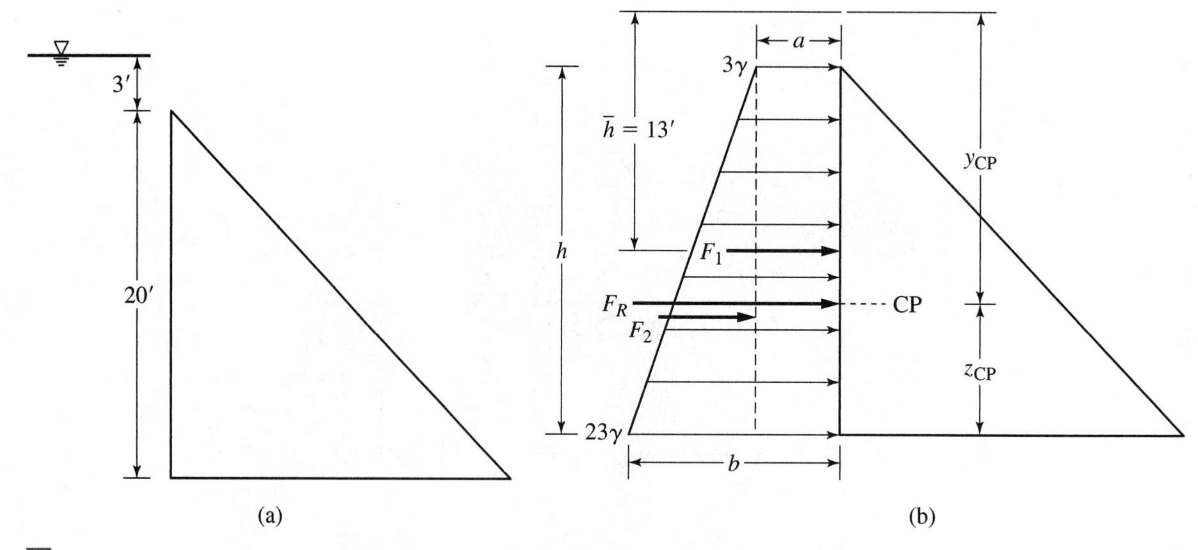

Figure 4.12 **(a) Example Problem 4.6. (b) Pressure schematic for Example 4.6.**

Example 4.7

Suppose the downstream face of the spillway in Example 4.6 is at an angle of 45° and the spillway is constructed of concrete with a specific gravity of 2.55. If the foundation material is fully permeable, what will be the resultant reaction force on the foundation of the spillway and where will it act?

Solution: The situation is shown in Figure 4.13. If the face angle is 45°, then the base of the spillway is equal to the height = 20 ft. The cross-sectional area $A = \frac{bh}{2} = \frac{20(20)}{2} = 200\ \text{ft}^2$. The weight of the spillway per unit width is $W = 2.55\gamma(200) = 31{,}824$ lb/ft. Now, for the permeable foundation, the uplift force will be given by $F_u = \frac{23\gamma}{2}A = \gamma(11.5)(20)(1) = 14{,}352$ lb/ft. The resultant vertical force is then $F_v = W - F_u = 31{,}824 - 14{,}352 = 17{,}472$ lb/ft. From the previous example we know that the horizontal force is $F_H = 16{,}224$ lb/ft. Then the resultant reaction force on the foundation is $F_R = \sqrt{F_H^2 + F_v^2} = 23{,}843$ lb/ft. Now, to find the point of action of this force on the base of the spillway, following Morris and Wiggert (1972), we sum the moments about the point of action of W and F_u. Thus $16{,}224\,(7.44) = 17{,}472\left(e + \frac{b}{6}\right)$, where e is the eccentricity. Then, solving for e, $120{,}706.5 - 17{,}472e - 58{,}240 = 0$, or $62{,}466.5 = 17{,}472e$. Thus, $e = 3.57$ ft. Then we find that $x_p = 10 + 3.57 = 13.57$ ft from the heel, or 6.43 ft from the toe of the spillway.

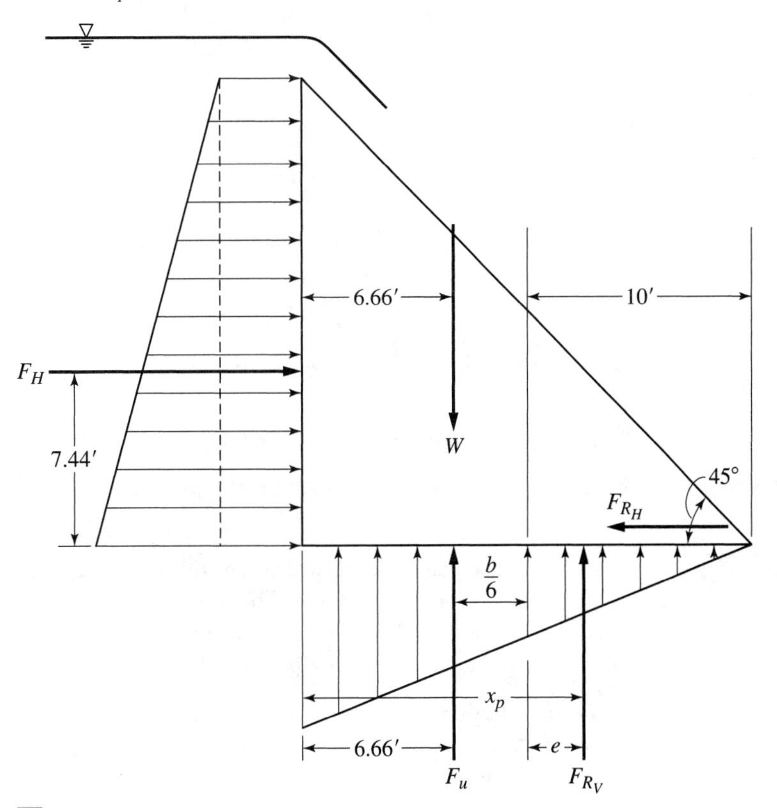

Figure 4.13 Solution schematic for Example 4.7.

Example 4.8

A bulkhead 5 m long divides a storage tank. One side of the tank is filled with oil of density 800 kg/m^3 to a depth of 2 m and the other side is filled with water to a depth of 1 m, as shown in Figure 4.14. Determine the resultant pressure force and its location.

Solution: Two pressure prisms are encountered, one on each side of the vertical plate, as shown in Figure 4.14. The magnitude of the pressure force on the oil side is equal to the volume of the pressure prism on that side, that is,

$$F_1 = 0.5 \times 800 \text{ (kg/m}^3) \times 9.81 \text{ (m/s}^2) \times 2 \text{ (m)} \times 2.0 \times 5.0 \text{ (m)}^2 = 78{,}480 \text{ kg-m/s}^2 \text{ (N)}$$

This force is located at $Z_1 = 0.67$ m from the bottom, and is directed from left to right. Likewise,

$$F_2 = 0.5 \times 1000 \text{ (kg/m}^3) \times 9.81 \text{ (m/s}^2) \text{ } 1.0 \text{ (m)} \times 1.0 \times 5.0 \text{ (m}^2) = 24{,}525 \text{ kg-m/s}^2 \text{ (N)}$$

This force is located at $Z_2 = 0.33$ m from the bottom and is directed from right to left. Note that F_1 and F_2 may also be calculated from equation (4.20). The resultant pressure force, F_R, is

$$F_R = 78{,}480 - 24{,}525 = 53{,}955 \text{ kg-m/s}^2 \text{ (N)}$$

Using equation (4.42), taking moments about point O,

$$F_R Z_R = F_1 Z_1 - F_2 Z_2$$

$$Z_R = (78{,}480 \text{ (N)} \times 0.67 \text{ (m)} - 24{,}525 \text{ (N)} \times 0.33 \text{ (m)})/53{,}955 \text{ (N)}$$
$$= 0.825 \text{ m from the bottom.}$$

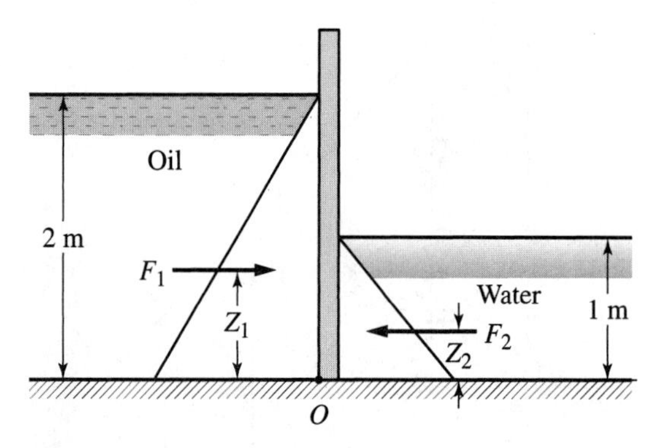

Figure 4.14 Representation of the bulkhead in Example 4.8.

Example 4.9

Consider a rectangular tank 2 m wide. The tank has water 2 m deep and oil on the top of water 3 m deep, as shown in Figure 4.15a. The specific gravity of oil is 0.8. Compute the two forces on the portions of the side wall that are in contact with oil and water and determine their locations. Compute the resultant force on the side wall of the tank and determine its location.

Solution: Referring to Figure 4.15b, the total force on the wall ABC (F_{ABC}) = force on the portion of the wall AB (F_{AB}) plus the force on the portion of the wall BC (F_{BC}). First, the force F_{AB} is computed.

$$F_{AB} = 9.81 \text{ (m/s}^2) \times 1{,}000 \text{ (kg/m}^3) \times 0.8 \times 1.5 \text{ (m)} \times 3 \text{ (m)} \times 2 \text{ (m)}$$
$$= 70.63 \times 10^3 \text{ kg-m/s}^2 = 70.63 \text{ kN}$$

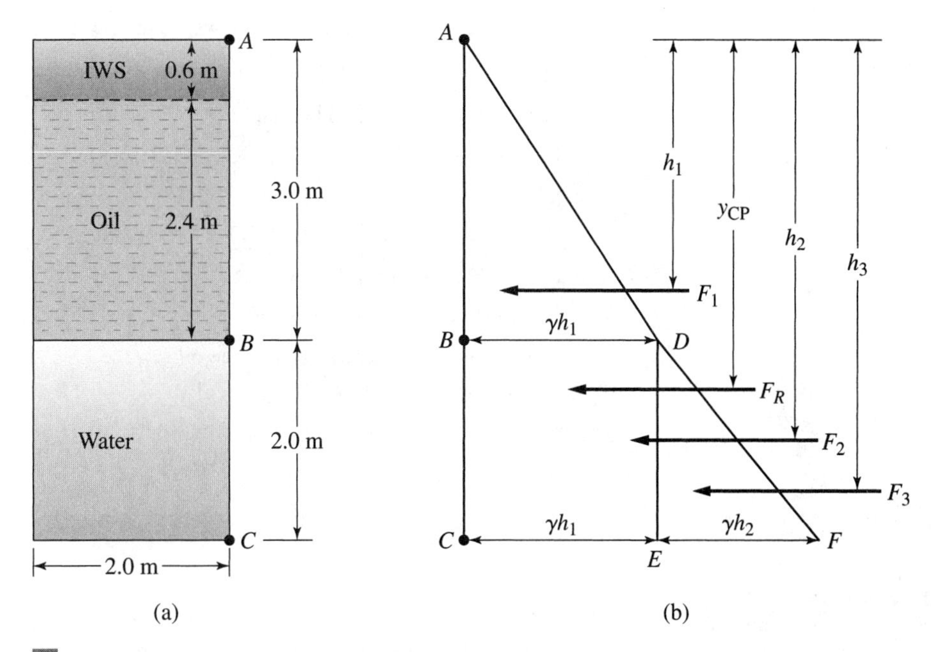

Figure 4.15 (a) A tank containing a layer of water and a layer of oil. (b) Hydrostatic forces and location of the resultant force.

This acts at $2/3 \times (h = 3) = 2$ m below the top. Now we compute the force F_{BC}. Water acts on the BC portion and the superimposed oil can be converted into an equivalent depth of water. To that end, consider an imaginary water surface (IWS) for this calculation. This is located at 0.8×3 (m) $= 2.4$ m above B or 0.6 m below the top layer or A.

$$F_{BC} = 9.81 \text{ (m/s}^2) \times 1{,}000 \text{ (kg/m}^3) \times (2.4 + 1) \text{ (m)} \times (2 \text{ m} \times 2 \text{ m})$$
$$= 133.42 \times 10^3 \text{ kg-m/s}^2 = 133.42 \text{ kN}.$$

This force acts at the center of pressure y_{CP}:

$$y_{CP} = \frac{I_c}{\bar{y}A} + \bar{y} = \frac{2(2)^3/12 \text{ m}^4}{(2.4 + 1)(\text{m})(2 \text{ m})(2 \text{ m})} + 3.4 \text{ (m)} = 3.5 \text{ m}$$

or $y_{CP} = 3.5 + 0.6 = 4.1$ m from A or the top.

The total resultant force $= 70.63 \times 10^3$ kg-m/s^2 $+ 133.42 \times 10^3$ kg-m/s^2 $= 204.05 \times 10^3$ kg-m/s^2 $= 204.05$ kN. This force acts at the center of pressure of the entire area.

Taking the moments around point A,

$$y_{CP} \times 204.05 \times 10^3 \text{ kg-m/s}^2 = 2 \text{ m} \times 70.63 \times 10^3 \text{ kg-m/s}^2 + 4.1 \text{ m} \times 133.42 \times 10^3 \text{ kg-m/s}^2$$

This yields $y_{CP} = [141.26 + 547.02]/204.05 = 3.37$ m measured from the top or from point A.

This problem can also be solved as follows. Referring to Figure 4.15b,

$$F_1 = (1/2)\gamma hbh = 0.5 \times 800 \times 9.81 \times 3.0 \times 3.0 \times 2.0 = 70{,}632 \text{ kg-m/s}^2 \text{ (N)}$$

F_1 is located at 2.0 m below the oil surface.

$$F_2 = 800 \times 9.81 \times 3.0 \times 2.0 \times 2.0 = 94{,}176 \text{ kg-m/s}^2 \text{ (N)}$$

F_2 is located at 4.0 m below the oil surface.

$$F_3 = 0.5 \times 1000 \times 9.81 \times 2.0 \times 2.0 \times 2.0 = 39{,}240 \text{ kg-m/s}^2 \text{ (N)}$$

F_3 acts at a distance of 4.333 m below the oil surface.

The resultant force, $F_R = F_1 + F_2 + F_3 = 204.048$ kN.

Taking the first moment of forces around the oil surface,

$$F_R y_{CP} = F_1 h_1 + F_2 h_2 = F_3 h_3$$

$$204.048\ y_{CP} = 70.632 \times 2.0 + 94.176 \times 4.0 + 39.24 \times 4.333$$

which yields $y_{CP} = 3.37$ m.

Example 4.10

A vertical wall separates sea water in an estuary from the fresh water upstream. If the depth on the freshwater side is 20 ft, and the sea water has a specific weight of 64 lb/ft^3, what must be the depth on the sea-water side for the resultant force to be zero?

Solution: Let the depth of water on the sea-water side be h. Then, the resultant pressure force acting on the wall on the sea-water side is $F = \gamma h(h \times 1)/2$, where $\gamma = 64$ lb/ft^3. The pressure force on the freshwater side = 62.4 (lb/ft^3) $\times$ 20 (ft) $\times$ 20 (ft)/2 = 12,480 lb/ft. In order for these two forces to be equal, $h^2 = 12{,}480$ lb/[32]. Therefore, $h = 19.75$ ft.

4.7 Buoyancy and Stability

When a body is immersed in a liquid, pressure forces act on its surface. Horizontal components of pressure force on a submerged body cancel each other. Upward and downward pressure forces on the upper and lower surfaces are not canceled because they are not equal. The lower surface is subject to higher pressure force, because the pressure increases with depth. The body is, thus, exposed to a net upward pressure force. This upward force is known as buoyancy force, F_b. The point of action of the buoyancy force is called the center of buoyancy, CB.

4.7.1 Buoyancy

Basic principles of buoyancy were discovered early by Archimedes over 2200 years ago. A submerged body in a liquid is subject to an uplift force which equals the weight of the liquid displaced by the body. This statement is known as Archimedes' principle. Thus, a submerged body that is in equilibrium below the liquid surface displaces a liquid weighing the same as the weight of the body. In other words, the submerged body displaces a liquid that just balances its own weight. The buoyant force acts upward through the centroid of the displaced volume.

Consider a submerged body in equilibrium conditions as shown in Figure 4.16. Horizontal pressure forces cancel each other (i.e., $\Sigma F_x = 0$). The upper surface of the body is

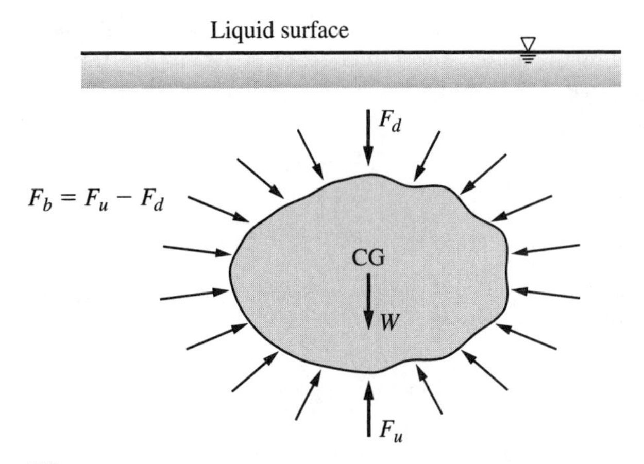

Figure 4.16 Submerged body in a equilibrium conditions.

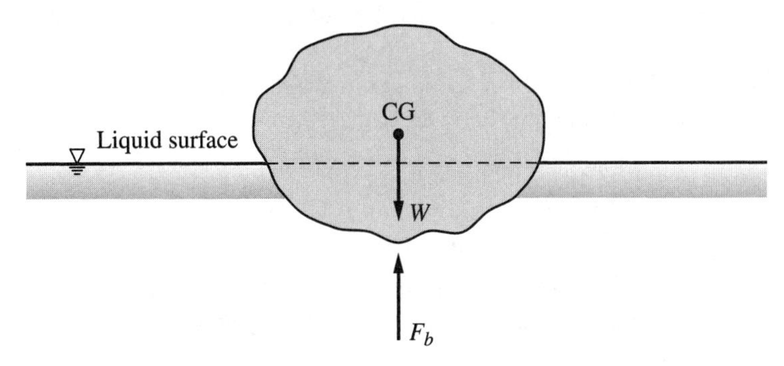

Figure 4.17 An object floating on the surface of a liquid.

subject to a downward vertical pressure force, F_d. The lower one is subject to an upward force, F_u, which is generally greater than F_d. The difference is the upward buoyant force, F_b. For a body in equilibrium, the buoyant force, F_b, must be equal to the weight of the body, W.

If the immersed body, shown in Figure 4.16, is in equilibrium, the densities of the body and the liquid must be the same. If the weight of the body is greater than the weight of the displaced liquid (i.e., the density of the body or object is greater than that of the liquid), the body will sink. If the weight of the object is less than the buoyant force, the object will rise to an equilibrium point and the object will float on the surface of the liquid as shown in Figure 4.17. In this case, no pressure force acts downward since there is no liquid above the object. The object will continue to rise and float to a stage where the buoyant force (in this case the upward pressure force) is equal to its weight. That is, a floating object displaces the liquid equal to its own weight in which it floats.

Example 4.11

An object weighs 490.5 kg-m/s^2 (N) in the air. When submerged in water it weighs 196.2 kg-m/s^2 (N). Find the volume, specific weight, and mass density of this object.

Solution: The buoyant force is equal to the difference between the weights of the object in the air and water. Therefore,

$$F_b = 490.5\text{ N} - 196.2\text{ N} = 294.3\text{ N}$$

According to the Archimedes principle, this buoyant force is equal to the weight of water displaced by the object, which is equal to the volume of water displaced (or the volume of the submerged object) multiplied by the specific weight of the water. Hence,

$$F_b = V_{\text{ol}} \cdot \gamma_w$$

Therefore,

$$V_{\text{ol}} = F_b/\gamma_w = 294.3/9{,}810 = 0.03 \text{ m}^3$$

Hence, the volume of the object is 0.03 m^3, and its specific weight, γ, is defined as the weight per unit volume, that is,

$$\gamma = \frac{490.5 \text{ kg-m/s}^2}{0.03 \text{ m}^3} = 16.35 \text{ kN/m}^3$$

The mass density ρ is defined as mass divided by the volume or $\rho = \gamma/g = 16{,}350 \text{ kg/m}^2\text{s}^2/9.81 \text{ m/s}^2 = 1667.67 \text{ kg/m}^3$.

Example 4.12

A 0.5 m cube of solid timber floats in water as shown in Figure 4.18. Find the submerged depth of the cube knowing that the density of the timber is 0.60 t/m^3.

Solution: The buoyant force is equal to the weight of the displaced liquid, i.e., $W = F_b$, where

$$W = \gamma V = 0.60 \text{ (t/m}^3) \times 1000 \text{ (kg/t)} \times 9.81 \text{ (m/s}^2) \times (0.5)^3 \text{ (m}^3)$$
$$= 735.75 \text{ kg-m/s}^2 \text{ (N)}$$

Also, F_b = weight of the liquid displaced

$$735.75 \text{ N} = \gamma_w (0.5 \times 0.5 \times d) = 2452.5\, d \text{ N}$$

Thus,

$$d = 735.75/2452.5 = 0.3 \text{ m}$$

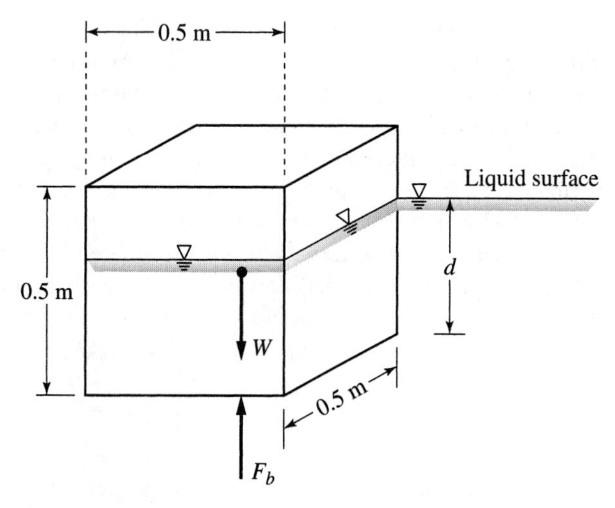

Figure 4.18 A cube of solid timber floats in water (Example 4.12).

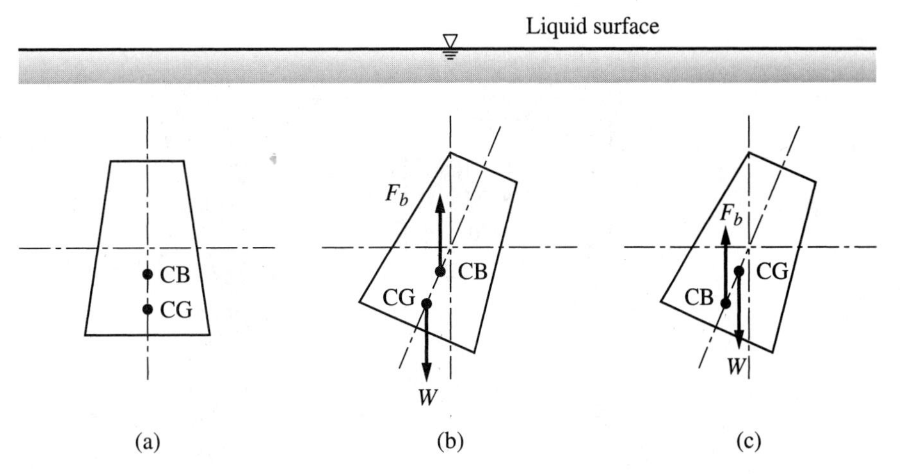

Figure 4.19 Stability of submerged bodies. (a) Center of gravity and center of buoyancy. (b) Stable object. (c) Unstable object.

4.7.2 Stability of Submerged Bodies

An object is said to be stable if it will return to its original position after being slightly displaced from its original position. This subject is of vital importance while designing submarines and sink chambers. Water currents tend to rotate objects about horizontal axes. Objects are in stable condition under water if they rotate back to their original position. Otherwise, they are not stable.

The stability of submerged objects can only be ensured if the center of buoyancy, CB (point of action of the buoyancy force), is kept always above the center of gravity, CG, of the object, as shown in Figure 4.19a. The center of buoyancy is located at the centroid of the displaced volume. Stability can generally be achieved by concentrating the weight of the object near its bottom to lower its center of gravity as much as possible. If some currents or any external forces cause this object to rotate slightly in clockwise direction, as shown in Figure 4.19b, the weight of the object itself and the buoyant force will create a moment that tends to produce an anticlockwise rotation. As the object gets back to its original position, this moment will vanish. On the contrary, for the object shown in Figure 4.18c, where the center of gravity is located above the center of buoyancy, if something causes a small rotation to the object in a clockwise direction, the created moment imposes an additional clockwise rotation and tends to overturn the object. This object is, thus, unstable.

4.7.3 Stability of Floating Bodies

Ships, boats, and buoys are examples of floating bodies. It is important to ensure their stability as they are used extensively in transportation. An analysis of stability of a floating body is somewhat different from that of a submerged body. Consider the floating cube shown in Figure 4.20a. The center of gravity, CG, is located over the center of buoyancy, CB, on the vertical axis *A-A*. As the cube is subject to some clockwise rotation, the center of gravity does not shift, however, the center of buoyancy shifts rightward as shown in Figure 4.20b. This is attributed to the change in the volume and shape of the displaced liquid. Therefore, the weight of the body, W, and the buoyant force, F_b, create a couple which produces counterclockwise rotation. The body is, thus, forced to go back to its original location. It is, therefore, a stable body.

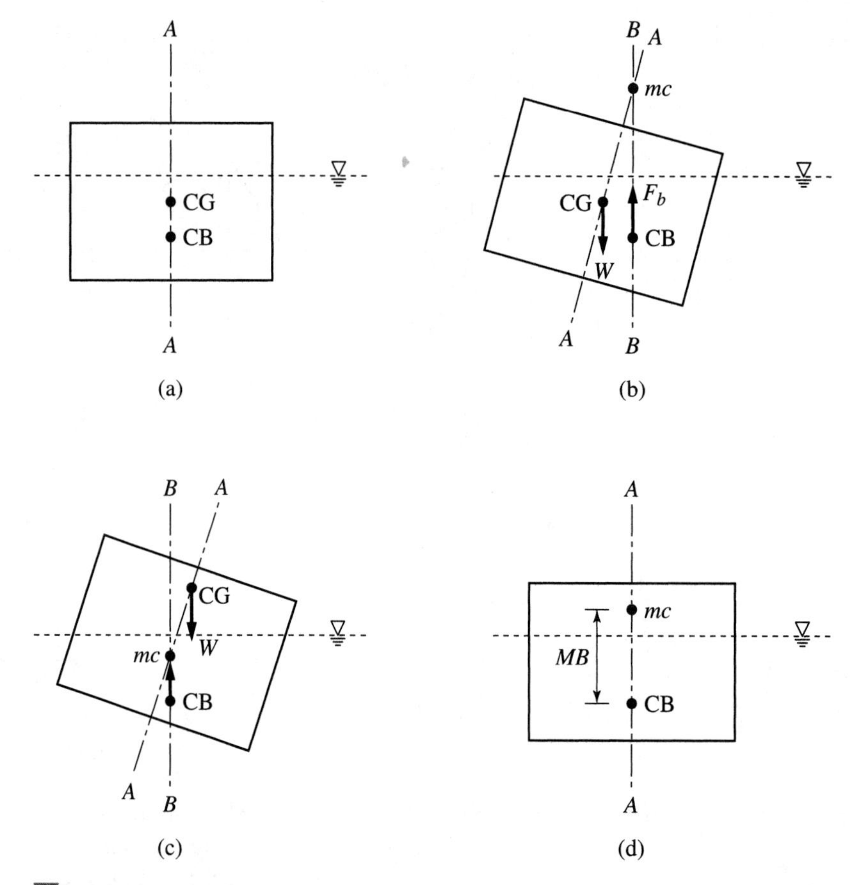

Figure 4.20 Stability of floating bodies: (a) Center of gravity and center of buoyancy. (b) Stable object. (c) Unstable object. (d) Distance between the center of buoyancy and the "metacenter."

In some cases, the center of gravity may be located right to the line of action of the buoyancy force, as illustrated in Figure 4.20c. In such cases, the couple resulting from the weight of the object and the buoyant force tends to overturn the object. The object is thus unstable.

The stability of a floating body is dependent on the relative location of the center of gravity and the center of buoyancy. For clockwise rotation, if the center of gravity is located left of the center of buoyancy, the body will be stable and vice versa.

The point of intersection between the axis *A-A* and the line of action of the buoyant force is defined as metacenter, mc (as shown in Figures 4.20b and 4.20c). A floating body is stable if its metacenter is located above its center of gravity, as shown in Figure 4.20b. If the metacenter is located below the center of gravity, as shown in Figure 4.20c, the body is unstable.

The distance *MB* (shown in Figure 4.20d) between the center of buoyancy, CB, and the metacenter, mc, can be obtained as

$$MB = \frac{I}{V_{ol\,d}} \tag{4.43}$$

where I is the moment of inertia of a horizontal section of the floating object calculated at the liquid surface while it is on an even keel, and $V_{ol\,d}$ is the volume of the displaced liquid.

Example 4.13

A solid wood cylinder has a diameter of 1.0 m and a height of 2.0 m. Check the stability of this cylinder when placed in water knowing that the specific density of the wood is 600.0 kg/m^3.

Solution: The problem is represented as shown in Figure 4.21. According to Archimedes' principle, the buoyant force, F_b, is equal to the cylinder weight, W, where

$$W = 600\,(\text{kg/m}^3) \times 9.81(\text{m/s}^2) \times \frac{\pi}{4}(1.0)^2\,(\text{m}^2) \times 2\ (\text{m}) = 2943\pi\ \text{kg-m/s}^2$$
$$= 9245.71\ \text{kg-m/s}^2(\text{N})$$

and

$$F_b = 1000\,(\text{kg/m}^3) \times 9.81(\text{m/s}^2) \times \frac{\pi}{4}(1.0)^2\,(\text{m}^2) \times D\,(\text{m}) = 7704.76D\ \text{kg/s}^2$$

where D is the submerged depth. Equating W and F_b, $9245.71 = 7704.76D$. Therefore, $D = 1.2$ m. The center of buoyancy, CB, is located at the middle of the submerged portion of the cylinder, that is, at 0.6 m measured above the bottom of the cylinder or below the water surface, as shown in Figure 4.21. The distance between the center of buoyancy and the metacenter, MB, is given as

$$MB = \frac{I}{V_{\text{ol}\,d}}$$

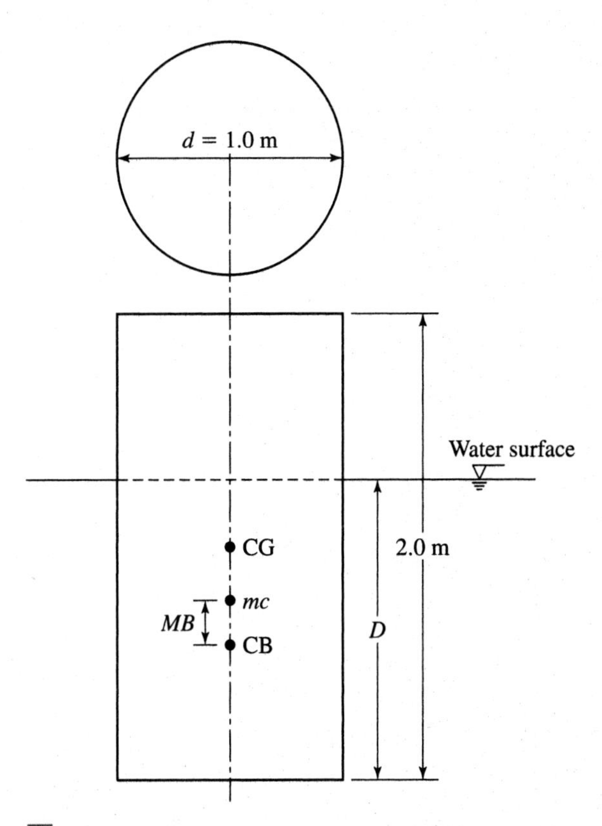

Figure 4.21 The solid wood cylinder given in Example 4.13.

where

$$I = (\pi d^4/64) = \pi(1.0)^4/64 = 0.04909 \text{ m}^4, \quad \text{and}$$

$$V_d = (\pi/4)d(1)^2 \times 1.2 = 0.9425 \text{ m}^3$$

Therefore, $MB = 0.04909/0.9425 = 0.05208$ m.

The metacenter is located 0.05208 m above the center of buoyancy. It is located at 0.6 + 0.05208 = 0.65208 m above the bottom of the cylinder. On the other hand, the center of gravity of the cylinder is located at its geometric center (i.e., at a distance of 1.0 m from the bottom). Therefore, the center of gravity is higher than the metacenter by about 0.348 m. The cylinder is thus unstable.

4.8 Measurement of Pressure

The pressure at a point in a fluid can be labeled as absolute pressure or gage pressure. Absolute pressure is measured relative to perfect vacuum or absolute zero pressure, and is therefore always positive. Gage pressure is measured relative to the local atmospheric pressure and can therefore be either positive or negative. If the pressure is above atmospheric pressure, the gage pressure takes on a positive value. If the pressure is below atmospheric pressure, the gage pressure takes on a negative value. Thus, a gage pressure of zero value corresponds to the local atmospheric pressure. A negative gage pressure refers to a suction or vacuum pressure.

The pressure is force per unit area. Its units in the BG system are lb/ft^2 or lb/in^2, commonly abbreviated as psf or psi. In the SI system, the units are N/m^2, commonly referred to as pascal abbreviated as Pa (1 N/m^2 = 1 Pa). Often, pressure is expressed as the height of a column of specified liquid [e.g., water, mercury (Hg), etc]. In that case, its units referring to the height of the column are in., ft., m, cm, mm, etc. The standard atmospheric pressure can be expressed as 760 mm Hg (absolute).

The atmospheric pressure is usually measured with a mercury barometer which consists of a glass tube closed at one end with the open end immersed in a container of mercury, as shown in Figure 4.22. The tube is initially filled with mercury. In the inverted tube, there

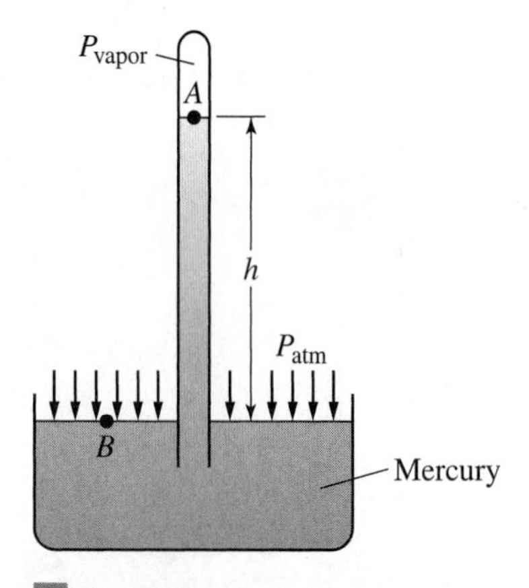

Figure 4.22 A mercury barometer.

will develop a space above the column of mercury. This column will attain an equilibrium position with the balance between the weight of mercury plus the force due to vapor pressure of mercury and the atmospheric pressure. Thus, one can write

$$p_{\text{atm}} = \gamma h + p_{\text{vapor}} \tag{4.44}$$

where γ is the specific weight of mercury. In general, the mercury vapor pressure is very small and can be neglected and therefore the atmospheric pressure is conventionally expressed in terms of the height h (in in. or mm) of mercury. The atmospheric pressure of 14.7 psi is about 29.9 in. of mercury or 34 ft of water.

4.9 Manometer

Standard pressure measuring devices employ liquid columns in vertical or inclined tubes and are called manometers. The mercury manometer is one type of manometer. The three common types of manometers are the piezometer tube, the U-tube, and the inclined-tube manometer. All three types of manometers are used in hydraulics. The first two are more common and are briefly discussed here.

4.9.1 Piezometer Tube

This type of manometer consists of a vertical tube which is open at the top and connected to the container in which the pressure is to be measured, as shown in Figure 4.23. Since the column of fluid is at rest, equation (4.13b) applies. Therefore, the pressure at point A, p_A is given by a measurement of h_1, since $p_A = \gamma h_1$, where γ is the specific weight of the liquid in the container. This device is suitable for measuring pressure, which is greater than the atmospheric pressure and is small to keep the height of the column tractable.

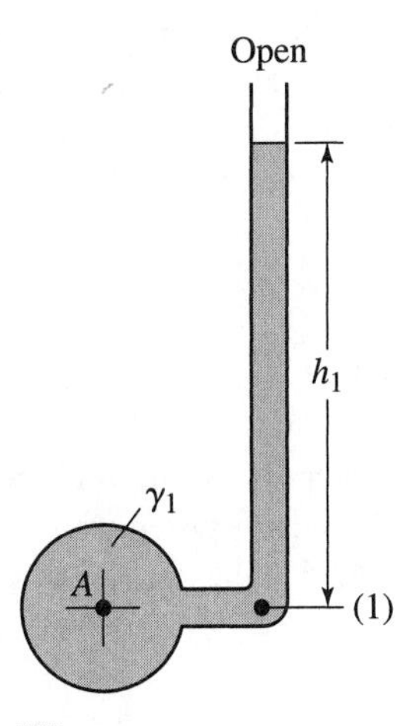

Figure 4.23 Piezometer tube.

Example 4.14

Container A is filled with water. The tube is filled with mercury (S.G. = 13.6) to a level of 0.5 m, followed by 0.2 m layer of water. Another 0.5 m layer of oil (S.G. = 0.8) is placed above the water as shown in Figure 4.24. Determine the pressure inside Container A.

Solution:

$$p_A = \gamma_o h_o + \gamma_w h_w + \gamma_m h_m$$

$$p_A = 0.8 \times 9810\ (\text{N/m}^3) \times 0.5\ (\text{m}) + 9810\ (\text{N/m}^3) \times 0.2\ (\text{m}) + 13.6 \times 9810\ (\text{N/m}^3) \times 0.5\ (\text{m}) = 72594\ \text{N/m}^2$$

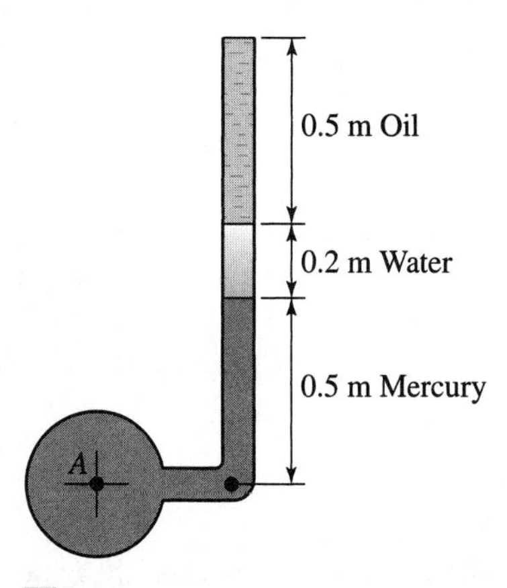

Figure 4.24 Piezometer tube, Example 4.14.

4.9.2 U-Tube Manometer

A U-shaped tube, as shown in Figure 4.25, is a widely used manometer. The fluid in the tube is known as the gage fluid. One end of the U-tube is open and the other end is attached to the container in which the pressure is to be measured. The fluids in the tube and the container can be the same or different. If different fluids are used, they must be immiscible. Since the fluid column is at rest, equation (4.13a) applies. In order to evaluate the pressure, it is advisable to start at one end of the system and work around to the other end. Choosing points 1, 2, and 3 as shown in Figure 4.25, one can express

$$p_A + \gamma_1 h_1 - \gamma_2 h_2 = 0 \tag{4.45}$$

which yields

$$p_A = \gamma_2 h_2 - \gamma_1 h_1 \tag{4.46}$$

The U-tube manometer can be used to measure the difference in pressure between two containers or two points in a system. Consider a system connecting two containers A and B by a manometer as shown in Figure 4.26. The difference in pressure can be evaluated by starting at one end of the system and working around to the other end. Choosing points A, 1, 2, 3, and 4, one can write

$$p_A + \gamma_1 h_1 - \gamma_2 h_2 - \gamma_3 h_3 = p_B \tag{4.47}$$

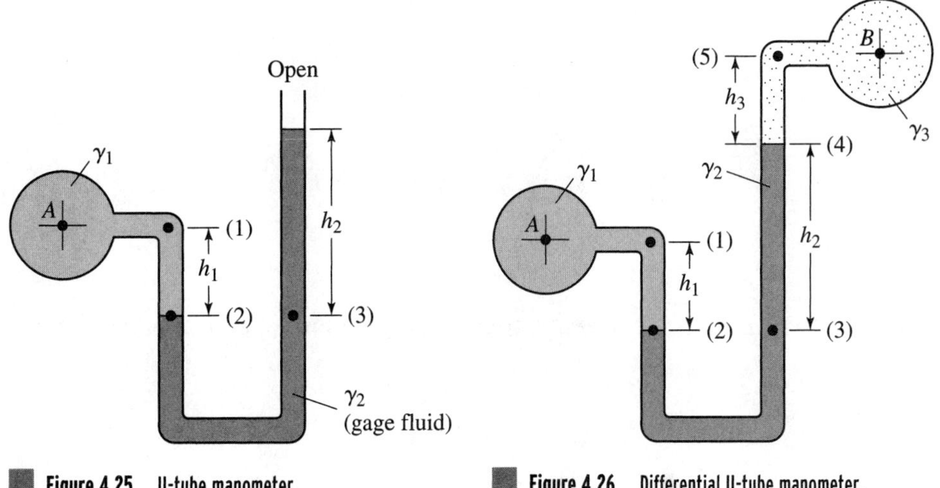

Figure 4.25 U-tube manometer.

Figure 4.26 Differential U-tube manometer.

which leads to the pressure difference:

$$p_A - p_B = \gamma_2 h_2 + \gamma_3 h_3 - \gamma_1 h_1 \tag{4.48}$$

Example 4.15

For the system shown in Figure 4.27, the pressure at B is 200 lb/ft^2. What is the pressure at A in psi knowing that the S.G. of mercury is 13.6?

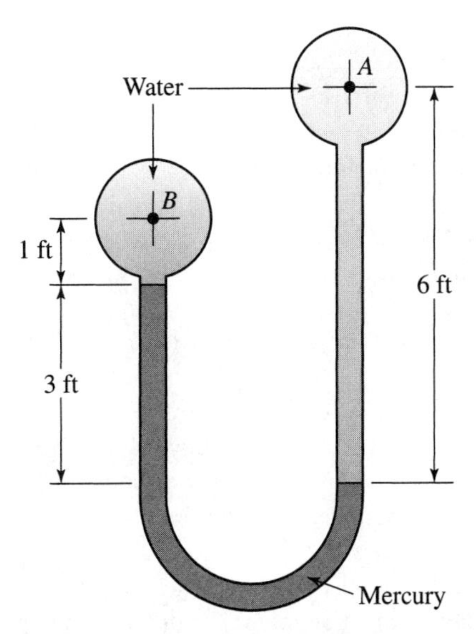

Figure 4.27 The system of a U-tube manometer given in Example 4.15.

Solution:

$$p_B + 62.4\ (\text{lb/ft}^3) \times 1\ (\text{ft}) + 13.6 \times 62.4\ (\text{lb/ft}^3) \times 3\ (\text{ft}) - 62.4\ (\text{lb/ft}^3) \times 6\ (\text{ft}) = P_A$$

Therefore,

$$p_A = 200\ (\text{lb/ft}^2) + 62.4\ (\text{lb/ft}^2) + 2545.92\ (\text{lb/ft}^2) - 374.4\ (\text{lb/ft}^2)$$

$$= 2433.92\ (\text{lb/ft}^2)$$

$$= 16.9\ \text{psi.}$$

READING AID

4.1. What is meant by hydrostatics? Write the fundamental equation for pressure variation with depth in a fluid at rest.
4.2. What is Pascal's principle? Define its limitations.
4.3. Explain why the pressure does not vary linearly with depth in compressible fluids?
4.4. How can you evaluate the vertical force acting on a horizontal surface immersed in a liquid?
4.5. Develop an equation to evaluate the total horizontal pressure acting on a vertical surface immersed in a liquid.
4.6. What is meant by the center of pressure? For a vertical surface immersed in a liquid, do you expect it to be located below or above the center of gravity and why?
4.7. For a vertically immersed surface, prove that the distance between the center of pressure and the center of gravity is equal to $I_G/[A\bar{x}]$, where A is the area of the surface, I_G is its moment of inertia around a horizontal axis passing through its center of gravity, and $\bar{x}$ is the distance between the liquid surface and the center of gravity.
4.8. Develop an equation to evaluate the pressure force acting on an inclined surface by an angle θ with the liquid surface.
4.9. For inclined surfaces immersed in water, as the depth of submergence increases do you expect the distance between the center of gravity and the center of pressure to increase or decrease? Why?
4.10. Explain the advantages of using the graphical method to evaluate the pressure force and its point of action on immersed bodies. Give two examples in which pressure prisms are triangular and trapezoidal, respectively.
4.11. Define the buoyant force and explain how it develops. Does the buoyant force vary with the depth of a submerged body? Why or why not?
4.12. Give three examples illustrating the importance of the study of buoyancy and stability.
4.13. What is meant by the Archimedes principle? Do floating bodies displace a liquid that balances their own volumes or their own weights?
4.14. Which parameter amongst weight, volume, and density determines whether a body will float or sink in water and why?
4.15. What is meant by the stability of a submerged body and how does one ensure it?
4.16. How does the stability of a floating object differ from that of a submerged one?
4.17. What is the main factor that determines whether a floating body is stable or unstable?

4.18. What is meant by metacenter and how can you evaluate the distance between the center of buoyancy and the metacenter?
4.19. Define *absolute pressure* and *gage pressure.* What are the units of pressure in the SI and BG systems?
4.20. What are the types of manometers commonly used in hydraulic applications?

Problems

4.1. A rectangular tank, 6 ft long and 3 ft wide, is filled with water up to a depth of 4 ft. Calculate the total force acting on the base of this tank.

4.2. A swimming pool, 20 m long and 8 m wide, is filled with water to a depth of 2.5 m. Determine the magnitude and location of the resultant force on each side of the pool.

4.3. A swimming pool, 20 m long and 8 wide, has a constant bed slope of 1:10. The depth of water at its shallow end is 0.5 m. Determine the magnitude and location of the resultant force on the sides of the pool.

4.4. A rectangular plate 10 ft long and 3 ft wide is immersed vertically in water in such a way that its longer side is parallel to the water surface and the 3 ft side is vertical. Find the total force acting on one side of the plate and its location.

4.5. A circular plate, 50 cm in diameter, is placed vertically in oil of specific gravity 0.9. The center of the plate is located 2 m below the oil surface. Calculate the total force acting on one side and its point of action.

4.6. A 10.0-m-long vertical wall separates seawater (specific gravity, S.G. = 1.025) from freshwater. If the depth of seawater is 8 m, what depth of freshwater is required to give zero resultant force? Calculate the moments at the toe of the wall.

4.7. An isosceles triangular plate of 2 m base and 2 m altitude is immersed vertically in an oil of specific gravity 0.85, as shown in Figure P4.7. Find the resultant force and its point of action.

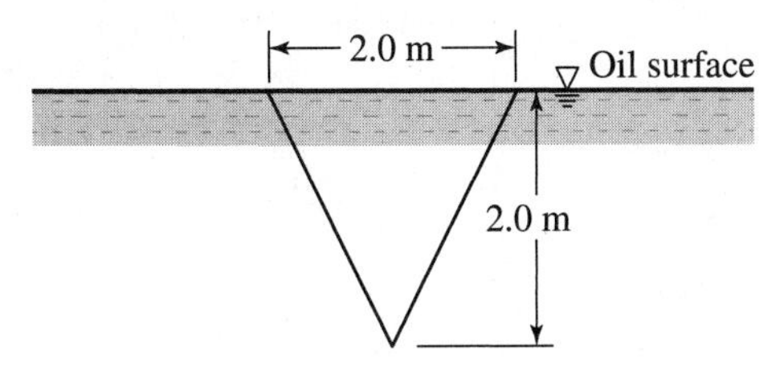

Figure P4.7 Isosceles triangular plate.

4.8. A circular plate 2 m in diameter is submerged in water with its higher and lower edges located at 2 m and 2.6 m, respectively, below the water surface. Find the resultant force on one side of the plate and the position of the center of pressure.

4.9. A square gate (4 ft by 4 ft) is located on the 45° face of a dam. The top edge of the gate lies 25 ft below the water surface. Determine the force of the water on the gate and its point of action.

4.10. The square gate, shown in Figure P4.10, is hinged at *A* and separates the water in the reservoir from the water in the tunnel. If the gate has dimensions of 3 m × 3 m and weighs 15 t-m/s^2, what is the maximum height, h, such that the gate will remain closed?

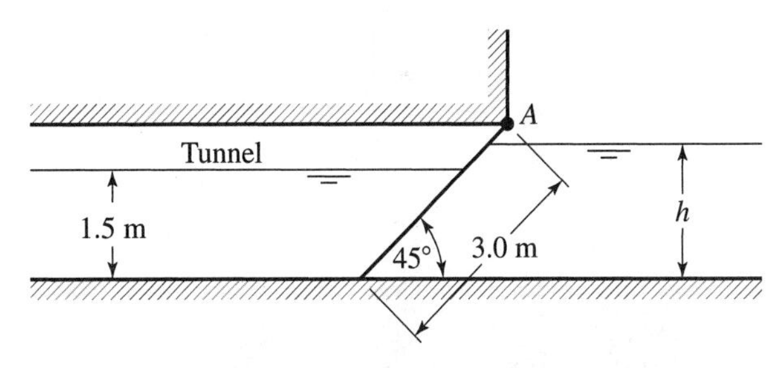

Figure P4.10 A square gate for water separation.

4.11. A gate having a triangular shape shown in Figure P4.11 is located in the vertical side of an open tank. The gate is hinged about the horizontal axis *AB*. Determine the moment exerted by the force of the water on the gate about the axis *AB*.

4.12. A 3-ft-diameter circular plate is located on the vertical side of an open tank containing gasoline. The force exerted by gasoline on the plate acts 3 in below the centroid of the plate. Determine the magnitude of this force.

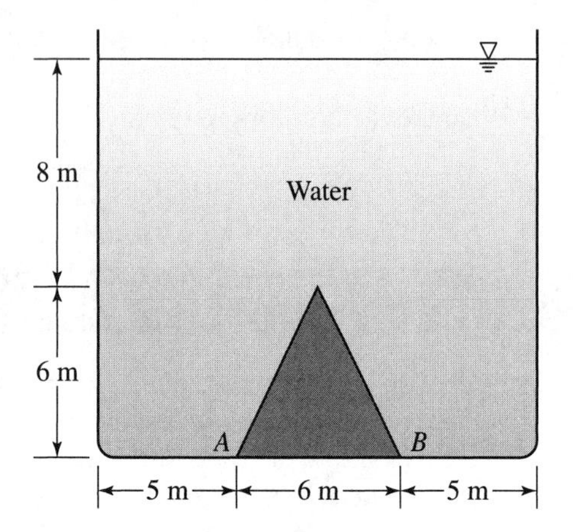

Figure P4.11 A triangular gate in a vertical side of a tank.

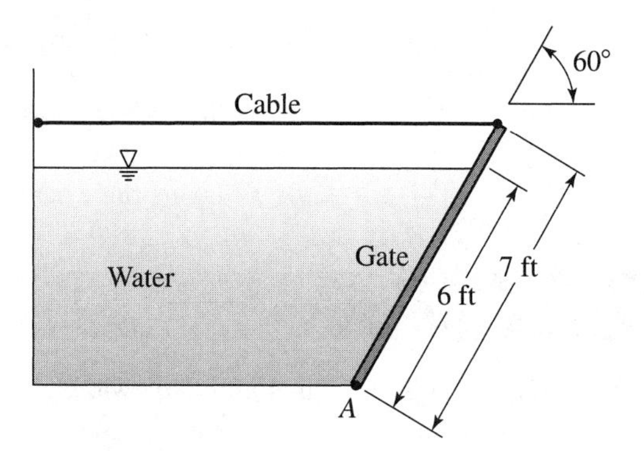

Figure P4.14 A rectangular gate held by a horizontal cable.

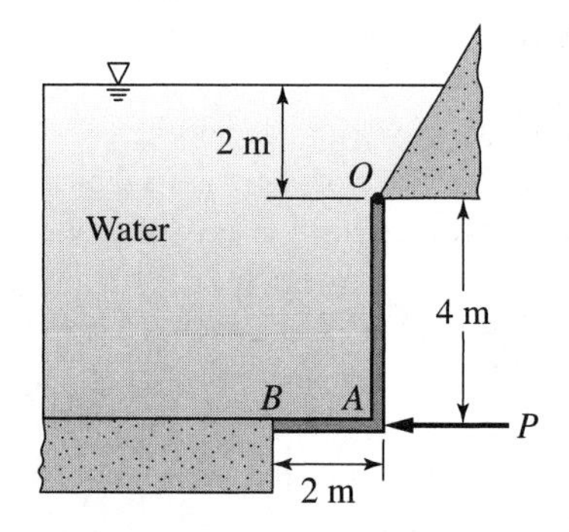

Figure P4.15 A 2-m-wide gate hinged at O.

4.13. The rectangular gate CD shown in Figure P4.13 is 2 m wide and 2.5 m long. Neglecting the friction at the hinge C, determine the weight of the gate necessary to keep it shut until the water level rises to 2 m above the hinge.

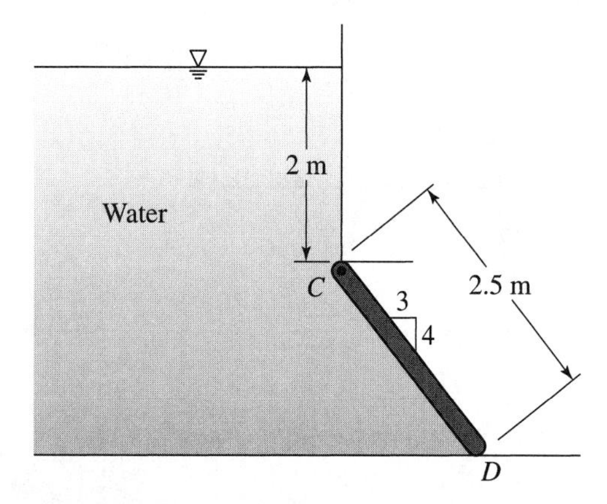

Figure P4.13 Rectangular gate CD.

4.14. A 3-ft-wide and 7-ft-long rectangular gate weighing 650 lb is held by a horizontal cable as shown in Figure P4.14. Neglecting the friction at the hinge A, determine the tension in the cable.

4.15. A 2-m-wide gate OAB is hinged at O and rests against a rigid support at B as shown in Figure P4.15. What minimum horizontal force, P, is required to hold the gate closed? Neglect the weight of the gate.

4.16. The gate shown in Figure P4.16 is mounted on a horizontal shaft. (a) Determine the force acting on the rectangular portion and that acting on the semicircular portion of the gate. (b) Determine the moment of the force acting on the semicircular portion about the shaft axis.

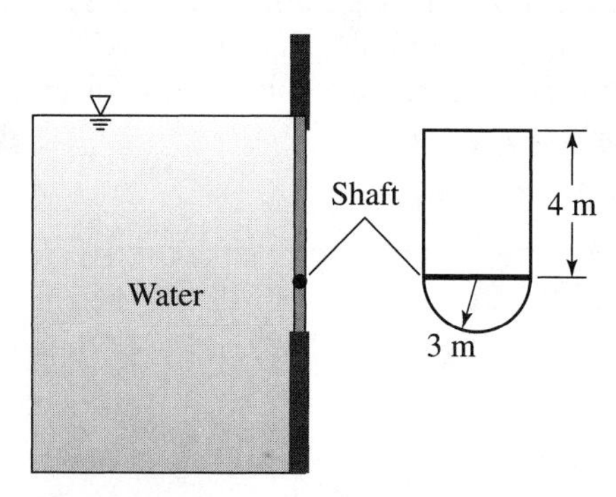

Figure P4.16 Gate setting.

4.17. The concrete dam shown in Figure P4.17 weighs 23.6 kN/m^3 and rests on a solid foundation. Determine the minimum coefficient of friction between the dam and the foundation required to keep the dam from sliding. Base your analysis on a unit length of the dam.

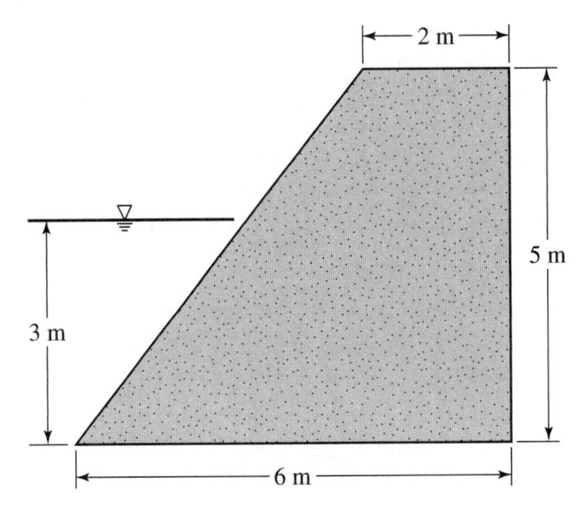

Figure P4.17 Physical setting of the dam.

4.18. A 2-m-diameter gate shown in Figure P4.18 swings around a horizontal pivot C located 50 mm below the center of the gate. To what depth, h, can the water rise without causing an unbalanced clockwise moment about pivot C?

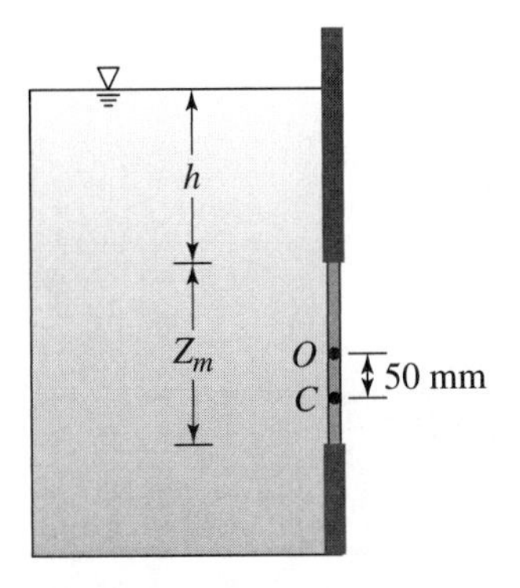

Figure P4.18 A 2-m-diameter gate.

4.19. A cubic block of 0.5 m on each side floats in water with 1/10 of its volume being out of water. Find the weight of the block.

4.20. An irregularly shaped piece of a solid material weighs 98.1 kg-m/s^2 in air and 58.86 kg-m/s^2 when fully submerged in water. Determine the specific weight of this material.

4.21. A hollow metallic sphere weighs 9.81 kg-m/s^2 and floats in water. What should be the external diameter of the sphere such that it can float half submerged?

4.22. A cylinder of specific gravity 0.95 and length 0.8 m floats with its axis vertical in a vessel containing water and oil of specific gravity 0.84. What is the minimum depth of oil required to ensure that the cylinder will be totally immersed?

4.23. A cube of steel 0.25 m on each side floats in mercury. The specific gravities of steel and mercury are 7.6 and 13.6, respectively. Determine the submerged depth of the cube.

4.24. What is the magnitude and direction of the force, F, required to hold the cube given in Problem 4.23 under mercury?

4.25. A metal cube, 50 cm on each side, has a specific gravity of 8. Determine the magnitude and direction of the force required to hold the cube in equilibrium (a) under water, and (b) under mercury. The specific gravity of mercury is 13.6.

4.26. A block of wood 15 × 10 × 4 m floats in oil of specific gravity 0.8. A couple holds the block in the position shown in Figure P4.26. Determine the buoyant force and the magnitude of the couple acting on the block and the position of the metacenter.

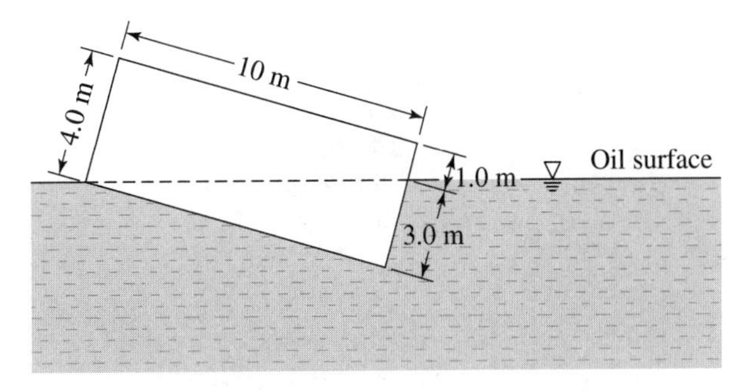

Figure P4.26 A floating block of wood.

4.27. A cube of solid wood, 1.0 m on each side, is placed in water. The specific gravity of the wood is 0.6. Check the stability of this cube if placed in (a) oil (specific gravity = 0.7) (b) water, and (c) mercury (specific gravity = 13.6).

4.28. A rectangular block 6 ft long, 2 ft wide, and 1 ft deep, floats in water. Determine the state of its equilibrium if it weighs 9.81 kN.

4.29. A solid wood cylinder has a diameter of 0.5 m and a height of 0.8 m. The specific gravity of the wood is 0.6. If the cylinder is placed vertically in oil (specific gravity = 0.8), would it be stable or not?

4.30. A stone weighs 100 N in air, and when immersed in water it weighs 60 N (see Figure P4.30). Compute the volume of the stone and its specific gravity.

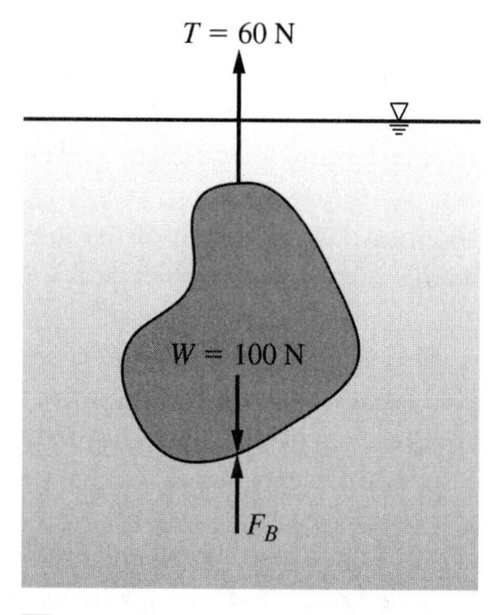

Figure P4.30 A stone immersed in water.

4.31. A prismatic object, 6 in. thick by 6 in. wide by 14 in. long, is weighed in water at a depth of 20 in and found to weigh 10 lb. What is its weight in air and its specific gravity?

4.32. A hydrometer weighs 0.022 N and has a stem at the upper end that is cylindrical and 2.5 mm in diameter. How much deeper will it float in oil of S.G. = 0.780 than in alcohol of S.G. = 0.821?

4.33. A 120 mm square piece of wood of S.G. = 0.75 is 1.5 m long. How many Newtons of lead weighing 110 kN/m^3 must be fastened at one end of the stick so that it will float upright with 0.3 m out of water?

4.34. What fraction of the volume of a solid piece of metal of S.G. = 6.50 floats above the surface of a container of mercury of S.G. = 13.57?

4.35. A rectangular open box, 6.5 m by 3.5 m in plan and 4 m deep, weighs 360 kN and is launched in fresh water. (a) How deep will it sink? (b) If the water is 4 m deep, what weight of stone placed in the box will cause it to rest on the bottom?

4.36. A regular block of wood floats in water with 60 mm projecting above the water surface. When placed in glycerin of S.G. = 1.35, the block projects 85 mm above the surface of that liquid. Determine the S.G. of the wood.

4.37. What length of 80 mm by 310 mm timber (S.G. = 0.50) will support a 445 N man in salt water (S.G. = 1.025) if he stands on the timber.

4.38. A barge (length = 40 ft) with a flat bottom and rectangular ends, as shown in Figure P4.38, has a draft of 6.0 ft when fully loaded and floating in an upright position. Is the barge stable? If the barge is stable, what is the righting moment in water when the angle of heel is 15°?

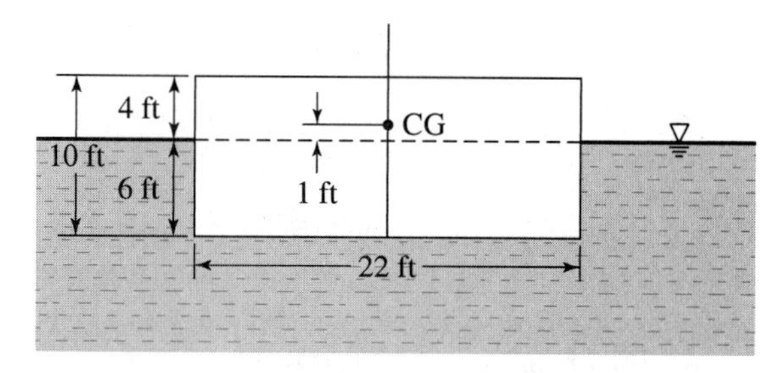

Figure P4.38 A barge with a flat bottom and rectangular ends.

4.39. Would the solid wood cylinder (Figure P4.39) be stable if placed vertically in oil? The specific gravity of wood is 0.60 and the specific gravity of oil is 0.857.

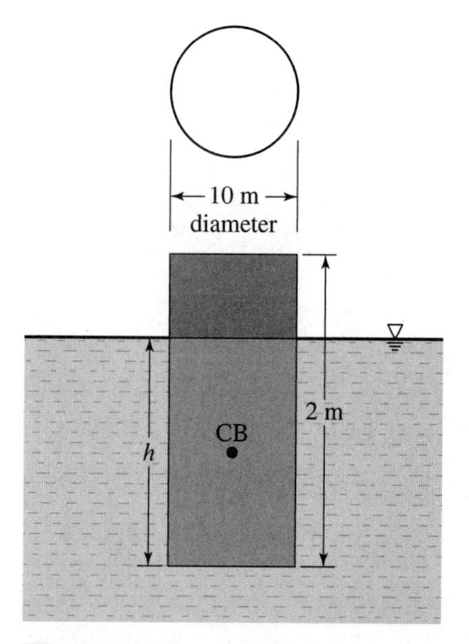

Figure P4.39 A solid cylinder of wood placed vertically in oil.

4.40. Determine the gage pressure at point A (see Figure P4.40).

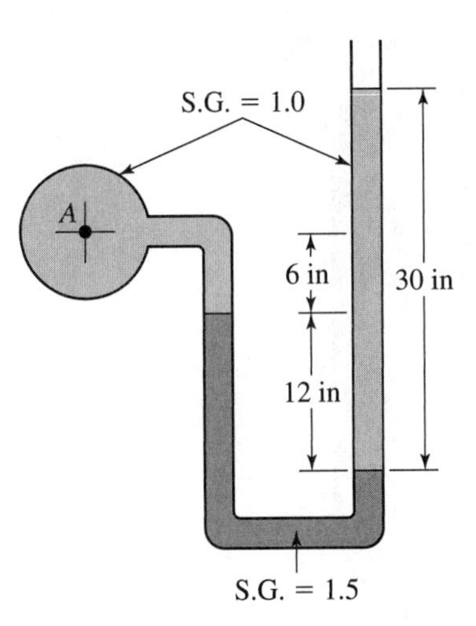

Figure P4.40 U-tube manometer.

4.41. Determine the gage pressure at point A (Figure P4.41) in pounds per square inch when the temperature is 70°F.

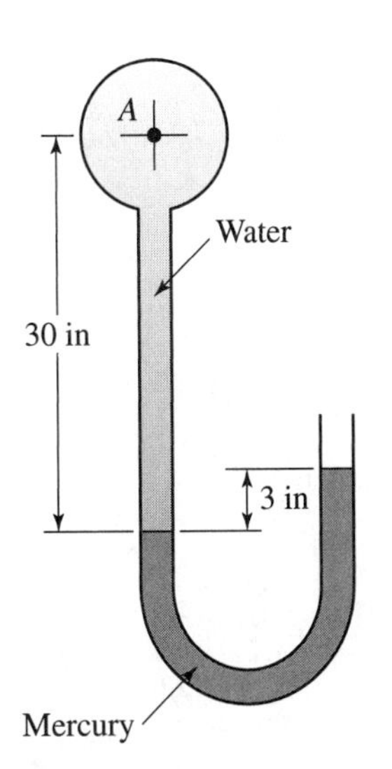

Figure P4.41 U tube manometer.

4.42. Determine the gage pressure at point A (Figure P4.42).

4.43. Assume all the dimensions of Problem 4.42 (Figure P4.42) are in feet instead of meters. Determine the gage pressure at point A.

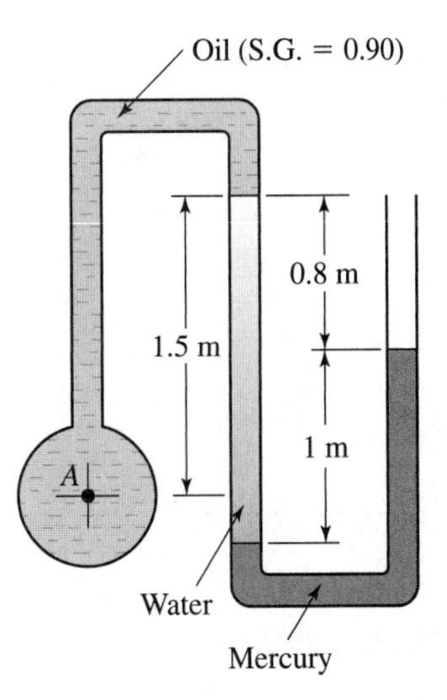

Figure P4.42 Differential U-tube manometer.

4.44. Determine the gage pressure at point A (Figure P4.44) in kilopascals and in pounds per square inch.

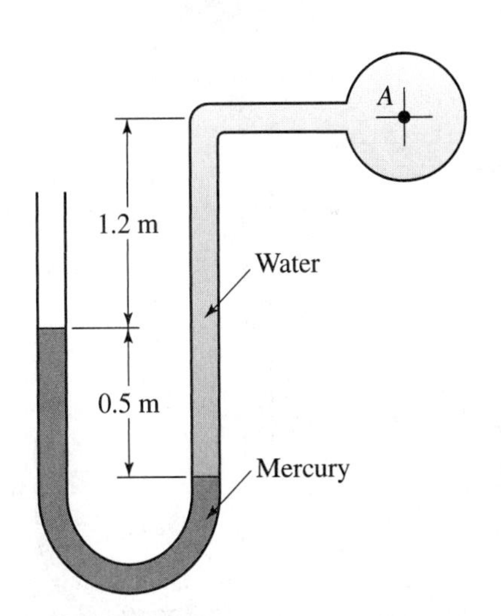

Figure P4.44 Differential U-tube manometer.

4.45. What is the gage pressure at point A (Figure P4.45)?

4.46. Water is poured into the tube shown in Figure P4.46 until it occupies 15 in. of the tube's length. Then oil (S.G. = 0.87) having a volume of 1/3 of water volume is poured in the right leg. Determine the difference between water and oil surfaces.

4.47. For the differential U-tube manometer shown in Figure P4.47, determine $(P_A - P_B)$.

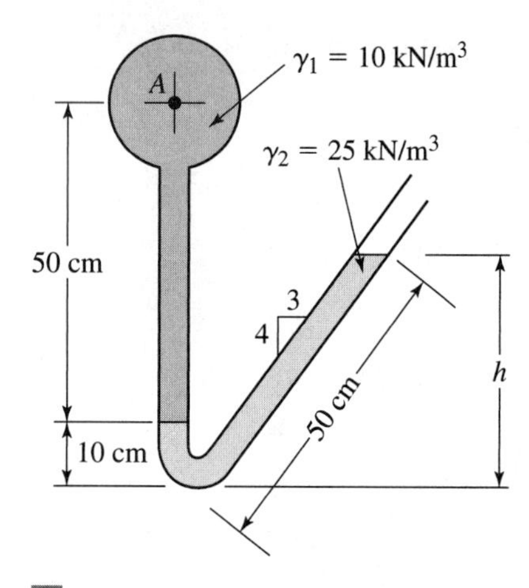

Figure P4.45 Inclined piezometer tube.

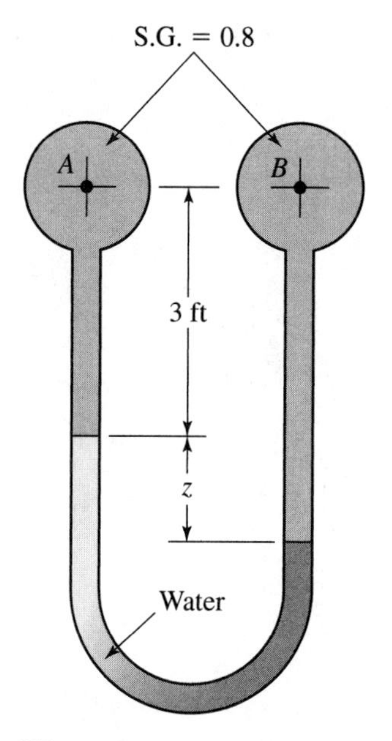

Figure P4.48 Differential U-tube manometer.

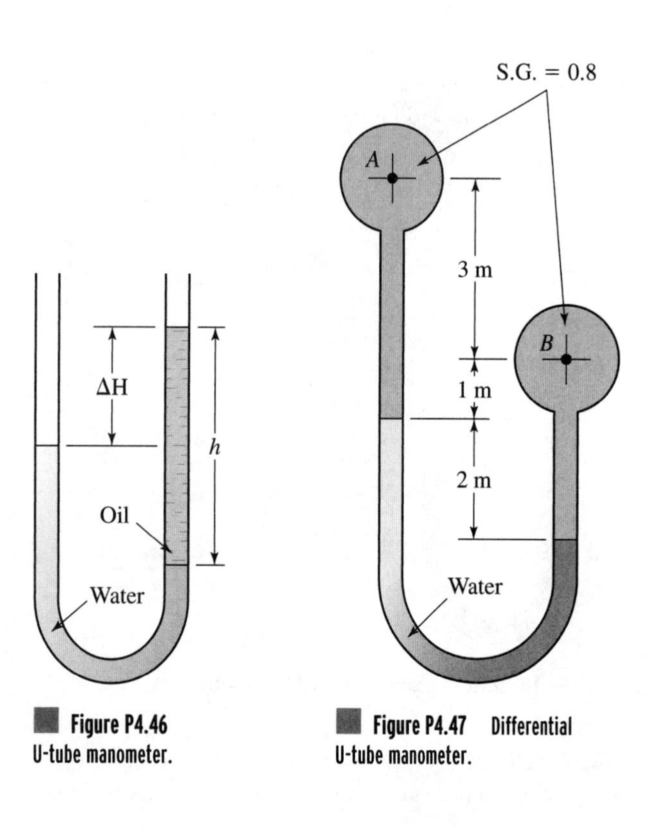

Figure P4.46 U-tube manometer.

Figure P4.47 Differential U-tube manometer.

4.48. For the differential U-tube manometer shown in Figure P4.48, determine z if $(P_B - P_A) = 0.30$ psi. Take the specific weight of water as 62.4 lb/ft³.

4.49. Determine the gage pressure at point A (Figure P4.49) in psi when the temperature is 10°C.

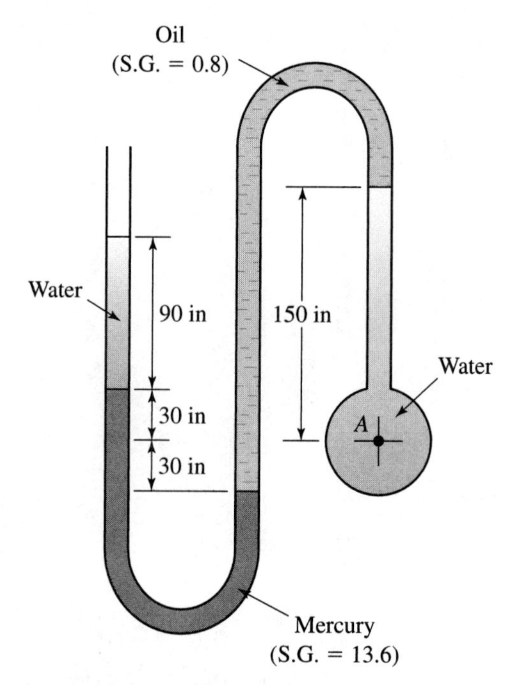

Figure P4.49 Differential U-tube manometer.

4.50. A U-tube manometer is connected to a water supply pipe in order to monitor the pressure in the line as shown in Figure P4.50. If the manometer fluid is mercury (S.G. = 13.6) and the readings are as shown in the figure, what is the pressure difference in the line between points *A* and *B*?

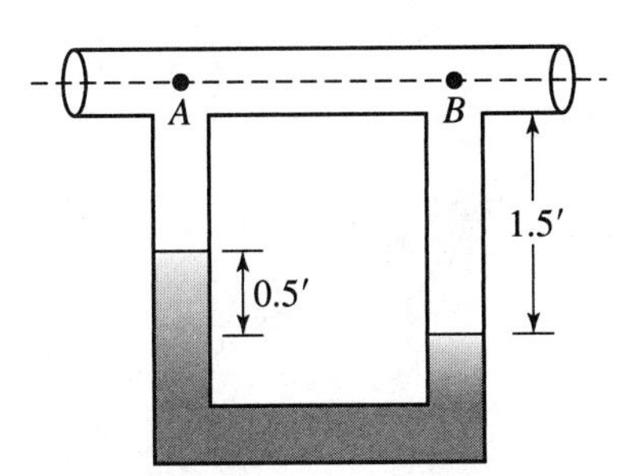

Figure P4.50 Schematic for Problem 4.50.

4.51. A differential manometer is connected across a Venturi meter as shown in Figure P4.51. The piezometer tube located upstream of the meter shows a reading of 5 ft as shown. The manometer fluid is mercury. If the manometer reading is 4 inches, what is the pressure at point *B* in psi?

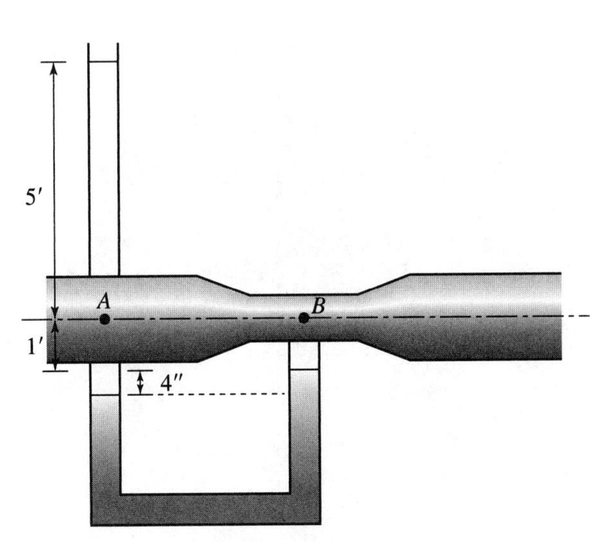

Figure P4.51 Schematic for Problem 4.51.

4.52. A differential manometer is connected across the vertical pipe as shown in Figure P4.52. If the manometer fluid is mercury and the readings and dimensions are as shown in the figure, what is the difference in pressure between points *A* and *B* in psi?

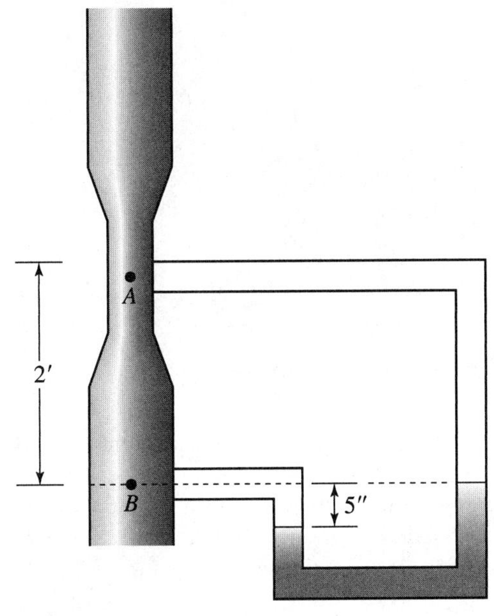

Figure P4.52 Schematic for Problem 4.52.

4.53. A paved levee embankment has a 45° slope as shown in Figure P4.53. A rectangular gate, 6 ft square, in the face of the embankment has its center 20 ft above the bed. When the river water surface is 40 ft above the bed, what is the total hydrostatic force on the gate and where is it applied?

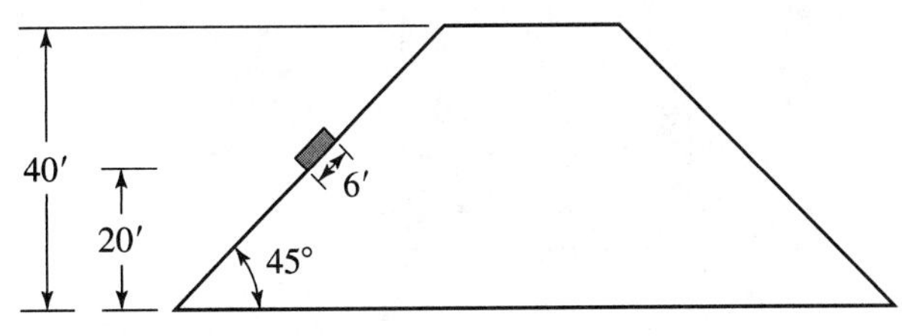

Figure P4.53 A 6-foot-square gate in a levee.

4.54. A spillway is constructed in a triangular shape with a height of 30 ft and a downstream face on a 45° angle as shown in Figure P4.54. The weight of the concrete is 150 lb/cubic feet. If the water level behind the spillway is 35 ft high, the foundation is impermeable, and tailwater can be ignored, what is the resultant reaction force and where is it located?

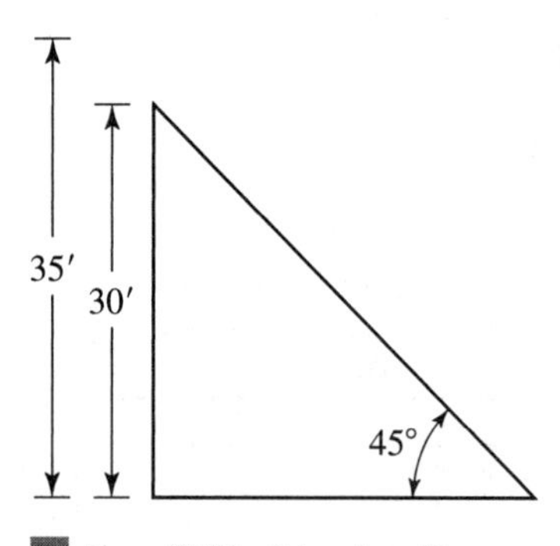

Figure P4.54 Triangular spillway.

4.55. Rework Problem 4.54 for a fully permeable foundation.

4.56. For the system of fluids shown in following Figure P4.56, determine the density of unknown oil, if atmospheric pressure is 101 kPa and pressure at the base of the tank is 169 kPa. $H = 0.5$ m.

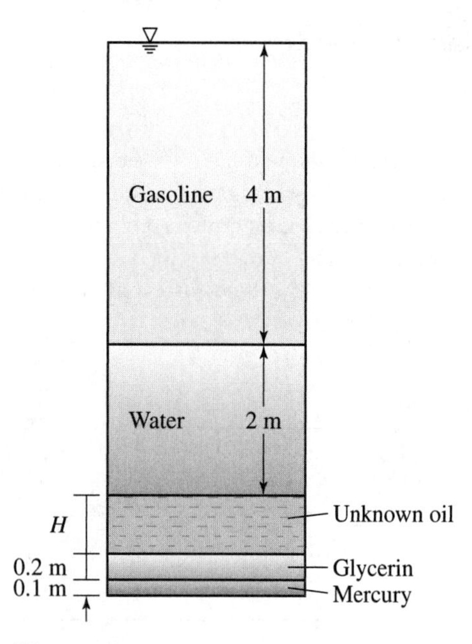

Figure P4.56 Tank with fluids.

4.57. In the figure of Problem 4.56, if the density of unknown oil is 1100 kg/m^3, determine the height H to maintain the same pressure at the base.

4.58. For Figure P4.58, liquid in the inclined pipe is water and that in the manometer is mercury. If the water is flowing downward, determine the total pressure drop between points 1 and 2.

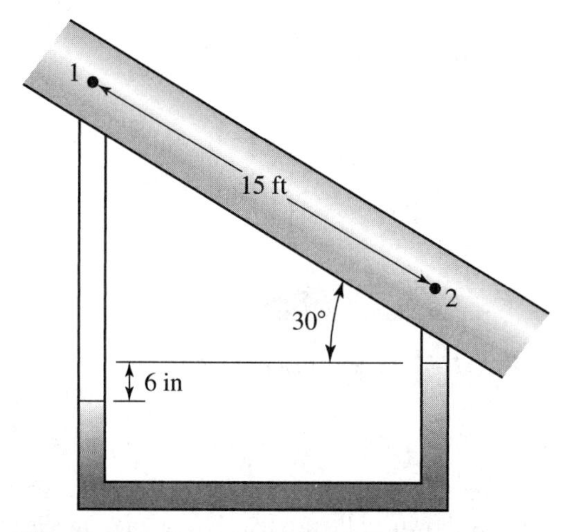

Figure P4.58 Inclined pipe drop manometer.

4.59. Figure P4.59 shows a slanted tube connected to a reservoir containing oil and water. If angle of inclination of the tube is 45°, determine the length L up to which water is filled in the tube.

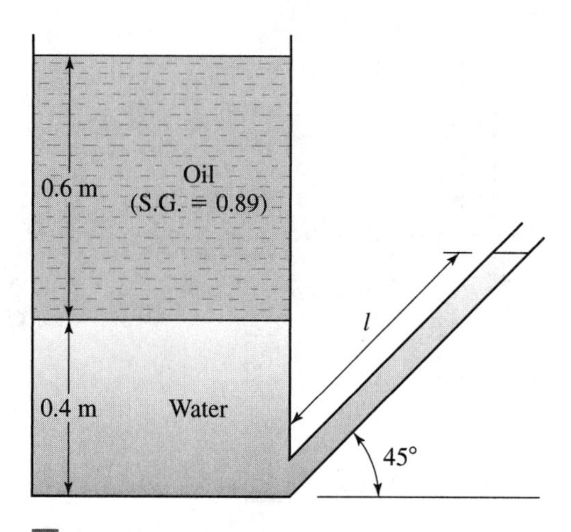

Figure P4.59 Inclined pipe from a tank.

4.60. Determine pressures at point A and point B in Figure P4.60. The vessel contains water at 20°C.

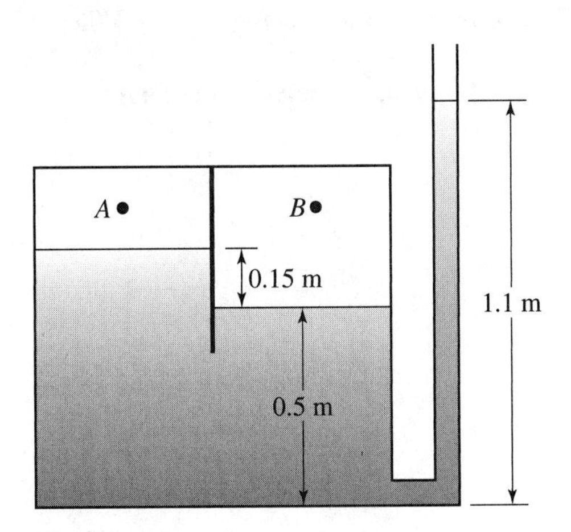

Figure P4.60 A tank with a stand pipe.

4.61. Solve Problem 4.60 if the water is replaced by glycerin (S.G. = 1.26).

4.62. For the problem shown Figure P4.62, determine the density of unknown fluid in the BG units.

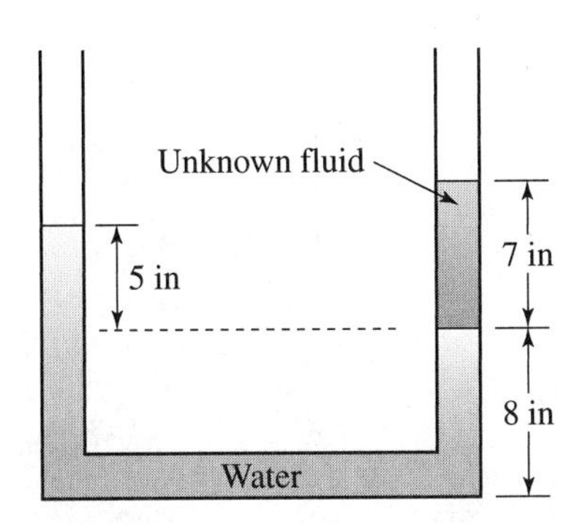

Figure P4.62 A manometer with unknown fluid.

4.63. A flood gate weighing 2000 lb is lying at an inclination of 5° as shown in Figure P4.63. The center of gravity of the gate is 45 in along the gate from the hinge point. Determine the depth of the water level at which the gate will start opening.

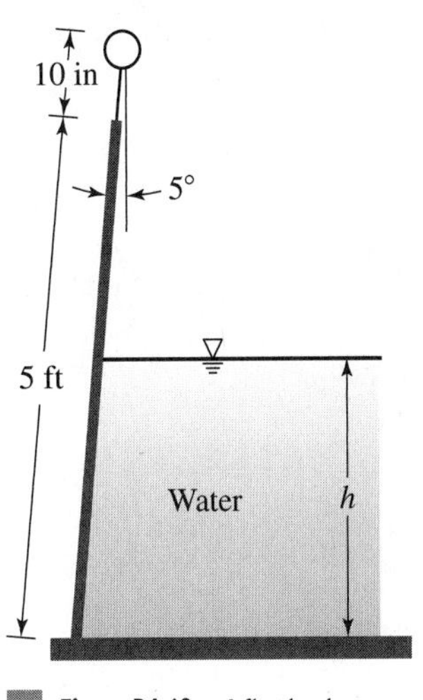

Figure P4.63 A flood gate.

4.64. A triangular dam of 50 m width into the paper is made up of concrete (S.G. = 2.4). Determine the hydrostatic force on the inclined plane and the moment due to this force on point A. Can this force over turn the dam?

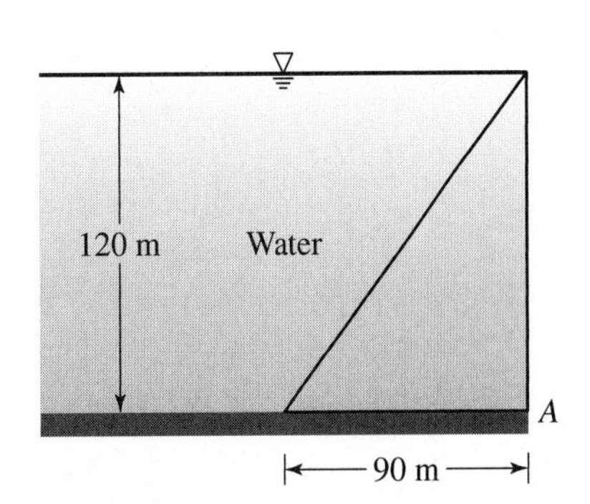

Figure P4.64 A triangular dam.

References

Fox, R. W. and McDonald, A. T., 1984. *Introduction to Fluid Mechanics*, Wiley, New York.

Morris, H. M. and Wiggert, J. M., 1972. *Applied Hydraulics in Engineering*, Wiley, New York.

Munson, B. R., Young, D. F., and Okiishi, T. H., 1990. *Fundamentals of Fluid Mechanics*, Wiley, New York.

CHAPTER 5

GOVERNING EQUATIONS

University Lake at the University of Alabama at Huntsville.

The flow of water in hydraulic conduits is governed by (1) conduit geometry, (2) fluid properties, and (3) two sets of constraints, external, and internal. The conduit geometric characteristics comprise area, wetted perimeter, hydraulic radius, and roughness of boundaries and will be discussed in Chapter 10. The fluid properties include density, viscosity, pressure, and compressibility (discussed in Chapter 2). External constraints, actually common to all physical phenomena, are the laws of conservation of mass, energy, and momentum. Internal constraints of hydraulic processes encompass the entropy increase law and the space-mass-time dimensional law and constitute the internal mechanics of the hydraulic systems.

The term entropy *is used to describe the disorder of a system. For a closed system, the entropy always tends to increase. In thermodynamics, entropy is usually defined in terms of the irreversible component of heat. In hydraulics, for a fluid to maintain its flow it has to perform work to overcome frictional forces resisting flow. This involves irreversible conversion of the flow energy into heat energy through friction. Thus the increase in entropy in a hydraulic system usually occurs due to the conversion of flow energy into heat through friction. The heat energy is conducted through the fluid and its boundary walls, if any, into the atmosphere and is eventually dissipated through space. The amount of energy so deployed is a function of the flow system geometry and the types of forces and energy affecting the system. The quantitative formulation of this function can be called the process equation, which is what leads to the resistance equation in hydraulics.*

Therefore, it can be concluded that every flow process is a conservation process, according to the energy conservation principle (the first law of thermodynamics) and a decay process according to the entropy-increase principle (the second law of thermodynamics). Because the flow process operates in the universe, which is a continuum of space and mass and time, the process equation must be expressible in space and mass and time dimensions (e.g., the meter-kilogram-second system).

The equations of continuity, energy, and momentum constitute the basic tools to address flow problems. It is, therefore, important to recall their underlying assumptions and the consequent limitations. These equations have been derived in different ways in one, two or three dimensions (Chiu et al., 1978; Singh, 1996; Taft and Reisman, 1965; Yen, 1973). In what follows, a brief discussion of these equations is presented.

5.1 Mass Conservation: The Continuity Equation

The law of mass conservation states that a matter can be neither created nor destroyed, though it can transform to some other form. Hence, the mass of water flowing in a hydraulic system (e.g., a pipe or an open channel) cannot increase or decrease unless water is added to or subtracted from the system. A mathematical statement of the mass conservation is the continuity equation.

The mass of a matter in a system, by definition, is composed of a collection of particles of fixed identity. The mass or quantity of this matter, m, does not change with time, t. Hence,

$$\left(\frac{dm}{dt}\right)_{\text{system}} = 0 \tag{5.1a}$$

where the mass in the system is

$$m_{\text{system}} = \int_{\text{mass (system)}} dm = \int_{V\text{(system)}} \rho \, dV_{\text{ol}} \tag{5.1b}$$

where ρ is the mass density, V_{ol} is the volume, dV_{ol} is the elemental volume, and dm is the elemental mass $= \rho dV_{\text{ol}}$.

For a control volume, the continuity equation is expressed as

$$\{\text{Rate of mass inflow}\} - \{\text{Rate of mass outflow}\} = \{\text{Rate of mass accumulation}\} \tag{5.2}$$

The mass conservation law is universally valid, except for processes involving mass to energy conservation. In such cases, a broader law of mass-energy conservation is applied. The law of mass conservation can be applied to various types of control volumes and different coordinate systems. Consider a control volume with dimensions of dx, dy, and dz in the three-dimensional fixed coordinate system x, y, and z, as shown in Figure 5.1. The mass of fluid flowing into or out of any face of the control volume per unit time is equal to the product of the density, the velocity, and the cross-sectional area normal to the direction of flow. Let v_x, v_y, and v_z be the average velocities of flow in the x, y, and z directions, respectively. The mass inflows into the control volume along the x axis, y axis, and z axis, respectively, are

$$\rho v_x \, dy \, dz \tag{5.3a}$$

$$\rho v_y \, dx \, dz \tag{5.3b}$$

$$\rho v_z \, dx \, dy \tag{5.3c}$$

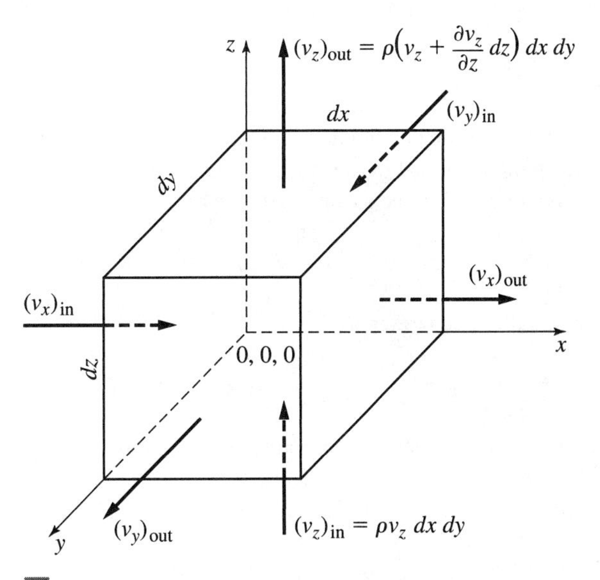

Figure 5.1 Control volume for mass conservation.

Using the Taylor series expansion, the approximate mass outflows from the control volume in the *x*, *y*, and *z* directions, respectively, are

$$\rho v_x \, dy \, dz + \frac{\partial}{\partial x}[\rho v_x \, dy \, dz] \, dx \tag{5.4a}$$

$$\rho v_y \, dx \, dz + \frac{\partial}{\partial y}[\rho v_y \, dx \, dz] \, dy \tag{5.4b}$$

$$\rho v_z \, dx \, dy + \frac{\partial}{\partial z}[\rho v_z \, dx \, dy] \, dz \tag{5.4c}$$

The approximate net mass inflow to the control volume (accumulation of mass) in the *x* direction is

$$-\frac{\partial}{\partial x}[\rho v_x \, dy \, dz] \, dx \tag{5.5a}$$

Similarly, the net mass inflow in the *y* direction is

$$-\frac{\partial}{\partial y}[\rho v_y \, dx \, dz] \, dy \tag{5.5b}$$

and the net mass flow in the *z* direction is

$$-\frac{\partial}{\partial z}[\rho v_z \, dx \, dy] \, dz \tag{5.5c}$$

Therefore, the approximate net mass accumulation in the control volume in the three directions is

$$-\left[\frac{\partial}{\partial x}\rho v_x + \frac{\partial}{\partial y}\rho v_y + \frac{\partial}{\partial z}\rho v_z\right] dx \, dy \, dz \tag{5.6}$$

The approximation becomes more accurate if the dimensions of the control volume *dx*, *dy*, and *dz* are diminished. The rate of change of mass within the control volume is

$$\frac{\partial \rho}{\partial t} \, dx \, dy \, dz \tag{5.7}$$

where $\partial\rho/\partial t$ is the time rate of change of density within the control volume. The net inflow to the control volume is equal to the rate of change of mass within the control volume. Therefore, equating equation (5.7) to equation (5.6), one obtains

$$-\left[\frac{\partial}{\partial x}\rho v_x + \frac{\partial}{\partial y}\rho v_y + \frac{\partial}{\partial z}\rho v_z\right] dx \, dy \, dz = \frac{\partial \rho}{\partial t} \, dx \, dy \, dz \tag{5.8a}$$

Equation (5.8a) can be written as

$$\frac{\partial \rho}{\partial t} = -\left[\frac{\partial}{\partial x}\rho v_x + \frac{\partial}{\partial y}\rho v_y + \frac{\partial}{\partial z}\rho v_z\right] \tag{5.8b}$$

Equation (5.8b) is the continuity equation in three dimensions and accounts for the unsteadiness of flow and compressibility of the fluid.

For two-dimensional flow (in the *x* and *y* directions), equation (5.8b) reduces to

$$\frac{\partial \rho}{\partial t} = -\left[\frac{\partial}{\partial x}\rho v_x + \frac{\partial}{\partial y}\rho v_y\right] \tag{5.9}$$

and for one-dimensional flow (in the x direction), it becomes

$$\frac{\partial \rho}{\partial t} = -\left[\frac{\partial}{\partial x}\rho v_x\right] \tag{5.10}$$

For steady-state conditions, equation (5.8b) becomes

$$\frac{\partial}{\partial x}\rho v_x + \frac{\partial}{\partial y}\rho v_y + \frac{\partial}{\partial z}\rho v_z = 0 \tag{5.11}$$

For incompressible flow, equation (5.11) reduces to

$$\frac{\partial v_x}{\partial x} + \frac{\partial v_y}{\partial y} + \frac{\partial v_z}{\partial z} = 0 \tag{5.12}$$

or simply

$$\nabla \cdot \mathbf{V} = 0 \tag{5.13}$$

where ∇ is the del operator expressed as

$$\nabla = \left(\frac{\partial}{\partial x}, \frac{\partial}{\partial y}, \frac{\partial}{\partial z}\right)$$

and $\mathbf{V}$ is the velocity vector $= \{v_x, v_y, v_z\}$.

If flow is considered along a stream tube or a stream line, where $dy\,dz$ in equation (5.8a) can be replaced by the differential area of the stream tube, δA, and the flow is one dimensional and incompressible then one can write equation (5.8a) for steady-state conditions as

$$\frac{\partial}{\partial x}(\rho v_x\,\delta A)\,dx = 0 \tag{5.14}$$

Referring to a differential stream tube shown within a finite stream tube in Figure 5.2, integration of equation (5.14) along the stream tube yields

$$\rho v_x\,\delta A = \text{Constant} \tag{5.15}$$

or

$$(\rho v_x\,\delta A)_1 = (\rho v_x\,\delta A)_2 \tag{5.16}$$

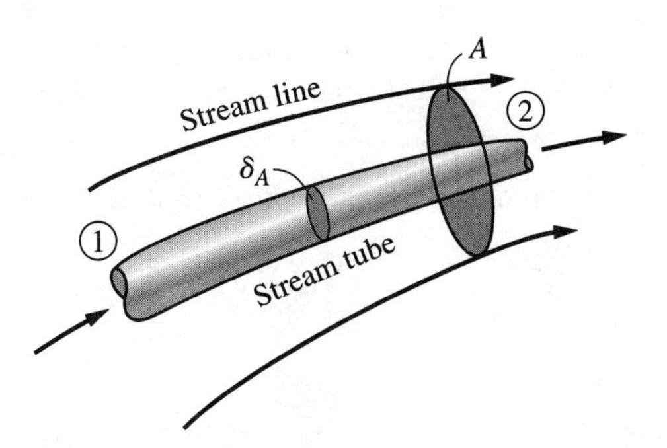

Figure 5.2 A stream tube.

where subscripts 1 and 2 denote two cross sections of the stream tube. Integrating equation (5.16) over the entire area, A, of the finite stream tube, and assuming that the velocity is normal to the cross section, one obtains

$$\rho_1 V_1 A_1 - \rho_2 V_2 A_2 = \frac{dm}{dt} \tag{5.17}$$

where V is the average velocity normal to the cross section, subscripts 1 and 2 denote the serial numbers of the cross section, and dm/dt is the rate of mass flow.

For incompressible flow, equation (5.17) can be written as

$$A_1 V_1 = A_2 V_2 = AV = \text{constant} = Q \tag{5.18}$$

where Q is the discharge ($L^3 T^{-1}$), which is the product of the cross-sectional area by the average velocity $= AV$.

The most common application of the stream tube concept in practice is to apply it to the whole region of flow. The boundaries of the stream tube are the physical boundaries of a hydraulic conduit (i.e., pipe, stream, etc.).

Example 5.1

A pipe of 20.0 cm diameter is carrying a discharge of 0.1 m^3/s. If the diameter of the pipe is reduced to 5.0 cm at its end, calculate the velocity and velocity head at the pipe outlet.

Solution: The flow is considered incompressible (i.e., the density is constant). Then

$$A_1 V_1 = A_2 V_2 = Q$$

$$A_2 = \pi d^2/4 = 0.0019635 \text{ m}^2$$

$$V_2 = Q/A_2 = 0.1/0.0019635 = 50.930 \text{ m/s}$$

The velocity at the outlet is 50.930 m/s. The velocity head at the pipe outlet is

$$V_2^2/2g = \frac{(50.930)^2}{2 \times 9.81} = 132.20 \text{ m}$$

Example 5.2

A river discharges at a rate of 500 ft^3/s into a lake, as shown in Figure 5.3. The water is drawn from the lake at a rate of 200 ft^3/s. The surface area of the lake is 60 acres. What is the rate of water rise in the lake in one day?

Solution: Applying the continuity equation for the control volume shown in Figure 5.3,

$$Q_{\text{in}} = Q_{\text{out}} + Q_{\text{rise}}$$

$500 = 200 + Q_{\text{rise}}$. Hence, $Q_{\text{rise}} = 300 \text{ ft}^3/\text{s}$.

$$Q_{\text{rise}} = A_{\text{lake}} \times v_{\text{rise}}$$

$300 = 60 \text{ acres} \times 43560 \text{ ft}^2/\text{acre} \times V_{\text{rise}}$, or $V_{\text{rise}} = 1.14784 = 10^{-4}$ ft/s.

Rate of rise in one day $= 1.14784 \times 10^{-4}$ ft/s $(24 \times 60 \times 60) = 9.92$ ft/day

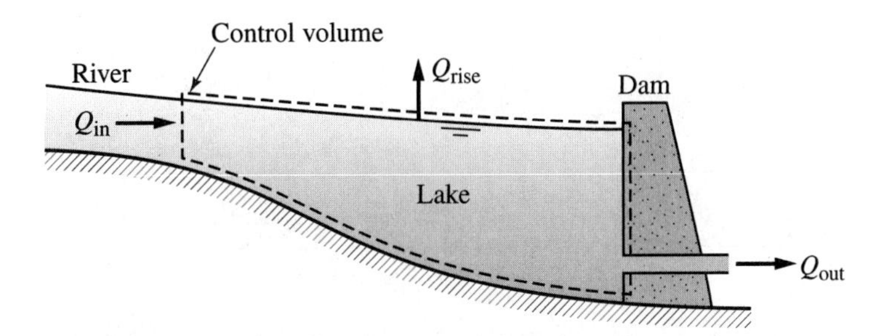

Figure 5.3 Control volume in a lake (Example 5.4).

Example 5.3

Air flows in a 5.0-cm diameter pipe at a velocity of 10.0 m/s under a pressure of 30.0 psi. Find the velocity at another section where the pipe diameter is 10.0 cm and the pressure is 15.0 psi. Consider the flow to be isothermal, where $\rho_1/\rho_2 = p_1/p_2$.

Solution: The mass conservation equation can be written as

$$\rho_1 A_1 V_1 = \rho_2 A_2 V_2$$

Therefore, $V_2 = (\rho_1 A_1 V_1)/(\rho_2 A_2)$.
For isothermal flow $\rho_1/\rho_2 = p_1/p_2$. Hence, the velocity at the other section is

$$V_2 = (p_1 A_1 V_1)/(p_2 A_2) = (p_1 d_1^2 V_1)/(p_2 d_2^2)$$

$$= (30.0 \times 5.0 \times 5.0 \times 10.0)/(15.0 \times 10.0 \times 10.0) = 5.0 \text{ m/s}$$

5.1.1 Continuity Equation in Open Channels

The continuity equation for steady flow in open channels can be expressed as

$$Q = Q_1 = Q_2 = \cdots = Q_n \tag{5.19a}$$

where Q is the discharge and subscripts 1, 2, . . . , n, denote sections of the channel. Equation (5.18) assumes incompressible flow as does equation (5.19a).

In open-channel flows, a complication, however, arises due to the changing water surface level. Consider the stream tube, of a short length Δx, to embrace the whole cross section of an open channel as shown in Figure 5.4. The discharges, Q_1 and Q_2, at the two ends may not be the same as water can accumulate in or evacuate from the control volume. Assuming that the volume of water within the control element changes by a rate of $\partial Q/\partial x$, one can write

$$\frac{\partial Q}{\partial x}\Delta x = Q_2 - Q_1 \tag{5.19b}$$

The discharge Q may vary with time and along the channel length. Let h be the height of the water surface above a specified datum. Then the rate of increase of the water volume, $\partial V_{ol}/\partial t$, between sections 1 and 2 can be expressed as

$$\frac{\partial V_{ol}}{\partial t} = B\frac{\partial h}{\partial t}\Delta x \tag{5.20a}$$

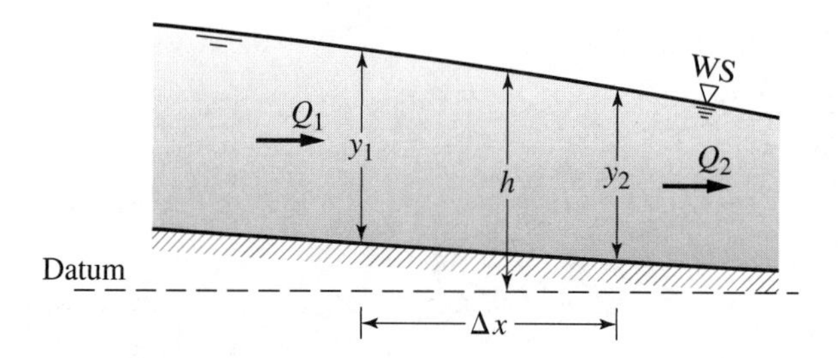

Figure 5.4 Definition sketch for the continuity equation in open channels.

where B is the width of the water surface (assumed constant) and $\partial V_{ol}/\partial t = \partial Q/\partial x$. The two terms evaluated from equations (5.19b) must be equal in magnitude but opposite in sign to equation (5.20a). Hence

$$\frac{\partial Q}{\partial x} + B\frac{\partial h}{\partial t} = 0 \tag{5.20b}$$

Equation (5.20b) is the continuity equation for unsteady flow in open channels.

In case of water withdrawal or addition from the lateral direction at a rate of q_x per unit length, equation (5.20a) is modified as

$$\frac{\partial Q}{\partial x} + B\frac{\partial h}{\partial t} = \pm q_x \tag{5.20c}$$

The positive sign for q_x refers to the case of additional lateral inflow to the channel and the negative sign refers to the lateral outflow from the channel.

5.1.2 Types of Flow

Flow in open channels may be classified into various types according to different criteria. A common way of classification is based on the variation of the depth of flow, velocity, or discharge with respect to time and space.

5.1.2.1 Steady and Unsteady Flow An open-channel flow is said to be steady if the depth of flow (y) at a certain location does not vary with time. Consequently, the velocity of flow (v) at any given point will not vary with time. Therefore, steady flow is the flow which satisfies the condition of zero local acceleration (i.e., $\partial v/\partial t = 0$). For three-dimensional steady flows the partial derivatives of all velocity components with respect to time are equal to zero. On the other hand, if the depth of flow and hence the velocity vary with time, then the flow is unsteady. Floods and surges are typical examples of unsteady flow. As waves advance, the stage of flow changes with time. The time variability of flow constitutes an important element in the design of hydraulic structures.

The discharge, Q, through an open channel, may be expressed as

$$Q = AV \tag{5.21}$$

where A is the cross-sectional area of flow normal to the direction of flow and v is the mean velocity. For constant discharge through different reaches of an open channel, one can write

$$Q = A_1V_1 = A_2V_2 = A_3V_3 = \cdots = A_iV_i \tag{5.22}$$

where subscripts 1, 2, . . . , i refer to the reach number. Equation (5.22) is the simple form of the continuity equation under steady-state conditions.

In some cases, it may be possible to transform an unsteady flow into a steady flow. This can be accomplished only if the stage and shape of the wave do not change while the wave propagates. A moving reference for the coordinates can be defined such that the reference is moved at the absolute velocity of the surge. This treatment offers many advantages, yet it is not possible to apply it in many cases. Waves generally deform as they propagate.

5.1.2.2 Uniform and Nonuniform (Varied) Flow If at a given time the depth of an open channel flow is constant throughout a channel reach, then the flow is uniform in this specified reach. Otherwise, it is nonuniform or varied. Hence, to define the uniformity of an open-channel flow, the depth of flow should be measured at different locations at the same time. If no differences are found, then the flow is uniform. This classification is, therefore, based on the variation of flow depth or flow velocity with space at a specified instant.

Partial derivatives of velocity components with respect to x, y, and z are all zero in uniform flow where convective acceleration does not exist. In many applications, it is assumed that $\partial v/\partial x = 0$, $\partial v/\partial y \neq 0$ and $\partial v/\partial z \neq 0$; this means the flow is considered uniform along the direction of flow, while allowing nonuniform velocity distribution in the lateral and vertical directions. In other words, the flow is considered uniform as long as no variations in depth and velocity along the flow direction are encountered at a certain instant of time. The pressure head, p/γ, and the velocity head, $V^2/2g$, do not vary throughout the reach of uniform flow, where p is pressure, γ is the specific weight, and g is acceleration due to gravity. Therefore, the bottom of the channel and the water surface or the hydraulic gradient line and the total energy line (defined in subsequent chapters) are parallel to each other (all have the same gradient).

Nonuniform or varied flow may be further classified according to the degree or rate of the depth or velocity variation. Gradually varied flow (GVF) is that one in which the depth and hence the velocity changes gradually and slightly with space. A backwater curve, as shown in Figure 5.5a, behind a hydraulic structure is an example of gradually varied flow. A 1-meter variation in the depth of flow may occur over a distance of tens of kilometers. On the other hand, if the variation occurs over relatively short distances, the flow is then rapidly varied flow (RVF). In many cases RVF is a local phenomenon. A hydraulic jump is an example of the

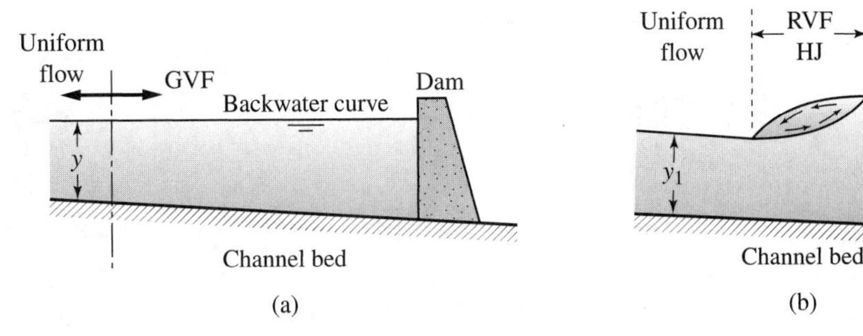

Note: GVF = gradually varied flow RVF = rapidly varied flow HJ = hydraulic jump

Figure 5.5 Nonuniform flow in open channels.

RVF, where a two meter variation in the depth of flow (or even more) may occur over a short distance of, say, several meters, as shown in Figure 5.5b.

5.1.2.3 Unsteady Nonuniform Flow Steady and unsteady flows are characterized by the variation of depth or velocity with respect to time at a specified location. Uniform and nonuniform flows are characterized by the variation of depth or velocity with respect to space at a specified time. An ideal case of an open-channel flow is that one which satisfies both the steady and the uniform flow conditions (i.e., $\partial v/\partial t = 0$ and $\partial v/\partial x = 0$). The flow is, therefore, steady uniform. Steady nonuniform open-channel flow is that flow in which the depth of flow does not change with time at any given location but does with location. The establishment of unsteady uniform flow would require that the water surface fluctuates as time elapses while remaining parallel to the channel bottom. Obviously, it is practically impossible to satisfy such a condition. Therefore, unsteady uniform flow does not exist in nature. Unsteady nonuniform or varied flow is the general case where the depth of flow is varied with respect to time and space. Figure 5.6a summarizes different types of flow in open channels.

5.1.2.4 Determination of Flow Type in Field The type of flow in the field can be determined by measuring the depth of flow in space and time. Suppose that at a distance $x = x_1$ and at a time $t = t_1$ the depth of flow in an open channel is measured as $y = y_1$. After a certain period of time where the time $t = t_2$ and at the same point $x = x_1$, the depth of flow is measured as $y = y_2$, as indicated in Figure 5.6b(a). If $y_1 = y_2$, the flow is steady. Otherwise, the flow is unsteady.

To determine whether the flow is uniform or nonuniform, one should measure the depth of flow at the same time but at two different locations. As before, the initial depth at $x = x_1$ and $t = t_1$ is measured as $y = y_1$. At the same point of time ($t = t_1$), the depth of flow is measured at another distance $x = x_2$, as y_1', as shown in Figure 5.6b(b). If $y_1' = y_1$, the flow is uniform. Otherwise, the flow is nonuniform.

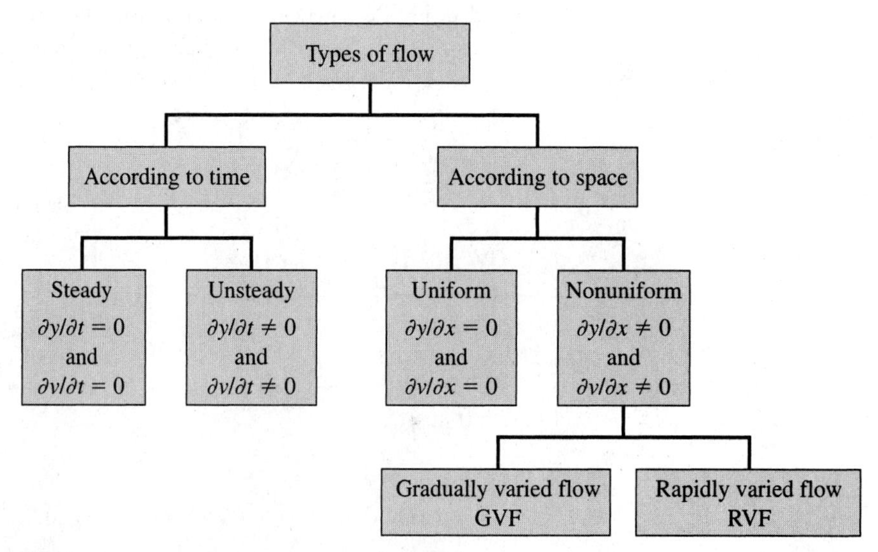

Figure 5.6(a) Different types of flow in open channels.

At the same point ($x = x_1$)
but different time

$y = y_1$ $y = y_2$

$t = t_1$ $t = t_2$ t

(a)

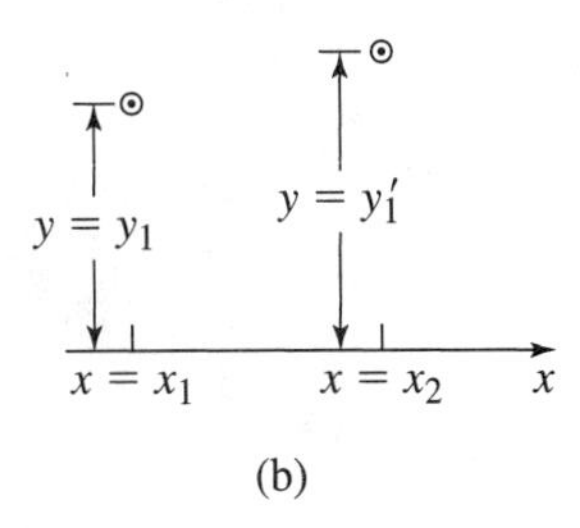

Figure 5.6(b) Determination of flow type.

Example 5.4

The continuity equation in open channels can be expressed in volumetric rate (or discharge) terms with lateral inflow. Using equation (5.20c), express the continuity equation in terms of discharge.

Solution: Consider a channel reach extending from $x = x_1$ to $x = x_2$, with a length $L = x_2 - x_1$, and width $B = 1$. Multiplying equation (5.20c) by dx and then integrating from x_1 to x_2 produces

$$B\frac{d}{dt}\int_{x_1}^{x_2} h\, dx + \int_{Q_i}^{Q_o} dQ = \int_{x_1}^{x_2} q_x\, dx$$

where Q_i is the inflow to the channel at $x = x_1$, Q_o is the outflow from the channel at $x = x_2$, and q_x is the lateral inflow per unit length. The quantity $h\,dx$ is ds = the change in storage per unit width of the channel. Therefore,

$$\frac{ds}{dt} = Q_i + q_x L - Q_o = Q_{i^*} - Q_o$$

where s is the storage per unit width, and

$$Q_{i^*} = Q_i + q_x L$$

5.2 Energy Conservation: The Bernoulli Equation

Energy is a scalar quantity and can be defined, in general terms, as the capacity for doing work. Work can be defined as the energy transfer by the action of force through a certain distance. The law of energy conservation states that energy can neither be created nor destroyed, but it may transform from one form to another. A mathematical statement of this law is the energy equation. For applications in hydraulics, some allowance must be made for energy losses, most commonly via dissipation of kinetic energy into heat energy. The energy equation deals with the conservation of various types of energy. There are five forms of intrinsic energy (Gerhart et al., 1992): kinetic energy, potential energy, internal energy, chemical energy, and nuclear energy. Kinetic energy is the energy inherent in the fluid due to its motion. Potential energy is attributed to the position of fluid particles with reference to a specified datum and fluid pressure. Internal energy is a system property and, as such, is related to its mass and temperature. Chemical and nuclear energies are generally absent (in most hydraulic applications) and are, therefore, neglected. Energy may be

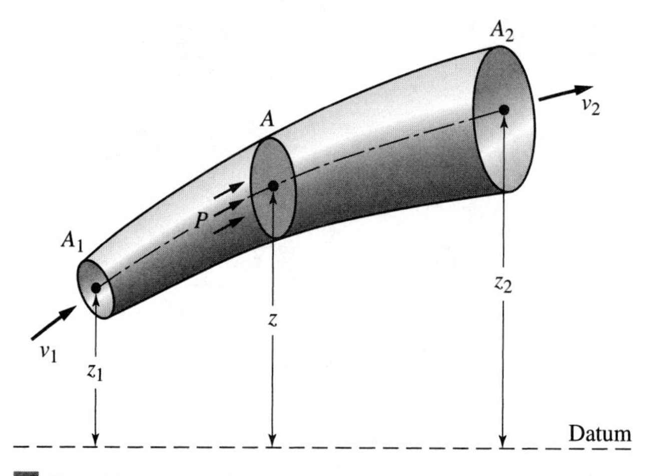

Figure 5.7 Control volume (stream tube).

transferred to a system by two different processes, heat and work. Heat transfer is attributed to the difference in temperature between a hydraulic system and its surroundings.

Consider a control volume (stream tube) shown in Figure 5.7. Let the cross-sectional area, A, of the stream tube be subjected to a pressure, p. Therefore, the force, F, exerted on the cross-sectional area is given as

$$F = pA \tag{5.23}$$

As the flowing fluid travels a distance, dx, along the stream tube during a time interval, dt, the work done, dW, within the time dt is

$$dW = F\,dx = pA\,dx \tag{5.24}$$

The mass, dm, entering the system during the same time interval is equal to $\rho A dx$. The kinetic energy, dKE, is thus defined as

$$dKE = \frac{1}{2}dmv^2 = \frac{1}{2}\rho A\,dx\,v^2 = \frac{1}{2}\rho A(v\,dt)v^2 = \frac{1}{2}\rho A\,dt\,v^3 \tag{5.25}$$

The potential energy of a mass $\rho A\,dx$ entering during the time interval dt is $\rho A\,dx\,gz$, where g is the acceleration of gravity, and z is the elevation above some arbitrary datum. The total energy, dE_T, during dt is the sum of the work done, kinetic energy, and potential energy. Therefore,

$$dE_T = pA\,dx + \frac{1}{2}\rho A\,dx\,v^2 + \rho A\,dx\,gz \tag{5.26}$$

The energy per unit weight of fluid (or the energy head, E) for the whole flow area is obtained by integrating equation (6.26) and dividing by $\rho g A\,dx$. The work done over the whole area will just be $pA\,dx$ (for constant dx) and the potential energy will remain $\rho A\,dx\,gz$. The kinetic energy will be $\int_A dKE = \frac{1}{2}\rho dx \int_A v^2 dA = \frac{1}{2}\alpha\rho\; dx\; V^2 A$, where V is the average cross sectional velocity. Hence, equation (5.26) becomes

$$E = \frac{p}{\rho g} + \frac{\alpha V^2}{2g} + z \tag{5.27}$$

where α is the energy connection factor to be discussed in the next section. It is often close to one and usually will be assumed so in this text.

At the inlet of the stream tube denoted by subscript 1, and noting that $\gamma = \rho g$, and letting $\alpha = 1$, the energy per unit weight is

$$E_1 = \frac{p_1}{\gamma} + \frac{V_1^2}{2g} + z_1 \tag{5.28}$$

Similarly, at the outlet denoted by subscript 2, the energy per unit weight will be

$$E_2 = \frac{p_2}{\gamma} + \frac{V_2^2}{2g} + z_2 \tag{5.29}$$

If no energy is lost or converted to some other form, the energy per unit weight of the fluid at the inlet must be equal to that at the outlet. Moreover, most of the applications in hydraulics are limited to incompressible fluids (i.e., $\rho_1 = \rho_2 =$ constant). Therefore, the energy equation is written as:

$$E = \frac{p_1}{\gamma} + \frac{V_1^2}{2g} + z_1 = \frac{p_2}{\gamma} + \frac{V_2^2}{2g} + z_2 = \text{Constant} \tag{5.30a}$$

Equation (5.30a) is the well-known Bernoulli equation, named after Daniel Bernoulli (1700–1782), the author of one of the early books in the science of fluid flow in 1738.

The dimension of each term in the Bernoulli equation is the length, L, and each term of the equation is accordingly called head. Thus, the energy per unit weight of fluid is the energy head and is the sum of the pressure head, the velocity head and the static head. The Bernoulli equation is a powerful equation because it relates the changes in pressure to changes in velocity and elevation along a stream line.

If there is no energy lost between two points of flow, equation (5.30a) simply states that

$$E_1 = E_2 = E \tag{5.30b}$$

However, in general there is energy loss. Therefore, equation (5.30b) becomes

$$E_1 = E_2 + h_L \tag{5.30c}$$

where h_L is the energy (head) loss.

Example 5.5

Water flows steadily through a horizontal nozzle, as shown in Figure 5.8. At the nozzle inlet the diameter is 6 in. At the nozzle outlet, where the pressure is atmospheric, the diameter is 2 in. Determine the pressure head at the nozzle inlet required to produce a jet velocity of 40 ft/s. Neglect the friction loss and assume incompressible flow.

Solution: At the nozzle entrance, the cross-sectional area is $\frac{\pi D^2}{4} = \frac{\pi(.5)^2}{4} = 0.196 \text{ ft}^2$.
Likewise, the area at the nozzle outlet is computed to be 0.0218 ft^2. Applying the continuity equation between sections 1 and 2,

$$A_1V_1 = A_2V_2, \quad \text{or} \quad v_1 = A_2V_2/A_1$$

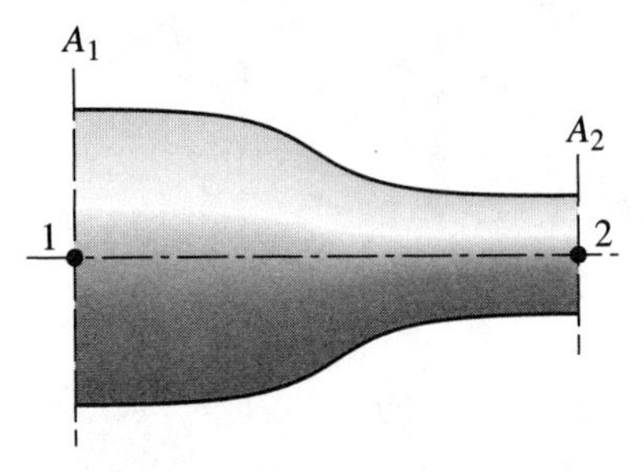

Figure 5.8 Horizontal nozzle (Example 5.5).

Then $V_1 = 0.0218 \times 40.0/0.196 = 4.44$ ft/s.
Applying the Bernoulli equation,

$$z_1 + \frac{p_1}{\gamma} + \frac{V_1^2}{2g} = z_2 + \frac{p_2}{\gamma} + \frac{V_2^2}{2g}$$

where $\gamma = \rho g$ is the specific weight and is equal to 62.4 lb/ft^3. Also, z_1 and z_2 have the same value and thus can be eliminated from the equation, and p_2/γ is the pressure head at the nozzle outlet and is equal to zero. Therefore,

$$\frac{p_1}{\gamma} = \frac{V_2^2 - V_1^2}{2g} = \frac{1600 - 19.71}{2 \times 32.2} = 24.54 \text{ ft}$$

Therefore, the pressure at the nozzle should be 1531 lb/ft^2 or 10.63 psi.

Example 5.6

A pipe 300 m long has a slope of 1:100 and converts from a 1-m diameter at its upper end, where the pressure is 6.87×10^4 kg/m-s^2 (68.7 kN/m^2) to 0.5 m at its lower end, as shown in Figure 5.9. If the water flows at a rate of 5400 l/min, determine the pressure at the lower end. Assume the energy loss due to friction inside the pipe is negligible.

Solution: The rate of flow is $Q = 5400$ l/min $= 0.09$ m^3/s. Assuming the datum line to intersect the center line of the pipe at its lower end, shown in Figure 5.9. The elevation of center line of the pipe at its upper end above this datum will be 3.0 m. Applying the continuity equation,

$$V_1 = \frac{Q}{A_1} = \frac{0.09}{\frac{\pi\, 0.5^2}{4}} = 0.46 \text{ m/s}$$

$$V_2 = \frac{Q}{A_2} = \frac{0.09}{\frac{\pi 1^2}{4}} = 0.115 \text{ m/s}$$

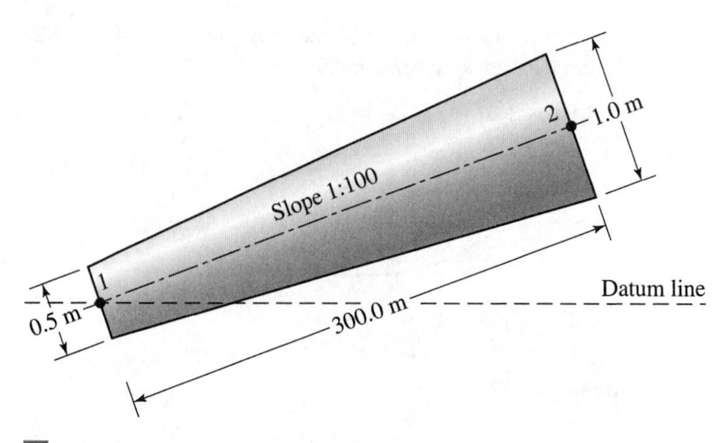

Figure 5.9 Sloping pipe (Example 5.6).

Applying the Bernoulli equation, one gets

$$\frac{P_1}{\gamma} + \frac{0.46^2}{2 \times 9.81} + 0 = \frac{6.87 \times 10^4}{9810} + \frac{0.115^2}{2 \times 9.81} + 3$$

$$\frac{P_1}{\gamma} = 7 + 0.674 \times 10^{-3} + 3 - 0.0108 = 9.99 \text{ m}$$

Therefore, the pressure, p_1, is equal to 98001 kg/m-s^2 or 98 kN/m^2. It should be noted that since the energy loss due to friction is neglected, the flow direction (either upward or downward) would not affect the results.

Example 5.7

Water discharges over a spillway as shown in Figure 5.10. The depth behind the spillway is 24 ft, the spillway and the downstream channel are both 50 ft wide, and the depth at the toe of the spillway is 0.41 ft. Assuming that the surface velocity behind the spillway is very small such that the velocity head can be neglected and ignoring friction losses on the spillway face, what is the discharge over the spillway?

Solution: Applying the Bernoulli equation to just upstream of the spillway and the downstream channel,

$$24 = 0.41 + \frac{V^2}{2g}$$

where V is the velocity in the downstream channel. This gives $V = 38.98$ or 39 ft/s. Thus, $Q = VA = 39 \times 0.41 \times 50 = 799$ cfs.

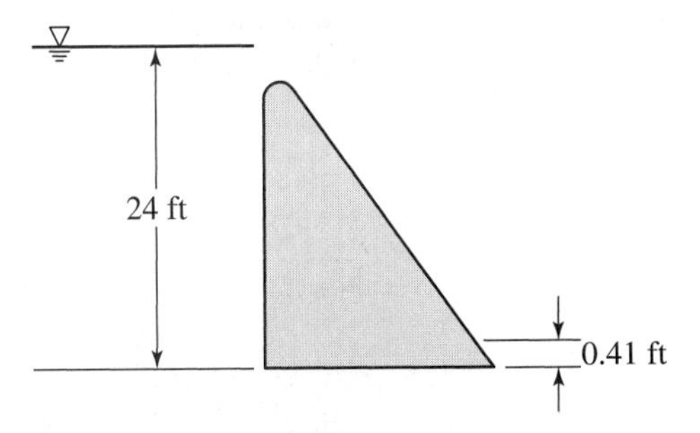

Figure 5.10 Flow over a spillway.

5.2.1 Energy Correction Factor

In the preceding discussion, it was assumed that during the movement of a body of mass (m) of water, all particles moved at the same average velocity, v. That is why the kinetic energy possessed by the body of m was expressed as $mV^2/2$ and the kinetic energy per unit weight as $V^2/(2g)$. In most situations, however, the velocities of the different water particles are not the same and the velocity variation must be accounted for. Consequently, in order to get the true value of the kinetic energy, one must integrate all portions of the stream. For practical purposes, it is more convenient to account for the effect of the velocity variation by introducing a correction factor α and the mean velocity V and then obtaining the true value of the kinetic energy (per unit weight), as shown earlier in equation (5.27), is

$$KE = \alpha \frac{V^2}{2g} \tag{5.31a}$$

An expression for α can be derived by considering the variation of velocity across a section. Consider that a stream is partitioned into small areas dA. If v is the local velocity at a point in the x direction, then the mass flow through an elementary area dA is $\rho\, dQ = \rho v\, dA$. Thus, the true flow of kinetic energy across area dA is $(\rho v\, dA)(v^2/2) = (\gamma/2g)v^3 dA$. The weight rate of flow through dA is $\gamma\, dQ = \gamma v\, dA$. Then, for the entire section, one can write

$$KE = \frac{\text{True } KE}{\text{Weight}} = \frac{\{\gamma/(2g)\}\int v^3\, dA}{\gamma \int v\, dA} = \frac{\int v^3\, dA}{2g\int v\, dA} \tag{5.31b}$$

Comparing equation (5.31b) with equation (5.31a), one obtains

$$\alpha = \frac{1}{V^2}\frac{\int v^3\, dA}{\int v\, dA} = \frac{1}{AV^3}\int v^3\, dA \tag{5.32}$$

The value of α is always greater than 1, but not greater than the cube of the average. Clearly, a greater variation of velocity across the section will give rise to a greater value of α. In most hydraulic applications, the value of α is assumed as 1, but in cases where α is significantly away from unity, a proper value of α must be used. For laminar flow in circular pipes, $\alpha = 2$, and for turbulent flow in pipes it ranges from 1.01 to 1.15, with a common range as 1.03 to 1.06.

5.2.2 Energy Equation in Open Channels

The total energy, E_T, in an open channel is expressed in terms of water head in meters. It is the sum of the bed elevation above a specified horizontal datum, the pressure head, and the velocity head. The total energy (head) at section 1, as shown in Figure 5.11, is given as

$$E_T = H = z_A + y_A + \frac{V_A^2}{2g} \tag{5.33a}$$

where z_A is the elevation of point A above a specified datum, y_A is the depth of point A below the water surface measured in the vertical direction, and V_A is the water velocity at point A. Note

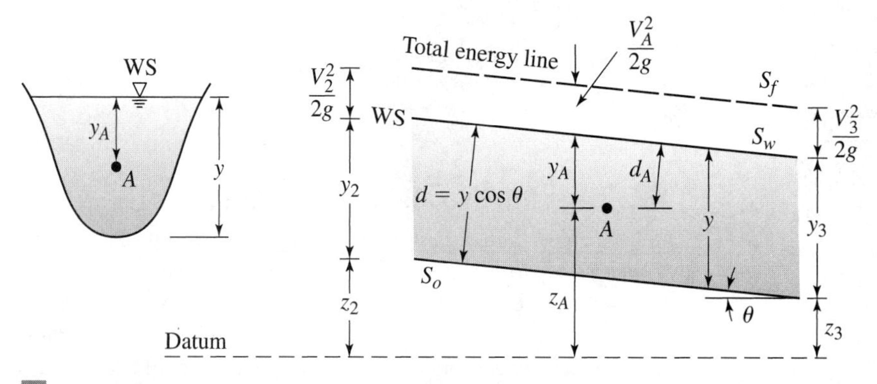

Figure 5.11 Definition sketch for energy equation in open channels.

that the depth normal to the channel bed, $d_A = y_A \cos\theta$, where θ is the bed slope angle and the total depth normal to the bed, $d = y\cos\theta$, where y is the total depth measured vertically.

Velocities vary from one point to another, not only in the lateral and longitudinal directions but also in the vertical direction. However, for practical purposes, velocity heads in the entire section are assumed equal and the energy distribution coefficient is introduced to account for the nonuniform velocity distribution. The total energy head, E_T, in a channel section is, thus, written as

$$E_T = z + y + \alpha\frac{V^2}{2g} \tag{5.33b}$$

The line representing the total energy, as shown in Figure 5.11, is the total energy line and its slope is known as the energy gradient, or friction slope S_f. As the pressure on the water surface is atmospheric, the hydraulic gradient line coincides with the water surface and its gradient is denoted by S_w. The bed slope is denoted by S_o. For uniform flow, where the depth of flow is constant with space, $S_f = S_w = S_o$.

According to the law of energy conservation, the total energy between any two sections (with consideration of losses) must be equal. For the two sections, 2 and 3, upstream and downstream of section 1 as shown in Figure 5.11, one can write

$$z_2 + y_2 + \alpha_2\frac{V_2^2}{2g} = z_3 + y_3 + \alpha_3\frac{V_3^2}{2g} + h_f \tag{5.34a}$$

where h_f is the head loss between sections 2 and 3. Note that if the bed slope is small ($\cos\theta \approx 1$), then the effects of bed slope can be ignored so that the depth normal to the bed d is approximately equal to the vertical depth y. In addition for regular cross sections, $\alpha \approx 1$, then equation (5.34a) can be written as

$$z_2 + y_2 + \frac{V_2^2}{2g} = z_3 + y_3 + \frac{V_3^2}{2g} + h_f \tag{5.34b}$$

Equation (5.34b) is discussed in more detail in Chapter 12.

5.2.3 Validity and Limitations of the Bernoulli Equation

Certain assumptions have been made in the derivation of the Bernoulli equation. Therefore, the following limitations should be noted while employing the Bernoulli equation: (1) The Bernoulli equation is applicable to only steady-state problems. (2) In many practical applications, there are some losses of energy due to frictional forces exerted between the flowing fluid and the conduit of the flow. Also, in turbulent flow some kinetic energy is converted into heat energy, while in viscous flow some energy is lost to overcome shear forces. Such losses should be integrated into the equation. (3) The velocity of every particle of the fluid was assumed constant across any cross-section of the hydraulic conduit (1-D flow). The Bernoulli equation does not account for velocity variation in the lateral direction through the cross section of flow. (4) The gravity force is considered to be the only external force acting on the flowing fluid. This is not always true in many problems where friction force as well as other external forces affect the flow pattern considerably. (5) Centrifugal forces due to curved paths are not considered. (6) The Bernoulli equation is not applicable through propeller machines. The equation was derived by integrating along a stream tube or streamline in the absence of moving surfaces such as vanes. However, it may be applied before and after the machines. In such cases, the energy loss inside these machines should be considered.

5.3 Momentum Conservation: The Momentum Equation

The law of momentum conservation states that a body in motion cannot gain or lose momentum unless some external force is imposed. The force acting on an element is equal to the rate of change of momentum (momentum flux). Newton's second law of motion defines force as the product of mass and acceleration. This statement is valid only for a fluid continuum in which all particles have the same rate of acceleration. According to Newton's second law of motion, the rate of change of momentum is proportional to the applied force and takes place in the direction in which the force acts. Expressed mathematically,

$$\mathbf{F} = \frac{d}{dt}(m\mathbf{V}) \tag{5.35}$$

Equation (5.35) is a vector relationship, true for both magnitude and direction. To cause the motion of a constant mass m or to change the state of an existing motion, the force $\mathbf{F}$ has to be applied, and this force causes the acceleration $d\mathbf{V}/dt$. The product $m\,d\mathbf{V}/dt$ is the inertial force, which depicts the natural resistance of fluid to any change in its state of motion. The mass in question is the mass of a unit volume of fluid [i.e., $m = \rho(\text{unit volume}) = \rho =$ fluid density]. Accordingly, mass times acceleration of an object equals the sum of forces acting on it. Therefore,

$$\mathbf{F} = m\mathbf{a} \tag{5.36a}$$

where a is the acceleration of the object. Integrating equation (5.36a) with respect to time yields

$$\int_{t_1}^{t_2} \mathbf{F}\,dt = m\int_{t_1}^{t_2} \mathbf{a}\,dt = m\int_{t_1}^{t_2} \frac{d\mathbf{V}}{dt}\,dt = m(\mathbf{V}_2 - \mathbf{V}_1) \tag{5.36b}$$

Equation (5.36b) is simply the impulse-momentum version of the momentum equation.

Consider a stream tube shown in Figure 5.12. In a time interval Δt, the momentum entering the stream tube is $\rho\Delta Q_1 \Delta t\, V_1$, whereas the momentum leaving it is $\rho\Delta Q_2 \Delta t\, V_2$. From the continuity principle, $\Delta Q_1 = \Delta Q_2 = \Delta Q$. Therefore, the force required to produce this change of momentum is

$$\Delta\mathbf{F} = \frac{\rho\,\Delta Q\,\Delta t\,(\mathbf{V}_2 - \mathbf{V}_1)}{\Delta t} = \rho\,\Delta Q(\mathbf{V}_2 - \mathbf{V}_1) \tag{5.37}$$

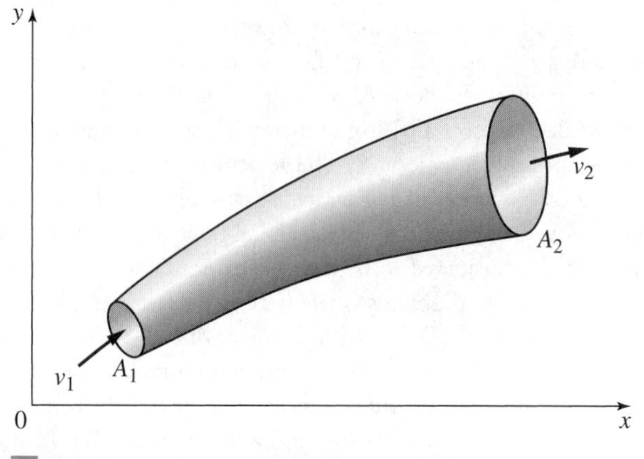

Figure 5.12 Momentum change in a stream tube.

Considering the force component in the x direction, F_x,

$$\Delta F_x = \rho\,\Delta Q\,(V_{2x} - V_{1x}) \tag{5.38}$$

Integration of equation (5.38) yields

$$F_x = \rho Q(V_{2x} - V_{1x}) \tag{5.39}$$

Equation (5.39) is the momentum flux equation in the x direction. Similarly, the other two force components F_y and F_z can be derived. Equation (5.39) is based on the assumption that the velocity is steady and uniform in the cross section. The momentum force is composed of the sum of all external forces acting on the control volume.

In many problems, the velocity varies from one point to the other along the cross section and the velocity distribution is not always known. Therefore, the mean velocity v (defined as the discharge divided by the cross-sectional area) must be used while solving both the energy and momentum equations. The energy and momentum coefficients, α and β, are thus introduced to account for the nonuniform velocity distribution. The effect of these coefficients on the accuracy of the final solution is rather limited and in many problems it can be ignored.

Numerous problems in hydraulics are solved using the momentum equation. Given the discharge and pressure in a pipe, the momentum equation is used to evaluate the forces on bends, divergent or convergent elements, and supporting elements of the system. In many problems both the momentum and Bernoulli equations are applied to obtain a solution.

Example 5.8

Calculate the magnitude and direction of the force exerted by the horizontal pipe bend, as shown in Figure 5.13, if the diameter is 20 cm and the upstream pressure head at the centroid is 5 m. The discharge through the pipe is 0.2 m^3/s.

Solution: Since the diameter of the pipe does not vary through the bend, the force exerted will be due to the change in the direction of flow. The x and y components of the momentum force are $F_{Mx} = (\gamma/g)Q(V_{2x} - V_{1x})$, and $F_{My} = (\gamma/g)Q(V_{2y} - V_{1y})$. The velocity through the bend is constant and is given from the continuity equation as $V = Q/A = 0.2/0.0314 = 6.366$ m/s.

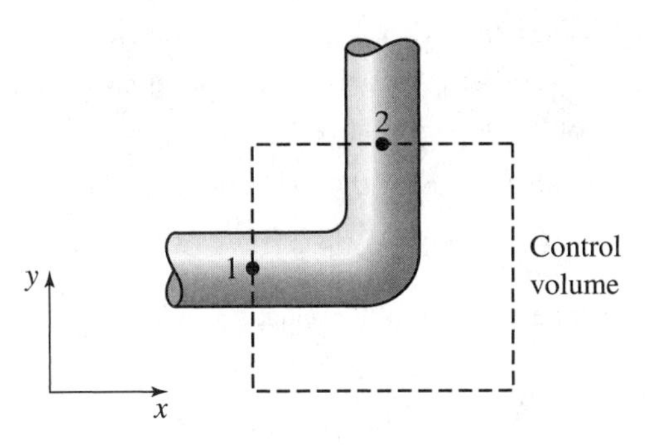

Figure 5.13 Horizontal pipe bend (Example 5.8).

Since the bend is horizontal, the static heads at the inlet and outlet must be the same (the Bernoulli equation). Therefore,

$$\frac{p_1}{\gamma} = \frac{p_2}{\gamma} = 5 \text{ m}$$

$$p_2 = p_1 = 9.81 \times 1000 \times 5 = 49050 \text{ N/m}^2$$

The pressure force, F_p, is equal to the pressure multiplied by the area. Therefore,

$$F_{px} = p_1A_1 - 0 = 49050 \times [\pi(0.2)^2/4] = 49050 \times 0.0314 = 1540 \text{ N}$$

$$F_{py} = 0 - p_2A_2 = -1540 \text{ N}$$

The momentum force can be calculated as

$$F_{Mx} = \rho Q(0 - V_1) = 1000 \times 0.2 \times (0 - 6.366) = -1273 \text{ N}$$

$$F_{My} = \rho Q(V_2 - 0) = 1273 \text{ N}$$

The resultant forces in the x and y directions, F_{Rx} and F_{Ry}, are given from

$$F_{Rx} + F_{px} = F_{Mx},$$

$$F_{Rx} = -1540 - 1273 = -2813 \text{ N, and}$$

$$F_{Ry} + F_{py} = F_{my},$$

$$F_{Ry} = 1540 + 1273 = 2813 \text{ N}$$

Hence,

$$F_R = \sqrt{F_{Rx}^2 + F_{Ry}^2} = 3978 \text{ N}$$

$$\theta = \tan^{-1}\frac{F_{Ry}}{F_{Rx}} = -45°$$

F_R is the force of the bend on the water. The force of the water on the bend is equal in magnitude and opposite in direction to F_R.

Example 5.9

A jet of water is deflected 60° by a frictionless vane moving with a constant velocity of 6 ft/s as shown in Figure 5.14. The jet velocity is 40 ft/s and its cross-sectional area is 0.0218 ft^2. Determine the force exerted by the jet on the vane.

Solution: The jet strikes the vane with a velocity of V_r, where

$$V_r = V_{\text{jet}} - V_{\text{vane}} = 40 - 6 = 34 \text{ ft/s}$$

The velocity at the outlet is equal to the velocity at inlet but differs in the direction.

$$F_x = \frac{\gamma}{g}Q(V_r \cos 60 - V_r)$$

$$F_x = 1.94(34 \times 0.0218)(34 \cos 60 - 34) = -24.45 \text{ lb}$$

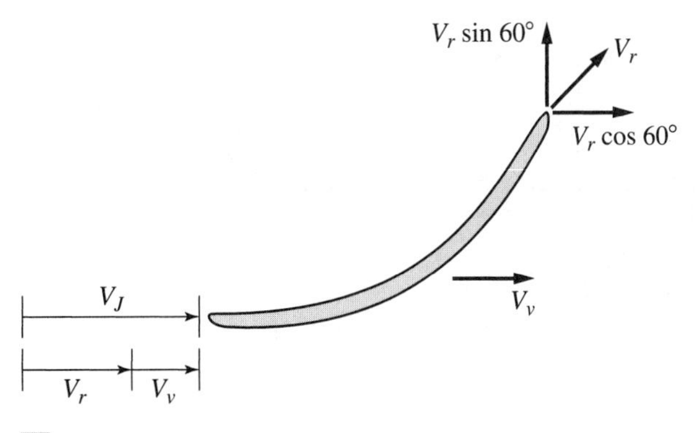

Figure 5.14 A moving vane (Example 5.9).

$$F_y = \frac{\gamma}{g} Q(V_r \sin 60 - 0)$$

$$F_y = 1.94(34 \times 0.0218)(34 \sin 60 - 0) = 42.34 \text{ lb}$$

The resultant force, F, is $\sqrt{F_x^2 + F_y^2} = 48.9$ lb and $\theta = \tan^{-1} \frac{F_y}{F_x} = -60°$.

5.3.1 Angular Momentum Equation

If the moment of momentum of flow leaving a control volume differs from that entering it, a torque which acts on the control volume will be produced. Assuming a steady discharge, Q, entering a control volume with an absolute velocity V_1, and leaving with an absolute velocity V_2, the momentum of the flow is $\rho Q V_1$ at the inlet and $\rho Q V_2$ at the outlet, respectively. If this flow has a curvature of radius r_1 at the inlet and of r_2 at the outlet, the moments of momentum at the inlet and the outlet are $\rho Q V_1 r_1$ and $\rho Q V_2 r_2$, respectively. According to the principle of moment of momentum, the resulting torque is

$$T = \rho Q(V_2 r_2 - V_1 r_1) \tag{5.40}$$

where T is the torque about the axis of rotation. Equation (5.40) is known as Euler's equation. Practical applications of this equation include problems with uniaxial moments such as flows in rotodynamics machines, garden sprinklers, and so on. In the absence of external moments, $T = 0$, then

$$V_1 r_1 = V_2 r_2 \tag{5.41}$$

where V_1 and V_2 are the absolute circumferential velocities of the fluid entering and leaving at radial locations r_1 and r_2, respectively.

Example 5.10

An asymmetrical sprinkler with a frictionless shaft has radii of 0.4 m and 0.6 m. The water leaves the sprinkler from two nozzles, as shown in Figure 5.15, with a velocity of 10 m/s relative to the nozzle. The sprinkler discharges 10 l/min. Determine the speed of the sprinkler in rpm. What torque will the sprinkler exert on hand if held stationary?

Solution: Since there is no external torque applied, the moment of momentum of the fluid leaving the system must be equal to zero, that is,

$$\frac{\gamma}{g} Q(V_1 r_1 - V_2 r_2) = 0$$

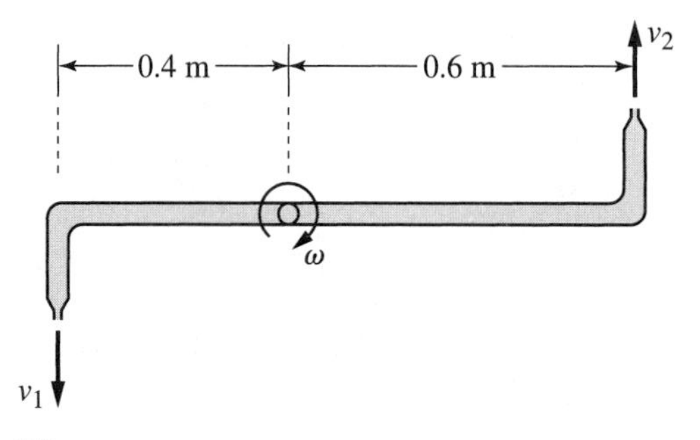

Figure 5.15 An unsymmetrical sprinkler (Example 5.10).

$0.4V_1 = 0.6V_2$, where $V_1 = 10 - 0.4\omega$, and $V_2 = 10 - 0.6\omega$. Hence, $0.4(10 - 0.4\omega) = 0.6(10 - 0.6\omega)$ then $\omega = 10$ rad/s:

$$N = \frac{60\omega}{2\pi} = \frac{600}{2\pi} = 95.5 \text{ rpm}$$

When stationary, the torque due to the nozzles' actions $= \rho QVr$ is

$$T_1 = 1000\left(\frac{10 \times 10^{-3}}{60}\right) \times 6 \times 0.4 = 0.40 \text{ N-m}$$

$$T_2 = 1000\left(\frac{10 \times 10^{-3}}{60}\right) \times 4 \times 0.6 = 0.40 \text{ N-m}$$

Torque required to hold the sprinkler $= T_1 + T_2 = 0.80$ N-m

5.3.2 Momentum Correction Factor

When water flows in a conduit, its velocity across any section is not uniform. This velocity variation influences the momentum flux and its computation. Normally, the average velocity is used in computation. But the momentum per unit time (momentum flux) transferred across a section is greater than that computed using the mean velocity. In order to account for the effect of velocity variation, one can compute the momentum flux over the section by integration. In practice, however, it is computed using a momentum correction factor, β, and the average velocity, just as in case of the energy correction factor. This correction factor can be derived as follows. Consider that a flow section is partitioned into small elementary areas dA. Let the local velocity in dA be designated as v. The rate of momentum transfer (or momentum flux) across an elementary area dA is $\rho v^2 dA$. Then the momentum flux across the entire section is

$$\text{Momentum flux} = \rho \int_A v^2 dA \tag{5.42}$$

The momentum flux using the average velocity is $\beta\rho QV = \beta\rho AV^2$. This momentum flux when equated with that of equation (5.42) leads to β as

$$\beta = \frac{1}{AV^2}\int_A v^2\, dA \tag{5.43}$$

The value of β is greater than one. In circular pipes, it is 4/3 for laminar flow and 1.005 to 1.05 for turbulent flow. In open channels its value can be larger. However, in most practical situations, the momentum correction factor, β, is frequently taken as 1.

5.3.3 Momentum Equation in Open Channels

Newton's second law of motion stipulates that for a fluid in continuum, in which all particles have the same acceleration, the force should be equal to the product of mass and acceleration. In other words, the change of momentum per unit time in the body of water in a flowing channel is equal to the resultant of all external forces acting on that body.

Consider the flow in an open channel, as shown in Figure 5.16, where a certain body of water is enclosed between sections 1 and 2. Let the channel be prismatic with no lateral flow between the two sections under consideration. The channel bed is sloping at an angle θ and the weight of the water body under consideration is W. Two forces resist the flow. The first one is the friction force along the surface of contact between the water and the channel bed and sides, F_f. The second one, F_a, is attributed to air resistance along the free water surface exposed to the atmosphere. F_a is generally small as compared to F_f. Referring to Figure 5.16,

$$\text{Mass inflow per unit time at section 1} = \rho Q \tag{5.44a}$$

in which ρ is the density of the flowing fluid. Let the mean velocity at section 1 to be V_1. Thus,

$$\text{Momentum inflow per unit time at section 1} = \rho Q \beta_1 V_1 \tag{5.44b}$$

in which β_1 is the momentum correction factor which accounts for the nonuniform velocity distribution. Similarly, one can write for section 2,

$$\text{Momentum inflow per unit time at section 2} = \rho Q \beta_2 V_2 \tag{5.44c}$$

The change of momentum per unit time between the two sections is, thus, equal to the difference between equations (5.44c) and (5.44b):

$$= \rho Q(\beta_2 V_2 - \beta_1 V_1) \tag{5.45}$$

$$\text{The pressure force at section 1} = F_{p1} = \gamma A_1 \bar{h}_1 \tag{5.46}$$

$$\text{The pressure force at section 2} = F_{p2} = \gamma A_2 \bar{h}_2 \tag{5.47}$$

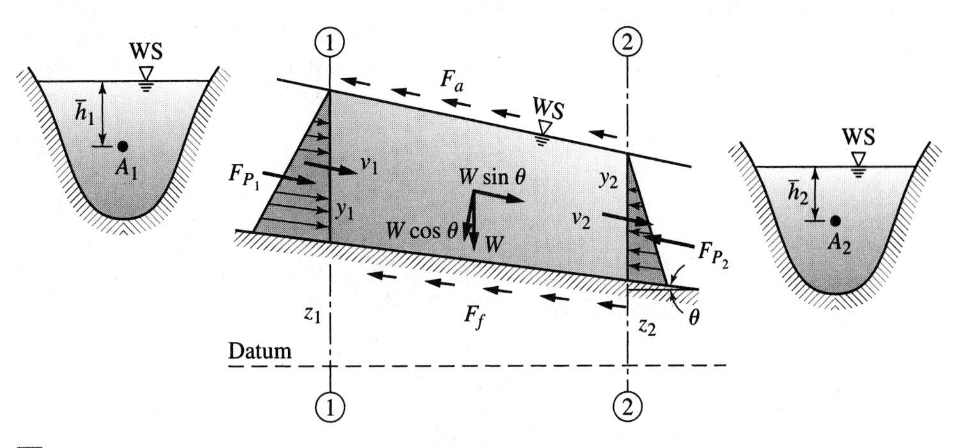

Figure 5.16 Definition sketch for the derivation of the momentum equation in open channels.

in which $\bar{h}$ is the depth of the area centroid below the water surface. A hydrostatic pressure distribution is assumed in equations (5.46) and (5.47), and note also that $\bar{h}$ is equal to $\bar{y}$.

The component of the fluid weight in the flow direction $= W \sin\theta$ **(5.48)**

The resultant external force, F_r, acting on the volume under consideration is the vector sum of all the aforementioned forces and is, thus,

$$F_r = \gamma A_1 \bar{h}_1 - \gamma A_2 \bar{h}_2 + W \sin\theta - F_f - F_a \tag{5.49}$$

According to Newton's second law, one can write noting that $\gamma = \rho g$,

$$\rho Q \beta_2 V_2 - \rho Q \beta_1 V_1 = \rho g A_1 \bar{h}_1 - \rho g A_2 \bar{h}_2 + W \sin\theta - F_f - F_a \tag{5.50}$$

Equation (5.50) constitutes the general form of the momentum equation for flow in open channels.

Neglecting F_a, and assuming a uniform velocity distribution, equation (5.50) is written as

$$\rho Q(V_2 - V_1) = \rho g(A_1 \bar{h}_1 - A_2 \bar{h}_2) + W \sin\theta - F_f \tag{5.51}$$

If the channel bed is assumed horizontal, the weight component will be zero. For frictionless channels, where the channel bottom and sides are smooth, F_f is also reduced to zero. Hence, equation (5.51) is reduced to

$$Q(V_2 - V_1) = g(A_1 \bar{h}_1 - A_2 \bar{h}_2) \tag{5.52}$$

which can be written as

$$\frac{Q^2}{gA_1} + A_1 \bar{h}_1 = \frac{Q^2}{gA_2} + A_2 \bar{h}_2 \tag{5.53}$$

Equation (5.53) is the momentum equation. The two sides of this equation are typical, except for the subscripts which refer to sections 1 and 2, respectively. Note that when θ is small, $y \cos\theta \sim y \sim h$.

Example 5.11

Water is stored behind a sluice gate in a rectangular open channel to a depth of 8 ft. If the depth of flow just after the gate is 0.5 ft, calculate the discharge neglecting the energy loss through the gate. Concrete blocks are placed, as shown in Figure 5.17, to create a 3.6-ft water depth. Determine the force exerted on these blocks.

Solution: Neglecting the friction losses between sections (1) and (2), then

$$8.0 + \frac{V_1^2}{2g} = 0.5 + \frac{V_2^2}{2g}$$

Also from the continuity equation, one can write $y_1 V_1 = y_2 V_2$ (i.e., $V_2 = 16 V_1$). Hence

$$8.0 + \frac{V_1^2}{2g} = 0.5 + \frac{256 V_1^2}{2g}$$

$$255 V_1^2 = 2g \times 7.5$$

Therefore, $V_1 = 1.37$ ft/s, $V_2 = 22$ ft/s.

The specific discharge, q, is thus $= y_1 V_1 = 10.96$ ft^2/s.

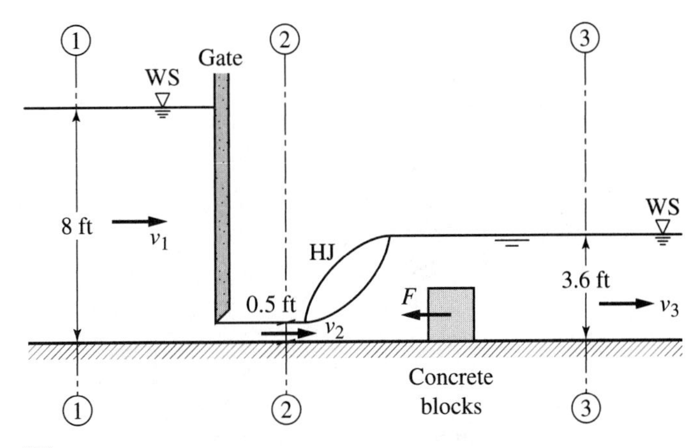

Figure 5.17 Open-channel setting given Example 5.11.

Applying the momentum principle (equation 5.51) between sections 2 and 3,

$$\rho q(V_3 - V_2) = \gamma(y_2\bar{h}_2 - y_3\bar{h}_3) - F$$

where $V_3 = q/y_3 = 3.04$ ft/s.

$$\bar{h}_2 = \frac{0.5}{2} = 0.25 \text{ ft}, \quad \bar{h}_3 = \frac{3.6}{2} = 1.8 \text{ ft; hence}$$

$$1.94 \text{ slugs/ft}^3 \times 10.96\ (3.04 - 22) = 62.4 \text{ lb/ft}^3\ (0.5 \times 0.25 - 3.6 \times 1.8) - F$$

or $F = 6.583$ lb/ft.

Therefore, the force on the blocks is 6.583 lb/ft and acts from right to left.

5.3.4 Impulse-Momentum Relationship to Energy Equation

Recall the application of the second law of motion in fluids, especially for flowing water. For simplicity and conformity with equations derived earlier in this chapter we will confine our discussion to the one-dimensional case. Consider a small differential element as shown in Figure 5.18. The direction of flow is s and the normal direction is denoted as n so that the dimensions of the element are $\Delta S\,\Delta N$ as shown in the figure. The element is suspended within the body of the flow, so that it is under a pressure differential as shown therein. The force per unit length (F_p) due to this differential pressure can be expressed from equation (5.36) as

$$F_p = p\,\Delta N - \left(p + \frac{\partial p}{\partial s}\Delta S\right)\Delta N \tag{5.54}$$

In the ideal case (neglecting friction and inertial viscous forces), the only other force per unit length acting on the element is gravity or the weight of the element, $W = \gamma\,\Delta S\,\Delta N$. Now, summing these forces in the direction of flow,

$$p\Delta N - p\Delta N - \frac{\partial p}{\partial s}\Delta S\Delta N + \gamma\Delta S\Delta N \sin\theta = -\frac{\partial p}{\partial s}\Delta S\,\Delta N + \gamma\,\Delta S\,\Delta N \sin\theta \tag{5.55}$$

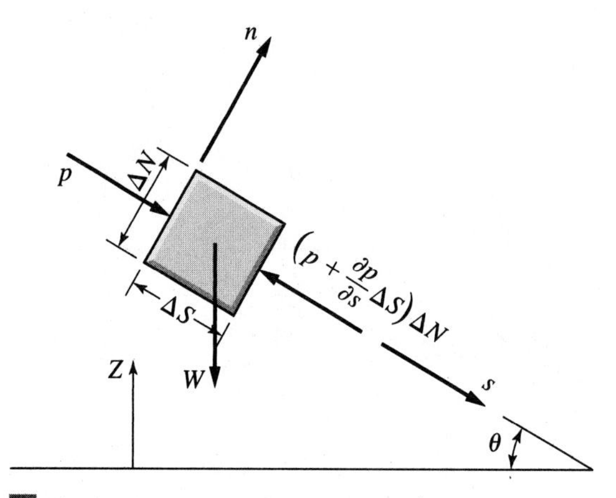

Figure 5.18 Forces on a differential element of flow.

Noting from Figure 5.18 that $\sin\theta = -\partial z/\partial s$, equation (5.55) becomes (Henderson, 1966)

$$\sum F_s = -\frac{\partial p}{\partial s}\Delta S \Delta N - \gamma \Delta S \Delta N \frac{\partial z}{\partial s} \tag{5.56}$$

The right side of equation (5.36a) can be written in the direction of flow as

$$ma = \rho \Delta S \Delta N a_s \tag{5.57}$$

where a_s represents the acceleration in the flow direction. Thus, equation (5.56) becomes

$$-\frac{\partial p}{\partial s}\Delta S \Delta N - \gamma \Delta S \Delta N \frac{\partial z}{\partial s} = \rho \Delta S \Delta N a_s$$

or

$$-\frac{\partial p}{\partial s} - \gamma \frac{\partial z}{\partial s} = \rho a_s$$

or

$$-\frac{\partial}{\partial s}(p + \gamma z) = \rho a_s$$

or

$$\frac{\partial}{\partial s}(p + \gamma z) + \rho a_s = 0 \tag{5.58}$$

Equation (5.58) is the most general one-dimensional form of the basic equation for flow of an ideal fluid known as the Euler equation after the Swiss mathematician who developed it in 1750. The potential energy term $(p + \gamma z)$ represents the piezometric pressure head. Now, the Euler equation can be further developed through the expansion of the acceleration term, a_s. Noting that acceleration is a function of both time and space, it can be written as (Henderson, 1966)

$$a_s = \frac{dv}{dt} = \frac{ds}{dt}\frac{\partial v}{\partial s} + \frac{\partial v}{\partial t} = v\frac{\partial v}{\partial s} + \frac{\partial v}{\partial t} \tag{5.59}$$

where v is local velocity, the spatial component $v(\partial v/\partial s)$ is known as the convective acceleration, and the temporal term $\partial v/\partial t$ is called the local acceleration. If we confine ourselves to the case of steady flow, then the local acceleration term is neglected and equation (5.58) becomes

$$\frac{\partial}{\partial s}(p + \gamma z) + \rho v \frac{\partial v}{\partial s} = 0 \tag{5.60}$$

Equation (5.60) represents the steady state one-dimensional governing equation of ideal flow and is, of course, another form of the Euler equation. As such, it represents a reduced and simplified form of the more complete set of equations of fluid flow known as the Navier-Stokes equations.

If we further integrate equation (5.60) with respect to space s, we simply obtain

$$p + \gamma z + \rho \frac{V^2}{2} = C \tag{5.61}$$

where V represents the average velocity integrated across the section and C is constant. Noting that $\gamma = \rho g$,

$$\frac{p}{\gamma} + z + \frac{V^2}{2g} = \frac{C}{\gamma} = H \tag{5.62}$$

which is just the Bernoulli equation.

The above derivation illustrates the essential one-dimensional, steady flow nature of the Bernoulli equation as it arises out of certain assumptions about the flow. Although this aspect of the equation does serve to limit the applicability of Bernoulli, it still does apply to a surprising number of practical problems. In many cases, even when the flow is in reality two- or even three-dimensional, the one-dimensional, steady approximation embodied in the Bernoulli equation can still give reasonable answers that are appropriate for engineering work. In all applications studied in this text, the Bernoulli approximation will be appropriate; nevertheless, the student should keep in mind that there are important cases in nature (e.g., storm surges, dam breaks, hydropower releases) for which it is not appropriate.

5.4 Choice between Energy and Momentum Equations

There is some confusion concerning the application of the energy and momentum principles to spatially varied flow in regard to the treatment of increasing and decreasing discharge (Yen and Wenzel, 1970; Yen et al., 1972). This can be attributed to the failure to recognize the difference between the friction slope, or the gradient associated with the average boundary shear, and the gradient of dissipated energy. The friction slope, the total energy (head) slope, and the dissipated energy gradient are different, and consequently, the corresponding friction resistance coefficient, total head loss coefficient, and dissipated energy coefficient differ from one another. Only for steady uniform flow without lateral inflow are the magnitudes of the three coefficients exactly equal and the slopes equal to the channel bed slope. In theory, the friction resistance should be used only in the momentum equation, the dissipated energy gradient in the energy equation, and the total head loss coefficient in the total head equation. In practice, however, for the steady flow case, the friction slope is usually close to the dissipated energy slope so that the difference is negligible. Thus in this text, the term "friction slope" is used to denote the dissipated energy gradient.

It is, therefore, concluded that for the case of no lateral inflow, there is no reason to distinguish between the energy and momentum equations. In their complete form valid for unsteady flow they are slightly different, but this difference is hardly significant in view of the uncertainties involved in expressing the boundary shear force (in the momentum equation) and the energy dissipation (in the energy equation). In the case of lateral inflow or outflow, the two equations show some difference in handling this effect, and a choice must be made. For lateral inflow the use of the momentum equation is appropriate, since an inflow of water into the main stream generally will be associated with considerable energy losses not accounted for by the assumed form of the energy dissipation, $S_f = \tau_o/\rho gR$. For lateral outflow one does not expect any significant energy dissipation, hence the energy equation seems to be appropriate. Henderson (1966) discussed the applicability of the momentum equation in cases of side-channel spillways and wash-water troughs both having lateral inflow where turbulence is much higher than most unsteady flow problems.

The difference between the energy and momentum equations in handling the lateral inflow is not surprising, since this essentially corresponds to a rapid variation. The two equations differ in that they are based on different conceptions and may be used to complement each other. Thus, in cases where physical insight suggests that only insignificant energy losses occur, the energy equation may be employed with advantage. In other cases, where energy losses are expected to be significant, or when the acting forces are to be estimated, the momentum equation may be used.

The energy principle is generally easy to apply, however, the momentum principle has certain advantages when applied to problems involving considerable energy losses, such as in the rapidly varied flow problems. Because the momentum equation deals with external forces only, application of the momentum equation would make the unknown energy losses avoidable. On the other hand, the term of frictional losses due to external forces can be omitted as such phenomena occur over short reaches.

The energy equation is applicable to a wider range of problems in closed conduit flow than is the momentum equation. The later is mostly employed whenever forces are to be evaluated. Such applications include the evaluation of forces on elbows, bends, divergents, convergents, and pipe supports.

READING AID

5.1. Define the laws of nature which are common to all physical processes. What are the other types of information needed to study hydraulic systems?
5.2. What is meant by external constraints in hydraulic processes? Define the elements of internal mechanics of such processes.
5.3. What is meant by entropy? Does entropy decrease or increase in closed systems? Why or why not?
5.4. What is meant by the process equation in hydraulic systems?
5.5. Every flow process can be considered either a conservation or a decay process. Explain this statement and define the domain in which the process equation is expressed.
5.6. Define the continuity principle. Derive the instantaneous three dimensional unsteady continuity equation for compressible fluids. Then, write it in one-dimensional form.
5.7. Write a three-dimensional form of the steady state continuity equation for compressible fluids. Then, deduce a two-dimensional steady form for incompressible fluids.
5.8. Derive the steady one-dimensional continuity equation along a finite stream tube.
5.9. Derive the continuity equation for unsteady incompressible flow in open channels. Then, write its general form where lateral flow is encountered. Check the homogeneity of the equation.

5.10. Define energy. Is it a scalar or vector quantity?
5.11. Discuss different types of energy. Define *work*.
5.12. Comment on the conditions under which the energy and momentum equations are applicable. What type of information can be obtained from each?
5.13. What is meant by energy conservation? Considering a stream tube as a control volume, derive the Bernoulli equation and check the homogeneity of the dimensions of its terms.
5.14. Discuss the validity and limitations of the Bernoulli equation. Give some examples in which the Bernoulli equation is not applicable.
5.15. Using a sketch, derive a formula for the energy equation in open channels.
5.16. What is meant by momentum conservation? Write Newton's second law of motion and express it mathematically.
5.17. Describe the forces normally included while applying the momentum equation.
5.18. A firefighter is holding a nozzle of a fire hose for extinguishing the fire. Will the nozzle be pushed toward him or away from him? Explain.
5.19. Derive a formula for the momentum force acting on a stream tube due to velocity variation.
5.20. What type of problems can be analyzed by the angular momentum equation? Give two examples.
5.21. Write an equation to evaluate the torque resulting from garden sprinkler and define its terms. What are the conditions which may create zero torque?
5.22. Discuss the main driving and resisting forces in open channels. Derive the momentum equation for open channels. Hence, deduce a form for horizontal and frictionless channels.
5.23. State the Euler equation for one-dimensional flow in an ideal fluid; hence derive the Bernoulli equation using the Euler equation.
5.24. Define *local velocity, convective acceleration,* and *local acceleration* in open-channel flow.
5.25. Why do the friction resistance coefficient, total head loss coefficient, and dissipated energy coefficient differ, in general, from each other? Can these coefficients be equal in magnitude?
5.26. Which of the energy and momentum equations is preferred in the steady spatially varied flow computation? Why?
5.27. Which equation is more appropriate in open-channel computations for the case of: (a) no lateral flow, and (b) lateral inflow? Why?
5.28. For computation of the conjugate depths of a hydraulic jump in an open channel, which of the energy and momentum equations is more accurate? Why?

Problems

5.1. Water flows in a 10-in-diameter pipe at a velocity of 6 ft/s. Determine the diameter of the nozzle fitted at the end of the pipe if the jet velocity is 50 ft/s.

5.2. If the diameter of a pipe is reduced by 50%, how much do you expect the velocity to change?

5.3. Water flows in a rectangular open channel of constant width through its entire length at a velocity of 1 m/s. The initial depth of flow was 2 m. After a certain hydraulic structure the depth was measured to be 1.2 m. Calculate the new velocity neglecting the head loss between the two sections.

5.4. A pipeline with a diameter of 12 inches carries a liquid with a specific gravity of 0.86 at a velocity of 10 ft/s. Determine the liquid velocity and the mass flux if the pipe diameter is reduced to 8 inches at another section.

5.5. A sheet of water issues from a wide rectangular opening in a side of a tank. The water emerges out horizontally with a velocity of 10 m/s and a sheet thickness of 2 cm. Under the gravity force the velocity and thickness of the jet vary, as shown in Figure P5.5. Determine the velocity and thickness of the jet at a section where it has a slope of 70° to the horizontal.

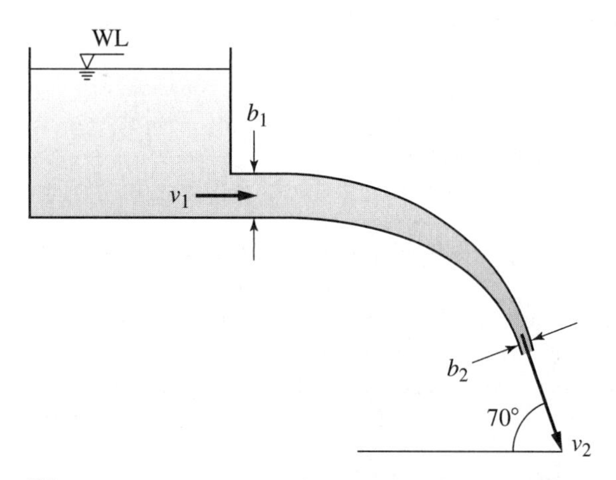

Figure P5.5 A wide rectangular opening on side of a tank.

5.6. Water flows through a pipeline whose diameter contracts from 45 cm at A to 30 cm at B after which it forks into two branches, as shown in Figure P5.6. The first branch is 15 cm in diameter discharging at C, and the second branch is 22.5 cm in diameter discharging at D. The velocities at A and D are 1.8 m/s and 3.6 m/s, respectively. What would be the velocities and discharges at sections B and C?

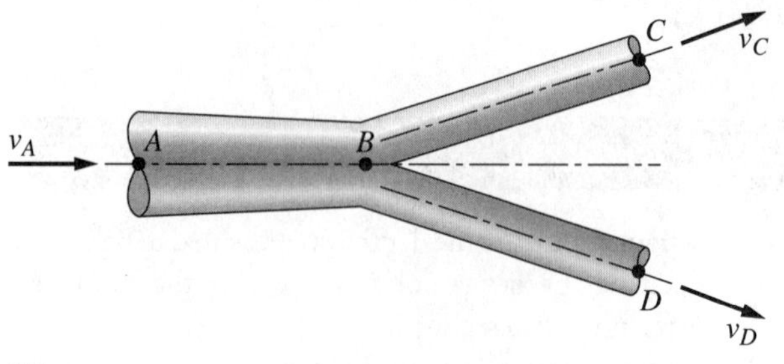

Figure P5.6 A pipe forking into two branches.

5.7. If the velocity components at a point in a compressible flow field are given by $V_x = 2xy + y^3 + b^{1/2}$ and $V_y = x^2 - y^2 + a^2$, where a and b are constants, prove that these components constitute a possible flow field.

5.8. The velocity components for an incompressible fluid, (i.e., the continuity equation is satisfied) in the x and y directions are given as

$$V_x = \frac{4xy^2}{3} - x^2y$$

$$V_y = xy^2 - \frac{4x^2y}{3}$$

Discuss whether the given velocity distribution satisfied the continuity equation or not.

5.9. The velocity components for an incompressible fluid in the x and y directions are given as

$$V_x = \frac{xy^2}{2} - x^2y$$

$$V_y = xy^2 - \frac{x^2y}{2}$$

Discuss whether the given velocity distribution is a possible field of flow.

5.10. The velocity components in a two-dimensional flow are $V_x = a(x^3 + xy^3)$ and $V_y = a(y^3 + yx^3)$, where a is a known constant. Is such a flow field possible?

5.11. Two pipes of diameters 1.5 ft and 0.8 ft, respectively, are set in series to convey a discharge of 10 ft^3/s. Find the velocity of flow in each pipe.

5.12. Water issues to an open tank through a jet 1 inch in diameter and a velocity of 30 ft/s. If the water leaves the tank at a rate of 0.1 ft^3/s, what is the rate of water accumulation in the tank and what is the total volume of water accumulated after 2 hours?

5.13. A water jet, 0.002 m^2 in cross section, issues with a velocity of 4 m/s, and is directed to an open tank, as shown in Figure P5.13. The water is extracted from

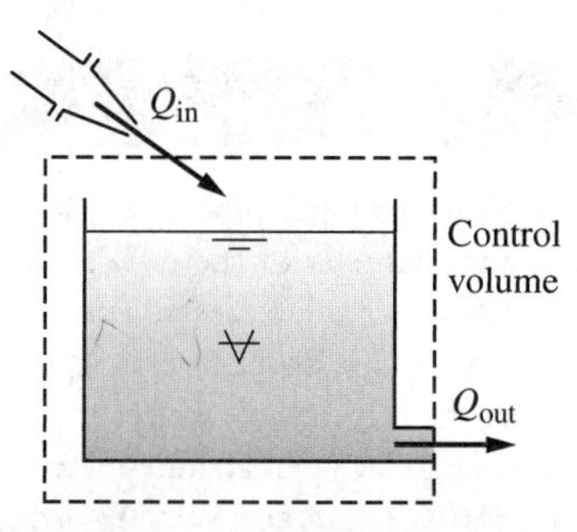

Figure P5.13 A water tank directed to an open tank.

the tank at a constant rate of 0.01 m^3/s. What time is needed to evacuate the tank if it contains 200 m^3 of water?

5.14. A cylindrical water tank with a diameter of 2.5 ft has a side hole of 4-in diameter and a bottom hole 6 inches in diameter. The water enters from the side hole and leaves from the bottom hole, as shown in Figure P5.14. The velocity of water at the outlet is $V = (2gh)^{0.5}$, where h is the height of the water surface above the outlet. At a certain time, where $h = 1.25$ ft, the water level was rising at a rate of 0.5 in/s. Assuming that the fluid is incompressible, determine the water velocity through the inlet.

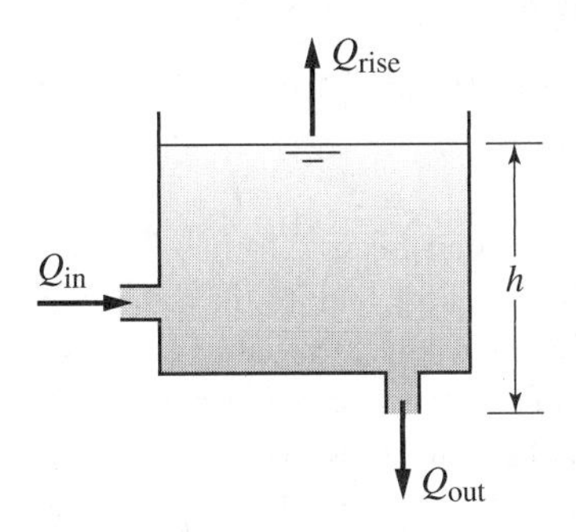

Figure P5.14 A water tank with two holes, one at the bottom and the other on a side.

5.15. Figure P5.15 shows a cylindrical tank of diameter 0.5 m connected to a horizontal pipe flow resulting in tank filling. The pipe inlet velocity is 5 m/s while the outlet velocity is 3.5 m/s. Determine

a. The rate of change of height of water inside the tank
b. The time required to fill the tank if the tank was initially empty

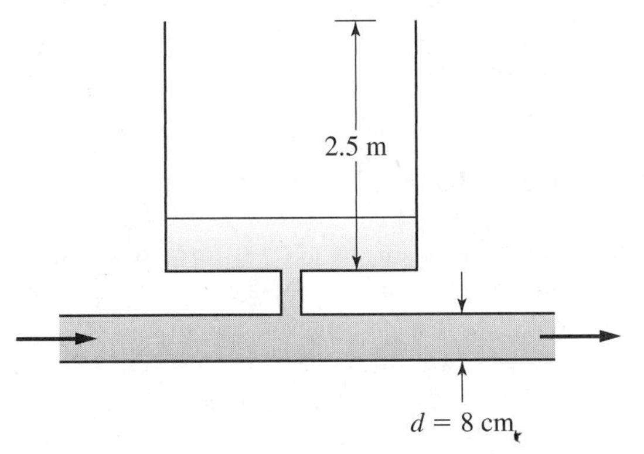

Figure P5.15 A cylindrical tank connected to a pipe.

5.16. Determine the flow rate in the tube shown in Figure P5.16 if the manometer fluid is mercury and the fluid in the tube is (a) water, (b) air.

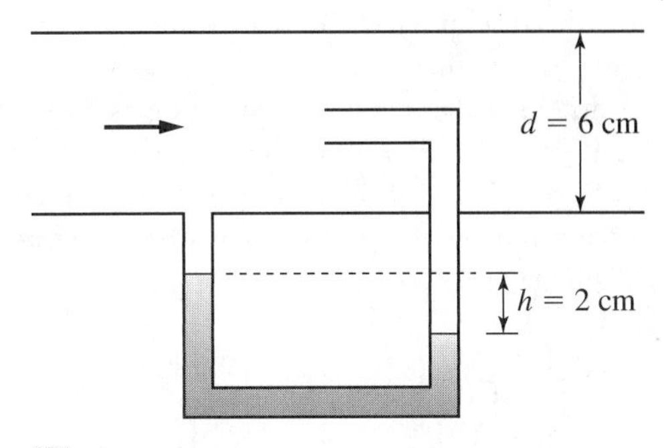

Figure P5.16 Manometer and flow tube for Problem 5.16.

5.17. Water at 20°C is being released from the nozzle into the open atmosphere as shown in Figure P5.17. If pressure at section 1 is 15.5 psi (abs), determine (a) flow rate, (b) height H in stagnation tube.

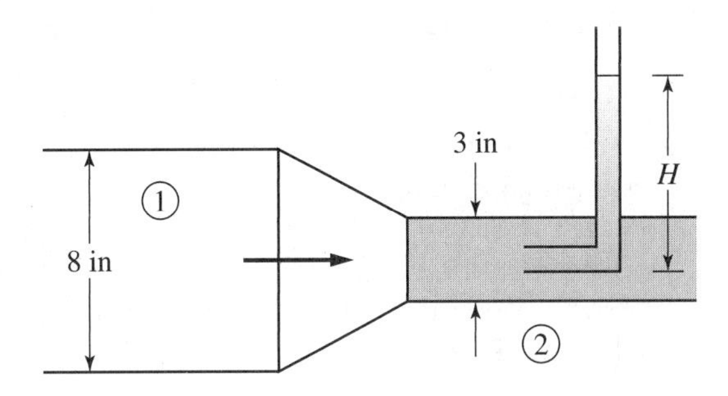

Figure P5.17 A nozzle discharging into the atmosphere.

5.18. Determine the total energy of water flowing in a pipe with a velocity of 10 m/s under a pressure of 294.3 kN/m^2. The center line of the pipe is 30 m above the datum line.

5.19. A fire hose nozzle has a diameter of 4 in, as shown in Figure P5.19. According to the fire code set by the state, the nozzle should be able to deliver 1000 gal/min. If the nozzle is connected to a 10-in-diameter hose, what pressure head must be maintained just upstream of the nozzle?

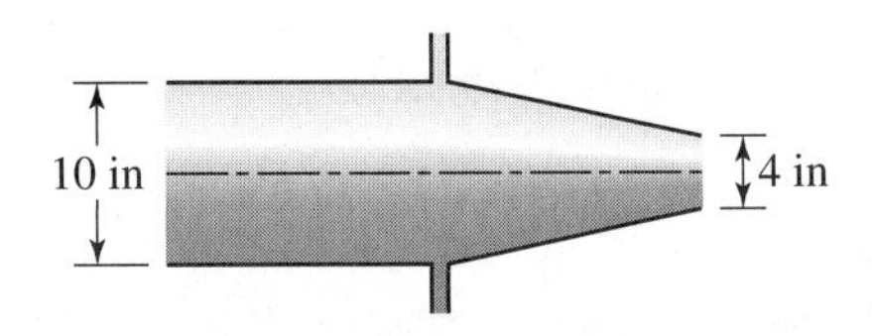

Figure P5.19 A fire hose nozzle.

5.20. A 5-cm-diameter pipe carries a discharge of 1 m^3/min of water at a pressure of 98.1 kN/m^2. Determine the pressure head in meter of water, the velocity head, and the total head with reference to a datum 10 m below the center line of the pipe.

5.21. Water flows through the pipe contraction as shown in Figure P5.21. The drop in the manometer level after contraction is 0.1 ft. If the manometer liquid is mercury, determine the flow rate as a function of the small pipe diameter, *D*.

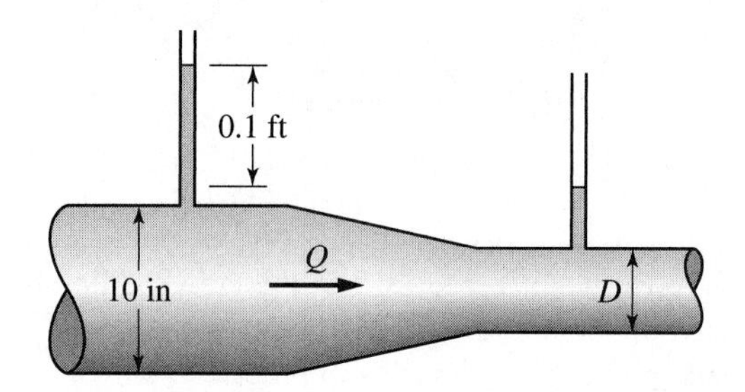

Figure P5.21 A pipe contraction.

5.22. A pipe 10 inches in diameter and 60 ft long is fixed in a vertical position, as shown in Figure P5.22. The pressure head at its upper end is 20.0 ft and the water flows at a velocity of 10 ft/s. The total friction loss in the pipe is 3 ft. Determine the pressure head at the lower end of the pipe when the direction of flow is (a) downward and (b) upward.

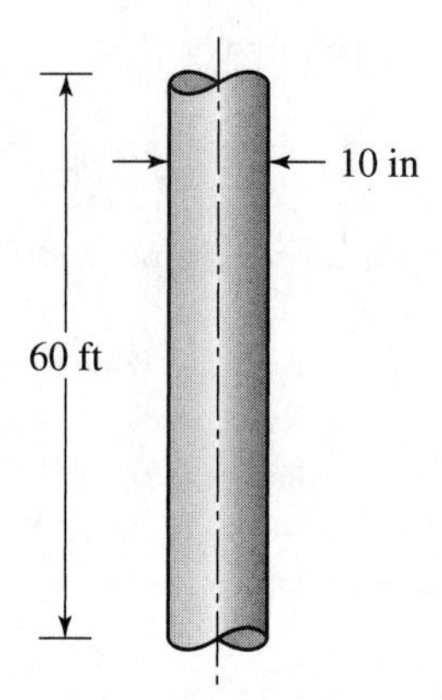

Figure P5.22 A pipe fixed in vertical position.

5.23. A convergent-divergent nozzle is fitted into the vertical side of a water tank, as shown in Figure P5.23. The pressure head at the throat is −1.9 m. The water level in the tank is 1.5 m above the center line of the nozzle and the nozzle discharges 4.5 l/s of water. Assuming that no losses take place in the convergent part of the nozzle and the losses in the divergent part are 0.18 times the velocity head at the exit, determine the throat and exit diameters of the nozzle.

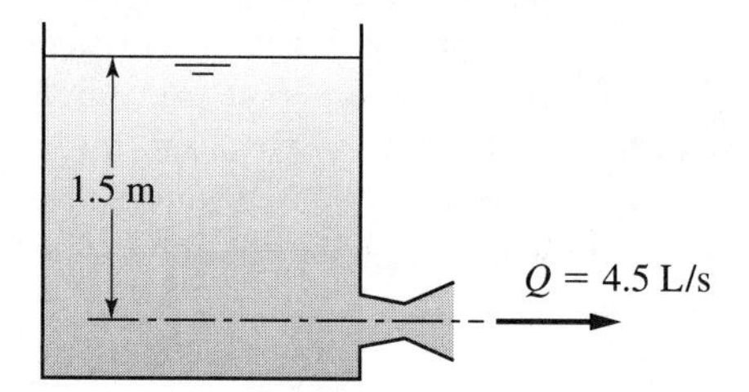

Figure P5.23 A convergent-divergent nozzle fitted to the vertical side of a tank.

5.24. A 0.2 m diameter siphon issues 220 l/s of oil of specific gravity 0.8 from a large tank to the atmosphere at an elevation of 3.0 m below the oil surface, as shown in Figure P5.24. The invert of the siphon is located 8.0 m above its exit to the atmosphere. Determine the losses in the siphon in terms of the velocity head. If two-thirds of the losses are encountered before the invert, what is the lowest pressure in that siphon?

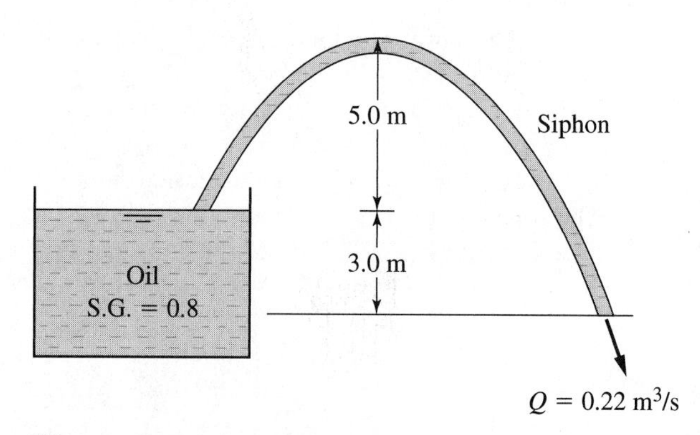

Figure P5.24 A siphon discharging oil from a tank.

5.25. A pipe 6 inches in diameter is fitted to a water tank as shown in Figure P5.25. The nozzle (2 inches in diameter) at the end of the pipe discharges into the atmosphere. Calculate the flow rate and the pressure at *A*, *B*, *C*, and *D*, neglecting all losses in the pipe and nozzle.

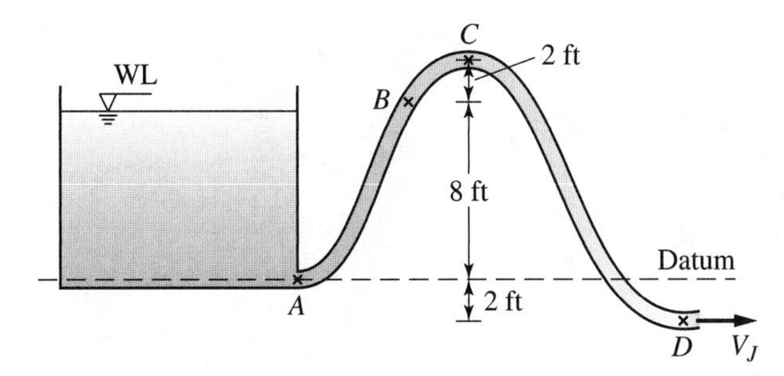

Figure P5.25 A pipe fitted to a water tank.

5.26. A horizontal pipe, 12 inches in diameter and 2500 ft in length, carries a discharge of 0.25 ft^3/s for a thermal power plant from a reservoir, as shown in Figure P5.26. Calculate the pressure in the pipe at a distance of 1000 ft from the reservoir. The head loss in the pipe is given as

$$h_L = \frac{0.025(L/d)V^2}{2g}$$

where L is the length of the pipe from the reservoir, d is the diameter of the pipe, V is the mean water velocity in the pipe, and g is the acceleration of gravity.

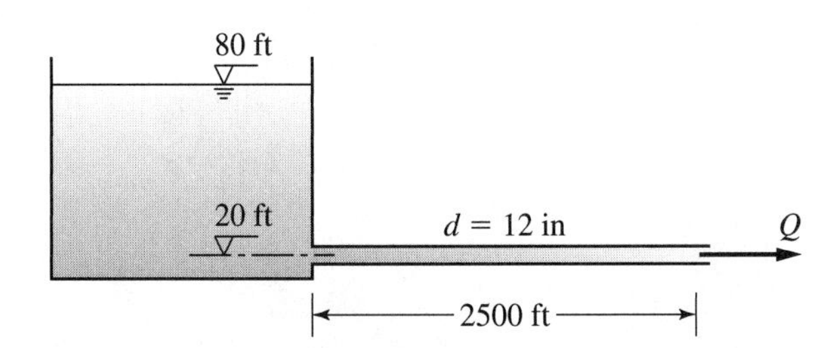

Figure P5.26 A horizontal pipe discharging water to a thermal power plant from a reservoir.

5.27. Carbon dioxide gas is allowed to pass through a diffuser of an inlet diameter 2 in and an outlet diameter of 4.5 in. The pressure at inlet is 46 psi and that at the outlet is 10 psi. The densities of the gas at the inlet and outlet are 0.356 lb/ft^3 and 0.0774 lb/ft^3, respectively. If the inlet velocity of the gas is 350 ft/s, determine (a) mass flow rate, and (b) exit velocity.

5.28. Water is coming into the fountain, shown in Figure P5.28, from a sump at a pressure $p_1 = 90$ kPa gage. Water exits from the space between two upper circular plates of diameter 60 cm as shown. Determine inlet and exit velocities and the flow rate of the water. Neglect frictional losses.

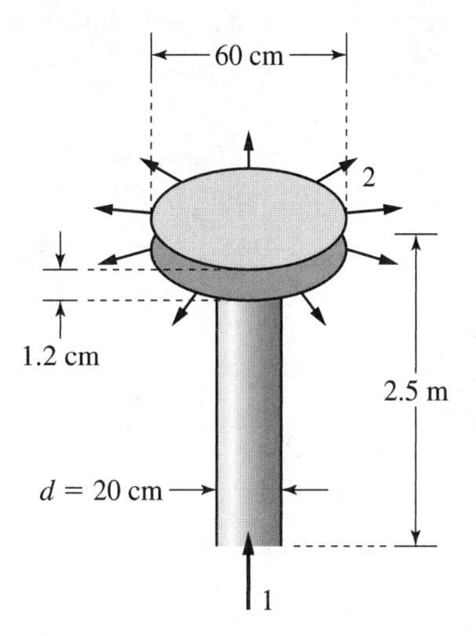

Figure P5.28 Fountain for Problem 5.28.

5.29. A 4.0-cm-diameter nozzle delivers a jet of water that strikes a flat plate normally. If the water speed is 50.0 m/s, calculate the force acting on the plate if (a) the plate is stationary, and if (b) the plate moves at 20 m/s in the same direction of the jet.

5.30. Determine the force exerted by a jet of water which has an area of 8 in^2 and a velocity of 30 ft/s on an inclined flat plate as a function of its angle of inclination to the horizontal, as shown in Figure P5.30. Assume the plate surface is smooth.

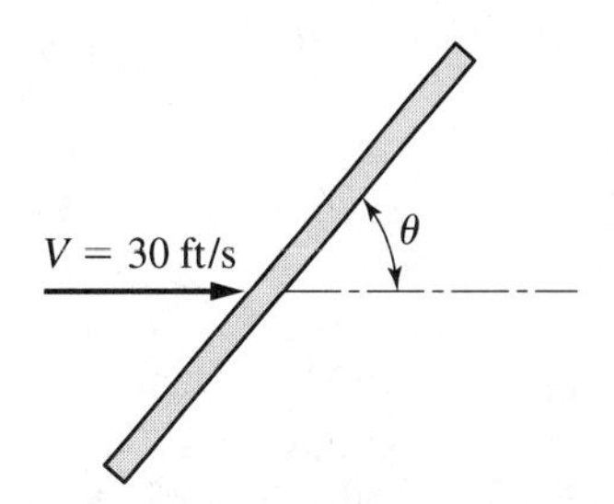

Figure P5.30 A water jet exerting force on an inclined plate.

5.31. A square plate of uniform thickness has a side length of 25 cm that hangs vertically from hinges at its top edge, as shown in Figure P5.31. When a horizontal jet strikes the plate at its center, the plate is deflected and comes to rest at an angle of 30° to the vertical. The jet is 2 cm in diameter and has a velocity of 8 m/s. Calculate the mass of the plate. What force should be applied at the lower edge of the plate to maintain its vertical position?

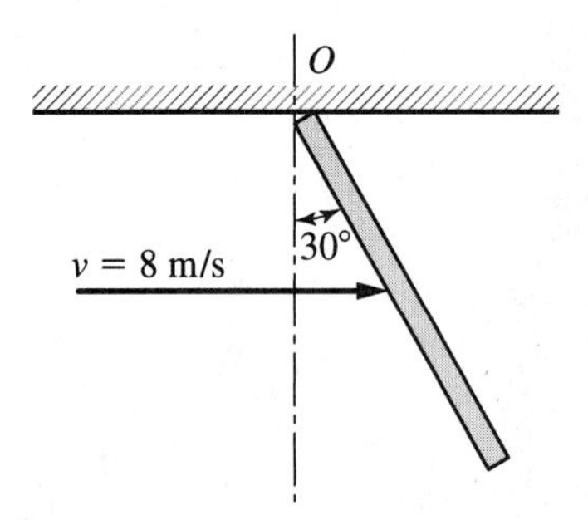

Figure P5.31 A vertically hanging square plate deflected by a horizontal jet.

5.32. A 4-in-diameter nozzle is fitted at the end of a 10-in-diameter pipeline, as shown in Figure P5.32. The pressure at the end of the pipe (nozzle inlet) is 20 psi and at the nozzle exit the pressure is atmospheric. Find the force exerted by the nozzle on the pipe line. Neglect the head loss in the nozzle.

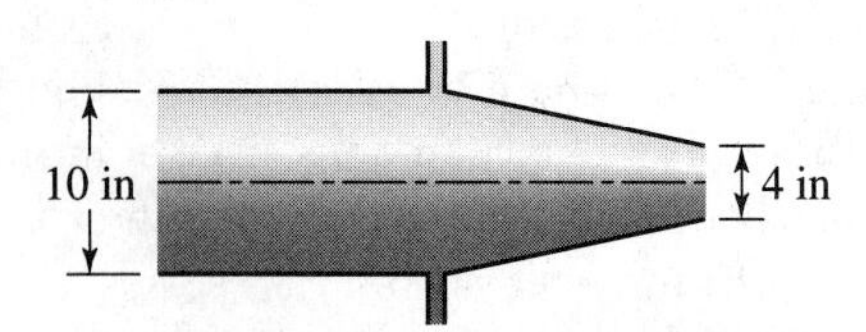

Figure P5.32 A nozzle fitted to a pipeline.

5.33. A horizontal reducer bend is used to connect a 20-cm-diameter pipe to 10-cm-diameter pipe, as shown in Figure P5.33. The flow, which has a velocity of 1.2 m/s in the first pipe, is deflected by an angle of 120°. The pressure at the bend entrance is 4905 N/m^2. Calculate the magnitude and direction of the force required to hold the bend in position.

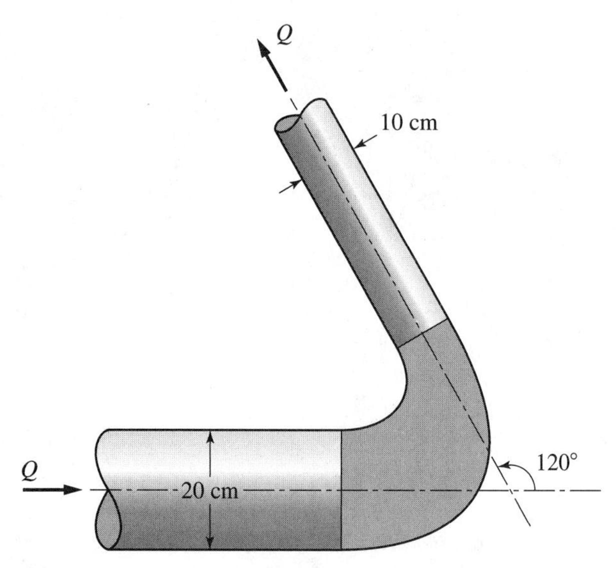

Figure P5.33 A horizontal reducer bend connecting two pipes.

5.34. Water flows through a Y-shaped horizontal pipe system consisting of two branches and a stem, as shown in Figure P5.34. The velocity in the stem is 3 m/s and the diameter is 20 cm. Each branch is 10.0 cm diameter and makes a 30° angle with the axis of the stem. The pressure in the stem is 196.2 kN/m^2 and the flow is divided equally between the two branches. Calculate the magnitude and direction of the force acting on the Y-shaped connection.

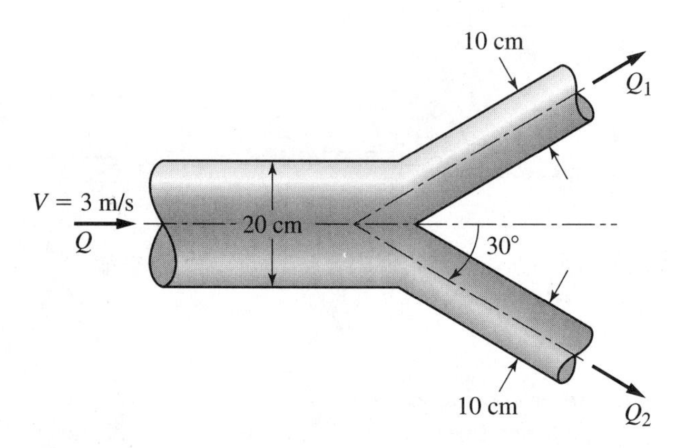

Figure P5.34 A horizontal pipe system consisting of two branches and a stem.

5.35. Water issues through a horizontal 180° pipe bend as shown in Figure P5.35. The pipe has the same diameter,

12 in, before and after the bend and the velocity through the bend is 10 ft/s. The pressures at the entrance and exist of the bend are 12 psi and 10 psi, respectively. Calculate the horizontal (x and y) components of the force exerted by the bend on the flowing water.

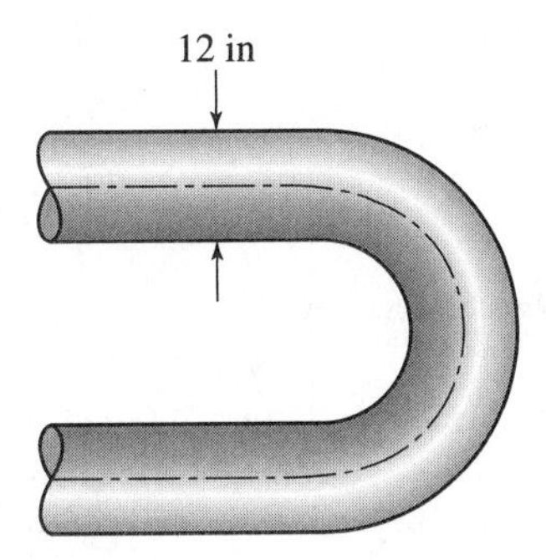

Figure P5.35 Water issuing through a pipe bend.

5.36. A horizontal nozzle with an outlet diameter of 5 in is fitted into a 20-in-diameter pipe, as shown in Figure P5.36. The pressure head at the nozzle inlet is 20 ft. Neglecting the head loss inside the nozzle and assuming that the weight of the nozzle and water in it is 250 lb, what force is needed to hold the nozzle in the pipe?

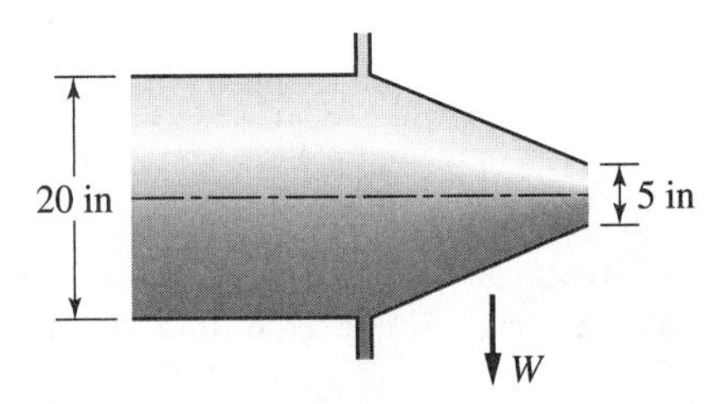

Figure P5.36 A horizontal nozzle fitted to a pipe.

5.37. A closed square tank with a base of 0.6 m and a height of 3 m weighing 6000 N has a side hole with an effective area of 8 cm^2, 10 cm above the tank bottom, as shown in Figure P5.37. The tank is filled with water to a depth of 2.5 m and the coefficient of friction between the wheels of the tank and the ground is 0.015. Determine the pressure needed on the water surface to move the tank if the wheels are locked.

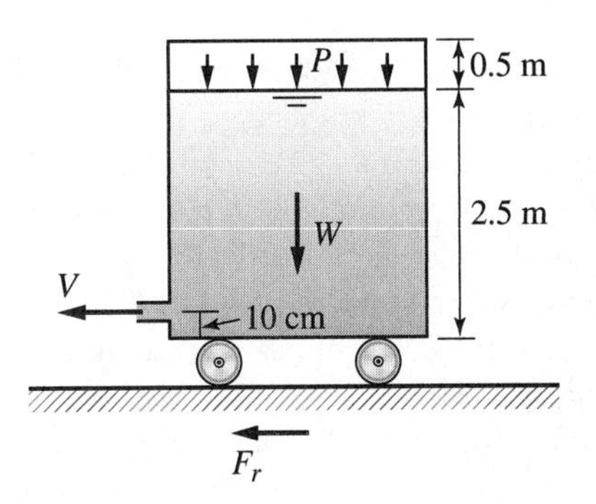

Figure P5.37 A closed tank on wheels.

5.38. A nozzle 8 cm in diameter issues a vertical jet of water with a velocity of 6 m/s. The jet strikes a horizontal and movable disc weighing 156.96 N. Determine the distance above the nozzle outlet at which the disc will be held in equilibrium.

5.39. A jet of water 8 cm in diameter strikes a vertical plate perpendicularly with a velocity of 10 m/s. The plate is moving toward the jet with a velocity of 5 m/s. Determine the force and power required to move this plate.

5.40. In Problem 5.39, if the plate moves away from the jet, what force and power will be required to move the plate?

5.41. The angle of a horizontal reducer bend with an initial diameter of 20 cm and a final diameter of 10 cm is 45° as shown in Figure P5.41. The discharge through the bend is 0.1 m^3/s and the pressure head at the bend inlet is 10 m of water. Assuming a friction loss inside the bend of 10% of the kinetic energy at the bend exit, determine the force required to hold the bend in place.

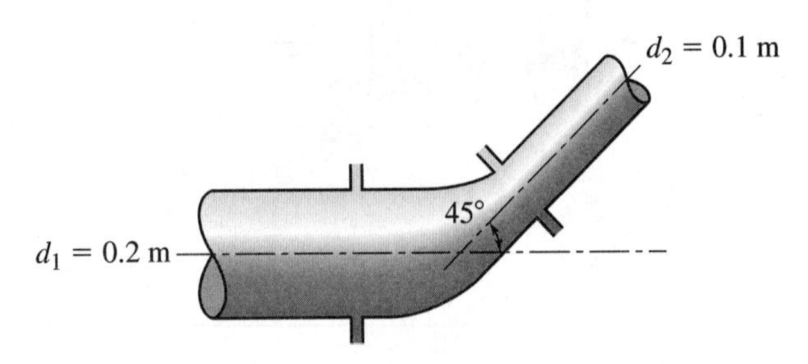

Figure P5.41 A horizontal reducer bend.

5.42. For Problem 5.41, if the angle of the reducing bend is 90°, what force will be required to hold the bend in place?

5.43. Water issues through a 180° vertical reducer bend with a discharge of 0.1 m^3/s, as shown in Figure P5.43. The diameter decreases from 20 cm at the inlet to 10 cm at the outlet of the reducer. The pressure head at the center line of the inlet of the bend is 11 m of water. If the volume of water inside the bend is 0.12 m^3 and assuming the validity of Bernoulli's equation, calculate the reaction needed to hold the bend in place. The weight of the bend metal is 900 kg-m/s^2.

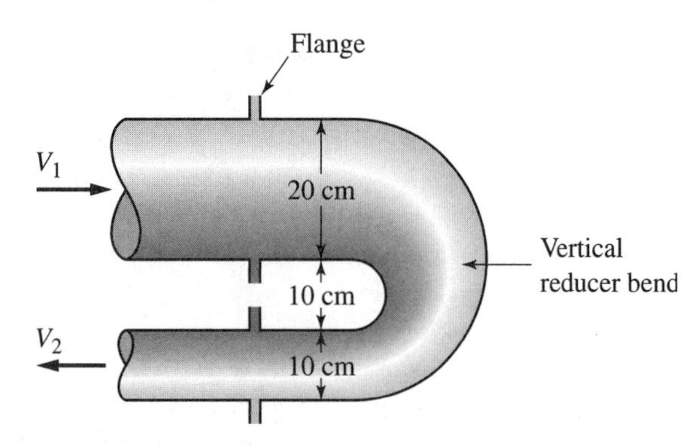

Figure P5.43 Water issuing from a 180° vertical reducer bend.

5.44. A 12-in-diameter pipe has a 60° horizontal bend as shown in Figure P5.44. The pipe carries a fluid with a specific gravity of 0.92 at a rate of 1.8 ft^3/s. The volume of liquid inside the bend is 2.5 ft^3 and the weight of the bend is 50 lb. What forces must be applied to hold the bend in place assuming a uniform pressure of 10 psi through the bend?

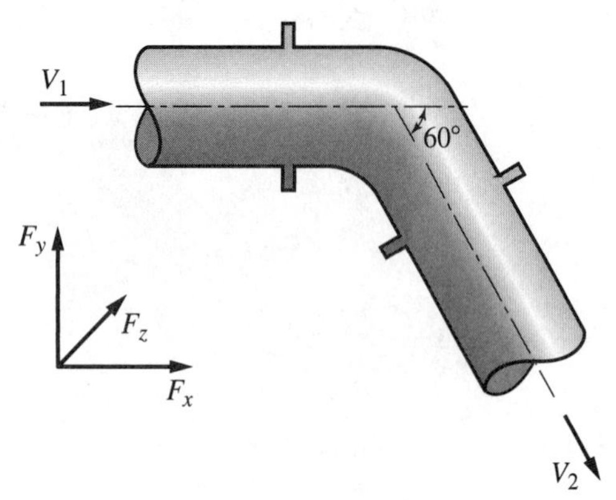

Figure P5.44 Water issuing from a 60° horizontal bend.

5.45. An asymmetrical sprinkler shown in Figure P5.45 has a frictionless shaft. Determine its speed of rotation in rpm if it has an equal flow velocity in the nozzles of 10 m/s. Determine the torque and power required to hold the sprinkler if the diameter of the nozzles is 2 cm.

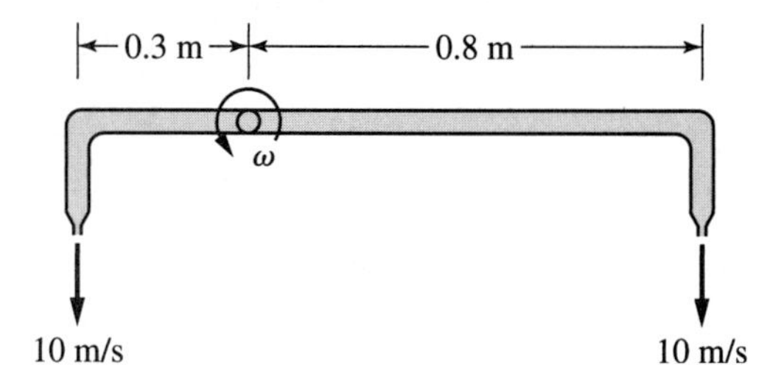

Figure P5.45 An asymmetrical sprinkler.

5.46. Referring to Problem 5.45, if the nozzle outlet in the short arm of the sprinkler is altered as shown in Figure P5.46, evaluate the torque and power required to hold the sprinkler.

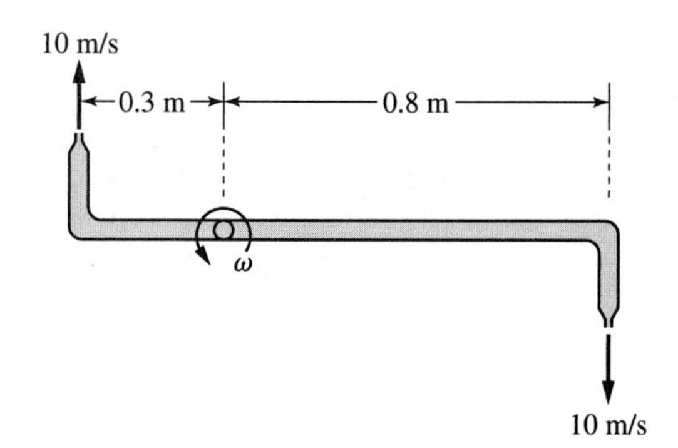

Figure P5.46 A sprinkler with an altered nozzle outlet.

5.47. The sprinkler shown in Figure P5.47 has three nozzles with jet velocities as given in the figure. Assuming frictionless shaft, determine the speed of rotation in rpm.

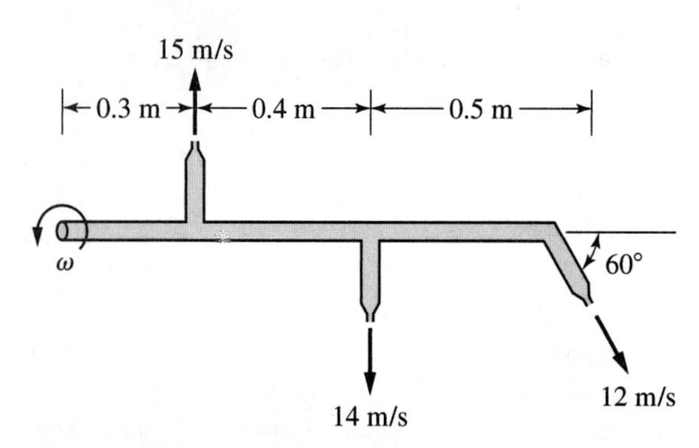

Figure P5.47 A sprinkler with three nozzles.

5.48. A sluice gate is placed across a rectangular channel where the water depth is 15 ft, as shown in Figure P5.48. Calculate the anchoring force required to hold the gate in place when (a) the gate is fully closed with no water downstream, and (b) when the gate is opened by 1 ft and operates under a free condition.

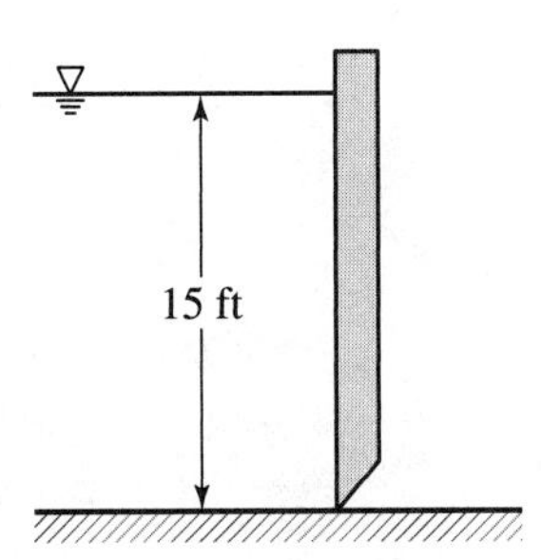

Figure P5.48 Sluice gate placed across a rectangular channel.

5.49. For a frictionless siphon shown in Figure P5.49, determine the discharge and the pressure head at *A*, given that the pipe diameter is 20 cm and the diameter of the nozzle outlet is 10 cm.

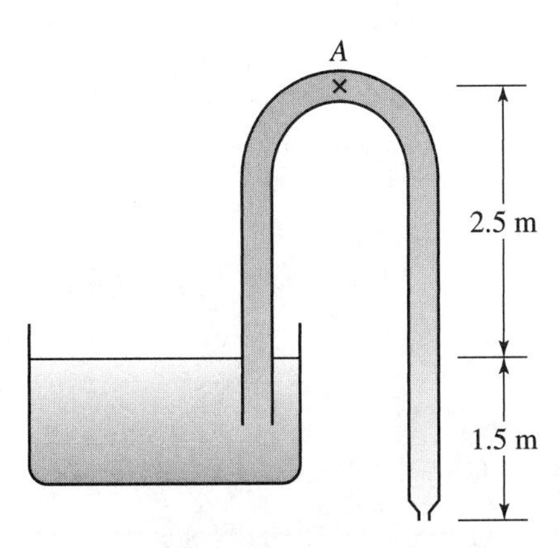

Figure P5.49 Frictionless syphon.

5.50. A pipe 15 cm in diameter is fitted to a water tank as shown in Figure P5.50. The nozzle (5 cm in diameter) at the end of the pipe discharges into the atmosphere. Calculate the flow rate and the pressure at *A*, *B*, *C*, and *D*, neglecting all losses in the pipe and nozzle.

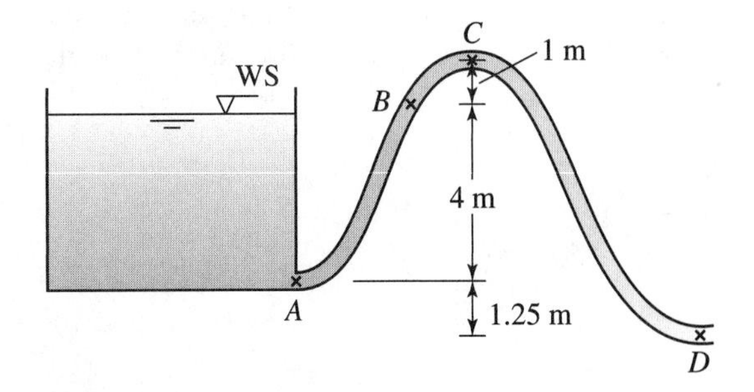

Figure P5.50 A siphon connected to a tank.

5.51. A horizontal pipeline 8 inches in diameter and 4000 feet long connects two reservoirs, with a 20-ft difference in surface elevations, Figure P5.51. Assuming the energy loss in the pipe to be 10 ft, what is the rate of flow in the pipe?

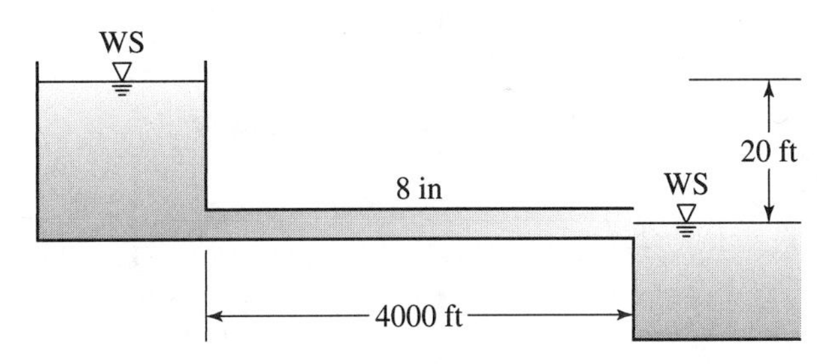

Figure P5.51 A pipe connecting two reservoirs.

5.52. Calculate the force required to hold a fire hose shown in Figure P5.52, with a discharge of 1000 gpm. The diameter of the nozzle at the inlet is 15 in and at outlet is 3 in. The pressure head at the inlet is 10 ft.

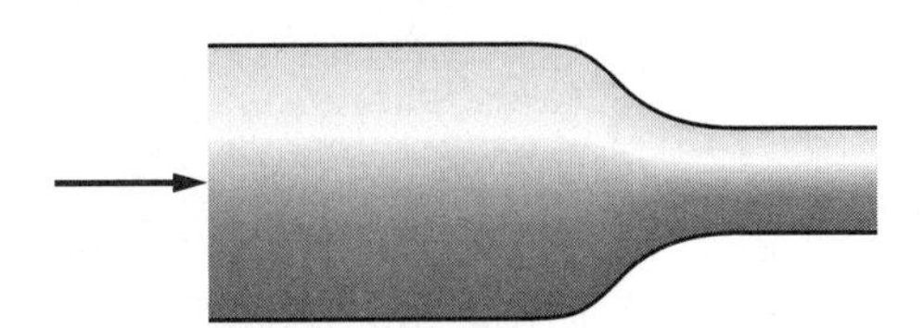

Figure P5.52 A firehouse.

5.53. A nozzle 10 cm in diameter issues a vertical jet of water with a velocity of 8 m/s. The jet strikes a horizontal and movable disc weighing 196.2 kg-m/s^2, as shown in Figure P5.53. Determine the distance above the nozzle out let at which the disc will be held in equilibrium.

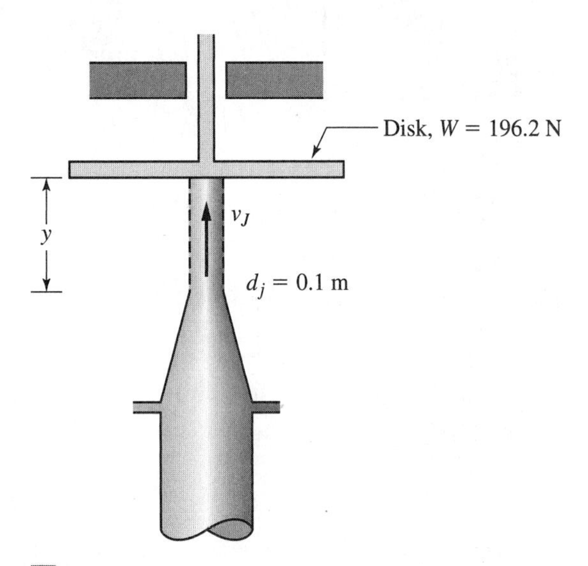

Figure P5.53 Movable disk subject to a water jet.

5.54. A jet of water 6 cm in diameter issues with a velocity of 12 m/s. The jet strikes a vertical plate perpendicularly as shown in Figure P5.54. The plate moves toward the jet with a velocity of 4 m/s. Determine the force and power required to move this plate.

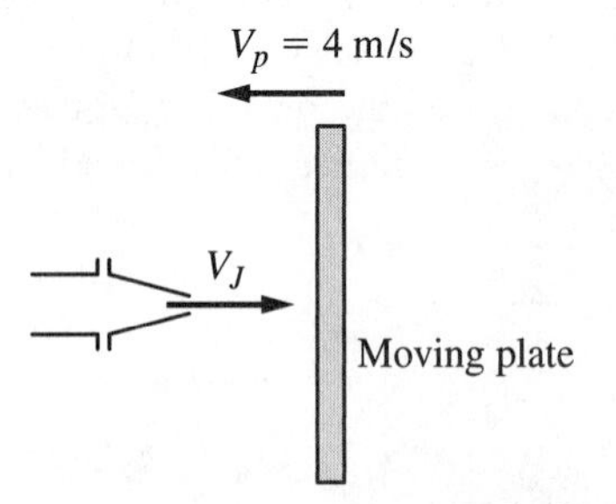

Figure P5.54 A vertical plate subject to a water jet.

5.55. A Venturi meter in a horizontal pipeline carrying water contracts from a normal section of 8 in diameter to a throat section 3 in diameter as shown in Figure P5.55. A differential manometer containing mercury (S.G. = 13.6) attached to the meter gives a reading of 2.5 in and a piezometer tube attached to the upstream full section shows a water level of 5 ft above the pipeline. Neglecting losses in the meter, calculate the pipe discharge and the resultant hydrodynamic force on the meter (lb).

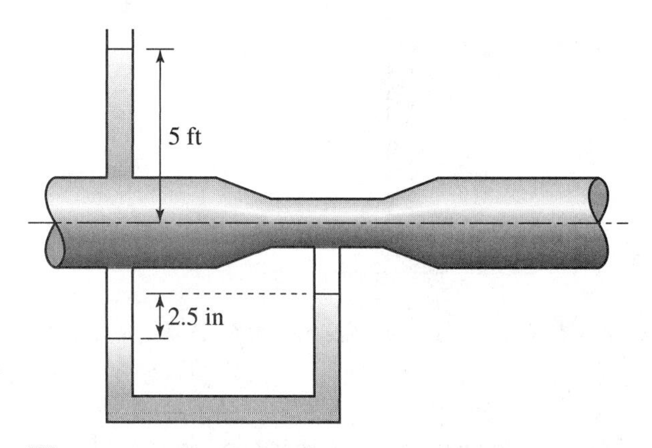

Figure P5.55 Venturi meter with a manometer attached.

5.56. A 10-ft-wide rectangular channel carries 500 cfs at a depth of 7 ft. Now, a sill, 3 ft high and 3 ft wide, is placed on the bottom of the channel, as shown in Figure P5.56. If the depth in the channel immediately downstream of the channel is 6.5 ft, what is the resultant hydrodynamic force acting on the sill?

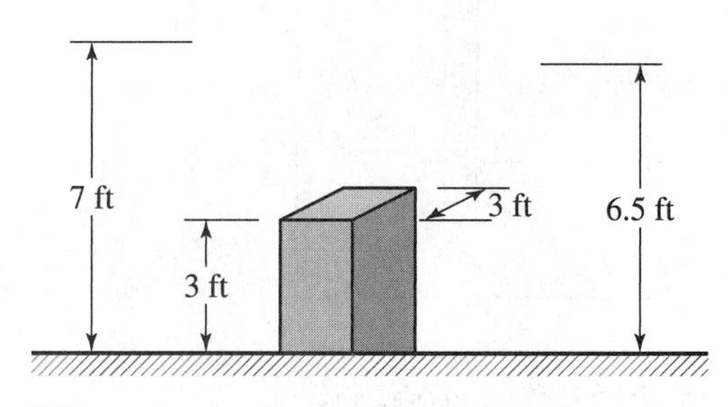

Figure P5.56 A sill in a rectangular channel.

5.57. A 20-ft-wide rectangular channel is crossed by a bridge which is supported by 3-ft square piers as shown in Figure P5.57. The discharge in the channel is 1000 cfs and the depth upstream of the piers is 10 ft. If the depth just downstream is 9.5 ft, what is the total resultant force acting on each pier? What is the resultant hydrostatic force on each pier?

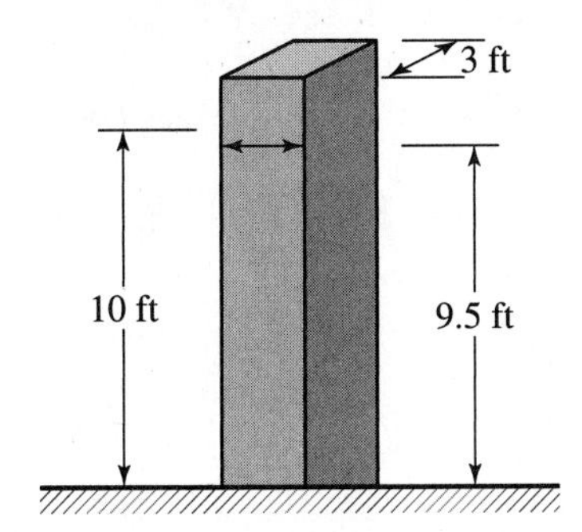

Figure P5.57 A bridge pier in a rectangular channel.

5.58. Water is fed through a fire hose from an open water tank located some distance from the building where the fire is to be extinguished, as shown in Figure P5.58. The water level in the tank is 20 ft above the invert of the floor, which is on fire. The nozzle of the hose is initially 12 inches tapering to a final diameter of 3 inches. If the flow rate is 1 cfs and there is 8 feet of head loss from the tank to the fire, what is the force required to hold the nozzle still? Ignore losses in the nozzle.

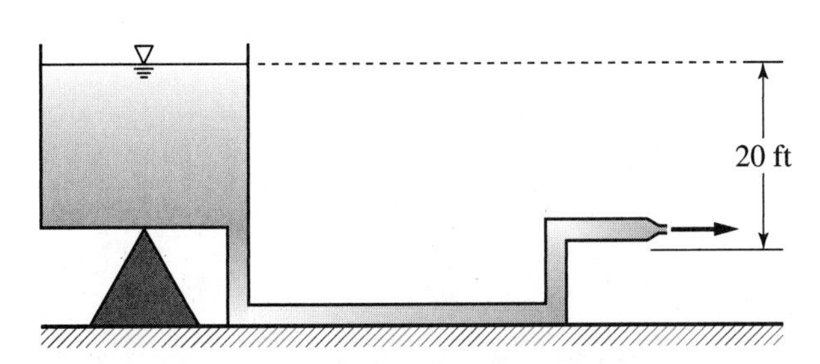

Figure P5.58 Water delivery from a tank through a hose.

5.59. Figure P5.59 shows a two-dimensional jet divided by a wedge. If the force required to hold the wedge is 20 lb per foot width of the wedge, determine the wedge angle θ. Assume the velocity of the jet remains constant at 15 ft/s.

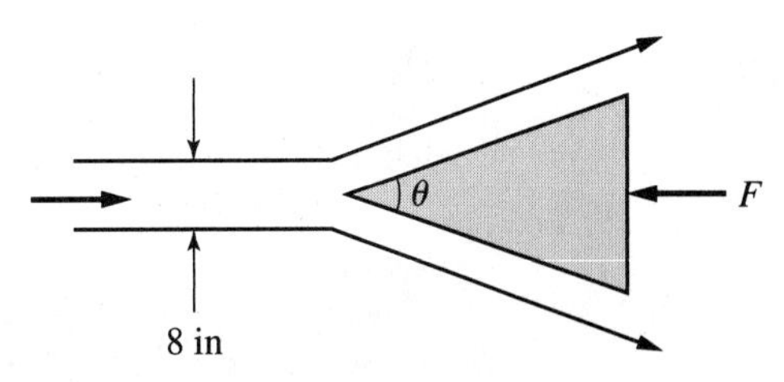

Figure P5.59 Jet for Problem 5.59.

5.60. A water jet from a nozzle of diameter 4 cm strikes a curved vane fixed on the wall as shown in Figure P5.60, with a velocity of 15 m/s. Determine the force exerted on the wall by this vane.

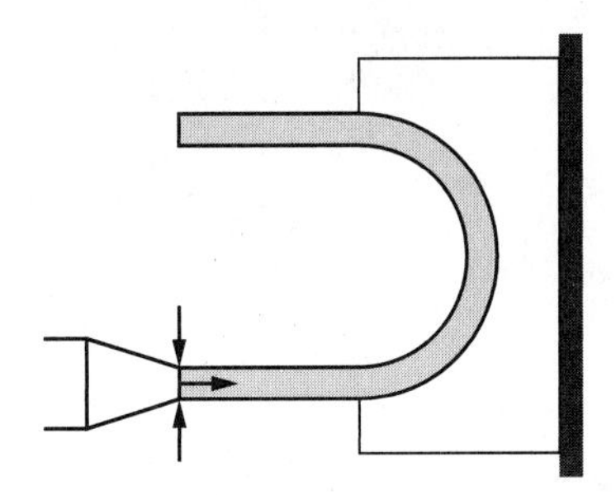

Figure P5.60 A nozzle striking a curved vane.

5.61. For the experimental setup shown in Figure P5.61, determine the theoretical mass required to balance the force due to the impact of a jet striking on a conical surface with an angle of cone $\theta = 60°$. The velocity of the water jet is 0.5 m/s.

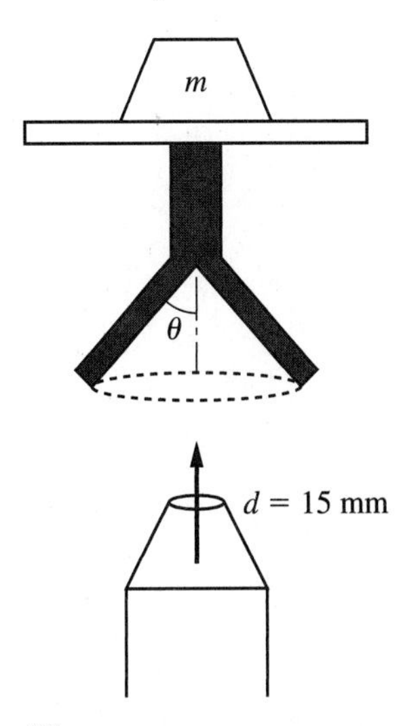

Figure P5.61 A jet striking a conical surface.

References

Chiu, C. L., Hsiung, D. E., and Lin, H. C., 1978. Three-dimensional open channel flow. *Journal of the Hydraulics Division.* ASCE, Vol. 104, No. HY8, pp. 1118–1136.

Gerhart, P. M., Gross, R. J., and Hochstein, J. I., 1992. *Fundamentals of Fluid Mechanics,* 2nd ed. Addison-Wesley Publishing Company, San Francisco, CA.

Henderson, F. M., 1966. *Open Channel Flow.* Macmillan, New York.

Singh, V. P., 1996. *Kinematic Wave Modeling in Water Resources: Surface Water Hydrology.* Wiley, New York.

Taft, M. I. and Reisman, A., 1965. The conservation equations: a systematic look. *Journal of Hydrology,* Vol. 3, pp. 161–179.

Yen, B. C., 1973. Open channel flow equation revisited. *Journal of the Engineering Mechanics Division,* ASCE, 99 (EMS): 979–1009.

Yen, B. C. and Wenzel, H. G., 1970. Dynamic equations for spatially varied flow, *Journal of the Hydraulics Division,* ASCE; 96(HY1): 801–813.

Yen, B. C., Wenzel, H. G., and Yoon, Y. N., 1972. Resistance coefficients for steady spatially varied flow. *Journal of the Hydraulics Division,* ASCE, Vol. 98, No. HY8, pp. 1395–1410.

CHAPTER 6

DIMENSIONAL ANALYSIS AND HYDRAULIC SIMILARITY

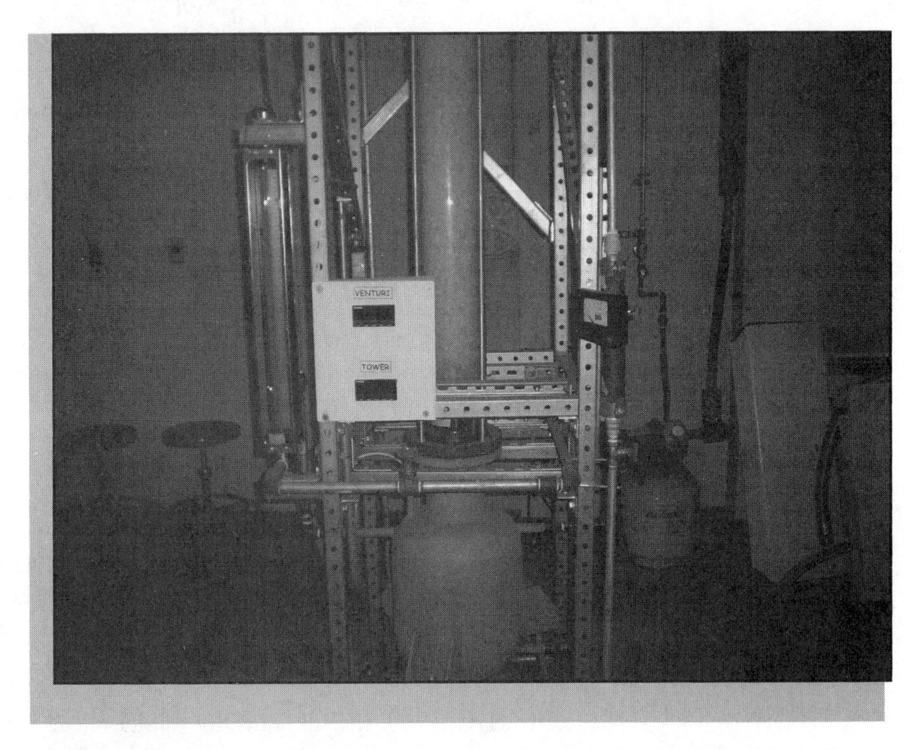

Porous media experiment at the Fluids Mechanics Lab at the University of Alabama at Huntsville.

Field problems and natural phenomena seldom have closed form or exact analytical solutions. A number of assumptions are usually made to obtain approximate solutions. In some cases we may need to develop empirical relations which should be tested and verified experimentally. Hydraulics encompasses more empirical and experimental relations than many other engineering fields. Hydraulic phenomena embody complex geometric, kinematic, and dynamic parameters and variables. The number of variables involved may be so large that it may be either impractical or highly expensive to experimentally examine them. However, for a specific problem some of the parameters predominate while others may be ignored without seriously affecting the accuracy of experimental results. In many cases, the theory of dimensional analysis is employed to determine the effective parameters and eliminate the less important parameters. Dimensional analysis describes various phenomena via dimensionally correct equations and reduces the cost of unnecessary testing. The number of effective parameters to be investigated can be significantly reduced by grouping them into a number of meaningful dimensionless terms. In this way, dimensional analysis maximizes data and information and minimizes experimental work, cost, and time of investigations.

6.1 Fundamental Dimensions, Systems of Units, and Hydraulic Variables

As discussed in Section 1.11 of Chapter 1, all variables can be expressed in terms of four fundamental quantities, length (L), time (T), mass (M), and temperature (θ). These quantities are independent of each other. Physical characteristics of various variables should be defined both qualitatively and quantitatively. The qualitative aspect defines the nature of a characteristic whether it is weight, velocity, pressure or whatever. The quantitative aspect provides a specific number on a standard universal scale through which various quantities can be compared to each other (Munson et al., 1990).

Two systems of units exist: the mass system encompassing L, T, M, and θ and the force system encompassing the same units but force (F) in place of mass (M). The two systems, based on mass and force, are related to each other by Newton's second law of motion: Force is equal to mass multiplied by acceleration. In dimensional form, Newton's law reads

$$[F] = [MLT^{-2}] \tag{6.1}$$

Here brackets denote that the quantities inside them are being represented by their corresponding dimensions and this convention will be followed in this chapter. The two systems of units are used alternatively to express the dimensions of various variables.

Dimensions of different variables can be easily determined either from their basic definitions or from physical laws describing them. In so doing, some variables are defined by the mass system while others are expressed using the force system. Equation (6.1) can be used to transform one system to the other.

The velocity, V, of an object, for example, is defined as the traveled distance divided by the corresponding time. In dimensional form one can write $[V] = [LT^{-1}]$. The surface tension, σ, is defined as force per unit length (i.e., $[\sigma] = [FL^{-1}]$). Using Newton's law, one can express σ using the mass system as $[\sigma] = [MLT^{-2}L^{-1}] = [MT^{-2}]$. Similarly, the density, ρ, is defined as the mass per unit volume (i.e., $[\rho] = [ML^{-3}]$), while the specific weight, γ, is defined as weight or force per unit volume (i.e., $[\gamma] = [FL^{-3}]$). Using the mass system, one can write $[\gamma] = [MLT^{-2}L^{-3}] = [ML^{-2}T^{-2}]$. Also, the momentum is defined as the mass multiplied by velocity, while both the work and energy are dimensionally defined as force multiplied by distance.

Some variables, unlike the ones above, do not have direct definitions. However, their dimensions can be easily deduced from the laws describing them. For example, the dimensions of the dynamic viscosity, μ, can be deduced from Newton's law of viscosity:

$$\tau = \mu \frac{dv}{dy} \tag{6.2a}$$

where τ is the shear stress defined as the force per unit area (FL^{-2}), dv is the variation in velocity (LT^{-1}), and dy is the distance in y direction (L). Using the force system, equation (6.2a) can be written in a dimensional form as

$$[FL^{-2}] = [\mu]\left[\left(\frac{LT^{-1}}{L}\right)\right] \tag{6.2b}$$

$$[\mu] = [FTL^{-2}] \tag{6.2c}$$

Using Newton's law, the dynamic viscosity can be expressed in the mass system as

$$[\mu] = [(MLT^{-2})\,TL^{-2}] = [ML^{-1}T^{-1}] \tag{6.2d}$$

The kinematic viscosity, ν, is defined as the dynamic viscosity, μ, divided by the fluid density, ρ. Hence,

$$[\nu] = \left[\frac{\mu}{\rho}\right] = \left[\frac{ML^{-1}T^{-1}}{ML^{-3}}\right] = [L^2T^{-1}] \tag{6.3}$$

Most parameters have dimensions, but a few do not. The slope (s) and the strain (ε) are examples of such dimensionless numbers. An angle θ is a dimensionless number. The efficiency of a hydraulic machine η is also a dimensionless number. The bulk modulus of elasticity, E_v, defined as the stress divided by the associated strain, has the same dimensions as the stress. The dimensions of most common physical parameters in fluid mechanics and hydraulics are given in Table 1.3.

The dimensions of all terms in any equation must be unified; otherwise, if all terms do not have the same dimensions, one can be assured that the equation is not correct or some important parameter is missing. In other words, one cannot add length to discharge or area to force. Only terms having the same dimensions can be added or subtracted from each other.

The variables encountered in hydraulics can be classified, according to their dimensions, into three groups: geometric, kinematic, and dynamic. The geometric group includes those variables describing the geometric conditions encountered in a certain problem, such as the length, area, or volume. They are only expressed in terms of L. Kinematic variables express the kinematics of the problem; they do not encompass either mass or force. Examples of kinematic variables include, among others, velocity, time, discharge, acceleration, angular velocity, and kinematic viscosity. Dynamic variables include, amongst others, force, density, specific weight, torque, power, pressure, surface tension, and dynamic viscosity.

6.2 Empirical Formulation of General Flow Equation

Dimensional analysis is a powerful tool through which empirical relations can be developed from any number of independent variables. The analysis is based on the fact that any physical phenomenon can be related through a dimensionally homogeneous equation. Consider, for example, a general case where a fluid is flowing in a hydraulic conduit under the impetus of a differential of potential energy. The shear stress, resulting from boundary roughness and fluid viscosity, tends to resist the flow, which is generally affected by the fluid density, surface tension, bulk modulus and other factors. The flow in open channels is dominated by gravity and friction forces, whereas the flow in pipes is dominated by pressure and viscous forces. Under certain pressure the fluid compressibility may have a significant effect on the flow pattern.

Let us consider a more general case, where all hydraulic parameters exist. The flow equation can generally be written as

$$f(D, V, \rho, L, \Delta(p+\gamma h), \mu, \gamma, \sigma, E, z, s, r, A, B, C, \ldots) = 0 \tag{6.4}$$

where f is a function; D is a parameter representing the geometry of the conduit such as the diameter of a pipe or the hydraulic depth or radius of an open channel; V is the average velocity; ρ is the fluid density; L is a parameter representing the longitudinal dimension of the conduit; $\Delta(p+\gamma h)$ is the differential of the unit potential energy over a length L; p is the pressure intensity; h is the static head above a specified datum; μ is the fluid dynamic viscosity; γ is the specific weight of the fluid; σ is the surface tension; E is the elastic modulus; z, s, r are dimensions of surface roughness elements such as height, longitudinal spacing, radius of rounding, and so on; and A, B, C are other dimensions as may be required to define the boundary geometry, such as radius of bends, width of channel transitions, weir height, and so on.

The variables in equation (6.4) can be grouped into a number of dimensionless groups or terms. Since there are three independent dimensional units (length, time, and force), the number of dimensionless groups will be equal to the number of variables minus three. Each dimensionless term in hydraulics is formed of three basic variables, generally chosen as D, v, and ρ, with exponents satisfying the dimensional requirements and one other variable whose exponent can be chosen arbitrarily, generally taken as minus one. Therefore, equation (6.4) can be written as

$$f\left(\frac{D^{a_1}V^{b_1}\rho^{c_1}}{L}, \frac{D^{a_2}V^{b_2}\rho^{c_2}}{\Delta(p+\gamma h)}, \frac{D^{a_3}V^{b_3}\rho^{c_3}}{\mu}, \frac{D^{a_4}V^{b_4}\rho^{c_4}}{\gamma}, \ldots\right) = 0 \tag{6.5}$$

Our objective now is to find the values of exponents to create dimensionless terms. For the first term in equation (6.5), it is clear that when $a_1 = 1$ and both b_1 and c_1 are equal to zero, then the first term will be D/L.

For the other terms we need to write the three basic equations for force or mass, time, and length to determine the numerical values of the exponents which lead to dimensionless terms. Therefore, the second term in equation (6.5) is written as

$$\frac{D^{a_2}V^{b_2}\rho^{c_2}}{\Delta(p+\gamma h)} = (L)^{a_2}(LT^{-1})^{b_2}(ML^{-3})^{c_2}(ML^{-1}T^{-2})^{-1} \tag{6.6}$$

To create a dimensionless term, the summation of exponents for each basic dimension should be equal to zero. Hence one can write

$$M^0L^0T^0 = (L)^{a_2}(LT^{-1})^{b_2}(ML^{-3})^{c_2}(ML^{-1}T^{-2})^{-1} \tag{6.7}$$

Then

for M, $0 = c_2 - 1$. This yields $c_2 = 1$.
for T, $0 = -b_2 + 2$. This yields $b_2 = 2$.
for L, $0 = a_2 + b_2 - 3c_2 + 1$. This yields $a_2 = 0$.

The resulting dimensionless term, therefore, is $\rho V^2/[\Delta(p + \gamma h)]$.

The same procedure can be repeated for other terms. Equation (6.5) can, thus, be recast in a dimensionless form as

$$f\left(\frac{D}{L}, \frac{\rho V^2}{\Delta(p + \gamma h)}, \frac{\rho VD}{\mu}, \frac{\rho V^2}{\gamma D}, \frac{\rho V^2 D}{\sigma}, \frac{\rho V^2}{E}, \frac{D}{z}, \frac{D}{s}, \frac{D}{r}, \ldots\right) = 0 \tag{6.8}$$

Equation (6.8) involves all of the parameters of equation (6.4) but is in a dimensionless form and with less terms. The geometric terms in equation (6.8) reveal that the flow pattern is, in fact, dependent on the ratio between geometric elements and not on the actual value of each element. All other nongeometric parameters in equation (6.8) involve some sort of forces, which also indicate that the flow pattern is governed by the relative values of different types of forces and not by the exact value of each force. The nongeometric parameters in equation (6.8) can be written as

$$\frac{\rho V^2}{\Delta(p + \gamma h)}, \frac{\rho V^2}{\mu v/D}, \frac{\rho V^2}{\gamma D}, \frac{\rho V^2}{\sigma/D}, \frac{\rho V^2}{E} \tag{6.9}$$

The numerators and denominators in equation (6.9) are each dimensionally equivalent to force per unit area or simply stress. The numerator is the inertia force inherent in fluid's motion. The denominators are dimensionally recognized as representative of the unit forces due to potential head, viscosity, gravity, surface tension, and elasticity, respectively. Each group, therefore, represents the ratio of inertial forces, which always exist in flowing fluids, to other forces which differ according to the type of conduit and boundary conditions.

The five dynamic terms (force ratios) can be modified into five kinematic terms (velocity ratios) by dividing both numerator and denominator by the density ρ and taking the square root. One should note that the terms, thus, obtained are still dimensionless. The dimensionless kinematic terms can, thus, be written as

$$\frac{V}{\sqrt{\Delta(p + \gamma h)/\rho}}, \frac{V}{\sqrt{\mu v/\rho D}}, \frac{V}{\sqrt{\gamma D/\rho}}, \frac{V}{\sqrt{\sigma/\rho D}}, \frac{V}{\sqrt{E/\rho}} \tag{6.10}$$

The dimensionless terms in equation (6.10) represent dimensionless numbers which are named after the investigators who first recognized their importance (Morris and Wiggert, 1972). These have been discussed in Chapter 3 and are rewritten here:

$$\textit{Euler number, } E_n = 2\left[\frac{V}{\sqrt{\Delta(p + \gamma h)/\rho}}\right]^{-2} = \frac{\Delta(p + \gamma h)}{\frac{1}{2}\rho V^2} \tag{6.11}$$

$$\textit{Reynolds number, } R_n = \left[\frac{V}{\sqrt{\mu v/\rho D}}\right]^2 = \frac{DV\rho}{\mu} = \frac{DV}{\nu} \tag{6.12}$$

where ν is the kinematic viscosity, μ/ρ.

$$\text{Froude number, } F_n = \frac{V}{\sqrt{\gamma D/\rho}} = \frac{V}{\sqrt{gD}} \tag{6.13}$$

$$\text{Weber number, } W_n = \left[\frac{V}{\sqrt{\sigma/\rho D}}\right]^2 = \frac{\rho D V^2}{\sigma} \tag{6.14}$$

$$\text{Mach number, } M_n = \frac{V}{\sqrt{E/\rho}} \tag{6.15}$$

The Cauchy number is the inverse of the square of the Mach number written as $\rho V^2/E$.

Equation (6.4) can, thus, be written as

$$f\left(E_n, R_n, F_n, W_n, M_n, \frac{D}{z}, \frac{D}{s}, \frac{D}{r}, \cdots\right) = 0 \tag{6.16}$$

The first step in doing dimensional analysis is to determine the dominating parameters, which should be taken into consideration, for the problem under investigation. For example, if we need to develop an equation for shear stress, τ, in a fluid flowing in a pipe, we may assume that the effective parameters are: the pipe diameter (D), the pipe roughness (e), the fluid density (ρ), the fluid velocity (V), and the dynamic viscosity (μ). We can thus write

$$\tau = \varphi(D, e, \rho, V, \mu) \tag{6.17}$$

It is obvious from the principle of homogeneity that the variables involved cannot be either added or subtracted since they have different dimensions. This constraint limits the application of dimensional analysis to the development of a combination of products of powers of the variables involved. After defining the parameters to be considered and expressing them in the form of equation (6.17) we can proceed with use of one of the two methods discussed below.

6.3 Methods of Dimensional Analysis

There are several methods for performing dimensional analysis. Two of these methods are popular in hydraulics and are discussed in what follows.

6.3.1 Rayleigh Method

The Rayleigh method is used to relate variables or develop dimensionless groups when there is a limited number of variables, generally five or six. As the number of variables exceeds this limit the Rayleigh method becomes more difficult to apply. The method is explained using the following examples.

Example 6.1

Assuming that the force acting on an object floating on a certain fluid depends on a characteristic length (L), velocity (V), fluid density (ρ), and dynamic viscosity (μ), find a relationship between the force and other parameters and define the dimensionless groups.

Solution: From the given information, one can write force F as a function φ of the given parameters:

$$F = \varphi[(L)^a(V)^b(\rho)^c(\mu)^d] \tag{6.18a}$$

In the above equation, it is assumed that the force depends on all the variables with different powers. Dimensional analysis reveals the values of a through d. For example, if F is proportional to V^2, then the value of b should be 2. On the other hand, if any of the variables does not have any effect on F, then its power will be zero. From a dimensional point of view, equation (6.18a) will only be correct if it satisfies the homogeneity condition. In dimensional form, equation (6.18a) is written as

$$MLT^{-2} = (L)^a(LT^{-1})^b(ML^{-3})^c(ML^{-1}T^{-1})^d \quad \textbf{(6.18b)}$$

Three fundamental equations for exponents of M, T, and L can be written as follows:

$$M: \quad 1 = c + d \quad \text{or} \quad c = 1 - d \quad \textbf{(6.18c)}$$

$$T: \quad -2 = -b - d \quad \text{or} \quad b = 2 - d \quad \textbf{(6.18d)}$$

$$L: \quad 1 = a + b - 3c - d \quad \textbf{(6.18e)}$$

Therefore, substituting equations (6.18c) and (6.18d) into equation (6.18e)

$$1 = a + (2 - d) - 3(1 - d) - d = a - 1 + d$$

or

$$a = 2 - d \quad \textbf{(6.18f)}$$

In the above steps, all powers are defined in terms of d. Equation (6.18a) can be written as

$$F = \varphi[(L)^{2-d}(V)^{2-d}(\rho)^{1-d}(\mu)^d] \quad \textbf{(6.18g)}$$

or

$$F = L^2V^2\rho\varphi\left[\left(\frac{\rho VL}{\mu}\right)^{-d}\right] \quad \textbf{(6.18h)}$$

where φ is some function. One may note that $(\rho VL/\mu)$ is a dimensionless number known as Reynolds number R_n. Removing the power of any dimensionless number will not disturb the homogeneity of the equation. Thus, equation (6.18h) may be written as

$$F = L^2V^2\rho\phi(R_n) \quad \textbf{(6.18i)}$$

or

$$\frac{F}{L^2V^2\rho} = \varphi(R_n) \quad \textbf{(6.18j)}$$

The two sides of the above equation are the required dimensionless numbers.

Example 6.2

Using the Rayleigh method of dimensional analysis develop an equation for the power delivered by a pump to lift a fluid of a specific weight γ with a rate of Q to a static level of H.

Solution: From the given data one can write the following equation of power as a function of the given variables:

$$P = \varphi[(\gamma)^a(Q)^b(H)^c] \quad \textbf{(6.19a)}$$

The power is defined as the force multiplied by velocity. In dimensional form, equation (6.19a) is written as

$$ML^2T^{-3} = \varphi[(ML^{-2}T^{-2})^a(L^3T^{-1})^b(L)^c] \quad \textbf{(6.19b)}$$

The fundamental equations are written as

$$M: \quad 1 = a \tag{6.19c}$$

$$T: \ -3 = -2 - b, \quad \text{or} \quad b = 1 \tag{6.19d}$$

$$L: \quad 2 = -2 + 3 + c, \quad \text{or} \quad c = 1 \tag{6.19e}$$

Equation (6.19a) can, thus, be written as

$$P = \varphi(\gamma QH) \tag{6.19f}$$

Example 6.3

Assuming that the distance (z) traveled by a freely falling body is a function of time, the weight of the body and the acceleration due to gravity, find the relation between z and other variables.

Solution: From the given information, one can write z as a function of the given variables:

$$z = \varphi[(T)^a (W)^b (g)^c] \tag{6.20a}$$

In dimensional form, equation (6.20a) can be written as

$$L = \varphi[(T)^a (F)^b (LT^{-2})^c] \tag{6.20b}$$

The fundamental equations are written as

$$F: \ 0 = b \tag{6.20c}$$

$$L: \ 1 = c \tag{6.20d}$$

$$T: \ 0 = a - 2c, \quad \text{or} \quad a = 2 \tag{6.20e}$$

Then, equation (6.20a) can be written as

$$z = \varphi(gT^2) \tag{6.20f}$$

In this example it is recognized that the traveled distance does not depend on the body weight. The Rayleigh method is ideal for application to problems where the number of variables is limited to four or five. Otherwise, the Buckingham method should be used.

6.3.2 Buckingham-Pi (π) Method

The Buckingham or Pi (π) theorem is widely used in dimensional analysis because of its advantage of dealing with any number of variables. It expresses the n number of variables into $(n - m)$ number of dimensionless groups, where m is the number of fundamental dimensions involved in the problem under consideration; m may be either 3 or 2. In most problems where three fundamental dimensions ($L, T,$ and M or F) exist, m is equal to 3. In a few cases, where the problem is kinematic in nature, m is equal to 2. In this method, not only the variables are arranged into dimensionless groups, but also the number of variables is reduced to $(n - m)$. According to the Pi (π) theorem, if a phenomenon involves n variables $a_1, a_2, a_3, \ldots, a_n$, and one of these variables, say a_1, is dependent on the remaining independent variables, we may express the relationship between the variables in a general form as

$$a_1 = \varphi(a_2, a_3, a_4, \ldots, a_n) \tag{6.21}$$

where φ is some function. Equation (6.21) can be expressed in an equivalent form as

$$\varphi_1(a_1, a_2, a_3, \ldots, a_n) = 0 \tag{6.22}$$

where φ_1 is an unknown function.

The n parameters can also be grouped into $(n - m)$ number of independent dimensionless groups (known as π-terms), and can be expressed as

$$\varphi_2(\pi_1, \pi_2, \pi_3, \ldots, \pi_{n-m}) = 0 \tag{6.23}$$

Equation (6.23) can be written as

$$\pi_1 = \varphi_3(\pi_2, \pi_3, \ldots, \pi_{n-m}) \tag{6.24}$$

Both φ_2 and φ_3 are unknown functions and cannot be predicted from dimensional analysis. A π-term is considered to be independent if it cannot be formed from a product or quotient of all other π-terms.

6.3.2.1 Steps for Solution The first step is to define the parameters which are believed to affect the phenomenon under consideration. Practical experience provides much help in this step. However, it is also advisable to consider the parameters about which there is some doubt as to whether they affect the problem or not. If an unimportant parameter is considered an extra π-term may be obtained, but experiments will reveal such a parameter. Adding unnecessary parameters will not cause any problem but missing a dominant parameter may adversely affect the results.

The following steps are thus recommended to identify relations and develop the π-terms:

1. Identify the parameters which are significant for the problem under consideration, and let the number of these parameters be n. Selection of the involved parameters is very important as the final result of analysis is totally dependent on the selected parameters.
2. Define the set of fundamental dimensions which are going to be employed in the solution. The set may consist of either M, L, and T or F, L, and T.
3. The number of recurring variables which is going to be chosen (m) is exactly equal to the number of fundamental dimensions encountered in the parameters selected in step 1 above. If M or F is found in any of the selected parameters, then m must be 3 to represent the three fundamental dimensions. For kinematic problems where M and F do not exist, then m must be 2. Therefore, m must take on the value of 2 or 3 according to the fundamental dimensions of the problem. In the majority of hydraulic problems m is equal to 3. However, one should always assure the assigned value.
4. After determining the number of recurring variables, select them from the entire set of parameters according to the following rules:
   (a) The first repeated variable should represent the geometric aspects of the problem. It must have the dimension of length, L, or its multiplication. Examples of the first repeated variable include length, width, depth, area, volume, and roughness height.
   (b) The second repeated variable should represent the kinematics of the problem. It should contain time, T, and should not contain either mass, M, or force, F. Examples of the second repeated variable include velocity, discharge, frequency, acceleration, angular velocity, and kinematic viscosity.
   (c) The third repeated variable should represent the dynamics of the problem. It should contain either mass or force. Examples of the third repeated variable include force, density, specific weight, pressure, torque, dynamic viscosity, and energy.

5. The number of π-terms is equal to $(n - m)$. Each π-term contains the repeated variables selected in step 4 (2 or 3) and another new independent variable. Set up the dimensional equations for the variables in each π-term. Solve the dimensional equations to determine the required π-terms.
6. Check the dimensions of each π-term obtained. Make sure that all π-terms are dimensionless. Then put the π-terms in the required form. Experimental work would be required to determine the functional relationship between the obtained π-terms.

Example 6.4

Using the Buckingham-π method, derive an expression for the shear stress, τ, in a fluid flowing in a pipe assuming that it is a function of the diameter, D, pipe roughness, e, fluid density, ρ, dynamic viscosity, μ, and fluid velocity, v.

Solution: The variables in this problem are τ, D, e, ρ, μ, and V. The number of variables $n = 6$. The variables contain F and M. Therefore, the number of repeated variables, m, is thus 3. The number of π-terms $= (n - m) = 6 - 3 = 3$. We select the following repeated variables:

1. D, representing the geometry of the problem
2. V, representing the kinematics of the problem (containing T)
3. ρ, representing the dynamics (containing M or F)

Each of the π-terms should contain the above three repeated variables plus another new variable. The first π-term can be expressed as

$$\pi_1 = (D)^{a_1}(V)^{b_1}(\rho)^{c_1}(\tau)^{-1}$$

$$M^0T^0L^0 = (L)^{a_1}(LT^{-1})^{b_1}(ML^{-3})^{c_1}(ML^{-1}T^{-2})^{-1}$$

For M, $0 = c_1 - 1$. This gives $c_1 = 1$.

For T, $0 = -b_1 + 2$. This gives $b_1 = 2$.

For L, $0 = a_1 + b_1 - 3c_1 + 1 = a_1 + 2 - 3 + 1$. This gives $a_1 = 0$.

Therefore,

$$\pi_1 = V^2\rho\tau^{-1} = \frac{\rho V^2}{\tau} \quad \text{or} \quad \pi_1 = \frac{\tau}{\rho v^2}$$

The second π-term can be expressed as

$$\pi_2 = (D)^{a_2}(V)^{b_2}(\rho)^{c_2}(e)^{-1}$$

$$M^0T^0L^0 = (L)^{a_2}(LT^{-1})^{b_2}(ML^{-3})^{c_2}(L)^{-1}$$

For M, $0 = c_2$. This yields $c_2 = 0$.

For T, $0 = -b_2$. This yields $b_2 = 0$.

For L, $0 = a_2 + b_2 - 3c_2 - 1$. This yields $a_2 = 1$.

Therefore,

$$\pi_2 = \frac{D}{e} \quad \text{or} \quad \pi_2 = \frac{e}{D}$$

The third π-term can be expressed as

$$\pi_3 = (D)^{a_3}(v)^{b_3}(\rho)^{c_3}(\mu)^{-1}$$
$$M^0T^0L^0 = (L)^{a_3}(LT^{-1})^{b_3}(ML^{-3})^{c_3}(ML^{-1}T^{-1})^{-1}$$

For M, $0 = c_3 - 1$. This gives $c_3 = 1$.
For T, $0 = -b_3 + 1$. This gives $b_3 = 1$.
For L, $0 = a_3 + b_3 - 3c_3 + 1$. This gives $a_3 = 1$.

Therefore,

$$\pi_3 = \frac{\rho VD}{\mu} = R_n$$

Now, we collect all the π-terms and write

$$\varphi(\pi_1, \pi_2, \pi_3) = 0$$

or

$$\pi_1 = \varphi_1(\pi_2, \pi_3)$$

Hence,

$$\frac{\tau}{\rho V^2} = \varphi_1\left(\frac{D}{e}, R_n\right)$$

or

$$\tau = \rho V^2 \varphi_1\left(\frac{D}{e}, R_n\right)$$

Example 6.5

The efficiency of a propeller, η, is believed to be dependent on its diameter, D, the fluid density, ρ, and dynamic viscosity, μ, angular velocity, ω, and discharge, Q. Using dimensional analysis, develop a relation between η and these variables.

Solution: The variables are η, D, ρ, μ, ω, and Q. The number of variables $n = 6$. The number of repeated variables $m = 3$. The number of π-terms $= 6 - 3 = 3$. The repeated variables are selected as D, ω, and ρ. The first π-term is

$$\pi_1 : (D)^{a_1}(\omega)^{b_1}(\rho)^{c_1}(\eta)^{-1}$$
$$M^0T^0L^0 : (L)^{a_1}(T^{-1})^{b_1}(ML^{-3})^{c_1}(M^0L^0T^0)^{-1}$$

For M, $0 = c_1$. This yields $c_1 = 0$.
For T, $0 = -b_1$. This yields $b_1 = 0$.
For L, $0 = a_1 - 3c_1$. This yields $a_1 = 0$.

Therefore,

$$\pi_1 = \frac{1}{\eta} \quad \text{or} \quad \pi_1 = \eta$$

The second π-term is

$$\pi_2: (D)^{a_2}(\omega)^{b_2}(\rho)^{c_2}(\mu)^{-1}$$
$$M^0T^0L^0: (L)^{a_2}(T^{-1})^{b_2}(ML^{-3})^{c_2}(ML^{-1}T^{-1})^{-1}$$

For M, $0 = c_2 - 1$. This gives $c_2 = 1$.
For T, $0 = -b_2 + 1$. This gives $b_2 = 1$.
For L, $0 = a_2 - 3c_2 + 1$. This gives $a_2 = 2$.

Therefore,

$$\pi_2 = \frac{D^2\omega\rho}{\mu}$$

The third π-term is

$$\pi_3: (D)^{a_3}(\omega)^{b_3}(\rho)^{c_3}(Q)^{-1}$$
$$M^0T^0L^0: (L)^{a_3}(T^{-1})^{b_3}(ML^{-3})^{c_3}(L^3T^{-1})^{-1}$$

For M, $0 = c_3$. This yields $c_3 = 0$.
For T, $0 = -b_3 + 1$. This yields $b_3 = 1$.
For L, $0 = a_3 - 3c_3 - 3$. This yields $a_3 = 3$.

Therefore,

$$\pi_3 = \frac{\omega D^3}{Q}$$

Relating the three π-terms, we obtain

$$\pi_1 = \varphi(\pi_2, \pi_3)$$

or

$$\eta = \varphi\left(\frac{D^2\omega\rho}{\mu}, \frac{\omega D^3}{Q}\right)$$

Example 6.6

The discharge over a V-notch in an open channel is investigated via dimensional analysis. The following parameters are considered to be effective: gravity acceleration (g), head over the V-notch (h), kinematic viscosity of the fluid (ν), fluid density (ρ), surface tension (σ), and angle of the notch (θ). Develop a relationship between discharge and other parameters.

Solution: The variables are (Q, g, h, ν, ρ, σ, θ). Thus, $n = 7$ and $m = 3$. The number of π-terms $= 7 - 3 = 4$. The repeated variables are selected as h, g, ρ. The first π-term is

$$\pi_1: (h)^{a_1}(g)^{b_1}(\rho)^{c_1}(Q)^{-1}$$
$$M^0T^0L^0: (L)^{a_1}(LT^{-2})^{b_1}(ML^{-3})^{c_1}(L^3T^{-1})^{-1}$$

For M, $0 = c_1$. This gives $c_1 = 0$.
For T, $0 = -2b_1 + 1$. This gives $b_1 = 1/2$.
For L, $0 = a_1 + b_1 - 3c_1 - 3$. This gives $a_1 = 5/2$.

Therefore,

$$\pi_1 = \frac{g^{1/2}h^{5/2}}{Q} \quad \text{or} \quad \pi_1 = \frac{Q}{g^{1/2}h^{5/2}}$$

The second π-term is

$$\pi_2 \colon (h)^{a_2}(g)^{b_2}(\rho)^{c_2}(\nu)^{-1}$$

$$M^0T^0L^0 \colon (L)^{a_2}(LT^{-2})^{b_2}(ML^{-3})^{c_2}(L^2T^{-1})^{-1}$$

For M, $0 = c_2$. This gives $c_2 = 0$.
For T, $0 = -2b_2 + 1$. This gives $b_2 = 1/2$.
For L, $0 = a_2 + b_2 - 3c_2 - 2$. This gives $a_2 = 3/2$.

Therefore,

$$\pi_2 = \frac{g^{1/2}h^{3/2}}{\nu}$$

The third π-term is

$$\pi_3 \colon (h)^{a_3}(g)^{b_3}(\rho)^{c_3}(\sigma)^{-1}$$

$$M^0T^0L^0 \colon (L)^{a_3}(LT^{-2})^{b_3}(ML^{-3})^{c_3}(MT^{-2})^{-1}$$

For M, $0 = c_3 - 1$. This yields $c_3 = 1$.
For T, $0 = -2b_3 + 2$. This yields $b_3 = 1$.
For L, $0 = a_3 + b_3 - 3c_3$. This yields $a_3 = 2$.

Therefore,

$$\pi_3 = \frac{\rho g h^2}{\sigma}$$

The fourth π-term is

$$\pi_4 \colon (h)^{a_4}(g)^{b_4}(\rho)^{c_4}(\theta)^{-1}$$

$$M^0T^0L^0 = (L)^{a_4}(LT^{-2})^{b_4}(ML^{-3})^{c_4}(M^0T^0L^0)^{-1}$$

For M, $0 = c_4$. This gives $c_4 = 0$.
For T, $0 = -2b_4$. This gives $b_4 = 0$.
For L, $0 = a_4 + b_4 - 3c_4$. This gives $a_4 = 0$.

Therefore,

$$\pi_4 = \frac{1}{\theta} \quad \text{or} \quad \pi_4 = \theta$$

Relating the 4 π-terms,

$$\pi_1 = \varphi(\pi_2, \pi_3, \pi_4)$$

or

$$\frac{Q}{g^{1/2}h^{5/2}} = \varphi\left(\frac{g^{1/2}h^{3/2}}{\nu}, \frac{\rho g h^2}{\sigma}, \theta\right)$$

or

$$Q = g^{1/2}h^{5/2}\,\varphi\left(\frac{g^{1/2}h^{3/2}}{\nu}, \frac{\rho g h^2}{\sigma}, \theta\right)$$

Example 6.7

The frequency of the surface wave (the number of waves passing a certain station per unit time) caused by the motion of a floating body is proportional to (V/l), where V is the velocity of the body and l is its characteristic length. The coefficient of proportionality is a function of Froude number. Using the Buckingham theory, prove the above statement.

Solution: The variables which can be deduced from the above statement are the frequency, f, the velocity, v, the length, L, and the acceleration due to gravity, g. Three of these variables are found in the Froude number. Therefore, the variables are f, V, L, and g. These variables do not contain either force or mass. Therefore, $n = 4$, and $m = 2$. The number of π-terms $= n - m = 4 - 2 = 2$. The repeated variables are chosen as L and V. The first π-term is

$$\pi_1: (L)^{a_1}(V)^{b_1}(f)^{-1}$$

$$T^0L^0: (L)^{a_1}(LT^{-1})^{b_1}(T^{-1})^{-1}$$

For T, $0 = -b_1 + 1$. This gives $b_1 = 1$.

For L, $0 = a_1 + b_1$. This gives $a_1 = -1$.

Therefore,

$$\pi_1 = \frac{V}{lf} = \frac{fl}{V}$$

The second π-term is

$$\pi_2: (L)^{a_2}(V)^{b_2}(g)^{-1}$$

$$T^0L^0 = (L)^{a_2}(LT^{-1})^{b_2}(LT^{-2})^{-1}$$

For T, $0 = -b_2 + 2$. This gives $b_2 = 2$.

For L, $0 = a_2 + b_2 - 1$. This gives $a_2 = -1$.

Therefore,

$$\pi_2 = \frac{V^2}{gL}$$

which also can be written as

$$\pi_2 = \frac{V}{\sqrt{gL}} = F_n$$

Relating the two π-terms, we get

$$\pi_1 = \varphi(\pi_2)$$

or

$$\frac{fl}{V} = \varphi(F_n)$$

or

$$f = \left(\frac{V}{l}\right)\varphi(F_n)$$

6.3.2.2 Useful Hints

1. In using dimensional analysis we are not concerned with the exact value of any dimensionless term but rather we are interested in maintaining its dimensionless character. Therefore, any π-term can be multiplied or divided by itself or any other π-term, that is,

$$\pi_1 = \varphi_1(\pi_1^2) = \varphi_2(\sqrt{\pi_1}) = \varphi_3\left(\pi_1\left(\frac{\pi_2}{\pi_3}\right)\right) = \varphi_4\left(\pi_1\sqrt{\frac{\pi_3}{\pi_1}}\right), \text{ and so on}$$

2. Any dimensionless number is a function of others, that is,

$$\pi_1 = \varphi_1(\pi_2, \pi_3, \pi_4)$$

   Similarly,

$$\pi_3 = \varphi_2(\pi_1, \pi_2, \pi_4)$$

   We can also write

$$\frac{\pi_3}{\pi_2} = \varphi_3(\pi_1, \pi_4)$$

   or

$$\frac{\pi_1}{\pi_3} = \varphi_4\left(\pi_2, \frac{\pi_4}{\pi_1}\right)$$

3. There are some other ways to select the repeated variables. One of them is to choose the first repeated variable to represent the geometry, the second repeated variable to represent the characteristics of the fluid (ρ, ν, γ, σ, μ), and the third repeated variable to represent the characteristics of flow (V, g, P, F). If the final formula is to be proved using dimensional analysis, the repeated variables should be deduced from the given formula. If any variable appears more than once in different π-terms, then it is a repeated variable. However, a repeated variable does not necessarily appear more than once.
4. The dependent variable (which is required to be on one side of the equation) should never be selected as a repeated variable. Otherwise, it may appear in different π-terms and could be difficult to separate on one side.
5. Never select a repeated variable if there is doubt as to its dimensions. Repeated variables are most likely to appear in all π-terms. Always ensure the dimensions of the repeated variables.
6. If any parameter has the dimensions of a repeated variable, the π-term is equal to the division of the latter by the former. Also, if any of the parameters is dimensionless, the π-term is that parameter.

Example 6.8

Prove that the power, P, developed by a Kaplan turbine can be expressed as

$$P = \rho D^5 N^3 \varphi\left(\frac{H}{D}, \frac{\rho D^2 N}{\mu}, \frac{DN^2}{g}\right)$$

where ρ is the fluid density, D is the average diameter of the vanes, N is the speed in rpm (revolutions per minute), H is the operation head, μ is the dynamic viscosity, and g is the acceleration of gravity.

Solution: From the given formula the variables are

$P, \rho, D, N, H, \mu, g.$ Then $n = 7$, $m = 3$

The number of π-terms $= n - m = 7 - 3 = 4$

The power, P, is excluded from being selected as a repeated variable because it is a dependent variable. From the given formula, we find that ρ, D, and N have appeared in more than one π-term. They are, thus, repeated variables. It may also be noted that each of the two π-terms, $P/[\rho D^5 N^3]$ and $[\rho D^2 N/\mu]$, contains four variables. Three of the four variables in each π-term are repeated. On the other hand, $P/[\rho D^5 N^3]$ is a π-term. Since P cannot be a repeated variable because it is a dependent variable, then the other variables must be repeated.

The first π-term can be written as

$$\pi_1: (\rho)^{a_1}(D)^{b_1}(N)^{c_1}(P)^{-1}$$

From dimensional analysis,

$$M^0T^0L^0: (ML^{-3})^{a_1}(L)^{b_1}(T^{-1})^{c_1}(ML^2T^{-3})^{-1}$$

For M, $0 = a_1 - 1$. This gives $a_1 = 1$.
For T, $0 = -c_1 + 3$. This gives $c_1 = 3$.
For L, $0 = -3a_1 + b_1 - 2$. This gives $b_1 = 5$.

Therefore,

$$\pi_1 = \frac{\rho D^5 N^3}{P} \quad \text{or} \quad \pi_1 = \frac{P}{\rho N^3 D^5}$$

The second π-term can be expressed as

$$\pi_2: (\rho)^{a_2}(D)^{b_2}(N)^{c_2}(H)^{-1}$$

Since D and H have the same dimensions,

$$\pi_2 = \frac{D}{H}$$

The third π-term can be expressed as

$$\pi_3: (\rho)^{a_3}(D)^{b_3}(N)^{c_3}(\mu)^{-1} \quad M^0T^0L^0: (ML^{-3})^{a_3}(L)^{b_3}(T^{-1})^{c_3}(ML^{-1}T^{-1})^{-1}$$

For M, $0 = a_3 - 1$. This gives $a_3 = 1$.
For T, $0 = -c_3 + 1$. This gives $c_3 = 1$.
For L, $0 = -3a_3 + b_3 + 1$. This gives $b_3 = 2$.

Therefore, $\pi_3 = \dfrac{\rho D^2 N}{\mu}$

The fourth π-term can be expressed as

$$\pi_4: (\rho)^{a_4}(D)^{b_4}(N)^{c_4}(g)^{-1} \quad M^0T^0L^0: (ML^{-3})^{a_4}(L)^{b_4}(T^{-1})^{c_4}(LT^{-2})^{-1}$$

For M, $0 = a_4$. This gives $a_4 = 0$.
For T, $0 = -c_4 + 2$. This gives $c_4 = 2$.
For L, $0 = -3a_4 + b_4 - 1$. This gives $b_4 = 1$.

Therefore, $\pi_4 = \dfrac{DN^2}{g}$

Relating the four π-terms, we write

$$\pi_1 = \varphi(\pi_2, \pi_3, \pi_4) \quad \text{or} \quad P = \rho D^5 N^3 \varphi\left(\frac{H}{D}, \frac{\rho D^2 N}{\mu}, \frac{DN^2}{g}\right)$$

6.4 Hydraulic Scale Models

Hydraulic projects can be massive, expensive, and complex. Examples of such projects include dams, levees, barrages, spillways, channel transitions, regulators, culverts, breakwaters, harbors, and so on. Even in relatively simple hydraulic projects it is often impossible to predict the exact flow patterns and actual dynamic forces and stresses on a hydraulic structure without conducting a model study. In more complex projects, such as those involving the release of pollutants from industrial or thermal outfalls, models are almost mandatory (Sharp, 1981). Hydraulic problems associated with such projects are too complex to be described and resolved by rigorous mathematical equations and numerical techniques. Furthermore, hydraulics involves many empirical relations and most hydraulic problems do not have exact analytical solutions. Therefore, their governing equations are generally simplified by introducing a number of assumptions, and boundary conditions are always idealized to allow approximate numerical solutions. For example, applications of mathematical equations to spillway flows can provide accurate solutions only if the geometry is fairly simple. If the spillway is curved in plan there is a possibility that most of the flow will not be evenly distributed across the channel and that most of the flow will be diverted to one side of the channel. If baffles or other devices are used to promote the formation of a jump in a stilling basin, mathematical techniques are not adequate to define the percentage of energy

Scale model of the Mississippi River at Louisiana State University.

dissipation or to predict the size, spacing and number of baffles necessary to provide the required amount of energy dissipation (Sharp, 1981).

A model may be thought of as a prediction tool in which the full-scale "prototype" is reproduced at another scale. A prototype and its model are two similar physical systems of different sizes. Normally, but not essentially, the model is smaller in size. Like dimensional analysis, models alone do not give the complete answer to a problem under consideration but certainly complement other techniques to more fully understand the complex natural phenomenon. Models are used alternatively or conjunctively with other techniques to obtain efficient and satisfactory designs. Hydraulic models are powerful tools in design and their use has long been practiced in various facets of hydraulics.

Fluids different from those in prototypes may be used in models. Models may also be subject to different pressures. Hydraulic models allow flexibility in testing prototypes under various conditions to which they might be exposed. The cost of building a model is relatively affordable, especially for the information and benefits they provide. The total cost of a hydraulic project may be increased by two or even five percent, yet better designs can be achieved and many problems can be avoided before the execution of prototype projects.

6.5 Types of Similarity

Three types of similarity should exist between a prototype and its model in order to ensure that the model is completely similar to and thus accurately represents the prototype. The main types of similarity are geometric, kinematic, and dynamic. Two other conditions could be employed: thermal similarity, which requires equality of temperature ratios at corresponding positions, and chemical similarity, which deals with concentration ratios. However, hydraulic models are mostly based on the first three types of similarity. Even in the models of cooling water outfall plumes it is not necessary, or sometimes economically possible, to achieve complete thermal similarity (Sharp, 1981). The three main types of similarity are explained in what follows.

6.5.1 Geometric Similarity

A model should be similar in shape to the prototype. Any unspecialized person can recognize that the model is just another version of the prototype but with a different scale, generally smaller. All dimensions in the prototype are reduced by a fixed factor called the scale ratio. A 1/50 model means that each dimension in the model is 1/50 of that in the prototype.

Consider, for example, a ship given in Figure 6.1a, with a length L_p, width B_p, and total depth of d_p. A model is to be built in the laboratory to test the performance and ability of the ship to accommodate lateral dynamic forces due to wave action. The length, width, and total depth of the proposed model are L_m, B_m, and d_m, respectively as shown in Figure 6.1b. In order to ensure the geometric similarity, the following conditions should hold:

$$L_r = \frac{L_m}{L_p} = \frac{B_m}{B_p} = \frac{d_m}{d_p} \tag{6.25}$$

where L_r is the length ratio generally less than 1 and may be small as 1/1000 or even less. In the relations to be used in this section all quantities with subscript p will correspond to prototype, those with subscript m to model, and those subscript r to the ratios of model to prototype quantities. If equation (6.25) is true, then one can deduce that

$$A_r = \frac{A_m}{A_p} = \frac{L_m^2}{L_p^2} = L_r^2 \tag{6.26}$$

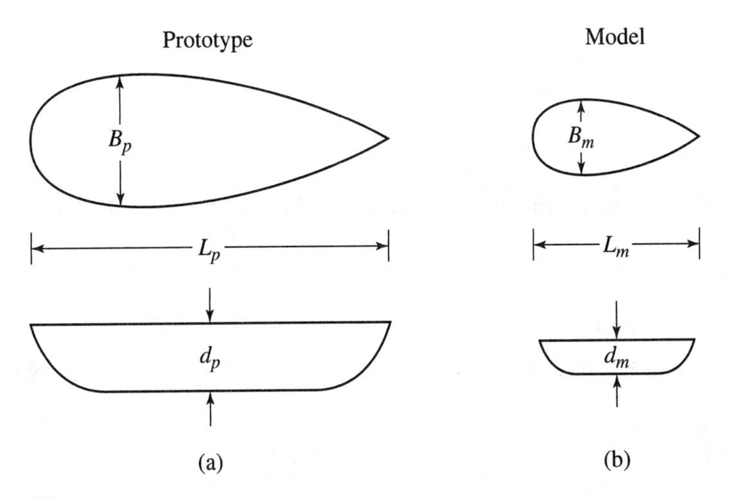

Figure 6.1 Geometric similarity between hydraulic systems.

where A_r is the area ratio, and A_m and A_p are the areas of model and prototype, respectively. For the horizontal area, $A_m = L_m B_m$ and $A_p = L_p B_p$, and for the area in the vertical section, $A_m = L_m d_m$ and $A_p = L_p d_p$.

In a similar way, one can write

$$\bar{V}_r = \frac{\bar{V}_m}{\bar{V}_p} = \frac{L_m^3}{L_p^3} = L_r^3 \tag{6.27}$$

where $\bar{V}_r$ is the volume ratio, and $\bar{V}_m$ and $\bar{V}_p$ are the volumes of the model and prototype, respectively. Also, $\bar{V}_m = L_m B_m d_m$ and $\bar{V}_p = L_p B_p d_p$. If a geometric similarity exists between two physical systems, say prototype and model, all parameters having the dimension of length to some power in the two systems can be related to each other through equations (6.25) through (6.27).

Example 6.9

A model for a huge circular water conduit is constructed. The prototype has a diameter of 5.0 m and the roughness height of its concrete wall, e, is 3.0 mm. The model is built to a scale ratio of 1/10; it has a length of 4.0 m. Estimate the diameter, cross-sectional area, and roughness height of the model and the length of the prototype.

Solution:

$$d_r = L_r = \frac{1}{10} = \frac{d_m}{d_p} = \frac{d_m}{5.0}$$

Therefore, $d_m = 0.5$ m.

$$A_p = \frac{\pi d_p^2}{4} = 19.635 \text{ m}^2$$

$$A_r = L_r^2 = \frac{1}{100} = \frac{A_m}{19.635}$$

Therefore, $A_m = 0.19635\ \text{m}^2$.

The roughness height, e, has the dimension of L:

$$e_r = L_r = \frac{e_m}{e_p} = \frac{e_m}{3.0}$$

Therefore, the model roughness height $e_m = 0.3$ mm.

$$L_r = \frac{1}{10} = \frac{4.0}{L_p}$$

Therefore, the prototype length $L_p = 40.0$ m.

Example 6.10

A prototype of a volume of 1000 m^3 is investigated by a hydraulic model 1.0 m^3 in volume. If the length of the model is 2.25 m, what is the length of the prototype?

Solution:

$$\bar{V}_r = L_r^3 = \frac{\bar{V}_m}{\bar{V}_p} = \frac{1}{1000}$$

where $\bar{V}_r$ is the volume ratio of the model volume to the prototype volume. Therefore,

$$L_r = \frac{1}{10}$$

$$L_r = \frac{L_m}{L_p} = \frac{2.25}{L_p}$$

Therefore, $L_p = 22.5$ m.

6.5.2 Kinematic Similarity

Kinematic similarity is defined as the similarity of motion between two geometrically similar systems (prototype and model). It requires that the shape of the stream lines at any time and for any two corresponding points is the same in both systems. The boundaries of the two systems should, thus, have exactly the same shape and deformation. Geometric similarity does not ensure kinematic similarity, but kinematic similarity does ensure geometric similarity.

The velocity ratio between any two corresponding points in two kinematically similar systems should be constant. The same is true for discharge, acceleration, time, kinematic viscosity, and all other parameters involving time in their dimensions. These terms do not compress force or mass. For kinematically similar systems, one can write

$$V_r = \frac{V_m}{V_p} = \text{constant} \tag{6.28}$$

Also,

$$Q_r = \frac{Q_m}{Q_p} = A_r V_r = L_r^2 V_r \tag{6.29}$$

$$T_r = \frac{T_m}{T_p} = L_r V_r^{-1} \tag{6.30}$$

$$a_r = \frac{a_m}{a_p} = V_r T_r^{-1} \tag{6.31}$$

where a is the acceleration. Substituting for T_r from equation (6.30) into equation (6.31),

$$a_r = V_r^2 L_r^{-1} \tag{6.32}$$

In a similar fashion, all ratios of other kinematic parameters can be determined from their basic dimensions in terms of L_r and V_r.

6.5.3 Dynamic Similarity

Dynamic similarity is defined as the similarity of forces of two kinematically similar systems. The two systems must therefore be geometrically similar. Motion is caused by the action of one or more forces. Forces accelerate or decelerate fluid particles or maintain their constant velocity against viscous and/or frictional forces which tend to bring moving particles to rest. This relation between forces and motion (discussed in Chapter 3) requires that the force action in the model must be identical to that in the prototype. The polygon of forces acting on a particle of fluid in the model must be geometrically similar to that in the prototype. In other words, the ratio of any two forces acting in the model must be equal to the ratio of the corresponding two forces acting in the prototype.

In general, three types of forces act on a hydraulic system: (1) external forces such as the gravity force or force due to pressure head in pipes and so on, (2) internal forces involving physical characteristics of the fluid itself such as viscous force and surface tension force, and (3) resultant forces, such as drag force on a moving body in a fluid or pressure force on hydraulic structures.

The inertial force is a hypothetical force equal in magnitude but opposite in direction to the resultant force. Every accelerating or decelerating fluid system must experience inertial forces because acceleration and deceleration can only result from an out-of-balance resultant force on the system.

The ratio between two different types of forces in two dynamically similar systems (model and prototype) must be constant. Assuming that F_I, F_g, F_v, F_σ, F_p, and F_e are the inertia force, gravity force, viscous force, surface tension force, pressure force, and elastic force, respectively, then one can write

$$F_r = \frac{(F_I)_m}{(F_I)_p} = \frac{(F_g)_m}{(F_g)_p} = \frac{(F_v)_m}{(F_v)_p} = \frac{(F_\sigma)_m}{(F_\sigma)_p} = \frac{(F_e)_m}{(F_e)_p} = \text{constant} \tag{6.33}$$

Regardless of the specific type of the force, its dimensions are equivalent to the dimensions of mass (M) multiplied by acceleration (a). Therefore,

$$F_r = M_r a_r = (\rho_r L_r^3)(V_r T_r^{-1}) = \rho_r L_r^2 V_r (L_r T_r^{-1}) = \rho_r L_r^2 V_r^2 \tag{6.34}$$

Based on equation (6.34), the ratios of all dynamic parameters can be identified in terms of the fluid density ratio (ρ_r), a geometric ratio (L_r) characterizing the geometry of the problem under consideration, and the velocity ratio (V_r) between the two systems. For example, the work (W) is dimensionally equal to the force multiplied by distance. Therefore,

$$W_r = F_r L_r = \rho_r L_r^2 V_r^2 \cdot L_r = \rho_r L_r^3 V_r^2 \tag{6.35}$$

The power (P) is dimensionally equal to the force multiplied by velocity. Therefore,

$$P_r = F_r V_r = \rho_r L_r^2 V_r^2 \cdot v_r = \rho_r L_r^2 V_r^3 \tag{6.36}$$

The pressure (p) is dimensionally equal to the force divided by the area. Therefore,

$$p_r = F_r L_r^{-2} = \rho_r L_r^2 V_r^2 L_r^{-2} = \rho_r V_r^2 \tag{6.37}$$

Also, the specific weight (γ) is dimensionally equal to the force or weight divided by volume. Therefore,

$$\gamma_r = F_r L_r^{-3} = \rho_r L_r^{-1} V_r^2 \tag{6.38}$$

It is, thus, clear that the ratios of all parameters in two dynamically similar systems can be expressed in terms of ρ_r, L_r, and V_r. Most models have the same type of fluid as the prototype (generally water) and hence ρ_r will be equal to unity. Otherwise, if the fluid in a model is different from that in the prototype, then

$$\rho_r = \frac{\rho_m}{\rho_P} \tag{6.39}$$

Because the fluid properties are known, ρ_r is a defined ratio. The length ratio, L_r, is selected according to the available facilities and space in the lab (discharge, head, material, and so on). In most cases, the length ratio is given or is defined prior to conducting modeling exercises. The only missing ratio is the velocity ratio, V_r. When this is defined, the ratios of all other parameters can be evaluated. To determine the velocity ratio, the dominating force(s) must be first defined.

6.6 Dominating Forces

Hydraulic models are subject to different forces according to the problem to be investigated. For a particular problem, a number of forces exist. However, only one or two forces dominate. For example, in pipes, viscous and elastic forces exist but the effect of viscous force is much more dominating than the effect of elastic force. To reduce the cost and simplify the solution, we generally consider one or two forces which dominate the phenomenon under investigation and neglect the effect of other forces.

6.6.1 Gravity Force

The gravity force is the most common force in the majority of hydraulic models. Examples of problems in which gravity force dominates include open channels, natural rivers, hydraulic structures, wave action, sediment transport, and so on. Other forces, such as surface tension force and viscous force, may exist as well in such problems but their effect is quite minimal.

To ensure the dynamic similarity the ratio between inertia force and gravity force in the two systems must be constant, that is,

$$\left(\frac{F_I}{F_g}\right)_m = \left(\frac{F_I}{F_g}\right)_P \tag{6.40}$$

or

$$\left(\frac{F_I}{F_g}\right)_r = 1 \tag{6.41}$$

where $F_I = \rho L^2 V^2$, and $F_g = \gamma L^3$.

Then

$$\left(\frac{\rho L^2 V^2}{\gamma L^3}\right)_r = 1 \tag{6.42}$$

which can be written as

$$\left(\frac{V^2}{gL}\right)_r = 1 = (F_n^2)_r \tag{6.43}$$

or

$$(F_n)_r = \left(\frac{V}{\sqrt{gL}}\right)_r = 1 \tag{6.44}$$

Equation (6.44) indicates that when gravity force dominates, the Froude number in the model must be exactly equal to that in the prototype. The velocity ratio for gravity problems is given from equation (6.44) as

$$V_r = \sqrt{g_r L_r} \tag{6.45}$$

In most cases both the model and prototype are exposed to the same gravity acceleration (i.e., g_r will be equal to 1). Therefore,

$$V_r = L_r^{1/2} \tag{6.46}$$

Otherwise, if the model and the prototype are not subject to the same gravity, g_r will be given as

$$g_r = \frac{g_m}{g_p} \tag{6.47}$$

Example 6.11

A 1:10 scale model is constructed for a hydraulic structure. Determine the ratios of force and pressure. If the force and pressure on the model are equal to 1.90 kN and 500 N/m^2, respectively, what are the corresponding values in the prototype? Assume the same fluid is used in both the model and prototype.

Solution:

$$L_r = \frac{1}{10}$$

Using equation (6.34),

$$F_r = \rho_r L_r^2 V_r^2$$

where

$$\rho_r = 1, \; V_r = L_r^{1/2}$$

$$F_r = L_r^3 = \frac{1}{1000}$$

$$p_r = \frac{F_r}{A_r} = \rho_r L_r^2 V_r^2 L_r^{-2} = \rho_r V_r^2$$

$$p_r = \frac{1}{10}$$

$$\frac{F_m}{F_p} = \frac{1}{1000} = \frac{1.9}{F_p}$$

Then $F_P = 1.9 \times 1000 = 1900$ kN.

$$p_r = \frac{p_m}{p_P} = \frac{1}{10} = \frac{500}{p_P}$$

$$p_P = 5 \text{ kN/m}^2$$

Example 6.12

The force required to tow a 1:25 model of a boat at a speed of 2.5 m/s is 0.6 N. What are the corresponding speed and force in the prototype? If 20 seconds are required by the model to fulfill certain maneuvers, what is the corresponding time in the prototype?

Solution:

$$L_r = \frac{1}{25}$$

$$V_r = L_r^{1/2} = \frac{1}{5} = \frac{V_m}{V_P}$$

$$V_P = 2.5 \times 5.0 = 12.5 \text{ m/s}$$

$$F_r = \rho_r L_r^2 V_r^2 = L_r^3 = \frac{F_m}{F_P} = \left(\frac{1}{25}\right)^3$$

$$F_P = 0.6 \times (25)^3 = 9.375 \text{ kN}$$

$$T_r = L_r V_r^{-1} = L_r^{1/2} = \frac{1}{5}$$

$$\frac{T_m}{T_P} = \frac{1}{5}, \quad T_P = 20 \times 5 = 100 \text{ seconds}$$

Example 6.13

A model is built for a spillway to a scale of 1:20. The discharge over the spillway is expected to be 1200 m^3/s under a head of 3.0 m. What are the corresponding values of discharge and head in the laboratory? If the horsepower dissipated by the hydraulic jump in the model is 0.75 HP, what would be the horsepower dissipated in the prototype?

Solution:

$$L_r = \frac{1}{20}$$

$$Q_r = A_r V_r = L_r^{5/2} = \left(\frac{1}{20}\right)^{5/2} = 0.00056$$

$$Q_r = \frac{Q_m}{Q_p}, \quad Q_m = 0.00056 \times 1200 = 0.672 \text{ m}^3\text{/s}$$

$$H_r = L_r = \frac{1}{20}$$

$$H_m = 3.0 \times \frac{1}{20} = 0.15 \text{ m}$$

$$(HP)_r = F_r V_r = \rho_r L_r^2 V_r^3 = L_r^{7/2} = \left(\frac{1}{20}\right)^{7/2}$$

where $\rho_r = 1$

$$(HP)_P = 0.75(20)^{7/2} = 26832\ HP$$

6.6.2 Viscous Force

Viscous forces are generally encountered in combination with gravity forces such as in flow in pipes and flow around partially submerged objects (ships and boats). Viscous forces are more appreciable in laminar flows. Flow around totally submerged bodies, such as submarines, also is dominated by viscous forces. Referring to Newton's law of viscosity,

$$\tau = \mu \frac{dv}{dy}$$

where τ is the shear stress due to the effect of viscous force, μ is the dynamic viscosity, and $\frac{dv}{dy}$ is the change in velocity with the change of distance, then

$$\frac{F_v}{L^2} = \mu \frac{v}{L} \tag{6.48}$$

The viscous force, F_v, can thus be given as

$$F_v = \mu v L \tag{6.49}$$

For dynamic similarity, the ratio between the inertial force and the viscous force in the two systems should be constant. Then

$$\left(\frac{F_I}{F_v}\right)_m = \left(\frac{F_I}{F_v}\right)_p = \text{constant} \tag{6.50}$$

or

$$\left(\frac{F_I}{F_v}\right)_r = 1 \tag{6.51}$$

but

$$\frac{F_I}{F_v} = \frac{\rho L^2 V^2}{\mu VL} = \frac{\rho VL}{\mu} = \frac{VL}{\nu} = R_n \tag{6.52}$$

Then it is concluded that for dynamically similar systems with a viscous force dominating the flow, we have

$$(R_n)_r = \left(\frac{\rho VL}{\mu}\right)_r = 1 \tag{6.53}$$

or

$$V_r = \mu_r \rho_r^{-1} L_r^{-1} = \nu_r L_r^{-1} \tag{6.54}$$

If the same fluid is used in both systems, then

$$V_r = L_r^{-1} \tag{6.55}$$

Equations (6.54) and (6.55) replace equations (6.45) and (6.46), respectively, when the flow is dominated by the viscous force. If different fluids are used, then μ_r and ρ_r can be evaluated from the physical properties of the fluids used.

Example 6.14

A 1:10 model for a submarine is tested in the laboratory. The submarine is required to be tested when it operates 50 m below the seawater level and moves at a velocity of 12 mph. What are the corresponding depth and velocity for the model knowing that ν_r between the water used in the model and seawater is 0.95?

Solution: The dominating force is the viscous force.

$$d_r = L_r = \frac{1}{10} = \frac{d_m}{d_p} = \frac{d_m}{50}$$

$$d_m = 5.0 \text{ m}$$

$$V_r = \nu_r L_r^{-1} = 0.95 \times 10 = 9.5$$

$$V_r = \frac{V_m}{V_P} = \frac{V_m}{12}$$

$$V_m = 9.5 \times 12 = 114 \text{ mph}$$

Example 6.15

A model is built to determine the energy loss in a pipeline 1.20 m in diameter and 10 km in length. The pipe conveys a liquid of a density of 0.8 g/cm^3 and a dynamic viscosity of 0.02 poise at a rate of 1.8 m^3/s. Tests are conducted on a 12-cm-diameter pipe using freshwater at 20°C (ρ = 0.998 g/cm^3, μ = 0.01 poise). The head loss in a length of 20 m of the model is 150 cm. Evaluate the model discharge and the head loss in the entire pipe. If the pressure at a certain point in the model is 700 kN/m^2, what is the corresponding pressure in the actual pipe?

Solution: The viscous force dominates and the Reynolds number should be the same in the two systems to ensure dynamic similarity. Therefore,

$$L_r = \frac{d_m}{d_p} = \frac{0.12}{1.2} = \frac{1}{10}$$

$$V_r = \mu_r \rho_r^{-1} L_r^{-1} = \left(\frac{0.01}{0.02}\right) \times \left(\frac{0.998}{0.8}\right)^{-1} \times \left(\frac{1}{10}\right)^{-1} = 4$$

$$Q_r = A_r V_r = L_r^2 V_r = \frac{1}{100} \times 4 = 0.04$$

$$Q_m = Q_r Q_P = 0.04 \times 1.8 = 0.072 \text{ m}^3\text{/s}$$

In pipes, the head loss (h_L) is evaluated as

$$h_L = \frac{fLV^2}{2gd}$$

where f is the friction coefficient (dimensionless). Therefore,

$$(h_L)_r = \frac{L_r V_r^2}{L_r} = V_r^2 = 16$$

where $g_r = 1.0$.

$$\text{The head loss in the prototype} = \frac{150}{16} = 9.375 \text{ cm.}$$

This head loss encountered in a length of $20 \times 10 = 200$ m.

$$\text{The head loss in the entire pipe} = 9.375 \times \frac{10000}{200} = 468.75 \text{ cm.} = 4.7 \text{ m}$$

The pressure ratio, p_r, is equal to the force ratio divided by the area ratio:

$$p_r = \frac{\rho_r L_r^2 V_r^2}{L_r^2} = \rho_r V_r^2 = \frac{0.998}{0.8}(4)^2 = 19.96$$

$$\text{The corresponding pressure in the prototype} = \frac{700}{19.96} = 35.07 \text{ kN/m}^2$$

6.6.3 Gravity and Viscous Forces

Drag forces on floating bodies like ships and boats are composed of two components: (a) the wave resistance and (b) the frictional or viscous resistance. Therefore, along with the inertia force, also exist the gravity and viscous forces and both of the latter forces affect the phenomenon under consideration in such a way that neglecting either one of them will undermine dynamic similarity. Considering that the variables are the drag force, F, the fluid velocity, V, a characteristic length of the moving body, L, density, ρ, viscosity of the liquid, μ, and the acceleration due to gravity, g, dimensional analysis would reveal that

$$F = \rho L^2 V^2 \varphi(F_n, R_n) \tag{6.56}$$

Therefore, to ensure the dynamic similarity, both F_n and R_n should have the same value in the two systems. That is,

$$(F_n)_m = (F_n)_P \quad \text{or} \quad (F_n)_r = 1 \tag{6.57a}$$

and

$$(R_n)_m = (R_n)_P \quad \text{or} \quad (R_n)_r = 1 \tag{6.57b}$$

The velocity ratio, ν_r, should have the same value when evaluated from equation (6.45) or from equation (6.54) assuming that $g_r = 1$. Therefore,

$$V_r = L_r^{1/2} = \nu_r L_r^{-1} \tag{6.58}$$

or

$$\nu_r = L_r^{3/2} \tag{6.59}$$

It is, therefore, possible to satisfy both of the gravity and viscous constraints with the condition that a different fluid is used in the model according to equation (6.59). Once L_r is determined and the physical properties of the fluid in prototype (μ, ρ, or ν) are known, the properties of the fluid to be used in the model can be defined. Using the same fluid in the two systems; that is, $\nu_r = 1$ in equation (6.59), would be possible only if the model has the same size of the prototype which is not practical. In some cases, the scale ratio, L_r, may be governed by the available fluids to be used in the model.

Example 6.16

A ship 400 ft long has a cruising speed of 30 mph in a river with a water kinematic viscosity of 1.217×10^{-5} ft²/s. A 1:100 model for the ship is tested in a towing basin containing a liquid with a specific gravity of 0.92. Determine the kinematic viscosity for the liquid to be used in the model. At what speed should the model be tested? If a force of 2.4-lb is imposed on the model, what is the corresponding force in the prototype?

Solution: Since both gravity and viscous laws are satisfied, the velocity ratio can be evaluated through either of them. It also should be noted that ρ_r is equal to γ_r. Therefore,

$$\nu_r = \frac{\nu_m}{\nu_p} = L_r^{3/2} = 0.001$$

$$\nu_m = 0.001 \times 1.217 \times 10^{-5} = 1.217 \times 10^{-8} \text{ ft}^2/\text{s}$$

The fluid to be used in the model should have a kinematic viscosity of 1.217×10^{-8} ft²/s.

$$V_r = L_r^{1/2} = \frac{1}{10} = 0.1$$

$$V_m - 0.1 \times 30 = 3 \text{ mph}$$

$$F_r = \rho_r L_r^2 V_r^2 = 0.92 L_r^3 = 0.92 \times 10^{-6}$$

$$F_P = \frac{2.4}{0.92 \times 10^{-6}} = 2.61 \times 10^6 \text{ lb}$$

6.6.4 Pressure Force

When the pressure force, F_p, dominates, the dynamic similarity can only be achieved when

$$\left(\frac{F_I}{F_p}\right)_m = \left(\frac{F_I}{F_p}\right)_P \tag{6.60}$$

where F_p is equal to the pressure intensity, p, multiplied by the area. Therefore,

$$\frac{F_I}{F_p} = \frac{\rho L^2 V^2}{p L^2} = \frac{\rho V^2}{p} = \frac{V^2}{p/\rho} \tag{6.61}$$

The Euler number, E_n, is defined as the square root of F_I/F_p. Then

$$E_n = \frac{V}{\sqrt{p/\rho}} \tag{6.62}$$

For a phenomenon controlled by the pressure force, the Euler number should have the same value in the two systems, that is,

$$(E_n)_r = \frac{V_r}{\sqrt{p_r/\rho_r}} = 1 \tag{6.63}$$

or

$$V_r = \sqrt{p_r/\rho_r} \tag{6.64}$$

Models for draft tubes in turbines or those built to simulate cavitation in pipes should satisfy both viscous and pressure constraints. The Reynolds and Euler numbers should have the same values in the two systems.

6.6.5 Elastic Force

In some cases where compressibility of fluids is important, the elastic force, F_E, must be considered. Examples for such situations include high speed flow and motion of objects through air at supersonic speeds. In this case, the dynamic similarity can be achieved when

$$\left(\frac{F_I}{F_E}\right)_m = \left(\frac{F_I}{F_E}\right)_p \tag{6.65}$$

F_E is equal to the bulk modulus, E_v, multiplied by the area. Therefore,

$$\frac{F_I}{F_E} = \frac{\rho L^2 V^2}{E_v L^2} = \frac{V^2}{E_v/\rho} \tag{6.66}$$

The Mach number, M_n, is defined as the square root of F_I/F_E; then

$$M_n = \frac{V}{\sqrt{E_v/\rho}} \tag{6.67}$$

To ensure dynamic similarity,

$$(M_n)_r = \frac{V_r}{\sqrt{E_{v_r}/\rho_r}} = 1 \tag{6.68}$$

or

$$V_r = \sqrt{E_{v_r}/\rho_r} \tag{6.69}$$

6.6.6 Surface Tension Force

The ratio between the inertia force and the surface tension force should be the same in the model and prototype when surface tension force is dominating. In general, the depths of flow in the models of rivers and open channels should not be very small to avoid appreciable effects of viscous and capillary phenomena in shallow flow regimes. To ensure dynamic similarity when surface tension dominates,

$$\left(\frac{F_I}{F_\sigma}\right)_m = \left(\frac{F_I}{F_\sigma}\right)_p \tag{6.70}$$

where F_σ is the surface tension force and is equal to σL.

$$\frac{F_I}{F_\sigma} = \frac{\rho L^2 V^2}{\sigma L} = \frac{V^2}{\sigma/\rho L} \tag{6.71}$$

The Weber number is defined as the square root of F_I/F_σ; then

$$W_n = \frac{V}{\sqrt{\sigma/\rho L}} \tag{6.72}$$

For dynamic similarity,

$$W_n = \frac{V_r}{\sqrt{\sigma_r/\rho_r L_r}} = 1 \tag{6.73}$$

or

$$V_r = \sqrt{\sigma_r/\rho_r L_r} \tag{6.74}$$

Table 6.1 summarizes the main forces acting on different hydraulic models and presents velocity ratios to ensure dynamic similarity under different types of dominating forces. If both gravity and viscous forces dominate, the velocity ratio can be evaluated from either gravity or viscous constraints; both will give the same value. However, the fluid in the model should differ from that in prototype to satisfy equation (6.59).

TABLE 6.1 Main Forces and Velocity Ratios in Different Hydraulic Models

Main Force	Dimensionless Number	Velocity Ratio	Application
Gravity	$F_n = \frac{V}{\sqrt{gL}}$	$V_r = g_r^{1/2} L_r^{1/2}$	Open channels, rivers, harbors, waves, spillways, weirs, stilling basins, channel transitions, sediment transport
Viscous	$R_n = \frac{\rho VL}{\mu}$	$V_r = \mu_r \rho_r^{-1} L_r^{-1}$ $= \nu_r L_r^{-1}$	Flow in pipes, hydraulic machines, submarines, flow around submerged bodies
Pressure	$E_n = \frac{V}{\sqrt{p/\rho}}$	$V_r = p_r^{1/2} \rho_r^{-1/2}$	Cavitation in pipes, draft tubes
Elastic	$M_n = \frac{V}{\sqrt{E_v/\rho}}$	$V_r = E_{v_r}^{1/2} \rho_r^{-1/2}$	High speed flow, airplanes, rockets
Surface tension	$W_n = \frac{V}{\sqrt{\sigma/\rho L}}$	$V_r = \sigma_r^{1/2} \rho_r^{-1/2} L_r^{-1/2}$	Very shallow flow in open channels and rivers, capillary waves, unsaturated groundwater studies

6.7 Distorted Models

The previous discussion focussed on undistorted models in which the model has only one scale ratio, L_r. In some cases, it is not possible to maintain this condition of complete geometric similarity. For example, if a model is to be built for a 10-km segment of a river with an average width of 1 km and an average depth of 4 m, we first investigate the available space in the laboratory to conduct the test. Assuming that the available space is 100 m × 30 m, then the 10-km distance can be reduced to 100 m or less and the 1-km distance can be reduced to 30 m or less. It is obvious that if we take L_r as 1:100, the above two constraints will be satisfied. The size of the model will be 100 m in length and 10 m in width. However, if the same ratio is applied to the vertical direction, the depth of flow in the model will be 4 cm. In some other cases the flow depth may be as small as a few millimeters. As a result, forces which may be entirely insignificant in the prototype may dominate in the model. Also, forces, which do not exist in the prototype, such as surface tension, would be introduced into the model. The model may, thus, not reflect the actual behavior of the prototype as different forces are introduced and sometimes dominate. To avoid creation of surface tension forces in models the Weber number should be kept at a high value. Another disadvantage of having very small flow depths in models is that small errors in measurements will be appreciable and would significantly affect the results.

Models are called distorted if we are not able to maintain full geometric similarity for economic, practical, or physical reasons. In distorted models two scale ratios are used: horizontal scale ratio, L_r, and vertical scale ratio, Z_r. All dimensions in the horizontal direction are reduced by L_r, while those in the vertical direction are reduced by Z_r. It is not possible to reduce two dimensions in the horizontal extension (say, length and width) by two different ratios. The vertical scale ratio, Z_r, applies only to the vertical direction.

Distorted models are used extensively in rivers and harbors as flow depths are quite small as compared to other areal dimensions. In such problems, for example, Z_r, is taken as 1:100, while L_r is taken as 1:200 or even 1:500.

Complete hydraulic similarity may be difficult to achieve in many cases. For example, even if a model is similar in size and shape to a prototype, this does not mean perfect similarity. The roughness of both surfaces in the two systems should also satisfy the condition of geometric similarity. Because roughness heights in prototypes are already very small, it could be impossible to reduce them with the same scale ratio as all other dimensions.

In distorted models, the discussion will be limited to the gravity force as a dominating force. Most distorted models are built for rivers, harbors, waves and other systems where flow is mainly controlled by the gravity force. Therefore,

$$V_r = g_r^{1/2} Z_r^{1/2} \tag{6.75}$$

Equation (6.75) is similar to equation (6.45), but L_r is replaced with Z_r. Because fluid velocities are created by vertical heads, we need to consider the vertical scale to evaluate the velocity ratio. This is also true for undistorted models, but because L_r is equal to Z_r, we simply write L_r.

If $g_r = 1$, then

$$V_r = Z_r^{1/2} \tag{6.76}$$

Dealing with distorted models, we always need to remember whether the considered ratios are horizontal or vertical. For example, the ratio for a horizontal area, A_{r_h}, is

$$A_{r_h} = L_r^2 \tag{6.77}$$

The ratio for the vertical area, A_{r_v}, is

$$A_{r_v} = L_r Z_r \tag{6.78}$$

Although we are dealing with the area in the vertical direction in equation (6.78), L_r has appeared because one side of any vertical area must be horizontal. However, V_r (regardless of its direction) is always evaluated from the vertical scale ratio, Z_r.

A similar procedure can be followed for all other ratios. For example, if the discharge ratio is required, we first need to know whether this discharge takes place in the horizontal direction like normal flow in open channel or in the vertical direction as in case of sudden fall in a channel bed. A horizontal flow takes place through a vertical area while vertical flow takes place through a horizontal area. Therefore,

$$Q_{r_h} = A_{r_v} \cdot V_r = L_r Z_r \cdot Z_r^{1/2} = L_r Z_r^{3/2} \tag{6.79}$$

while

$$Q_{r_v} = A_{r_h} \cdot V_r = L_r^2 Z_r^{1/2} \tag{6.80}$$

The time ratio for a particle moving in the horizontal direction is

$$T_{r_h} = \frac{L_r}{V_r} = L_r Z_r^{-1/2} \tag{6.81}$$

and for a particle moving in the vertical direction,

$$T_{r_v} = \frac{Z_r}{V_r} = Z_r^{1/2} \tag{6.82}$$

The force ratio in the horizontal direction, F_{r_h}, is

$$F_{r_h} = \rho_r Q_{r_h} V_r = \rho_r A_{r_v} V_r^2 = \rho_r (L_r Z_r) Z_r = \rho_r L_r Z_r^2 \tag{6.83}$$

The force ratio in the vertical direction, F_{r_v}, is

$$F_{r_v} = \rho_r Q_{r_v} V_r = \rho_r A_{r_h} V_r^2 = \rho_r L_r^2 Z_r \tag{6.84}$$

The work ratio in the horizontal direction, W_{r_h}, is

$$W_{r_h} = F_{r_h} \cdot L_r \tag{6.85}$$

Substituting for F_{r_h} from equation (6.83), we obtain

$$W_{r_h} = \rho_r L_r^2 Z_r^2 \tag{6.86}$$

The power ratio in the vertical direction, P_{r_v}, is

$$P_{r_v} = F_{r_v} \cdot V_r \tag{6.87}$$

Substituting for F_{r_v} from equation (6.84), we get

$$P_{r_v} = \rho_r L_r^2 Z_r^{3/2} \tag{6.88}$$

The former procedure can be applied to all other parameters.

Example 6.17

A distorted model for a 5 km × 5 km harbor is to be built in a laboratory space 7.5 m × 7.5 m. The larger pump available in the laboratory provides a discharge of 60 l/s. The depth of flow in the harbor is 5.0 m and the flow is 6×10^5 l/s. Find the horizontal and vertical scales, time scale in the horizontal direction, and power scale in the horizontal direction. What velocity in the laboratory corresponds to a velocity of 10 km/hr in the harbor? Assume that the same seawater is used in the model.

Solution:

$$L_r = \frac{7.5}{5000} = 0.0015$$

$$Q_{r_h} = \frac{Q_m}{Q_p} = \frac{60}{6 \times 10^5} = 1 \times 10^{-4}$$

$$Q_{r_h} = A_{r_v} \cdot V_r = L_r Z_r^{3/2}$$

$$1 \times 10^{-4} = L_r Z_r^{3/2} = \frac{7.5}{5000} \times Z_r^{3/2}$$

$$Z_r = (0.0667)^{2/3} = 0.164$$

$$T_{r_h} = \frac{L_r}{V_r} = \frac{L_r}{Z_r^{1/2}} = \frac{0.0015}{(0.164)^{1/2}} = 0.0037$$

$$P_{r_h} = \rho_r Q_{r_h} V_r^2 = A_{r_v} V_r^3 = L_r Z_r^{5/2} = 0.0015(0.164)^{5/2} = 16.34 \times 10^{-6}$$

$$V_r = \frac{V_m}{V_p} = \frac{V_m}{10} = Z_r^{1/2} = (0.164)^{1/2}$$

Therefore, $V_m = 4.05$ km/h.

Example 6.18

A distorted model was built for a regulator to a horizontal scale of 1:200 and a vertical scale of 1:50. If the prototype discharge was 12.0 m^3/s, find the required discharge in the model. If a fluid particle traveled a certain horizontal distance in 44 seconds, what is the corresponding time in the prototype? What force is acting on the gates of the prototype if the corresponding force in the model is 45.0 kg-m/s^2.

Solution:

$$L_r = \frac{1}{200}, \quad Z_r = \frac{1}{50}$$

$$Q_{r_h} = A_{r_v} V_r = L_r Z_r^{3/2} = \frac{1}{200}\left(\frac{1}{50}\right)^{3/2} = 14.1 \times 10^{-6}$$

$$14.1 \times 10^{-6} = \frac{Q_m}{12 \times 10^3}, \text{ then } Q_m = 0.169 \text{ l/s}$$

$$T_{r_h} = \frac{L_r}{V_r} = L_r Z_r^{-1/2} = \frac{1}{200}\left(\frac{1}{50}\right)^{-1/2} = 35.35 \times 10^{-3}$$

$$35.35 \times 10^{-3} = \frac{44}{T_P}, \text{ then } T_P = 1.24 \times 10^3 \text{ s}$$

$$F_{r_h} = \rho_r Q_{r_h} V_r = L_r Z_r V_r^2 = L_r Z_r^2 = \left(\frac{1}{200}\right)\left(\frac{1}{50}\right)^2 = 2 \times 10^{-6}$$

Therefore,

$$F_P = \frac{F_m}{2 \times 10^{-6}} = \frac{45}{2 \times 10^{-6}} = 22.5 \times 10^6 \text{ kg-m/s}^2 = 22.5 \times 10^3 \text{ kN}$$

Example 6.19

A model is constructed for a wide river whose maximum discharge is 3×10^4 m^3/s and Manning roughness coefficient, n, is 0.04. The maximum discharge in the laboratory is 0.105 m^3/s. Assuming that the vertical scale is 5 times the horizontal scale, determine the horizontal and vertical scales. If a velocity of 0.03 m/s is observed in the model, find the corresponding velocity in the river. What is the value of the Manning roughness coefficient in the model? The Manning equation is given as

$$V = \frac{1}{n} R^{2/3} S_o^{1/2}$$

where v is the flow velocity, n is the Manning roughness coefficient, R is the hydraulic radius which is equal to the depth y for wide channels, and S_o is the bed slope.

Solution: For wide channels, the hydraulic radius, R, is equal to the flow depth, y. Therefore, R_r is equal to Z_r.

$$Z_r = 5L_r$$

$$Q_{r_h} = \frac{Q_m}{Q_P} = \frac{0.105}{3 \times 10^4} = A_{r_h} V_r = L_r Z_r^{3/2} = \frac{Z_r^{5/2}}{5}$$

$$Z_r = \frac{1}{80}, \text{ and } L_r = \frac{Z_r}{5} = \frac{1}{400}$$

$$V_r = Z_r^{1/2} = \frac{V_m}{V_P}$$

$$V_P = \frac{0.03}{\left(\frac{1}{80}\right)^{1/2}} = 0.268 \text{ m/s}$$

Using Manning's equation,

$$V_r = \frac{1}{n_r} Z_r^{3/2} S_r^{1/2} = \frac{1}{n_r} Z_r^{3/2} \frac{Z_r^{1/2}}{L_r^{1/2}}$$

Substituting for $V_r = Z_r^{1/2}$, then

$$n_r = \frac{Z_r^{3/2}}{L_r^{1/2}} = \frac{(400)^{1/2}}{(80)^{3/2}} = 0.02795$$

$$n_m = n_r n_P = 1.118 \times 10^{-3}$$

READING AID

6.1. What are the main advantages of applying dimensional analysis in fluid mechanics and hydraulics?

6.2. Define the two systems of units. How are these two systems related to each other?

6.3. Using the mass system, write the dimensions of density, surface tension, viscosity, torque, power, and work.

6.4. What is meant by dimensionless numbers? Give some examples of such numbers.

6.5. What is the main constraint of the Rayleigh method in preforming dimensional analysis?

6.6. Why is the Buckingham method used widely? How many π-terms can be developed from 11 parameters with 2 of them involving mass?

6.7. Explain how the repeated variables are chosen in the Buckingham method? Can you select two variables having the same dimensions as repeated variables?

6.8. Explain the usefulness of physical models in hydraulic applications. Are models used independently or used with other techniques?

6.9. What is meant by geometric, kinematic, and dynamic similarities? If dynamic similarity exists, do you think that the geometric similarity is satisfied or not?

6.10. Write the ratio of the following variables in terms of the length ratio, L_r, and the velocity ratio, V_r, assuming that the same fluid is used in the model and prototype: discharge, time, angular acceleration, energy, surface tension force, and pressure force. Consider the gravity force to be dominating.

6.11. Discuss the types of forces acting on hydraulic systems. What is the role of the inertia force?

6.12. Give some examples of models for problems where the gravity force dominates. Deduce the velocity and pressure ratios in terms of the length ratio for such models.

6.13. Give some examples of models for problems where the viscous force dominates. Deduce the velocity ratio for such models.

6.14. If similarity exists between a prototype and its model, evaluate the scale ratio in terms of the kinematic viscosity of fluids in both systems assuming that Reynolds and Froude laws are satisfied.

6.15. Give examples for problems in which the pressure force dominates and hence determines the velocity ratio in terms of pressure and density ratios.

6.16. If the surface tension force dominates, evaluate the velocity ratio in terms of the surface tension ratio, density ratio, and length ratio.

6.17. What is meant by distorted models? Give some examples where such models are useful.

6.18. When evaluating the velocity ratio in distorted models, do you employ the horizontal scale ratio or the vertical scale ratio? Why?

6.19. For a distorted model, evaluate the discharge ratio in the horizontal and vertical directions in terms of horizontal and vertical scales L_r and Z_r, respectively.

6.20. Explain the advantages and disadvantages of distorted models.

Problems

6.1. Define the dimensions of the following groups:

(a) $\dfrac{\partial p}{\partial x}\cdot\dfrac{D^4}{\mu Q}$; (b) $\dfrac{T}{\rho N^2 D^5}$; (c) $\dfrac{N\sqrt{P}}{H^{5/4}}$; (d) $\dfrac{V^2\rho L}{\sigma}$

where p is the pressure, D is the diameter of a pipe, μ is the dynamic viscosity, Q is the discharge, T is the torque, ρ is the fluid density, N is the speed in rpm, P is the power, H is the head of water, V is the velocity, L is a characteristic length, and σ is the surface tension.

6.2. Using the Rayleigh method, find an expression for the drag force (D) of a sphere of diameter (d_o) falling slowly with a velocity (V) in a large vessel containing a liquid of density (ρ) and viscosity (μ).

6.3. Using the Rayleigh method, prove that the frequency of a wave caused by the motion of a yacht is proportional to (V/l), where V is the yacht velocity, l is its characteristics length, and the coefficient of proportionality is the Froude number.

6.4. In the metric system, the Chezy equation for a certain open channel is written as

$$V = 50\sqrt{RS}$$

where V is the water velocity, R is the hydraulic radius, and S is the bed slope. Find the corresponding equation in the BG system.

6.5. Using the Rayleigh method, determine the rational formula for discharge through a sharp-edged orifice flowing freely into the atmosphere in terms of constant head H, orifice diameter D, fluid density ρ, dynamic viscosity μ, and gravity acceleration g.

6.6. The variables in a Saybolt viscometer are the time (t) required to empty a certain volume of oil of density (ρ) and viscosity (μ), a length (L) representing the dimensions of the viscometer, and the gravitational acceleration (g). Using the π-theorem, show that

$$\frac{t}{\sqrt{L/g}} = \varphi\left(\frac{\mu}{\rho\sqrt{gL^3}}\right)$$

6.7. The drag resistance (D) of a projectile depends on its diameter (d), its velocity (V), fluid density (ρ) and viscosity (μ), and bulk modulus (k). Prove that

$$D = C_D A \frac{1}{2}\rho V^2$$

where C_D is the coefficient of drag and A is the biggest cross-sectional area of the projectile perpendicular to the flow direction.

6.8. To study the drop of pressure in pipes, several variables are considered. These include the pressure gradient (dp/dx), the diameter of the pipe (d), the density (ρ) and viscosity (μ) of the fluid, the fluid velocity (V), and the roughness of the pipe wall (e). Relate the pressure gradient to other parameters using the π-theorem.

6.9. Using dimensional analysis, show that the power P required to operate a test tunnel is given by $P = \rho l^2 V^3 \varphi(R_n)$, where ρ is the density, l is a characteristic linear dimension of the tunnel, V is the velocity of the fluid relative to the tunnel, and R_n is the Reynolds number.

6.10. Using dimensional analysis, find the relation between the shear stress, τ, exerted on a pipe wall and the fluid density ρ, viscosity μ, velocity V, and the pipe diameter d.

6.11. The thrust force F of a propeller running in a fluid depends on its diameter d, speed V, fluid density ρ, speed of the propeller N, and fluid dynamic viscosity, μ. Show that

$$F = \rho d^2 V^2 \phi\left[R_n, \frac{dN}{V}\right]$$

6.12. Derive a relation to evaluate the discharge, Q, over a spillway as a function of the water head, H, the depth at the throat, D, the velocity of flow, V, and the gravity acceleration, g.

6.13. Show that the flow over a rectangular notch can be expressed as

$$Q = bg^{1/2}h^{3/2}\varphi\left(\frac{Q\rho}{b\mu}, \frac{\sigma}{\rho h^2 g}\right)$$

where Q is the discharge, b is the width of the notch, g is the acceleration due to gravity, ρ is the density, μ is the fluid viscosity, and σ is the surface tension.

6.14. In rotating a thin circular disc submerged in a fluid, a torque has to be applied to the disc to overcome the frictional resistance of the fluid. Using dimensional analysis, show that

$$T = \rho N^2 D^5 \varphi\left(\frac{\mu}{D^2 N \rho}\right)$$

where T is the applied torque and D is the diameter of a disc rotating at a speed N in fluid of viscosity μ and density ρ.

6.15. A flow from an injector is leading to the formation of a spray of liquid drops. The drops formed in the spray have an average diameter d, which depends on the diameter of the pipe to which the injector is fitted D, the liquid velocity V, the acceleration of gravity g, and the liquid properties including density ρ, viscosity μ, and surface tension σ. Using dimensional analysis, show that

$$\frac{d}{D} = \phi\left(\frac{\rho VD}{\mu}, \frac{\rho V^2 D}{\sigma}, \frac{gD}{V^2}\right)$$

6.16. The capillary rise h, to which a liquid will rise in a small-bore tube depends on the tube radius, r, the acceleration of gravity g, and the fluid properties including fluid viscosity μ, density ρ, and surface tension σ. Develop an equation encompassing dimensionless groups to relate h/r to the other variables.

6.17. Show that the rate of flow Q of a fluid over a triangular notch can be expressed as

$$Q = H^2\sqrt{gH}\,\varphi\left(\rho H\frac{\sqrt{gH}}{\mu}, \theta\right)$$

where H is the head over the notch, g is the gravity acceleration, ρ and μ are the density and viscosity of the fluid, respectively, and θ is the angle of the notch.

6.18. The Shield relation in open channels commonly used to evaluate the sediment transport is written as

$$\frac{\tau}{D(\gamma_s - \gamma)} = \varphi\left(\frac{DU_*}{\nu}\right)$$

where τ is the shear stress between water and soil, D is a representative diameter for the soil particles, $(\gamma_s - \gamma)$ is the difference between the specific weight of solid particles and water, U_* is the shear velocity, and ν is the water kinematic viscosity. Prove this relation using dimensional analysis.

6.19. A cylinder of given length/diameter ratio is kept in steady rotation of N rpm in a uniform stream with a fluid velocity V. The power P required to maintain the motion depends on the fluid density ρ, kinematic viscosity of the fluid ν, diameter of the cylinder D, and V and N. Using dimensional analysis, derive the following relation:

$$P = \rho \nu^3 D^{-1} \varphi\left(\frac{VD}{\nu}, \frac{ND^2}{\nu}\right)$$

where φ is an arbitrary function.

6.20. Show that the discharge Q passing through a pipe of diameter D, as measured by an orifice meter of diameter d, can be expressed as

$$Q = cd^2\sqrt{\frac{p}{\rho}}$$

where c is a function of Reynolds number and the ratio between the pipe and orifice diameters, p is the pressure difference on the orifice, and ρ and ν (found in Reynolds number) are the density and kinematic viscosity of the fluid.

6.21. To investigate the effect of roughness on the shear stress in open channel flow, the following variables are considered: the average velocity, V, the normal depth, d, the fluid density and viscosity, ρ and μ, the acceleration due to gravity, g, the boundary shear stress, τ, and the roughness height, e. Use the Buckingham method to show that (g/C^2) is a function of the Reynolds number, Froude number, and relative roughness, where C is the Chezy coefficient.

6.22. Using dimensional analysis, derive a relation to evaluate the headloss, h_f, in closed conduits assuming that the headloss depends on the conduit diameter, d, water velocity, V, gravity acceleration, g, and the length of the pipe under consideration, l.

6.23. The discharge, Q, over a notch depends on the head, H, the acceleration due to gravity, g, the density, ρ, and the dynamic viscosity μ. Prove that

$$Q = g^{1/2} H^{5/2} \varphi\left(\frac{\rho g^{1/2} H^{3/2}}{\mu}\right)$$

Find the flow of water over the notch which corresponds to an oil flow over the notch with a head of 0.5 m. The specific gravity of oil is 0.8 and its dynamic viscosity is 8 times that of the water. The discharge of the oil over the notch is evaluated as $Q = 1.5H^{5/2}$.

6.24. A dam 15 m long is to discharge water at a rate of 114 m^3/s under a head of 3 m. Find the corresponding length and head of its model if the discharge in the laboratory is fixed at 30 l/s.

6.25. A model for a ship is built to a scale ratio of 1/12 to test its performance in a seawater of density 1030 kg/m^3. Calculate the velocity ratio, the force ratio, and the power ratios between the model and the prototype.

6.26. A tidal basin is built to a scale of 1:100 to study the waves and forces in a harbor. If the velocity in the model is 0.15 m, what is the corresponding velocity in the prototype? What is the time period in the model corresponding to 1 day in the prototype? If the breakwater in the model is subjected to a force of 20 N, determine the corresponding force in the prototype.

6.27. The Machio dam in Japan was modeled to a scale of 1/60. The prototype is an ogee spillway designed to carry a flood of 3200 m^3/s. Estimate the corresponding value for the model in m^3/s. What is the time in the model which represents one day in the prototype?

6.28. The depth and velocity at a point in a river are measured as 3.25 m and 1.5 m/s, respectively. The flow in the river is simulated by a 1/50 model in the laboratory. What are the corresponding depth and velocity in the model?

6.29. The drag characteristics of a blimp 4 m in diameter and 50 m in length are to be examined in a wind tunnel with a 1/10 scale model. If the speed of the blimp through still air is 8 m/s, what airspeed in the wind tunnel is needed to satisfy the conditions of similarity? Consider the air pressure and temperature to be the same in both the model and prototype.

6.30. A 1/30 model is built for a spillway. The actual discharge in the prototype is 20.0 m^3/s under a head of 2.0 m. Determine the head and discharge in the model. If the model dissipates 0.05 horsepower, what energy will be dissipated in the prototype?

6.31. The flow around a bridge pier is investigated using a 1/10 scale model. When the velocity in the model is 0.8 m/s, the height of the standing wave at the pier nose is observed to be 4.0 cm. Determine the corresponding values and wave height in the prototype.

6.32. A tidal estuary is to be modeled at a scale of 1/200. The maximum water velocity in the actual estuary is evaluated as 4 m/s, while the tidal period is estimated at 14 hours. What are the corresponding values in the model?

6.33. A dam is to be modeled in a laboratory to determine the adequacy of its design. The length of the river to be modeled is 1500 m, while its largest width (also the dam width) is 300 m. The expected maximum flow over the spillway of the dam is 3600 m^3/s, while the maximum discharge in the laboratory is 1.0 m^3/s. The available area in the laboratory to conduct this project is 50 m $\times$ 25 m. Determine the largest feasible scale for the model.

6.34. For Problem 6.33, determine the water velocity and water depth in the prototype if the corresponding values in the model were 0.5 m/s and 8 cm, respectively. What are the time and pressure ratios? Assume $\rho_r = 1$.

6.35. A 1/20 scale model of a spillway requires a discharge of 5.0 m^3/s in the laboratory. A hydraulic jump is formed in the model setup where the power lost was measured as 0.2 HP. Calculate (a) the discharge in the prototype neglecting viscous and surface tension effects, (b) the velocity in the prototype corresponding to a model velocity of 0.5 m/s, and (c) the horsepower lost in the prototype.

6.36. A 1/25 model of a ship with a wetted area of 800 m^2 is towed in a water basin at a speed of 2.0 m/s. The model experiences a resistance of 20 N. Calculate (a) the speed of the prototype, (b) the drag on the ship due to wave, (c) the skin friction, assuming that the skin-drag coefficient for the model is 0.004 and for prototype 0.015, (d) the total drag on the ship, and (e) the power to propel the ship.

6.37. A 20-cm-diameter pipe conveys water with a kinematic viscosity of 1×10^{-6} m^2/s. Determine the velocity of water in the pipe to ensure dynamic similarity with an oil flow of a kinematic viscosity of 1×10^{-5} m^2/s flowing with a velocity of 1.5 m/s through the same pipe.

6.38. For Problem 6.37, if the oil is flowing in a pipe 10 cm in diameter, evaluate the water velocity. Hence, determine the discharge ratio, the time ratio, the force ratio, and the pressure ratio. Let $\rho_o/\rho_w = 0.8$.

6.39. Oil with a kinematic viscosity of 4.8×10^{-6} m^2/s flows through a pipe 5.0 cm in diameter with a velocity of 3.2 m/s. A different oil with a kinematic viscosity of 4.0×10^{-6} m^2/s is flowing in a 2.5-cm-diameter pipe. At what velocity should the oil in the second pipe run to satisfy the conditions of similarity?

6.40. The moment acting on the rudder of a submarine is investigated by a 1/50 scale model. The model is tested in a tunnel where the moment is found to be 2 Nm at a speed of 8 m/s. What are the corresponding moment and speed of the prototype assuming both operate in the seawater?

6.41. A large valve has an inlet diameter of 75 cm conveying a discharge at the rate of 2.0 m^3/s. A model for this valve is tested in the laboratory, where the same water was used at the same temperature as in the prototype. The inlet diameter for the tested valve is 5 cm. Determine the required discharge in the model.

6.42. The drag of a small submarine hull is to be investigated when moving deep below the water surface using a 1/10 scale model. At what speed should the model be tested if the submarine is to run at 10 km/h? What drag force on the prototype corresponds to a drag force of 3 kN in the model?

6.43. A 1/50 scale model is built for a ship 90 m long that is cruising in a river with a water of kinematic viscosity of 1×10^{-6} m^2/s at a speed of 40 km/h. The model is tested in a towing basin containing a liquid of specific gravity of 0.94. Determine the kinematic viscosity of the liquid to be used in the model to satisfy both the gravity and viscous constraints. At what speed should the model be tested and what force in the prototype corresponds to a force of 64 kg-m/s^2 in the model?

6.44. The gravity and viscous forces are to be satisfied in a modeling exercise. The kinematic viscosities of the liquids used in the model and prototype are 1 stoke (cm^2/s) and 8 stokes, respectively. Determine the scale ratio, velocity ratio, and discharge ratio. Find the power dissipated due to friction through the prototype if the power dissipated in the model is 0.02 HP. Take $\rho_r = 1.2$.

6.45. For Problem 6.28, if the vertical scale is reduced to 1/20 while the horizontal scale is kept unchanged, what are the corresponding depth and velocity for this case in the model?

6.46. A distorted model is built for a turbine with a horizontal scale of 1/10. The speed of the prototype is 400 rpm under a head of 50 m. Find the speed of the model when running under a head of 10 m.

6.47. A quarter-scale turbine model is tested under a head of 15 m. The full-scale turbine is required to work under a head of 30 m and to run at a speed of 420 rpm. At what speed should the model be tested? If the model develops 1177.20 HP and uses 0.8 m^3/s, what power will be developed from the prototype assuming that the overall efficiency of the prototype is 5% better than that of the model?

6.48. A distorted model is built for a harbor to scales of 1:900 and 1:100 in horizontal and the vertical directions, respectively. Find (a) time in model corresponding to 5 s in prototype for a particle to fall down in water, (b) time in model corresponding to 10 s in prototype for a wave to travel across the harbor, (c) power ratio in the vertical direction, and (d) power ratio in the horizontal direction.

6.49. A distorted model is constructed for a wide river whose maximum discharge is 8000 m^3/s. The maximum discharge in the laboratory is 0.27 m^3/s. The vertical scale in the model is six times the horizontal one. The Chezy coefficient for the river is 50. Determine the horizontal and vertical scale ratios and the value of the Chezy coefficient in the model. If a velocity of 0.2 m/s is observed in the model, what is the corresponding velocity in the prototype?

6.50. For Problem 6.49, if the bed slope in the river is 20 cm/km, what will be the bed slope in the model? Determine the time ratio for a particle to move in the flow direction and that for a particle to move in the vertical direction. Determine the force ratio in the horizontal and vertical directions. Assume $\rho_r = 1$.

6.51. A distorted model is to be built for a harbor of 6×6 km^2 in area, 15 m average depth, and flow rate of 3000 m^3/s. The available basin in the laboratory is 30 m $\times$ 25 m in area and 0.6 m in depth. Determine the scale ratio in the horizontal and vertical directions and the discharge required to run the experiment. What is the time in the model that corresponds to 10 min in the prototype for a particle to travel both in the horizontal and vertical directions?

References

Morris, H. M. and Wiggert, J. M., 1972. *Applied Hydraulics in Engineering.* Wiley, New York.

Munson, B. R., Young, D. F., and Okiishi T. H., 1990. *Fundamentals of Fluid Mechanics.* Wiley, New York.

Sharp, J. J., 1981. *Hydraulic Modeling.* Butterworth Publishers Inc., London.

CHAPTER 7

Flow Resistance and Velocity Distributions

A natural channel in poor condition.

The study of resistance and velocity distributions in the flow of real fluids plays a vital role in the transition from the study of theoretical fluid mechanics principles to the solution of practical real-world problems. Were it not for the resistance imparted to the fluid by the boundary, the velocity in the flow field would remain uniform and velocity profiles would not develop. The formulation of the resistance and velocity principles developed in this chapter makes possible the estimation of energy losses in the flow of real fluids and allows for the application of the energy conservation (Bernoulli) equation in practical problems. The study of the relationship between resistance and velocity is so important to engineers that the development of the theory and observations on the subject by Prandtl, von Karman, and Nikuradse is recognized as the founding of the science of fluid mechanics even though many of the basic principles of fluid flow were known for centuries before. In succeeding chapters the foundation laid in this chapter will be used as the basis for the development of methodologies for solution of many problems in a variety of contexts.

The resistance to the flow of water provided by the cross-sectional boundary, as well as the internal properties of the fluid, is vital to the flow behavior. In most practical cases, the dominant resistance term is due to the shear stress imparted to the fluid by the boundary. The shear stress creates friction that results in a translation of some of the energy in the water into heat, which is then dissipated through the mass of water, or possibly through the boundary to the atmosphere. Either way, the net result is a loss of useful energy in terms of work which the water would be capable of doing. Of practical concern, enough energy must exist in the flow at any particular point to overcome the resistance in the cross sectional length ahead of it if the water is to continue to flow. Therefore, resistance losses must be computed in the design of pump systems, pressure tanks, stand tanks, etc. In addition, unequal resistance forces are the reason for the development of nonuniform water surface profiles in free surface flow.

The analysis of the resistance issue presents a difficult problem for hydraulic engineers. Even though it has been systematically studied for over a hundred years, a definitive solution still has not been found. Much of the body of knowledge which we use is due to a combination of theoretical advancement over the past 60 years and pure empiricism. The theoretical advancement and some focused empirical experiments on pipe flow are in the area of applied fluid mechanics. For that reason, this chapter, as well as Chapter 8, is intended to provide a vital link between the more fundamental and theoretical material in the previous chapters and the applied and practical material in later chapters.

7.1 Factors Affecting Flow Resistance

Flow resistance in pipes and channels is affected by a combination of fluid properties, flow conditions, and other factors related to the cross-sectional geometry and nature of the boundary material. Obviously, in some cases, one or more of these factors may be subject to significant spatial and temporal variability. It is this complexity of the issue which makes the resistance problem so intractable. In the present text, we will not be so concerned with temporal variability, and spatial variability is simplified by either discretizing the boundary into relatively homogeneous subunits or averaging conditions over some spatial area. Spatial variability presents a much greater difficulty in the cases of natural channels and watershed surfaces than in the case of closed conduit flow. In these cases, it is necessary to make some simplifying assumptions in order to deal with the problem.

The fluid property of most importance to resistance is the viscosity of the water. One needs to recall Newton's viscosity relationship to understand the importance of this property to the resistance issue. Viscosity enters into the analysis primarily through its role in determination of the internal shear or deformation of the flow on a microscale level as evidenced by the intensity and degree of turbulence.

Similarly, flow conditions such as velocity, depth, uniformity, and degree of turbulence play a vital role in determining the energy lost through resistance. Flow resistance increases as the uniformity of the flow decreases and intense turbulence can result in very large energy losses. In addition, the depth of flow relative to the representative roughness elements of the boundary plays an important role in determining energy losses due to resistance. The role of velocity is self-evident, higher velocities result in greater losses through friction and in fact, some energy losses are even expressed in units of velocity head.

The last set of factors affecting resistance losses is related to the geometry of the flow cross section and nature of the boundary. High roughness of the boundary, such as large stones or thick vegetation, will obviously provide obstruction to the flow, resulting in higher resistance. Elements, such as corrugations in metal pipes, will also result in greater energy losses than those occurring in smooth pipes, such as steel. Similarly, irregularities in the boundary surface may induce greater turbulence and increase resistance losses. Lastly, nonuniformity in the geometry of the flow cross section will result in instantaneous losses of energy due to sudden pressure changes in pipes and sudden changes in hydraulic control and associated depth in channels. These losses are usually not associated with resistance, however. Features such as pipe bends, valves, junctions, sudden channel constrictions or expansions, and gates are examples of geometric nonuniformities which affect energy losses.

Energy losses in the flow of water due to the resistance factors discussed above occur through the conversion of some of the available energy to heat through a mechanism which is a function of the particular flow regime which prevails. In laminar flow, energy losses occur due to the shear stress developed by the slippage of adjacent laminae of fluid over one another. This condition is characterized by relatively small values of the Reynolds numbers ($R_n \leq 2000$). When relatively fast moving fluid comes in contact with slower moving fluid particles, internal vortices are generated which are characteristic of turbulent flow conditions. Turbulent flow is characterized by the continuous multidimensional generation and dissipation of these vortices. The greater part of the energy losses in this case is caused by the internal mechanism of vortex generation. If combinations of the laminar and turbulent mechanisms of energy dissipation are at work, then the flow is classified as transitional. This state exists in a relatively confined range of Reynolds numbers between the upper limit for laminar flow and the lower limit for turbulent flow.

Were it not for the role of resistance, velocity profiles in steady flow in conduits and channels would be uniform. The resistance provided to the fluid particles adjacent to the boundary results in these particles adhering to the boundary and creates a velocity differential throughout the flow field. This velocity differential results in the development of shear stresses throughout the flow field that are the ultimate cause of turbulence and energy losses. The two most common velocity distributions encountered in hydraulics are the logarithmic and power law distributions. In the following sections these distributions will be derived and discussed with respect to their particular functions in the analysis of resistance losses.

7.2 Steady Uniform Flow

Before beginning our discussion of velocity distributions and shear stresses, we must first define the spatial and temporal flow properties which we will be concerned with. In the previous section, the effects of unsteadiness and nonuniformity of flow properties on resistance were mentioned. In our initial discussions in this chapter we will only be concerned with steady flow conditions and, primarily, with uniform spatial flow. So, to begin the development of the framework for the analysis of velocity distributions and resistance we must first develop the conditions of steady, uniform flow. Steady, uniform flow is primarily a resistance condition based upon the concept of dynamic equilibrium within the system. Of course, this condition can exist within any medium; however, our initial development will deal primarily with closed conduit flow as an example. It is important to note that the general relationships developed are applicable to flow in any media or conduits.

Consider the situation of flow through a section of uniform straight pipe as shown in Figure 7.1. As discussed in Chapter 5, the total energy available at any point in the pipe relative to the datum is given by the Bernoulli equation

$$E_T = z + \frac{p}{\gamma} + \alpha\frac{V^2}{2g} \tag{7.1}$$

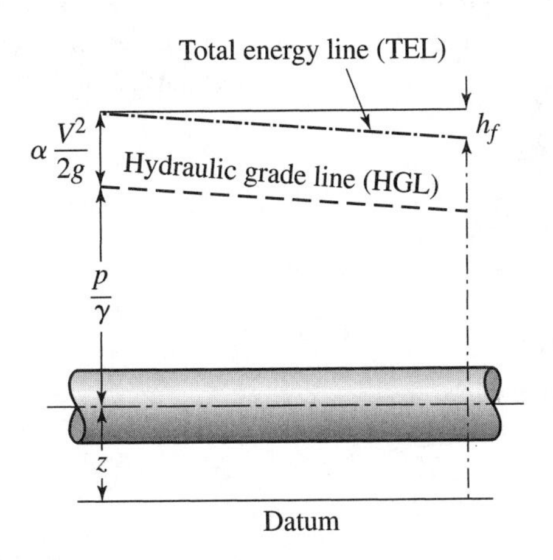

Figure 7.1 Steady, uniform flow in closed conduits.

Inside a large steel pipe. (Copyright Terry Vine/CORBIS.)

where z is the position head, p/γ is the pressure head (p is the fluid pressure and γ is the specific weight), $V^2/2g$ is the velocity head (V is the average flow velocity and g is the acceleration due to gravity), and α is the velocity coefficient. For regular cross sections such as rectangles, circles, and trapezoids, the velocity coefficient is usually close enough to unity that it can be considered so without appreciable error. The sum of the position head and pressure head is called the piezometric head. The piezometric gradient across the pipe length is also known as the hydraulic grade line. If the pipe diameter is uniform, then the average velocity is constant across the pipe length, and thus the velocity heads are equal across the pipe. In this case, the slopes of the hydraulic grade line and the energy grade line are equal, and the energy loss is manifested as a change in piezometric head across the pipe length. If the pipe slope is known, then the energy loss can be computed from the pressure drop across the pipe measured by piezometers. This flow condition is known as (steady) *uniform flow* and is always identifiable by the equality of the slopes of the energy and hydraulic grade lines. Obviously, the same condition can also occur within a uniform channel reach and thus one might expect uniform flow to represent an important general flow condition.

Now we wish to examine the conditions under which this situation occurs. From an engineering perspective, it is perhaps most satisfying to approach the problem from the force-momentum relationship. From this perspective, we consider the length of pipe to be a control volume (CV) and realize that all of the forces acting on the CV must be in dynamic equilibrium. Referring to Figure 7.2, we can see that the applicable forces are the hydrostatic (pressure) forces, the gravity force, and the boundary shear stress that is acting to retard the flow. The engineering approach to the problem, then, would be to relate the energy loss across the pipe to the shear stress at the boundary. Thus, in the direction of flow the force balance yields

$$p_1A - p_2A + \gamma LA \sin\alpha - \tau_o PL = 0 \tag{7.2}$$

where τ_o = boundary shear stress (N/m^2), p_1 = average pressure at section 1 (N/m^2), p_2 = average pressure at section 2 (N/m^2), γ = specific weight of water (N/m^3), P = wetted perimeter (m) = πd, d = pipe diameter (m), L = pipe length (m), and A = cross-sectional area (m^2).

Dividing equation (7.2) by A, we get

$$p_1 - p_2 + \gamma L \sin\alpha - \tau_o(P/A)L = 0 \tag{7.3}$$

The ratio A/P is a factor that represents a length reflecting the area and geometry of a particular cross section. In hydraulics, as discussed in Chapter 3, it is called the *hydraulic radius* (R) and in some sense can be thought of as a cross-sectional resistance length. Dividing equation (7.3) through by γ, we obtain

$$\frac{p_1 - p_2}{\gamma} + L\sin\alpha - \frac{\tau_o L}{\gamma R} = 0 \tag{7.4}$$

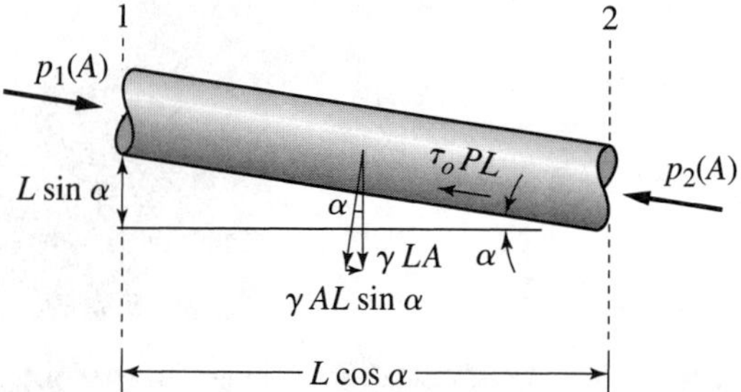

Figure 7.2 Forces acting on steady closed conduit flow.

Note that $\tan\alpha = (\sin\alpha/\cos\alpha) = S_o$, where S_o = pipe slope relative to the horizontal. So, $\sin\alpha = S_o \cos\alpha$. Then equation (7.4) becomes

$$\frac{p_1 - p_2}{\gamma} + S_o L \cos\alpha = \frac{\tau_o L}{\gamma R} \tag{7.5}$$

Because $L \sin\alpha = S_o L \cos\alpha = \Delta z$, the change in the pipe elevation with respect to the horizontal datum across the length L, equation (7.5) becomes

$$\frac{\Delta p}{\gamma} + \Delta z = \frac{\tau_o L}{\gamma R} \tag{7.6a}$$

or

$$\Delta\left(\frac{p}{\gamma} + z\right) = \frac{\tau_o L}{\gamma R} \tag{7.6b}$$

From Figure 7.1 it can be seen that the quantity $\Delta(p/\gamma + z)$ is the change in the hydraulic grade line across the pipe with respect to the datum, which is also equal to the energy loss across the pipe. Thus, this loss, which we will now refer to as the friction loss (h_f), is

$$h_f = \frac{\tau_o L}{\gamma R} \tag{7.7}$$

For circular pipes, $R = A/P = \dfrac{\pi d^2}{4\pi d} = d/4$, where d is the pipe diameter. Then

$$h_f = \frac{4\tau_o L}{\gamma d} \tag{7.8}$$

Dividing equation (7.7) by the length, and writing the resulting slope of the energy grade line (h_f/L) as S_f, equation (7.7) is often written as

$$\tau_o = \gamma R S_f \tag{7.9}$$

In this form, equation (7.9) has many practical design applications, particularly in the case of channels, as we shall see in later chapters.

In the case of open channels, a similar analysis can be made and will lead to the same general conclusion [equations (7.7) or (7.9)]. The open channel case is shown figuratively in Figure 7.3. It is important to note that in the case of free surface flow, the pressure head term ($\Delta p/\gamma$) in equation (7.6a) is zero, thus the piezometric head change must be equal to the

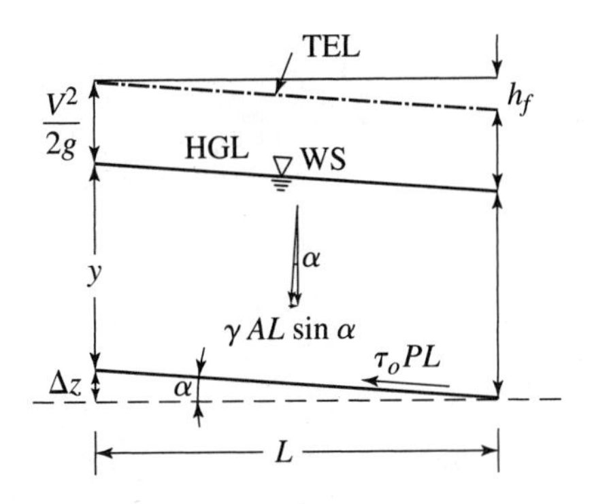

Figure 7.3 Steady, uniform flow in open channels.

change in bed elevation relative to the datum (Δz). Thus, in the case of steady, uniform flow in open channels, the slope of the channel bed, the slope of the water surface, and the slope of the energy grade line are all equal. An obvious implication of this fact is that steady uniform flow in channels must exhibit a constant depth throughout. In hydraulics, this depth is known as the *normal depth* (y_o). The conditions for steady uniform flow in open channels can also be derived from the consideration that in this case the pressure force terms in equation (7.2) fall out, and thus the force driving the flow (gravity) must be exactly countered by the force tending to oppose the flow (shear).

Applying Newton's law to the channel reach shown in Figure 7.3,

$$\sum \mathbf{F} = M\mathbf{a}$$

where **F** represents forces in the flow direction, M is the fluid mass, and a is the acceleration of flow. However, since the dynamic equilibrium is assumed, the acceleration $a = 0$. This yields $\Sigma\mathbf{F} = 0$. Then, setting the two forces equal in the direction of flow as discussed above leads to the following expression:

$$A \cos\alpha \; L \sec\alpha \gamma \sin\alpha = \tau_o P \cos\alpha \; L \sec\alpha \tag{7.10}$$

for small α, $\cos\alpha \approx \sec\alpha$. Thus, $A\gamma \sin\alpha = \tau_o P$, or

$$\tau_o = \gamma R \sin\alpha \tag{7.11}$$

where R is the hydraulic radius $= A/P$. Note that $S_f = h_f/L = S_o = \Delta z/L = \tan\alpha$, where h is the energy head, and that $\tan\alpha = \sin\alpha/\cos\alpha$, then $\sin\alpha = S_o \cos\alpha$. Thus,

$$\tau_o = \gamma R S_o \cos\alpha \tag{7.12}$$

since, for small values of α, $\cos\alpha \simeq 1$, one can write

$$\tau_o = \gamma R S_o \tag{7.13}$$

which is equivalent to equation (7.9). Since $S_o = S_f$, it is customary to use the bottom slope in the expression when applied to uniform flow in channels.

Example 7.1

A horizontal pipe 0.3 m in diameter has an available pressure drop of 100 N/m^2 to drive a flow rate of 0.5 m^3/s across a length of 100 m. What shear stress will be developed on the pipe boundary by this flow condition?

Solution: For a horizontal pipe,

$$\frac{p_1}{\gamma} + \frac{V_1^2}{2g} = \frac{p_2}{\gamma} + \frac{V_2^2}{2g} + h_f$$

Since $V_1 = V_2$, then $\Delta p/\gamma = h_f = S_f L$. Then

$$\frac{\Delta p}{\gamma} = \frac{100 \text{ N/m}^2}{9810 \text{ N/m}^3} = 0.0102 \text{ m}$$

So

$$S_f = \frac{0.0102 \text{ m}}{100 \text{ m}} = 0.000102 \text{ m/m}$$

Then

$$\tau_o = \gamma R S_f = 9810 \text{ N/m}^3 (0.3 \text{ m}/4)(0.000102 \text{ m/m}) = 0.075 \text{ N/m}^2$$

Example 7.2

A trapezoidal channel is carrying water at a normal depth of 5.2 ft. The channel has a bottom width of 10 ft and side slopes of $2H$ to $1V$. It is running on a slope of 0.005 ft/ft. What will be the shear stress in lb/ft^2 exerted on the channel bed by this flow?

Solution: The cross-sectional area of the flow is given by $A = by + ty^2 = 10(5.2) + 2(5.2)^2 = 106.08$ ft^2. The wetted perimeter would be given by $P = b + 2y\sqrt{1 + t^2} = 10 + 2(5.2)(2.236) = 33.255$ ft. Then the hydraulic radius is given by $R = A/P = 106.08/33.255 = 3.19$ ft. Then the shear stress applied to the bed is given by $\tau_o = \gamma R S_o = 62.4(3.19)(0.005) = 0.995$ lb/ft^2.

Example 7.3

Water is flowing in a 6-inch-diameter pipe inclined (upward) at a slope of 0.1 ft/ft. If the observed shear stress on the pipe is 0.975 lb/ft^2, what is the slope of the hydraulic grade line? What is the pressure drop per foot of the pipe length?

Solution: The situation is shown in the sketch below.

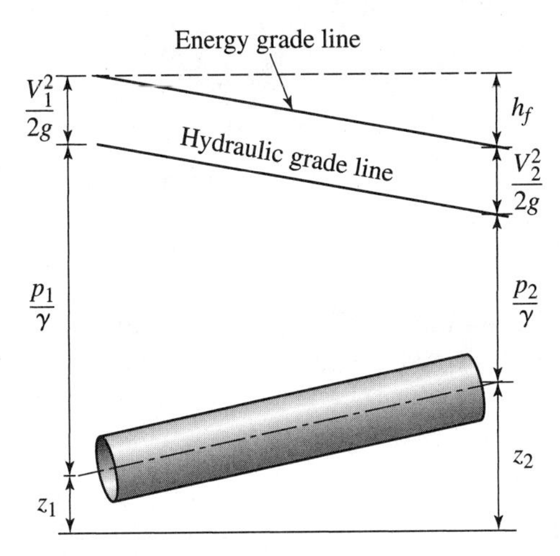

Figure 7.4 Solution for Example 7.3.

For the pipe, $A = \pi d^2/4 = 0.196$ ft^2, $R = A/P = 0.125$ ft. Then $\tau_o = \gamma R S_f$, or $0.975 = \gamma(0.125)S_f$, or $S_f = 0.125$ ft/ft. The angle of the pipe is $\alpha = \arctan(0.1) = 5.71°$, so $\cos\alpha = 0.995$. Then the head loss across 1 foot of pipe length is $h_f = 0.125/0.995 = 0.125$ ft/ft. Now, applying the Bernoulli equation across the pipe (positive upward) between sections 1 and 2, and noting that $V_1 = V_2$, then

$$(z_2 - z_1) + \left(\frac{p_2}{\gamma} - \frac{p_1}{\gamma}\right) = -h_f$$

$$\Delta z + \frac{\Delta p}{\gamma} = -h_f = 0.1 + \frac{\Delta p}{\gamma} = -0.125$$

or

$$\frac{\Delta p}{\gamma} = -0.225 \text{ ft}$$

This yields $\Delta p = -14.04$ psf or -0.0975 psi.

7.3 Resistance Equations for Steady Uniform Flow

A general shear stress relationship has been developed in the previous section that is valid in all cases of steady uniform flow. Now a method is needed which will relate the shear stress to the velocity in these situations. This can be accomplished by equating the physically derived equation (7.9) with an expression that involves the relevant properties and is dimensionally correct. From dimensional considerations, one can deduce that the relationship of the boundary shear stress to the fluid density and velocity must be of the form (Henderson, 1966):

$$\tau_o = a\rho V^2 \tag{7.14}$$

where a is a dimensionless constant of proportionality related to the boundary roughness characteristics and V is the average cross-sectional velocity (m/s). Then, relating equations (7.9) and (7.14),

$$a\rho V^2 = \gamma R S_f \tag{7.15}$$

Solving for V (noting that $\gamma = \rho g$),

$$V = \sqrt{\frac{g}{a}}\sqrt{RS_f} \tag{7.16}$$

Letting $\sqrt{\dfrac{g}{a}}$ equal C, equation (7.16) is usually written as

$$V = C\sqrt{RS_f} \tag{7.17}$$

Equation (7.17) is known as the Chezy equation after the French engineer Antoine Chezy (1718–1798), who first derived it around 1770 using the reasoning outlined above.

Now, from dimensional analysis considerations discussed in Chapter 6 (see Example 6.15), a general flow function relating flow parameters to the change in piezometric head in pipes can be developed:

$$\Delta\left(\frac{p}{\gamma} + z\right) = f\frac{L}{d}\frac{V^2}{2g} \tag{7.18}$$

where the factor f is a dimensionless parameter interpreted as a function of fluid properties and intrinsic boundary roughness. If we note that in the case of steady, uniform flow $h_f = \Delta(p/\gamma + z)$, then equation (7.18) becomes

$$h_f = f\frac{L}{d}\frac{V^2}{2g} \tag{7.19}$$

An equation similar in principle to (7.19) was first developed empirically in the mid-nineteenth century by Henri Darcy (1803–1858), while the equation was given in its present form by Julius Weisbach (1806–1871) in 1845 and is usually known as the Darcy equation or the Darcy-Weisbach equation.

Noting the fact that for circular pipes, $d = 4R$, equation (7.19) can also be written in terms of the velocity as

$$V = \sqrt{\frac{8g}{f}}\sqrt{RS_f} \tag{7.20}$$

Comparing equations (7.17) and (7.20), one can see that if the Chezy constant $C = \sqrt{8g/f}$, then the Darcy equation can be considered as a special case of the more general Chezy formula. In this case, the Chezy proportionality constant, a, would be equal to $f/8$.

The Chezy equation provides a means of relating the energy loss to the flow velocity if the resistance factors C or f can be determined. One reasonable interpretation would be that they are related to the boundary shear stress developed by the fluid movement. In fact, one might reasonably interpret the Chezy formula as an alternate form of the general shear relationship [equation (7.9)] expressed in terms of the velocity. Finally, it needs hardly to be pointed out that the Chezy and Darcy equations, like the general shear stress formulae that are their progenitors, are applicable in all cases of steady uniform flow regardless of the flow medium. However, the Darcy equation is most often applied in closed conduit situations for reasons that will be discussed in the next chapter. It should also be noted that one important difference between the Chezy and Darcy formulations is that the Chezy C coefficient necessarily possesses units ($L^{1/2}/T$) while the Darcy f factor is truly dimensionless. Thus, the Chezy coefficient would exhibit different values when employed in different units systems for example. This fact mitigates in favor of the use of the Darcy formulation wherever practicable.

Thus far, we have derived a number of functions relating the resistance loss to the flow parameters in the case of uniform flow. Equation (7.8) was derived based on physical force-momentum principles, while equations (7.17) and (7.19) were derived primarily from dimensional considerations, although the evaluation of the resistance factors C and f makes them quasi-empirical. Therefore, a possibility exists whereby one might relate the wall shear stress τ_o to the velocity through the friction factor f or the Chezy coefficient and thereby obtain the head loss for any given situation. In order to accomplish this task, three issues need to be addressed: (1) development of an analytical relationship between shear stress and velocity; (2) interpretation of the velocity distribution in the relationship derived in (1); and (3) final synthesis and application of the method to specific flow media. In this chapter, we will only be concerned with issues (1) and (2) and will leave the discussion of issue (3) for later chapters.

Taking issue (1) first, we relate equation (7.7) to equation (7.19) using the diameter $d = 4R$ for circular cross sections in order to remove d and keep the development general:

$$\frac{\tau_o L}{\gamma R} = f\frac{L}{4R}\frac{V^2}{2g} \tag{7.21}$$

Noting that $\gamma = \rho g$, multiplying both sides by 8, and then eliminating L, R, and g from both sides, we obtain

$$\frac{8\tau_o}{\rho} = fV^2 \tag{7.22}$$

Equation (7.22) illustrates that the friction factor is directly proportional to the boundary shear stress. Analysis of the ratio τ_o/ρ reveals that it has dimensions of velocity squared, so it is appropriately denoted as the *shear velocity* squared (v^{*2}). Thus, equation (7.22) can be rewritten as

$$\frac{V}{v^*} = \sqrt{\frac{8}{f}} \tag{7.23}$$

Equation (7.23) defines the relationship between the average flow velocity and the shear stress in dimensionless form through the friction factor f. It only remains now to develop velocity distributions for specific cases and to interpret them in the context of this relationship.

Equation (7.23) plays a vital role in the development of the resistance formulation not only for its physical significance, but because of its dimensionless form. It provides a framework whereby a whole range of velocity distributions applicable to a wide variety of circumstances can be placed on equal footing within the engineering force-momentum context. These velocity distributions can be analyzed within the shear stress framework and the energy losses due to boundary shear and friction can be fairly estimated.

Example 7.4

A rectangular channel 5 m wide carries a discharge of 10 m^3/s at a uniform depth of 3 m. The channel is running on a slope of 0.0025 m/m. Estimate the Darcy friction factor and the Chezy coefficient for this case. Assume the density of the flowing water as 998.2 kg/m^3.

Solution:

$$A = By = 5(3) = 15 \text{ m}^2;\ v = Q/A = 10/15 = 0.667 \text{ m/s};\ v^* = (\tau_o/\rho)^{0.5};\text{ and}$$

$$\tau_o = \gamma R S_f = 9810 \text{ N/m}^3\ [15/(2(3) + 5)] \text{ m } (0.0025) \text{ m/m} = 33.44 \text{ N/m}^2$$

$$v^* = (33.44/998.2)^{1/2} = 0.183 \text{ m/s}$$

Solving equation (7.23) for f; $f = (8v^{*2})/V^2 = 8(0.183)^2/(0.667)^2 = 0.602$ and

$$C = (8g/f)^{1/2} = (8(9.81)/0.602)^{1/2} = 11.418 \text{ m}^{0.5}/\text{s}$$

7.4 Velocity Distributions in Steady, Uniform Flow

As mentioned previously, in the text we will confine our discussion chiefly to logarithmic and power law velocity distributions. Typical laminar and turbulent logarithmic velocity distributions for pipes are given in Figure 7.5. An examination of these profiles, even in an elementary sense, will reveal that the shear stress imparted by the wall to a particle of water will depend on the distance from the particle to the wall. It is also obvious that an infinite number of possible velocity distributions exist ranging from fully laminar to fully turbulent conditions.

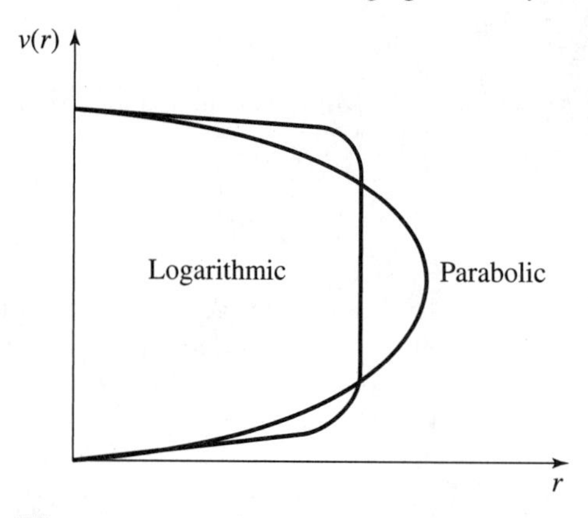

Figure 7.5 Parabolic and logarithmic velocity distributions in pipes.

Newton's law of viscosity can be used as the starting point for the examination of velocity profiles in the logarithmic family:

$$\tau = \mu \frac{dv}{dy} \tag{7.24a}$$

or, in polar coordinates,

$$\tau = -\mu \frac{dv}{dr} \tag{7.24b}$$

One may recall that the Newtonian viscosity principle directly governs flow in the laminar region. An equation similar in form can be written for turbulent flow:

$$\tau = \eta \frac{dv}{dy} \tag{7.24c}$$

where η is a quantity known as the *eddy viscosity,* which, unlike the dynamic viscosity in equation (7.24a), is not constant throughout the flow field. The eddy viscosity is not strictly a fluid property but also depends on the turbulence dynamics and fluid motion. It is normally evaluated in a probabilistic framework.

7.4.1 Laminar Flow

Since the laminar flow case is governed directly by the Newtonian viscosity principle, a simple analytical solution is possible for that case. Solving equation (7.8) for τ_o for unit length dL gives

$$\tau_o = \left(\frac{dE}{dL}\right)\frac{\gamma r}{2} \tag{7.25}$$

Note that $dE/dL = h_f$ for element dL and that $d = 2r$. Note also that equation (7.25) is valid throughout steady uniform flow and not just for the laminar case. Then, taking the shear stress at the wall so that $\tau = \tau_o$, substituting equation (7.25) into equation (7.24b), and solving for dv, we obtain the differential velocity for the laminar case:

$$dv = -\frac{dE}{dL}\frac{\gamma}{2\mu} r\,dr \tag{7.26}$$

Then, merely integrating equation (7.26) gives the velocity as a function of the radial position r:

$$v = \int_r^{r_0} dv = \frac{dE}{dL}\frac{\gamma}{4\mu}(r_0^2 - r^2) \tag{7.27}$$

where r_0 is the specific pipe radius. Noting from equation (7.27) that the laminar velocity distribution is a paraboloid, and that its maximum value is at the centerline of the pipe (i.e., $r = 0$), and that for a paraboloid the average value of the function is given by one half the maximum functional value, one can substitute to obtain an expression in terms of the average velocity (v):

$$\frac{dE}{dL} = \frac{32\mu V}{\gamma d^2} \tag{7.28}$$

Equation (7.28) gives the head loss in a pipe element in terms of the average velocity thus negating the necessity to deal directly with the velocity distribution in the case of laminar flow. The energy loss gradient (dE/dL) is also known as the *friction slope* since it represents the rate of energy dissipation due to boundary shear stress or friction. Recall also that in the case of uniform flow, the friction slope is equal to the slope of the piezometric grade line or the hydraulic gradient.

7.4.2 Velocity Distributions in Turbulent Flow

We have examined the relationship among energy loss, shear stress, and velocity for the case of steady, uniform laminar flow in channels and conduits. Unfortunately, the situation is not so simple in the case of turbulent velocity distributions since the relationships between wall shear stress and velocity distributions are more complex than in the laminar case. Solution of this complex case relies on the pioneering works of Ludwig Prandtl (1875–1953), Theodor von Karman (1881–1963), and Johann Nikuradse (1894–1979) at the Gottingen Fluid Mechanics Laboratory in Germany. The great contribution of Prandtl was the introduction of the boundary layer concept in 1901. For this he is considered to be the father of the modern discipline of fluid mechanics. Prandtl determined that in all flow cases there exists a thin layer of laminar flow near the boundary which is known as the laminar sublayer. In most practical cases of flow in both closed conduits and channels, the flow is turbulent outside of the boundary layer. The thickness of the laminar boundary layer decreases as the Reynolds number of the flow increases.

Figure 7.6 shows the behavior of the shear stress (τ_l) within the laminar sublayer region compared to that in turbulent flow above the sublayer (τ_t). In the case of turbulent flow, statistical and dimensional considerations led von Karman to develop the following relationship between shear stress and velocity distribution (Henderson, 1966):

$$\tau = \rho k^2 \frac{(dv/dy)^4}{(d^2v/dy^2)^2} \tag{7.29}$$

where k is the von Karman constant for turbulent flow which is equal to approximately 0.40 for clear water. Equation (7.29) was developed following a statistical analysis of internal turbulence dynamics related to equation (7.24b) that is beyond the scope of the present text. The equation is difficult, if not impossible, to solve analytically; however, an approximate solution yields the dimensionless velocity distribution given by von Karman as

$$\frac{v}{v^*} = \frac{1}{k} \ln \frac{y}{y_o} + C \tag{7.30}$$

where y is the distance from the boundary, y_o is the hydraulic depth, and C is a constant. In the case of regular channels, y_o is the normal depth and for pipes y_o is the radius. Equation (7.30) is

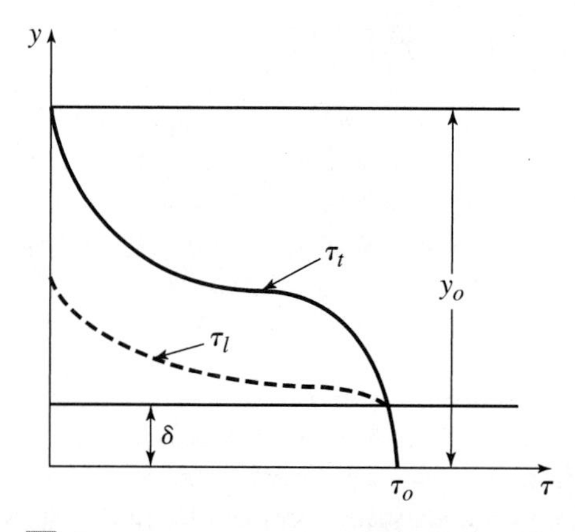

Figure 7.6 Shear stress near the wall in turbulent flow (Singh, 1996).

TABLE 7.1 Equivalent Roughness Elements for Various Materials

Material	ε (mm)	ε (ft)
Glass, drawn tubing (copper, brass)	0.0015	0.000005
Wrought iron, steel (new)	0.045	0.00015
Galvanized iron (new)	0.15	0.0005
Cast iron (new)	0.26	0.00085
Wood stave (smooth surface)	0.18	0.0006
Wood stave (rough surface)	0.9	0.003
Concrete (smooth surface)	0.3	0.001
Concrete (rough surface)	3.0	0.01
Riveted steel (few rivets)	0.9	0.003
Riveted steel (many rivets)	9	0.03
Corrugated metal	45	0.15

the famous logarithmic velocity distribution for turbulent flow shown in Figure 7.5. It will also be shown that the logarithmic velocity profile given by equation (7.30) is applicable to open channel conditions as well.

The thickness of the laminar sublayer depends on the intensity of turbulence of the flow (Reynolds number) as well as the intrinsic roughness of the boundary. The boundary roughness can be expressed as a *relative roughness* term y_o/ε (or ε/y_o), where ε was taken to be the diameter of a sand grain of the same size as the roughness elements of the boundary material. The behavior of the energy loss function was found to depend upon the relative thicknesses of the boundary layer (δ) and the equivalent sand roughness of the wall (ε). Equivalent roughness values for commonly encountered pipes are given in Table 7.1. According to Hwang and Houghtalen (1996), if $\delta > 1.7\varepsilon$, then the laminar sublayer dominates and the wall functions as a *hydraulically smooth* surface. In this situation, the wall roughness elements are well submerged beneath the sublayer so that a smooth interface is provided between the water particles outside the layer and the wall. Conversely, if $\delta < 0.08\varepsilon$, then the significance of the laminar sublayer is decreased (i.e., ε dominates, and the wall is said to be *hydraulically rough*). In these situations, the roughness elements protrude well above the laminar sublayer and the outside water particles encounter a very rough interface. Another factor in these cases is that the rough surface will not only provide greater intrinsic resistance to the flow but will also increase the turbulence of the flow field.

7.4.2.1 Turbulent Flow with Smooth Walls As stated above, a surface is hydraulically smooth if the equivalent roughness element of the wall is significantly less than the thickness of the laminar sublayer for the particular flow situation. Of course, since the boundary layer thickness depends on the Reynolds number, a surface might be smooth for some flow cases and not smooth for others.

In this case it is advantageous to be able to express the velocity distribution [equation (7.30)] as a function of the laminar sublayer properties. In the laminar sublayer the velocity gradient is nearly constant and the shear stress τ is assumed constant equal to $\mu v/y$ (i.e., $\tau = \mu v/y$). Thus, rearranging Newton's viscosity law [equation (7.24a)], recalling the definition of the shear velocity, and dividing by v^* yields the dimensionless velocity distribution for the laminar sublayer:

$$\frac{v}{v^*} = \frac{yv^*}{\nu} \tag{7.31}$$

where ν is the kinematic viscosity. The term (yv^*/ν) has the form of a Reynolds number and in fact is known as the *wall Reynolds number.* Then, by this analogy, there should exist some critical value at which the flow changes from laminar to turbulent, a point to which we will return later. Placing equation (7.31) in the context of equation (7.30), at the point where $y = \delta$, it can be shown that

$$\frac{v}{v^*} = \frac{1}{k} \ln \frac{yv^*}{\nu} + C_1 \tag{7.32}$$

Nikuradse conducted a series of experiments using sand grains attached to pipe walls to simulate various roughness conditions and determined the constant C_1 to be 5.5, thus making the distribution for turbulent flow in smooth pipes with $k = 0.4$ to be given by

$$\frac{v}{v^*} = 2.5 \ln \frac{yv^*}{\nu} + 5.5 \tag{7.33a}$$

and

$$\frac{v}{v^*} = 5.75 \log \frac{yv^*}{\nu} + 5.5 \tag{7.33b}$$

Singh (1996) states that the shear-velocity relationship in the laminar sublayer [equation (7.31)] holds for values of yv^*/ν up to 4 or 5 and that there is a gradual transition from fully laminar to fully turbulent flow. Morris and Wiggert (1972) conclude that, based upon Nikuradse's experiments and the work of Schlichting (1968), equation (7.31) should be preferred up to a value of the wall Reynolds number of 11.6. Based on this analysis, they show that at the point where $y = \delta$:

$$\frac{\delta v^*}{\nu} = 11.6, \quad \text{or} \quad \delta = \frac{11.6\nu}{v^*} \tag{7.34}$$

Noting from equation (7.23) that $v^* = V\sqrt{f/8}$, equation (7.34) is then rewritten to show the dependence of the boundary layer on the Reynolds number (R_n):

$$\frac{\delta}{d} = \frac{11.6\nu}{dV\sqrt{f/8}} = \frac{11.6\sqrt{8}}{R_n\sqrt{f}} \tag{7.35}$$

On the other hand, Singh (1996) concludes that there is no definitive boundary where the flow changes from entirely laminar to entirely turbulent and proposes the following division:

Laminar sublayer: $0 \leq yv^*/\nu \leq 4$

Overlap zone: $4 \leq yv^*/\nu \leq 30$ to 70

Turbulent zone: $yv^*/\nu > 30$ to 70

Schlichting (1968) plotted the experimental data from various sources against equations (7.31) and (7.33) as shown in Figure 7.7. The figure demonstrates the goodness-of-fit of equation (7.33) to the fully turbulent zone as well as the ambiguity of the transition from laminar to turbulent flow.

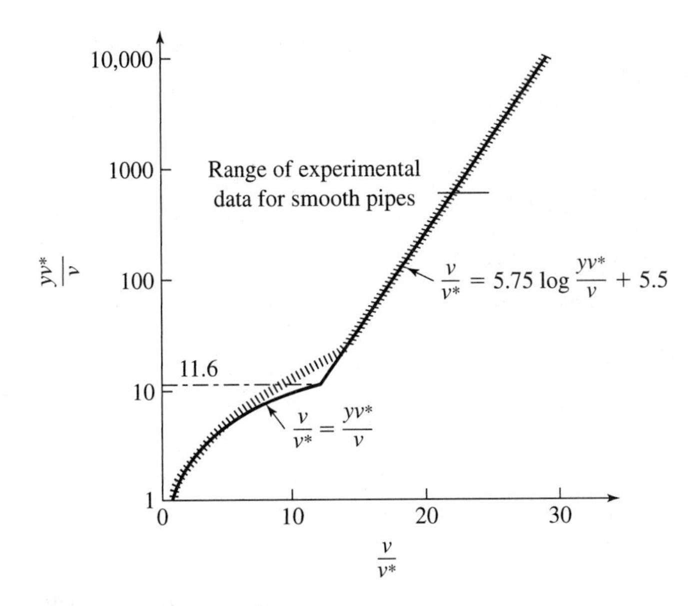

Figure 7.7 Velocity distribution in smooth pipes.

A smooth glass pipe.

Example 7.5

Water flows in a pipe which is running on a slope of 10% as shown in Figure 7.8 and is carrying a discharge of 0.3 m^3/s. The pipe diameter is 0.2 m and the length is 100 m. The temperature of the water is 20°C. At point 1 the pipe centerline is located 20 m above the datum and a piezometer reading there is 0.5 m as shown in the figure. At the end of the pipe (100 m from point 1) the piezometer reading is 0.35 m. Using the Morris and Wiggert criterion, what is the approximate thickness of the boundary layer in this situation? Under the Singh criterion, what would be the sublayer thickness? If the pipe in question is riveted steel, can it be considered a smooth pipe in this case?

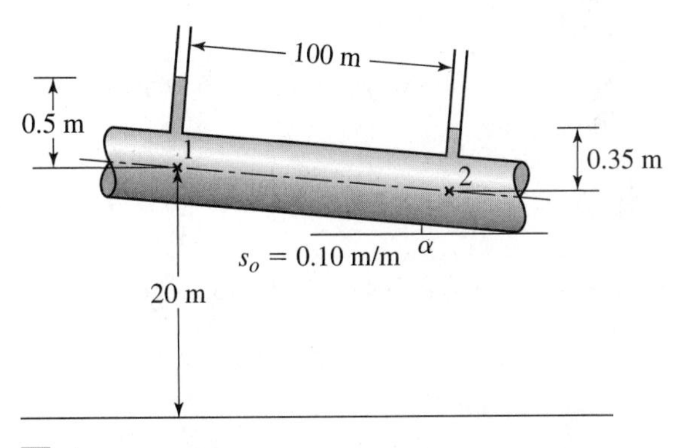

Figure 7.8. Sloping pipe for Exmaple 7.5.

Solution: The friction slope, $S_f = [\Delta(p/\gamma + h)]/[L \cos\alpha]$, where α is the angle of incline of the pipe. Since $\tan\alpha = 0.1$, then $\alpha = 5.71°$; and $\cos\alpha = 0.995$. Thus the pipe falls $0.1(99.5) = 9.95$ ft. Then

$$(p_1/\gamma) + z_1 = 0.5 + 20 = 20.5 \text{ m}$$

$$(p_2/\gamma) + z_2 = 0.35 + 10.05 = 10.4 \text{ m}$$

So $\Delta(p/\gamma + z) = 10.1$ m; then $S_f = 10.1/99.5 = 0.102$ m/m

$$\tau_o = \gamma R S_f = 9789(0.2/4)(0.102) = 49.92 \text{ N/m}^2$$

$$v^* = (\tau_o/\rho)^{0.5} = (49.92/998.2)^{0.5} = 0.223 \text{ m/s}$$

With the Morris and Wiggert criterion, $\delta = [11.6\nu/v^*]$:

$$\delta = 11.6(1.003 \times 10^{-6})/0.223 = 0.0000522 \text{ m}$$

From the Singh criterion, $yv^*/\nu = 4$; so $y(0.223)/(1.003 \times 10^{-6}) = 4$:

$$y = 0.000018 \text{ m}$$

Now, from Table 7.1, for riveted steel the minimum ε value is 0.9 mm = 0.0009 m. So $\delta/\varepsilon = 0.0000522/0.0009 = 0.058$.

Since $0.058 \ll 1.7$, the pipe cannot be considered smooth in this case.

7.4.2.2 Turbulent Flow with Rough Walls As stated previously, if the thickness of the laminar sublayer δ is significantly less than the representative roughness element of the boundary material ε, then the effects of the sublayer become negligible and the turbulent elements of the flow dominate. This is the case of fully turbulent flow against rough surfaces. In this case, the general equations (7.31) and (7.32) still hold for the shear stress-velocity relationship and the form of the velocity distribution respectively. However, in this case, it is advantageous to express equation (7.32) as a function of the roughness element rather than the thickness of the boundary layer (Roberson et al., 1988):

$$\frac{v}{v^*} = \frac{1}{k} \ln \frac{y}{\varepsilon} + B \qquad \text{(7.36a)}$$

where B is a coefficient. With the von Karman constant $k = 0.4$, and changing to the common base of logarithm,

$$\frac{v}{v^*} = 5.75 \log \frac{y}{\varepsilon} + B \qquad \text{(7.36b)}$$

The rough pipe experiments of Nikuradse led to the conclusion that the coefficient B should be equal to 8.48. Thus,

$$\frac{v}{v^*} = \frac{1}{k} \ln \frac{y}{\varepsilon} + 8.48 \qquad \text{(7.37)}$$

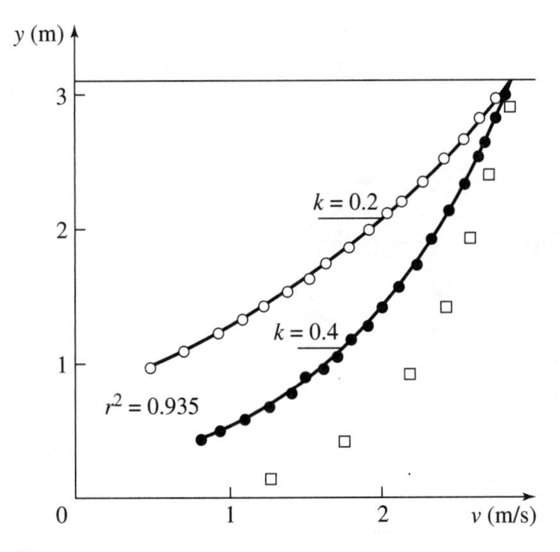

Figure 7.9 Logarithmic velocity distribution in open channels. (From Barbé, 1990)

Section of rough wrought iron pipe. (Courtesy of Jon Schladweiler, www.sewerhistory.org.)

The case of fully turbulent flow in rough surfaces is also the normal case in open-channel situations. In these cases, the Nikuradse–von Karman logarithmic velocity profile also applies. The profile is normally written as

$$\frac{v}{v^*} = \frac{1}{k}\ln\frac{h}{y} + \frac{v_m}{v^*} \tag{7.38}$$

where h is vertical distance from the channel bed, y is the flow depth, and v_m is the maximum velocity. Figure 7.9 shows this profile fitted to an observed velocity profile given by Barbé (1990).

Example 7.6

Water flows in a wrought iron pipe of diameter 0.25 m. The discharge is 0.15 m^3/s and piezometers stationed along the pipe show a piezometric grade line of 0.0343 m/m. The temperature of the water is 20°C. Calculate the velocity distribution in this case using both the smooth and rough pipe assumptions. In this case, are either of these assumptions substantially correct?

Solution: In this case, the piezometric gradient is equal to the friction slope, so $S_f = 0.0343$. Then

$$\tau_o = 9789(0.25/4)(0.0343) = 20.98 \text{ N/m}^2$$

So $v^* = (20.98/998.2)^{0.5} = 0.145$ m/s.

For the smooth law,

$$\frac{v}{v^*} = 5.75 \log\frac{yv^*}{\nu} + 5.5$$

and for the rough law,

$$\frac{v}{v^*} = 5.75 \log\frac{y}{\varepsilon} + 8.48$$

For $T = 20°C$, $\nu = 1.003 \times 10^{-6}$, and for wrought iron, from Table 7.1, $\varepsilon = 0.045$ mm = 0.000045 m. Inserting these values into the smooth and rough laws, respectively, and tabulating typical results leads to the following:

y (m)	Smooth Law v (m/s)	Rough Law v (m/s)
0.01	3.43	3.19
0.05	4.01	3.77
0.10	4.27	4.02
0.125	4.35	4.10

Now, the thickness of the boundary layer, δ, can be computed from the Morris and Wiggert criterion, $\delta = \dfrac{11.6\nu}{v^*} = \dfrac{11.6(1.003 \times 10^{-6})}{0.145} = 0.000802$ ft. Alternatively, using the Singh criterion, $\dfrac{yv^*}{\nu} = 4$, or $\dfrac{4(.145)}{1.003 \times 10^{-6}} = 4$. Thus, $y(\delta) = 0.0000276$ ft. Comparing to the ε for wrought iron (0.000045), we see that if we use the first criterion, then δ is greater than 1.7ε (0.0000765 ft), so we could say that the wall is (barely) smooth. However, the second criterion does not meet this requirement, so we would probably be safer by saying that it is neither smooth nor rough.

Example 7.7

Water is flowing in a semicircular aquaduct of 2-ft radius running on a slope of 0.0025 ft/ft and functioning as an open channel throughout. The discharge is 25 cfs with a depth 1.5 ft and an observed maximum velocity of 9 ft/s. Compute the vertical velocity profile for this situation assuming the rough law.

Solution: The situation is depicted in Figure 7.10.

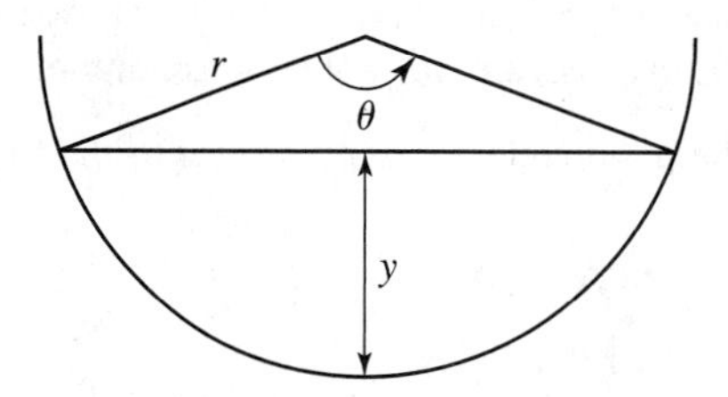

Figure 7.10 Solution for Example 7.7.

With a depth of 1.5 ft on a diameter of 4 ft, we find the angle θ from the expression given in Table 10.1 as $\theta = 2\cos^{-1}(1 - 2(y/d)) = 2\cos^{-1}(1 - 2(1.5/4)) = 2.636$ radians. From Table 10.1, we also get that the area of the depth of flow is given by

$$A = (d^2/8)(\theta - \sin\theta) = ((4)^2/8)(2.636 - 0.484) = 4.30 \text{ ft}^2$$

Also from Table 10.1, we find that the wetted perimeter would be given by $P = (d/2)\theta$. Then $P = 2(2.636) = 5.27$ ft. Therefore, $R = A/P = 4.30/5.27 = 0.815$ ft. Then the average velocity would simply be $V = Q/A = 25 \text{ cfs}/4.30 \text{ ft}^2 = 5.81$ ft/s. Then

$$\tau_o = \gamma R S_o = 62.4(0.815)(0.0025) = 0.127 \text{ lb/ft}^2$$

So $v^* = (\tau_o/\rho)^{0.5} = 0.256$ ft/s. Then, from the rough velocity law,

$$\frac{v}{v^*} = \frac{1}{k}\ln\frac{h}{y} + \frac{v_{max}}{v^*} = 2.5\ln(h/1.5) + 35.15$$

The computed velocity profile is given below for different values of h.

h (ft)	v (ft/s)
0.1	7.26
0.25	7.85
0.50	8.25
0.75	8.55
1.00	8.67
1.25	8.88
1.45	8.97

7.5 Power Law Velocity Distributions

The general form of the power law velocity distributions for turbulent flow is (Singh, 1996)

$$\frac{v}{v_m} = \left(\frac{y}{D}\right)^{m/(2-m)} \tag{7.39}$$

where v_m is the maximum velocity and D is the hydraulic depth. The power law distribution is valid only in the turbulent flow region (i.e., at significant distances from the boundary surface). Clearly, it has the weakness that if one assumes that m is a constant, then all profiles would be alike. For this reason, the law only makes sense from a physical standpoint if m is allowed to vary with the Reynolds number. The exponent is generally considered to be a function of the bed frictional resistance and thus of R_n. Experiments have shown that throughout a range of Reynolds numbers defined by $3000 \leq R_n \leq 70{,}000$, that $m = 1/4$, and so equation (7.39) becomes

$$\frac{v}{v_m} = \left(\frac{y}{D}\right)^{1/7} \tag{7.40}$$

The 1/7th exponent formulation is by far the most common power law formula encountered in hydraulic engineering. Because of its applicability to flow with fairly small Reynolds numbers ($<10^5$), the power law is generally only considered applicable for the case of smooth walls. In view of this fact, Morris and Wiggert (1972) expressed the power law in the smooth pipe formulation:

$$\frac{v}{v^*} = 8.7\left(\frac{yv^*}{\nu}\right)^{1/7} \tag{7.41}$$

In fact, Wooding et al. (1973) showed that the power law with a small exponent is essentially identical to the logarithmic law. Schlichting (1968) stated that the logarithmic law should be considered as an asymptotic distribution applicable to very large Reynolds numbers, while the power law is only valid for a range of small R_n as discussed above.

Example 7.8

Using the data from Example 7.6, determine the velocity profile using the 1/7 power law and compare with the results obtained using the logarithmic smooth wall law.

Solution:

$$R_n = \frac{Vd}{\nu} = \frac{3.06(0.25)}{1.003 \times 10^{-6}} = 7.627 \times 10^5$$

The Reynolds number in this case is greater than the range specified for application of the power law (1×10^{-5}); however, we will continue.

Using equation (7.33b) for the smooth wall and the maximum velocity from the smooth pipe law for v_m in equation (7.40), we obtain the following:

y (m)	Power Law v (m/s)	Smooth Pipe Law v (m/s)
0.01	3.03	3.43
0.05	3.82	4.01
0.10	4.21	4.27
0.125	4.35	4.35

The above results, shown in Figure 7.11, indicate that as the depth of flow increases, the two laws converge to each other. The differences in their velocities are significant only near the bed.

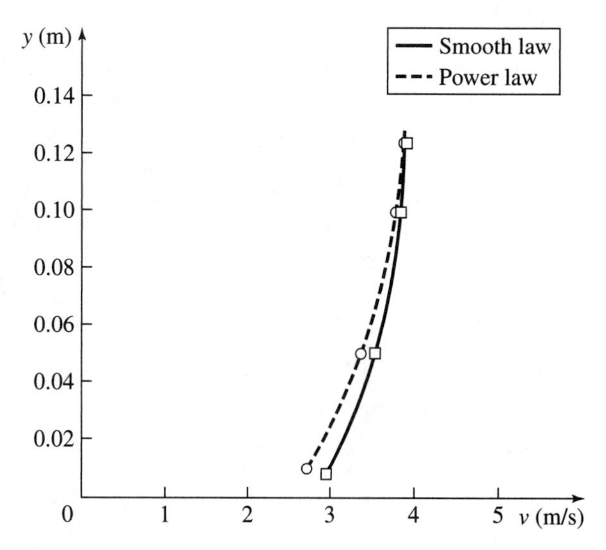

Figure 7.11 Velocity distribution using both the logarithmic smooth and 1/7 power law assumptions.

READING AID

7.1. Discuss the effect of flow resistance on the velocity distribution. In real-life problems, could the velocity be uniform in the entire cross section of the flow?
7.2. What is the dominant resistance force in most of the flow problems?
7.3. Explain how the flow resistance is converted into loss of energy. What is the importance of maintaining energy in the flow?
7.4. Is the study and analysis of flow resistance in hydraulic systems based on theoretical or empirical concepts?
7.5. Discuss the main factors affecting the flow resistance in hydraulic systems. Could these factors change in space and time within the same system?
7.6. What simplification could be made to avoid dealing with spatial and temporal variability of the factors affecting the flow resistance?
7.7. Discuss the mechanism of flow resistance and energy loss under laminar and turbulent flow conditions.
7.8. State the two most common velocity distributions encountered in hydraulic systems.
7.9. Under what condition will the hydraulic grade line and the energy grade line be parallel to each other in a pipe line?
7.10. Using the notations given in Figure 7.2, deduce an expression for the boundary shear stress in pipes in terms of the specific weight of the fluid, the hydraulic radius and the slope of the total energy line.
7.11. Using the notations given in Figure 7.3, deduce an equation to calculate the boundary shear stress in open channel flow.
7.12. Write down the Chezy equation and the Darcy-Weisbach equation and define all the parameters in the two equations. Is the Darcy equation more general than the Chezy equation?
7.13. State the dimensions of the Chezy coefficient, C, and the Darcy factor, f. Write an equation relating C to f and check its dimensional homogeneity.
7.14. Define *shear velocity* and state the relationship between the average flow velocity and the shear velocity.
7.15. What is the main difference between the dynamic viscosity and the eddy viscosity? Can the eddy viscosity be measured by the viscometer?
7.16. Deduce an equation for the head loss in a pipe element in terms of the average velocity under steady, uniform and laminar flow conditions.
7.17. Does the laminar boundary layer exist in all types of flow in pipes and open channels? How is the thickness of this layer affected by the Reynolds number of the flow?
7.18. Write the von Karman equation relating the shear stress to the velocity distribution. Define the parameters of the logarithmic velocity distribution equation under turbulent flow conditions for the case of regular channels and pipes.
7.19. What is meant by relative roughness? State the condition under which the boundary of the flow would function as hydraulically smooth surface.
7.20. Define the wall Reynolds number. Identify the limits of the wall Reynolds number for the fully laminar and fully turbulent flow based on Singh (1996).
7.21. Under what conditions can the effect of the laminar sublayer be ignored, and the turbulent effects assumed to dominate the flow?
7.22. Comment on the applicability of the power law velocity distribution with regard to the flow region and state its main weakness.

Problems

7.1. A pressure drop of 10 psi is available to carry water across a 1000-ft-long horizontal pipe with a diameter of 1 ft and a Darcy f of 0.018. (a) What is the discharge in the pipe under these conditions? (b) What shear stress will be applied to the pipe boundary by this flow?

7.2. A 0.25-m-diameter, 150-m-long horizontal pipe carries a discharge of 0.5 m^3/s. What pressure drop will cause a shear stress of 0.15 N/m^2 on the pipe boundary?

7.3. Repeat Problem 7.2 if the pipe is inclined upward so that there is 1 m difference in the elevation between the pipe ends.

7.4. Water is flowing in a 12-inch pipe inclined downward on a slope of 10% as shown in Figure P7.4. The discharge is 1000 gpm and piezometers located 30 ft apart measure a difference in piezometric head of 0.1 ft. (a) What is the pressure drop across the pipe? (b) What is the shear stress exerted on the pipe boundary? (c) What is the shear velocity? (d) What is the Darcy friction factor?

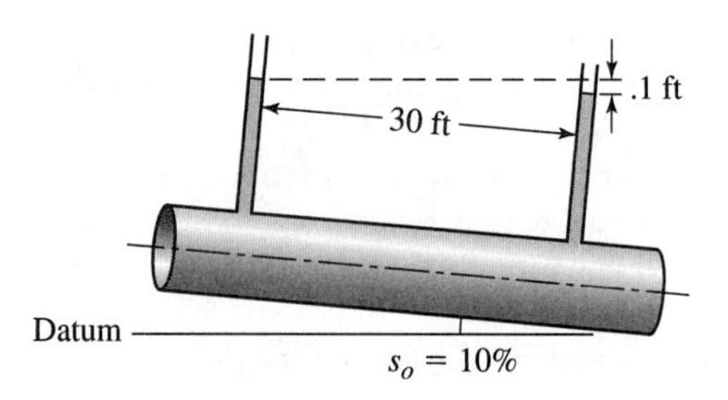

Figure P7.4 Sloping pipe setting for Problem 7.4.

7.5. Assume that the pipe in Problem 7.4 has a very smooth surface and that the water temperature is 65°F, compute (a) velocity at the center of the pipe, and (b) thickness of the boundary layer.

7.6. Water is flowing in a trapezoidal channel with 10 ft bottom width and $2H/1V$ side slopes. The channel is running on a slope of .00778 ft/ft and the flow exerts a shear stress of 1.5 lb/ft^2 on the channel bed. What is the depth of flow in this case?

7.7. If the Darcy friction factor for the channel in Problem 7.6 is estimated to be 0.3, what would be the Chezy C value for this channel? What would be the discharge for the flow conditions given in Problem 7.6?

7.8. A 3-m-wide rectangular channel is carrying water at a normal depth of 2 m. Calculate the shear stress exerted on the channel bed if the bed slope is 0.004 m/m.

7.9. A trapezoidal channel with 2 m bed width, 1.2 m normal depth, 1:1 side slopes, and bed slope of 0.0040 m/m carries a water discharge of 8.5 m^3/s. Estimate the Darcy friction factor and the Chezy coefficient for this case.

7.10. Water is flowing in a 6 in-diameter pipe line inclined (upward) at a slope of 0.01 ft/ft as shown in Figure P7.10. If the observed shear stress on the pipe boundary is 0.975 lb/ft^2, (a) what is the slope of the hydraulic grade line (ft/ft)? (b) If the friction factor, f, is calculated to be 0.088, what is the discharge in the pipe? (c) What is the pressure drop across a 1-foot section of this pipe?

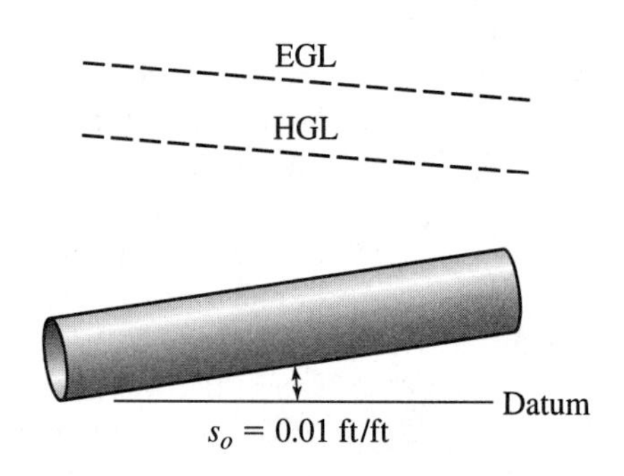

Figure P7.10 Inclined pipe for Problem 7.10.

7.11. Water is flowing in a 3-m-wide rectangular channel at a normal depth of 2 m. The channel is running on a slope of 0.0025 m/m and has a Chezy coefficient of 90. (a) What is the discharge in the channel in m^3/s (b) What shear stress will be imparted to the channel bottom by this flow?

7.12. Water is flowing in a 0.25-m-diameter pipe inclined (upward) at a slope of 15% with a length of 50 m as shown in Figure P7.12. The difference in piezometric head across the pipe is 0.1 m and the temperature of the water is 25°C. (a) Determine the thickness of the

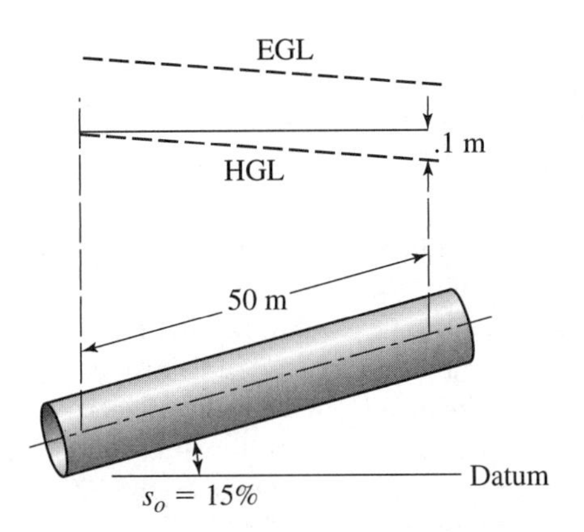

Figure P7.12 Inclined pipe setting for Problem 7.12.

boundary layer by both the Morris and Wiggert and Singh criteria. (b) If the pipe material is new cast iron, would the pipe be considered smooth, rough, or neither in this case? (c) Determine the pressure drop.

7.13. Water at 20°C is flowing in a 0.25-m-diameter and 120-m-long pipe which is inclined downward at a slope of 5%. The piezometric heads at the two ends are 1.6 m and 1.25 m as shown in Figure P7.13. Under the Morris and Wiggert criterion, what is the approximate thickness of the boundary layer in this case? Under the Singh criterion, what is the sublayer thickness? If this pipe is made of galvanized iron, can it be considered a smooth pipe in this case?

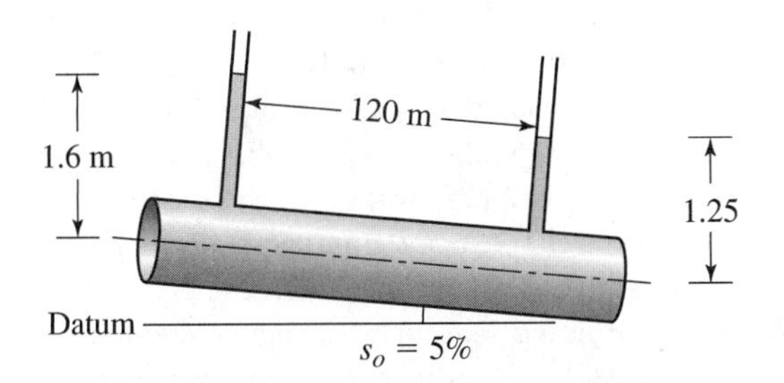

Figure P7.13 Sloping pipe setting for Problem 7.13.

7.14. Water is flowing in a trapezoidal channel 8 ft wide and side slopes of $2H/1V$. The discharge is 200 cfs, the flow depth is 4 ft, and the channel is running on a slope of 0.005 ft/ft. What are the Chezy C and Darcy f values in this case?

7.15. If the discharge in Problem 7.12 is 0.05 m^3/s, what would be the value of the Darcy f in this case? Compute the Chezy coefficient for this case.

7.16. Water flows in a 75-m-long wrought iron pipe with a 0.20 m diameter and carries a discharge of 0.2 m^3/s. The pipe is running horizontally and experiences a pressure drop of 120 N/m^2 across its length. The temperature of the water is 25°C. (a) Compute and plot the velocity profile for this flow under the smooth pipe assumption. (b) Compute and plot the velocity profile using the rough pipe assumption. (c) Which profile (if either) is more appropriate in this case?

7.17. Water flows in a 200-ft-long cast iron pipe 6 inches in diameter inclined on a slope of 10% as shown in Figure P7.17. The flow rate is 300 gpm (U.S.) and the observed piezometric head loss across the pipe is 1 foot. The temperature of the water is 65°F. (a) Using the Morris and Wiggert criterion, calculate the boundary layer thickness. (b) If the pipe material is cast iron, can this be considered a smooth pipe? (c) If the pipe material is wrought iron, can this be considered a smooth pipe case?

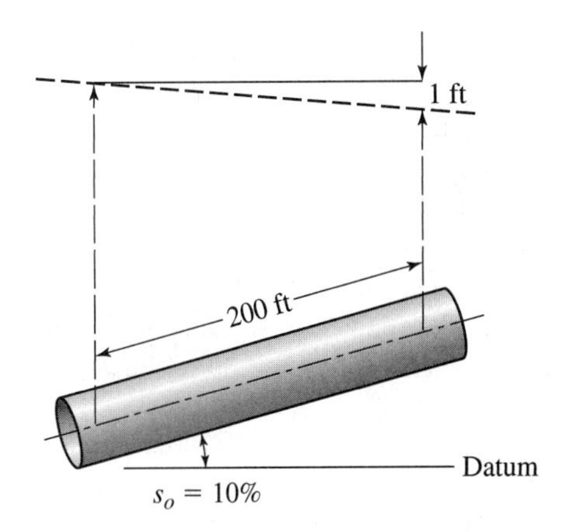

Figure P7.17 Inclined pipe setting for Problem 7.17.

7.18. Compute the velocity profile for the flow conditions of Problem 7.17 under both the smooth and rough pipe assumptions. How do these results compare with the conclusions of Problem 7.17?

7.19. Compute the velocity profile for the conditions of Problem 7.17 using the Morris and Wiggert formulation of the 1/7th power law.

7.20. A 0.30-m-diameter cast iron pipe is carrying 0.20 m^3/s of water at 20°C. If the observed shear stress on the pipe boundary is 30 N/m^2, calculate and plot the velocity distribution using both the smooth and rough pipe assumptions. Which assumption is more accurate for this case?

7.21. Using the data in Problem 7.20, determine the velocity profile using the 1/7 power law and compare the logarithmic rough wall law.

7.22. Water is flowing in a rectangular channel 5 m wide at a normal depth of 3 m. The channel is running on a slope of 0.0005 m/m and carries a discharge of 22.5 m^3/s. The representative roughness element of the bed material is 0.01 m. (a) Determine the Chezy C value for this channel. (b) Compute the value of the shear stress imparted to the channel boundary. (c) Using the velocity equation for fully rough flow, with the representative bed element as ε, compute the velocity profile for this condition. (d) Using the velocity obtained at a distance of 2.5 m from the bed as v_m, compute the velocity profile by the 1/7 power law and compare with the results from part (c).

7.23. Water is flowing at a rate of 15 m^3/s in a triangular channel with a 2.5 m water depth, 1:1 side slope, and 0.003 m/m bed slope. The representative roughness element of the boundary is 0.25 in. If the observed maximum velocity is 3.8 m/s, compute the vertical velocity profile using the rough law and the 1/7 power law.

7.24. An elevated tank is used to supply water to a building located 4800 ft away as shown in Figure P7.24. The water will be carried through a 12-inch cast iron pipe and the specified flow rate is 1000 gpm. The fire code requires that a pressure of 30 psi be available in the line at the building site. What should be the height of the water in the tank above the invert of the building to get the job done? Assume additional length of 200 ft for pipe connections and water temperature of 60°F.

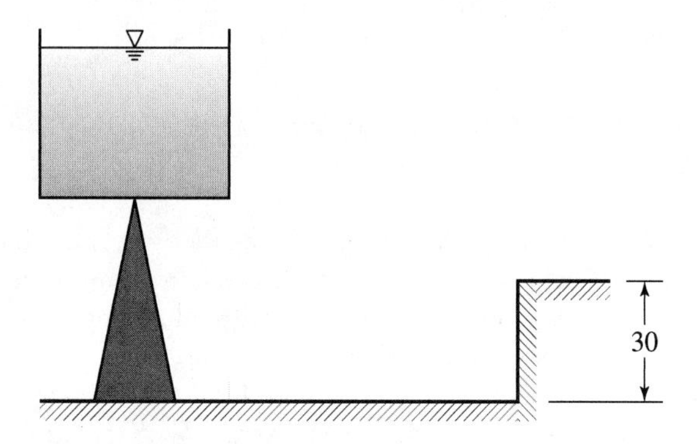

Figure P7.24 Elevated tank for Problem 7.24.

7.25. Water at 20°C is flowing in a pipe of length 10 m and diameter 10 mm. Inlet pressure is 500 kPa and outlet is at 1 atm. Determine the flow rate in liters per second if the pipe is (a) horizontal, (b) inclined at 20° with flow upward, (c) inclined at 20° with downward flow.

7.26. Find the flow rate in l/s between two tanks through a pipe of diameter 3 mm as shown in Figure P7.26. The liquid in the tanks is water at 20°C.

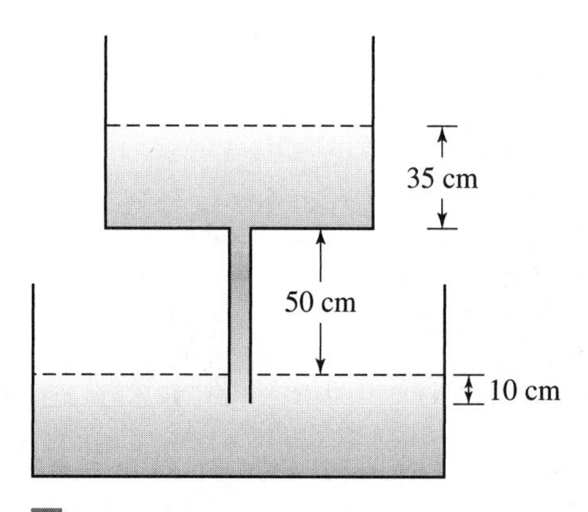

Figure P7.26 Tanks connected by a pipe.

7.27. Referring to Problem 7.26, calculate the pipe diameter to maintain a critical Reynolds number of 2000.

7.28. In an experimental setup for a pipe flow, velocity is measured at two different points. Point 1 is 1 cm away and point 2 is 2 cm away from the pipe wall. If the increase in the velocity from point 1 to point 2 is 16%, estimate the roughness ε for the pipe surface.

7.29. A 2-m-wide and 1-m-high rectangular conduit of concrete ($\varepsilon = 2$ mm) is used to supply water to a town population by gravity. To meet the supply requirement, the flow rate is required to be 8000 gal/min. Determine the inclination of the conduit to meet this requirement.

7.30. Derive a relation between velocity in radial direction $v(r)$ and centerline velocity v_c for a fully developed laminar flow inside a horizontal pipe with a head loss h_f for a pipe length L.

References

Barbé, D. E., 1990. *Probabilistic Analysis of Bridge Scour Using the Principle of Maximum Entropy*, Ph.D. dissertation, Louisiana State University, Baton Rouge, LA.

Henderson, F. M., 1966. *Open Channel Flow.* Macmillan, New York.

Hwang, N. H. C. and Houghtalen, R. J. 1996. *Fundamentals of Hydraulic Engineering Systems,* 3rd edition. Prentice-Hall, Upper Saddle River, N.J.

Morris, H. M. and Wiggert, J. M., 1972. *Applied Hydraulics in Engineering,* 2nd edition. Wiley, New York.

Roberson, J. A., Cassidy, J. J., and Chaudhry, M. H., 1988. *Hydraulic Engineering.* Houghton Mifflin, Boston, Mass.

Schlichting, H., 1968. *Boundary Layer Theory,* 6th edition, McGraw-Hill, New York.

Singh, V. P., 1996. *Kinematic Wave Modeling in Water Resources, Surface-Water Hydrology.* Wiley, New York.

Wooding, R. A., Bradley, E. F., and Marshall, J. K., 1973. Drag due to regular arrays of roughness elements of varying geometry, *Boundary Layer Meteorology,* 5(3): 285–308.

CHAPTER 8

Closed Conduit Flow

A large slurry pipe. (Copyright Terry Vine/CORBIS.)

In dealing with closed conduit flow, the main consideration is the ability to deliver the required flow rate to the appointed location under the prescribed circumstances. The conservation principles of mass, momentum and energy apply. In determination of discharges and pipe diameters the continuity (mass) and energy principles are usually applied. Estimation of forces resulting from flow in closed pipes involves application of the momentum principle and has been dealt with in previous chapters, so it will not be discussed here. In this chapter our focus will be devoted to the applications of the energy, velocity, and resistance principles developed in Chapters 5 and 7 to the case of flow in closed conduits. Some practical applications of the theory developed in the previous chapter for the case of closed conduit flow are demonstrated. We will show how the theory and experimental results obtained by the German school of investigators in the early twentieth century can be used to solve most practical pipe flow problems. However, our results are generally limited to the case of steady, uniform flow in pipes. The uniform flow restriction is normally accommodated by breaking the pipe system into segments of uniform diameter pipe and handling the transitions as minor losses as discussed in Section 8.4. The case of unsteady flow in pipes, which can occur either through the initial conditions, or through the boundary conditions (i.e., rapid closing of a valve or gate), requires the application of dynamic principles that are generally beyond the scope of this text. However, one case of unsteady flow, the water hammer phenomenon is treated in Section 8.6.

8.1 General Energy Considerations

In dealing with the flow of water in pipes, the dominant consideration is usually associated with energy development and loss in the flow system. Some of the principles developed in earlier chapters can be reiterated with the help of Figure 8.1. The energy developed in the flow at any

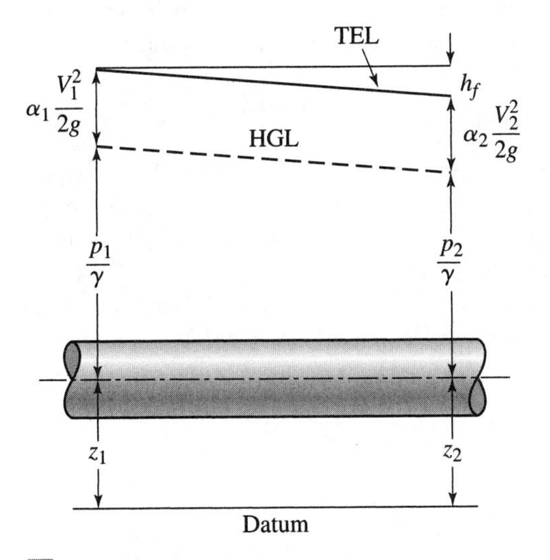

Figure 8.1 Steady, uniform flow in closed conduits.

point is the sum of the position, pressure and velocity heads as shown in the figure, that is,

$$E_1 = z_1 + \frac{p_1}{\gamma} + \frac{\alpha V_1^2}{2g} \tag{8.1}$$

where z is the elevation of the pipe center line above a specified horizontal datum, p is the pressure, V is the average water velocity, γ is the unit weight of water, α is the velocity coefficient (see Chapter 5), and subscript 1 refers to point 1. As discussed in Chapter 7, for regular and symmetrical cross sections such as pipes, the velocity coefficient can be set to unity. Then, writing the energy equation at points 1 and 2 along the pipe, the well- known Bernoulli equation results as discussed in Chapter 5.

$$z_1 + \frac{p_1}{\gamma} + \frac{V_1^2}{2g} = z_2 + \frac{p_2}{\gamma} + \frac{V_2^2}{2g} + h_f \tag{8.2}$$

where h_f is the energy or head loss due to resistance as discussed in Chapter 7. As discussed in the previous chapter, the sum $z + (p/\gamma)$ is known as the Piezometric (or hydraulic) head and, as shown in Figure 8.1, the gradient of this quantity is termed the hydraulic grade line, or the Piezometric grade line. The gradient of the total energy given in equation (8.1) is the energy grade line, or, because it represents the gradient of energy losses due primarily to boundary friction, it is sometimes termed the friction slope. Figure 8.1 also demonstrates the fact, that in pipes of constant diameter, the velocity remains constant for steady flow, and thus the hydraulic and energy gradients must be equal. In Chapter 7, this condition was defined as steady, uniform flow. Because of this equality, the head loss can be computed from the drop in hydraulic grade line ($\Delta(z + (p/\gamma))$) and there is a direct relationship between the pressure drop across a length of pipe and the resulting head loss.

Typically, conditions at point 1 are known as well as the pipe length, the discharge, and z_2. The requirement is to select the appropriate pipe material and diameter to deliver the discharge against the design head with the greatest efficiency and least cost. Occasionally, the conditions at 1 and the pipe diameter might be known and the discharge is to be computed; or alternatively, conditions at point 2 might be specified and a device must be designed to supply the required energy at point 1. Either way, most practical problems reduce to application of the Bernoulli equation and estimation of the head loss.

Example 8.1

It is desired to use an open air elevated tank to facilitate delivery of water to a building located some distance away. The required flow rate is 0.15 m^3/s and it is desired to have an available pressure of 20 kN/m^2 in the line at the building location. The ground surface at the building is 20 m higher than the ground surface at the tank as shown in Figure 8.2. The connecting pipe is 250 m long and 25 cm in diameter. The head loss through the pipe is 7.5 m. What must be the elevation of the water surface in the stand tank?

Solution: $d = 0.25$ m, $A = \dfrac{\pi d^2}{4}$, $V = Q/A = 0.15/0.049 = 3.06$ m/s. So

$$\frac{p_1}{\gamma} + \frac{V_1^2}{2g} + z_1 = \frac{p_2}{\gamma} + \frac{V_2^2}{2g} + z_2 + h_f$$

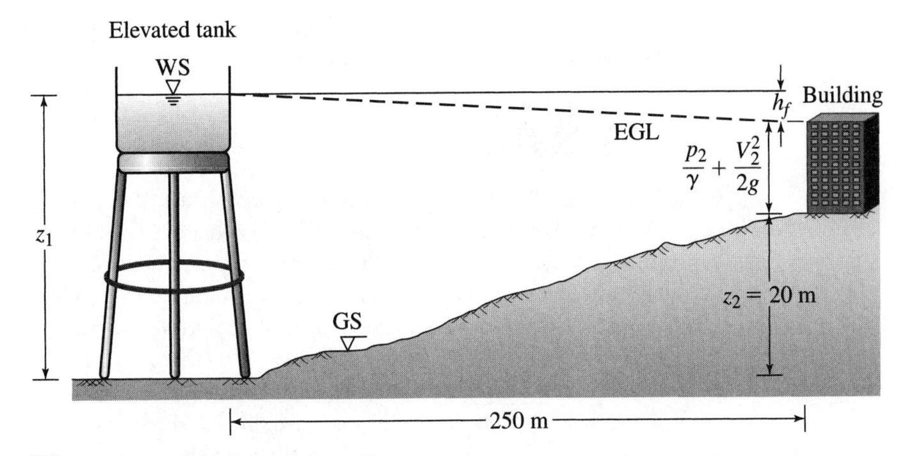

Figure 8.2 Layout and hydraulic heads for Example 8.1.

Letting the datum be at the invert of the tank, and since for the stand tank p_1 and V_1 are 0, we can simply write

$$z_1 = \frac{p_2}{\gamma} + \frac{V_2^2}{2g} + z_2 + h_f$$

$$z_1 = \frac{20000 \text{ N/m}^2}{9789 \text{ N/m}^3} + \frac{(3.06 \text{ m/s})^2}{2 \times 9.81 \text{ m/s}^2} + 20 \text{ m} + 7.5 \text{ m}$$

$$z_1 = 30.0 \text{ m}$$

8.2 Resistance Applications and Friction Losses in Pipes

The reader may recall from the formulation of the previous chapter that a general resistance equation for steady uniform flow was developed from computing the shear stress of a system in dynamic equilibrium:

$$\tau_o = \gamma R S_f \tag{8.3}$$

where τ_o is the boundary shear stress (N/m^2), γ is the specific weight of water (N/m^3), R is the hydraulic radius of the flow conduit and is equal to the ratio of the area to the wetted perimeter (A/P), and S_f is the friction slope which is equal to the change in energy divided by the flow length. It was also shown through dimensional analysis that the resistance equation could be written in terms of the average velocity:

$$V = C\sqrt{RS_f} \tag{8.4}$$

where C is a resistance coefficient. Equation (8.4), known as the Chezy equation, is applicable to all cases of steady uniform flow and can be interpreted merely as an alternate form of the boundary shear formula. However, in most cases of closed conduit flow, it is customary to compute energy losses due to resistance by use of the Darcy-Weisbach equation, which can be regarded as a special case of the Chezy formula:

$$h_f = \frac{fL}{d}\frac{V^2}{2g} \tag{8.5}$$

It was also shown in Section 7.3 that if we assume $C = (8g/f)^{0.5}$, then the Chezy and Darcy-Weisbach equations are the same. The fact that the Darcy form of the equation is preferred is due, in large part, to the influence of the experiments conducted by Prandtl, von Karman, and Nikuradse in the early part of the twentieth century. The results of these experiments on pipes led to the development of several analytical expressions for the Darcy factor f based upon the velocity distributions discussed in Chapter 7. Henderson (1966) showed that the results of these experiments can just as easily be converted to show the behavior of the Chezy C; however, that formulation for resistance in closed conduits has not caught on with most engineers. In addition, as mentioned in the previous chapter, the fact that the Chezy coefficient has units while the Darcy f does not is another factor in favor of Equation (8.5).

Example 8.2

For the situation in Example 8.1, what was the Darcy friction factor and the Chezy coefficient for the pipe connecting the stand tank and the building?

Solution: The given head loss (h_f) was 7.5 m. Then

$$h_f = \frac{fL}{d}\frac{V^2}{2g} \quad \text{or} \quad 7.5 = \frac{f(250)(3.06)^2}{0.25(2)(9.81)}$$

$$7.5 = 477.2477f$$

This yields $f = 0.0157$ and $C = 70.7\ \mathrm{m^{0.5}/s}$.

Example 8.3

Water is fed from a pressurized tank through a 1-ft-diameter pipe 1000 ft long to a point located at an elevation of 50 ft above the point where the pipe exits the tank. The pressure in the tank is 30 psi and the pipe has a Darcy friction factor value of 0.035. If the required pressure at the end of the pipe is 5 psi, what will be the flow rate in the line?

Solution: The situation is sketched in Figure 8.3. The energy head at the tank is constituted by the pressure head and equals $30(144)/62.4 = 69.23$ ft. Equating this energy head to that at the pipe end gives the head loss. At the end of the pipe, the elevation head is 50 ft and the pressure head is 11.54 ft. Now, ignoring the velocity head for now, the head loss = 7.69 ft. Applying the Darcy-Weisbach relation, the flow velocity is computed as

$$7.69 = \frac{0.035(1000)V^2}{1(2 \times 32.2)}$$

This gives $V = 3.76$ ft/s. Then $(V^2/2g) = 0.22$ ft. Subtracting this quantity from our initial estimate of head loss leaves $h_f = 7.47$ ft. Recomputing V as above leads to a velocity estimate of 3.71 ft/s. Then the new velocity head is 0.21 ft, close enough to the initial estimate of 0.22. Thus, the discharge $Q = VA = 3.71\left(\frac{\pi(1)^2}{4}\right) = 2.91$ cfs.

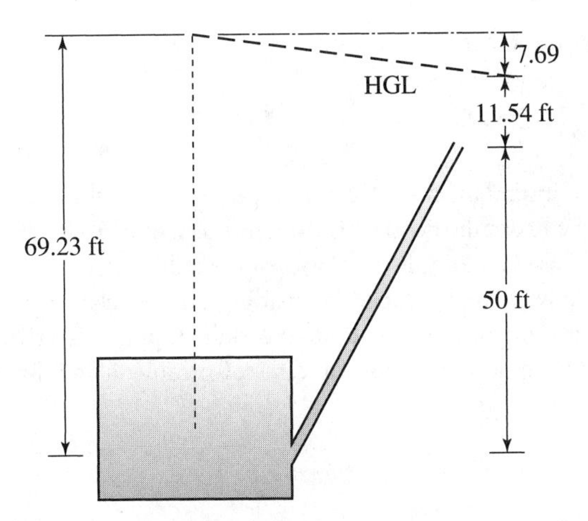

Figure 8.3 Pressurized tank connected to a pipe (Example 8.3).

8.2.1 Laminar Flow in Pipes

In the case of laminar flow a simple relationship was developed in Chapter 7 between energy loss and the average flow velocity based upon Newton's law of viscosity. Rewriting equation (7.28) for a pipe length L,

$$h_f = \frac{32\mu VL}{\gamma d^2} \tag{8.6}$$

Equation (8.6), which is known as the Poiseuille equation, can also be rewritten as

$$h_f = \frac{64V^2L}{R_n 2gd} \tag{8.7}$$

where $R_n = [\rho Vd]/\mu$ is the Reynolds number of the flow.

Comparing equation (8.7) with the Darcy equation (8.5), we see that in terms of the Darcy friction factor, $f = 64/R_n$ for the case of laminar flow, thus indicating that it is independent of the roughness of the pipe surface. This phenomenon should not be surprising in view of our discussion of the laminar sublayer development in the last chapter. In this case, the laminar sublayer merely extends across the entire pipe diameter. Note also that since the Darcy equation can be derived from dimensional considerations, the calculation of shear stresses and energy losses in the case of laminar flow involves no empirical relationships at all and is completely general relative to the flow medium.

8.2.2 Turbulent Flow in Pipes

As discussed in Chapter 7, most practical problems involve turbulent flow, at least in the civil and environmental engineering fields. In dealing with turbulent flow in pipes, the work of the Prandtl school of experimentalists is relied upon to a large extent. Recall that a general expression for turbulent velocity distributions is given as

$$\frac{v}{v^*} = \frac{1}{k}\ln\frac{y}{D} + C \tag{8.8}$$

where v is the instantaneous velocity at location y from the boundary, v^* is the shear velocity (m/s), D is the hydraulic depth, k is the von Karman constant (0.4 for clear water), and C is a constant. In closed conduit flow, D is the pipe radius. Recall also that two distinct conditions of turbulent flow were developed, depending on the relative thicknesses of the laminar sublayer and the roughness elements of the wall. Using this criterion, dimensionless velocity distributions were developed for the cases of turbulent flow in smooth-walled conduits and rough-walled conduits:

Smooth:

$$\frac{v}{v^*} = 5.75\log\frac{yv^*}{\nu} + 5.5 \tag{8.9}$$

Rough:

$$\frac{v}{v^*} = \frac{1}{k}\ln\frac{y}{\varepsilon} + 8.48 \tag{8.10}$$

where the velocity distribution in the smooth walled case is a function of the laminar sublayer properties (wall Reynolds number) and in the rough wall case it is a function of the wall roughness element ε. Following up on this idea, these distributions can be used to determine the behavior of the Darcy friction factor under the prescribed conditions of turbulence.

8.2.2.1 Turbulent Flow in Smooth Pipes Using the velocity distribution given in equations (8.9) and (8.10) with the shear stress/velocity relationship embodied in equation (7.23), it is possible to simply solve for the friction factor as follows:

$$\frac{V}{v^*} = \sqrt{8/f} \tag{8.11a}$$

$$f = 8\left(\frac{v^*}{V}\right)^2 = 8\left[\frac{Av^*}{\int v\,dA}\right]^2 \tag{8.11b}$$

Inserting the velocity distribution from equation (8.9) and integrating across the pipe radius, von Karman and Prandtl (Prandtl, 1926; von Karman, 1934) derived an expression for the friction factor f for the case of turbulent flow in smooth pipes:

$$\frac{1}{\sqrt{f}} = 2\log(R_n\sqrt{f}) - 0.80 = 2\log\left(R_n\frac{\sqrt{f}}{0.627}\right) \tag{8.12}$$

It is not surprising to note from equation (8.12) that energy loss for turbulent flow in smooth pipes is a function of the flow Reynolds number only and not of the pipe roughness since in this case the laminar sublayer dominates. An analysis of the equation reveals that the friction factor decreases as the Reynolds number increases. The use of the average cross sectional velocity (as embodied in the calculation of R_n) is made possible by the integration of the velocity distribution across the section. This was one of the desired effects of the operation since it relieves the user of having to directly deal with the velocity profile.

Example 8.4

Water is flowing in a pipe with a diameter of 6 in. at a discharge of 1000 gpm. The temperature of the water is 70°F and the pipe can be considered hydraulically smooth. If the pipe length is 75 ft, estimate the Darcy friction factor and the resultant head loss in this situation.

Solution: $d = 0.5$ ft; $A = \pi d^2/4 = 0.196$ ft^2, $Q = 1000$ gpm $= 2.23$ cfs, $V = Q/A = 2.23/0.196 = 11.38$ ft/s. For a temperature of 70°F, $v = 1.059 \times 10^{-5}$ ft^2/s.

$$R_n = \frac{Vd}{v} = \frac{11.38(0.5)}{1.059 \times 10^{-5}} = 5.37 \times 10^5$$

Therefore

$$\frac{1}{\sqrt{f}} = 2\log(R_n\sqrt{f}) - 0.8$$

By trial and error, $f = 0.013$. Then

$$h_f = \frac{fL}{d}\frac{V^2}{2g} = \frac{0.013(75)(11.38)^2}{0.5(2)(32.2)} = 3.92 \text{ ft}$$

8.2.2.2 Turbulent Flow in Rough Pipes Using equation (8.11b) with the velocity distribution for rough walls [equation (8.10)] and integrating across the pipe radius leads to the well-known Nikuradse rough pipe equation for the friction factor:

$$\frac{1}{\sqrt{f}} - 2\log\frac{r}{\varepsilon} = 1.74 \tag{8.13a}$$

In the integration the relative distance term y has been replaced with its limiting value (r), and the natural logarithm has been converted to the common base. Equation (8.13a) is often expressed in terms of the pipe diameter:

$$\frac{1}{\sqrt{f}} - 2\log\frac{d}{\varepsilon} = 1.14 \tag{8.13b}$$

Equations (8.13a and 8.13b) demonstrate that in the case of turbulent flow in rough pipes the friction factor is a function of the pipe roughness only (expressed as a relative roughness term) and not of the intensity or degree of turbulence (Reynolds number). This follows from the fact that in this case the pipe roughness elements dominate and the laminar sublayer effects are negligible. It is important to reiterate that, as in the smooth pipe case, since the sublayer thickness is itself a function of the Reynolds number, whether or not a pipe behaves as a smooth or rough pipe (or neither) depends on the flow conditions of the particular situation.

Example 8.5

An experiment is conducted such that water is discharged at the rate of 0.1 m^3/s through a pipe of diameter 0.15 m. The temperature of the water is 20°C. The slope of the hydraulic grade line (piezometric slope) is measured to be 0.217 m/m. Determine the Darcy friction factor, and estimate whether this pipe is better represented by the smooth or rough pipe formulation.

Solution: $d = 0.15$ m; $A = \pi d^2/4 = 0.0177$ m^2, $V = Q/A = 0.1/0.0177 = 5.65$ m/s. For $T = 20$°C, $v = 1.003 \times 10^{-6}$. Therefore,

$$R_n = \frac{Vd}{v} = \frac{5.65(0.15)}{1.003 \times 10^{-6}} = 844965 \quad \tau_o = \gamma R S_f = 9789(0.15/4)(0.217) = 79.66 \text{ N/m}^2$$

and $$v^* = \sqrt{\frac{\tau_o}{\rho}} = \sqrt{\frac{79.65}{998.2}} = 0.282 \text{ m/s} \qquad \frac{V}{v^*} = \sqrt{\frac{8}{f}} \Rightarrow \frac{5.65}{0.282} = \sqrt{\frac{8}{f}} \Rightarrow f = 0.02$$

Using the conservative Morris and Wiggert criterion for the thickness of the boundary layer as discussed in Chapter 7,

$$\delta = \frac{11.6\nu}{v^*} = \frac{11.6(1.003 \times 10)^{-6}}{0.282} = 0.0000413 \text{ m}$$

From the rough pipe formulation:

$$\frac{1}{\sqrt{f}} - 2 \log\left(\frac{d}{\varepsilon}\right) = 1.14$$

Therefore

$$\frac{1}{\sqrt{0.02}} - 2 \log\left(\frac{0.15}{\varepsilon}\right) = 1.14$$

from which we obtain $\varepsilon = 0.000162$ m. Then, $\delta/\varepsilon = 0.25$. According to the Hwang and Houghtalen criterion discussed in Chapter 7, if the ratio $\delta/\varepsilon < 0.08$ the pipe can be considered rough. However, if we use the less conservative Singh criterion for the sublayer thickness, that is,

$$\delta = \frac{4\nu}{v^*} = \frac{4(1.003 \times 10)^{-6}}{0.282} = 0.0000142 \text{ m}$$

Then $\delta/\varepsilon = 0.0877$, which is much closer to the rough criterion and the pipe might fairly be considered rough.

8.2.2.3 Transition Region The previous example shows that it is quite possible that the flow regime which prevails in a particular case would follow neither the smooth nor rough pipe formulations. In fact, one might imagine that this would be the case in the majority of practical situations. In addition, observations have revealed that the head loss in commercially available pipe fails to follow the relationships developed by von Karman and Nikuradse precisely. This is caused by the definition of the roughness element in terms of equivalent sand grain diameters used in their experiments which does not correspond to the roughness evident in real pipes. In these cases, the British researchers Colebrook and White (1937) proposed a semiempirical function which would be asymptotic to both the rough and smooth pipe equations.

Subtracting 2 log r/ε from the smooth pipe equation (8.12) yields an expression comparable to equation (8.13a):

$$\frac{1}{\sqrt{f}} - 2 \log \frac{r}{\varepsilon} = 2 \log\left(\frac{R_n\sqrt{f}}{r/\varepsilon}\right) - 0.80 \tag{8.14}$$

Combining the smooth pipe relationship of equation (8.14) with the expression for rough pipes given in equation (8.13a) leads to the desired expression:

$$\frac{1}{\sqrt{f}} - 2 \log \frac{r}{\varepsilon} = 1.74 - 2 \log\left(1 + 18.7 \frac{r/\varepsilon}{R_n\sqrt{f}}\right) \tag{8.15}$$

The final term in this equation represents the empirically derived addition to the von Karman-Nikuradse expressions to account for the deviation of observations from the original equations (Singh, 1996). An analysis reveals that the transition function approaches the smooth pipe equation for low values of R_n and the rough pipe formula for high values of Reynolds number.

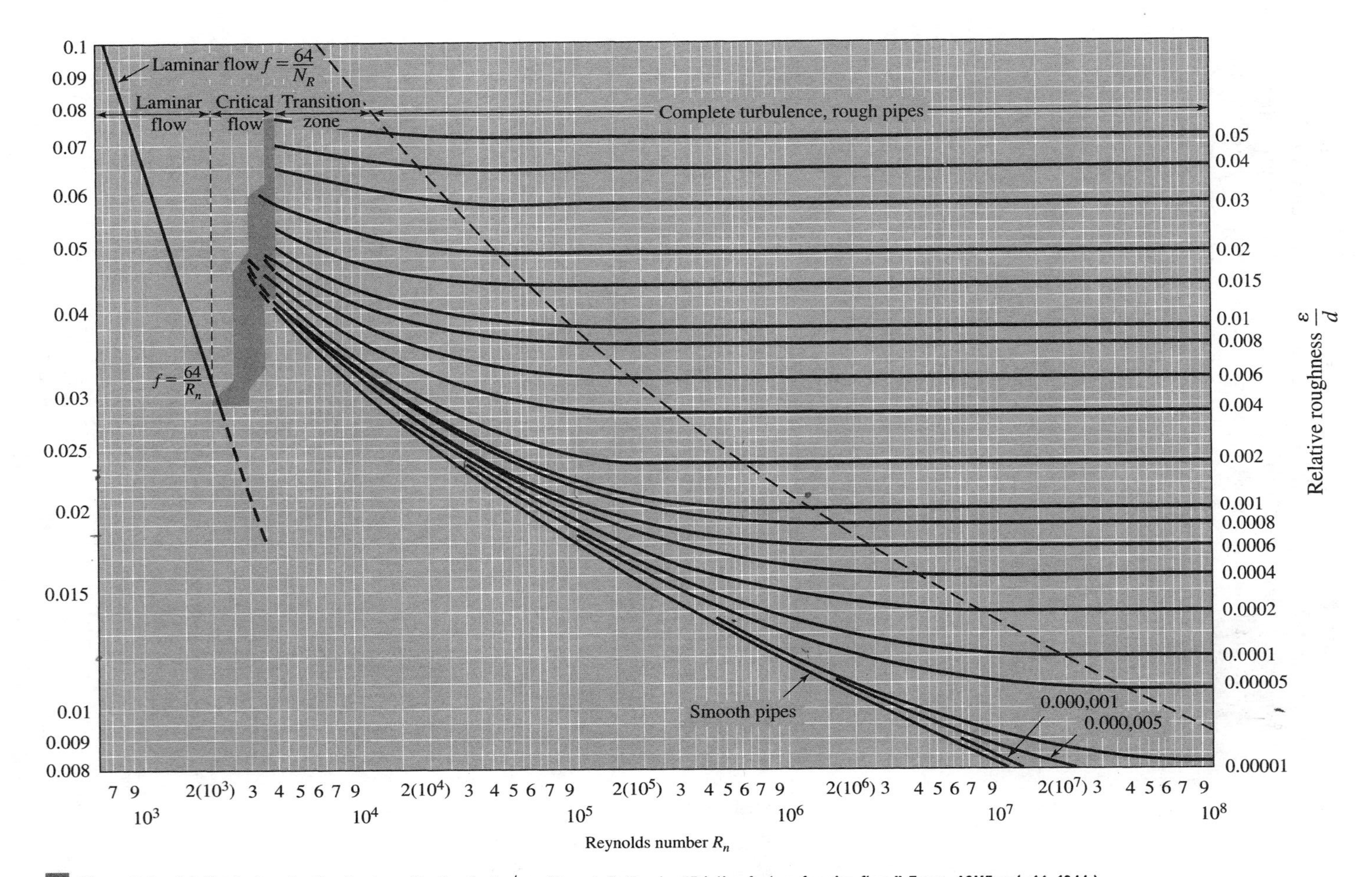

Figure 8.4 Friction factors for flow in pipes, the Moody diagram (From L. F. Moody, "Friction factors for pipe flow," *Trans. ASME*, vol. 66, 1944.)

Many of the analytical expressions that have been developed in the preceding sections for determining the friction factor are nonlinear. For this reason it would be very awkward and cumbersome to have to solve them in every case for which they are applied. This is particularly true, since many applied problems involve a trial and error solution to obtain the optimum pipe diameter and material to deliver a specified flow rate against a given head. To facilitate the determination of the friction factor to be used in practical situations, Lewis Moody (see Henderson, 1966) developed graphical plots of the behavior of f as given in the preceding expressions. The so-called Moody diagram, shown in Figure 8.4, gives f as a function of both R_n and a relative roughness measure (ε/d), thus resulting in a family of curves as shown. Note that the diagram divides into three broad regions: laminar flow which prevails up to a R_n of about 2000 and is bounded by the straight line given by $f = 64/R_n$; the smooth pipe and transition zone in which f is a function of both R_n and ε/d and which prevails up to a R_n of roughly 10^6; and the fully turbulent rough zone where the linear Nikuradse rough pipe equation applies and f is a function of ε/d only.

If we reevaluate the results of Example 8.4 in light of the Moody diagram we can see that the f value of 0.02 in conjunction with a Reynolds number of about 8.5×10^5 would place the flow just on the boundary between the transition region and the rough pipe zone, thus confirming our earlier conclusion. The Moody diagram is very useful for solving many practical design problems in closed conduit flow.

Example 8.6

It is desired to deliver a discharge of 0.1 m^3/s through 500 m of wrought iron pipe. If the allowable head loss is 1.5 m, and the water temperature is 25°C, what is the minimum size of the pipe to accomplish the task?

Solution:

$$h_f = \frac{fLV^2}{d2g} = \frac{fL[Q/(\pi d^2/4)]^2}{d2g} = \frac{8fLQ^2}{g\pi^2 d^5} = 1.5 \text{ m}$$

The solution proceeds in an iterative fashion as follows: Assume $f = 0.015$ for instance. Then, with $Q = 0.1$ m^3/s and $L = 500$ m, solve for d: $1.5 = \dfrac{8(0.015)(500)(0.1)^2}{g\pi^2 d^5}$.

This yields $d = 0.33$ m.

$A = \pi(0.33)^2/4 = 0.085$ m^2; $V = Q/A = 1.18$ m/s. For $T = 25$°C, $\nu = 0.893 \times 10^{-6}$ m^2/s. Therefore, $R_n = \dfrac{Vd}{\nu} = \dfrac{1.18(0.33)}{0.893 \times 10^{-6}} = 4.36 \times 10^5$.

For wrought iron, from Table 7.1, we find $\varepsilon = 0.26$ mm $= 0.00026$ m, so $\varepsilon/d = 0.00079$. Then, from the Moody diagram, with $R_n = 4.36 \times 10^5$ and $\varepsilon/d = 0.00079$, we read $f = 0.0195$. Then, using this f, we recalculate a new $d = 0.35$ m, which leads to the following iteration: $A = 0.096$ m^2, $V = 1.04$ m/s, $R_n = 4.07 \times 10^5$, $\varepsilon/d = 0.000743$. Then, from the Moody diagram we read a new $f = 0.019$, which results in $d = 0.35$ m, which approximately equals the previous value.

Example 8.7

What is the minimum diameter of a galvanized iron pipe to carry a discharge of 0.15 m^3/s a distance of 750 m if the allowable head loss is 1 m and the water temperature is 20°C?

Solution: For galvanized iron pipe from Table 7.1, $\varepsilon = 0.00015$ m. Here, $\nu = 1 \times 10^{-6}$ m^2/s, and the head loss = 1 m. Applying the Darcy-Weisbach relation, we compute the pipe diameter as $1 = \frac{8fLQ^2}{g\pi^2 d^5} = \frac{8f(750)(0.15)^2}{9.81\pi^2 d^5} = \frac{1.394f}{d^5}$. This gives $d^5 = 1.394f$. Now, assume that $f = 0.02$. Then, $d = 0.49$ m, $A = 0.1885$ m^2, $V = 0.795$ m/s, $R_n = 3.89 \times 10^5$, $\varepsilon/d = 0.000306$ and from Figure 8.4, $f = 0.0165$, which is less than the assumed value. Assuming $d = 0.47$ m, $A = 0.173$ m^2, $V = 0.867$ m/s, and $R_n = 4 \times 10^5$, then $f = 0.0165$, which is okay. Thus, the pipe diameter is 0.47 m. It may be noted that the final answer depends somewhat on number of significant digits carried during the computations.

8.3 Empirical Resistance Equations

8.3.1 Blasius Equation

One of the earliest, simplest and most accurate of empirical expressions for f was that developed by another associate of Prandtl, Paul Blasius (1913), for the case of smooth-walled pipes:

$$f = \frac{0.316}{R_n^{0.25}} \tag{8.16}$$

The Blasius equation corresponds to the one-seventh power law for velocity distribution as discussed previously. The equation is considered to be very accurate for smooth walled pipes for Reynolds numbers up to 100,000. In fact, Singh (1996) cites evidence to show that the Blasius expression is more accurate for the region $R_n < 100{,}000$ than is the Colebrook-White expression embedded in the Moody diagram.

Singh (1996) has shown how the Blasius equation can be used in a more general sense and related directly to the power law velocity distributions. Generalizing equation (8.16),

$$f = aR_n^b \tag{8.17}$$

where a and b are coefficients to be determined empirically. It was shown that the exponent b could be determined from

$$b = -\frac{2}{n + 1} \tag{8.18}$$

where n is the exponent in the power velocity law. If $R_n = 10^5$ and $n = 7$, then $b = -0.25$ as in the Blasius equation. Experimental data on smooth pipes cited by Singh (1996) found the coefficient a to vary from 0.192 to 0.687. With b fixed at a value of -0.25, a varied from only 0.281 to 0.345; thus an average value of 0.316 provided an excellent fit to the experimental data.

8.3.2 Manning Equation

Like the Darcy equation, the Manning formula is a special case of the Chezy. For many years after its introduction by Chezy in 1769, research was focused on the empirical evaluation of the resistance factor C. In recent history, the Chezy equation has most often been employed in free surface situations. In that context, several rather cumbersome empirical formulae have been proposed for the determination of C. In 1890 Robert Manning proposed a very simple relation which had been in use in some parts of the world for several years at that time. The Manning relation, which was originally proposed for use in open channels, is an empirical formula based upon observations from many rivers and canals from various parts of world. The relation is (in SI units)

$$C = \frac{1}{n} R^{1/6} \tag{8.19}$$

TABLE 8.1 Manning n Values for Various Pipe Materials

Pipe Material	Manning n
Smooth glass, brass, copper	0.01
Vitrified clay	0.01–0.017
Cast iron	0.015–0.02
Wrought iron	0.01–0.017
Riveted steel	0.015–0.02
Reinforced concrete	0.012–0.014
Corrugated metal	0.022–0.026

where n is a coefficient which is a function primarily of the boundary roughness and to a certain extent the hydraulic radius. Substitution of this relation into the Chezy formula leads to what is known in the United States as the Manning equation:

$$V = \frac{1}{n} R^{2/3} \sqrt{S_f} \tag{8.20a}$$

or in BG units [since $(3.28)^{1/3} = 1.485$]

$$V = \frac{1.485}{n} R^{2/3} \sqrt{S_f} \tag{8.20b}$$

Since $Q = VA$, from the Manning equation (SI units)

$$Q = \frac{1}{n} A R^{2/3} \sqrt{S_f} \tag{8.21a}$$

or (BG units)

$$Q = \frac{1.485}{n} A R^{2/3} \sqrt{S_f} \tag{8.21b}$$

As stated above, the so-called Manning n is usually considered to be a constant for a given situation, although like the Chezy C before it, it must possess units, and has been shown to be a slight function of R as well. That is the reason for the conversion of equation (8.20a) to (8.20b). Henderson (1966) showed that the Manning relation for C does, in fact, fit the data from the Nikuradse rough pipe experiments to a remarkable degree. In light of this fact, it is easy to understand why the roughness coefficient n would not be a function of turbulence characteristics or the Reynolds number but would vary slightly with the flow depth (through the hydraulic radius). In view of this, therefore one can say that the Manning equation would be strictly applicable to rough pipes only, although it has frequently been employed as a general resistance formula for pipes. However, it is employed far more frequently in open channel situations. For a more complete discussion of the Manning Equation, see Section 11.1. Typical values of the Manning n are given in Table 8.1 for pipes which are commonly encountered in practice. Further discussion of the Manning equation and a table of values for channels will be provided in Chapter 11.

8.3.3 Hazen-Williams Equation

Williams and Hazen presented an empirically derived equation in 1905 which is widely used in water supply and irrigation work (Williams and Hazen, 1933). The equation, which is only valid for water flow under turbulent conditions, is written as

$$V = C_{HW} R^{0.63} (S_f)^{0.54} \tag{8.22}$$

where R is the hydraulic radius of the pipe (i.e., $d/4$ for circular sections), S_f is the slope of the energy grade line or friction slope, and C_{HW} is the resistance coefficient related to the pipe material and roughness condition. In time, it was seen as advantageous to make the Hazen-Williams coefficient conform to the older, more established Chezy coefficient previously discussed (Fair et al., 1971). Recall that the Chezy equation is given by

$$V = C\sqrt{RS_f} \tag{8.23}$$

where C is the Chezy constant. Then in order to make equation (8.22) conform with equation (8.23), it must be set such that the Hazen-Williams V equals the Chezy V. Then

$$C_{HW}\, R^{0.63} S_f^{0.54} = CR^{0.5} S_f^{0.50} \tag{8.24}$$

Since the objective is to set $C_{HW} = C$, there must exist some factor b for which the two coefficients will be equal, that is,

$$R^{0.5} S_f^{0.5} = bR^{0.63} S_f^{0.54}$$

This yields

$$b = \frac{R^{0.5} S_f^{0.5}}{R^{0.63} S_f^{0.54}} = R^{-0.13} S_f^{-0.04} \tag{8.25}$$

Then, using values for R and S_f encountered in typical water supply situations of $R = 1$ ft (0.3 m) and $S_f = 1/1000$ results in a value of $b = 1.318$. Thus, the Hazen-Williams equation for V (in BG units) is

$$V = 1.318\, CR^{0.63} S_f^{0.54} \tag{8.26a}$$

or in SI units

$$V = 0.849\, CR^{0.63} S_f^{0.54} \tag{8.26b}$$

In equations (8.26a) and (8.26b) the parameter C is the Chezy coefficient, although in this context it is usually referred to as the Hazen-Williams coefficient (C_{HW}). Typical values of this coefficient are given in Table 8.2. Although the Hazen-Williams equation is considered to be independent of pipe diameter, it is generally considered to be valid for larger pipe diameters ($R \geq 1$).

TABLE 8.2 Values of the Hazen-Williams Coefficient for Various Pipe Materials

Pipe Material	C_{HW}
Cast iron (new)	130
Cast iron (lined)	100–130
Cast iron (10 yr old)	107–113
Cast iron (20–30 yr old)	75–100
Riveted steel (coated)	90–110
Riveted steel (welded)	100–140
Concrete	130–140
Wood stave	130
Plastic	130–140
Vitrified clay	110–140

As a final note, it should be stated that the Hazen-Williams equation, as well as the Manning relation when applied to pipes, is usually employed in the context of the design and operation of sewers and culverts. In these situations pipes of other than circular geometry are frequently employed. Arch pipe as well as elliptical pipe are common noncircular geometries encountered. The use and hydraulic analysis of these pipes are discussed in Chapter 16 in the sections on storm sewer and culvert design.

8.4 Minor Losses in Pipes

As previously mentioned, all energy losses which occur in pipes are not due to boundary friction. Instantaneous losses can occur due to changes in pipe diameter or geometry, or due to control devices such as valves and fittings. These appurtenances cause nonuniformities in the flow regime of the conduit and result in small energy losses. Minor losses also occur at the entrances and exits of pipe sections.

Minor losses are normally expressed in units of velocity head, that is,

$$h_l = k_l \frac{V^2}{2g} \tag{8.27}$$

where k_l is a loss coefficient associated with a particular type of minor loss. It is usually assumed that $k_l = f(R_n, R/D$, bend angle, type of valve, etc.). The types of minor losses which are usually dealt with are contractions, expansions, bends, valves, fittings, entrances and exits. In the case of expansions of cross-sectional area, the loss function is sometimes written in terms of the difference between the velocity heads in the original and expanded section due to momentum considerations:

$$h_l = k_e \frac{(V_1 - V_2)^2}{2g} \tag{8.28}$$

where k_e is the expansion coefficient.

In the case of contractions or expansions in pipe diameter, the head loss is caused by a rapid decrease or increase in the pressure head in the pipe. Naturally, the magnitude of the energy loss is a function of the degree and abruptness of the transition as measured by the ratio of the two diameters and its angle to the horizontal. Table 8.3 provides typical contraction and expansion coefficients as functions of the ratio of the smaller diameter to the larger diameter and the transition angle. Pipe fittings tend to behave like very abrupt transitions and so have loss coefficients which are much higher than ordinary transition coefficients. Loss coefficients for some typical fittings are also given in Table 8.3.

Entrance and exit losses are also expressed in terms of the velocity head in the pipe as given in equation (8.27). The entrance loss coefficient is strongly influenced by the nature of the entrance. In particular, entrances which are flush with the surrounding environment (wall or embankment) perform much more efficiently than do protruding entrances. Rounded entrances are also more efficient than are square entrances. In rounded entrances, the loss coefficient is a function of the radius of the entrance and the pipe diameter. Typical values of pipe entrance coefficients are given in Figure 8.5. The exit loss coefficient is normally taken as 1 (i.e., exit losses are equal to 1 velocity head).

Bend losses are functions of the angle and radius of the bend, r and the diameter of the pipe, d. The most common bend angles are 45°, 60°, and 90°. Typical coefficients for bend losses are given in Table 8.4.

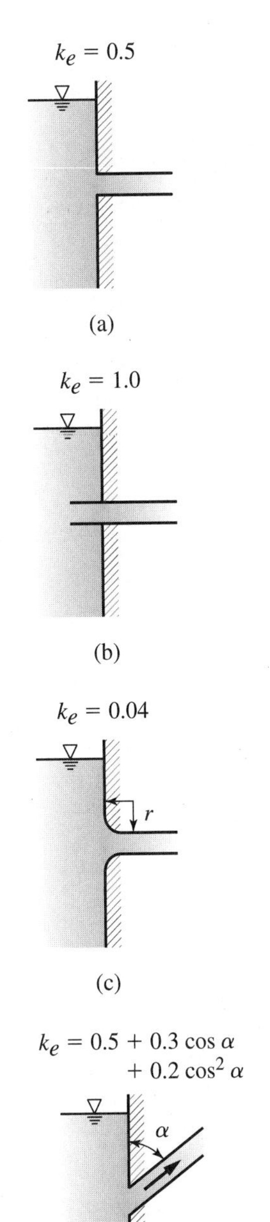

Figure 8.5 Coefficients of entrance loss for pipes (after Wu et al., 1979).

TABLE 8.3 Expansion and Contraction Coefficients and Coefficients for Threaded Fittings

Contraction	d_2/d_1	k_c, $\theta = 60°$	k_c, $\theta = 180°$
$h_l = k_c V_2^2/2g$	0.0	0.08	0.50
	0.20	0.08	0.49
	0.40	0.07	0.42
	0.60	0.06	0.32
	0.80	0.05	0.18
	0.90	0.04	0.10
Expansion	d_1/d_2	k_e, $\theta = 10°$	k_e, $\theta = 180°$
$h_l = k_e V_1^2/2g$	0.0		1.00
	0.20	0.13	0.92
	0.40	0.11	0.72
	0.60	0.06	0.42
	0.80	0.03	0.16
Threaded pipe fittings	Globe valve—wide open		$k_v = 10.0$
	Angle valve—wide open		$k_v = 5.0$
	Gate valve—wide open		$k_v = 0.2$
	Gate valve—half open		$k_v = 5.6$
	Return bend		$k_b = 2.2$
	Tee		$k_t = 1.8$
	90° elbow		$k_e = 0.9$
	45° elbow		$k_e = 0.4$

TABLE 8.4 Bend Loss Coefficients (Beij, 1938)

r/d	1	2	4	6	10	16	20
k_b	0.35	0.19	0.17	0.22	0.32	0.38	0.42

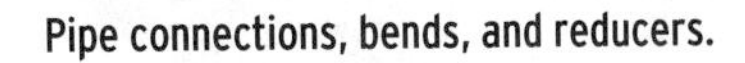

Pipe connections, bends, and reducers.

Sleeve valve. (Courtesy TVA.)

Valve loss coefficients vary with the geometry and opening of the valve. Obviously, valves which are wide open function more efficiently than do those which are partly closed. Likewise, valves with a straight geometric design are more efficient than are those that spiral. Loss coefficients for some typical valve designs are given in Table 8.5.

TABLE 8.5 Values of k_v for Some Common Hydraulic Valves

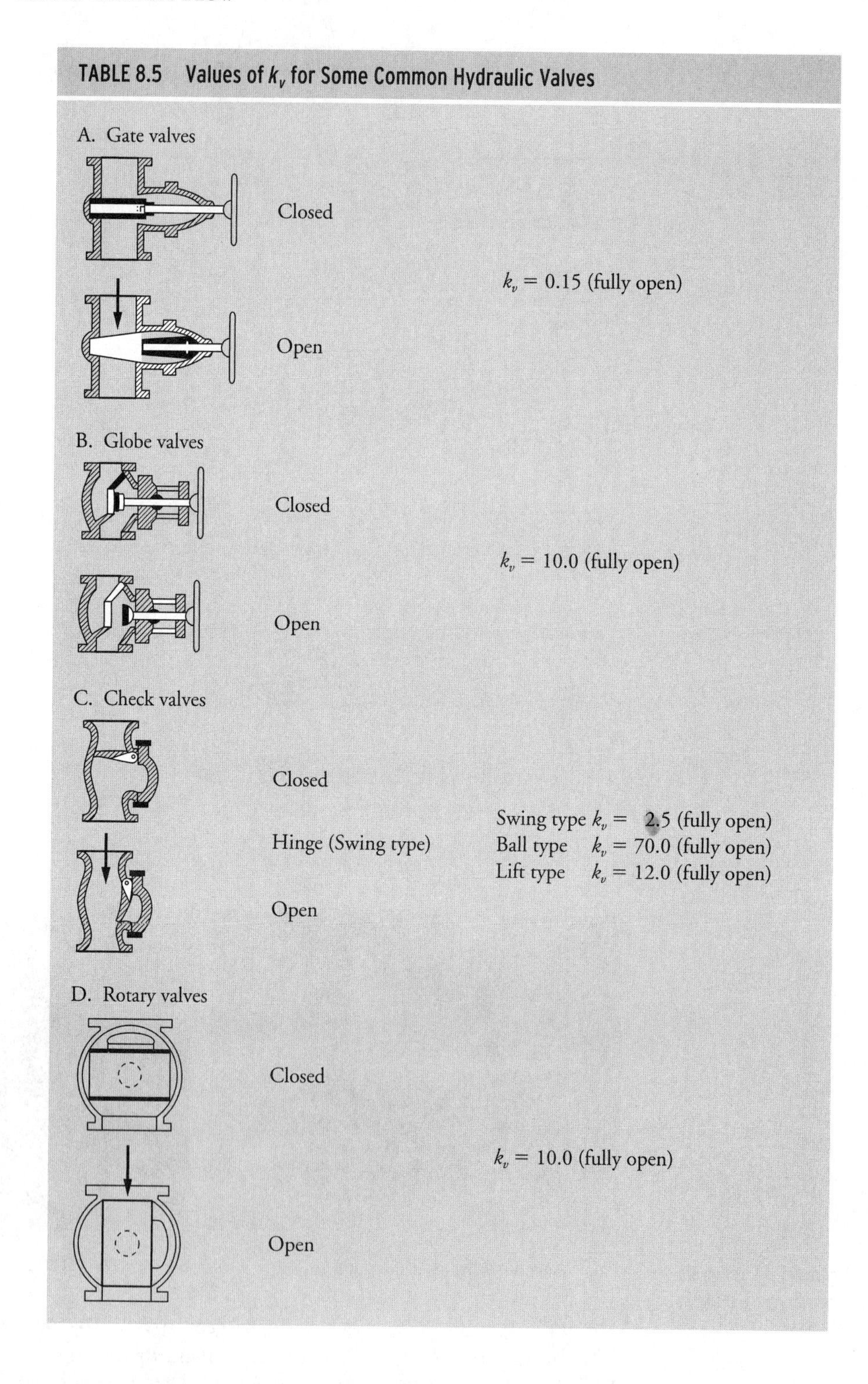

Example 8.8

A pump is used to deliver a discharge of 0.05 m^3/s from a river to a storage reservoir. The system consists of 500 m of 0.15-m cast iron pipe with a 90° bend with a radius of 0.75 m, which then transitions to a 0.1-m cast iron pipe 100 m long, as shown in Figure 8.6. If the difference between the water levels in the river and the reservoir is 20 m and the water discharges into the atmosphere at the reservoir through a fully open gate valve, what horsepower pump would be required? Assume the water temperature is 20°C.

Solution: $Q = 0.05\ m^3/s$, $T = 20°C$, so $\nu = 1.003 \times 10^{-6}\ m^2/s$. $A_1 = \pi(0.15)^2/4 = 0.0177\ m^2$; $A_2 = \pi(0.1)^2/4 = 0.00785\ m^2$; $V_1 = Q/A_1 = 2.82$ m/s; $V_2 = Q/A_2 = 6.36$ m/s.

$$R_{n1} = \frac{2.82(0.15)}{1.003 \times 10^{-6}} = 4.22 \times 10^5; \quad R_{n2} = \frac{6.36(0.1)}{1.003 \times 10^{-6}} = 6.34 \times 10^5$$

For cast iron, from Table 7.1, $\varepsilon = 0.00026$ m; $\varepsilon/d_1 = 0.0017$; $\varepsilon/d_2 = 0.0026$. From Figure 8.4, $f_1 \approx 0.022$ and $f_2 \approx 0.025$.

Friction Losses

$$h_{f1} = \frac{f_1 L_1 V_1^2}{d_1 2g} = 29.7\ \text{m}, \quad h_{f2} = \frac{f_2 L_2 V_2^2}{d_2 2g} = 51.5\ \text{m}$$

Minor Losses

Contraction: Ratio of smaller to larger pipe diameters: $d_2/d_1 = 0.1/0.15 = 0.67$, velocity in the 0.1 m pipe = 6.36 m/s. From Table 8.3, $k_c \approx 0.054$ for $\theta = 60°$, $h_{lc} = k_c V^2/(2g) = 0.11$ m.

Bend: Ratio of bend radius to pipe diameter = 0.75/0.15 = 5. From Table 8.4, $k_b = 0.20$, $h_{lb} = k_b V^2/(2g) = 0.08$ m.

Valve: From Table 8.5, for a gate valve (fully open), $k_v = 0.15$, and $h_{lv} = k_v V^2/(2g) = 0.31$ m

So total pumping head $H_p = \Delta Z + h_{l1} + h_{l2} + h_{lc} + h_{lb} + h_{lv}$

$$= 20 + 29.7 + 51.5 + 0.11 + 0.08 + 0.31 = 101.7\ \text{m}$$

Power $P_o = \gamma Q h_p = 9790\ N/m^3 \times 0.05\ m^3/s \times 101.7$ m

$= 49782$ N-m/s ≈ 49800 watts (1 N-m/s = 1 watt)

$= 49.8$ kw $= 66.7$ hp (1 kw = 1.34 hp)

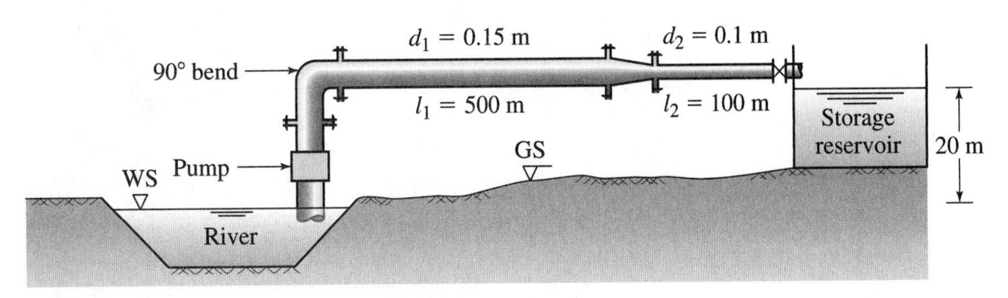

Figure 8.6 Water delivery system given in Example 8.8.

8.5 Water Distribution Systems

An important problem faced by managers of water resources systems, particularly those engaged in municipal work, is the distribution of water to all of the users, or customers of the agency. The computations and analyses involved in this work rely on the general principles developed in earlier sections of the text. However, the actual application of these principles may be quite different than anything illustrated to this point. The material will begin with analysis of the classic three reservoir problem and then proceed to development of the method for pipelines in parallel and then to fully developed pipe networks as found in municipalities.

8.5.1 Three-Reservoir Problem

Consider the situation shown in Figure 8.7. The three reservoirs are connected by pipelines running to the common node at *D*. The problem is to determine the discharge and direction of flow in each line given the elevations of the water in each reservoir and the length, diameter and material of each pipe. Note that the reservoirs merely represent sources of water that are applying a certain fixed head to the system. It is obvious that the flow will be from the reservoirs with the higher elevations, or heads, toward those of lower heads. Thus, in Figure 8.7, the flow will ultimately be toward reservoir *C*. Therefore, one can write

$$Q_C = Q_A + Q_B \tag{8.29}$$

The solution is one of trial and error using the principles of energy and continuity. The solution proceeds by assuming a pressure head at node *D*, which then leads to the computation of the allowable head loss between each reservoir and the node. Then, for each pipe the Darcy head loss law applies:

$$h_f = \frac{fL}{d}\frac{V^2}{2g} \tag{8.30}$$

The procedure then reduces to a trial solution for V by assuming a value for the Darcy f in the manner previously done in the chapter. This is then repeated for each reservoir and continuity is checked using equation (8.29). If continuity is not achieved, then another head is assumed at *D* and the entire procedure is repeated.

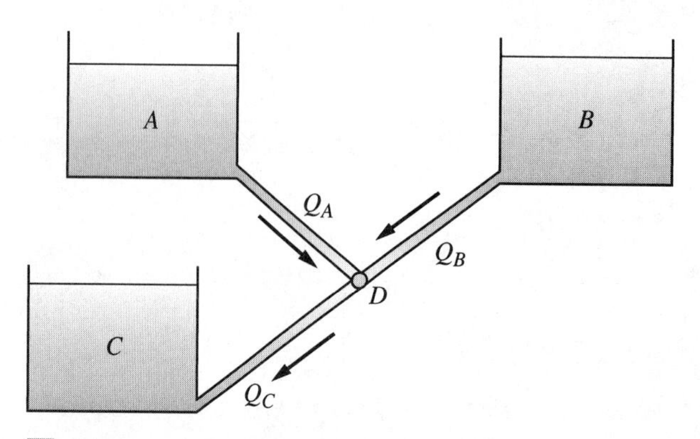

Figure 8.7 Schematic of three reservoir problem.

Example 8.9

Three reservoirs are connected as shown in Figure 8.8. All related information are as shown in the figure. Determine the discharge and flow direction in each pipe.

Solution: A technique that is commonly employed is to assume that the head at D is the same as the middle reservoir, thereby cutting off flow from that direction. The other two discharges are computed and compared and then the appropriate adjustment to D can be made. So, we begin by assuming that the head at $D = 110$ ft. Then the allowable head losses are

$$h_{fB-D} = 120 - 110 = 10 \text{ ft}$$

$$h_{fD-C} = 110 - 70 = 40 \text{ ft}$$

Then, for the pipe connecting reservoir B to D,

$$\frac{f(1000)}{1}\frac{V^2}{2g} = 10$$

Or $fV^2 = 0.644$. Similarly, for the line connecting node D to reservoir C, $\frac{f(500)}{0.667}\frac{V^2}{2g} = 40$, or $fV^2 = 3.43$. We will assume that for the 12-in cast iron pipe the Darcy friction factor will be 0.02. Then $(0.02)V^2 = 0.644$, or $V = 5.67$ ft/s. Checking this assumption, for a water temperature of 65°F, we find from Table 2.2 that the kinematic viscosity, $\nu = 1.138 \times 10^{-5}$ ft²/s. For cast iron, we find from Table 7.1 that the roughness element, $\varepsilon = 0.00085$ ft; thus $\varepsilon/d = 0.00085$. The Reynolds number is $R_n = \frac{(5.67)(1)}{1.138 \times 10^{-5}} = 4.982 \times 10^5$.

From the Moody diagram (Figure 8.4), we find that $f \sim 0.02$ so our assumption is valid. Then $Q = VA = 5.67(0.785) = 4.45$ cfs. Turning to the line, D-C, again assuming for the galvanized iron pipe that the friction factor will be 0.02, $(0.02)V^2 = 3.43$, or $V = 13.09$ ft/s. Then $R_n = \frac{Vd}{\nu}$, or $R_n = \frac{(13.09)(0.667)}{1.138 \times 10^{-5}} = 7.67 \times 10^5$. From Table 7.1 we find that for galvanized iron $\varepsilon = 0.0005$ ft, thus $\varepsilon/d = 0.00075$. From the Moody diagram, we find that $f = 0.019$. Thus, $(0.019)V^2 = 3.43$, or $V = 13.43$ ft/s and $Q = 13.43(.349) = 4.69$ cfs. Comparing, we see that the $Q_{DC} > Q_{BD}$ by 0.24 cfs, so we need to lower the head at D only slightly in order to increase the flow from B to D and decrease the flow from D to C. Of course, this will also bring in some flow from A to D as well.

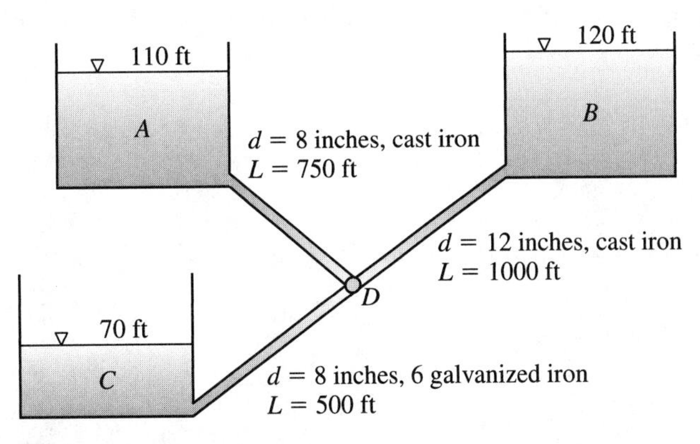

Figure 8.8 Sketch for Example 8.9.

Next, try lowering the head at D to 109.5 ft. Then

$h_{fB-D} = 120 - 109.5 = 10.5$ ft

$h_{fA-D} = 110 - 109.5 = 0.5$ ft

$h_{fD-C} = 109.5 - 70 = 39.5$ ft

Then, following the same procedure as before, and using what we have already learned, we compute the following by initially assuming the f values that we found to be correct in the previous trials:

for B-D: $(0.02)V^2 = 10.5/15.52 = 0.676$; $V = 5.82$ ft/s; $Q = 4.56$ cfs, $R_n = 5.11 \times 10^5$

for A-D: $(0.02)V^2 = 0.5/17.46 = 0.0286$; $V = 1.19$ ft/s; $Q = 0.41$ cfs, $R_n = 7.0 \times 10^4$

for D-C: $(0.019)V^2 = 39.5/11.64 = 3.39$; $V = 13.36$ ft/s; $Q = 4.66$ cfs; $R_n = 7.8 \times 10^5$

Checking our assumptions for f, we find that all are correct except for A-D, which we find from the Moody chart should be 0.022 for $\varepsilon/d = 0.00075$. Using this value for f, we obtain for line A-D: $V = 1.14$ ft/s; $Q = 0.39$ cfs, and $R_e = 6.68 \times 10^4$, which checks. Then, summing the discharges, we find

$$4.56 + 0.39 - 4.66 = 0.29 \text{ cfs}$$

This computation shows us that there is too much flow coming into line D-C, so we have lowered the head at D too much. So we adjust the head at D to a value of 109.9 ft. Then

$h_{fB-D} = 120 - 109.9 = 10.1$ ft

$h_{fA-D} = 110 - 109.9 = 0.1$ ft

$h_{fD-C} = 109.1 - 70 = 39.9$ ft

Using these values, we find the following:

for B-D: $(0.02)V^2 = 10.1/15.52 = 0.650$; $V = 5.7$ ft/s; $Q = 4.47$ cfs, $R_n = 5.0 \times 10^5$

for A-D: $(0.022)V^2 = 0.1/17.46 = 0.0057$; $V = 0.51$ ft/s; $Q = 0.18$ cfs, $R_n = 2.98 \times 10^4$

for D-C: $(0.019)V^2 = 39.9/11.64 = 3.42$; $V = 13.43$ ft/s; $Q = 4.68$ cfs; $R_n = 7.87 \times 10^5$

Checking our assumptions for f, we again find that A-D is incorrect. From the Moody chart, we find that $f = 0.025$. Using this value we compute for line A-D; $V = 0.48$ ft/s; $Q = 0.16$ cfs, $R_n = 2.8 \times 10^4$, which checks. Summing the discharges, $4.47 + 0.16 - 4.68 = -0.05$ cfs. We conclude that this is close enough.

Actually, solutions of tedious iterative problems such as these are facilitated by use of a computer spreadsheet program such as Microsoft Excel or Corel Quattro Pro. A spreadsheet solution for the above example is as follows.

Line	L (ft)	d (ft)	A (sq ft)	L/d	f	$fL/2gd$	h (ft)	V^2	V (ft/s)	R_n	Q (cfs)	Error
A-D	750	0.667	0.349	1124.43			0	0	0	0	0	
B-D	1000	1	0.785	1000	0.02	0.31	10	32.2	5.6745	498638	4.454	0.239
C-D	500	0.667	0.349	749.625	0.02	0.22	40	180.9	13.449	767225	4.694	

Line	L (ft)	d (ft)	A (sq ft)	L/d	f	$fL/2gd$	h (ft)	V^2	V (ft/s)	R_n	Q (cfs)	Error
A-D	750	0.667	0.349	1124.43	0.025	0.44	0.1	0.229	0.4786	28054	0.167	
B-D	1000	1	0.785	1000	0.02	0.31	10.1	32.52	5.7028	501125	4.477	
C-D	500	0.667	0.349	749.625	0.019	0.22	39.9	180.4	13.432	787253	4.688	0.0439

In this technique, the computations for the first trial were set as shown in the first four lines of the table to arrive at the first error term of 0.239 cfs. Then the four lines were merely copied below as shown and the solution was arrived at by merely toggling the available head losses in column 8 until an acceptable error term was obtained. It just must be noted that each time the head loss is changed and a new velocity and Reynolds number are computed, the Moody diagram must be checked to determine if the assumed f value was correct or not. Then, when a new f value is substituted in the table, all other computations will automatically be corrected by the program.

8.5.2 Pipes in Parallel

Consider the situation of two pipes running parallel as shown in Figure 8.9. Given the inflow to the node A, and the diameter, lengths and material properties of the two pipes, the problem is to distribute the discharge among the two lines. In some ways the problem is similar to the one just discussed above, as the head loss between the pipes must be balanced in the same way, and, of course, continuity must be maintained. Since, from the figure one can see that the pressure at point B must be independent of the pipes in the branch, then the head loss in each pipe must be equal in order to obtain the same pressure difference between A and B. Therefore,

$$h_{fC} = h_{fCC} \tag{8.31}$$

where h_{fC} = head loss in the clockwise direction and

h_{fCC} = head loss in the counterclockwise direction

Thus

$$\frac{fL_c}{d_c}\frac{V_c^2}{2g} = \frac{fL_{cc}}{d_{cc}}\frac{V_{cc}^2}{2g} \tag{8.32}$$

where d_c = clockwise pipe diameter, L_c = clockwise pipe length, d_{cc} = counterclockwise pipe diameter, and L_{cc} = counterclockwise pipe length.

The procedure is then one of trial and error, assuming values of f for each pipe and computing the associated velocities and discharges. As before, when the sum of the discharges agrees to within an acceptable tolerance, the problem is solved.

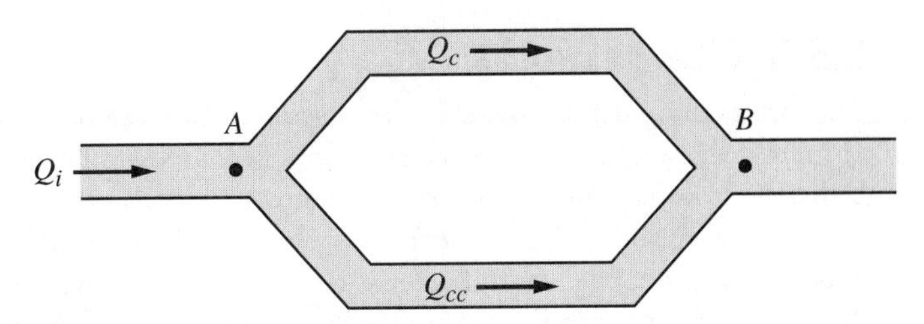

Figure 8.9 Pipes in parallel.

Example 8.10

Two pipes are laid in parallel as shown in Figure 8.10. The pipe characteristics and dimensions are as shown in the figure. If the total discharge is 0.35 m^3/s and the water temperature is 25°C, what will be the discharge in each line?

Solution: From the discussion above we know that

$$\frac{f_1(100)}{0.2}\frac{V_1^2}{2g} = \frac{f(100)}{0.3}\frac{V_2^2}{2g}$$

$$25.484 f_1 V_1^2 = 16.989 f_2 V_2^2$$

Assume that $f_1 = f_2 = 0.02$. Then we can write

$$V_1^2 = 0.666 V_2^2 \quad \text{or} \quad V_1 = 0.816 V_2$$

Then, from continuity, $V_1A_1 + V_2A_2 = 0.35$ m^3/s, or $0.816V_2(0.0314) + 0.0707V_2 = 0.35$. Solving, we find that $V_2 = 3.63$ m/s and thus $V_1 = 2.96$ m/s. Now, checking our assumed f values, $R_{n1} = \dfrac{2.96(0.2)}{0.893 \times 10^{-6}} = 6.629 \times 10^5$ and $\varepsilon/d = 0.0013$, so from the Moody chart we find that $f = 0.021$. Likewise, $R_{n2} = \dfrac{3.63(0.3)}{0.893 \times 10^{-6}} = 1.219 \times 10^6$ and for $\varepsilon/d = 0.00015$ we find that $f = 0.0145$.

Using these values for f, we recompute the velocities in each line as above and find that $V_2 = 3.80$ m/s and $V_1 = 2.57$ m/s. Then $Q_2 = 0.269$ m^3/s and $Q_1 = 0.081$ m^3/s. Checking our new values of f, we compute $R_{n1} = 8.5 \times 10^5$ and $R_{n2} = 8.63 \times 10^5$, and using the Moody chart we find that the friction factors do approximately check out.

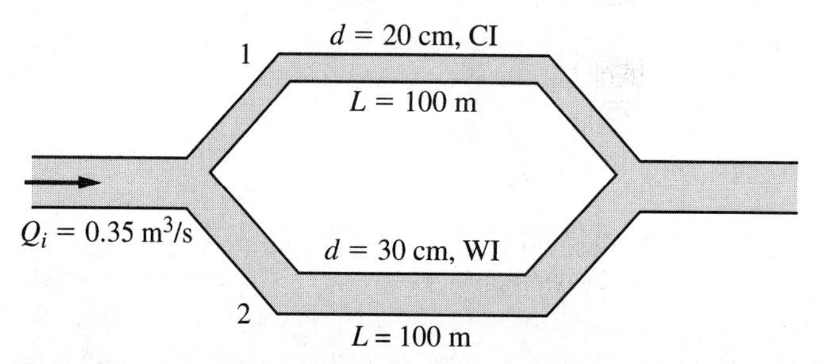

Figure 8.10 Schematic for Example 8.10.

8.5.3 Pipe Networks

The methodology demonstrated in the previous section can be extended to networks consisting of more than two pipes; however, the process becomes significantly more tedious if performed by hand. However, a number of standard software packages are available for the design of water distribution networks. The problem may be to design the original network, or to add additional nodes to an existing network. If an existing water distribution system is to be extended to additional users, then the problem is to determine the required flow and pressure at each node to meet the demand on the system. Figure 8.11 illustrates an example of this type of system. The

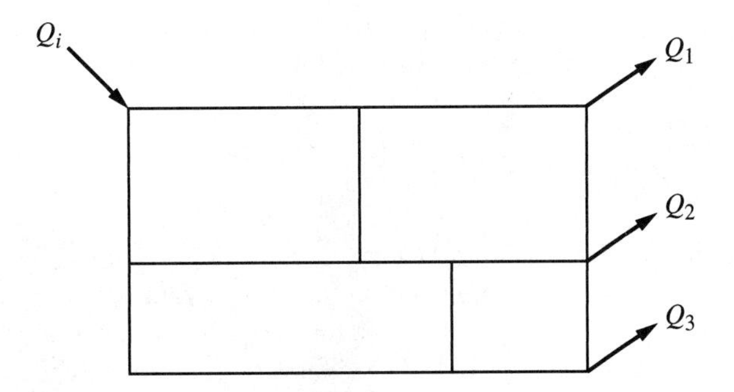

Figure 8.11 A pipe network.

technique is to follow some systematic trial procedure to arrive at the flow rates and pressure loss in each line.

With the availability of computers, any one of many trial procedures can be employed. However, the oldest and most well known procedure was presented by Hardy Cross in 1936 and is based on a similar method developed by him for distribution of stresses in indeterminate beams. It is known as the Hardy Cross method and is presented here.

The procedure is principally an extension of that given in the previous section in that the two guiding principles are the same: continuity must be maintained and the head loss between any two nodes must be independent of the route taken to get there. These conditions are summarized by the following governing equations for each loop:

$$\sum Q_i = 0 \tag{8.33}$$

$$\sum h_{fC} = \sum h_{fCC} \tag{8.34}$$

where the definitions are as given previously. The method is one of successive approximation, taking one loop at a time.

Consider the simple two-loop network shown in Figure 8.12 with only one inflow point and one discharge node. The inflow and outflow, as well as pipe diameters, lengths and material are all known. Taking loop *ABCD* first, the procedure is to first assume discharges in each line based on the inflow Q_i, taking care to maintain continuity at each node. The head losses

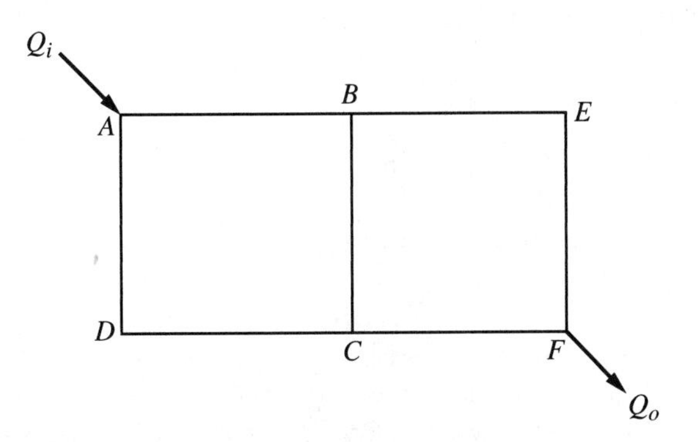

Figure 8.12 A two-loop network.

in each pipe are then computed from the Darcy equation or some other head loss function. It is common practice to express the head loss in terms of the Q only:

$$h_f = \frac{fL}{d}\frac{V^2}{2g} = \frac{fL}{d}\frac{Q^2}{2gA^2} = \frac{fL}{2g\,dA^2}Q^2 = kQ^2 \tag{8.35}$$

and again, remember that for the loop: $\sum h_{fC} = \sum h_{fCC}$, or $\sum k_c Q_c^2 = \sum k_{cc} Q_{cc}^2$. In the above equation the k value is computed from the Darcy equation but any appropriate resistance equation will work. Using the known data for the situation, the head loss in each line can be computed and the totals checked. The difference is known as the closure error (Hwang and Houghtalen, 1996):

$$\text{Closure error} = \sum k_c Q_c^2 - \sum k_{cc} Q_{cc}^2 \tag{8.36}$$

Assuming that the first assumptions for Q were incorrect, it is then necessary to compute a correction to the assumed flows to be added to one side of the loop and subtracted from the other. This is where the Hardy Cross method really comes in. Suppose we need to subtract a ΔQ term from the clockwise side and add it to the counterclockwise side in order to balance the head losses. Then we can write

$$\sum k_c(Q_c - \Delta Q)^2 = \sum k_{cc}(Q_{cc} + \Delta Q)^2 \tag{8.37}$$

A Taylor's series expansion of the above expression leads to

$$\sum k_c(Q_c^2 - 2Q_c\Delta Q + \Delta Q^2 - \cdots) = \sum k_{cc}(Q_{cc}^2 + 2Q_{cc}\Delta Q + \Delta Q^2 + \cdots). \tag{8.38}$$

Now, assuming that the higher order terms are much smaller than are the other terms and can be ignored, we can thus solve for ΔQ:

$$\Delta Q = \frac{\sum k_c Q_c^2 - \sum k_{cc} Q_{cc}^2}{2(\sum k_c Q_c + \sum k_{cc} Q_{cc})} \tag{8.39}$$

Hwang and Houghtalen (1996) suggested writing the correction factor in terms of the head loss directly by dividing equation (8.35) through by Q to obtain the expression $kQ = \frac{h_f}{Q}$, which is then used to rewrite equation (8.39) as:

$$\Delta Q = \frac{\sum h_{fc} - \sum h_{fcc}}{2\left(\sum \frac{h_{fc}}{Q_c} + \sum \frac{h_{fcc}}{Q_{cc}}\right)} \tag{8.40}$$

Once the heads and discharges are balanced for the first loop ($ABCD$), then we move on to the next one. However, here we see that we have one line (BC) that is common to both loops. Therefore, any changes we may have to make to that pipe in order to balance loop $BEFC$ will also require us to go back and recompute loop $ABCD$. Once this is done, we must then re-check the original loop and so on. When both loops are in balance, we move on to the next loop if there is one and so on. The procedure is demonstrated through the following example.

Example 8.11

A pipe network is represented in Figure 8.13. All information is as shown in the figure. The water temperature is 70°F. With the data shown, determine the flow rate in each member of the system.

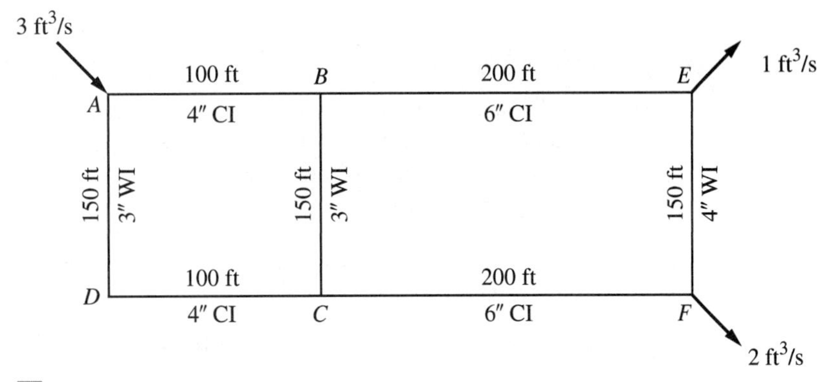

Figure 8.13 Pipe network for Example 8.11.

Solution: As in the case of Example 8.9, the solution is greatly facilitated by the use of a spreadsheet program. In the following solution, the Excel spreadsheets for each individual step are shown for illustrative purposes. In reality, once the initial table for each loop is formed, the solution can be arrived at by merely toggling the Q value for that loop similarly to the way that the H value was toggled in Example 8.9.

First we make a table of all relevant physical data needed to solve the problem:

Pipe	d (in)	Type	ε (ft)	ε/d	A (sq. ft)
AB	4	CI	0.00085	0.00255	0.087
BC	3	WI	0.00015	0.0006	0.049
CD	4	CI	0.00085	0.00255	0.087
AD	3	WI	0.00015	0.0006	0.049
BE	6	CI	0.00085	0.0017	0.196
EF	4	WI	0.00015	0.00045	0.087
CF	6	CI	0.00085	0.0017	0.196

Next, our solution begins with loop *ABCD*. We will assume flows in this loop, carefully maintaining continuity in each direction. We will begin by assuming that 2 cfs will flow in *AB* (leaving 1 cfs for *AD* and *DC*) and 1 cfs in *BC*. Using these assumptions, the following sheet is developed:

Pipe	Q (cfs)	V (ft/s)	$R_n \times 10^5$	f	h_f (ft)
AB	2	22.99	7.23	0.0245	60.93
AD	1	20.4	4.81	0.018	69.79
DC	1	11.49	3.61	0.025	15.39
BC	1	20.4	4.81	0.018	69.79

Of course, the head loss was computed from the Darcy equation with the friction factor obtained from the Moody diagram (Figure 8.4). In computing the Reynolds number, a kinematic viscosity of 1.059×10^{-5} ft²/s was obtained from Table 2.2. From the above table we can then find

$$\sum h_{fC} = 60.93 + 69.79 = 130.72 \text{ ft}$$

$$\sum h_{fCC} = 69.79 + 15.39 = 85.18 \text{ ft}$$

Thus, the closure error is 45.54 ft with the clockwise head loss being the highest. Thus our correction must be subtracted from the clockwise side of the loop and added to the counterclockwise side. We then calculate our correction term:

$$\Delta Q = \frac{\sum h_{fc} - \sum h_{fcc}}{2\left(\sum \frac{h_{fc}}{Q_c} + \sum \frac{h_{fcc}}{Q_{cc}}\right)} = \frac{45.54}{2(100.255 + 85.18)} = 0.123 \text{ cfs}$$

Applying this correction, we find our new Q values in each pipe as shown in the following table:

Pipe	Q (cfs)	V (ft/s)	$R_n \times 10^5$	f	h (ft)
AB	1.877	21.57	6.78	0.0245	53.15
BC	0.877	17.9	4.22	0.0182	54.33
AD	1.123	22.92	5.41	0.018	88.1
DC	1.123	12.9	4.05	0.025	19.4

Summing each side of the loop, we find that $\sum h_{fC} = 107.48$ ft and $\sum h_{fCC} = 107.5$ ft and thus loop $ABCD$ is in balance.

Next we move on to loop $BEFC$ using the flows we have developed in the previous loop and carefully maintaining continuity so that the outflows at E and F are achieved. This leads us to the distribution of flows as given in the following table:

Pipe	Q (cfs)	V (ft/s)	$R_n \times 10^5$	f	h (ft)
BC	0.877	17.9	4.22	0.0182	54.33
BE	1	5.1	2.4	0.023	3.71
CF	2	10.2	4.81	0.022	14.21
EF	0				0

Summing the head losses in each direction, we find that $\sum h_{fC} = 3.71$ ft and $\sum h_{fCC} = 68.54$ ft, and thus the closure error is -64.83 ft. Thus the correction term must be subtracted from the counterclockwise side and added to the clockwise side of the loop. Computing ΔQ, we find that $\Delta Q = \dfrac{64.83}{2(3.71 + 69.05)} = 0.445$ cfs. Applying this figure to the appropriate discharges, we compute the new flows as given below:

Pipe	Q (cfs)	V (ft/s)	$R_n \times 10^5$	f	h (ft)
BC	0.432	8.81	2.08	0.0195	14.12
BE	1.445	7.37	3.48	0.022	7.42
CF	1.555	7.93	3.74	0.022	8.59
EF	0.445	5.11	1.61	0.0195	3.56

If we sum the sides of the loop, we find that $\sum h_{fC} = 10.98$ ft and $\sum h_{fCC} = 22.71$ ft, thus leading to a correction term of $\Delta Q = \frac{11.73}{2(37.81 + 13.52)} = 0.114$ cfs, which must be subtracted from the counterclockwise side and added to the clockwise side of the loop. This results in the following distribution of flows:

Pipe	Q (cfs)	V (ft/s)	$R_n \times 10^5$	f	h (ft)
BC	0.318	6.49	1.53	0.02	7.84
BE	1.559	7.95	3.75	0.022	8.63
CF	1.441	7.35	3.47	0.022	7.38
EF	0.559	6.42	2.02	0.0195	5.62

Summing as before, we find that $\sum h_{fC} = 14.25$ ft and $\sum h_{fCC} = 15.22$ ft, leading to a closure error of only 0.97 ft. This difference leads to a correction factor ΔQ of 0.01 cfs. Based upon these results, we decide to stop here and go back to check the previous loop. Based on our latest flows, then the distribution for loop *ABCD* would be the following:

Pipe	Q (cfs)	V (ft/s)	$R_n \times 10^5$	f	h (ft)
AB	1.877	21.57	6.78	0.0245	53.15
BC	0.318	6.49	1.53	0.02	7.84
AD	1.123	22.92	5.41	0.018	88.1
DC	1.123	12.9	4.05	0.025	19.4

Summing as usual, we find that $\sum h_{fC} = 60.99$ ft and $\sum h_{fCC} = 107.5$, leading to a closure error of 46.51 ft. Using this error, we compute the correction factor $\Delta Q = 0.156$ cfs, which is subtracted from the counterclockwise and added to the clockwise side of the loop, leading to the distribution of flows given in the following table:

Pipe	Q (cfs)	V (ft/s)	$R_n \times 10^5$	f	h (ft)
AB	2.03	23.33	7.33	0.025	63.39
BC	0.474	9.67	2.28	0.019	16.55
AD	0.967	19.73	4.65	0.018	65.28
DC	0.967	11.11	3.49	0.0245	14.10

Summing the sides in each direction, we determine that $\sum h_{fC} = 79.94$ ft and $\sum h_{fCC} = 79.38$ ft and thus the loop is in balance.

Next, we must check loop *BEFC* again. From our latest results the following table is developed:

Pipe	Q (cfs)	V (ft/s)	$R_n \times 10^5$	f	h (ft)
BC	0.474	9.67	2.28	0.019	16.55
BE	1.556	7.95	3.75	0.022	8.63
CF	1.441	7.35	3.47	0.022	7.38
EF	0.556	6.39	2.00	0.0195	5.57

Summing around this loop, we find that $\sum h_{fC} = 14.2$ ft and $\sum h_{fCC} = 23.93$ ft; thus the closure error is 9.73 ft. Employing this value, we can compute a correction factor of $\Delta Q = 0.087$ cfs, which must be subtracted from the counterclockwise side and added to the clockwise side of the loop. This operation leads to the following distribution of flows:

Pipe	Q (cfs)	V (ft/s)	$R_n \times 10^5$	f	h (ft)
BC	0.387	7.89	1.86	0.02	11.6
BE	1.643	8.38	3.95	0.022	9.59
CF	1.354	6.90	3.25	0.022	6.5
EF	0.643	7.39	2.32	0.019	7.25

Summing the sides of the loop, we find that $\sum h_{fC} = 16.84$ ft and $\sum h_{fCC} = 18.1$ ft, thus leading to a closure error of 1.26 ft. Using this difference, we can calculate the correction term to be $\Delta Q = 0.012$ cfs. Considering this to be a small error, we now arrive at the final flow distribution for the network:

Pipe	Q (cfs)
AB	2.03
BC	0.375
AD	0.97
DC	0.97
BE	1.655
CF	1.342
EF	0.655

We notice that all of these flows changed only marginally from their previous values; thus we conclude that the network is in balance.

We should note that it is also possible to perform the analysis using pressure heads rather than discharges. In this process, initial values of pressure head are assumed at each node and then the head loss is computed across each pipe. With this head loss, the discharge can then be calculated from the Darcy equation, or some other friction equation such as Hazen-Williams. The correction term for the heads is then computed from the closure error on the computed discharges and applied to each node as appropriate. Then the process is repeated to successive loops as described above.

8.6 Transient Flow in Closed Conduits

Although this text deals primarily with steady flow conditions, there is one case of unsteady, or transient, flow with which the practicing engineer must be familiar. At times operations such as valve closure or cessation of pumping can cause the flow in a pipeline to become highly unstable for a short period of time. For example, suppose that water is flowing freely in a pipe when a valve is suddenly closed. The velocity near the valve will rapidly decrease from the initial value V_0 toward a value of 0. A simple application of the momentum equation will show what must happen:

$$\sum F = \rho Q(V_2 - V_1) \tag{8.41}$$

As the magnitude of V_2 drops below that of V_1, then a negative force is developed. This force must take the form of a wave of increased pressure that propagates back up the pipe toward the source of the flow. When it reaches the source, then a wave of reduced pressure must propagate back down the pipe towards the valve, and so on. This process constitutes the well-known *water hammer* phenomenon.

Consider the pipe line shown in Figure 8.14. In Figure 8.14(a) the water is flowing freely from the reservoir to the destination and the pressure and energy relationship is as expected.

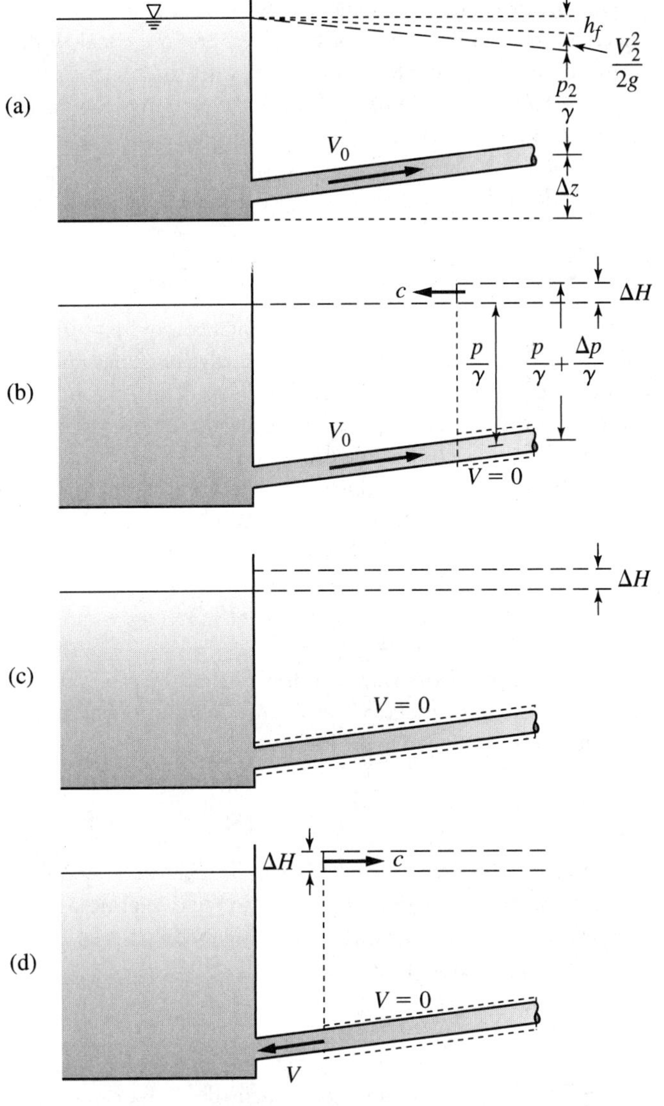

Figure 8.14 Schematic of the water hammer process.

However, when the valve is closed suddenly as in Figure 8.14(b), the water in the region near the valve comes to a stop, and a zone of increased pressure emanates from the valve and starts to propagate back towards the reservoir. This wave travels at a speed that is called its *celerity* (c). Now, it must be realized that neither the water in the pipe nor the conduit itself is completely rigid, so there is a slight compression of the water column and corresponding expansion of the pipe walls as shown in Figure 8.14(b). This behavior has been termed "elastic water column theory" (Parmakian, 1963), and the elasticity of the water-pipe system tends to somewhat lessen the effects of the water hammer. Now, when the wave reaches the reservoir, as shown in Figure 8.14(c), then the velocity throughout the pipe has come to 0, and the entire pipe has expanded as shown in the figure. However, the energy grade line in the pipe is now greater than that in the reservoir, so water will now begin to flow from the pipe into the reservoir Figure 8.14(d). Then, since the energy grade line just inside the pipe must drop to that of the reservoir, necessarily, a zone of decreased pressure equal to the previous increase will start to move down the pipe back towards the valve as shown in the figure. This process is known as *wave reflection.* Once the reflected wave reaches the valve, the process begins again and continues until it is damped due to energy losses from friction in the pipe.

A simple momentum analysis will reveal the importance of valve closure time on the magnitude of the pressure wave (Hwang and Houghtalen, 1996):

$$\sum F = ma = m\frac{dv}{dt} \tag{8.42}$$

Obviously, if the valve was closed instantaneously ($dt = 0$), the resulting force would be infinite; however, instantaneous closure of a valve or gate is not possible due to mechanical restrictions, even if the operator wanted to do so. Thus, the magnitudes of the pressures that will develop in the pipe are dependent on the closure time as well as the celerity of the wave. The issue is whether or not the initial wave travels the length of the pipe and back before or after the valve or gate is completely closed. If the valve is completely closed by the time the shock wave returns, then the water hammer pressure will reach its maximum. However, if the valve is still partially open when the initial wave returns, then some relief is afforded due to some continuing velocity in the initial direction of flow. If the valve closure is faster than the travel time of the reflected wave, the operation is termed *rapid closure,* while the reverse is known as *slow closure* (Simon and Korom, 1997).

The celerity of the wave is a function of the modulus of elasticity of the water and the pipe material. Theoretically, in a fluid of infinite extent, a shock wave would travel at a velocity given by Newton as (Simon and Korom, 1997):

$$c' = \sqrt{\frac{E_v}{\rho}} \tag{8.43}$$

where c' = theoretical wave celerity (L/T); E_v = bulk modulus of elasticity of the fluid (F/L^2); and ρ = density of the fluid. However, in the case of flow in a confined space, the elasticity of the boundary must also be considered. The modulus of elasticity of the fluid-pipe system can be calculated from (Hwang and Houghtalen, 1996)

$$\frac{1}{E_s} = \frac{1}{E_v} + \frac{d}{E_p \varepsilon} \tag{8.44}$$

where E_s = composite modulus of elasticity fluid-pipe system; d = pipe diameter; E_p = modulus of elasticity of the pipe; and ε = thickness of the pipe walls. Values of E_v are given in Table 2.3

TABLE 8.6 Modulus of Elasticity of Various Pipe Materials

Pipe Material	E_p (N/m²)	E_p (psi)
Cast iron	1.1×10^{11}	16×10^6
Steel	1.9×10^{11}	28×10^6
Lead	3.1×10^8	4.5×10^4
Reinforced concrete	1.6×10^{11}	25×10^6
Glass	7.0×10^{10}	10×10^6
Copper	9.7×10^{10}	14×10^6

and typical values of E_p for some common pipe materials are given in Table 8.6. Using equation (8.44) with equation (8.43) leads to the following expression for the celerity of a pressure wave in a confined (but elastic) pipe (Roberson et al., 1998):

$$c = \sqrt{\frac{c'^2}{1 + \left(\frac{E_v d}{\varepsilon E_p}\right)}} \tag{8.45}$$

Now, the time necessary for the wave to travel from the valve back up the pipe to the source would be simply given by $t_L = L/c$, where $L =$ pipe length. Then the time required for the wave to travel from the valve to the reservoir and back would simply be $t_{2L} = 2L/c$. So the pressure levels in the pipe will be governed by the relationship of the time of valve closure (T_v) to t_{2L}.

It is very important for the engineer to calculate the maximum pressure that will develop in a pipe due to a sudden interruption of flow. From the discussion given above, one can see that the maximum possible pressure for the given conditions will be if $T_v < t_{2L}$. The computation of these pressure magnitudes involves a straightforward application of the momentum-impulse principle. Following the reasoning given by Hwang and Houghtalen (1996), consider the pipe shown in Figure 8.15. Assume the water is flowing freely when the valve is closed. As discussed previously, the mass of water in the immediate vicinity of valve will decelerate to $V = 0$; however until the initial wave propagates all the way back to the source (t_L), some water will still be entering the upstream end of the pipe. This is shown as the condition at time $t + \Delta t$ in the figure. Since the water at the downstream end of the pipe has nowhere to go, then a compression of the water column on this end of the pipe must ensue

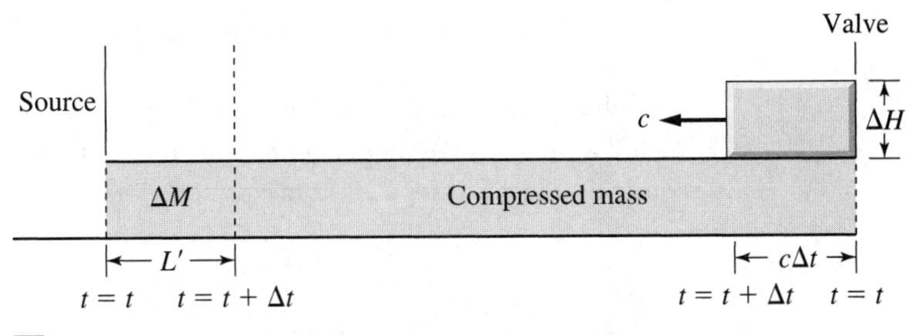

Figure 8.15 Compression of water column during water hammer.

with a corresponding increase in pressure. Writing the momentum equation for this situation leads to an expression for the pressure increase (Δp) as follows:

$$\Delta F = \Delta M \frac{dV}{dt} = \rho A L' \left(\frac{dV}{dt} \right) \tag{8.46}$$

where ΔF = resultant force created by the pressure increase Δp above the initial pressure in the pipe; ΔM = increased mass of water that enters the pipe in time $\Delta t < L/c$; and L' = length of pipe containing the increased mass at time $t = t + \Delta t$. Now, substituting that $L' = c\Delta t$, $\Delta F = \Delta p\, A$, and $V_2 = 0$ into equation (8.46), we determine that

$$\Delta p A = \rho A c \Delta t \left(\frac{V_0}{\Delta t} \right) = \rho A c V_0. \tag{8.47}$$

or

$$\Delta p = \rho c V_0 \tag{8.48}$$

Note that equation (8.48) gives the maximum pressure surge that will occur for a rapid interruption of the flow. The maximum pressure would then be

$$p_{\max} = p_0 + \rho c V_0 \tag{8.49}$$

Also, as pointed out by Simon and Korom (1997), the pressure will fluctuate in each cycle as $p = p_0 \pm \Delta p$ at times $t = (2L/c)$. However, the pressure will gradually decrease with each cycle due to pipe friction.

A solution for the pressure situation for the slow closure case was given by Lorenzo Allievi (1929) as

$$\Delta p = p_0 \left(\frac{N}{2} + \sqrt{\frac{N^2}{4} + N} \right) \tag{8.50}$$

where $N = (\rho L V_0 / p_0 T_v)$. Then, as before, the total pressure is $p = p_0 + \Delta p$.

Example 8.12

A reinforced cast iron pipe 2000 m long is carrying a discharge of 100 l/s. The pipe diameter is 25 cm and the wall thickness is 1 cm while the initial pressure in the line is 250 kN/m^2. If a valve in the line is closed in 2 s, what will be the pressure surge in the line? If the valve closure time is increased to 3.5 sec, what pressure surge would develop. Assume the water temperature to be 25°C.

Solution: The following data are developed: $d = 0.25$ m; $A = 0.049$ m^2, $V_0 = \dfrac{Q}{A} = \dfrac{0.1}{0.049} =$ 2.04 m/s.

From Table 2.3, $E_v = 2.2 \times 10^9$ N/m^2 and from Table 8.6, $E_p = 1.6 \times 10^{11}$ N/m^2. Then the theoretical wave celerity is computed from equation (8.43) as $c' = \sqrt{E_v/\rho} = 1483.23$ m/s. Then the actual celerity is computed from equation (8.45):

$$c = \sqrt{\frac{c'^2}{1 + \dfrac{E_v d}{\varepsilon E_p}}} = \sqrt{\frac{(1483.23)^2}{1 + \dfrac{2.2 \times 10^9 (0.25)}{(0.01)(1.6 \times 10^{11})}}} = 1279.5 \text{ m/s}$$

The time for the wave to return to the valve is $t = (2L/c) = 3.12$ s, which is greater than the valve closure time of 2 s, so the maximum pressure surge will develop in the line as given by equation (8.48):

$$\Delta p = \rho c V_0 = 1000 \text{ kg/m}^3 \text{ (1279.5 m/s)(2.04 m/s)} = 2.61 \text{ kN/m}^2$$

This would result in an increased pressure head $(\Delta p/\gamma) = 266.07$ m.

Now, if the valve closure time was increased to 3.5 s, this time would exceed the wave travel time of 3.12 s, so the pressure would be calculated by the Allievi procedure.

$$N = \left(\frac{\rho L V_0}{p_0 T_v}\right) = 4.66$$

Then $\Delta p = p_0\left(\frac{N}{2} + \sqrt{\frac{N^2}{4} + N}\right) = 1.375 \text{ kN/m}^2$ for the initial pressure p_0 of 250 kN/m^2.

This would result in an increase of pressure head of 140.16 m, or about 53% of the maximum value computed previously.

8.7 Surge Tanks

As the above calculations demonstrate, water hammer forces can produce tremendous pressure surges in closed pipe lines. Various measures have been devised to relieve water hammer pressures including slow valve closure, installation of relief valves, and incorporation of surge tanks in the pipe system. The surge tank has the advantage that water is not allowed to escape from the system, in contrast to the relief valve. A surge tank is normally installed at the downstream end of the pipe, just upstream of the valve or gate that will be operated as shown in Figure 8.16. As the surge forms, water is merely allowed to escape from the pipe into the tank and then can flow back into the pipe system once the valve is re-opened. A simple method to estimate the maximum level to which the water will rise in the tank due to a pressure surge has been given by Morris and Wiggert (1972). The relevant energy and pressure terms are shown in Figure 8.16 as the initial piezometric grade line is at elevation p_0/γ and the water level in the tank (which is

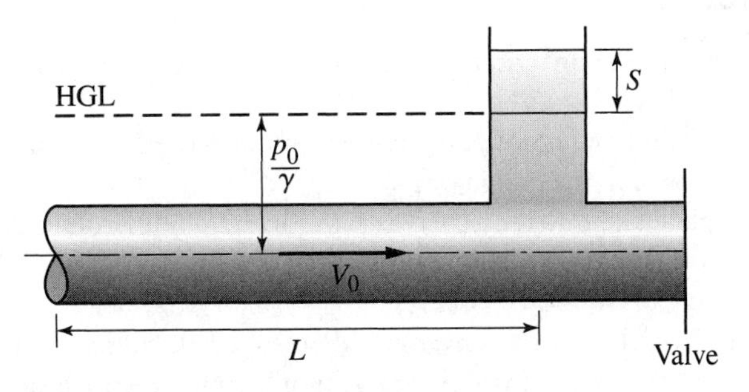

Figure 8.16 Mechanics of the surge tank.

acting as a piezometer) is at the same level. Note that friction losses between the reservoir and the tank are ignored in this analysis. Now, after the surge develops and the water has risen to level S above the initial level, writing the momentum equation for the pipe with cross-sectional area A, for the additional force developed by the surge, we get,

$$\Delta \mathbf{F} = m\mathbf{a}, \quad \text{or} \quad -\gamma SA = \rho AL\frac{dV}{dt} \tag{8.51}$$

or

$$\frac{dV}{dt} = \frac{dv}{dS}\frac{dS}{dt} = -\frac{gS}{L} \tag{8.52}$$

Now, to satisfy continuity between the pipe and the tank, we must have

$$\frac{dS}{dt} = \frac{AV}{A_t}, \text{ where } A_t = \text{area of the tank.} \tag{8.53}$$

Then, combining equations (8.52) and (8.53), we can get

$$\frac{dV}{dS} = -\frac{A_t}{AV}\frac{gS}{L}, \quad \text{or} \quad V\frac{dV}{dS} = -\frac{A_t}{A}\frac{g}{L}S \tag{8.54}$$

Integrating equation (8.54) with respect to S, with boundary condition that $V = V_0$ at $t = t_0$, we obtain

$$V^2 = -\frac{A_t}{A}\frac{g}{L}S^2 + V_0^2 \tag{8.55}$$

Now, realizing that the maximum surge (S) will occur when the water velocity in the pipe (V) is 0, we can solve equation (8.55) for S_{max}:

$$S_{max} = V_0\sqrt{\frac{AL}{A_t g}} \tag{8.56}$$

READING AID

8.1. State the main criteria in the design of closed conduits. Do the conservation laws apply?
8.2. Explain how non-uniform flow conditions can be solved using the uniform flow approach in the design of pipelines.
8.3. Explain the difference between the hydraulic (piezometric) head and the total energy. Under uniform flow condition will the hydraulic grade line be parallel to the total energy line or not? Why?
8.4. In a pipeline with converging diameter, will the gradient of the total energy line be steeper or milder than the gradient of the piezometric head line? Why?

8.5. In computing the head loss in closed conduits, would you prefer to use Darcy equation or Chezy equation? Why?
8.6. Write Poiseuille equation incorporating Reynolds number. Does the Darcy friction factor, f, depend on the roughness of the pipe material for the case of laminar flow?
8.7. Why does the energy loss for turbulent flow in smooth pipes depend on Reynolds number only? Will the friction factor, f, decrease or increase with the increase of Reynolds number?
8.8. For the case of turbulent flow in rough pipes, does the friction factor depend on the Reynolds number? Why?
8.9. Explain how the transition function approaches both the smooth and rough pipes formula under different values of Reynolds number.
8.10. Explain the Moody diagram. For which type of flow can it be applied?
8.11. For Reynolds numbers up to 100,000, would it be more accurate to use Moody diagram (Colebrook-White expression) or the Blasius expression?
8.12. Using the SI units, write an expression relating Manning Coefficient, n, to the Chezy Coefficient, C. State the dimensions of both.
8.13. Does the Manning coefficient depend on the Reynolds number of the flow or on the hydraulic radius of the conduit?
8.14. Express the Hazen-Williams equation (in English and SI units) using the Chezy coefficient.
8.15. Give examples for the different types of minor losses and express the loss function in terms of the velocity head.
8.16. Explain the approach that is commonly used to solve the three reservoir problem.
8.17. For any two pipes of different diameters connected in parallel, which of the following is true? (a) The two pipes have the same discharge, (b) the two pipes have the same head loss, (c) the two pipes have the same discharge and head loss.
8.18. In a horizontal pipeline, if the diameter of the pipe is increasing, do yo expect the velocity head to increase or to decrease?
8.19. What are the main constrains that should be satisfied in the design of pipe networks among any two nodes?
8.20. Explain how to calculate the closure error in pipe networks when using the Hardy Cross method. Express the correction factor in terms of the head loss.
8.21. Discuss the main reasons for the occurrence of water hammer in pipelines.
8.22. Define the celerity of wave which might be developed due to sudden closure of valves.
8.23. What is meant by the elastic water column theory? Explain the wave reflection process inside closed conduits.
8.24. State the factors affecting the wave celerity. Would you recommend slow or rapid closure of valves in pipelines?
8.25. What is the minimum time recommended for the closure of valves in pipe lines?
8.26. Write an expression to estimate the composite modulus of elasticity of a fluid-pipe system.
8.27. State some of the measures which could be used to reduce the risks associated with the water hammer phenomenon in closed hydraulic systems.
8.28. Identify two main advantages for using surge tanks in pipelines.
8.29. Explain how the volume of the surge tank is determined.

Problems

8.1. Two reservoirs containing water at 20°C are connected with a smooth pipe of length 5000 m and diameter 10 mm, as shown in Figure P8.1. What should be the minimum height H to maintain a flow rate of 0.5 L/min? Is this flow laminar?

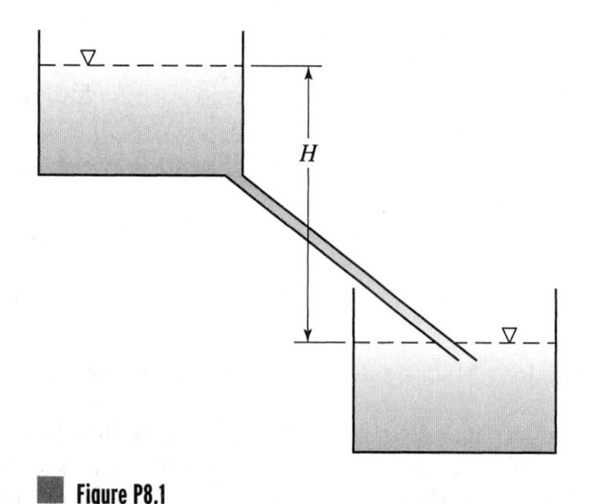

Figure P8.1

8.2. For Problem 8.1, find the minimum height H required to maintain a flow rate of 5 L/min if (a) the pipe is smooth, (b) pipe is made up of cast iron.

8.3. A swimming pool filled with water at 65°F is being emptied by gravity using a galvanized iron pipe of 3/4 in diameter and 20 ft length. Determine the depth of water level if the flow rate is 12 gal/min.

8.4. If 1.5 cfs of water flows through 1200 ft of 6 in steel pipe at a temperature of 70°F, what is the friction loss in ft?

8.5. An elevated open air water tank is used to supply water to a building located 5000 ft away as shown in Figure P8.5. The water is carried through a 12-inch cast iron pipe with a flow rate of 1000 gpm. Fire codes require that a pressure of 30 psi be available on the second floor of the building which is at an elevation of 20 ft above a reference datum. The water temperature is 70°F. What is the required elevation of the water in the tank relative to the reference datum?

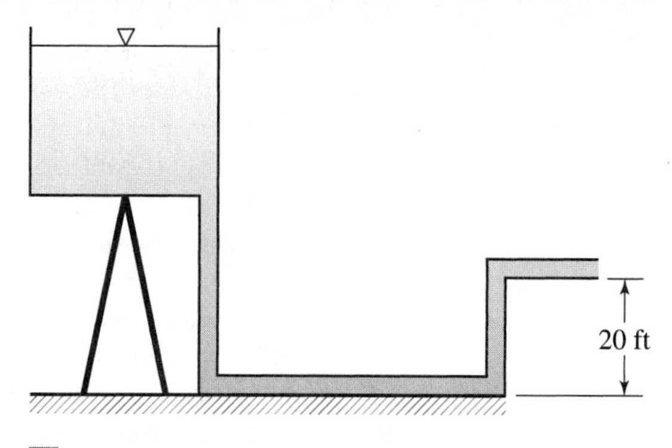

Figure P8.5 Elevated water tank for Problem 8.5.

8.6. Water flows to a building from an elevated water tank by gravity through 2000 ft of 10-inch cast iron pipe. The water temperature is 65°F and the flow rate is 2500 gpm. If the water level in the tank is 250 ft above the slab of the building, what is the pressure at the slab in psi?

8.7. Points A and B are separated by 3000 ft of new 6 inch steel pipe. A discharge of 750 gpm of water at a temperature of 60°F flows from A to B. Point B is 60 ft in elevation above Point A. If the minimum pressure at B is 50 psi, what should be the pressure at A?

8.8. Calculate the head loss due to friction and the power required to maintain flow in a horizontal circular pipe of 40 mm diameter and 750 m long. The pipe roughness is 0.08 mm. Use $\rho = 1000$ kg/m^3 and $\mu = 1.14 \times 10^{-3}$ N-s/m^2.

(a) when water flows at a rate of 66.7 cm^3/s

(b) when water flows at a rate of 500 cm^3/s

8.9. A garden hose of roughness 0.2 mm is used to irrigate a small farm, with water at 20°C. It is required to supply a flow rate of 20 m^3/hr, to irrigate the farm in 6 hr. If the length and diameter of the hose are 50 m and 5 cm, respectively, determine the appropriate pressure that must be supplied at the inlet of the hose, to maintain the required flow rate.

8.10. A 10-hp pump is used to deliver water through an 18 inch wrought iron pipe to a location 2000 ft away and 10 ft higher in elevation. If the desired flow rate is 1500 gpm and the water temperature is 70°F, what will be the pressure in the line in psi at the final destination?

8.11. Water is fed from an open air tank through a 1-ft-diameter pipe as shown in Figure P8.11. The pipe is 1000 ft long and is made of reveted steel. The water temperature is 65°F. If the flow experiences 5 ft of head loss across the 1000 ft of pipe, (a) what is the discharge in the pipe, (b) what shear stress is exerted on the pipe boundary, (c) what is the pressure in the end of the line, and (d) can this pipe be considered smooth, rough, or neither?

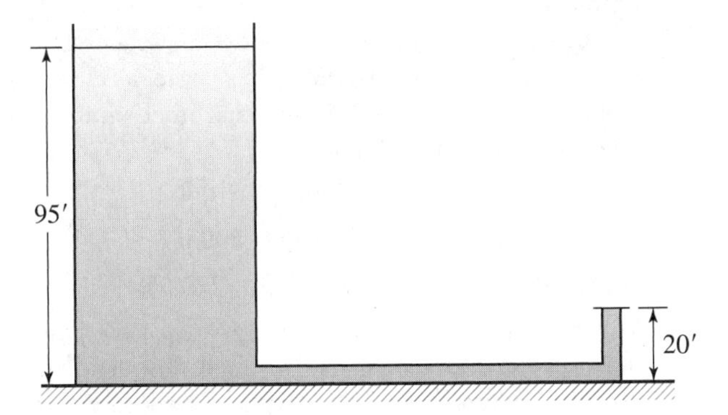

Figure P8.11 Water distribution system for Problem 8.11.

8.12. Water is flowing in a pipe with a diameter of 0.25 m at a discharge of 0.2 m^3/s. The temperature of the water is 25°C and the pipe is a 300-m-long section of wrought iron. (a) What is the head loss in this situation? (b) Can this pipe be considered smooth, rough, or neither?

8.13. A 0.3-m-diameter pipe inclines at a slope of 10% and carries a discharge of 0.35 m^3/s. Piezometers located along the pipe measure a hydraulic gradient of 0.125 m/m. The water temperature is 25°C. (a) What is the shear stress exerted on the boundary? (b) What is the shear velocity of this situation? (c) Compute the Darcy friction factor theoretically (without using the Darcy equation).

8.14. Now, assuming that the pipe in Problem 8.13 has a rough surface and the flow is fully turbulent, compute (a) the equivalent sand roughness diameter, ε; and (b) the thickness of the boundary layer using the Singh criterion for fully turbulent flow from Chapter 7. (c) Is our rough pipe assumption correct?

8.15. Water is flowing in a 6-inch-diameter pipe inclined downward on a slope of 5% at a flow rate of 1000 gpm. There is an observed pressure drop of 1 psi across 150 feet of this pipe. The temperature of the water is 70°F. (a) Compute the wall shear stress. (b) Compute the shear velocity. (c) Compute the Darcy friction factor theoretically (without using the Darcy equation). (d) Compute the Reynolds number of the flow.

8.16. Assume the pipe in Problem 8.15 is smooth walled. (a) Compute the velocity at the centerline of the pipe. (b) Compute the thickness of the boundary layer using the Morris and Wiggert criterion. (c) Compute the Darcy friction factor from the smooth pipe formula. (d) Is our assumption of a smooth-walled pipe justified?

8.17. A mercury differential manometer is connected between 2 points 10 m apart, on a straight horizontal pipe of 20 mm diameter. Calculate the Darcy friction factor if water is flowing at 2.0 L/s and the height difference in manometer is 6.0 cm.

8.18. What discharge can be delivered through a 0.25 m diameter pipe 500 m long running on a slope of 0.001 m/m with an available pressure drop of 0.01 N/cm^2 per meter of pipe if the pipe can be considered hydraulically smooth? The water temperature is 20°C.

8.19. What discharge can be delivered under the conditions specified in Problem 8.18 if the pipe is galvanized iron and can be considered hydraulically rough?

8.20. A 2-m-wide and 1-m-high rectangular conduit of concrete (ε = 2.0 mm) is used to supply water to a town population, by gravity. To meet the supply demands the flow rate is required to be 8000 gpm. Determine the elevation of the supply tank if the length of the conduit is 2.0 km.

8.21. What discharge can be transmitted through 2000 ft of 6 inch diameter hydraulically rough pipe if the allowable head loss is 10 ft? Assume the equivalent sand diameter for the roughness height of the pipe wall is 0.002 ft.

8.22. Determine the head loss in 10,000 m of 1 m diameter galvanized iron pipe running horizontally and carrying a discharge of 1.5 m^3/s. The water temperature is 25°C.

8.23. What will be the total energy loss generated in a 24-inch cast iron pipe running one mile on a slope of 0.001 ft/ft and carrying a discharge of 5 cfs if the water temperature is 70°F?

8.24. A wrought iron pipe is laid on a slope of 0.005 m/m and runs for a distance of 1000 m. If the pipe is to carry a discharge of 1 m^3/s with a pressure drop of no more than 5000 N/m^2, what must be the minimum diameter of the pipe? Assume the water temperature is 20°C.

8.25. Water is to be delivered from an open air stand tank through a horizontal galvanized iron pipe to a building located 5000 ft away from the tank as shown in Figure P8.25. The water level in the tank is 50 ft above the ground level. The required flow rate is 1500 gpm and the required pressure in the line at the building is 15 psi. If the water temperature is 65°F, what is the minimum diameter of the pipe to meet these requirements?

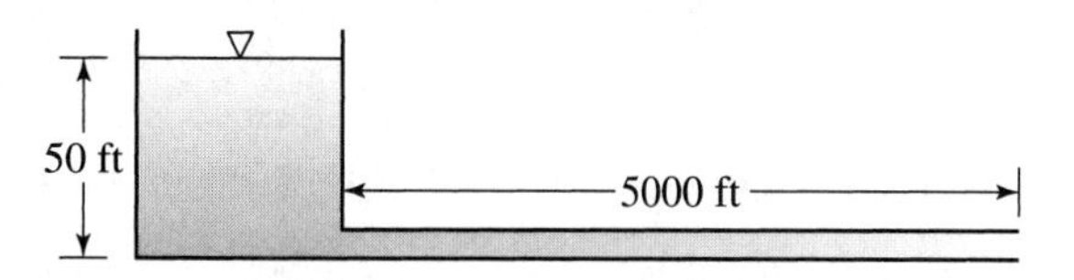

Figure P8.25 Water delivery system for Problem 8.25.

8.26. A 20-HP pump is employed to deliver water from a stream to an open air holding tank located 1500 ft away where the water level is to be maintained at an elevation 75 ft above the stream level (Figure P8.26). The pipe line is cast iron and the water temperature is 60°F. If the required flow rate is 1000 gpm, what is the minimum diameter of the pipe to meet these requirements?

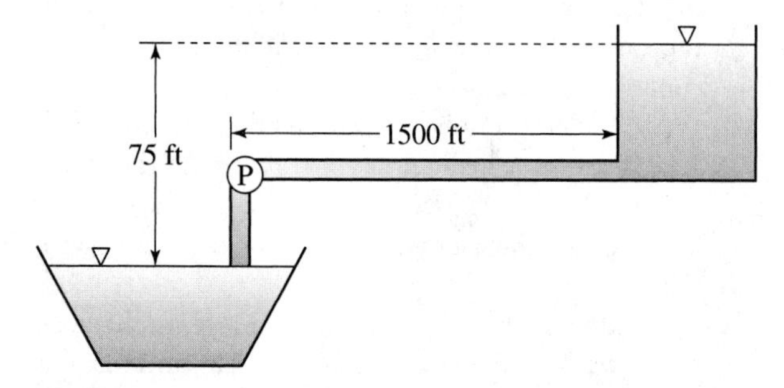

Figure P8.26 Water distribution system for Problem 8.26.

8.27. A smooth pipe of 15 cm diameter and 150 m length has a flush entrance and sharp exit. If the flow velocity is 5 m/s for the water at 20°C, what will be the total head loss? Use Blausius equation to find f.

8.28. Using the Hazen-Williams equation, determine the discharge capacity of a 0.5 m diameter reveted steel storm sewer line running on a slope of 0.001 m/m and flowing by gravity.

8.29. A gravity driven new cast iron storm sewer line is to carry 5 cfs on a slope of 0.005 ft/ft. Using the Hazen-Williams equation, determine the minimum diameter pipe to meet these requirements.

8.30. After a 10-year service period, what will be the discharge capacity of the line that was designed in Problem 8.29?

8.31. A pipe network connects junctions A, B, C, and D as shown in Figure P8.31. All pipes have a Hazen-Williams coefficient of 110 and the other pipe and flow parameters are as follows:

A-B: $L = 1000$ m, $d = 0.5$ m, $Q = 1000$ l/s;

B-C: $L = 500$ m, $d = 0.4$ m, $Q = 500$ l/s;

C-D: $L = 2000$ m, $d = 0.3$ m, $Q = 220$ l/s.

Water can be added or subtracted at any of the junctions in order to achieve the required flows and the flow direction is from A to D with no backward flow at any junction. If minor losses are negligible, and the minimum allowable pressure anywhere in the line is 100 kN/m^2, what must be the pressure at point A and what is the elevation of the hydraulic grade line there?

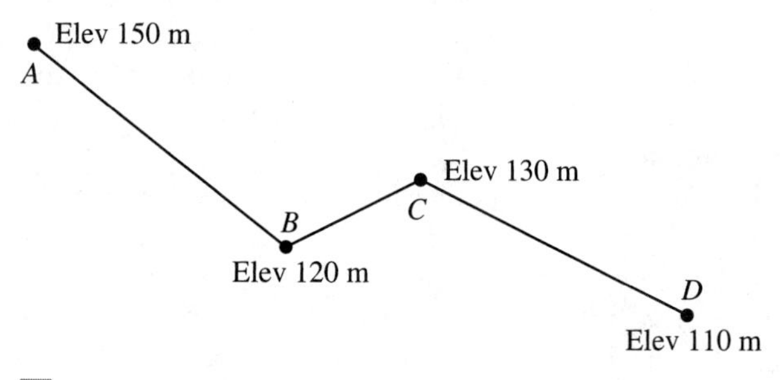

Figure P8.31 Pipe system for Problem 8.31.

8.32. Water is delivered from a pressure tank through a 1000-m-long cast iron pipe of diameter 0.5 m at a discharge rate of 0.4 m^3/s (see Figure P8.32). If the line has a flush inlet with the tank and goes through two 90° bends each of radius 0.5 m, and then exits into the atmosphere through a fully open gate valve, what will be the total head lost in the pipe? What is the pressure in the tank (N/m^2) for this situation? Assume the water temperature is 20°C.

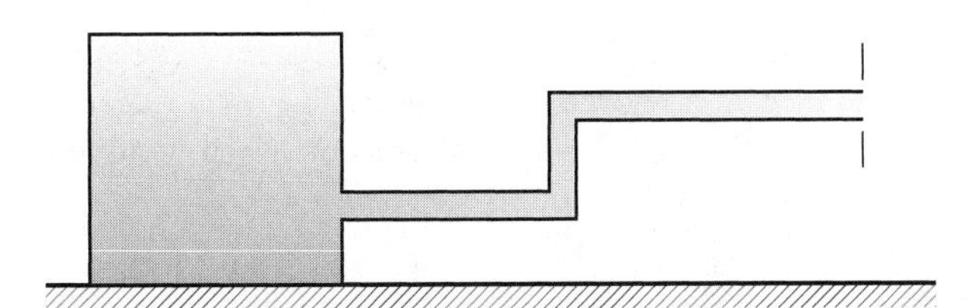

Figure P8.32 Distribution network for Problem 8.32.

8.33. A smooth pipe of a total length of 20 m is used to supply water into the house from an overhead tank. The pipe line has a protruded entrance, 4 standard 90° elbows, one tee (straight flow), and one fully open globe valve. Determine the size of the pipe to maintain a flow rate of 30 L/min for water at 20°C, if the available head is 5 m.

8.34. A discharge of 500 gpm of water flows through 300 ft of 6 in steel pipe at a temperature of 100°F. The pipe contains two 6 in steel elbows, two fully open gate valves, and a swing check valve. If the discharge end of the pipe is located 20 ft higher than the entrance, what is the pressure difference between the two ends of the pipe?

8.35. Two pipes are connected in series with a diameter ratio of 2:5. If the flow velocity in the smaller diameter pipe is 12 m/s, determine the head loss due to (a) sudden contraction, (b) sudden enlargement.

8.36. Two reservoirs with elevation difference of 10 ft are connected by a 6-in-diameter pipe ($f = 0.032$). The pipe is having a flush entrance and submerged exit. Determine the flow rate if the pipe length is 110 ft. If the last 10 ft of pipe length is replaced by a 10° conical diffuser, compute the flow rate.

8.37. Three reservoirs are connected as shown in Figure P8.37. All information is as shown in the figure. The water temperature is 20°C. What is the discharge and flow direction in each line?

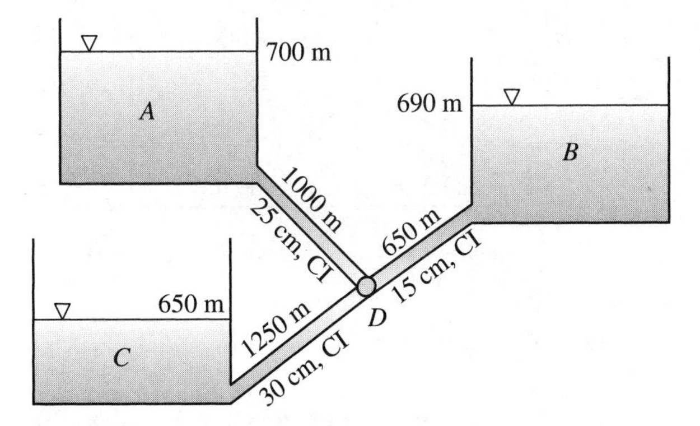

Figure P8.37 Schematic for Problem 8.37.

8.38. Three reservoirs are connected as shown in Figure P8.38. All quantities are as shown in the figure. What is the discharge and flow direction in each line? Assume the water temperature is 70°F.

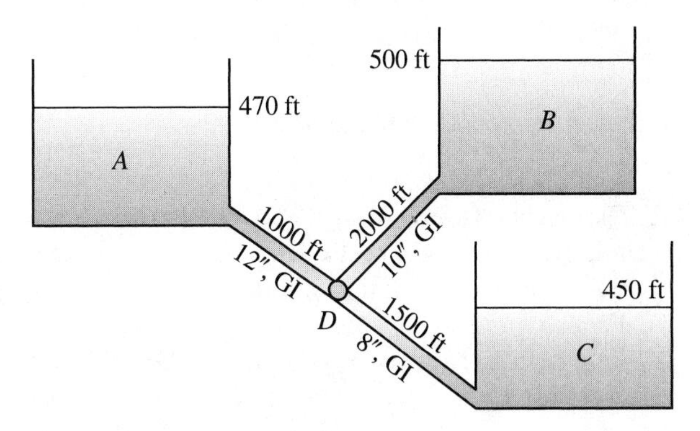

Figure P8.38. Three reservoir schematic for Problem 8.38.

8.39. Three reservoirs are interconnected as shown in Figure P8.39. All relevant data are as shown in the figure. If minor losses can be ignored, and all pipes have a friction factor of 0.02, what is the direction of flow and pressure at point *D*?

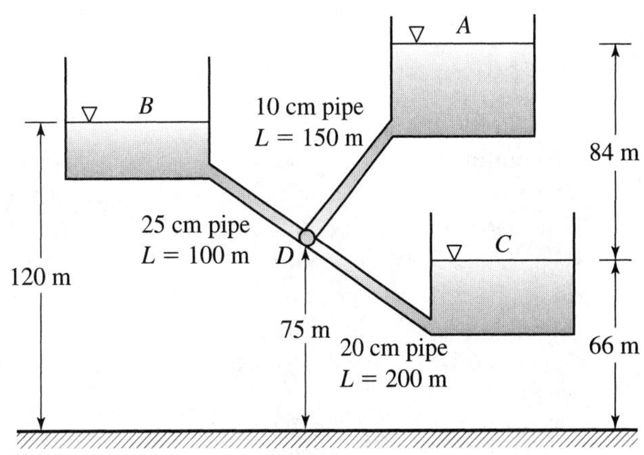

Figure P8.39 Schematic of three reservoirs for Problem 8.39.

8.40. A pipeline branches into two parallel lines as shown in Figure P8.40. The pipe diameters, lengths and materials are as shown in the figure. If the flow entering the branch is 0.5 m^3/s, what is the flow in each branch? Assume the water temperature is 20°C.

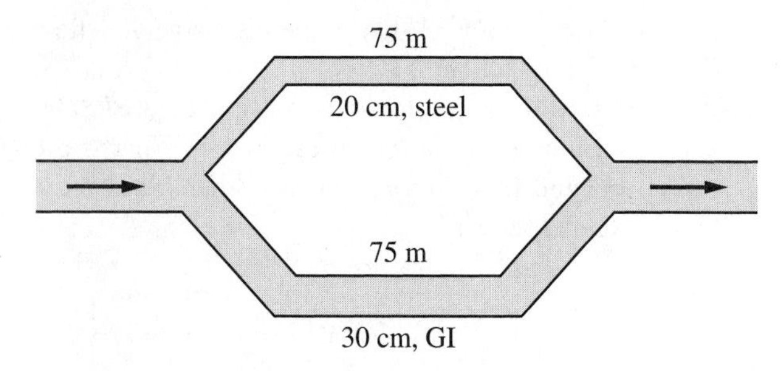

Figure P8.40 Branching pipeline for Problem 8.40.

8.41. The pipeline shown in Figure P8.41 branches into two lines as shown. The pipe diameters, lengths, and materials are as given in the figure. If the discharge entering the branch is 1000 gpm, what is the flow in each line of the branch? Let the water temperature be 70°F.

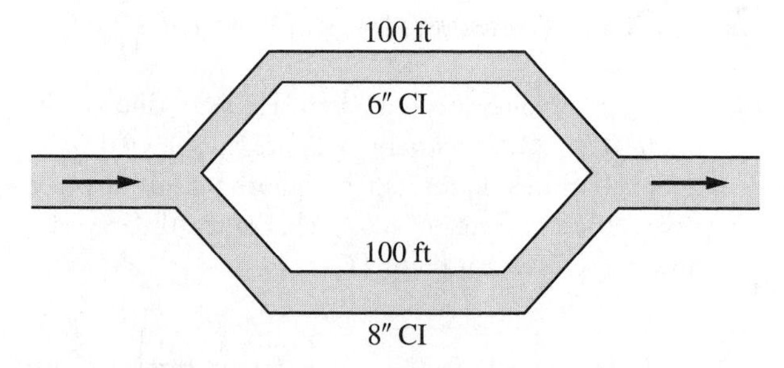

Figure P8.41 Branching pipeline for Problem 8.41.

8.42. Smooth brass pipes 1, 2 and 3 are used in an element of sugar refinery as shown below. The lengths of the three pipes are 550 ft, 350 ft and 600 ft, respectively. The diameters of the 3 pipes are 2 in, 3 in, and 4 in, respectively. If the net flow of water ($\nu = 1.08 \times 10^{-5}$ ft²/s) is 0.7 ft³/s, determine the head loss from inlet to outlet and the flow rate in each pipe.

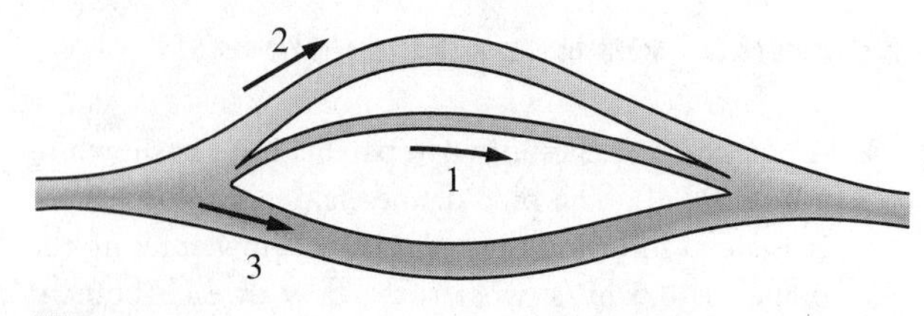

Figure P8.42 Three branching pipes for Problem 8.42.

8.43. A pipe network is shown in Figure P8.43. All relevant information is given on the figure. The water temperature is 20°C. With the information given, compute the flow and direction in each line.

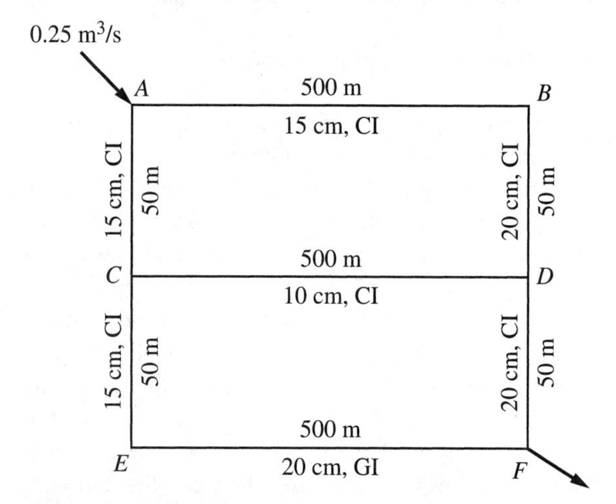

Figure P8.43 Pipe network for Problem 8.43.

8.44. A pipe network is represented in Figure P8.44. All pertinent data are given on the figure. Compute the flow and direction in each line. Use the Chezy equation for head loss.

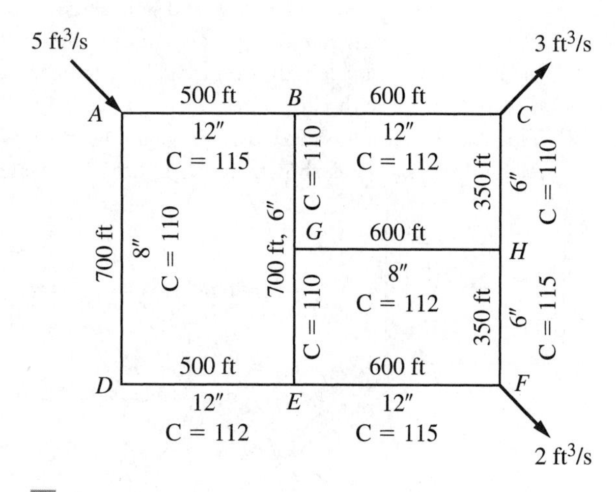

Figure P8.44 Pipe network for Problem 8.44.

8.45. Water at 20°C is flowing through the system shown in Figure P8.45. Length AB = 2000 ft and BD = 1500 ft.

All the pipes have equal diameter of 6.0 in each. The friction factor is considered to be 0.022 for all pipes. If the gage pressure at A is 90 lb/in^2, determine the flow rate in all the pipes and pressure at points B, C, and D.

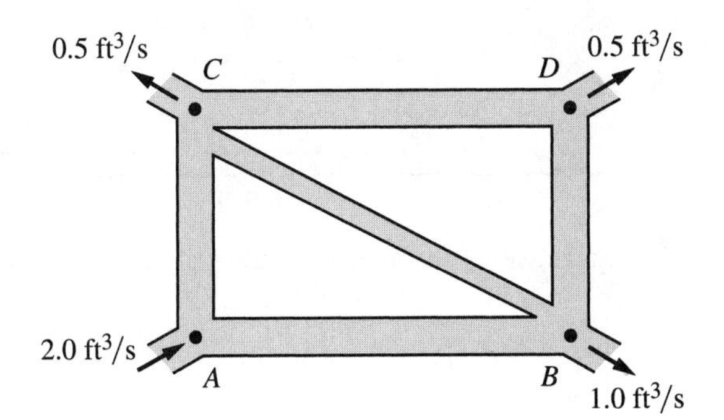

Figure P8.45 Pipe network system for Problem 8.45.

8.46. For the pipe network shown in Figure P8.46, determine the flow in each pipe. For the head loss $h_f = kQ^2$, k values for each pipe flow are as follows:

$$k_{AB} = k_{CD} = 1.0,\ k_{AC} = k_{BD} = 2.0,\ k_{BC} = 3.0$$

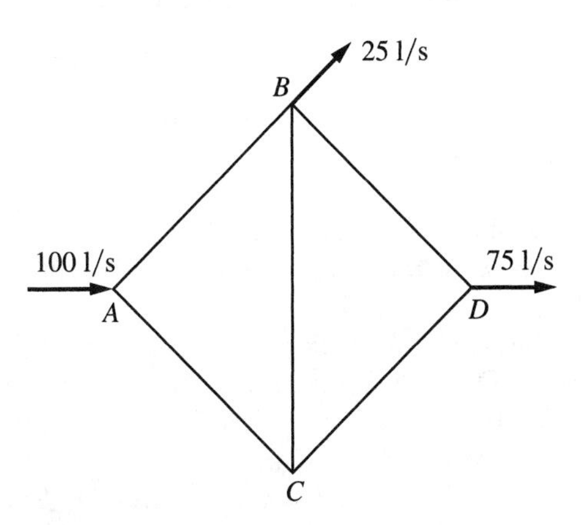

Figure P8.46 Pipe network system for Problem 8.46.

8.47. A single-loop pipe network is shown in Figure P8.47. The distance between each junction is 1000 ft and all junctions are at the same elevation. If the C value of all pipes is 100, what is the flow rate and direction in each line of the network? Determine the flow rates to within an accuracy of 2 gpm.

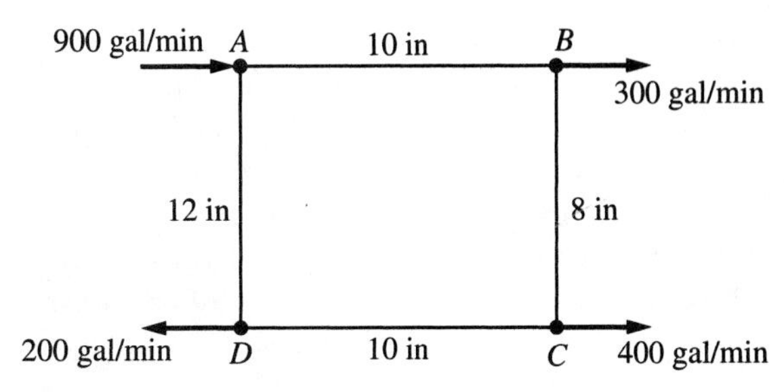

Figure P8.47 Pipe network for Problem 8.47.

8.48. A double-loop pipe network is shown in Figure P8.48. The lengths of all pipes are 700 ft and the C value for each pipe is 110. The elevations of the nodes are as follows: A (200 ft); B (150 ft); C (300 ft), D (150 ft), E (200 ft), and F (150 ft). If the pressure at node C is 40 psi, what is the flow rate and direction of flow in each line, and what is the pressure at each node?

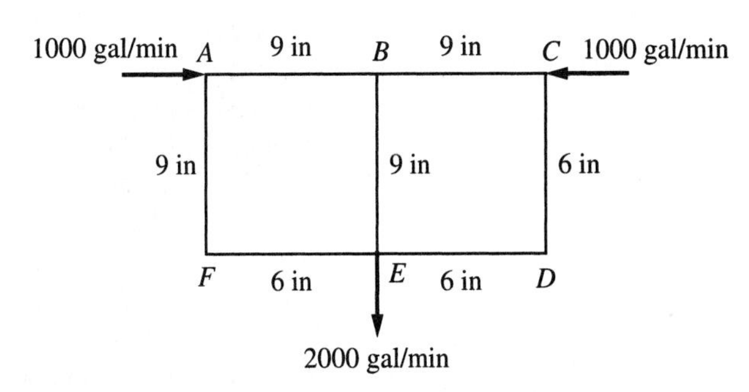

Figure P8.48 Pipe network for Problem 8.48.

8.49. A cast iron pipe has an inside diameter of 18 in and a wall thickness of 0.50 in. The pipeline is 1000 ft long and the flow velocity is 5 ft/s with a water temperature of 70°F.

a. If a valve at the end of the line is closed instantaneously, what will be the pressure increase experienced in the line?
b. Over what length of time must the valve be closed in order to create a pressure increase equivalent to that associated with instantaneous closure?

8.50. A 250 m long steel pipe issues from a reservoir as shown in Figure P8.50. The pipe runs horizontally with a diameter of 0.5 m and 20 mm wall thickness. The discharge in the pipe is 1 m^3/s. The water surface in the reservoir is 50 m above the pipe outlet as shown and the temperature is 20°C.

a. If a valve is located at the end of the line, what is the minium closure time in order to avoid a maximum pressure surge?
b. If the valve is closed faster than the time computed in part (a), what will be the maximum pressure that will develop in the line?
c. If the valve is closed in twice the time computed in part (a), what pressure surge will develop in the line?

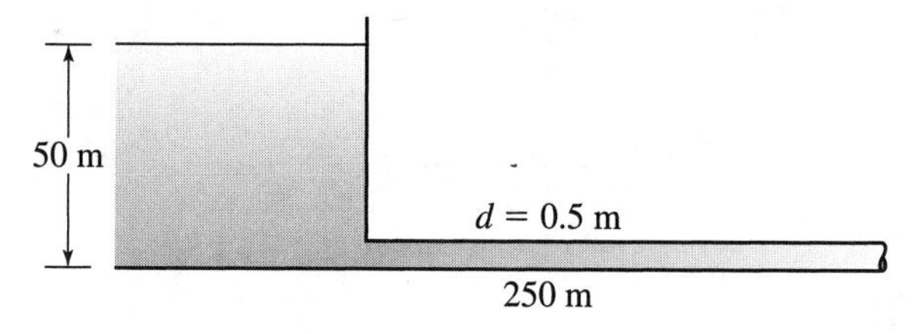

Figure P8.50 A pipe connected to a reservoir for Problem 8.50.

References

Allievi, L. Theory of Water hammer (translated by E.E. Halmos), *Transactions of the ASME:* 1929.

Beij, K. H., 1938. Pressure losses for fluid flow in 90° pipe bends, *Journal of Research of National Bureau of Standards,* 21.

Blasius, H., 1913. Das Aehnlichkeitsgessetz bei Reibungsvorgangen in Flussigkeiten, Forschung im Ingenieurwesen, No. 131.

Colebrook, C. F. and White, C. M., 1937. Experiments with fluid friction in roughened pipes. *Proceedings, Royal Society of London,* Series A, Vol. 161, pp. 367–81.

Fair, G. M., Geyer, J. C., and Okun, D. A., 1971. *Elements of Water Supply and Waste Water Disposal,* Wiley, New York.

Henderson, F. M., 1966. *Open Channel Flow.* Macmillan, New York.

Hwang, N. H. C. and Houghtalen, R. J., 1996. *Fundamentals of Hydraulic Engineering Systems,* Third Edition. Prentice Hall, Upper Saddle River, N.J.

Manning, R., 1891. On the flow of water in open channels and pipes, *Transactions of the Institution of Civil Engineers of Ireland,* Dublin, Vol. 20, pp. 161–207.

Moody, L. F., 1944. Friction factors for pipe flow. *Transactions, American Society of Mechanical Engineers,* Vol. 66, pp. 671–84.

Morris, H. M. and Wiggert, J. M., 1972. *Applied Hydraulics in Engineering.* Wiley, New York.

Parmakian, John, 1963. *Waterhammer Analysis.* Dover Publications, Inc., New York.

Prandtl, L., 1926. Uber die ausgebildeter Turbulenz. *Proceedings of the Second International Congress on Applied Mechanics,* Zurich, Switzerland, pp. 62–74.

Roberson, J. A., Cassidy, J. J., and Chaudhry, M. H., 1998. *Hydraulic Engineering,* Second Edition. Wiley, New York.

Simon, A. L. and Korom, S. F., 1997. *Hydraulics,* Fourth Edition, Prentice Hall, Upper Saddle River, N.J.

Singh, V. P., 1996. *Kinematic Wave Modeling in Water Resources: Surface Water Hydrology.* Wiley, New York.

von Karman, T., 1934. Turbulence and skin friction. *Journal of Aeronautical Science* 1(1): 1–20.

Williams, G. S. and Hazen, A., 1933. *Hydraulic Tables,* Third Edition, Wiley, New York.

Wu, C., et al., 1979. *Hydraulics.* The Chengdu University of Science and Technology Press, Chengdu, Sichuan, China.

CHAPTER 9

PUMPS

Pumps operating in a chilled water handling unit at the University of Alabama at Huntsville.

Pumps are used everywhere, as, for example, in water distribution networks and drainage systems, wastewater treatment and disposal, stormwater control, and so on. In this chapter, the focus is devoted to the basic concepts of the most common types of pumps as well as the methodology for selecting the proper pump for a given set of design conditions.

9.1 Introduction

Hydraulic machines are built to convert between mechanical and hydraulic energies. Pumps convert the mechanical energy, often developed from an electrical source, into hydraulic energy (position, pressure and kinetic energies). Special motors are designed to transform the electrical energy into mechanical energy that drives the pumps. Unlike pumps, water turbines are built to convert the hydraulic energy into mechanical energy. The mechanical energy is then used to drive generators that develop electricity. Figure 9.1 presents the setting of the main components of hydraulic machines.

Water turbines are manufactured upon demands according to the specific characteristics of the water power stations under construction and are not readily available in the market. Because each power station has its own conditions and parameters (water head, discharge, water and power demands, and so on), turbines are tailored and designed specifically according to the requirements of the power stations under consideration. Water power stations are generally of national character, large scale, and require very high investments and thus are mostly carried out by governments.

Pumps are available everywhere, and many individuals have dealt with them in some way or the other. Pumps are installed in buildings to elevate water to higher levels. A typical car includes a water pump, an oil pump, and a gas pump. Pumps are installed in gas stations for car washing and gas refilling. Pumps are installed in groundwater wells to deliver the subsurface water to the ground surface. The applications of pumps in our daily life are numerous.

All pumps operate almost under the same principle. A vacuum pressure is created in the working chamber of the pump by expelling the air. The pressure (atmospheric) on the fluid surface in the sump will then be higher than that in the working chamber of the pump; thus the fluid will be lifted to the chamber under the pressure difference. The fluid will then be pushed to the delivery pipe either by displacement or under a pressure head.

The clarity and characteristics of the water or fluid to be pumped should be considered while selecting the suitable pump. A pump that would be used to deliver wastewater should

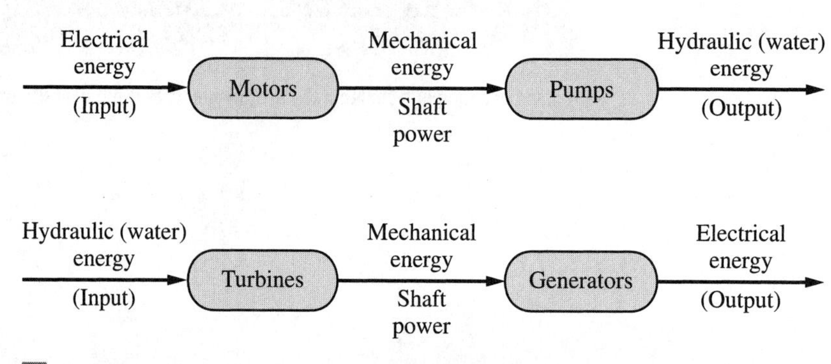

Figure 9.1 Main components of hydraulic machines.

be able to handle sediment particles, debris, corrosive products, and otherwise. Civil engineers should be able to select a pump (according to its characteristics) that would suit the required application. Mechanical engineers are responsible for the mechanical design of the pump itself, including, for example, the shape of the casing, runner dimensions, guide blade and vane angles, diffuser rings, and mechanical and overall efficiencies. In this chapter, the discussion will mostly be limited to the aspects related to civil engineers. The selection and operation of pumps and pumping systems can represent an important aspect of an engineer's career. Pumps are used in water distribution and drainage systems, desalination plants, wastewater treatment, and disposal and stormwater control.

9.2 Overall Efficiency of Hydraulic Machines

The overall efficiency, e_o, of any machine can be defined as the ratio between the output power and input power of the machine. For turbines, as shown in Figure 9.1, the input is water horsepower (WHP) and the output is shaft horsepower (SHP). For the pumps the situation is vice versa. Likewise, the efficiency of the motors, e_m, that drive the pumps is evaluated as the shaft power (SP) divided by the electrical power (EP),

$$e_m = \frac{\text{output}}{\text{input}} = \frac{\text{SP}}{\text{EP}} = \frac{T\omega}{V_e I} \tag{9.1}$$

where T is the torque exerted by the shaft, ω is the angular speed of the shaft ($2\pi N/60$) and N is the speed in revolutions per minute (rpm), V_e is the electrical voltage supplied to the motor, and I is the electrical current in amperes.

The overall efficiency of a pump can be evaluated as the water power (WP) divided by the shaft power,

$$e_p = \frac{\text{output}}{\text{input}} = \frac{\text{WP}}{\text{SP}} = \frac{\gamma QH}{T\omega} \tag{9.2}$$

where γ is the specific weight of the pumped fluid, Q is the discharge or flow rate of the pump, and H is the total head of the pump. The total head includes the static head to be lifted and all other losses that might be encountered in the system. The overall efficiency of the pump depends on several factors, including the hydraulic losses within the system, the mechanical friction at the bearings and other moving mechanical parts, eddy losses due to flow separation and secondary currents, and volumetric efficiency, which accounts for the slip of fluid through the clearances between the runner and the housing chamber. In a general form, the overall efficiency can be expressed as

$$e_o = e_h e_v e_m \tag{9.3}$$

where e_h is the hydraulic efficiency, e_v is the volumetric efficiency, and e_m is the mechanical efficiency.

9.3 Classifications of Pumps

Pumps are classified into positive (displacement) pumps and dynamic pressure (rotodynamic) pumps. In the positive pumps the fluid is guided from the suction to the delivery side by displacement. The pump has one or more chambers that are filled and emptied alternatively with the fluid to be pumped. The air in these chambers is displaced first. Such pumps have parts that

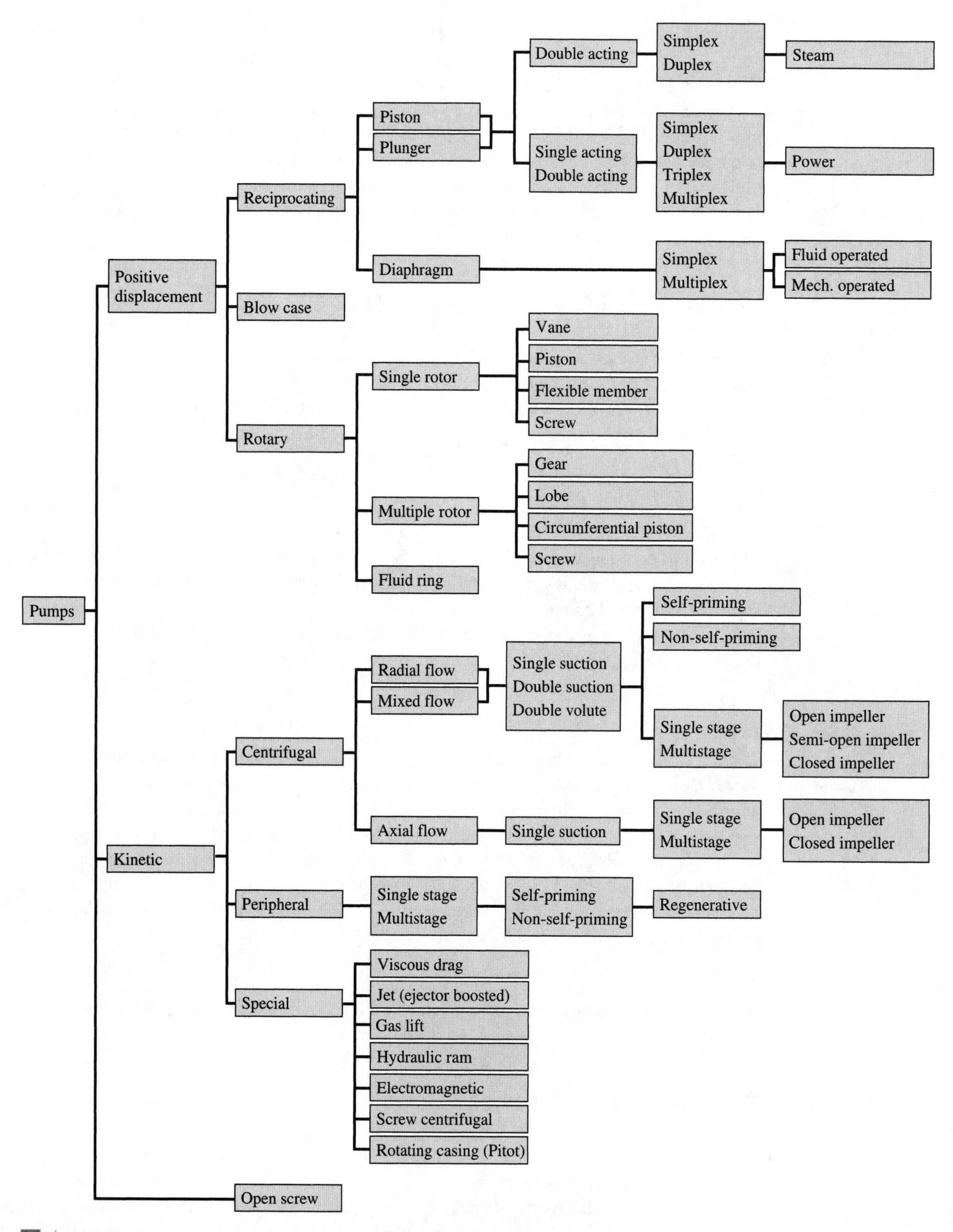

Figure 9.2 General classification of pumps (Courtesy of Hydraulic Institute).

interact in such a manner that definite volumes of fluids are displaced in the desired directions. The discharge in such pumps is mostly dependent on the speed of rotation and nominally on the working pressure. Various kinds of positive pumps exist. Reciprocating and gear pumps are the most common of the positive pumps. Displacement pumps, as compared to rotodynamic pumps, have limited but important applications.

In the dynamic pressure pumps, the fluid is guided from the suction to the delivery side by creating a pressure difference between the two sides. The mechanical energy of the shaft is transformed into pressure energy. Due to their characteristics, dynamic pressure pumps have more practical and diverse applications than the positive pumps. According to the flow path, pressure pumps are categorized into three main types: (1) radial or centrifugal pumps, (2) axial or propeller pumps, and (3) mixed-flow pumps, which represent a combination of radial and axial pumps. Figure 9.2 presents a general classification for pumps.

9.4 Positive (Displacement) Pumps

9.4.1 Reciprocating Pumps

The layout and the main elements of a typical single-acting reciprocating pump are shown in Figure 9.3. The pump consists of a ram, or plunger, that is connected to a rod which moves in a circular motion at one end, thus imparting a linear alternating motion to the ram at the other end of the rod. A crank with a radius R is connected to the rod. The ram moves forward and backward inside a cylinder (chamber) that is fitted with two self-acting lift valves (suction valve, S_v and delivery valve, D_v) to allow for flow only from the suction to the delivery side. The suction and delivery pipes are connected to one side of the cylinder as shown in Figure 9.3.

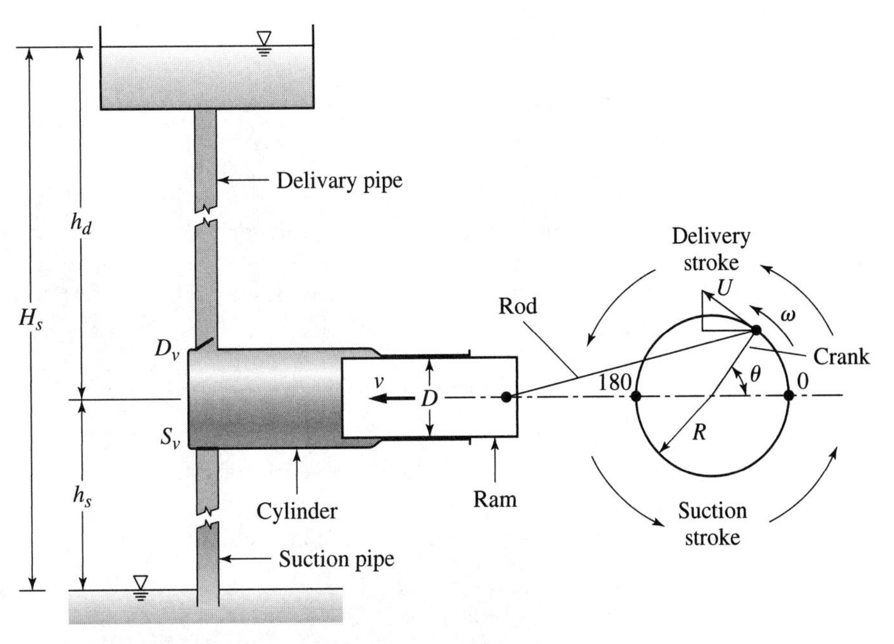

Figure 9.3 Main element of single-acting reciprocating pump.

Assuming that the crank is rotating in an anticlockwise manner, as the angle of rotation changes from $\theta = 0°$ to $\theta = 180°$, the ram moves forward inside the cylinder and this half-cycle is called the delivery stroke. During the delivery stroke the delivery valve, D_v, is self-lifted and the suction valve, S_v, is self-shut by the water. The water in the cylinder will therefore be pushed to the delivery pipe. As the crank continues to rotate from an angle of $\theta = 180°$ to $\theta = 360°$ (suction stroke), the ram reverses the motion and D_v is shut while S_v is lifted. Therefore, during the suction stroke, where the ram moves from left to right, the fluid fills the cylinder, whereas during the delivery stroke, the ram moves from right to left and the fluid is expelled from the cylinder to the delivery pipe. This process is repeated and the suction and delivery valves do not allow for return flow.

According to the notations given in Figure 9.3, h_s is the suction head, h_d is the delivery head, $H_s = h_s + h_d$ is the static head, R is the crank radius, N is the crank speed in rpm, and D is the ram diameter (approximately equal to the inner diameter of the cylinder).

Then

$$\text{Volume of water delivered in one revolution} = \frac{\pi D^2}{4} 2R \tag{9.4}$$

Assuming that no leakage (slip) of water take place from any of the valves due to lag of closure, then the theoretical discharge is given as

$$Q_{\text{th}} = \frac{\pi D^2}{4} 2R \frac{N}{60} \tag{9.5}$$

Due to leakage from the valves, the actual discharge, Q_a, is somewhat less than the theoretical discharge. Therefore,

$$Q_a = e_v Q_{\text{th}} = e_v \frac{\pi D^2}{4} 2R \frac{N}{60} \tag{9.6}$$

where e_v is the volumetric efficiency expressed as

$$e_v = \frac{Q_a}{Q_{\text{th}}} \tag{9.7}$$

Alternatively, e_v can be written as

$$e_v = \frac{100 - \text{slip}}{100} \tag{9.8}$$

The percentage of slip is generally small (less than 2%). However, high delivery pressure heads may increase the slip. According to the suction and delivery heads, h_s and h_d, the force on the ram during the suction and delivery strokes differs. These forces are given assuming that no losses take place:

$$\text{Force on the ram during suction stroke} = \gamma h_s \frac{\pi D^2}{4} \tag{9.9}$$

$$\text{Force on the ram during delivery stroke} = \gamma h_d \frac{\pi D^2}{4} \tag{9.10}$$

$$\text{The power delivered by the pump (output power)} = \gamma Q H_s \tag{9.11}$$

Note that equation (9.11) does not account for the head loss in the suction and delivery lines and other minor losses.

Example 9.1

A single-acting reciprocating pump of 20-cm ram diameter and 30-cm crank radius operates at a speed of 40 rpm. Assuming that the water slip is 2.5%, calculate the actual discharge of the pump.

Solution:

$$Q_{th} = \frac{\pi D^2}{4} 2R \frac{N}{60} = \frac{\pi (0.2)^2}{4} \times 2 \times 0.3 \times \frac{40}{60} = 0.012566 \text{ m}^3/\text{s} = 12.566 \text{ l/s}$$

$$e_v = \frac{100 - 2.5}{100} = 97.5\%$$

$$Q_a = 0.975 \times 12.566 = 12.25 \text{ l/s}$$

Example 9.2

A single-acting reciprocating pump delivers a discharge of 15 l/s when running at a speed of 55 rpm. The diameter of the ram is 25 cm and the volumetric efficiency is 96%. Determine the crank radius and the force on the ram during the delivery stroke knowing that the delivery head is 5.2 m. If the suction head is 3.8 m and the overall efficiency of the pump is 85%, calculate the power required to run the pump.

Solution:

$$Q_a = e_v Q_{th} = e_v \frac{\pi D^2}{4} 2R \frac{N}{60}$$

Hence

$$0.015 = 0.96 \frac{\pi (0.25)^2}{4} 2R \frac{55}{60}$$

Therefore, $R = 0.1736 \text{ m} = 17.36 \text{ cm}$

Force on the ram during the delivery stroke $= 9810 \times 5.2 \times (\pi/4)(0.25)^2 = 2504$ N

Water horse power delivered by the pump $= (\gamma Q H_s)/745$

$= \{9810 \times 0.015 \times (5.2 + 3.8)\}/745$

$= 1.78$ hp

Power required to run the pump $= 1.78/0.85 = 2.09$ hp

9.4.1.1 Instantaneous Rate of Discharge The actual rate of discharge in the delivery pipe at any point of time varies from zero during the entire suction stroke to a maximum value at the middle of the deliver stroke. Assuming that q and V are the instantaneous discharge and velocity, respectively, in the delivery pipe at any instant, v is the ram velocity at any instant, θ is the angle through which the crank has rotated from the outer dead center, U is the uniform linear speed of the crank-pin (moving on the circumference of a circle with radius R), and d is the diameter of the delivery pipe (Figure 9.3). Then the instaneous linear velocity of the ram (piston) is given as (Karassik, et al. 2001):

$$v = R\left[1 - \cos\theta + L\left(1 - \sqrt{1 - \frac{R^2}{L^2}}\right)(\sin\theta)^2\right] \tag{9.12a}$$

If the length of the rod, L, is relatively large as compared to R, then the instaneous linear velocity of the ram can be approximated as:

$$v = U \sin \theta \tag{9.12b}$$

Equation (9.12*b*) implies that the ram moves in a periodic motion or approximately in a simple harmonic motion. Applying the mass balance concept, the amount of water displaced by the ram should be equal to the amount of water flowing through the delivery pipe assuming no losses. Therefore,

$$\frac{\pi D^2}{4} v = \frac{\pi d^2}{4} V \tag{9.13}$$

or

$$V = v\left(\frac{D}{d}\right)^2 = U \sin \theta \left(\frac{D}{d}\right)^2 = \omega R \sin \theta \left(\frac{D}{d}\right)^2$$
$$= \frac{2\pi NR}{60} \sin \theta \left(\frac{D}{d}\right)^2 \tag{9.14}$$

The discharge through the delivery pipe, Q_p, is given as

$$Q_p = A_p V = \frac{\pi}{4} d^2 \frac{2\pi NR}{60} \sin \theta \left(\frac{D}{d}\right)^2 = \frac{\pi}{4} D^2 \frac{2\pi NR}{60} \sin \theta \tag{9.15}$$

where A_p is the cross-sectional area of the delivery pipe.

The instantaneous velocity and discharge will be a maximum when θ is set equal to 90° in equations (9.14) and (9.15), respectively. Therefore,

$$V_{\text{max}} = \frac{2\pi NR}{60}\left(\frac{D}{d}\right)^2 \tag{9.16}$$

and

$$(Q_p)_{\text{max}} = \frac{\pi D^2}{4} \frac{2\pi NR}{60} \tag{9.17}$$

The mean pipe velocity is evaluated as

$$V_{\text{mean}} = \frac{Q_{\text{th}}}{A_p} = \frac{\frac{\pi D^2}{4} 2R\left(\frac{N}{60}\right)}{\frac{\pi d^2}{4}} = \left(\frac{D}{d}\right)^2 2R\left(\frac{N}{60}\right) \tag{9.18}$$

Comparing equations (9.16) and (9.18), it is deduced that

$$\frac{V_{\text{max}}}{V_{\text{mean}}} = \pi \tag{9.19}$$

Likewise, comparing equations (9.5) and (9.17), one gets

$$\frac{(Q_p)_{\text{max}}}{Q_{\text{th}}} = \pi \tag{9.20}$$

Figure 9.4 presents the variation of the water velocity in the delivery pipe with the crank rotation angle, θ. As the crank rotates from $\theta = 0°$ to $\theta = 180°$ (delivery stroke), the ram moves forward and the water is pushed through the delivery valve, D_v, to the delivery pipe. Referring

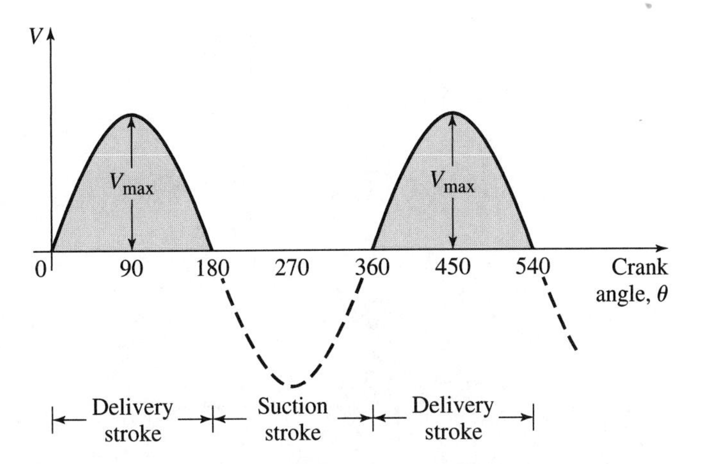

Figure 9.4 Velocity variation in the delivery pipe of a single-acting reciprocating pump.

to equation (9.14), the water velocity will increase from $V = 0$ at $\theta = 0°$ to a maximum value at $\theta = 90°$ and will then start to reduce again to reach zero at $\theta = 180°$. The suction stroke is encountered between $\theta = 180°$ and $\theta = 360°$. During the entire suction stroke no flow takes place in the delivery pipe and the water velocity in the delivery pipe is equal to zero. The same velocity cycle encountered between $\theta = 0°$ and $\theta = 360°$ is repeated again every one complete crank rotation. It should be noted that the velocity pattern in the suction pipe is a reverse of the velocity pattern in the delivery pipe.

9.4.1.2 Air Vessels An air vessel is a closed chamber that may be connected to both the suction and delivery pipes of reciprocating pumps. Air vessels provide the following advantages:

1. Continuous discharge and almost constant velocity are maintained in the suction and deliver pipes. This will also reduce acceleration and deceleration of flow inside the pipe system.
2. Considerable reduction in the friction losses inside the suction and delivery pipes is achieved.
3. Fitting an air vessel on the suction pipe would allow the pump to run at a higher speed and be set at a higher elevation above the water level in the sump.

Figure 9.5 presents a single-acting reciprocating pump with air vessels installed at the section and delivery pipes. The air vessels contain compressed air above the water stored in the vessels. At the suction side, the air must have a pressure that is lower than the atmospheric pressure to allow water to be lifted from the sump. At the delivery side, the air in the vessel is compressed at a certain pressure. According to the pressure in the suction and delivery pipes, the air vessel would take water from or release water to the flow in the pipes. During the delivery stroke, the excess of discharge from the pump, as shown in Figure 9.6a, will be diverted and stored in the air vessel at the delivery pipe. During the suction stroke, the stored water will be released to the delivery pipe providing a continuity of flow. The same concept applies to the air vessel at the suction side. The pressure of the compressed air in the air vessels should be adjusted in such a way as to provide uniform flow as much as possible at the delivery side.

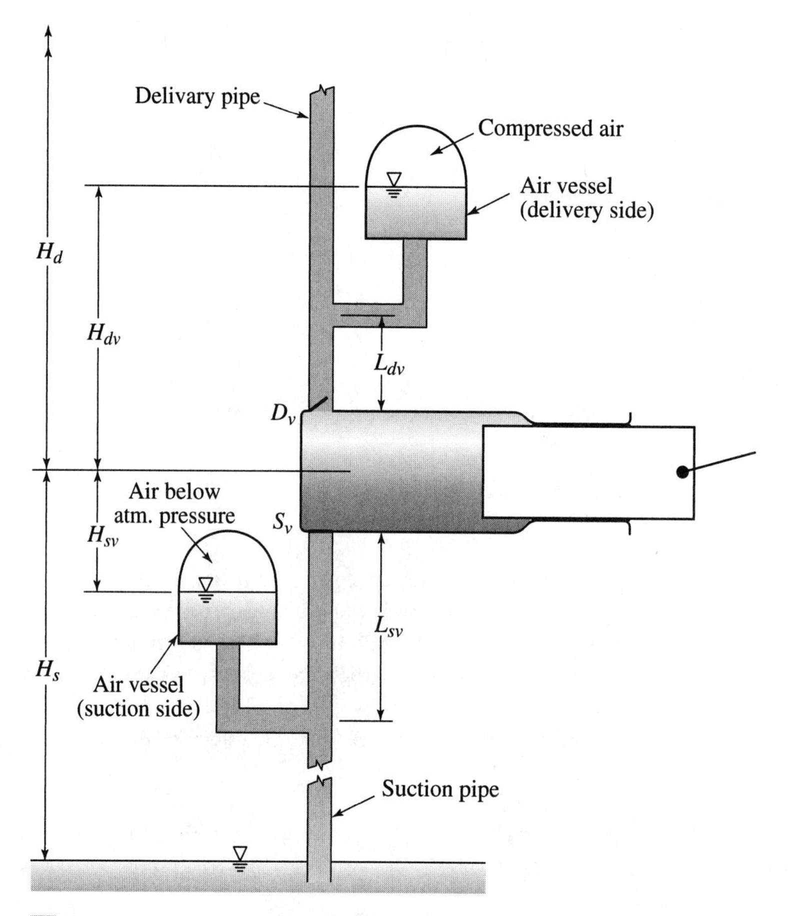

Figure 9.5 Air vessels on suction and delivery sides.

The water level and air pressure inside the air vessels will naturally fluctuate; however, these fluctuations can be controlled within certain limits by increasing the capacity of the vessels. For a single-acting reciprocating pump, the volume of the air vessel is generally six to nine times the pump displacement volume. Figure 9.6b presents flow patterns with and without air vessels.

9.4.2 Double-Acting Reciprocating Pump

In the single-acting reciprocating pump the water is encountered at one side of the ram (or piston). In the double-acting reciprocating pump the water is encountered at both sides of the piston. Each stroke of the piston carries out a suction process at one side of the piston and a delivery process at the other side of the piston as shown in Figure 9.7. The main suction and delivery pipes are diverted into two suction pipes and two delivery pipes each connecting to one side of the cylinder. Four self-opening suction and delivery valves are also fitted to control the flow direction. As the piston moves from left to right, the suction valve S_{v1} and the delivery valve D_{v1} are opened (Figure 9.7), while the other valves are kept shut. Therefore, the water is

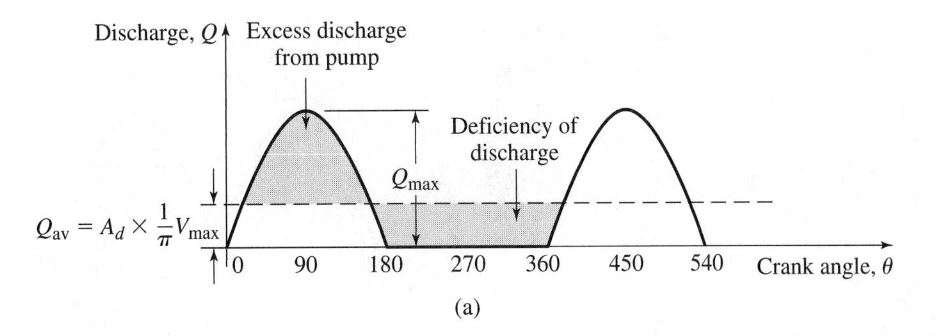

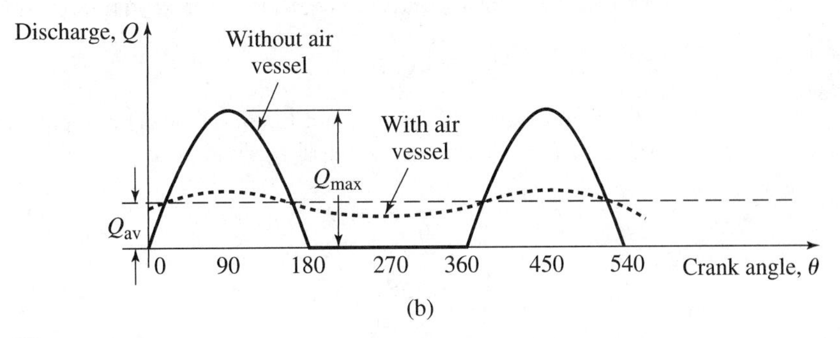

Figure 9.6 (a) Discharge variation in a single-acting reciprocating pump (without air vessels). (b) Fluctuations in discharge in delivery pipe.

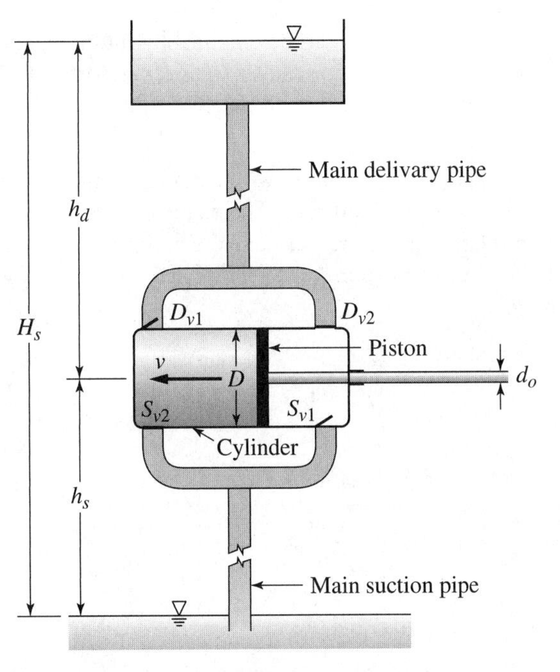

Figure 9.7 Double-acting reciprocating pump.

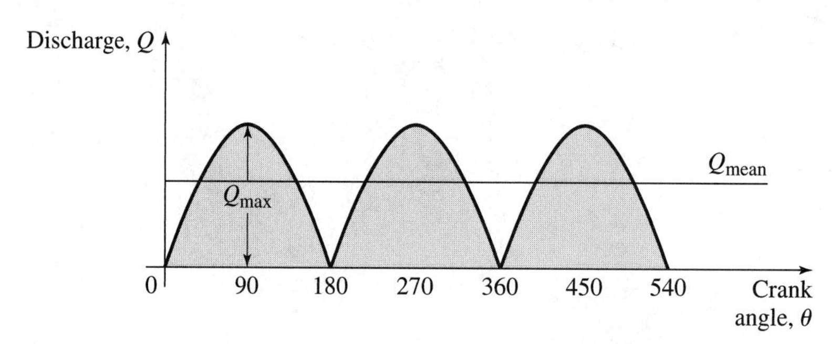

Figure 9.8 Discharge pattern versus crank angle in a double-acting reciprocating pump.

filled at one side of the piston and forced to the delivery pipe at the other side of the piston simultaneously. The theoretical discharge is given as

$$Q_{\text{th}} = \frac{\pi D^2}{4} 2R\frac{N}{60} + \frac{\pi}{4}(D^2 - d_o^2)2R\frac{N}{60}$$

$$= \frac{\pi}{4} 2R\frac{N}{60}\{2D^2 - d_o^2\} \tag{9.21}$$

where D is the diameter of the piston and d_o is the diameter of the piston rod. Assuming that D is much bigger than d_o, then

$$Q_{\text{th}} \cong 2\frac{\pi D^2}{4} 2R\frac{N}{60} \tag{9.22}$$

The actual discharge, Q_a, can be obtained by multiplying equation (9.22) by the volumetric efficiency, e_v. Figure 9.8 presents the discharge pattern versus the crank rotation angle. Air vessels can be used to provide a more uniform flow.

Example 9.3

A double-acting reciprocating pump with a crank radius of 25 cm and a piston diameter of 20 cm operates at a speed of 80 rpm. The diameter of the piston rod is 3 cm. Determine the discharge and the power required to drive the pump knowing that the static head is 8 m and the overall efficiency of the pump is 84%. Neglect the effect of water slip, and other losses in the pipeline.

Solution: Using equation (9.21),

$$Q_{\text{th}} = \frac{\pi}{4} \times 2 \times 0.25 \times \frac{80}{60}\{2(0.2)^2 - (0.03)^2\}$$

$$= 0.04146 \text{ m}^3/\text{s} = 41.46 \text{ l/s}$$

$$\text{Output water power} = \frac{\gamma QH}{745} = \frac{9810 \times 0.04146 \times 8}{745} = 4.36 \text{ hp}$$

$$\text{Power required to drive the pump} = \frac{4.36}{0.84} = 5.19 \text{ hp}$$

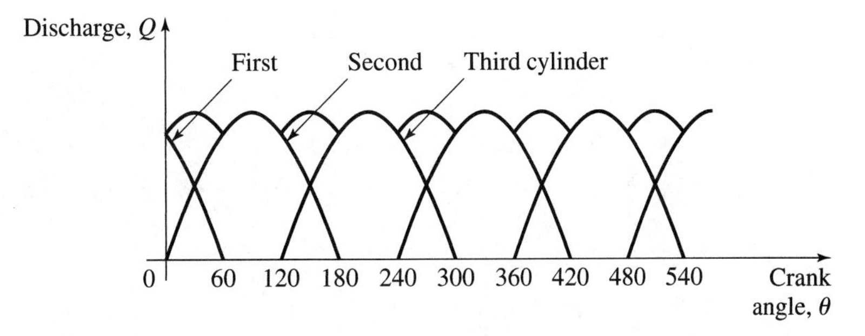

Figure 9.9 Discharge pattern versus crank angle: three cylinders set with an angle difference of 120°.

9.4.3 Multicylinder Reciprocating Pumps

One crank can be designed to drive rams of two or more cylinders. For the two-cylinder system, called a single-acting duplex pump, a difference of 180° is set between strokes of the two rams. Therefore, as the first cylinder caries out a suction stroke, the second carries out a delivery stroke. In this case, the two cylinders will act as a double-acting reciprocating pump.

For the case of a reciprocating pump with three cylinders running on one crank, called a triple pump, a difference of 120° is set between the strokes of the three cylinders. A single-acting pump is a reciprocating unit with one compression and one intake per cycle. A double-acting pump has two compression and two intake strokes per cycle. On the other hand, the number of cylinders determines whether the pump is simplex (one cylinder), duplex (two cylinders), triplex (three cylinders), or other. Figure 9.9 presents the flow characteristics for a reciprocating pump with three cylinders set with a crank angle difference of 120°. Figure 9.10 presents the flow characteristics of reciprocating pumps with different cylinder settings. Air vessels can be installed at the suction and delivery pipes to provide uniform discharge.

9.5 Dynamic Pressure Pumps

Dynamic pressure pumps move fluid through the system by creating a pressure differential between the inflow and outflow sides of the pump. The pressure differential is created through the addition of an external force applied to the liquid by means of vanes or propeller blades. This force is converted to pressure head within the pressurized housing of the pump assembly. In the case of axial flow pumps, the force is applied in the same plane and direction as the inflow, while in radial or centrifugal pumps the pressure differential is magnified through the addition of a centrifugal or radial force. Mixed flow pumps employ both concepts (i.e., forces applied both in the same direction as the flow and a centrifugal force). Mixed flow pumps are usually classified as centrifugal pumps; however, these pumps do not take full advantage of the centrifugal force concept and thus are not capable of generating the same head differential that radial pumps may generate. On the other hand, mixed flow pumps are capable of pumping larger flow rates than do fully radial pumps. In the following sections, centrifugal radial flow and axial flow pumps will be discussed in detail. Obviously, the same concepts apply to the mixed flow pumps.

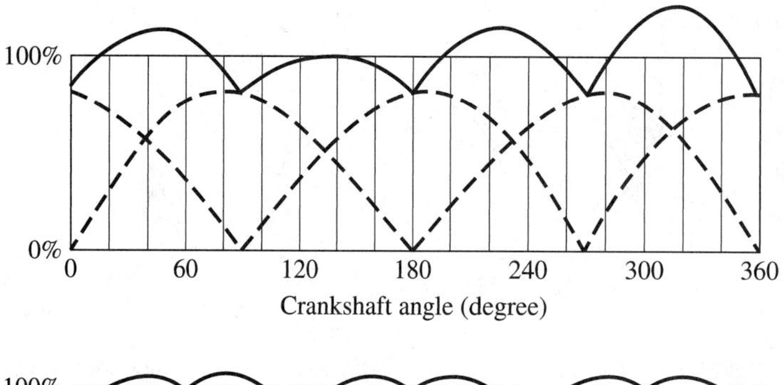

Duplex Double Acting
Average flow—100%
Maximum flow—100% + 24%
Minimum flow—100% − 22%
Total flow var.—46%

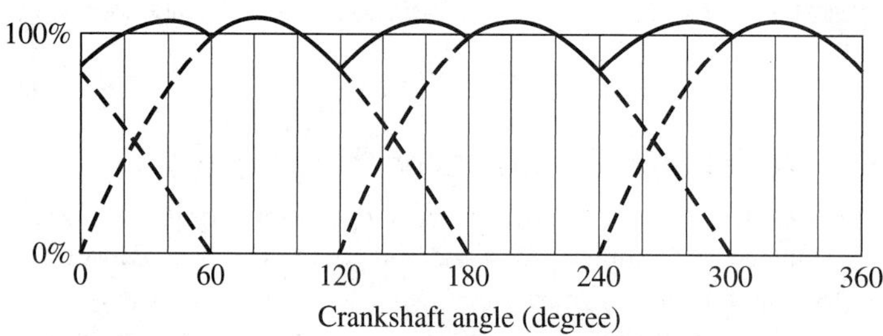

Triplex Single Acting
Average flow—100%
Maximum flow—100% + 6%
Minimum flow—100% − 17%
Total flow var.—23%

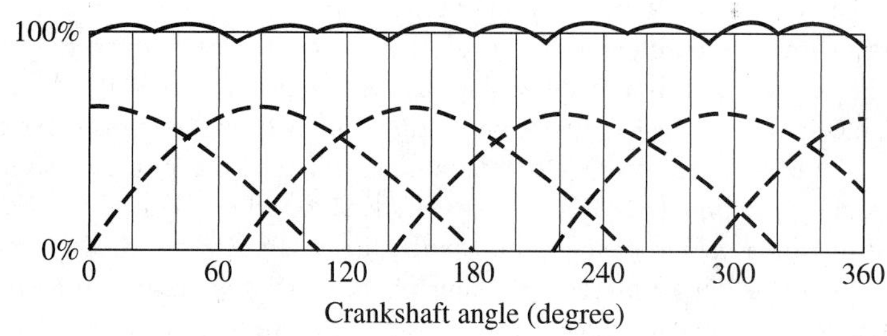

Quintuplex Single Acting
Average flow—100%
Maximum flow—100% + 2%
Minimum flow—100% − 5%
Total flow var.—7%

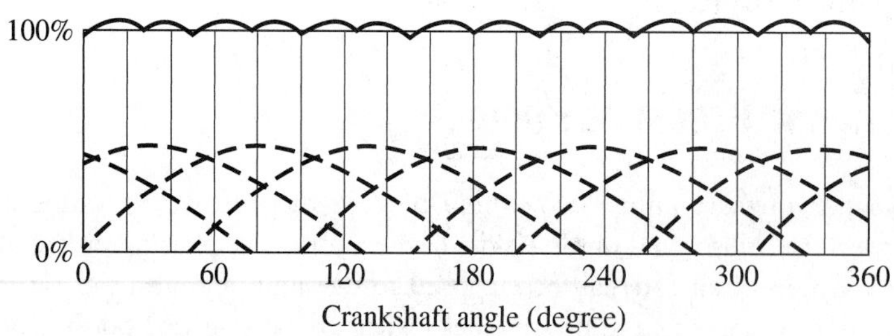

Septuplex Single Acting
Average flow—100%
Maximum flow—100% + 1.2%
Minimum flow—100% − 2.6%
Total flow var.—3.8%

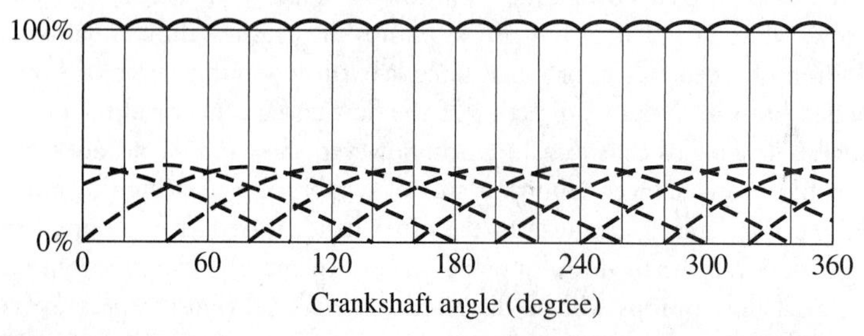

Nonuplex Single Acting
Average flow—100%
Maximum flow—100% + 0.6%
Minimum flow—100% − 1.5%
Total flow var.—2.1%

Figure 9.10 Reciprocating pump flow characteristics (Rayner, R., *Pump User Handbook*, 1995).

9.5.1 Centrifugal (Radial) Pumps

The concept of the centrifugal pump has been known since 1730, when it was first demonstrated by Demour (Hwang and Houghtalen, 1996). A simple tee made of pipes is submerged in water and the arms rotated to generate a centrifugal force. When this force is sufficient to counter the gravity forces holding the water in place, the liquid will rise and flow out of the horizontal arms of the tee. This basic principle of the centrifugal pump is still in use today with some refinements. A cross-sectional diagram of the modern pump is shown in Figure 9.11. As shown in the figure, the pump simply consists of the rotating element (known as the impeller) and a closed housing that seals the pressurized water inside. The housing may also include vanes that direct and guide the flow through the pump. Power is supplied to the impeller by a motor connected to the shaft. The resulting hydraulic operations are sketched in Figure 9.12. The

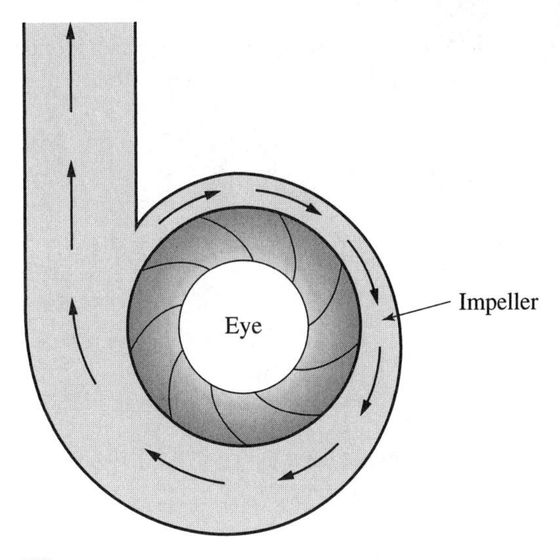

Figure 9.11 Cross section of a modern centrifugal pump.

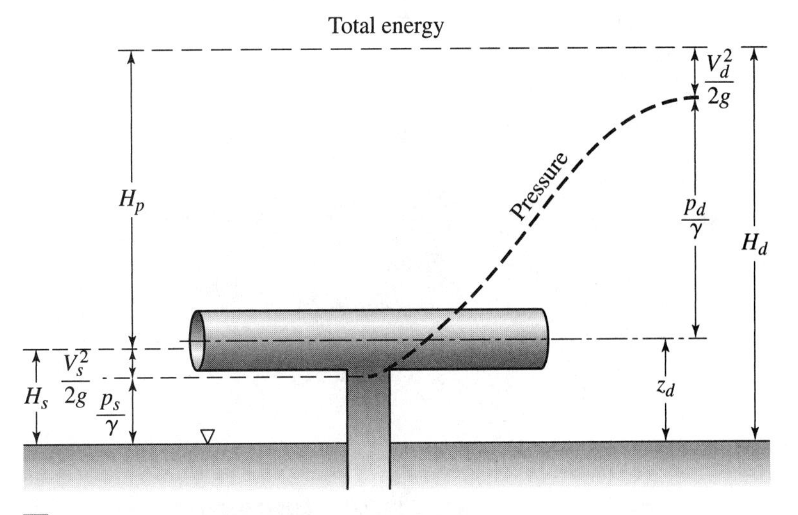

Figure 9.12 Hydraulic diagram of a centrifugal pump.

rotation of the impeller draws the water in through the eye, where the pressure is low (below atmospheric, in fact) from whence it is directed toward the higher-pressure region near the outside of the housing. The housing is designed with a gradually expanding cross section to minimize losses and increase the efficiency of the conversion of energy from velocity head (in the vicinity of the eye) to pressure head at the outlet.

The principle of energy conversion at work in the centrifugal pump is one of angular momentum. The reader may recall that linear momentum is defined by the following:

$$\text{Mom} = mV$$

where Mom = momentum, m = mass, and V = average velocity. Now, the angular momentum is just the moment of the linear momentum about a fixed axis. So

$$\text{Angular Mom} = r \times \text{Mom} = r \times m \times V \tag{9.23}$$

where r is the radial distance. Conservation of angular momentum merely means that the time rate of change of angular momentum of the fluid must equal the external moments applied, called *torque.* Then

$$T = \frac{rmV}{t} = r\rho QV \tag{9.24}$$

where T = torque applied, Q = discharge, and ρ = density. The geometry of the situation as the flow is directed from the vicinity of the eye toward the outer housing is shown in Figure 9.13.

From Chapter 5 we recall that in the linear case conservation of momentum is expressed as

$$\sum \mathbf{F} = \rho Q(\mathbf{V}_2 - \mathbf{V}_1) \tag{9.25}$$

Now, in the angular case, and referring to Figure 9.13, the torque applied to the shaft is given by

$$T = F_{t(2)}r_2 - F_{t(1)}r_1 = \rho Q(V_2 \cos\alpha_2 r_2 - V_1 \cos\alpha_1 r_1) \tag{9.26}$$

where F_t = force in the tangential direction, which is a function of the tangential velocities (i.e., $V_t = V\cos\alpha$, where the angle α refers to the angle with the tangent at the pump inlet or in the housing as shown in the figure).

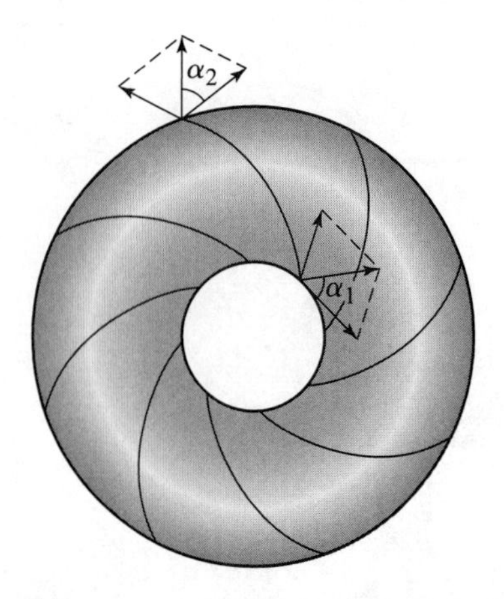

Figure 9.13 Geometric relationships of angular momentum.

9.5.1.1 Power, Head, and Efficiency Now the power *input* to the pump is just the shaft speed (ω) in radians per second times the torque, or

$$P_i = \omega T = \rho Q \omega (V_2 \cos \alpha_2 r_2 - V_1 \cos \alpha_1 r_1) \tag{9.27}$$

The angular velocity of the shaft can converted from revolutions per minute (rpm) to radians per second by

$$\omega = \frac{2\pi}{60} \times (\text{rpm}) \tag{9.28}$$

The output of the pump is described in terms of the amount of discharge that can be pumped against the amount of head generated. Figure 9.12 shows the energy relationship associated with the centrifugal pump. From the figure, one can see that the net head generated by the pump is given by the difference in the energy levels at the inlet and outlet locations, that is,

$$H_p = H_d - H_s = \frac{V_d^2 - V_s^2}{2g} + \frac{p_d - p_s}{\gamma} + z_d \tag{9.29}$$

This value represents the amount of energy (relative to the datum) that is imparted to the liquid by the operation of the pump.

The *output* power can be expressed as

$$P_o = \gamma Q H_p \text{ (ft-lb/s)} \tag{9.30a}$$

or

$$= \frac{\gamma Q H_p}{550} \text{ (horsepower)} \tag{9.30b}$$

$$\text{In the SI system, } P_o = \frac{\gamma Q H_p}{745} \tag{9.30c}$$

In the form of equations (9.30b) and (9.30c) the power is usually referred to as the *water horsepower*.

The *efficiency* of the pump is just the ratio of the output power to the input power,

$$e_p = P_o/P_i = \frac{\gamma Q H_p}{\omega T} \tag{9.31}$$

The power needed by the motor to drive the pump is called the *brake horsepower* and is computed as (Simon and Korom, 1997)

$$\text{Brake HP} = \frac{\text{water HP}}{e_p} \tag{9.32}$$

Also of interest is the efficiency of the motor that drives the pump. This efficiency is given by the ratio of the power supplied to the pump by the operation of the motor (P_i) to the power input to the motor through either electricity or fossil fuel (P_m). Thus, the overall efficiency of the pump-motor system is given by

$$e = e_p e_m = \frac{P_o}{P_i} \cdot \frac{P_i}{P_m} = \frac{P_o}{P_m} \tag{9.33}$$

Example 9.4

A pump must move a discharge of 1.1 cfs against a head of 30 ft. If the pump is driven by a 5-kW motor with an efficiency of 75%, what is the overall efficiency and the brake horsepower?

Solution: The output power generated by the pump is given by equation (9.30b), $P_o = \gamma Q H_p/550 = \frac{\gamma(1.1)(30)}{550} = 3.744$ hp. Since 1 kW = 1.34 hp, then $P_o = 2.79$ kW. Then, from equation (9.33), $e = P_o/P_m = 2.79/5 = 56\%$. Then, also from equation (9.33), $e = e_p e_m$, whence we find $e_p = 74.5\%$. Then, from equation (9.32), Brake HP $= 3.744/0.745 = 5.025$ hp.

9.5.1.2 Pump Affinity Relationships In Example 6.5, two functional relationships involving pump characteristics were derived through dimensional analysis:

$$e_p = \varphi\left(\frac{D^2\omega\rho}{\mu}, \frac{Q}{\omega D^3}\right) \tag{9.34}$$

where D is the impeller diameter and the other terms are as previously defined.

Now, a third term can be developed by using the pressure added to the system by the pump:

$$\eta = \varphi(\omega, D, \Delta P, \rho^{-1}) \tag{9.35}$$

Using the methodology employed in Example 6.5, we can derive the third term as

$$e_p = \varphi\left(\frac{\Delta P}{\rho\omega^2 D^2}\right) \tag{9.36}$$

Then, converting the pressure term to head by employing the relation $\rho = \gamma/g$, we obtain

$$e_p = \varphi\left(\frac{\Delta Hg}{\omega^2 D^2}\right) \tag{9.37}$$

It is obvious that the first term in equation (9.34) represents the effects of viscous forces (Reynolds number) and for turbulent flow conditions (high Reynolds numbers) these effects can be effectively ignored. Thus, in this case, we can write

$$e_p = \varphi\left(\frac{\Delta Hg}{\omega^2 D^2}, \frac{Q}{\omega D^3}\right) \tag{9.38}$$

So, for two geometrically similar pumps, we can write (Wurbs and James, 2002)

$$\left(\frac{\Delta Hg}{\omega^2 D^2}\right)_1 = \left(\frac{\Delta Hg}{\omega^2 D^2}\right)_2 \tag{9.39a}$$

and

$$\left(\frac{Q}{\omega D^3}\right)_1 = \left(\frac{Q}{\omega D^3}\right)_2 \tag{9.39b}$$

The expression given in equation (9.39a) is known as the *head coefficient,* while that given in equation (9.39b) is called the *discharge coefficient.*

It is also possible to combine the head, discharge, and shaft speed into one factor, known as the *specific speed* of the pump, that is,

$$\eta_s = \frac{\omega Q^{0.5}}{(H_p g)^{0.75}} \tag{9.40}$$

where η_s = specific speed, ω = shaft speed (rad/sec), Q = discharge (m^3/sec), and H_p = pumping head (m). The specific speed is a dimensionless parameter that has many uses in pump analysis. It is widely used to compare operating conditions of different pumps and to compare performance characteristics of the same pump operating under different head and discharge conditions.

For a given centrifugal pump, there is a shaft speed at which the pump operates with maximum efficiency. Although the pump may (and usually is) able to operate at other speeds, the

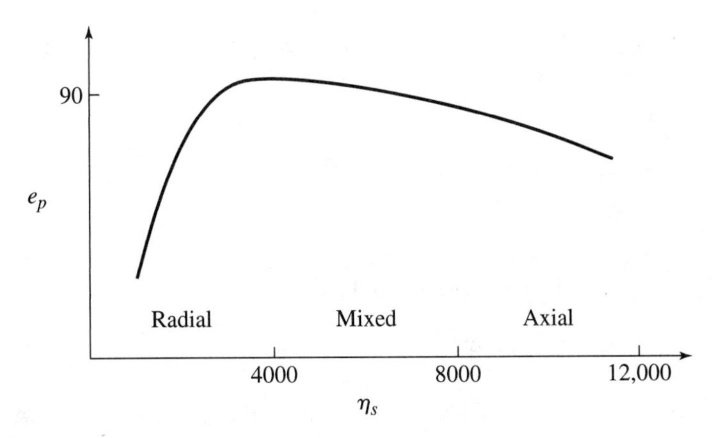

Figure 9.14 Relationship between specific speed and efficiency.

efficiency will decline at any speed other than the optimum. Figure 9.14 shows a typical relationship between pump efficiency and specific speed. Pumps with low specific speeds are usually employed to deliver small flows against high heads, while those with high specific speeds do the reverse. Centrifugal radial flow pumps operate at specific speeds of between 500 and 3500 and are capable of pumping relatively small discharge amounts ($Q < 1000$ gpm) while generating large heads, axial flow pumps operate at much larger specific speeds (7500–15,000) and deliver relatively large flow rates ($Q > 5000$ gpm), but generate dynamic heads of less than 12 m (Gupta, 1989; Hwang and Houghtalen, 1996). Mixed flow pumps are used to meet requirements between radial and axial flow pumps and generally operate at specific speeds between 3500 and 7500 with flow rates greater than 1000 gpm. Gupta (1989) noted that, in general, pumping efficiency decreases significantly when the specific speed drops below 1000, and that for all pumps and all specific speeds the efficiency is lower for smaller pump capacities. The relationships among pumping head, discharge, efficiency, and horsepower are supplied by the manufacturer and are known as *pump characteristic charts* or curves.

Example 9.5

A pump is turning 1200 rpm to pump a discharge of 500 l/s against a total head of 10 m. Now, if the rpm's are increased to 1500, what discharge would be pumped against the same head? What is the specific speed of this pump?

Solution: From equation (9.39b) we know that $\left(\frac{Q}{\omega D^3}\right)_1 = \left(\frac{Q}{\omega D^3}\right)_2$. Now, since we are using the same pump in both instances, the impeller diameter, D, remains the same so that the equation can be rearranged such that $\frac{Q_1}{Q_2} = \frac{\omega_1}{\omega_2} = \frac{500}{Q_2} = \frac{1200}{1500} = 0.8$. Then $Q_2 = 625$ l/s. The specific speed is given by equation (9.40). First converting the specific speed to rad/sec:

$$\omega = \frac{2\pi}{60} \times 1200 = 125.66 \text{ rad/sec}$$

$$\eta_s = \frac{\omega \times Q^{0.5}}{(H_p \times g)^{0.75}} = \frac{125.66 \times (0.5)^{0.5}}{(10 \times 9.81)^{0.75}} = 2.85$$

Example 9.6

Suppose the head that the pump in Example 9.5 is operating against is increased to 15 m. What rpm would the pump turn in order to move the same discharge? What would be the specific speed in this case?

Solution: From equation (9.39a) we know that $\left(\frac{\Delta Hg}{\omega^2 D^2}\right)_1 = \left(\frac{\Delta Hg}{\omega^2 D^2}\right)_2$. Again, since the D remains the same, the expression can be reduced to $\frac{\Delta H_1}{\Delta H_2} = \frac{\omega_1^2}{\omega_2^2}$. Then $\frac{10}{15} = \frac{(1200)^2}{\omega_2^2}$, from which we find that $\omega_2 = 153.9$ rad/sec.

Again, from equation (9.40), $\eta_s = \dfrac{153.9 \times (0.5)^{0.5}}{(15 \times 9.81)^{0.75}} = 2.57.$

It can be observed from the previous examples that the specific speed (and thus from Figure 9.14, the efficiency) of the pump decreased when the pumping head was increased. Thus, one can use the pump affinity relations and specific speeds to help select the pump that will operate at the optimum efficiency.

9.5.1.3 Pumping Head In practical problems, the engineer wishes to select the pump that will move the desired discharge (or range of discharges) and result in a specified set of conditions at the point of destination with the greatest efficiency. A typical example is shown in Figure 9.15. As shown in the figure, the pump must supply the energy to meet the destination requirements, including increased station head (elevation), required line pressure, velocity head associated with the flow rate, and friction losses in the line (h_f). Minor losses associated with

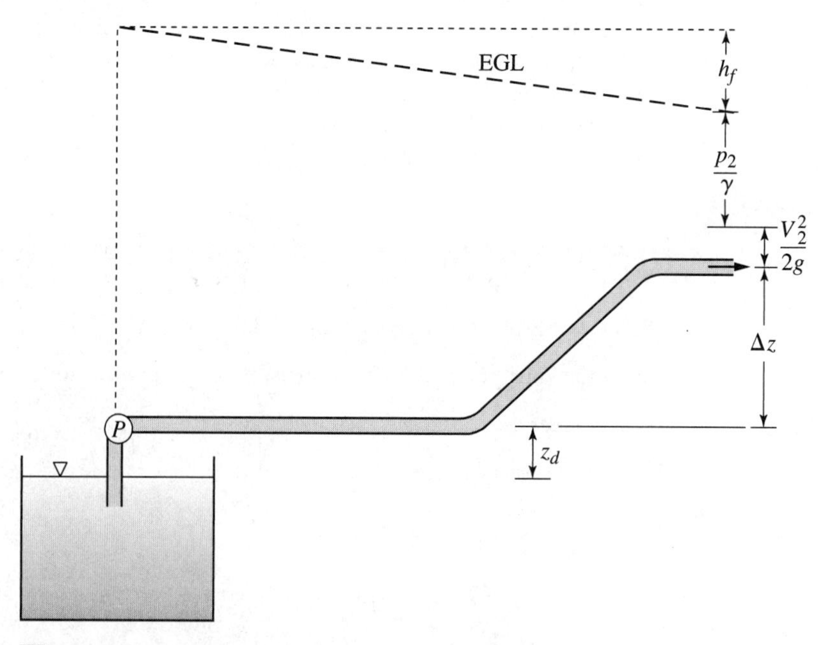

Figure 9.15 A water distribution system.

bends, contractions, valves, and so on should also be added where necessary. The normal situation is that some required flow rate be pumped against a certain change in station head and that some required pressure be available in the line at the destination point. Of course, the flow of water through the pipe system generates some head loss that must also be overcome by the pump.

The computation of the required pumping head involves many of the concepts and methods discussed in Chapter 8. The following example will serve to demonstrate the methodology for computing pumping head.

Example 9.7

A 12-inch-diameter cast iron pipe line 1000 ft long is used to carry 1000 gpm to an elevation 10 ft above the source of the water. In addition, it is necessary to have a pressure of 10 psi in the line at the destination location. Determine the required pumping head for the pump.

Solution: The situation is shown in Figure 9.16.

The required pumping head will be computed as

$$H_p = \Delta z + \frac{p_2}{\gamma} + \frac{V_2^2}{2g} + h_L$$

So $p_2 = 10 \text{ psi} = \dfrac{10(144)}{62.4} = 23.07$ ft of pressure head (energy) and a discharge of 1000 gpm $= \dfrac{1000}{60(7.5)} = 2.22$ cfs. The area of the pipe is $\pi d^2/4 = 0.785 \text{ ft}^2$. So the resulting velocity in the pipe is $Q/A = 2.83$ ft/s. Now, we must find the head loss that will be generated by this flow rate. From Table 7.1, we find that the ε value for cast iron is 0.00085 ft, so that $\varepsilon/d = 0.00085$. Now, the Reynolds number is

$$R_n = \frac{\rho V d}{\mu}$$

Now assuming a water temperature of 65°F, we find that the dynamic viscosity (μ) is 2.196×10^{-5} lb-s/ft^2. Using these values, we determine the Reynolds number to be 2.50×10^5. Then, using the Moody diagram given in Figure 8.4, we determine the Darcy f to be 0.0205. Then,

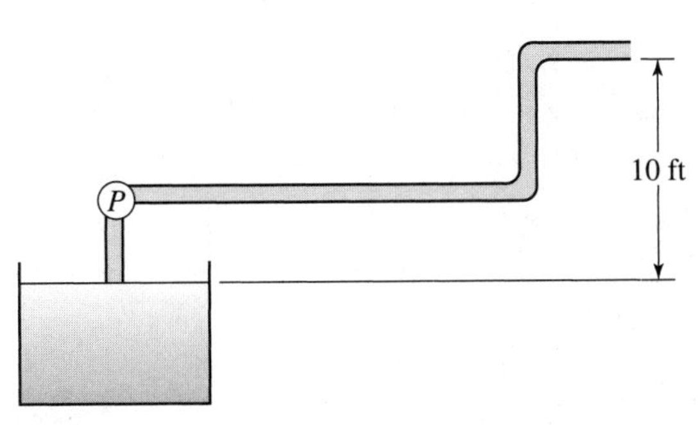

Figure 9.16 Schematic for Example 9.7.

from the Darcy equation, we find the projected friction loss to be

$$h_f = \frac{fL}{d}\frac{V^2}{2g} = \frac{0.0205(1000)}{1}\frac{(2.83)^2}{2g} = 2.55 \text{ ft}$$

Ignoring the minor losses associated with the bends, we determine the pumping head to be

$$H_p = 10 + 23.07 + 0.12 + 2.55 = 35.74 \text{ ft}$$

9.5.1.4 Pump Selection The selection of the proper and most efficient pump to meet specified flow and head requirements is often an important part of the civil engineering profession. Civil engineers often work with water supply systems, water or sewage treatment plants, or drainage networks. All of these commonly encountered situations involve analysis of flow requirements and selection of pumps.

The first step in pump selection is to develop the requirements that must be met by the pumping system. The system includes the pipe(s), appurtenances (such as valves, gates, etc.), the static head requirements, and the requirements that must be met at the discharge end (such as pressure or velocity). If these items are known, then a unique head-discharge (*H-Q*) curve can be obtained for the system by computing and plotting the required head (including losses) versus several selected discharges. Such a curve is shown in Figure 9.17.

Pump manufacturers also provide sets of performance (sometimes called characteristic) curves (or charts) for the pumps in their inventory. These curves are available in various formats, but in some manner they show the complete information about the pumps, including pumping head, discharge, horsepower, shaft speed (rpm) and efficiency. The performance characteristics of the pump operating under a range of conditions are determined experimentally in order to produce these charts. In some cases, a single chart may show all of the relevant information. In other situations, a separate chart is given for each pump based either on the horsepower of the pump or its shaft speed in rpm. Figure 9.18 shows a typical curve of the first type, while typical curves of the second type are shown in Figures 9.19 and 9.20. The *H-Q* curve developed for a given situation can then be superimposed on the pump performance charts and the proper pump can be selected. This will normally be the pump that meets the minimum requirements with the greatest overall efficiency and thus the minimum operating costs.

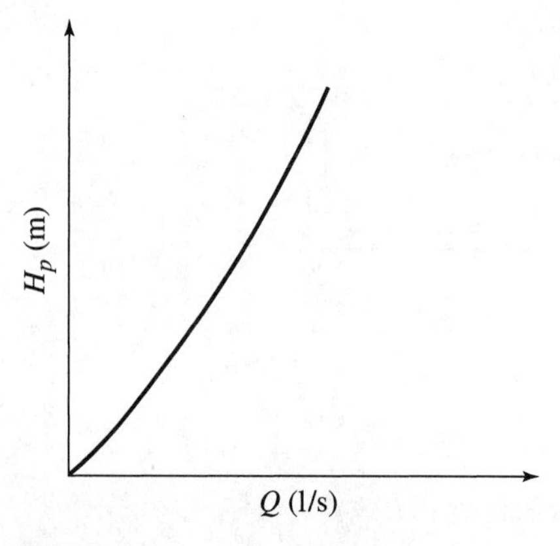

Figure 9.17 Typical head-discharge curve.

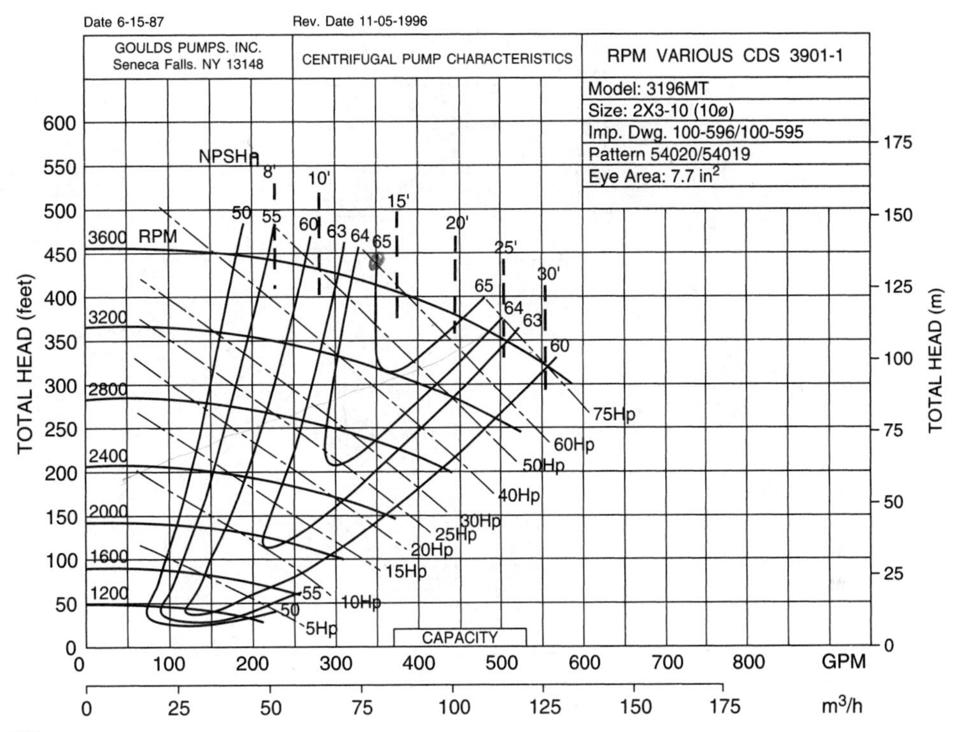

Figure 9.18 Performance curves for various pumps.

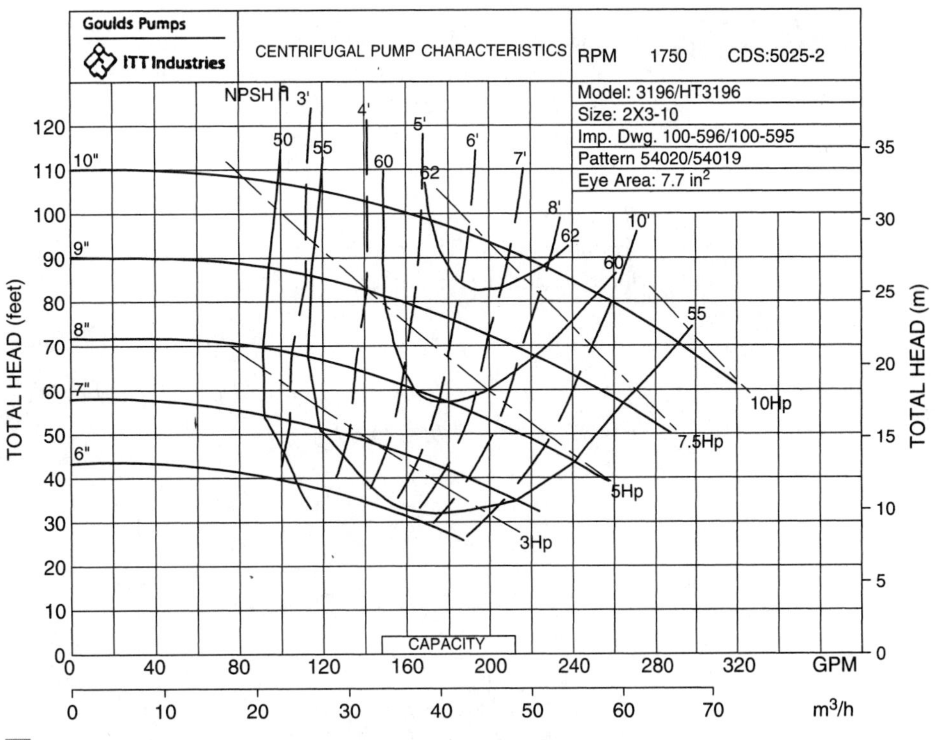

Figure 9.19 Performance curves for 1750-rpm pump.

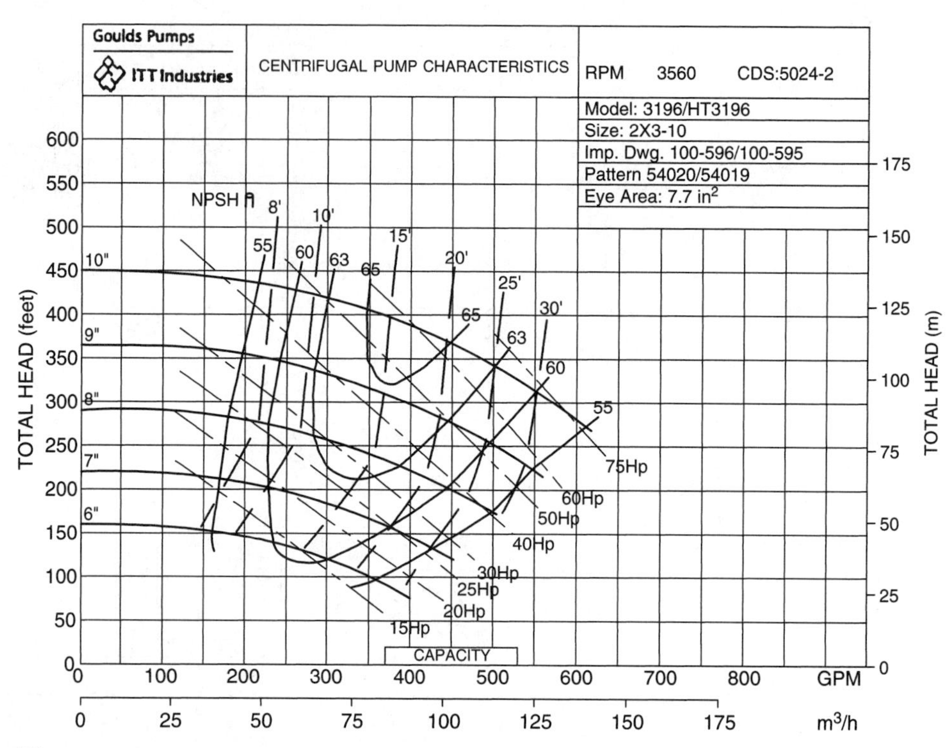

Figure 9.20 Performance curves for 3560-rpm pump.

Example 9.8

It is necessary to select a pump to deliver water from a source to a location 25 m higher in elevation. A 0.25-m cast iron pipe 1000 m long will be used. At the destination end of the pipe it is required that a pressure of 50,000 N/m^2 be available. The required flow rate is 25 l/s and the water temperature is 20°C. Using the pump characteristic curves given in Figures 9.18, determine the optimum pump (hp) for this situation.

Solution: The situation is depicted in Figure 9.21. The first step is to develop the *H-Q* curve for this situation. The static head requirements are the difference in elevation between the two locations and the required pressure. Thus

$$H_s = 25 + \frac{p}{\gamma} = 25 + \frac{50000}{9810} = 30.1 \text{ m}$$

Then, ignoring any minor losses, the total pumping head will be

$$H_p = 30.1 + h_f + (V^2/2g)$$

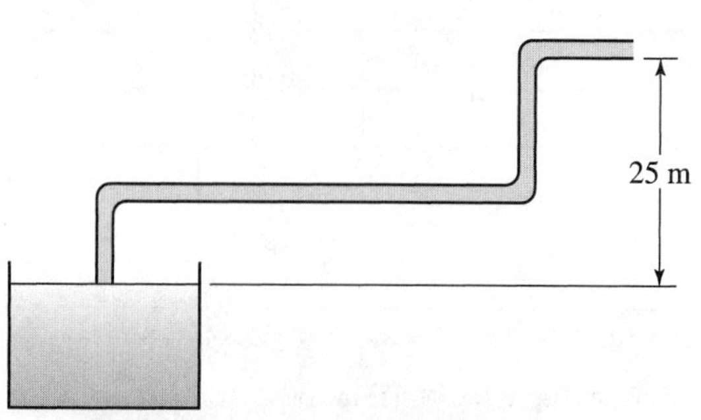

Figure 9.21 Schematic for Example 9.8.

The process proceeds as follows:

Water temperature = 20°C; $\mu = 1.002 \times 10^{-3}$ N-s/m^2; $d = 0.25$ m, $A = 0.049$ m^2

So let $Q = 36$ m^3/hr = 0.01 m^3/s; then $V = Q/A = 0.01/0.049 = 0.2$ m/s

$$\text{So } R_n = \frac{\rho V d}{\mu} = \frac{1000(0.2)(0.25)}{1 \times 10^{-3}} = 5 \times 10^4$$

Using Table 7.1, we find that the ε value for cast iron is 0.26 mm, or 0.00026 m; thus $\varepsilon/D =$ 0.00104. Using the Moody diagram (Figure 8.4), we find that for this situation the Darcy friction factor would be 0.0245. Then $h_f = \frac{0.0245(1000)}{(0.25)}\frac{(0.2)^2}{2g} = 0.2$ m. The total pumping head for this flow rate is 30.1 + 0.2 + 0.002 = 30.3 m.

Following this process, the following table was generated

Q (l/s)	Q (m^3/hr)	H (m)
10	36	30.3
25	90	31.2
40	144	32.9

From the table, one can see that the minimum requirements are for a discharge of 25 l/s against a total head of 31.2 m. Then the H-Q curve is plotted on the characteristic curves given in Figure 9.18 as shown in Figure 9.22. Looking at the results, it appears that, all else being equal, the proper selection would be the 25-hp pump, with a shaft speed of about 2400 rpm whose efficiency is approximately 54%.

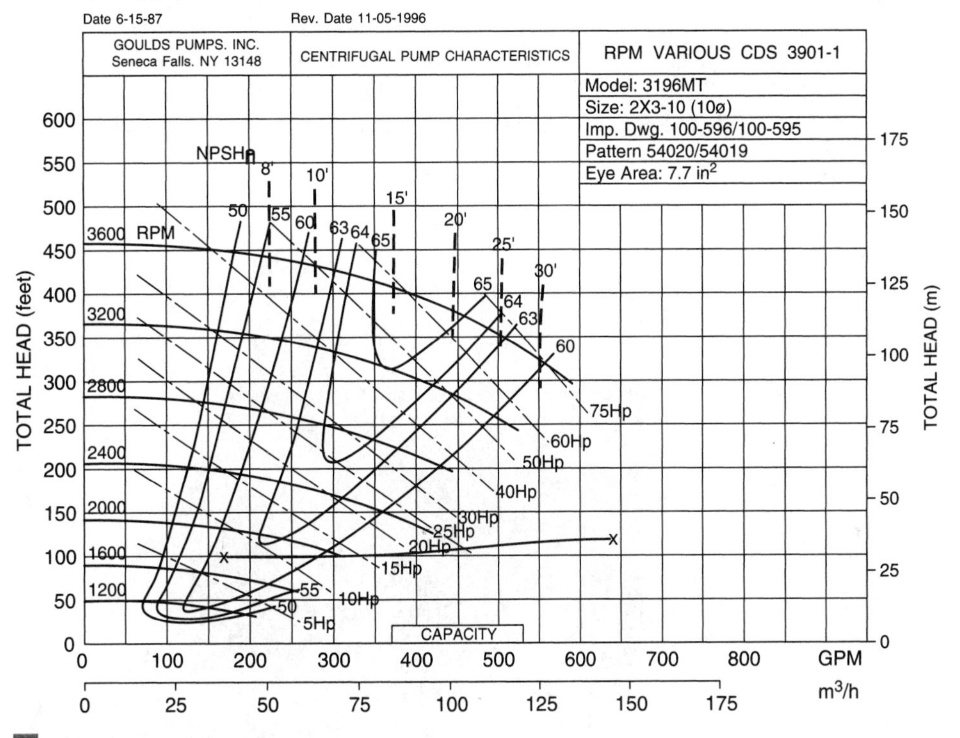

Figure 9.22 Solution for Example 9.8.

9.5.1.5 Siting or Locating Pumps In the examples considered above, the pump was actually located at some unspecified elevation above the water level in the source. This is an important consideration in pump design. It should be noted that the pressure in the inlet side of the pump is less than atmospheric or is in suction. Of course, this is necessary in order for the water to rise to the inlet of the pump. The action is one of a *siphon*, or a negative pressure causing water to rise above its initial station. There is also some energy loss in the inlet pipe due to friction. The total head at the suction side of the pump (i.e., the location of the pump above the water level + the velocity head + the energy loss) is called the total suction head (H_s). If the total head at the suction side of the pump is less than the vapor pressure of water for the particular temperature of the operating conditions, then the water will turn to vapor and a condition known as *cavitation* will occur. In this condition, vapor bubbles form in the fluid and then burst when they reach the high-pressure region of the pump casing. The resulting force can cause excessive vibration of the pump housing and can cause significant damage to the pump.

When cavitation occurs, the actual, or absolute, pressure of the water at the inlet has been reduced to the vapor pressure. The difference between the total absolute head at the suction side of the pump and the vapor pressure head is known as the *net positive suction head* (NPSH). Note that the head at the suction side of the pump includes all the energy terms at that point. Thus

$$\text{NPSH} = \left(\frac{p_s}{\gamma} + \frac{V_s^2}{2g}\right) - \frac{p_v}{\gamma} \tag{9.41}$$

Note that in equation (9.41) the pump elevation term z_s cancelled because it would appear in both the suction and vapor pressure head terms. The terms p_s and p_v are the pressures at the suction side of the pump and the vapor pressure of the fluid, respectively. The fluid vapor pressure is a function of its temperature and is given in Tables 2.1 and 2.2. Thus, with the NPSH known, the maximum safe elevation of the pump above the source can be determined. NPSH curves are also given on pump performance curves as shown in Figures 9.18–9.20.

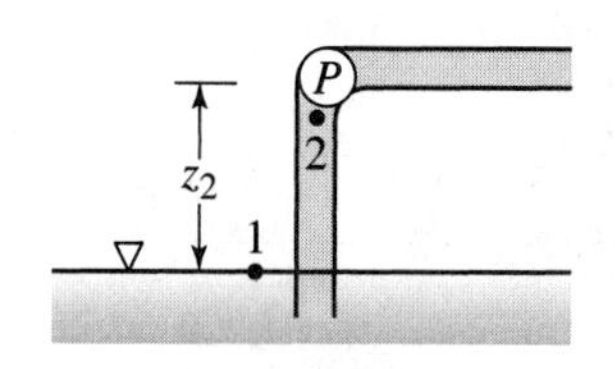

Figure 9.23 Site location for a pump.

The calculations are simply a matter of applying Bernoulli's equation from the surface of the source to the suction side of the pump as shown in Figure 9.23. Thus

$$z_1 + \frac{p_1}{\gamma} + \frac{V_1^2}{2g} = z_2 + \frac{p_s}{\gamma} + \frac{V_s^2}{2g} + h_L \tag{9.42}$$

Then, assuming that the velocity there is zero, as given by Chin (2000):

$$\text{NPSH} = \frac{p_1}{\gamma} - \left(\Delta z + h_L + \frac{p_v}{\gamma}\right) \tag{9.43}$$

where h_L is the cumulative head losses in the line between the source and the pump.

A submersible pump.

Another allied issue is the one of pump priming. Due to the energy relationship on the suction side of the pump, the suction pipe and the entire housing and impeller of the pump must be filled with water in order for the pump to function. If water is allowed to drain out when the pump is not in operation, then the priming will be lost and the pump will have to be primed before operation can begin. For this reason, a nonreturn valve is usually fixed at the end of the suction pipe to hold water in the system unless the pump is self-priming.

Example 9.9

Water is to be pumped from a source under the following conditions: $p_{atm} = 14.1$ psia; water temperature $= 68°$F; flow rate $= 1000$ gpm. Now, suppose that given the load requirements of the pump (elevation head, destination pressure, etc.) we have selected a pump from one of our pump characteristic charts and find that the NPSH of our pump is 20 ft. Let the diameter of the suction line be 6 inches. Under the given conditions, what is the maximum height that the pump can be installed above the surface of the source?

Solution: From the given conditions, we find $Q = 1000$ gpm $= 2.22$ cfs; $A = \frac{\pi(0.5)^2}{4} = 0.196$ ft^2; $p_v = 0.34$ psia for $T = 68°$F (Table 2.2); $V_1 = 0$; $V_2 = 2.22/0.196 = 11.37$ ft/s. Then, writing the Bernoulli equation from the source to the intake side of the pump,

$$z_1 + \frac{14.1(144)}{\gamma} + 0 = z_2 + \frac{p_2}{\gamma} + \frac{(11.37)^2}{2g} + h_L$$

and, from equation (9.41),

$$\text{NPSH} = 20 = \frac{p_2}{\gamma} + \frac{V_2^2}{2g} - \frac{p_v}{\gamma} = \frac{p_2}{\gamma} + \frac{(11.37)^2}{2g} - \frac{(0.34)(144)}{\gamma}$$

from whence we find the limiting pressure head at 2; $\frac{p_2}{\gamma} = 18.785$ ft. So $z_1 + 32.54 = z_2 + 18.785 + 2 + h_L$.

Then, from the above we find that $(z_2 - z_1) + h_L = 11.75$ ft. Since we do not know the material of the intake pipe, we cannot say with certainty what the total head losses might be. However, assuming $f = 0.02$ (a fairly reasonable assumption), and taking $L = 11$ ft, we find that $h_f = \frac{(0.02)(11)}{0.5}(2) = 0.88$ ft. Assuming an entrance loss coefficient of 0.1, then the total losses in this situation would be $0.88 + 0.1(2) = 1.08$ ft. Thus, Δz is approximately 10 to 11 ft. The NPSH curves indicate that if the pump is placed at a level higher than this value above the surface of the source, then cavitation will occur.

9.5.2 Axial Flow (Propeller) Pumps

The basic principle of the axial flow pump is the same as in the centrifugal pumps; however, the actual operation of the pump is somewhat different. Although the pump operates according to the principle of momentum conservation, in the axial flow case, the momentum is conserved along the plane of the axis of the pump. Figure 9.24 shows the hydraulic principles of the axial flow pump.

The principle at work in this case is simple linear momentum conservation:

$$\sum \mathbf{F} = \rho Q(\mathbf{V}_2 - \mathbf{V}_1) \tag{9.44}$$

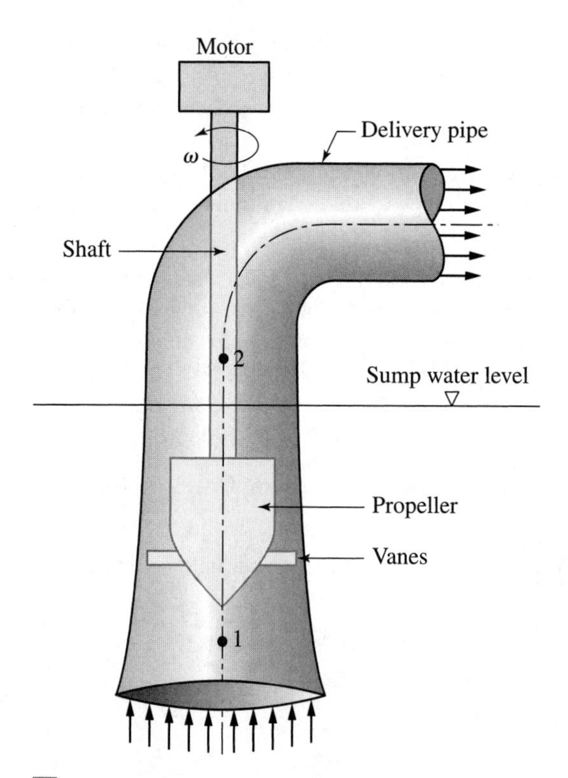

Figure 9.24 Schematic of axial flow pump.

where the terms are as previously defined. Now, referring to Figure 9.24, and treating the pump section as a control volume, one can write (Hwang and Houghtalen, 1996)

$$p_1A_1 - p_2A_2 + F_p = \rho Q(V_2 - V_1) \tag{9.45}$$

where F_p represents the force imparted to the water by the propeller. Now, assuming that $A_1 = A_2$, then equation (9.45) becomes

$$F_p/A = p_2 - p_1 \tag{9.46}$$

Equation (9.46) implies that the force exerted on the water by the propeller takes the form of a pressure increase from point 1 to point 2 (i.e., before and after the propeller). Then the total energy (head) developed by the pump would be given by

$$H_p = \frac{p_2}{\gamma} - \frac{p_1}{\gamma} \tag{9.47}$$

Of course, the output power generated by the pump would be computed in the usual way (i.e., $P_o = \gamma Q H_p$) and the efficiency of the pump would just be the ratio of the output and input powers as usual.

As stated in Section 9.5.1.2, axial flow pumps are employed in situations where fairly large flow rates are required to be pumped against relatively low heads. Axial flow pumps operate at large specific speeds (7500–15,000) and pump relatively large flow rates ($Q > 5000$ gpm) but generate fairly low dynamic heads (less than 12 m).

9.6 Pumps Operating in Combination

At times, it may be more efficient to have multiple pumps operating in some combination in a line than it would be to have one very large pump. This situation often arises in cases were the conditions (required discharge or pumping head) may vary dramatically with time. In these cases, only those pumps that are required to meet the particular discharge and head requirements of the moment can be employed and the others can be shut off to save power. The discussion from the previous section also implies that if the specific speed of the pumps can be maintained at a high level, then the efficiency will be increased.

The most common combination for pumping systems is in a parallel configuration as shown in Figure 9.25a. The pumps are connected to a manifold that is then connected to the

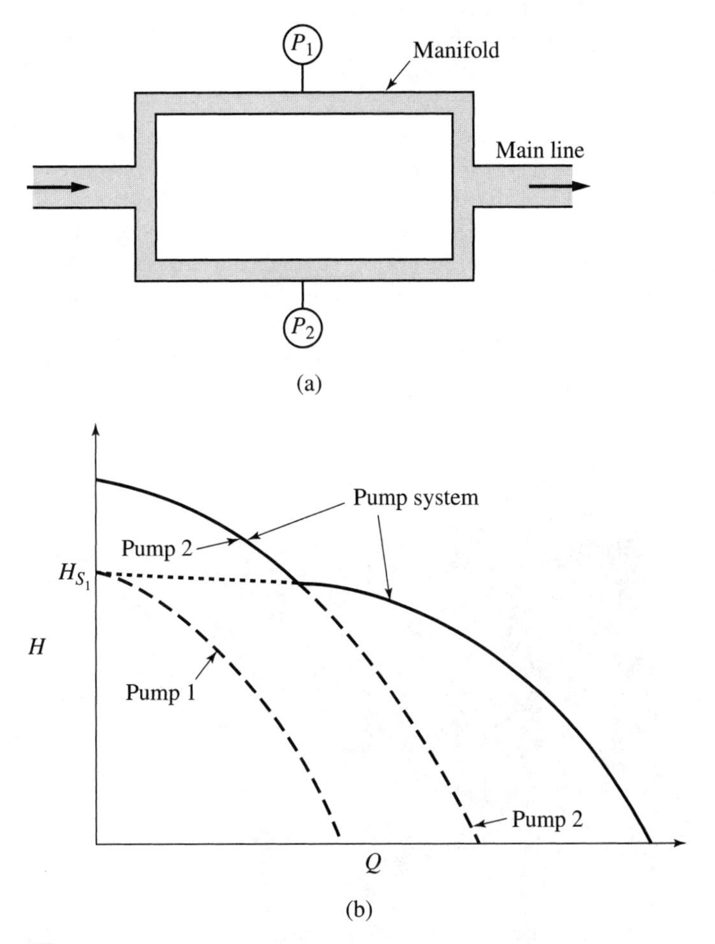

Figure 9.25 (a) Two pumps operating in parallel. (b) Characteristic curve for two pumps operating in parallel.

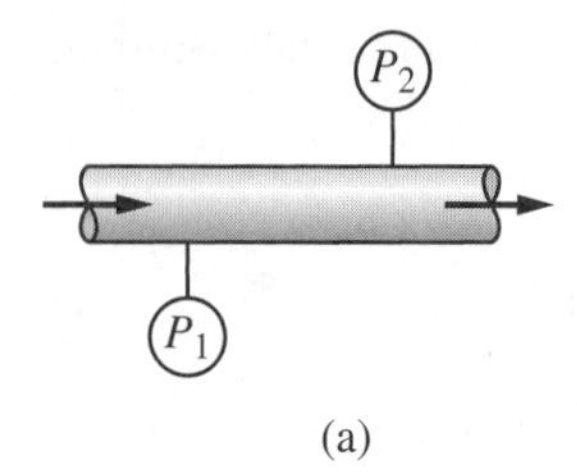

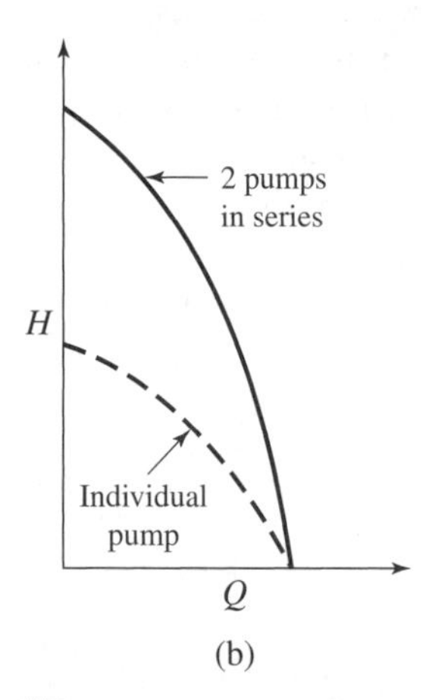

Figure 9.26 (a) Two pumps operating in series. (b) Characteristic curve of two pumps operating in series.

main distribution line as shown. The effect of this configuration is that the discharge is increased but the head remains the same as it would be for one pump since the pressure across each wing of the manifold must be equal. The discharge is computed by merely summing the discharge for each pump at the particular operating head obtained from its individual characteristic curve. However, the increased head loss due to the separate members of the manifold must also be taken into account, so that two identical pumps operating in parallel would not exactly double the discharge for a given head. A typical characteristic curve for two pumps operating in parallel is shown in Figure 9.25b. The schematic depicts a case of two nonidentical pumps in parallel, with pump No. 2 operating during the time that the head requirements are above the operating limit of pump No. 1 (H_{s_1}, or "shut-off head") and then both pumps are in operation for heads less than H_{s_1}.

Conversely, two pumps operating in a series alignment, as shown in Figure 9.26a, would result in a system that could operate in higher heads with the discharge remaining essentially unchanged from that of the single pump. The pressure head that can be developed by the series configuration is again computed by just summing the heads of each pump at the particular pumping discharge obtained from its individual characteristic curve [Figure 9.26b]. It should be noted that the two pumps in Figure 9.26 have the same H-Q characteristic curve. Again, however, two identical pumps operating in series will not develop exactly twice the pressure head of a single pump due to increased resistance in the line.

The effects of pumps in combination can also be obtained within a single pump through unique pump construction. A *double-suction* pump has a two-sided impeller that allows fluid to enter both sides, thus increasing the discharge while the head remains constant as in the case of pumps in parallel. The advantage is that the energy losses are nearly the same as for a single pump; thus the discharge is effectively doubled. Conversely, a *multistage pump* consists of two or more impellers within a single housing. The flow from one impeller feeds the suction side of the next one and so on. This operation is similar to pumps in a series configuration; thus the discharge remains fixed while the head is increased by each impeller. As shown in Figure 9.14, efficiencies of single pumps are normally small for low specific speeds. With multistage pumps, the specific speed can be kept high, thus maintaining optimum efficiencies against high heads.

Two centrifugal pumps operating in parallel.

The procedure for analysis of pumps in combination is essentially the same as for single pumps. A characteristic curve of the pump system is obtained (as in Figures 9.25 and 9.26) and then an operating curve is computed from the requirements of the pipe system as in Example 9.8. The operating curve is superimposed on the characteristic curve and the operating conditions of the pump combination can be read from the graph.

READING AID

9.1. What is the main function of hydraulic machines? Which machine is used to convert the hydraulic energy to mechanical energy?
9.2. Gives some examples for the uses of pumps in our daily life. Can the human heart be regarded as a pump?
9.3. Describe the main process in the operation of all kinds of pumps. Would the characteristics of the pumped water or fluid affect the selection of the pump?
9.4. Define the overall efficiency of hydraulic machines. Write the equations for the overall efficiency of pumps and turbines.
9.5. Discuss the factors affecting the overall efficiency of the pumps.
9.6. What are the main differences between displacement and rotodynamic pumps? Classify the rotodynamic pumps according to the flow path.
9.7. Sketch a single-acting reciprocating pump and describe the operation mechanism of the pump. What is the function of the suction and delivery valves in reciprocating pumps?
9.8. Using the notations in Figure 9.3, derive an equation for the theoretical discharge from reciprocating pumps.
9.9. Why is the discharge of the single-acting reciprocating pump not uniform and not continuous? Sketch the theoretical discharge of the pump against the angle of rotation of the crank. What is the relation between the maximum discharge and the theoretical discharge of this pump?
9.10. Explain the advantages of including air vessels in reciprocating pump systems. Are they connected to the suction or delivery pipes?
9.11. Sketch the flow pattern of a single-acting reciprocating pump connected to an air vessel. What is the approximate size of the vessel as compared to the pump displacement volume?
9.12. Is the flow of double-acting reciprocating pumps continuous and uniform? How many suction valves are fitted in such a pump?
9.13. What is the main benefit of the multicylinder reciprocating pumps? Sketch the flow characteristics of a triplex reciprocating pump.
9.14. How is the pressure difference between inflow and outflow sides of dynamic pressure pumps created?
9.15. Explain the concepts of operation of axial, radial, and mixed flow pumps. Discuss the advantages of the three different types of the rotodynamic pumps.
9.16. Why is the casing (housing) of the centrifugal pumps generally designed with gradual increase in the flow area as the fluid moves from the eye of the impeller to its outer edge?
9.17. Using the momentum principles and the notations given in Figure 9.13, derive an expression for the torque applied to the shaft of a centrifugal pump.

9.18. What is meant by the *water horsepower* and the *brake horsepower?* Explain how both are calculated.
9.19. Why is the output power (shaft power) of the motor called *brake horsepower?*
9.20. Using the concepts of dimensional analysis, find the head coefficient and the discharge coefficient for two geometrically similar pumps.
9.21. Can the similarity condition [equation (9.39)] be used for the same pump under different head and discharge?
9.22. Write an expression for the specific speed of the pump and check its dimensions.
9.23. Discuss the range of the specific speed for radial (centrifugal), mixed, and axial flow pumps. Which pump would you select to deliver large volumes of water against a head of 10 feet?
9.24. Explain how the total pumping head is calculated accurately for a pump delivering water from an underground tank to an elevated tank.
9.25. Define the term *net positive suction head* (NPSH) of a centrifugal pump. From your point of view, would it be better to lower the pump as much the possible near the sump water level? Why?
9.26. Would you prefer to fix throating valves and flow meters at the suction or delivery pipes of a centrifugal pump. Why or why not?
9.27. Explain how to avoid cavitation in the suction pipes of centrifugal pumps. Which point in the suction pipe has the least pressure?
9.28. Can axial flow pumps operate without a suction pipe? Explain.
9.29. What is the main benefit of operating a number of pumps in a parallel configuration? If the required lifting head is higher than the total head of the available pumps, would you recommend a parallel or series configuration in the pumping system?
9.30. What are the advantages of double-suction pumps and multistage pumps. Which of them would you recommend for pumping groundwater from deep aquifers?

Problems

9.1. A single-acting reciprocating pump of 25 cm ram diameter and 40 cm crank radius operates at a speed of 50 rpm. Calculate the volumetric efficiency of the pump if the actual discharge is 32.07 l/s.

9.2. A single-acting reciprocating pump of 25 cm ram diameter and 20 cm crank radius delivers a discharge of 200 l/s with a volumetric efficiency of 97%. What is the operating speed of the pump? Knowing that when the suction head is 3.0 m and the overall efficiency of the pump is 85%, the power required to run the pump is 25 hp, find the maximum delivery head of the pump. What is the force on the ram during the delivery stroke?

9.3. A double-acting reciprocating pump with a crank radius of 25 cm and a piston diameter of 22 cm operates with an overall efficiency of 85%. If the power required to run the pump is 50 HP and the piston rod diameter is 3 cm, what is the operating speed of the pump knowing that the static head is 7.5 m? Neglect the effect of water slip.

9.4. A double-acting reciprocating pump with a crank radius of 30 cm and a piston diameter of 20 cm operates at a speed of 75 rpm. The piston rod diameter is 2.5 cm. Determine the discharge and the power required to run the pump if the static head is 8.5 m and the overall efficiency is 87%. Neglect the effect of water slip.

9.5. The head-discharge relationship for a certain pump can be represented by the equation $H = 28 - 9Q^2$. The pump is fixed 2.5 m above the water surface in a river and it forces the water to a level 7.8 m above the pump. The suction and delivery pipes are 16 m and 764 m long, respectively, and each pipe is 0.8 m in diameter. The pipe coefficient of friction is 0.005. Estimate the time required to lift 1500 tons of water from the river by using this pump. Draw the hydraulic grade line and the total energy line for the pumping system.

9.6. A 750-kW motor drives a pump with an efficiency of 70% and the pump operates with an efficiency of 80%. Determine the flow rate of the pump if it must overcome an energy head of 75 ft.

9.7. If the pump in Problem 9.6 must overcome an energy head of only 50 ft, what would be the efficiency for the same flow rate? What if the required head was increased to 85 ft?

9.8. A 100-kW motor with an efficiency of 75% drives a centrifugal pump delivering a flow of 100 l/s against a head of 30 m. What is the pump efficiency?

9.9. A submersible pump, whose characteristics are given in the following table, has been lowered to the bottom of a circular deep water pond 6 m in diameter. The pump is fitted with a discharge line 100 m in length and 20.5 cm in diameter. The line rises vertically from the pump and then lies on a horizontal surface; the coefficient of pipe friction is 0.0075. The water level in the pond was 4 m below the ground surface. Find the time required to reduce the water level by 16 m, assuming there is no inflow into the water pond.

Head (m)	30	25	20	15	10	5
Discharge (l/s)	10	70	105	125	135	140

9.10. A centrifugal pump with four stages in parallel delivers 180 l/s of water against a manometric head of 25 m, the diameter of the impellers is 23 cm, and the pump's speed is 1700 rpm. Another pump is to be designed with a number of stages in series, of similar construction to those of the first pump (same specific speed), to run at 1250 rpm and to deliver 240 l/s against a manometric head of 245 m. Find the diameter of the impellers and the number of stages required.

9.11. A pump is required to operate at 2000 rpm with a head of 8 m and a discharge rate of 0.12 m^3/s. Determine the type of the required pump.

9.12. A pump was intended to run at 1750 rpm when driven by a 0.5-hp motor. What is the required power rating of a motor that will turn the pump at 2000 rpm?

9.13. A pump delivering 1100 l/s of water is turning at 900 rpm. The pump adds 7 ft of head to the water.

a. What is the specific speed of this pump?

b. If the pump is a double-suction pump operating under the same conditions, what would be the specific speed?

9.14. At a level of 480 m below the ground surface it is required to pump groundwater with $Q = 50.0$ l/s to a tank whose water level is 14 m above ground surface using a multistage pump running at a speed of 1500 rpm. The pump has a specific speed of 85 m and overall efficiency of 75%. The pump is connected with a pipeline 500 m long, 22 cm diameter, and 0.008 pipe friction factor. Calculate the required motor power. If the pump works 10 hours/day, calculate the energy cost during one month if energy cost is 0.01 $/hr for each 1 hp.

9.15. 1.25 cfs of water at 70°F are pumped from the bottom of a tank through 700 ft of 4-in steel pipe. The line includes a 50-ft rise in elevation, two 90° elbows, a wide-open gate valve, and a swing check valve. The inlet pressure is 50 psi and a working pressure of 20 psi is needed at the end of the line. What is the necessary power developed by the pump (kW) to accomplish this task?

9.16. 80 gpm of water at 80°F are lifted 12 ft vertically by a pump through a total length of 50 ft of a 2-in-diameter copper line as shown in Figure P9.16. The discharge end of the line is submerged in 8 ft of water as shown. What head must be added by the pump?

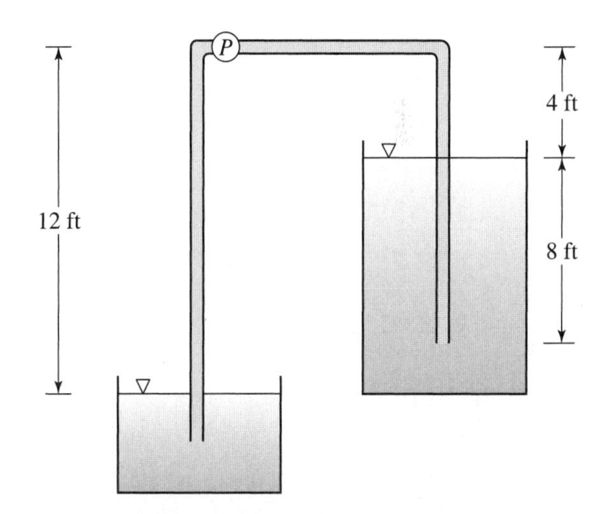

Figure P9.16 Pumping system for Problem 9.16.

9.17. A 20-hp motor drives a centrifugal pump whose efficiency is 70%. The pump discharges water at 60°F with a velocity of 12 ft/s into a 6 in steel line. The inlet is 8 in steel pipe. The pump suction is 5 psi below atmospheric and the friction loss in the system is 10 ft. The suction and dischage lines are at the same elevation. What is the maximum height above the pump inlet that water is available at standard atmospheric pressure?

9.18. A pump station is used to fill a tank on a hill above from a lake below as shown in Figure P9.18. The flow rate is 10.5 l/s at a temperature of 16°C. The pump is 4 m above the lake and the tank water level is 115 m above the pump. The suction and discharge lines are 10.2-cm-diameter steel pipe. The equivalent length of the inlet line between the lake and the pump is 100 m while the total equivalent length between the lake and the tank 2300 m. The overall efficiency of the pump and motor is 70%.

a. What is the required power of the motor in kW?
b. What is the NPSH for this application?

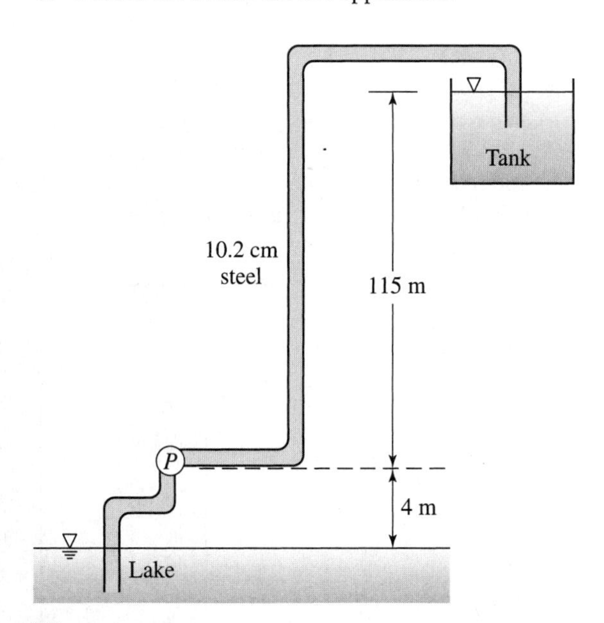

Figure P9.18 Pumping system for Problem 9.18.

9.19. A pump transfers 3.5 mgd of water from the clear well of a 10-ft-by-20-ft sand filter to a higher elevation as shown in Figure P9.19. All related data are as shown in the figure. The pump efficiency is 85% and the motor efficiency is 90%. Friction and minor losses are insignificant.

a. What is the static suction head in this case?
b. What is the static discharge head?

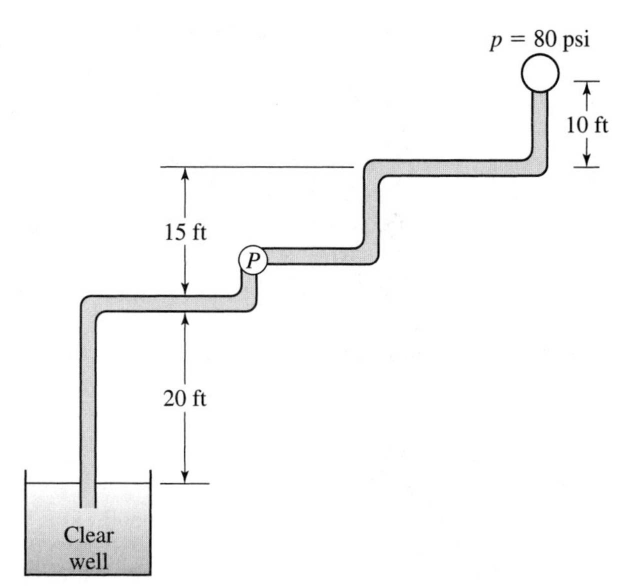

Figure P9.19 Water pumped from a clear well for Problem 9.19.

c. What is the total dynamic head required by the pump?
d. What motor power is required?

9.20. It is desired to pump a flow rate of 20 l/s through a 10-cm-diameter pipe 1000 m long to a location whose invert is 30 m above the water source from which the pump is drawing. Assume that the water temperature is 20°C and the pipe is new cast iron. Select the "best" pump in terms of efficiency and horsepower using Figure 9.18.

9.21. It is necessary to pump a flow rate of 10 l/s from a source to a destination that is located at an elevation 15 m above the source. The water will flow through a 10-cm-diameter wrought iron pipe that is 100 m long. The water temperature is 20°C. Using Figure 9.19, determine the best pump for this job in terms of horsepower and efficiency.

9.22. Using Figure 9.18, do the preliminary selection of a pump to deliver water to a reservoir at a flow rate of 30 l/s. The difference in station head between the source and destination is 20 m and the water is delivered through a 15-cm galvanized iron pipe 100 m long. The flow is controlled by a check valve in the pipeline.

9.23. A centrifugal pump is used to facilitate flow to a building located at an elevation 50 ft above the source of water. A 12-inch-diameter pipe 1000 ft long carries the flow. Now, for a flow rate of 500 gpm, it is desired to have a pressure of 30 psi at the end of the line. Using the appropriate figure in the text, select the proper pump for this task. Give the pump speed, horsepower, and efficiency of your selection.

9.24. A centrifugal pump running at 1400 rpm has the characteristic curve shown in Figure P9.24. The pump will be installed in an existing line with known head requirements given by the formula $H = 30 + 2Q^2$, where H is the system head in ft and Q is the flow rate in cfs. The pump efficiency is 86%.

a. What is the flow rate if the pump is turning at 1400 rpm?
b. What power is required to drive the pump?
c. What is the flow rate if the pump is turned at 1200 rpm?

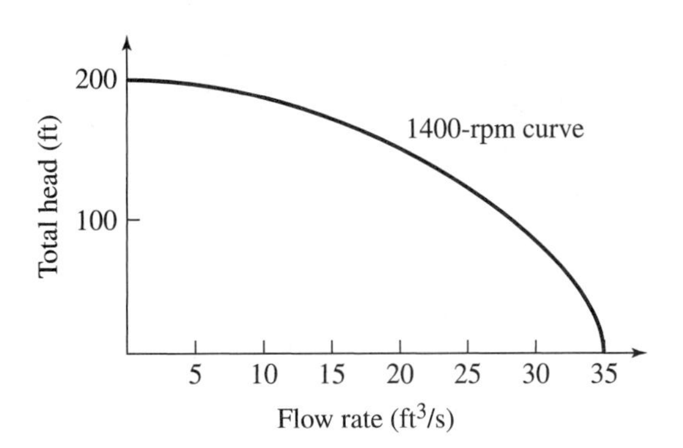

Figure P9.24 Performance curve of a pump running at 1400 rpm for Problem 9.24.

9.25. If the NPSH for the pump operating under the conditions specified in Problem 9.20 is 10 ft, how far off of the water surface could this pump be placed without engendering cavitation? Assume a water temperature of 20°C.

9.26. If the NPSH for the pump operating under the conditions as in Problem 9.21 is 7 ft, at what height above the water surface could this pump be placed without risking cavitation?

9.27. If the NPSH for the pump selected in Problem 9.23 is 15 ft, how far above the water surface could this pump be located without risking cavitation? Assume a water temperature of 68°F.

9.28. A centrifugal pump, whose characteristics are given in the following table, is installed in the system given in Figure P9.28.

Q (l/s)	18.9	25.5	31.5	37.8	44.0
H_m (m)	39.0	37.8	36.0	33.2	28.0
NPSH required (m)	0	1.5	3.4	5.5	8.2

a. What discharge will be produced by the system if the friction coefficient is $f = 0.005$?
b. Will the pump experience cavitation? Given that the vapour pressure head = 0.715 m and the atmospheric pressure head (H_a) = 9.52 m of water, if the answer is yes, specify changes that might be made to avoid cavitation.
c. What is the maximum allowable value for z_s?

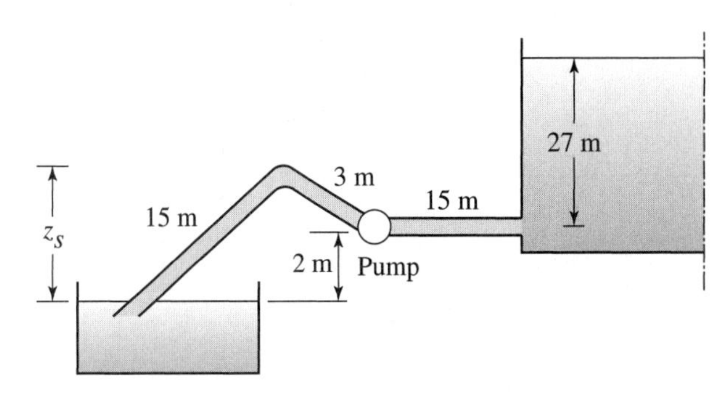

Figure P9.28 Pumping system for Problem 9.28.

9.29. A pump operates at speed of 1160 rpm and delivers water through a 300 m pipe with a diameter of 15.0 cm and a friction coefficient of 0.005 to an elevated tank 20 m above the water level at the suction side. The pump has the following characteristics:

Head (m)	0	14	19	23	26	29	30
Discharge (l/s)	142	127	113	85	57	28	0
NPSH required (m)		9.1	6.7	4.3	3.6	3.0	

The pump is located 2.75 m above the intake level and the length of the suction pipe is 55 m (which is included in total length of the pipe, 300 m).

a. Determine the discharge and head at which the pump operates.

b. If, during a power failure, the pump is operated by a stand-by diesel engine, which runs at a speed of 1250 rpm, what is the discharge in this case?
c. Will the pump be safe from cavitation when operated by the diesel engine, assuming that the atmospheric pressure is 9.705 m and the vapor pressure is 0.357 m?
d. Is this pump running near its point of best efficiency?

9.30. 100 gpm of water at 70°F under a pressure of 80 psi is drawn through 30 ft of 1.5-in-diameter steel pipe into a 2-psi tank. The inlet and outlet are both 20 ft below the surface of the water when the tank is full. The inlet line contains a square mouth inlet and two wide-open gate valves. The pump's NPSH is 10 ft for this application. Will the pump cavitate?

References

Chin, David A., 2000. *Water-Resources Engineering.* Prentice Hall, Upper Saddle River, N.J.

Gupta, Ram S., 1989. *Hydrologic and Hydraulic Systems.* Prentice Hall, Englewood Cliffs, N.J.

Hwang, N. H. C. and R. J. Houghtalen, 1996. *Fundamentals of Hydraulic Engineering Systems,* 3rd edition. Prentice Hall, Upper Saddle River, N.J.

Karassik, I. J., Messina, J. P., Cooper, P. and Heald, C. C. (editors), 2001. *Pump Handbook,* 3rd Edition. McGraw-Hill, New York.

Rayner, R., 1995. *Pump User Handbook.* Elsevier Science Publishers LTD.

Simon, Andrew L. and Scott F. Korom, 1997. *Hydraulics,* 4th edition. Wiley, New York.

Wurbs, Ralph and Wesley P. James, 2002. *Water Resources Engineering,* Prentice Hall, Upper Saddle River, N.J.

CHAPTER 10

CHANNEL GEOMETRY

A channelized stream in poor condition.

Flow of water, sediment, and pollutants occurs in natural or human-made systems, called conduits. Hydraulic conduits can be defined as conveyance structures or flow passages which transfer water from one location to the other under pressure or gravity forces. A conduit may be either a pipe (pressure flow) or an open channel where gravity forces dominate. Flows in open channels are similar to flows in closed pipes in many ways but differ in some respects. The characteristics of flow are greatly influenced by the geometric characteristics of these conduits. A hydraulic study of flow, therefore, requires consideration of these characteristics. Since pipes are of well-defined shapes and are usually circular, this chapter focuses on open channels and their geometric characteristics.

10.1 Channel Flow

Open channels are not completely enclosed by solid boundaries, and their water surfaces are subject to atmospheric pressure. Due to the bed slope, gravity force constitutes the main driving force for flow in open channels. There must be a net change in the elevation of the channel bottom to generate flow. The fluid flows from the sections of higher bed elevations to the sections of lower ones. Unlike flow in pipes, no external head or force is needed to create flow in open channels. Such a flow is often referred to as free-surface flow or gravity flow. In pipes flowing full, the flow is mainly under the pressure force; therefore, no direct interaction with the atmosphere may exist. The flow is bounded at all points in a closed conduit. The cross-sectional area of flow in an open channel may vary with respect to space and time. Any variation in the discharge will be associated with the corresponding variation in the cross-sectional area of flow. The cross-sectional area of flow in a pipe flowing full does not vary with time or discharge; however, it may vary with space as the pipe diameter may change from one section to another.

Open-channel flow is much more complicated to deal with than is pipe flow. Practically, almost all pressurized systems have a circular section or at least a section of well-defined geometry (square, horseshoe, etc.). Most open channels, such as in natural rivers, have irregular sections. Therefore, an infinite number of sections may exist. Roughness coefficients in open channels have a much wider range of variation than those in pipes. Even for the same type of boundaries, they vary with the depth of flow, and they are, therefore, time dependent. This time dependency also holds for most of the other parameters of open channels. The complexity of open-channel flow arises from the fact that almost all of the geometric, hydraulic, and physical parameters are interdependent. Moreover, these parameters exhibit a certain degree of uncertainty. Even for the simplest parameters, such as the depth of flow, measurement errors may exist. On the other hand, no one can miss a diameter of a pipe.

A flow in a closed conduit may be classified as an open-channel flow if it has a free surface at which atmospheric pressure holds. At two different times a certain section in a closed conduit system may act as an open channel and as a pipe. The corresponding two types of flow may occur at the same instant at two different locations. Under such circumstances, the system should be designed to match all flow possibilities. A good example of such a case is encountered in storm sewer systems, which are generally designed as open channels. However, under flash flood conditions, such systems should be able to act as pressurized systems. Generally, a closed conduit system may have free surface conditions in a portion of its length and pressurized conditions in other portions. Such cases are usually encountered when the downstream end of the closed system is submerged.

10.2 Types of Open Channels

The term *open channel* is used in hydraulics in a broad sense. Open channels encompass any system conveying a fluid under atmospheric pressure. Rivers, bayous, creeks, streams, brooks, canals, sewers, ditches, streets, airports runways, parking lots, wadis, floodplains, golf courses, agricultural fields, gardens, lawns, parks, and others are examples of open channels. The cross-sectional area of flow may take any regular or irregular shape.

Open channels can be classified into two main types: natural channels and artificial channels. Natural channels or natural water courses have been developed throughout history by natural processes without major human interference. Most of the rivers around the globe belong to this category. Creeks, bayous, streams, brooks, and tidal estuaries are other kinds of such channels.

Artificial channels are designed and constructed, according to specifications, to meet certain water conveyance requirements. Irrigation canals, drainage ditches, gutters, culverts, flood control channels, and navigation channels are mainly developed by human efforts and hence are artificial. A natural river with levees on both sides can fit both categories. Canals are generally long channels with mild slope and they may be either unlined or lined with any material or membrane to prevent seepage. Channels may be either prismatic or nonprismatic. Prismatic channels are those that have a constant cross section and bed slope with distance (Chow, 1959). In nonprismatic channels the cross section and bed slope (at least one of them) vary with distance. Flumes are open channels built above the ground surface to convey water through depressions. Chutes are open channels of steep bed slope. Drops are short chutes. Culverts are channels used to convey water under highways, railways, and so on. Open-channel tunnels are long covered channels designed to convey water through a hill or any other obstruction on the ground surface.

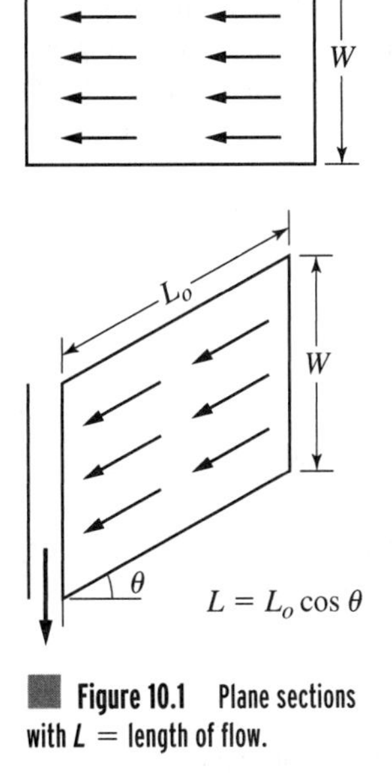

Figure 10.1 Plane sections with L = length of flow.

Planes on the land surface are also embraced by the hydraulic connotation of an open channel. Exemplifying such planes are parking lots, lawns, runways, golf courses, gardens, orchards, agricultural fields, wetlands, road surfaces, overland flow areas, coastal plains, floodplains, and so on. Flow on a plane is different from that in, say, a river or canal, although hydraulic principles remain the same. On planes the depth of flow may sometimes be less than 1 cm, while the width may be as large as several kilometers. A plane is defined by length L and width W as shown in Figure 10.1 (Singh, 1996). It can be either a horizontal or inclined plane, rectangular or triangular, convex, concave, converging or diverging. Its length is not necessarily greater than width. The length of flow is taken to represent the length of the plane along the direction of flow. The land slope, length of flow, and area are sufficient to define a plane section.

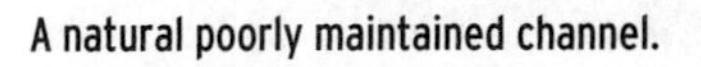
A natural poorly maintained channel.

A well-maintained channel.

Floodplains are typically low areas adjacent to a river, ocean, or other water body and are subject to flooding. The amount of land inundated by a flood depends on the magnitude of the flood. The floodplain of a river is the area of the valley floor adjacent to the incised channel. As water rises in the river over the bankful stage, this area gets flooded. There is, however, some difficulty in precisely defining the bankful stage and the floodplain. It is well documented that the floodplain is subject to frequent flooding. Floodplains in the eastern and central United States are flooded, on average, twice in 3 years. Approximately 7% of the United States is subject to flooding by the 100-year flood. This area is almost as big as Texas (Singh, 1996).

On the basis of the capacity at peak flow, the floodplains land can be classified into floodway and pondage areas. Floodway, or flowage land, is the principal flow carrying part of the natural cross section of the stream. Any encroachment upon the floodway will increase flood heights. Pondage area is the land on which water is stored as dead water during flooding. These dead areas are not a part of a flow-carrying area and are often referred to as flood fringes. The boundary between floodway and pondage lands is not a fixed one but changes with the magnitude of the flood.

Although open channel flow in natural channels is discussed in different locations throughout the text, a comprehensive treatment of flow in natural channels is outside the scope of this book. The reader is referred to Chow (1959), Henderson (1966), among others for a more complete treatment of this subject.

10.3 Channel Geometry

The term *channel geometry* encompasses all characteristics that define the channel geometry. Because the geometry of a channel varies in most cases in space as well as in time when the time horizon is long, the characteristics should be defined at a given point as well as along the length. At a point, the characteristics include cross-sectional area, flow depth, wetted perimeter, top width, bottom width, side slope, and friction. For the entire channel reach, these include length, the average width both at the top and bottom, slope, average friction, average cross section, and average flow depth.

The term *channel cross section* is used to denote the cross section of a channel taken normal to the direction of flow. The vertical section is the one that passes through the lowest point of the channel cross section. As mentioned above, a channel built with constant cross section and constant bed slope throughout its length is defined as prismatic channel. Otherwise, it is nonprismatic. A spillway of variable width and curved alignment is an example of a nonprismatic channel. Sections of natural channels are usually very irregular. Such sections not only vary with respect to space but may also vary with respect to time. Sections of a natural stream may

A concrete trapezoidal channel. (Courtesy of Dr. Mohammad Al-Hamdan.)

vary from one year to another due to deposition of sediment and/or erosion of bed and boundaries. Usually these sections vary from an approximate parabola to an approximate trapezoid.

Natural rivers are subject to floods, and the discharge, therefore, increases dramatically during flood stage. In such cases a channel consists of a main channel section which conveys the normal discharge and one or more side channel sections which accommodate overflows.

Artificial channels are generally designed with sections of specified geometric shapes. Table 10.1 lists different cross sections that are in common use. Because of its stability and

TABLE 10.1 Properties and Geometric Elements of Typical Channel Cross Sections

Section	Area, A	Wetted Perimeter, P	Hydraulic Radius, R	Top Width, B	Hydraulic Depth, D	Cross Section
Rectangular	by	$b + 2y$	$by/(b + 2y)$	b	y	
Trapezoidal	$(b + ty)y$	$b + 2yw$, $w = (1 + t^2)^{0.5}$	A/P	$b + 2ty$	A/B	
Triangular	ty^2	$2yw$	$ty/(2w)$	$2ty$	A/B	
Circular	$(\theta - \sin\theta)\dfrac{d^2}{8}$	$r\theta$	$\left(1 - \dfrac{\sin\theta}{\theta}\right)d/4$	$2r\sin(\theta/2)$	A/B	$\theta = 2\cos^{-1}\left(1 - 2\dfrac{y}{d}\right)$
Semicircular	$\pi r^2/2$	πr	$r/2$	$2r$	$\pi r/4$	
Parabolic Section	$2/3By$	$B + (8/3)y^2/B$*	$2B^2y/(3B^2 + 8y^2)$	$3A/(2y)$	$2/3y$	

*Approximation for the interval $0 < \dfrac{4y}{B} < 1$.

feasibility, the trapezoidal section is used extensively with unlined earth banks. Rectangular and triangular sections are special cases of the trapezoidal section. Rectangular sections are usually built with stable materials such as rocks and metals to stabilize the vertical sides. Triangular sections are rare and are generally used in small ditches, gutters, and laboratory experimental channels. Sewers and culverts are constructed below the ground surface. Therefore, because of its ability to resist soil pressure, circular section is popular for these types of water conduits.

10.4 Geometric Elements

The cross-sectional shape of a channel governs the pattern of the velocity distribution, boundary shear stress, and secondary circulation in the channel, and, in turn, channel discharge capacity. Also of interest are wetted perimeter, top width, and conveyance factor. These geometric elements are functions of stage which vary with discharge.

The parameters which define various types of channel cross sections are called geometric elements. Since these parameters describe the geometry of the sections, they all have a dimension of length (L) or its multiplication. Geometric elements are extremely important for flow computation. For a simple rectangular section, the geometric elements may be expressed in terms of the depth of flow and the width of the section. Each section has its own elements which define its geometry entirely. Natural sections have no simple rule or elements to define their geometry. Curves and equations are, therefore, developed to represent the geometric elements of such sections. Several terminologies and definitions are given hereafter for the geometric elements. Consolidation of these terminologies is important at this stage.

The depth of flow, y, is the vertical distance of the lowest point of a channel section from the free surface. Depths of flow are measured vertically from the channel bed to free surfaces. The depth of a flow section, d, is the depth of flow measured normal to the bed, as shown in Figure 10.2. For a channel with a longitudinal bed slope φ the depth of flow

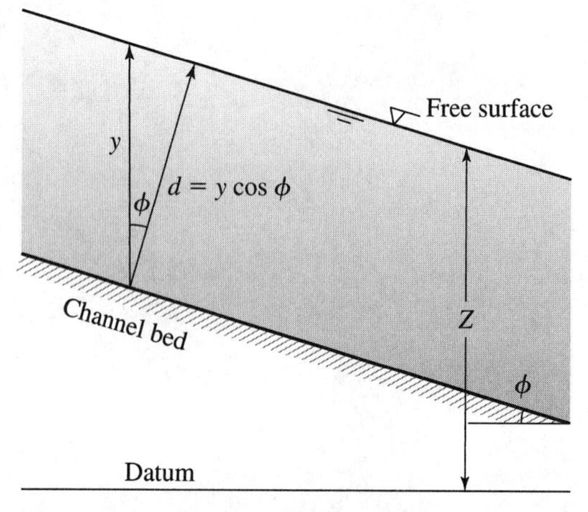

Figure 10.2 Depth of flow and depth of flow section.

section is equal to the depth of flow multiplied by $\cos\varphi$ (i.e., $d = y\cos\phi$). If the channel bed slope is small, then d is approximately equal to y. The depth of flow, y, is much more common in use for open-channel calculations. The stage, Z, is the elevation of the free surface above a specified datum. If the datum is the channel bed, then the stage equals the depth of flow, y.

The top width, B, is the width of the channel section at the free surface. The bottom width, b, is the width of the channel section at the channel bottom. In rectangular sections the top width is equal to the bottom width (i.e., $B = b$). For triangular and circular sections the bottom width is equal to zero. In trapezoidal sections, the top width is related to the bottom width and the side slopes of the channel boundaries.

The cross-sectional area of flow, A, is the area normal to the direction of flow. In natural channels, areas are generally calculated with integration or summation of subareas. Cross-sections are plotted from field measurements, and then areas can be calculated. For regular channels, areas are calculated from simple equations.

The wetted perimeter, P, is the length of the line of intersection of the channel wetted surface with a cross-sectional plane normal to the direction of flow. The hydraulic radius, R, is the ratio of the cross-sectional area, A, to its wetted perimeter, P,

$$R = A/P \tag{10.1}$$

The hydraulic depth, D, is the ratio of the cross-sectional area, A, to the top width, B:

$$D = A/B \tag{10.2}$$

For rectangular sections, the cross-sectional area is equal to By. Therefore, the hydraulic depth, D, is equal to the depth of flow, y, in this case. The hydraulic depth is the depth of a rectangular section having the same area of the section under consideration. In other words, different cross sections are transformed to equivalent rectangular sections; therefore, comparisons can be made on the same basis.

The section factor, M, is defined as the product of the cross-sectional area, A, and the square root of the hydraulic depth, D:

$$M = A\sqrt{D} = A\sqrt{\frac{A}{B}} \tag{10.3}$$

The main geometric elements for different sections are given in Table 10.1.

If the width of a section is very large as compared to the depth of flow, rectangular and sometimes trapezoidal and natural sections are approximated as wide sections. The top width, in this case, is set equal to the bottom width. The cross-sectional area of flow is set equal to by, where b is the bottom width. The wetted perimeter is set equal to the bottom width. The wetted length of the two sides of the channel is neglected while calculating the wetted perimeter of the channel. The hydraulic radius is therefore equal to the depth of flow. A wide section implies that $B = b = P$, where $R = D = y$. This approximation significantly reduces the computations. The assumption of a wide section is valid only when B is greater than $10y$. Accurate results are obtained as B/y increases. However, this assumption should be employed only when the side slopes or the channel width are not known. In this case, the discharge may be estimated per unit width of the channel.

Some key geometric parameters of open channels are often expressed as power functions of the flow depth y (Ackers, 1988; Allen et al., 1994; Cahoon, 1995; Lane and Foster, 1980; Singh, 1996) as

$$X = \alpha y^{\beta}, \quad X = A, P, R \tag{10.4}$$

where A is the area, P is the wetted perimeter, R is the hydraulic radius, and α and β are coefficients. If the cross-sectional area is expressed by equation (10.4), then for a rectangular channel $\beta = 1$ and $\alpha =$ bottom width, b. If the channel is triangular and the cross-sectional area is expressed by equation (10.4), then $\beta = 1$ and $\alpha =$ half the top width or $\beta = 2$ and $\alpha =$ horizontal component of the side slope for one vertical t (t:1). For regular concave sections that display smooth changes in the top width with flow depth, equation (10.4) is generally satisfactory. From a computational standpoint, this is highly efficient, because it permits the use of an analytical solution. In the case of flood routing, repeated evaluation of the geometric parameters is needed as flow depth varies with discharge, and equation (10.4) is very handy. The same is true for backwater profile computations.

Example 10.1

Compute the hydraulic radius, hydraulic depth, and section factor for a trapezoidal channel given in Figure 10.3. The depth of flow is 5 m, the bottom width is 15 m, and the side slope is 2 (horizontal): 1 (vertical).

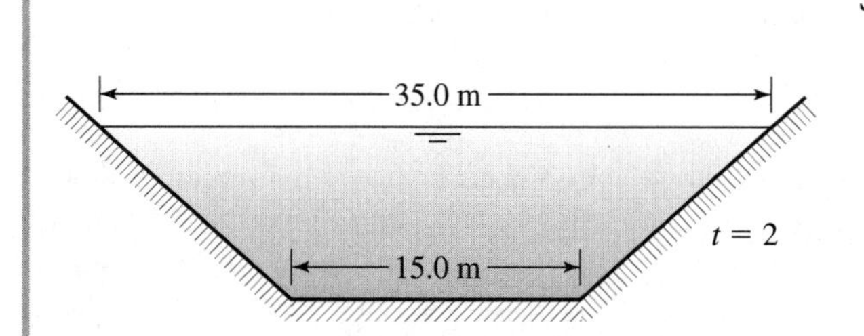

Figure 10.3 A trapezoidal channel section (Example 10.1).

Solution: $B = b + 2ty = 15 + 2 \times 2 \times 5 = 35.0$ m.

$$P = b + 2y\sqrt{1 + t^2} = 15.0 + 2 \times 5\sqrt{5} = 37.36 \text{ m}$$

$$A = \tfrac{1}{2}(15 + 35) \times 5 = 125 \text{ m}^2$$

$$R = \frac{A}{P} = \frac{125}{37.36} = 3.35 \text{ m}$$

$$D = \frac{A}{B} = \frac{125}{35} = 3.57 \text{ m}$$

$$M = A\sqrt{D} = 125\sqrt{3.57} = 236.23 \text{ m}^{2.5}$$

Example 10.2

A rectangular channel 4.0 ft wide has a bed slope of 10 ft/mi. The bed and water surface levels at section 1 are 120 ft and 124.2 ft, respectively. The flow is uniform with a velocity of 3 ft/s. What is the water level at section 2 (5.0 miles downstream of section 1)? Calculate the flow rate and draw the total energy line.

Solution: Since the flow is uniform, the depth of flow will not vary with distance.

Recall from fluid mechanics that the total energy head is the sum of the elevation head, the pressure head, and the kinetic energy head.

Depth of flow, $y = 124.2 - 120.0 = 4.2$ ft

Bed level at section 2 $= 120 - ((10/5280) \times 5(5280)) = 70$ ft.

Water level at section 2 $= 70 + 4.2 = 74.2$ ft

The flow rate $= A \times V = by \times V$
$= 4.0 \times 4.2 \times 3.0 = 50.4$ ft³/s

Velocity head $= V^2/2g = 0.14$ ft

The total energy line is located 0.14 ft above the water surface as shown in Figure 10.4.

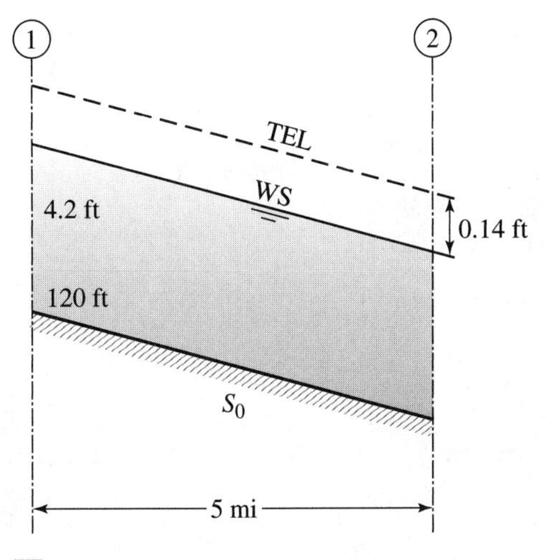

Figure 10.4 Water surface and total energy lines for Example 10.2.

Example 10.3

A triangular channel has 5.0-ft top width and 3:2 side slope. Express the cross-sectional area, the wetted perimeter, and the hydraulic radius as power functions of the flow depth.

Solution:

$$A = ty^2 = 1.5y^2;\ P = 2y\sqrt{1 + t^2} = 3.61y;\ R = \frac{A}{P} = \frac{1.5y^2}{3.61y} = 0.416y$$

10.5 Cross-Sectional Asymmetry

A common characteristic of natural rivers is their tendency to manifest varying degrees of asymmetry in their cross-sectional form, especially in, but not limited to, meandering reaches. According to Leopold and Wolman (1960), about nine-tenths of a meandering channel is asymmetric in cross section. Knighton (1981) proposed three measures to describe that asymmetry:

$$A^* = \frac{A_R - A_L}{A} \tag{10.5}$$

$$A_1 = \frac{2xh_{\max}}{A} \tag{10.6}$$

$$A_2 = \frac{2x(h_{\max} - h)}{A} \tag{10.7}$$

where A_R and A_L are, respectively, the cross-sectional areas to the right and left (looking downstream) of the channel centerline; $A = A_R + A_L$ is the total area of the channel; $h_{\max}$

is the maximum depth; x is the horizontal distance (measured positive to the right and negative to the left) from the center line to the vertical of the maximum depth; and h is the mean depth.

The index given by equation (10.5) considers differences in area between two parts of the channel. When $A^* = -1$, it represents an extreme left asymmetry, and $A^* = +1$ corresponds to an extreme right asymmetry. When A^* is plotted against A_R/A_L, the resulting curve has the form of a logistic function:

$$A^* = \frac{A_R - A_L}{A_R + A_L} = +1 - \frac{2}{1 + (A_R/A_L)} \tag{10.8}$$

A realistic range of A^* is $(-0.65, 0.65)$ for natural channels (Knighton, 1981).

The indices given by equations (10.6) and (10.7) are based on the relative displacement of depth instead of area. Both horizontal and vertical asymmetry types are explicitly included in these indices. However, if $A_1 = 0$, it represents symmetry only in the horizontal direction, without implying uniformity of depth either across the entire channel or between the two parts of the channel separated by the centerline. On the other hand, a value of $A_2 = 0$ implies either horizontal symmetry ($x = 0$) or vertical symmetry ($h_{max} = h$) or both, although $h_{max} = h$ is a sufficient condition to imply that $x = 0$, since it is possible for only a perfectly rectangular channel. According to Knighton (1981), realistic ranges for A_1 and A_2, respectively, are $(-2, 2)$ and $(-1, 1)$.

Channel cross-sectional asymmetry is one of the factors that led to the development of alternating scour holes and meandering flow. This reinforces the belief that bed deformation is the fundamental cause of meandering. Therefore, asymmetry provides a link between two adjustable elements of channel geometry, cross-sectional form and channel pattern.

Example 10.4

At a certain section on a 304-m-wide river the depth of flow is found to be a maximum of 7.63 m at a distance of 19.0 m from the center line (to the right). The cross-sectional areas to the right and left are 912 m^2 and 560 m^2, respectively. Compute A^*, A_1, and A_2 to reflect the section asymmetry.

Solution: $A_R = 912\ m^2$, $A_L = 560\ m^2$, $x = 19.0$ m, $A = 912 + 560 = 1472\ m^2$; then

$$A^* = \frac{912 - 560}{1472} = 0.239; \quad A_1 = \frac{2 \times 19 \times 7.63}{1472} = 0.197$$

$$A_2 = \frac{2 \times 19(7.63 - 4.84)}{1472} = 0.072; \quad \text{Mean depth, } h = \frac{A}{B} = \frac{1472}{304} = 4.84 \text{ m}$$

10.6 Compound Sections

Natural channels possess compound or irregular cross-sectional forms. When a simple power function, such as equation (10.4), is employed to represent an irregular or compound cross section, it performs poorly (Cunge, 1975a,b; Cunge et al., 1980). By employing two sets of coefficients (α and β)—one for main-channel flow and the other for overbank flow

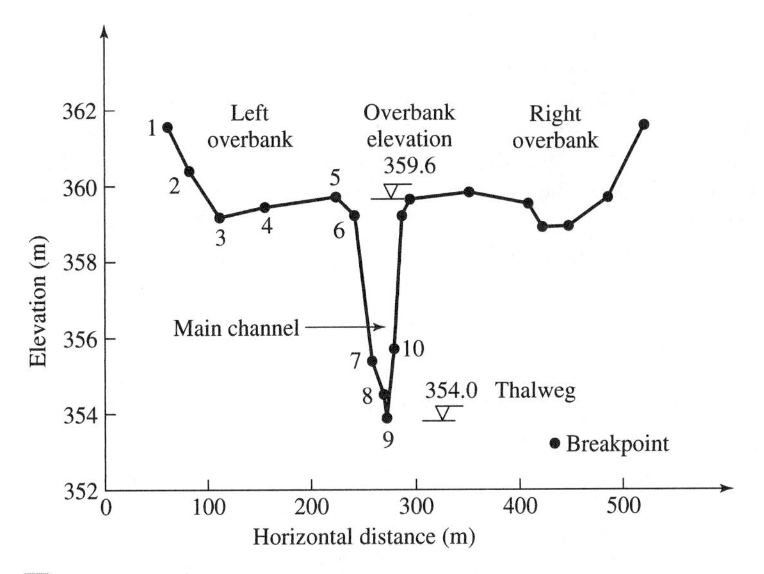

Figure 10.5 Compound channel cross section (adapted from Garbrecht, 1990).

An asymmetrical natural channel.

conditions—this simple representation can be adapted to account for overbank flow. However, a discontinuity occurs when coefficients are changed as free-water surface reaches and exceeds overbank elevation. This causes instability in numerical solutions of flow routing equations. To alleviate this problem, Garbrecht (1990) proposed an alternative power-function representation for channels with compound cross sections. The discussion in what follows is based on his work. A compound cross section is assumed to be composed of a single main-channel subsection and an optional right or left bank subsection, as shown in Figure 10.5. Overbank flow occurs when the free-water surface in the main channel exceeds overbank or levee elevation. This representation is aimed at natural channel sections with smooth changes in top width with the flow depth as opposed to human-made rectangular or trapezoidal channels.

The power-function method for compound cross sections represents the geometric properties by two power functions: one power function for main channel flow conditions and a separate power function for overbank flow conditions. The power function for main-channel flow conditions can be expressed as

$$X = m_1 y^{p_1} \qquad \text{for } h < h_o \tag{10.9}$$

where m_1 and p_1 are coefficients derived from actual data on X for main-channel flow conditions only. Note that equation (10.9) is equivalent to equation (10.4). The power function for overbank flow only is expressed as

$$X = m_2(h - h_o)^{p_2} + X_o \qquad \text{for } h > h_o \tag{10.10}$$

where m_2 and p_2 are coefficients to be derived from data on X for overbank flow conditions only, h_o is the overbank elevation (depth), and X_o is the value of X (geometric property) at h_o.

Equations (10.9) and (10.10) have four parameters, m_1, m_2, p_1, and p_2. Because they are derived from field data, computation of the X values is in order. For an effective evaluation of the X values, a compound section is discretized into simple elements. An element is the portion of the cross section between two consecutive breakpoints. As shown in Figure 10.5, a pair of (x, y) coordinates defines the breakpoint, where x is the horizontal distance from an arbitrary reference point (or origin) and y is the vertical distance from an arbitrary datum. The commencement of overbank channels is defined by breakpoints separating the main-channel and overbank subsections. Frequently, the overbank elevations on both sides of the channel are not equal; in such cases, the lower of the two elevations is used to indicate the beginning of the overbank flow.

The cross-sectional area of flow A, normal to the direction of flow, is computed for a discretized section as

$$A = \sum_{i=1}^{N} A_i \tag{10.11}$$

Likewise the wetted perimeter and hydraulic radius are calculated as:

$$P = \sum_{i=1}^{N} P_i \tag{10.12}$$

$$R = \frac{A}{P} \tag{10.13}$$

Subscript i refers to the ith section element, A_i and P_i are A and P for the ith element, and N is the number of sections at or below the free-water surface.

The conveyance K, a measure of a cross section's discharge capacity, can be defined with Manning's uniform flow formula as

$$K = \frac{1}{n} A R^{2/3} \tag{10.14}$$

where n is Manning's roughness coefficient. It must be noted that K is evaluated for an entire subsection and not as a summation over section elements. For compound cross sections, the total conveyance K_c is the sum of the conveyance of the main channel K_m and the overbank channel K_o (Chow, 1959; Posey, 1950). This is based on the observation that for a river in flood stage with overbank flow, the flow in the main (deep) channel is quite different from that in the overbank (shallow) channel and should therefore be computed for the two portions separately. Hence,

$$K_c = K_m + K_o \tag{10.15}$$

The hydraulic radius for the compound section, R_c, is taken as the conveyance weighted sum over the subsections:

$$R_c = \frac{K_m R_m + K_o R_o}{K_c} \tag{10.16}$$

where subscript c stands for the compound section, m stands for the main-channel subsection, and o stands for the overbank subsection.

The corresponding value of the equivalent roughness is obtained from equation (10.14) as

$$n_e = \frac{A R_c^{2/3}}{K_c} \tag{10.17}$$

It may be noted that the equivalent hydraulic radius for compound sections may not be a monotonically increasing function of the flow depth. As shown in Figure 10.6, R_c may decrease as flow switches from predominantly main-channel flow (higher hydraulic radius) to predominantly overbank flow (smaller hydraulic radius). With transition of flow, R_c goes

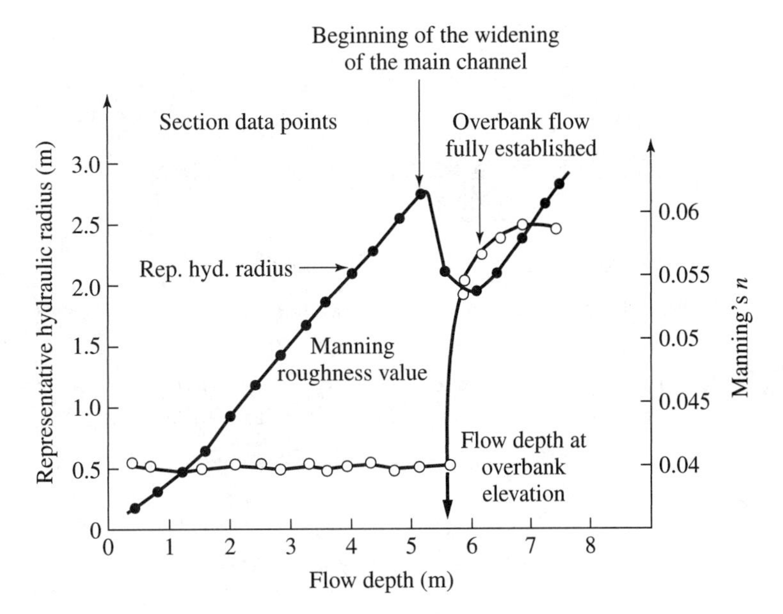

Figure 10.6 Representative hydraulic radius as a function of flow depth (adapted from Garbrecht, 1990).

through a gradual transition from a higher value to a lower value. However, this decrease in R_c is followed by a renewed increase with continuing increase in the flow depth. The equivalent roughness for compound sections remains unaltered for main-channel flow conditions. With occurrence and dominance of overbank flow, it tends to approach the value given by equation (10.17), as shown in Figure 10.6.

For flow routing calculations, the channel cross-sectional properties are interpolated, as in the case with the U.S. Army Corps of Engineers River Analysis System (HEC-RAS; Hydrologic Engineering Center, 1998). Interpolated data between cross sections are used to refine estimations of steady open-channel water surface profiles. Traver and Miller (1993) found that the error included in the geometric representation of the stream was of the same magnitude as the current methods used to measure cross sections. Hence, they confirmed the use of interpolated data, as long as the change in the geometric properties between measured cross sections was uniform, and concluded that the overall benefits of interpolation outweighed any added numerical error.

Example 10.5

Express the area of flow for the compound section given in Figure 10.7 as a power function of the depth of flow.

Solution: For the main-channel flow condition,

$$A = 10y; \quad y \leq 3.0 \text{ ft}$$

The power function for overbank flow is expressed as

$$A = 16(y - 3) + 30; \quad y > 3.0 \text{ ft, or}$$

$$A = 16y - 18; \quad y > 3.0 \text{ ft}$$

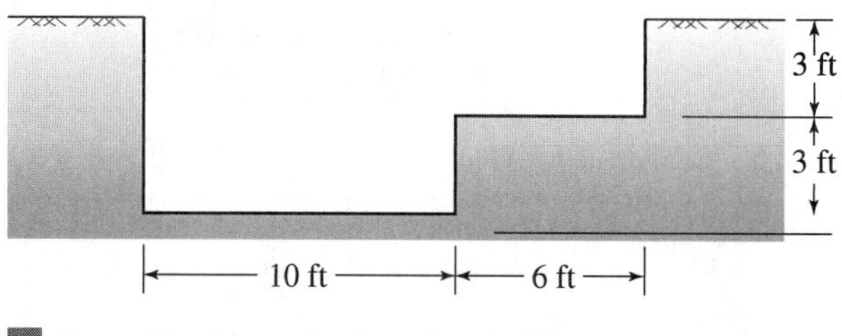

Figure 10.7 Compound section in Example 10.5.

Example 10.6

Evaluate the conveyance factor, equivalent hydraulic radius, and equivalent roughness of the compound section given in Figure 10.8. Take the Manning roughness coefficient, n, as 0.02.

Solution: The section is divided into four subsections as shown in Figure 10.8. The following table is developed:

Section	A (m^2)	P (m)	R (m)	K
1	0.5	1.41	0.355	12.53
2	6.0	4.0	1.5	393.12
3	16.0	5.66	2.857	1610.76
4	9.0	6.0	1.5	589.66
Total	31.5	17.07		2606.07

$$K_c = \sum K_i$$

$K_c = 2606.07$. Applying equation (10.16),

$$R_c = \frac{4.45 + 589.65 + 4601.94 + 884.51}{2606.07} = 2.33$$

$$n_e = \frac{31.5(2.33)^{2/3}}{2606.07} = 0.0212$$

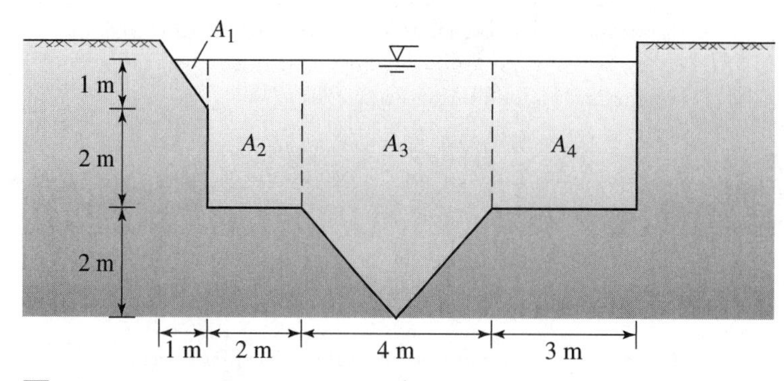

Figure 10.8 Compound section in Example 10.6.

10.7 Channel Slope

The channel slope or slope of the bed of a channel, S_o, has a profound effect on the velocity of flow in the channel and, consequently, on the flow characteristics of runoff from a drainage basin. The channel slope must be defined first in order to determine the discharge and velocity using the Manning or Chezy equation. Because the slope varies longitudinally, an average value of slope must be determined for the channel reach under consideration.

Method 1: For use in either the Manning or Chezy equation, the channel slope is a local measurement to approximate the energy slope, assuming uniform flow. For this purpose, the channel slope, S_o, is computed from the equation:

$$S_o = h/L \tag{10.18}$$

where h is the fall of the channel bed in meters and L is the horizontal distance (length) over which the fall occurs. The fall over the channel reach of interest is measured. However, this measurement is a local measurement and cannot hold for other channel reaches in the drainage basin. The cross-sectional area, channel roughness, and cross-sectional shape also affect the flow velocity. Therefore, there is not as much difference in the flow velocity between various reaches of the drainage basin as one might think.

Method 2: The arithmetic slope may also be evaluated by computing the fall from the head of the uppermost beginning of the channel to the channel outlet and dividing this fall by its horizontal length. In this method the local slope for the segment under consideration is not evaluated. Only one value is calculated for the entire channel.

Method 3: A geometric slope is sometimes used. This slope is determined by locating the median channel-profile elevation on the main channel and computing the fall from this point to the outlet. The length is the horizontal distance between the point of median elevation to the outlet. Slope is then computed using equation (10.18).

Method 4: Benson (1962) found that the "85-10" slope factor was the most satisfactory in his study of floods in New England. This factor is the slope between 85% (excluding the upper 15%) and 10% (excluding the lower 10%) of the distance along the stream channel from the basin outlet to the divide. It should be noted that the distance is measured to the divide and not to the end of this defined stream channel. The fall and horizontal length are thus used to compute the slope using equation (10.18).

Method 5: This method is from Johstone and Cross (1949). The channel can be divided into N number of reaches, each having a uniform slope, S_i. Then the equivalent uniform slope S_m is

$$S_m = \left[\frac{\sum_{i=1}^{N} L_i S_i^{1/2}}{\sum_{i=1}^{N} L_i} \right]^2 \tag{10.19}$$

Equation (10.19) estimates the slope that would result in the same total time of travel as for the actual stream if the length, roughness, channel cross section, and any other pertinent factors other than slope were unchanged.

Method 6: This is due to Laurenson (1962). Again, the stream is divided into N reaches, each of uniform slope. Furthermore, it is assumed that the effects of roughness and hydraulic radius on the velocity are the same for all reaches. This assumption is questionable but has been used by others (Taylor and Schwarz, 1952). This assumption is also implied in the first method.

The velocity V_i of flow through any reach i can be written as

$$V_i = aS_i^{0.5} \tag{10.20}$$

where a is a constant. Then the time of flow t_i is

$$t_i = \frac{L_i}{V_i} \tag{10.21}$$

Therefore, the total time of travel T_c down the main channel is

$$T_c = \frac{1}{a}\sum_{i=1}^{N}\frac{L_i}{S_i^{0.5}} \tag{10.22}$$

The mean velocity of flow V_m can be written as

$$V_m = \sum_{i=1}^{N}\frac{L_i}{T_c} = \frac{a\sum_{i=1}^{N}L_i}{\sum_{i=1}^{N}L_i/S_i^{0.5}} \tag{10.23}$$

Furthermore,

$$V_m = aS_m^{0.5} \tag{10.24}$$

Hence,

$$S_m = \left[\frac{\sum_{i=1}^{N}L_i}{\sum_{i=1}^{N}L_i/S_i^{0.5}}\right]^2 \tag{10.25}$$

Example 10.7

Figure 10.9 shows a segment of a natural river 10,000 ft in length. To determine the bed slope of the river, several sections, as indicated in the figure, along the river were investigated. The horizontal distance and bed level at the centerline of the river are given in the following table:

Section	1	2	3	4	5	6	7	8	9	10	11
Horizontal distance (ft)	0	700	1500	2300	3500	5000	6150	7050	7900	9000	10,000
Bed level (ft)	79.2	78.15	76.5	74.3	72.3	71.6	70.3	68.9	67.3	66.5	65.4

1. Calculate the bed slope of each reach.
2. Calculate the arithmetic bed slope.
3. Calculate the bed slope using methods 4 and 5.

Solution: The bed slope for different reaches is evaluated from equation (10.18) and is given in the following table:

Reach	1–2	2–3	3–4	4–5	5–6	6–7	7–8	8–9	9–10	10–11
Drop (ft)	1.05	1.65	2.2	2.0	0.7	1.3	1.4	1.6	0.8	1.1
Length (ft)	700	800	800	1200	1500	1150	900	850	1100	1000
Bed slope ($\times 10^{-4}$)	15	20.6	27.5	16.7	4.7	11.3	15.6	18.8	7.27	11
$LS^{1/2}$	27.11	36.31	41.95	49.04	32.55	38.66	35.55	36.86	29.66	33.17

$$\sum L_i S_i^{1/2} = 360.83$$

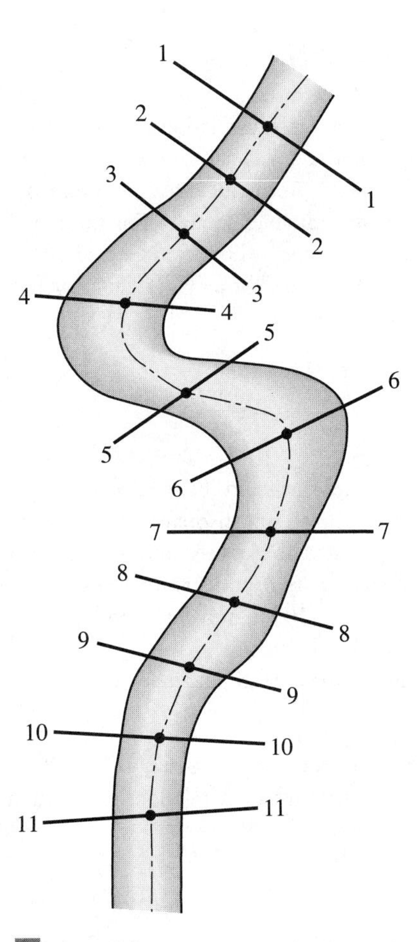

Figure 10.9 A river segment given in Example 10.7.

The arithmetic bed slope is evaluated over the entire river segment and equals

$$\frac{79.2 - 65.4}{10{,}000} = 13.8 \times 10^{-4}$$

Using method 4, the bed slope equals

$$\frac{76.5 - 66.5}{7500} = 13.3 \times 10^{-4}$$

Using method 5, the bed slope is evaluated from equation (10.19) as

$$S_m = \left(\frac{360.83}{10{,}000}\right)^2 = 13.0 \times 10^{-4}$$

In this example, the values of bed slope obtained from different methods are quite similar. This is mainly attributed to the relative uniformity of the bed slope throughout the entire channel length. Different values may be obtained from different methods if the channel bed slope varies significantly from one reach to the other.

10.8 River Hydraulic Geometry

The term *hydraulic geometry* connotes the relationships between the mean stream channel form and discharge both at a station and downstream along a stream network in a hydrologically homogeneous basin. The channel form includes the mean cross-sectional geometry (width, depth, etc.) and the hydraulic variables include the mean slope, mean friction, and mean velocity for a given influx of water and sediment and the specified channel boundary conditions. Leopold and Maddock (1953) expressed the hydraulic geometry relations for a channel in the form of power functions of discharge as

$$B = aQ^b \tag{10.26}$$

$$y = cQ^f \tag{10.27}$$

$$V = kQ^m \tag{10.28}$$

where B is the width of the water surface in the stream in meters or feet at a given cross section; y is the mean water depth in the stream in meters or feet at the cross section; V is the mean velocity of flow in m/s or ft/s at that cross section; a, c, and k are parameters or constants of proportionality; and b, f, and m are exponents, or slopes on a semilog graph. The relations expressed by equations (10.26) to (10.28) are plotted in Figure 10.10. Similar equations are also found for channel friction and bed slope.

The hydraulic variables, width, depth and velocity, satisfy the continuity equations:

$$Q = ByV \tag{10.29}$$

Therefore, the coefficients and exponents in equations (10.26) to (10.28) satisfy certain constraints derived as follows. Substituting these relations in equation (10.29), we get

$$Q = aQ^b \times cQ^f \times kQ^m \tag{10.30}$$

or

$$Q = ackQ^{b+f+m} \tag{10.31}$$

from which

$$ack = 1 \tag{10.32}$$

and

$$b + f + m = 1 \tag{10.33}$$

Equations (10.32) and (10.33) help verify the mathematical relation between the geometry of any stream channel and its discharge. The relations for a particular stream channel are illustrated in Figure 10.10. The slopes of the lines in Figure 10.10 are believed to be determined by the nature of the soil and rock material in which the stream channel is incised. Slopes b, f, and m determine the ordinate intercepts for each set of relations for any stream.

The hydraulic geometry relation holds for any given ralations up to the mean annual discharge on any natural stream and specifies a unique relation between that annual discharge and the stream width, depth, and mean velocity of flow. There is a different relation for discharges exceeding the mean annual discharge. The hydraulic geometry of streams presumes that an equilibrium condition exists between the streams draining a watershed and the water and sediment supplied to those streams (Singh, 1992).

The hydraulic geometry of a stream remains constant for any stream as long as its drainage environment is not altered. Urbanization, agricultural development, or any other factors that change the drainage environment will change the hydraulic geometry of the stream. If such changes are made, then a new relation is established among discharge and the

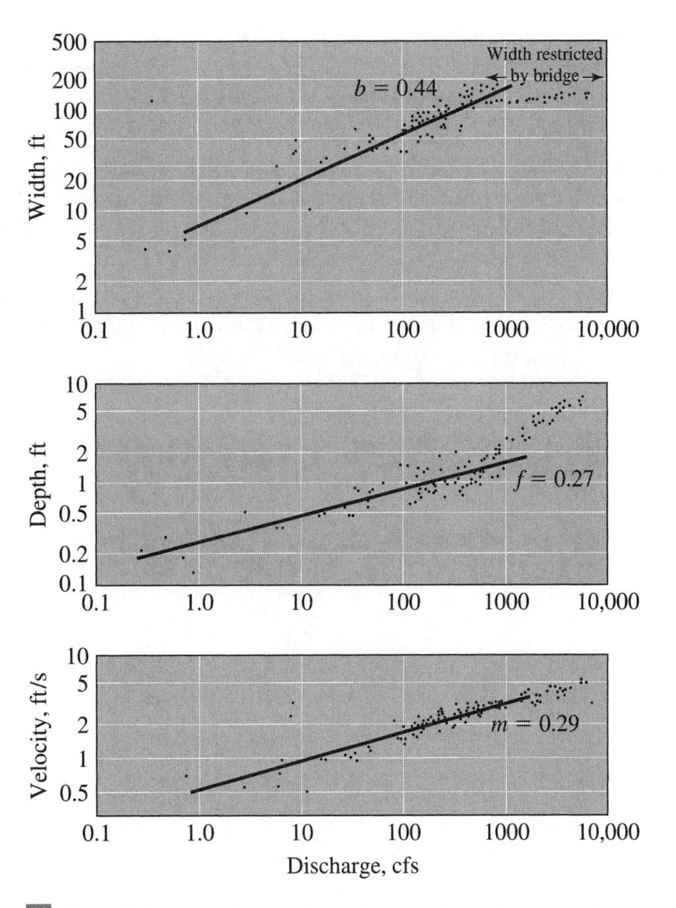

Figure 10.10 Hydraulic geometry for the power River at Arvada, Wyoming.

width, depth, and velocity of the channel. When development, whether urban or agricultural, is taking place on the drainage basin, the hydraulic geometry resumes a state of flux until static conditions are reestablished.

In Figure 10.10, the slopes for the width, depth, and velocity are $b = 0.44$, $f = 0.27$, and $m = 0.29$, respectively, and their sum is equal to unity in accordance with equation (10.33). The product of intercepts a, c, and k are also unity.

The hydraulic geometry relations are of great practical value in prediction of alluvial channel behavior, such as scour and fill; channel deformation; layout of river training works; design of stable channels, canals, and intakes; river flow control works; irrigation schemes and river improvement works; channel management; river restoration; modeling aquatic biota production systems; flow and sediment routing; flood estimation; and drainage net configurations. Through their exponents these relations can also be employed to discriminate between different types of river sections as well as in planning for resource and impact assessment. Therefore, at-a-station hydraulic geometry relations have been a subject of much interest and discussion in hydraulic and hydrologic literature. The mean values of the hydraulic variables of equations (10.26) to (10.28) are known to follow necessary hydraulic laws and the principle of the minimum energy dissipation rate. These mean values correspond to the equilibrium state of the channel. The implication is that an alluvial channel adjusts its width, depth, slope, velocity, and friction to achieve a stable condition, which is regarded as the one corresponding to the maximum sediment transporting capacity. The average river system tends to develop in a manner that produces an approximate equilibrium between the channel and the water and sediment it must transport. One can determine stable width of an alluvial channel using the hypothesis that an alluvial channel attains a stable width when the rate of change of unit stream power with respect to its width is a minimum. This means that an alluvial channel with stable cross section has the ability to vary its width at a minimum consumption of energy per unit width per unit time.

The hydraulic geometry of streams is a complex subject and involves sediment transport and other factors that are beyond the scope of this text. The geometric characteristics of the stream itself yield information related to the hydraulic geometry that assist in predicting the expected runoff from a drainage basin.

Example 10.8

The following data are available on the cross-section of a natural stream. Construct curves showing the relationships among the depth y and the section elements A, R, D, and Z. Determine the geometric elements for $y = 1.2$ m from the curves.

Distance from a Reference Point near Left Bank (m)	Stage (m)
−1.5	1.67
−1.2	1.38
−0.6	1.20
0.0	0.57
0.3	0.24
0.6	0.06
0.9	0.09
1.5	0.06
2.1	−0.03
2.7	−0.03
3.3	−0.15
3.9	−0.03
4.5	0.21
5.1	0.78
5.7	0.96
6.0 (right bank)	1.23

Solution: First, the stage is plotted against the distance from the reference point on an arithmetic paper, as shown in Figure 10.11. This plot produces the cross-section of the stream. Using the formulas,

$$R = \frac{A}{P}; \quad D = \frac{A}{B}; \quad Z = A\sqrt{D}$$

the following quantities are obtained:

Stage Height (m)	A (m^2)	P (m)	B (m)	R (m)	D (m)	Z ($m^{5/2}$)
0.00	0.11	2.12	2.1	0.05	0.05	0.02
0.25	0.99	4.39	4.25	0.23	0.23	0.47
0.50	2.11	5.11	4.75	0.41	0.44	1.40
0.75	3.36	5.79	5.25	0.58	0.64	2.69
1.00	4.78	6.88	6.15	0.69	0.78	4.22
1.25	6.38	7.67	6.80	0.83	0.94	6.19

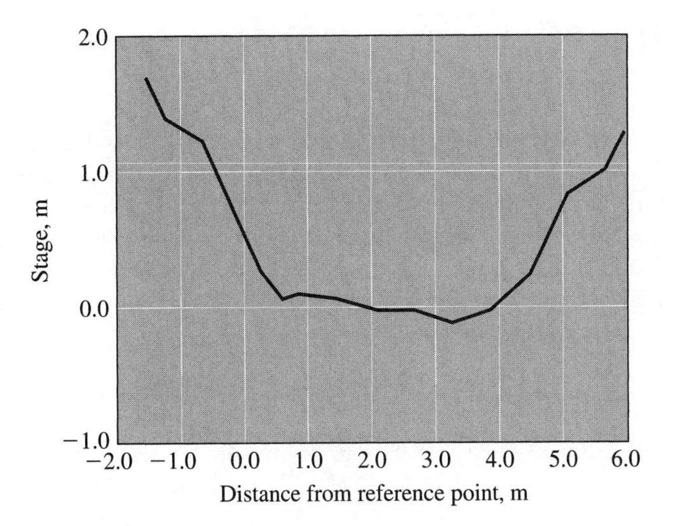

Figure 10.11 Cross section of a natural stream (Example 10.8).

The values of A, R, D, and Z versus the stage height and y are plotted on the graph paper as shown in Figure 10.12. From this figure, the following values are obtained for $y = 1.2$ m: $A = 5.3$ m^2, $R = 0.74$ m, $D = 0.84$ m, $Z = 4.85$ m$^{5/2}$.

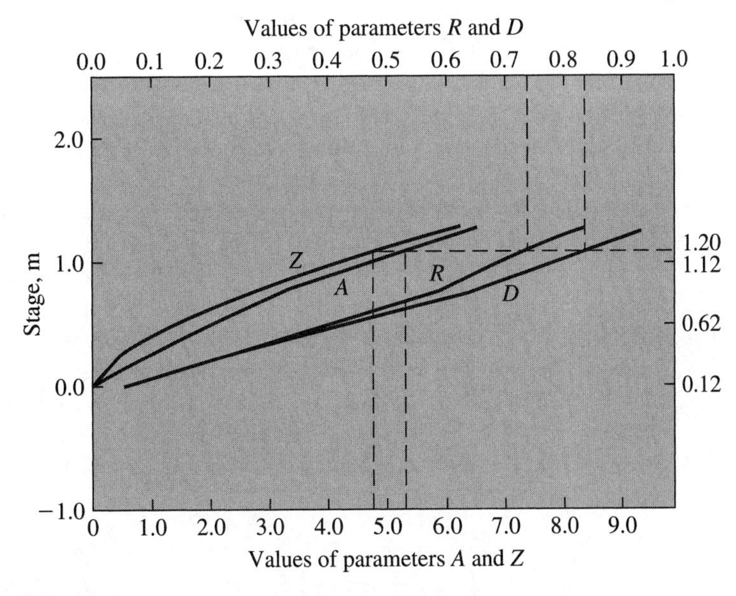

Figure 10.12 Stage versus hydraulic elements of a natural stream (Example 10.8).

10.9 Hydraulic Geometry of Basins

The average values of B, y, V, and A obtained for a channel reach are not sufficient to describe flow conditions in the entire stream for aquatic habitat evaluation or to evaluate the relationship between habitat suitability and streamflow variability for other but unmeasured reaches in the entire stream network. Singh and Broeren (1989) and McConkey and Singh (1992) developed basin hydraulic geometry relations from data measured near 14 stations in the Sangamon and South Fork Sangamon River basins in central Illinois. The relations derived for a given stream network in a hydrologically homogeneous basin from stream gaging stations representing a range of drainage areas are

$$\log x = a + bF + c \log A_d \quad (10.34)$$

$$\log x = A_f + B_f \log A_d \quad (10.35)$$

where $x = B$, y, or V, F is the flow duration in decimal form, A_d the drainage area in mi^2, and a, b, c, A_f, and B_f are coefficients. The values of these coefficients were found to be as follows:

	Sangamon River Basin			South Fork Sangamon River Basin		
x	a	b	c	a	b	c
B	0.55	−0.77	0.58	0.68	−0.93	0.55
y	−0.32	−1.17	0.41	−0.31	−1.22	0.45
V	−0.0054	−0.53	0.13	−0.025	−0.59	0.067

These basin hydraulic geometry relations define the average values of B, y, and V for a given Q or for a given F and A_d. The geometric variables increase with A_d for a fixed F. Implied in equations (10.34) and (10.35) is that B, y, and V change with A_d at the same rate for all values of F. Field data tended to support this assumption.

Bhowmik (1984) used data of 13 river basins (drainage area varying from 1828 to 28,275 km^5) in the United States (Bhowmik and Stall, 1979) to establish hydraulic geometry relationships for floodplains. The floodplain width, cross-sectional area, surface area, depth, sinuosity, and incision were related to stream order. A number of cross sections (10 to 60) were chosen for each stream segment to provide a representative value of the computed cross-sectional area and the depth of floodplain for that particular stream. The cross sections of the floodplain were then determined by calculating the difference in elevations of different contour lines intersecting the cross-sectional line. The cross-sectional area of the floodplain was calculated at each cross section and the average value of the cross-sectional area, A, for the entire reach was determined. The average width, B, of the floodplain for a reach was computed by the ratio of the planimetered floodplain area in the plan view to the down-valley length (the distance that the river could have traveled had it flowed at the middle of the valley in each reach of the river). The average depth of the floodplain, Y_F, was then computed from the average area, A, and the average width, B. The basic equation postulated by Bhowmik (1984) for a floodplain was of the form

$$\ln Y_F = a + bU \quad (10.36)$$

where a and b are coefficients, U is the stream order, and $Y = f(A, D, B, V, Q, A_D)$, where V is the velocity and Q is the 100-year flood discharge. The drainage A_D was found to be related to the stream order U for all streams studied by Stall and Fok (1968) and Stall and Yang (1970). Thus, U in equation (10.36) can be replaced by A_D. The average values of a and b, as cited by Bhowmik

TABLE 10.2 Values of Coefficients a and b in Equation (10.36) for 13 River Basins in the United States

Definition of Y_F	a	b
B	4.680	0.52
A	5.440	0.66
D	0.898	0.14
Q_{100}	1.835	0.85
A_D	−2.378	1.45

(1984), for all the streams are given in Table 10.2. Klein (1981) related the variation of channel geometry downstream with drainage area and found that variation of channel width and channel depth did not behave as a simple power function of discharge (or area) over a wide range of discharges. For both small and large basins, the exponent value of 0.5 was a good average.

Example 10.9

Using the staff gage to determine the geometric elements of a natural river, the following measurements were taken:

Distance from the Left Bank (m)	Water Depth (m)	Distance from the Left Bank (m)	Water Depth (m)
0.0	0.00	140	4.20
20	1.25	160	5.10
40	1.82	180	3.10
60	2.30	200	1.90
80	3.12	220	0.70
100	3.90	240	0.30
120	4.10	252	0.00

Calculate the area and hydraulic radius of the river.

Solution: The section is plotted in Figure 10.13. According to the given data, the section is divided into 13 subareas. The area and wetted perimeter for each subarea are

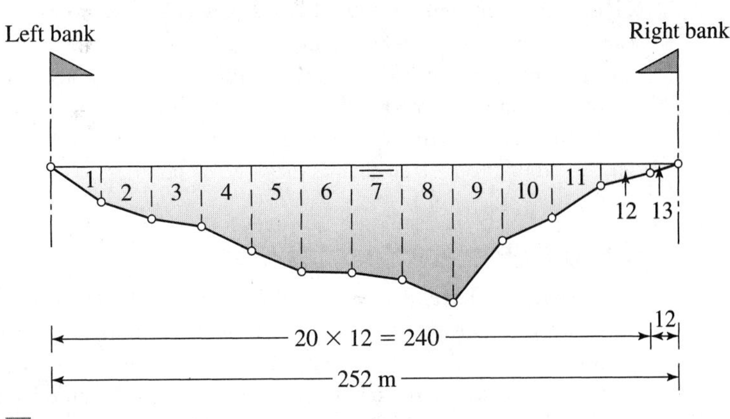

Figure 10.13 The river section given in Example 10.9.

as follows:

No. of Subarea	Area (m^2)	Wetted Perimeter (m)
1	12.5	20.04
2	30.7	20.01
3	40.2	20.00
4	43.2	20.05
5	70.2	20.02
6	80.0	20.00
7	83.0	20.00
8	93.0	20.02
9	82.0	20.10
10	50.0	20.04
11	16.0	20.04
12	10.0	20.00
13 (right bank)	1.80	12.00
Total	612.6	252.29

$$R = \frac{A}{P} = \frac{612.6}{252.29} = 2.428 \text{ m}$$

10.10 Measurement of Geometric Elements of Natural Rivers

As discussed in the preceding chapters, the study of open-channel and river hydraulics requires a determination of many parameters related to geometry and length of rivers, slope of channels beds, channels routes, characteristics of drainage basins, land use, catchment areas, topographic forms, and others. Natural streams and rivers have irregular boundaries as shown in Figure 10.13. The boundaries of natural rivers may change with time according to various developmental activities in drainage areas and associated flow conditions. A high stage flow may cause erosion in some localities while a low stage flow will generally cause deposition of suspended matters. Hydraulic parameters of natural rivers vary according to flow conditions. A flash flood may cause considerable geometric changes in a specific channel section. Calculations of velocity, discharge, and water-surface profiles in natural rivers cannot be done without accurate measurements or data of geometric elements. Sources of data can be classified as primary sources and secondary sources. Primary sources include direct measurements in the field. Secondary sources are related to analog data, such as maps, aerial photographs, and existing digital databases.

10.10.1 Primary Sources

The geometric elements of a natural river vary from one section to the other along its length. Field investigations are generally required to determine the geometry of the section under consideration. The objective of such an investigation is to determine the shape of the bed across the river width at specific sections, known as the cross section. Traditional methods include the use of staff gages and wire gages, which are manually operated. A staff gage, as shown in Figure 10.14, is the

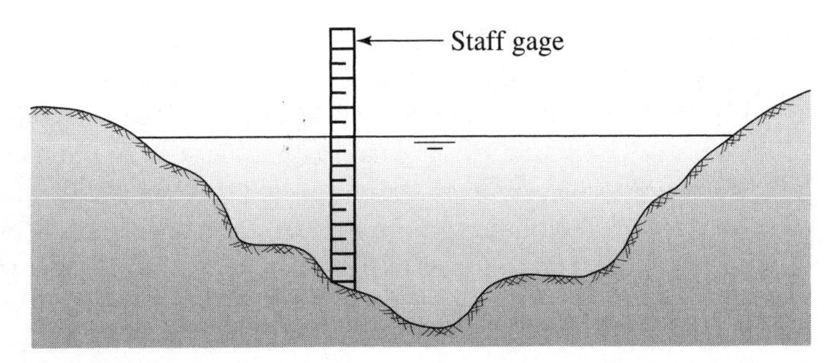

Figure 10.14 A general cross section in a natural river and a staff gage.

simplest way to determine bed shape. The gage indicates the bed level at the point of measurement, as compared to the water level. The difference between the water level and the bed level indicates the depth of flow at the point of measurement. Repeating this measurement at several points across the river width will lead to a good identification of the river bed. The river section is then plotted on graph paper. All other geometric parameters, including the side slope, bottom width, top width, wetted perimeter, hydraulic radius, and hydraulic depth, can thus be determined.

A wire gage, or lead line, can also be used from above such as from a bridge or any other overhead structure, or from a boat. A weight is lowered from the structure until it reaches the water surface. It is then lowered again until it reaches the bottom of the river. The gage has a drum with a circumference equal to 1 m or 1 ft. The number of revolutions of the drum as the weight is lowered from the water surface to the river bed is recorded by a mechanical counter, which, in turn, measures the depth of flow in the river. The operating range of a wire-weight gage is about 25 m (75 ft).

In modern times, hydrographic sounding techniques, such as sonar, are more often employed to determine channel cross-sectional geometry. Channel bottom bathymetry is obtained from sounding the channel bed from a boat using a radar or sonar device and the data are automatically recorded and processed by the receiver. Bathymetric maps can be produced in near real time. The operation can be made even more accurate and timely through the use of some of the systems discussed below, particularly Differential Global Positioning Systems.

10.10.2 Secondary Sources

The process of data collection from the field is not only a tedious and time-consuming task but is also costly and requires special arrangements. On the other hand, the availability and accuracy of data represent the most important element in any design problem or related decision. In addition, for large-scale projects, such as the Three Gorges River Project in China or the Mississippi River in the United States, in which open channels run for hundreds of kilometres, it might be impossible to collect the required data manually.

Recent techniques have emerged for data collection and processing that are powerful, accurate and efficient tools. Global Positioning Systems (GPS), remote sensing images (RSI), digital elevation models (DEM), digital terrain models (DTM), and geographical information systems (GIS) have been used extensively over the last two decades. Most current work and research related to river hydraulics involve one or more of the above tools. These techniques are virtually indispensable when dealing with large-scale projects but can also be useful on smaller projects.

10.10.2.1 Global Positioning Systems The GPS was developed by the U.S. Department of Defense (DOD) but now has many thousands of civilian users worldwide, although the system is still operated by the U.S. military. The operational system consists of 24 satellites that orbit the earth in a 12-hour cycle. The receiver must lock on to a minimum of four GPS satellite signals at a given time in order to compute *x, y, z* positions and the time offset in the receiver clock (Dana, 1999). Positioning is simply based on triangulation between the satellites, knowing their position at a given time and the speed of the signal. Under ideal conditions, simple GPS can provide reliable and accurate position and elevation data. Users of GPS receivers can plot their positions on earth to within a few meters of their true locations.

Position in the *x, y, z* plane is converted within the receiver to geodetic latitude, longitude, and height above the ellipsoid (earth surface). Horizontal and vertical positions are provided in the geodetic datum on which GPS is based (Universal Transverse Mercator, for example). Receivers can be set to convert to other user-required datums. Offsets of hundreds of meters in position can result from the use of the wrong datum. The users of GPS should be aware of the sources of errors that might degrade the quality of the obtained information (August et al., 1994).

GPS data can be made even more accurate through the use of Differential Global Positioning Systems (DGPS). A reference receiver, or base station, is set up at a location whose position is known with a high degree of accuracy, such as a United States Geological Survey benchmark. The base station computes corrections for each satellite signal from the known location. These errors are then transmitted to the roving GPS receivers for their use in correcting their own data. Because errors in identifying the locations and altitudes within an area of a few kilometers (along or across a channel, for example) will most likely be equal, DGPS can be used to determine the geometric characteristics of a channel under consideration with reasonable accuracy. The range of accuracy of DGPS is on the order of less than a meter horizontally and a few centimeters vertically.

10.10.2.2 Remote Sensing Images Remote sensing images (RSI) provide spatial and periodic information for large areas and are powerful tools for gathering regional information. The term *remote sensing* refers to the methods that employ electromagnetic energy observed in various band widths (visible, near-infrared [IR], thermal, etc.) as the means of detecting and measuring characteristics of objects of interest. Aerial photography is the original form of remote sensing and remains a widely used method. Recent developments in the science of remote sensing and its capabilities have significantly improved our understanding of river characteristics.

The resolution of useful RSI may vary from 1 m to more than 1 km. Based on the purpose of the study and the required accuracy, appropriate RSI should be acquired. Typical information that could be gathered through RSI includes, among others, general soil type, surface roughness, and channel slope. The drainage patterns, characteristics, land cover, and land use can also be identified from RSI. Figure 10.15 presents an RS image for Wadi Tawiyean, United Arab Emirates. The green circle indicates an area of a surface water body which can be characterized. The drainage pattern of the main tributaries can also be identified from the image. The image was used to identify hydrological parameters such as drainage network, land cover, and soil type for simulation of surface water runoff in the drainage system using the Corps of Engineers Hydrologic Modeling System HEL-HMS (Sherif et al., 2005). Remote sensing is also useful in estimating water levels and discharges in rivers (Bradley et al., 2005).

Satellite data can also be used to delineate flood-affected areas, and this information provides timely help to relief agencies. In India, a large number of irrigation projects were constructed in the postindependence period for improving agricultural production and

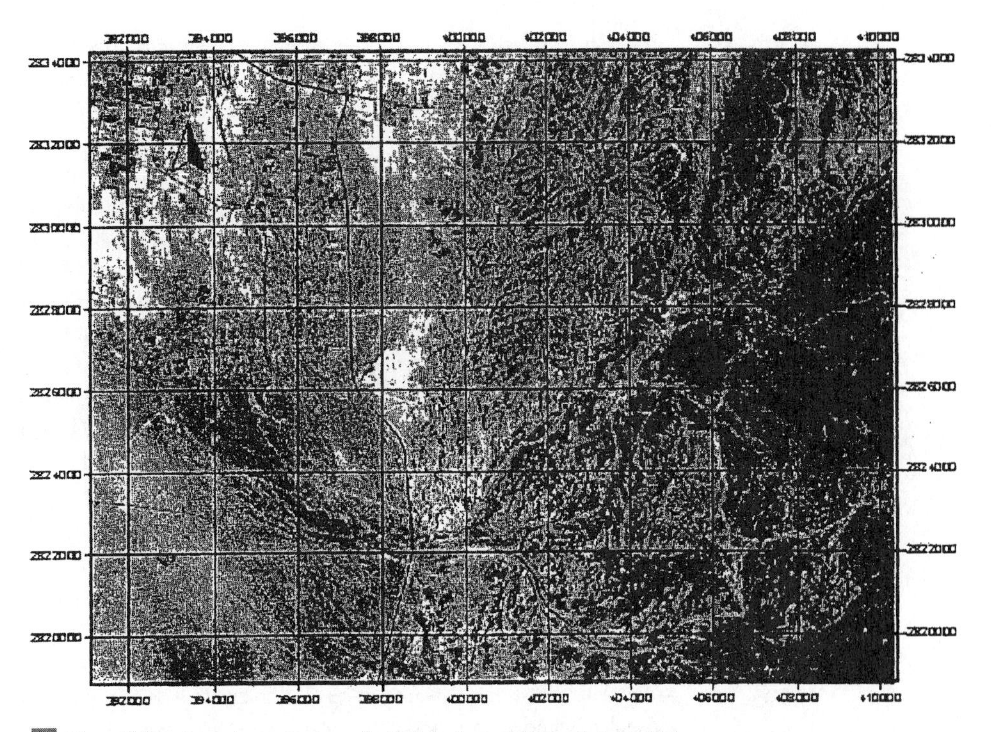

Figure 10.15 A remote sensing image for Wadi Tawiyean, UAE (Sherif et al., 2005).

productivity. Under the Command Area Development (CAD) program, the National Remote Sensing Agency (NRSA) has carried out an evaluation study of 14 irrigation commands in Andhra Pradesh, Assam, Maharashtra, Rajasthan, and West Bengal, the objective being the assessment of land utilization in terms of crops and soil conditions and to identify the performance of canals and distributaries during the past decade.

10.10.2.3 Digital Elevation Models Topographic data can be obtained in digital form in two basic formats. Digital line graph (DLG) files are simply representations of normal topographic maps with the features superimposed in vector format. On the other hand, digital elevation model (DEM) files present the topographic data in a gridded, or raster format. There are many kinds of DEM generation methods, such as stereo-matching from aerial photographs or satellite images, or interpolation of topographic maps. DEM has become a popular source for digital terrain modeling and watershed characterization due to its simple data structure and widespread availability.

DEMs are available in various grid sizes and elevation resolutions. Grid sizes of 30 m are common, but smaller (or larger) grid sizes are possible. The accuracy of DEM-generated data depends on the source and resolution of the data samples from which it was derived. The DEM data accuracy is derived by comparing linear interpolation elevations in the DEM with corresponding benchmark location elevations and computing the statistical standard deviation or root-mean-square error (RMSE). The RMSE is normally used to describe the DEM accuracy.

A runoff analysis or drainage pattern extraction is a popular application of DEM. The slope aspect and gradient are calculated from the DEM using clusters of pixels or grid cells. The objective is to determine the most likely path that water will flow through the basin on a cell-by-cell basis. Flow directions are assigned to each cell based on the computed aspect and gradient. Grid clusters of various sizes have been employed in this analysis in order to gain a clearer picture of the true flow gradient. The effective rainfall or runoff must flow to the next cell downstream according to the slope aspect and gradient.

The flow direction grid can be computed from the active DEM region using a Geographical Information System such as Geographic Resources Analysis Support System (GRASS), ARC/INFO, ArcView Spatial Analyst, or any other program that supports American Standard Code for Information Interchange (ASCII) formats. With the flow directions assigned for each grid, the flow accumulation at each cell can be computed. The flow accumulation for a given cell is defined as the number of cells whose flow paths eventually pass through that point. For example, cells that are part of a stream have high flow accumulation values since the flow paths of all "upstream" points will pass through them. Streams are easily identified by displaying all cells with a flow accumulation value greater than a user-defined threshold.

The cross sections of rivers and the bed slopes can also be generated from imported DEM data through the GIS interface. The generated slope and cross sections can be used in any hydraulic model, such as HEC-RAS.

10.10.3 Geographic Information Systems

A geographic information system (GIS) is a computer technology that has the ability to store, arrange, retrieve, classify, manipulate, analyze, and present large data sets and information in a simple manner. A GIS database contains both coordinate (location) data and attribute (descriptive) data. The GIS software can be used to create new products from the existing information for visualizing and analyzing the data. It is also an *information system.* It is comprised of four sections: spatial data input, spatial data manipulation, spatial data analysis, and spatial data visualization. The visual and analytical abilities of the GIS technology, along with developments in different spatial data input devices, such as Global Positioning Systems, have allowed GIS to be used not just in hydraulics but in virtually all areas of human endeavor. Field data can also be integrated into GIS, since most of the river geometric data would be available from hydrographic surveys, and can be interpolated to create spatial data (i.e., each grid cell would have valid data without any gaps). Point data could be water level, flow depth, flow width, porosity of soil, sediment size, and others. Depending on the objectives of the study, important layers of informational data and other related variables that would be stored in different layers can be identified. A GIS database, including various layers, should be developed in such a manner as to allow for efficient access, retrieval, organization, manipulation, and analysis of the available information. Contour or zonation maps can be developed from the input information to provide a better understanding of the areal distribution of variables in the study domain. Leading GIS software vendors in recent years have made extra efforts to improve the analytical and modeling capabilities of their products. Pioneered by HEC-RAS, developed by the Army Corps of Engineers, several commercial software vendors have developed stand-alone GIS modules with functions that can be used for a variety of hydraulic modeling needs. Certain hydraulic modeling functions have been embedded in leading generic GIS software packages, such as Environmental Systems Research Institute (ESRI) ArcStorm and ArcGrid and Integraph's InRoads. This approach builds on the top of a commercial GIS software package and takes full advantage of built-in GIS functionalities.

READING AID

10.1. What is meant by hydraulic conduits? What are the main driving forces for pipe flow and for open-channel flow?

10.2 Explain why problems of open-channel flow are more complicated than those of pipes.

10.3. State the main features of flow in open channels. When can flow in a pipe be an open-channel flow?
10.4. Give several examples of open channels. What are the two main types of open channels?
10.5. Explain the difference between flowage land and pondage areas.
10.6. What is meant by prismatic and nonprismatic channels? Explain the difference between flumes and culverts.
10.7. State three types of cross sections commonly used in artificial channels. Give the geometric elements of each.
10.8. Define the depth of flow, wetted perimeter, hydraulic radius, hydraulic depth, and section factor in open channels.
10.9. Write an equation to express the area, wetted perimeter, and hydraulic radius as a power function of flow depth.
10.10. What is meant by the cross-sectional asymmetry? What factors are used to reflect the section asymmetry?
10.11. What is meant by compound sections? How can the area be represented as a power function of flow depth in compound sections?
10.12. What is meant by the conveyance of a compound section? How do you evaluate the equivalent hydraulic radius and equivalent roughness of compound sections?
10.13. State three different methods to calculate the channel bed slope. What is the importance of defining bed slopes in open channels?
10.14. How can you measure the geometric elements of natural rivers?
10.15. Verify the mathematical relation between the geometry of any stream channel and its discharge.
10.16. Discuss whether the average values of B, y, V, and A are sufficient to describe flow conditions in a stream for aquatic habitats or not. What are the other factors which should be incorporated?
10.17. Provide some examples of primary and secondary sources of data related to geometric elements of natural rivers.
10.18. Discuss the benefits of using global positioning systems, remote sensing images, digital elevation models, and geographic information systems in water resources with emphasis on river hydraulics.

Problems

10.1. A 10-ft^2 hydraulic conduit issued 10 ft^3/s of flow. When the discharge was doubled, the flow velocity was also doubled. What is the type of this conduit? Calculate the velocity of flow.

10.2. The bottom of an open channel at a specified section is 2.0 m above a horizontal datum. The area of the cross section (rectangular section) of flow is 4.0 m^2, which accommodates a discharge of 8.0 m^3/s. The total energy is 3.0 m (measured from the horizontal datum). Calculate the depth of flow in the channel.

10.3. A rectangular channel 10.0 ft wide has a 4-ft water depth. Calculate the hydraulic radius, the hydraulic depth, and the section factor. What is the depth of the flow in a triangular section having the same cross sectional area and 1:1 side slope? Calculate the geometric elements of that section.

10.4. A trapezoidal section has a 5.0-ft bed width, 2.5-ft depth, and 1:1 side slope. Evaluate its geometric elements.

10.5. Calculate the geometric elements of a parabolic section with a cross-sectional area of 10.0 m^2 and a top width of 4 m. If the section is replaced by a trapezoidal section with a 4.0-m bed width and 3:2 side slope, calculate the depth of flow, top width, hydraulic radius, and section factor.

10.6. What will be the depth of flow in a trapezoidal channel that has a bottom width of 2.5 m and side slopes of 1:1? The channel carries a flow rate of 4 m^3/s with a velocity of 1.44 m/s.

10.7. A rectangular channel 6 ft wide has a 1.5-ft water depth. What is the depth of flow in a square channel with one diagonal vertical having the same cross-sectional area of the rectangular channel? The side of the square channel is 4 ft.

10.8. Evaluate the geometric elements for the above cross-section if the area of flow is 30 ft^2 and the side of the square is 6 ft.

10.9. A triangular channel with side slopes of 1:1 carries a discharge of 4 m^3/s with a velocity of 0.74 m/s. Evaluate the geometric elements of the channel.

10.10. Calculate the geometric elements of a circular cross section with a cross-sectional area of 0.15 m^2 and a diameter of 1.0 m.

10.11. It is required to cut a rectangular canal in a rock formation to carry a discharge of 475 cfs with a velocity of 5 ft/s. Determine the cross-sectional elements if the bottom width is twice the water depth.

10.12. A rectangular channel is 4 m wide and 2.5 m deep. The water in the channel is 1.75 m deep and is flowing at a rate of 21 m^3/s. Determine the area of flow, wetted perimeter, and hydraulic radius. Is the flow laminar or turbulent?

10.13. An equilateral triangular channel has a surface water width of 40 ft. Determine the hydraulic radius of flow and the flow rate if the water is moving at an average speed of 7 ft/s.

10.14. In an equilateral triangular conduit, a certain increase in the flow rate causes the maximum depth to increase by 1.5 times. Determine the percentage change in area, hydraulic depth, and discharge if the velocity of the flow is increased by 20%.

10.15. A trapezoidal channel has a side slope of 20° to the vertical and has a base width of 150 m. If the depth of flow is 6 m, determine the hydraulic radius, hydraulic depth, and section factor.

10.16. Determine the Reynolds number and the Froude number for the flow in Problem 10.15 if the discharge rate is 800 m^3/s. What is the flow regime?

10.17. The cross-sectional profile of a natural river can be approximated as parabolic. The distance between left and right banks is 100 m and the maximum water depth is 8 m. If the average velocity of the river is 0.5 m/s, determine the flow rate. Also determine the flow regime.

10.18. In a sewer pipe of circular section, flow is allowed up to 85% of the maximum depth. Determine the required pipe diameter to carry a flow rate of 2.25 m^3/s if the flow velocity is 0.8 m/s.

10.19. The channel shown in Figure P10.19 has a symmetric cross section with left and right overbanks. If a, b, c are 4 m, 2 m, and 5 m, respectively, determine the hydraulic radius and flow rate of the channel to maintain an average velocity of 2 m/s.

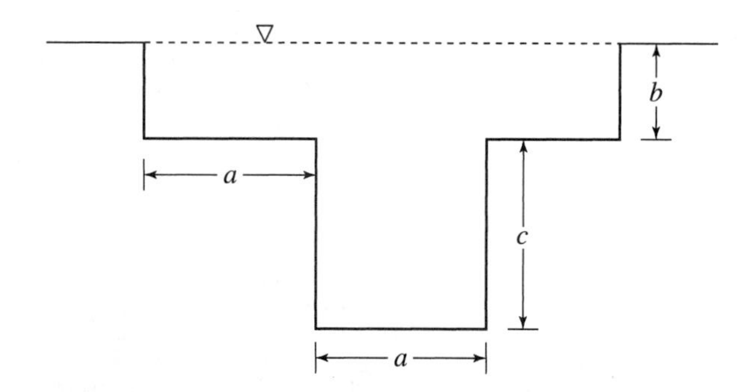

Figure P10.19 Channel cross section for Problem 10.19.

10.20. A symmetric artificial channel is made up of 5 steps of equal size as shown in Figure P10.20. If b and y for each step is 20 in and 16 in, respectively, determine the hydraulic depth and the hydraulic radius of the channel assuming that the channel section is fully occupied with flow.

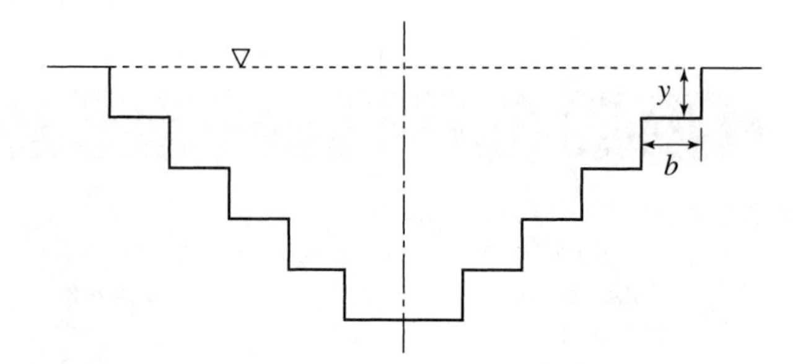

Figure P10.20 Channel cross section for Problem 10.20.

10.21. Starting from upstream to downstream, the following data have been collected for a 1.5-km-long portion of a natural river. Determine the equivalent bed slope and the discharge in the river if constant a in equation (10.30) is equal to 12.6. Also determine the time required by a floating object to cross this portion of the river.

Section	1	2	3	4	5	6	7	8
Horizontal distance (m)	0	150	350	520	700	1000	1230	1500
Bed level (m)	38	35	33.8	33.5	31.3	29.8	28	26

10.22. The following data have been collected for a section of natural river:

Section	1	2	3	4	5	6
Distance (km)	0	0.35	0.57	0.8	1.1	1.25
Bed level (m)	86	83.2	82	79.5	76	75.8

Calculate the arithmetic bed slope and the equivalent bed slope.

10.23. The following data have been collected for a river section starting from the left bank and reaching the right bank, taking readings for the water depth every 40 m. Plot the river section and determine the hydraulic radius and hydraulic depth of the river.

Distance from Left Bank (m)	Water Depth (m)
0	0
40	1.2
80	4.9
120	6.9
160	9.2
200	11.7
240	14
280	16.5
320	18.8
360	21
400	17
440	15.8
480	14
520	12.5
560	10.2
600	7.3
640	2.3
680	1
705	0

References

Ackers, P., 1988. Alluvial channel hydraulics. *Journal of Hydrology,* Vol. 100, pp. 177–204.

Allen, P. M., Arnold, J. G., and Byars, B. W., 1994. Downstream channel geometry for use in planning-level models. *Water Resources Bulletin.* Vol. 30, No.4, pp. 663–671.

August, P., Michaud, J., Labash, C., and Smith, C., 1994. GPS for environmental applications: Accuracy and precision of location data. Photogram. *Engrg. and Remote Sensing,* Vol. 60, No. 1, pp. 41–45.

Benson, M. A., 1962. Evolution of methods for evaluating the occurrence of floods. Water Supply Paper 180-A. 30 pp. U.S. Geological Survey, Washington, D.C.

Bhowmik, N. G., 1984. Hydraulic geometry of flood plains. *Journal of Hydrology,* Vol. 68, pp. 369–401.

Bhowmik, N. and Stall, J. B., 1979. Hydraulic geometry and carrying capacity of flood plains. Contribution Report 225, 147 pp., Illinois State Water Survey, Champaign, Ill.

Bradley, A. A., Holly, F. M., Lakshmi, V., Kruger, A., and Birkett, C., 2005. Coupling remote sensing and unsteady flow modeling for discharge estimation. Research Project funded by NASA Land Surface Hydrology Program, The University of Iowa, Iowa City, Iowa.

Cahoon, J. E., 1995. Defining furrow cross-section. *Journal of Irrigation and Drainage Engineering,* Vol. 121, No.1, pp. 114–119.

Chow, V. T., 1959. *Open Channel Hydraulics.* Macmillan, New York.

Cunge, J. A., 1975a. Rapidly varying flow in power and pumping canals. In: *Unsteady Flow in Open Channels,* Vol. II, edited by K. Mahmood and V. Yevjevich, Chapter 14, pp. 529–586. Water Resources Publications, Fort Collins, Colo.

Cunge, J. A., 1975b. Applied mathematical modeling of open channel flow. In: *Unsteady Flow in Open Channels,* Vol. I, edited by K. Mahmood and V. Yevjevich, Chapter 14, pp. 539–586. Water Resources Publications, Fort Collins, Colo.

Cunge, J., Holley, F. M., and Verwey, A., 1980. *Practical Aspects of Computational River Hydraulics.* Pitman, London.

Dana, P., 1999. Global positioning system overview. Department of Geogaphy, University of Texas at Austin, Tex.

Garbrecht, J., 1990. Analytical representation of cross-section hydraulic properties. *Journal of Hydrology,* Vol. 119, pp. 43–56.

Henderson, F. M., 1966. *Open Channel Flow.* Macmillan, New York.

Hydrologic Engineering Center, 1998. HEC-RAS: River Analysis System: Hydraulic Reference Manual. U.S. Army Crops of Engineers, Davis, Calif.

Johnstone, D. and Cross, W. P., 1949. *Elements of Applied Hydrology.* Ronald Press, New York.

Klein, M., 1981. Drainage area and the variation of channel geometry downstream. *Earth Surface Processes and Landforms,* Vol. 6, pp. 589–593.

Knighton, A. D., 1981. Asymmetry of river channel-cross sections. I. Quantitative Indices. *Earth Surface Processes and Landforms,* Vol. 6, pp. 581–588.

Lane L. J. and Foster, G. R., 1980. Modeling channel processes with changing land use. Proceedings of the ASCE Symposium on Watershed Management, Vol. 1, Boise, Idaho.

Laurenson, E. M., 1962. Hydrograph synthesis by runoff routing. Report 66. Water Research Laboratory, the University of New South Wales, Manly Vale, Kensington, New South Wales, Australia.

Leopold, L. B. and Maddock, T., 1953. The hydraulic geometry of stream channels and some physiographic implications. Professional Paper 252, pp. 1–57. U.S. Geological Survey, Washington, D.C.

Leopold, L. B. and Wolman, M. G., 1960. River meanders. *Bulletin, Geological Society of America,* Vol. 71, pp. 769–794.

McConkey, S. A. and Singh, K. P., 1992. Alternative approach to the formulation of basin hydraulics geometry equations. *Water Resources Bulletin,* Vol. 58, No.2, pp. 305–312.

Posey, C. J., 1950. Gradually varied flow. In: *Engineering Hydraulics,* edited by H. Rouse, Wiley, New York.

Sherif, M. M. (Project Leader), et al., "Assessment of the effectiveness of Al Bih, Al Tawiyean, and Ham Dams in Groundwater Recharge using Numerical Models," Final Report Vol. 1: Main Report, Ministry of Agriculture and Fisheries, Dubai, UAE, June 2005.

Shih, S. F., 1996. Integration of remote sensing and GIS for hydrological studies. Chapter 2 in *Geographical Information Systems in Hydrology,* edited by V. P. Singh and M. Fiorentino, Kluwer Academic Publishers, Dordrecht, the Netherlands.

Stall, J. B. and Fok, Y. S., 1968. Hydraulic geometry of Illinois streams. Research Report 15, Water Resources Research Center, University of Illinois, Urbana, Ill.

Stall, J. B. and Yang, C. T., 1970. Hydraulic geometry of 12 selected stream systems of the United States. Report 32. Water Resources Research Center, University of Illinois, Urbana, Ill.

Singh, K. P. and Broeren, S. M., 1989. Hydrologic geometry of streams and stream habitat assessment. *Journal of Water Resources Planning and Management,* Vol. 115, No. 5, pp. 583–597.

Singh, V. P., 1996. *Kinematic Wave Modeling in Water Resources: Surface-Water Hydrology.* Wiley, New York.

Singh, V. P., 1992. *Elementary Hydrology,* Prentice Hall.

Taylor, A. B. and Schwarz, H. E., 1952. Unit Hydrograph lag and peak low related to basin characteristics. *Transactions, American Geophysical Union,* Vol. 33, pp. 235–246.

Traver, R. G. and Miller, A. C., 1993. Open channel interpolation of cross sectional properties. *Water Resources Bulletin,* Vol. 29, No. 6, pp. 767–776.

CHAPTER 11

RESISTANCE IN OPEN CHANNELS

Steady, uniform flow in a well-maintained channel. (Courtesy of Dr. Mohammad Al-Hamdan.)

Resistance in open channels was discussed briefly in Chapter 7; however, that chapter dealt with general concepts and the specifics of resistance applications to open-channel situations were not covered. The same general concepts that govern resistance in closed conduits also apply to open channels or free surface cases. However, the specific applications are quite different. The quantification of resistance factors in channels is usually much more problematic than in the case of pipes. As discussed in Section 7.4.2.2 of Chapter 7, flow in open channels almost always represents the case of turbulent flow in rough surfaces. Channel boundaries are not as uniform as are pipe boundaries, and the cross-sectional geometry may not be as regular. Because of these factors, the turbulent components of the velocity profiles may exert influences that are more complex than in the closed conduit case. Thus, it is difficult to arrive at a simple resistance factor that can properly represent the complexity of the resistance forces in open-channel flow. As in the case of pipes, empiricism plays an important role in the applications of resistance theory to real-world open-channel cases.

11.1 Steady, Uniform Flow in Open Channels

The case of steady, uniform flow in open channels was briefly discussed in Section 7.2 of Chapter 7. As discussed in that section, uniform flow in channels represents a state of dynamic equilibrium. This situation is depicted in Figure 11.1. Note that since the depth is uniform throughout the channel length, the hydrostatic forces are equal and opposite. Thus, the gravitational forces tending to drive the flow are exactly countered by the resistance forces tending to retard the flow. This situation is also one of equilibrium of specific energy, since the energy lost through resistance is exactly balanced by the energy gained through the change in the channel invert. Thus, uniform flow in channels always exhibits equality in the slopes of the energy grade line, the channel bottom, and the water surface as shown in the figure. As mentioned in Section 7.2, the constant depth that occurs throughout the channel length is called *normal* depth.

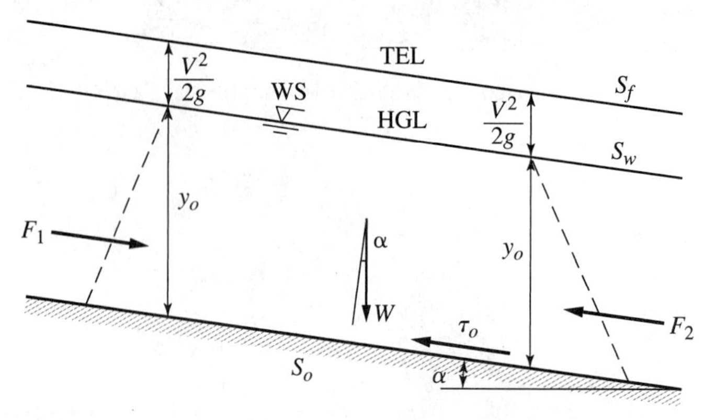

Figure 11.1 Forces in dynamic equilibrium in steady uniform flow.

Applying Newton's second law of motion to the channel reach shown in Figure 11.1, we get

$$\sum \mathbf{F} = m\mathbf{a} \tag{11.1}$$

where $\mathbf{F}$ is the force vector, $\Sigma\mathbf{F}$ is the sum of forces, m is the mass, and $\mathbf{a}$ is acceleration. However, since we have dynamic equilibrium, the acceleration $\mathbf{a} = 0$, so that $\Sigma\mathbf{F} = 0$. Then, setting the two forces equal in the direction of flow as discussed above leads to the following expression:

$$A \cos\alpha L \sec\alpha\gamma \sin\alpha = \tau_o P \cos\alpha \; L \sec\alpha \tag{11.2}$$

where A is the cross-sectional area, L is the reach length, P is the wetted perimeter, α is the slope angle of the channel bed, γ is the specific weight of water, and τ_o is the boundary shear stress. Thus,

$$A\gamma \sin\alpha = \tau_o P \tag{11.3a}$$

or

$$\tau_o = \gamma R \sin\alpha \tag{11.3b}$$

where R is the hydraulic radius $= A/P$. Now, noting that $S_f = \Delta h/L = S_o = \Delta z/L = \tan\alpha$, and that $\tan\alpha = \sin\alpha/\cos\alpha$, then $\sin\alpha = S_o \cos\alpha$. Thus,

$$\tau_o = \gamma R S_o \cos\alpha \tag{11.3c}$$

Since, for small values of α, $\cos\alpha \approx 1$, we can write equation (11.3c) as

$$\tau_o = \gamma R S_o \tag{11.4}$$

which is equivalent to equation (7.9) of Chapter 7. Since $S_o = S_f$, it is customary to use the bottom slope in the expression when applied to uniform flow in channels. Thus we see that the general expressions developed in Section 7.2 for steady uniform flow are applicable in the open-channel case, and since the resistance equations developed in Section 7.3 of Chapter 7 are based upon these expressions, then these equations are also valid.

The resistance equation most useful in governing uniform flow in channels is the Chezy equation (7.17):

$$V = C\sqrt{RS_o} \tag{11.5}$$

where C is the Chezy resistance coefficient ($L^{1/2}/T$). In Section 7.3 of Chapter 7, it was shown that the Chezy formula can be considered the most general form of other resistance equations such as the Darcy formula depending on how the Chezy C is defined. In particular, it was shown that if $C = \sqrt{8g/f}$, then the Chezy formula is essentially equal to the Darcy formula. The f-to-C transformation can be employed to examine the behavior of the Chezy C by developing a sort of Moody curve for C as demonstrated by Henderson (1966). This curve is shown in Figure 11.2. The figure demonstrates that for the region beyond Reynolds numbers values of about 10^5 the Chezy C, like the Darcy f factor, becomes constant. Of course, this is the range of Nikuradse's rough pipe experiments discussed in Chapter 7. Flow in free surfaces, such as channels or floodplains, almost always falls in the category of high Reynolds numbers and rough boundary surfaces. Thus the Chezy C coefficient is not usually considered to be a function of the Reynolds number or of fluid characteristics.

The C-f relationship points up the fact that the resistance factor C must be obtained through empiricism just as the Darcy f factor was by Prantl and his associates. The Chezy equation has a long history of use in channel and canal design, where it is desired to design the channel to carry a given discharge at uniform flow. A few investigators made early

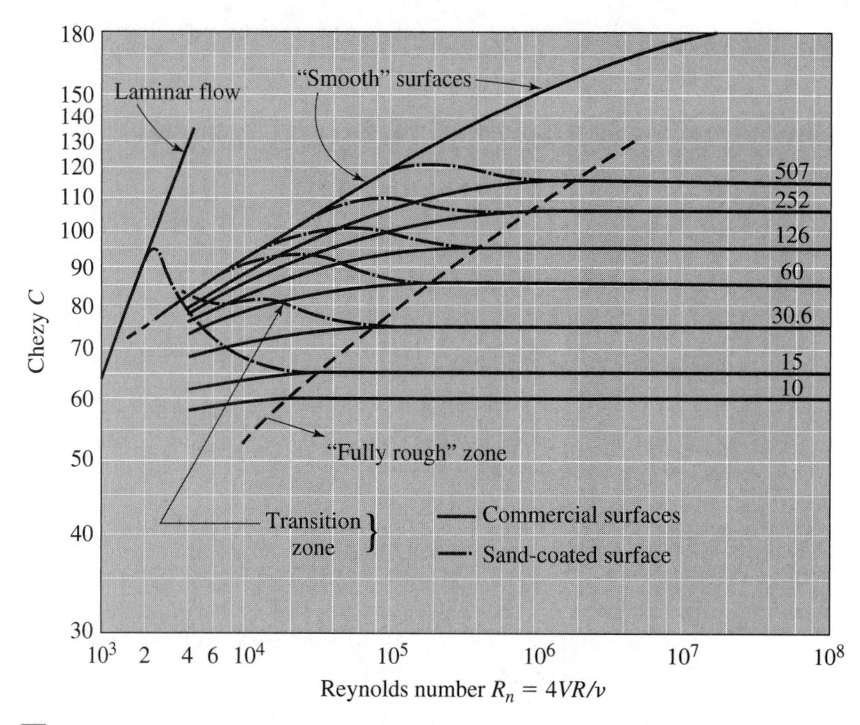

Figure 11.2 Modified Moody diagram showing the behavior of the Chezy C (after Henderson, 1966).

attempts to examine the behavior of C in channels, notably Henry Darcy and H. E. Bazin between 1855 and 1860 (Sturm, 2001). The Chezy equation was extensively used in channel design by British engineers engaged in the design of canals in the Indian subcontinent during the nineteenth century. In these efforts several empirical relationships for the resistance factor C were developed. One popular equation was the Kutter formula (or Ganguillet and Kutter formula), actually based primarily on observations made on the Mississippi River in the United States during the middle part of the nineteenth century:

$$C = \frac{[1.811/n] + \{41.65 + [0.00281/S_o]\}}{1 + [n/R^{1/2}]\{41.65 + [0.00281/S_o]\}} \tag{11.6}$$

The factor n that appears in this equation is an empirical constant that relates the resistance of the channel to the roughness of the boundary. In 1890 the Irish engineer Robert Manning (1816–1897) synthesized much of the extant information and data pertinent to resistance in open channels and proposed that the velocity would vary as the product of the hydraulic radius to the 2/3 power and the square root of the slope multiplied, of course, by a resistance coefficient. Subsequently, it became common to assume that the Manning resistance coefficient could be expressed as the reciprocal of the Kutter n, leading to a simple expression for the Chezy C:

$$C = \frac{1}{n}R^{1/6} \quad \text{(SI units)} \tag{11.7a}$$

or

$$C = \frac{1.49}{n}R^{1/6} \quad \text{(BG units)} \tag{11.7b}$$

The same expression had been in use in various parts of the world before its recommendation by Manning. It is a matter of some controversy whether or not the n factor that appears in equations (11.7a and b) really has exactly the same meaning as the n that appears in the Kutter formula (Morris and Wiggert, 1972). Substitution of equations (11.7a and b) into the Chezy equation leads to what is called the Manning equation:

$$V = \frac{1}{n} R^{2/3} S_o^{1/2} \quad \text{(SI units)} \tag{11.8a}$$

$$V = \frac{1.49}{n} R^{2/3} S_o^{1/2} \quad \text{(BG units)} \tag{11.8b}$$

Also, since $Q = AV$,

$$Q = \frac{1}{n} A R^{2/3} S_o^{1/2} \quad \text{(SI units)} \tag{11.9a}$$

$$Q = \frac{1.49}{n} A R^{2/3} S_o^{1/2} \quad \text{(BG units)} \tag{11.9b}$$

As mentioned above, the so-called Manning n is considered to be a factor relating the resistance to the roughness of the channel boundary. However, one can note from the above equations that the Manning n factor must possess units of $T/L^{1/3}$ in order for the equation to be dimensionally consistent. This fact represents one of the great weaknesses of the Manning equation (and the Chezy equation directly for that matter), since it implies that the Manning "constant" is not really constant after all. Thus, just as is the case of the Chezy coefficient, the Manning n values will differ depending on the system of units employed. A great deal of guidance has been provided to aid in the selection of n values for various situations. The Manning equation has been shown to be appropriate for the case of fully turbulent rough flow (Henderson, 1966) so that the coefficient is not thought to be a function of fluid properties. It has traditionally been related to the roughness properties of the boundary and recommended values of Manning's n for various channel surfaces have been presented as least as far back as Horton (1916). As implied by the units discussion above, the coefficient has also been shown to be a slight function of the hydraulic radius itself, and curves of n versus velocity $\times$ hydraulic radius have been produced for guidance with various grassed surfaces (Coyle, 1975, Palmer, 1945; see Chapter 16). In addition, guidance in the form of photographs of conditions corresponding to given n values has been provided by Chow (1959) as well as the U.S. Geological Survey (1967, 1989). These publications are often used by professionals for guidance in the field.

In recent history, the problem of resistance in free surface flow has been identified with the work on fully turbulent, rough flow performed by Nikuradse in the first third of the twentieth century that was discussed in Chapters 7 and 8 (see Henderson, 1966, for example). Thus, the energy losses are related to the relative roughness of the surface (k_s/y) and the resistance effect lessens as the depth increases (Figure 11.3). In the previous expression, k_s is the representative roughness height and y is the depth. A notable example of this work was Strickler's (1923) development of the relationship between the Manning n and the median particle size of channel beds:

$$n = 0.034 d_m^{1/6} \tag{11.10}$$

where d_m is the median grain diameter of the bed material in feet (k_s in Figure 11.3). Henderson (1966) showed that this equation is almost identical to one that can be derived from the Nikuradse rough pipe experiment data discussed in Chapter 7. This analogy was further

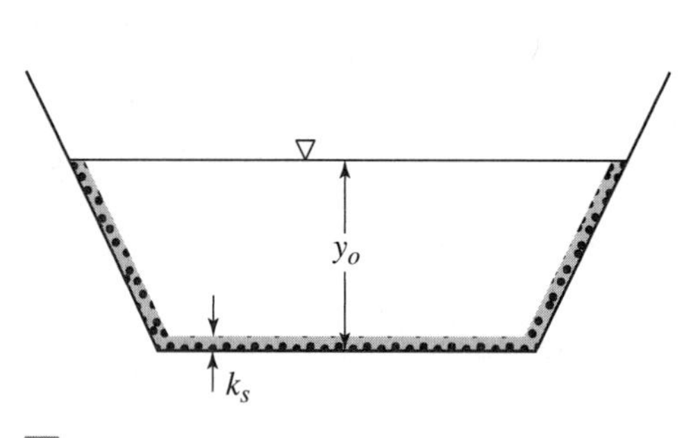

Figure 11.3 Roughness elements for open channels.

A natural channel.

enhanced by Keulegan (1938), who even developed a form of the Nikuradse rough pipe equation for trapezoidal channels (Sturm, 2001). This relative roughness analogy has led to the general acceptance of the assumption that the Manning factor would not be a function of depth in most channel situations, and thus it is common practice to use a constant n value for a given channel reach regardless of flow depth. Recent work cited by Sturm (2001) has given the limits of constancy of the Manning n versus the R/k_s, ratio as $4 < R/k_s < 500$ (Yen, 1992), or $3.6 < R/k_s < 360$ (Hager, 1999).

The work cited above has led to the conclusion that the Manning equation is consistent with the turbulent rough pipe formula developed by Nikuradse and thus, strictly speaking, should only be employed in rough boundary situations. However, experience has shown that most open channel situations meet this criterion. Table 11.1 gives representative roughness coefficients in BG units for some commonly encountered channel conditions.

TABLE 11.1 Manning's Roughness Coefficients for Channels

Channel Type	Condition	Manning n
Timber lined	Planed, carefully laid	0.010–0.012
Timber lined	Unplanned	0.012–0.014
Concrete lined	Unfinished	0.014–0.020
Concrete lined	Finished	0.011–0.016
Metal lined	Steel, painted	0.012–0.017
Metal lined	Steel, riveted	0.015–0.020
Metal lined	Corrugated	0.022–0.026
Natural earth	Straight, good condition	0.018–0.020
Natural earth	Regular, weathered	0.022–0.025
Natural earth	Regular, stones and weeds	0.028–0.040
Natural earth	Winding, irregular	0.035–0.050
Natural earth	Sluggish, deep pools	0.050–0.080
Natural earth	Weedy, obstructed with debris	0.050–0.15
Overbanks	Short grass	0.025–0.035
Overbanks	High grass	0.030–0.050
Overbanks	Mature crops	0.035–0.050
Overbanks	Light brush and trees	0.040–0.080
Overbanks	Medium to dense brush	0.070–0.16

11.2 Calculation of Normal Depth

Flood control channels and water delivery canals are usually designed to carry a specified discharge at normal depth. Thus, the relationship among discharge, depth, and channel geometry embodied in equation (11.9) is central to the practice of open-channel hydraulics. Design situations are normally presented in one of two cases: (a) The channel geometry and slope are known and the normal depth is to be calculated; or (b) channel dimensions are to be determined to carry a given discharge at a specified normal depth. The Manning equation (11.9) is employed in either case. In discussions of the Manning equation, it is sometimes convenient to separate the terms representing channel geometry and roughness from the slope term, that is,

$$Q = KS_o^{1/2} \tag{11.11}$$

where

$$K = \frac{1}{n}AR^{2/3} \quad \text{(SI units)} \tag{11.12}$$

and the factor K is known as the *conveyance*. It is convenient to discuss channel capacity in this manner since channel slopes are usually fairly fixed by the natural topography of the land surface. On the other hand, engineers have some control over the factors that make up the conveyance. Channel capacities for various discharges can be computed by plotting K versus depth for given cross sections. The conveyance can also be used in cases of compound channel sections to balance the discharge between the different channel components, since the slopes of all components will be the same (U.S. Army Corps of Engineers, 1993).

To solve equation (11.9) for normal depth, we formulate it as follows:

$$AR^{2/3} = \frac{nQ}{S_o^{1/2}} \quad \text{(SI units)} \tag{11.13a}$$

or

$$AR^{2/3} = \frac{nQ}{1.49S_o^{1/2}} \quad \text{(BG units)} \tag{11.13b}$$

The normal situation is that everything on the right side of equation(s) (11.13) is known and the only unknown is either the channel width or depth on the left side of the equation.

A concrete trapezoidal channel. (Courtesy of Dr. Mohammad Al-Hamdan.)

Example 11.1

A rectangular channel 5 m wide carries a discharge of 100 m^3/s at normal depth. The channel is running on a slope of 0.005 m/m and has a Manning n value of 0.035 (metric). Determine the normal depth for this situation.

Solution: Employing equation (11.13a) and letting b = channel width and y_o = normal depth,

$$AR^{2/3} = \frac{nQ}{S_o^{1/2}}; \quad by_o\left[\frac{by_o}{2y_o + b}\right]^{2/3} = \frac{0.035 \times 100}{(0.005)^{1/2}} = 49.5;$$

$$5y_o\left[\frac{5y_o}{2y_o + 5}\right]^{2/3} = 49.5; \quad y_o = 6.65 \text{ m}.$$

Equation (11.13) can be solved by any numerical technique or by trial and error. The relationship is very smooth and regular with only one real positive root, so that it is amenable to solution by most numerical techniques such as bisection, secants, and so on. Many modern calculators have root finders for the solution of nonlinear equations and, furthermore, the equation is ideal for solution using spreadsheet software such as Microsoft Excel or Corel Quattro Pro. A spreadsheet solution is as follows:

y (m)	A (m^2)	P (m)	R (m)	$AR^{2/3}$	Error
2	10	9	1.111111	10.7277	−38.7693
6.65	33.25	18.3	1.81694	49.51012	0.013123

The error term in the last column of the spreadsheet is just the value in the previous column minus the right side of the equation $\left(\text{i.e., } \frac{nQ}{S_o^{1/2}}\right)$. Note that once the calculations were set up for the first assumption, it was just copied down a line; then it was a matter of merely toggling through other assumed values of y on the second line until an acceptable error term was obtained. In reality, it was not even necessary to copy the first line down, as the toggling could proceed on that line until the solution was achieved.

Example 11.2

A trapezoidal channel with a bottom width of 10 ft and side slopes of $2H/1V$ carries a discharge of 500 cfs at normal depth. If the channel is laid on a slope of 0.0025 ft/ft and has a roughness coefficient of 0.030, what is the normal depth?

Solution: Employing equation (11.13b) with b = bottom width, t = horizontal component of side slope,

$$AR^{2/3} = \frac{nQ}{1.49S_o^{1/2}} \quad \text{or}$$

$$(by_o + ty_o^2)\left[\frac{by_o + ty_o^2}{b + 2y_o(1 + t^2)^{1/2}}\right]^{2/3} = \frac{0.03 \times 500}{1.49 \times (0.0025)^{1/2}} = 201.34$$

$$(10y_o + 2y_o^2)\left[\frac{10y_o + 2y_o^2}{10 + 4.47y_o}\right]^{2/3} = 201.34$$

by spreadsheet, or trail, y_o = 4.87 ft.

Example 11.3

A 3-ft-diameter circular culvert is laid on a slope of 0.005 ft/ft and is carrying a discharge of 25 cfs (Figure 11.4). If the roughness coefficient of the culvert is 0.012, what is the normal depth for this case?

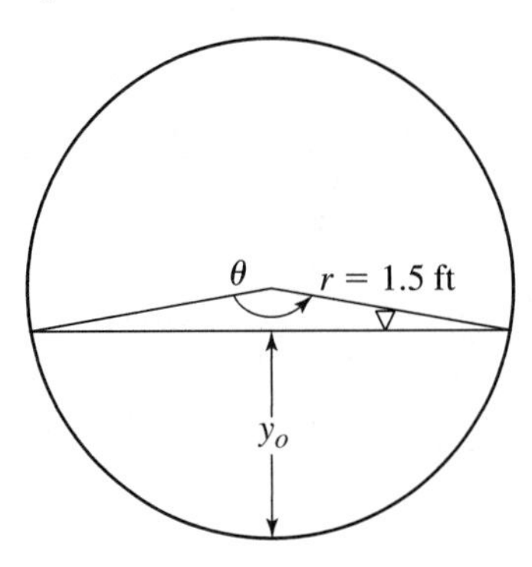

Figure 11.4 Culvert for Example 11.3.

Solution: The Manning equation is written as

$$AR^{2/3} = \frac{nQ}{1.49S_o^{1/2}} = \frac{0.012(25)}{1.49(0.005)^{1/2}} = 2.847$$

From Table 10.1, we find that for circular cross sections

$$\theta = 2\cos^{-1}\left(1 - 2\frac{y}{d}\right) \text{ and then } A = \frac{d^2}{8}(\theta - \sin\theta)$$

while $P = \frac{d}{2}\theta$. The solution proceeds by spreadsheet, setting up the following table and iterating on θ until the Manning relation is satisfied:

θ (rad)	A (ft²)	P (ft)	R (ft)	$AR^{2/3}$
2.5	2.13	3.75	0.57	1.47
3.0	3.21	4.50	0.71	2.56
3.1	3.44	4.65	0.74	2.81
3.11	3.46	4.66	0.74	2.84

Thus, $\theta = 3.11$ rad and from the equation given above:

$$\theta = 2\cos^{-1}\left(1 - 2\frac{y}{d}\right) = 3.11 = 2\cos^{-1}\left(1 - 2\frac{y}{d}\right)$$

Then $1.555 = \cos^{-1}(1 - 2(y/d))$, or $0.0158 = 1 - 2(y/d)$, or $0.492 = y/d$, from whence we find that $y = 0.492d = 1.48$ ft.

Example 11.4

A trapezoidal channel is to be designed to carry a discharge of 100 m³/s at a normal depth of 2 m. Geotechnical considerations require that the channel side slopes be no steeper than $2H/1V$ and the channel must be laid on a slope of 0.001 m/m. The estimated roughness coefficient is 0.035. What must be the bottom width of the channel to accomplish this purpose?

Solution:

$$(by_o + ty_o^2)\left[\frac{by_o + ty_o^2}{b + 2y_o(1 + t^2)^{1/2}}\right]^{2/3} = \frac{0.035 \times 100}{(0.001)^{1/2}} = 110.68$$

$$(b(2) + 8)\left[\frac{b(2) + 8}{b + 8.94}\right]^{2/3} = 110.68$$

$$b = 33.8 \text{ m}$$

11.3 Other Applications

It is quite common for some restrictions to be placed on the channel geometry in design cases. For instance, soils and geotechnical considerations may limit the allowable shear stress that the bed material can resist and thus limit the maximum velocity allowable to reduce erosion of the channel boundary. Then, since $\tau_o = \gamma R S_o$, the required hydraulic radius can be determined if the bed slope is known. With the hydraulic radius and slope known, then the Manning equation is employed to determine the allowable velocity. Therefore, through continuity, this effectively limits the cross-sectional area available for a specified discharge. Thus, the Manning formula is still the governing resistance equation employed in the design subject to the area restriction (Problems 11.15, 11.16, 11.18, 11.20, and 11.28).

In other cases, the required channel slope associated with a given normal depth may need to be computed for a channel. In this case, solution of the Manning formula for the slope term leads to the following formulation:

$$S = \left[\frac{nQ}{AR^{2/3}}\right]^2 \quad \text{(SI units)} \tag{11.14}$$

In some cases it is desirable to use the velocity in the computation, in which cases the formula becomes

$$S = \left[\frac{nV}{R^{2/3}}\right]^2 \tag{11.15}$$

The use of the Manning resistance formula to compute slopes associated with a given channel configuration for normal depths corresponding to specified discharges is an important aspect of open-channel hydraulics. In Chapter 3 (Section 3.8.2) flow conditions have previously been classified as either subcritical or supercritical depending on whether the dimensionless Froude number is less than or greater than unity (i.e., if the flow velocity is less than or greater than the velocity of a shallow gravity wave of the same depth). The situation is illustrated in Figure 11.5. In the case of uniform flow, the normal depth can only be supercritical if the slope is steep enough to maintain the velocities necessary for that case. Consequently, if the normal depth is supercritical, then the channel slope is said to be "steep" as shown in the figure. Conversely, if supercritical velocities cannot be maintained in a particular channel, then the channel slope is designated as "mild" and the normal depth is subcritical as in

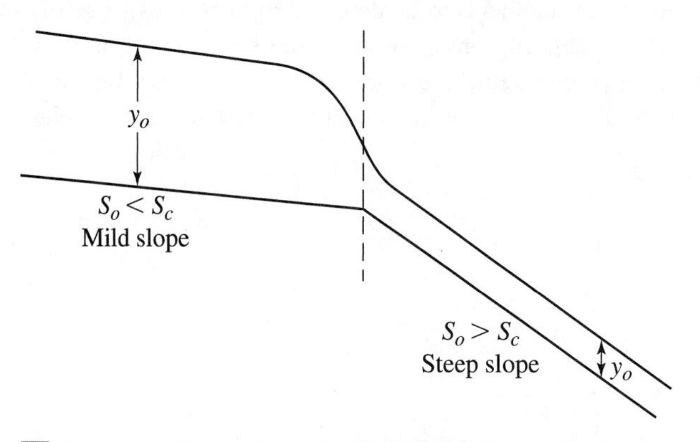

Figure 11.5 Steep and mild slopes for channels.

Figure 11.5. The designations "steep" or "mild" are always used in hydraulics in the context of the potential for supercritical or subcritical normal depth in a channel. If a channel slope is such that the normal depth would be exactly critical, then the slope is designated as "critical."

From the preceding analysis it can be seen that, in the case of steady uniform flow, the state of the flow, supercritical or subcritical, can actually be considered a resistance condition. The slope as determined from the Manning (or Chezy) equation is a function of both channel roughness and cross-sectional geometry, as well as discharge. Thus the capability of a particular channel to be steep or mild is not fixed (i.e., the same channel may support supercritical flow for one discharge and may not for another flow). Likewise, a channel may be in one state for a particular flow rate, and a change in either geometry or roughness may change it to another state. These issues will be fully explored in the next chapter.

The preceding discussions lead to another point; that is, since uniform flow is a state of dynamic equilibrium of both forces and energy, then the slope calculated by equation (11.14) also represents the rate of energy dissipation per unit length of channel in that case. This is demonstrated in Figure 11.5 by the parallel nature of the bed slope (S_o) and the friction slope (S_f). That is the reason that no subscript was used with the slope term in equations (11.14) and (11.15).

11.4 Channel Efficiency

The hydraulic efficiency of an open channel can be judged based on the resistance, or energy lost, in carrying a given discharge. Alternatively, it can be regarded in terms of the discharge capacity of the channel for a given head loss. From either perspective, it appears that the efficiency of a channel is a function of the channel geometry and roughness condition. The Manning equation shows that for a given slope, roughness, and cross-sectional area, the capacity of a channel varies directly with the hydraulic radius. Since the hydraulic radius is equal to the ratio A/P, it is clear that the capacity or efficiency of a channel varies inversely with the wetted perimeter. This observation would also follow from the fact that the resistance losses in a channel result from frictional contact between the water and the channel boundary. Thus energy loss will be less in channels with smaller wetted perimeters and greater in channels with larger perimeters.

An open channel with a cross-sectional area of 20 m^2, for example, can be excavated in a number of ways. Some of the options for construction of a rectangular section are given in Figure 11.6. All sections have the same area but differ in the length of their wetted perimeter. The wetted perimeters for cases (a), (b), (c), and (d) are 22 m, 14 m, 13 m, and 14 m, respectively. Therefore, of the given rectangular sections, the section in case (c) has the minimum wetted perimeter and hence will encounter less energy loss.

From the above discussion it is clear that the channel with maximum hydraulic efficiency will have the minimum wetted perimeter. Such a section is called the most efficient section or the best hydraulic section. In the case of lined channels, it may also be the most economical section, as it gives, for a given area, the maximum discharge and the minium wetted perimeter. This reduces the cost of lining. For a given cross-sectional area, the best hydraulic section has the minimum wetted perimeter. For a given perimeter, it has the maximum area. In the case of trapezoidal sections it is a simple matter to develop geometric relationships to minimize the wetted perimeter. Recall the basic equations for area and wetted perimeter for trapezoids:

$$A = by + ty^2 \tag{11.16}$$

$$P = b + 2y(1 + t^2)^{1/2} \tag{11.17}$$

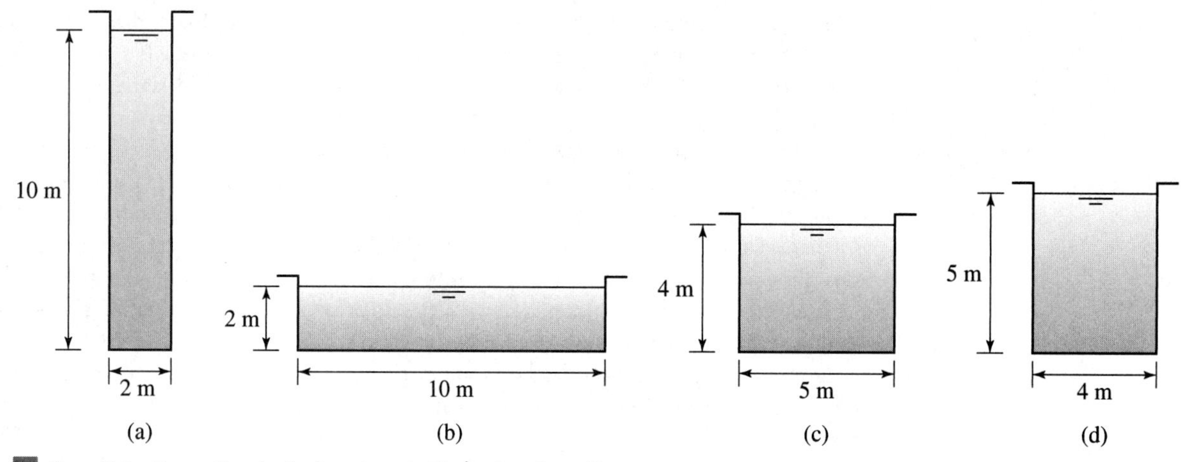

Figure 11.6 Some options for the dimensions of a 20 m² rectangular section.

We can determine the relationship between b and y to minimize P for a fixed cross-sectional area and side slope. Solving equation (11.16) for b yields

$$b = \frac{A - ty^2}{y} \tag{11.18}$$

Substitution of equation (11.18) into equation (11.17) gives

$$P = \frac{A - ty^2}{y} + 2y(1 + t^2)^{1/2} \tag{11.19}$$

Then we minimize P with respect to y for fixed A and t:

$$\frac{dP}{dy} = 0$$

$$\frac{d}{dy}[Ay^{-1} - ty + 2y(1 + t^2)^{1/2}] = -\frac{b}{y} - 2t + 2(1 + t^2)^{1/2} = 0$$

or

$$b = 2y[(1 + t^2)^{1/2} - t] \tag{11.20}$$

Equation (11.20) gives the relationship between the width and depth for the most hydraulically efficient trapezoidal section. However, the use of this equation frequently leads to impractical channel designs, such as very deep, narrow channels, so that it is not often employed in practice.

Alternatively, equation (11.20) can be written as

$$b + 2ty = 2y\sqrt{1 + t^2} \tag{11.21}$$

which implies that the top width is equal to twice the side length of the channel L (i.e., $B = 2L$, Figure 11.7). Referring to Figure 11.7, one can write

$$OQ = OP \sin\theta$$

$$= \frac{B}{2}\frac{1}{\sqrt{1 + t^2}} = \frac{b + 2ty}{2}\frac{1}{\sqrt{1 + t^2}}$$

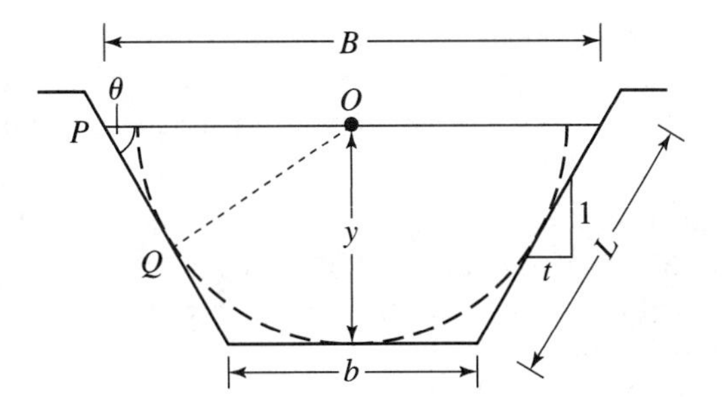

Figure 11.7 A best hydraulic trapezoidal section.

Substituting from equation (11.20) for b, one obtains

$$OQ = \frac{2y[\sqrt{1 + t^2} - t] + 2ty}{2} \frac{1}{\sqrt{1 + t^2}} = y$$

Thus, a semicircle with its center at O coinciding with the channel axis and of radius y can be drawn tangential to the bed and sides as shown in Figure 11.7.

Minimization of the wetted perimeter with respect to the horizontal side slope element t entails a similar operation as above (Problem 11.21) and leads to the result that for the most efficient trapezoidal section the side slope is

$$t = \frac{1}{\sqrt{3}} \tag{11.22}$$

which corresponds to a slope angle of 60°. The most efficient section in this case is one-half of a hexagon. This slope is often too steep for use in natural channels depending on the angle of repose of the bank material. Substituting equation (11.22) into equation (11.20) leads to the relationship that defines the trapezoidal section of greatest possible efficiency:

$$b = 1.155y \tag{11.23}$$

Recall that the hydraulic radius of the trapezoidal section for side slope t is given by

$$R = \frac{A}{P} = \frac{by + ty^2}{b + 2y(1 + t^2)^{1/2}} \tag{11.24}$$

Substitution of equation (11.23) into equation (11.24) will lead to an expression for the hydraulic radius of the most efficient trapezoidal section. It can be shown (Problem 11.22) that this expression will reduce to the simple relation

$$R = \frac{y}{2} \tag{11.25}$$

Equation (11.25), that is, that the most efficient trapezoidal section will be the one with hydraulic radius equal to one-half the normal depth, appears to have been known in many cultures for several centuries. For instance, archaeological remains from South American Indian cultures have revealed the presence of water supply and irrigation canals designed on this principle.

In the case of rectangular channels ($t = 0$), equation (11.20) reduces to

$$b = 2y \tag{11.26}$$

which implies that the most efficient rectangular channel is one in which the depth is one-half the width. This is a very nice result; however, it is only practicable in cases in which the channel is lined with concrete or some other nonerodible and stable material.

Example 11.5

A trapezoidal channel is to be designed to carry a discharge of 150 m^3/s and run on a slope of 0.0025 m/m with side slopes of $2H/1V$. If the channel is to be designed for maximum hydraulic efficiency (subject to the side slope restriction), what would be the depth and width? Let the Manning n value be 0.035.

Solution: Employing equation (11.20), we obtain

$$b = 2y[(1 + t^2)^{1/2} - t]$$

For $t = 2$, $b = 0.472y$. Then, from Manning's equation,

$$AR^{2/3} = \frac{nQ}{S_o^{1/2}} = \frac{0.035 \times 150}{(0.0025)^{1/2}} = 105$$

or

$$(by + ty^2)\left[\frac{by + ty^2}{b + 2y(1 + t^2)^{1/2}}\right]^{2/3} = 105$$

but since $b = 0.472y$,

$$2.472y^2(0.5y)^{2/3} = 105$$

$$y = 4.85 \text{ m}$$

and

$$b = 0.472y = 2.29 \text{ m}$$

The same concept can be applied for triangular sections. For the most efficient triangular section, the wetted perimeter should be minimum for a given area. Referring to Table 10.1 for triangular sections,

$$A = ty^2$$

$$P = 2y\sqrt{1 + t^2} = 2\sqrt{\frac{A}{t}}\sqrt{1 + t^2}$$

Hence,

$$P^2 = 4A\left(t + \frac{1}{t}\right) \tag{11.27}$$

Differentiating equation (11.27) with respect to t with A assumed as constant, one gets

$$2P\frac{dP}{dt} = 4A\left(1 - \frac{1}{t^2}\right) = 0 \tag{11.28}$$

or

$$\frac{1}{t^2} = 1 \quad \text{or} \quad t = 1 \tag{11.29}$$

Therefore, any triangular section must have a right apex to be most efficient. Referring to Figure 11.8, $OQ = y/\sqrt{2}$. It can also be shown that a semicircle with its center at O and a radius

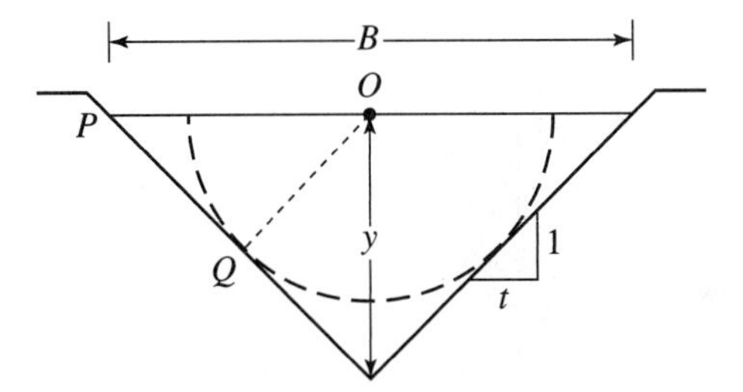

Figure 11.8 A best hydraulic triangular section.

of $y/\sqrt{2}$ will be tangential to the two sides of the triangle. The hydraulic radius for the most efficient triangular section is

$$R = \frac{A}{P} = \frac{ty^2}{2y\sqrt{1 + t^2}} = \frac{y}{2\sqrt{2}} \tag{11.30}$$

It should be noted that circular and semicircular sections have the least wetted perimeter for a given area, as compared with all other sections. For a given wetted perimeter, the circular section encloses the largest cross-sectional area. On that basis, the circular section is hydraulically the most efficient, as compared to all other sections. Yet it is rare to observe such circular sections in nature (except in culverts, of course) due to the difficulties associated with its construction and maintenance. Trapezoidal sections are much more practical.

11.5 Resistance in Steady Nonuniform Flow

So far, the resistance relationships have only been discussed in the context of steady, uniform flow in open channels. Of course, the Chezy and Manning equations have great practical utility in the design of channels and canals to carry specified discharges at specified depths and velocities. However, it can be shown that these equations can also be generalized to the more prevalent cases of steady, nonuniform flow. The importance of this generalization is that the Manning (or Chezy) equation can be used to compute head loss or energy gradients in most steady flow situations.

The nonuniform flow situation is shown in Figure 11.9. Now, in the nonuniform case to be considered here, the assumption is made that the stage is changing only gradually with respect to the longitudinal distance along the channel. This assumption is the reason that this situation is usually referred to as *gradually varied* flow. However, the acceleration is not considered to be negligible, and this is what accounts for the major difference between the uniform and nonuniform cases. Note that in the nonuniform flow case, the slopes of the channel bed, the water surface (hydraulic gradient), and the energy gradient are not equal.

Applying Newton's second law to the gradually varied flow case, one can write

$$\sum F_s = ma_s$$

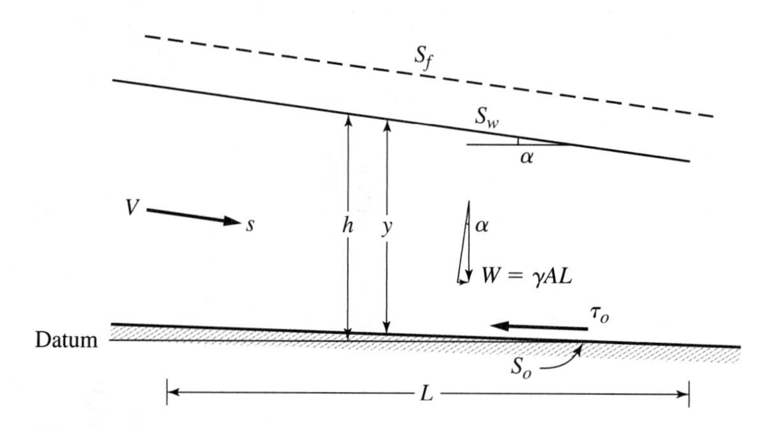

Figure 11.9 Steady, nonuniform flow in channels.

where the subscript s denotes the direction of flow. For purposes of this derivation we will consider the line of action of the flow as along the water surface. Then, following the procedure given in Section 11.1 and referring to Figure 11.9, one writes

$$\gamma AL \sin\alpha - \tau_o LP = \rho ALa_s \tag{11.31}$$

Note that in the formulation of equation (11.31) the two forces on the left are considered to be acting in parallel planes. This is an assumption that is consistent in the case of small slopes as encountered in actual practice (Henderson, 1966). Now, in handling the acceleration term, we note, as discussed in Chapter 5 (Section 5.4), that in the case of steady flow, only the convective term $v\frac{\partial v}{\partial s}$ needs to be considered. Dividing equation (11.31) through by AL, one obtains

$$\gamma \sin\alpha - \frac{\tau_o}{R} = \rho v\frac{\partial v}{\partial s} \tag{11.32}$$

or

$$\tau_o = \gamma R \sin\alpha - \rho R v\frac{\partial v}{\partial s} \tag{11.33}$$

Noting that $\rho = \gamma/g$, one writes

$$\tau_o = \gamma R \sin\alpha - \gamma R\left(\frac{v}{g}\right)\frac{\partial v}{\partial s} \tag{11.34}$$

Now, it is noted that with respect to the flow direction s, the angle of the line of action (α) is the angle of the water surface with the horizontal. Then one can write

$$\tan\alpha = \frac{\sin\alpha}{\cos\alpha} = S_w$$

where S_w is the slope of the water surface. Then $\sin\alpha = S_w \cos\alpha$. Hence, equation (11.34) can be written as

$$\tau_o = \gamma R S_w \cos\alpha - \gamma R\left(\frac{v}{g}\right)\frac{\partial v}{\partial s} \tag{11.35}$$

Now, noting that $S_w = -dh/ds$, and the cos of small angles approaches 1, one can further write

$$\tau_o = -\gamma R\left(\frac{dh}{ds} + \frac{v}{g}\frac{\partial v}{\partial s}\right) \tag{11.36}$$

or

$$\tau_o = -\gamma R\frac{d}{ds}\left(h + \frac{V^2}{2g}\right) \tag{11.37}$$

Note that the velocity coefficient α, which would normally be included with the velocity head term in equation (11.37) after the integration of equation (11.36), is assumed equal to 1 as usual.

One now simply notes that $\frac{d}{ds}\left(h + \frac{V^2}{2g}\right) = S_f$ or the friction slope. Thus, one has

$$\tau_o - \gamma R S_f \tag{11.38}$$

The operation that leads to equation (11.36) (i.e., substitution of dh/ds for $\sin\alpha$) illustrates the fundamental principle that the flow is actually being driven by the hydraulic gradient, which is represented in equation (11.31) by the component of the gravity force in the flow direction.

Since it has been shown that the velocity-based resistance equations (Darcy-Weisbach, Chezy, Manning, etc.) are based upon, and derived from, equation (11.38), these equations must also be valid for the case of steady, nonuniform flow as well as long as the friction slope is used as the slope term.

11.6 Clarifying Remarks

In this chapter it was demonstrated that the basic resistance concepts developed in Chapter 7 and employed in Chapter 8 are equally valid in the case of open channels, or free surface flow as in the case of pipes. However, the actual applications of the principles differ somewhat. While the general relationship relating velocities to resistance losses is still the Chezy equation, the specific formula used for computational purposes in channel cases is the Manning formula rather than the Darcy equation. The equivalency of the Darcy and Manning equations, through their relationship with the Chezy formula and the resultant relationship of the Manning n and the Darcy f, was demonstrated. The relationship for the Chezy C that resulted in the Manning equation was developed specifically for channel flow and predated the work on pipe resistance performed in Germany early in the twentieth century. Thus, the Manning formula was already well established for channel flow by that time.

In the present chapter the Manning resistance formula was applied in the case of steady, uniform flow in open channels and thus is the governing equation in that case. The basic resistance principles were rederived for that case and it was demonstrated that it represents a case of dynamic equilibrium both in resistance and energy terms. Various applications of the principle were demonstrated, primarily in design situations. However, it was shown that, in the special case of uniform flow, the Manning equation can be used to compute the rate of energy dissipation since the slope of the energy grade line and the channel bottom are the same. Lastly, it was proved that the Manning resistance equation can also be used to compute the rate of energy losses in the case of nonuniform flow as well. It is this generalization of the principle that gives the equation its great strength and usefulness in hydraulic engineering beyond its use for design.

READING AID

11.1. Explain why the quantification of resistance factors in open channels is more complicated than the case of pipes.

11.2. Define uniform and nonuniform flow in open channels. Does the definition holds for the case of pipes?

11.3. Under uniform flow conditions in open channels, will the gravity force (driving force) be bigger or smaller than the resistance force? Explain.

11.4. If the depth of flow in an open channel is increasing along the flow direction, would you expect the gravity force to be bigger or smaller than the resistance force?

11.5. For a channel reach under dynamic equilibrium conditions, what is the value of the water acceleration?

11.6. Explain whether the Chezy coefficient, C, in open channels is a function of Reynolds number, or not.

11.7. Write two equations to express the relationship between the Chezy coefficient and the Manning resistance coefficient in terms of the hydraulic radius using both SI and BG systems. Check the dimensions of the equations.

11.8. Can the Manning coefficient be regarded as a constant, or does it differ according to the employed system of units?

11.9. For a certain pipe with a pressurized flow, will the hydraulic radius depend on the flow rate? Explain.

11.10. For an open channel, will the hydraulic radius and the hydraulic depth vary with the change of the flow rate? Explain.

11.11. Write an expression to estimate the Manning n from the mean particle size of channel beds.

11.12. Write an equation for the channel conveyance, K. Check its dimensions.

11.13. What are the main parameters to be considered in the design of a channel section?

11.14. For erodible soil, would you recommend steep or mild bed slope for an open channel? Why?

11.15. For an open channel with a steep slope, would the depth of flow be greater or smaller than the critical depth?

11.16. What is the relationship between the bed slope and flow depth in an open channel of given cross section? Are they directly or adversely proportional?

11.17. Explain how to ensure high hydraulic efficiency for a given area of a channel section. Define the best hydraulic section.

11.18. Derive an expression for the best hydraulic section of a trapezoidal section. Why it is not always possible to design open-channel sections as best hydraulic sections?

11.19. What is side slope of the most efficient trapezoidal sections?. Express the bottom width, b, and the hydraulic radius, R, of such sections in terms of the flow depth, y.

11.20. For most efficient rectangular open channels, what is the relation between the channel width, b, and the flow depth, y?

11.21. Derive an expression for the most efficient triangular section.

11.22. Why are the circular and semicircular sections regarded as the best hydraulic section? Comment on the practicality and feasibility of such sections.

11.23. What is the relation among the slopes of total energy line, water surface, and channel bed for the cases of uniform and nonuniform flow conditions?

11.24. Derive an equation for the shear stress for gradually varied flow conditions.

Problems

11.1. Water is flowing in a trapezoidal earthen channel with bottom width 2 m and side slopes 1.5 $H/1V$. The channel is carrying a discharge of 50 m^3/s and is running on a slope of 0.0025 m/m. If the roughness coefficient is 0.030, what is the normal depth in the channel?

11.2. A 3-m-wide rectangular irrigation channel carries a discharge of 25.3 m^3/s at a uniform depth of 1.2 m. Determine the slope of the channel with $n = 0.022$. If the discharge is increased to 40 m^3/s, what is the normal depth of flow?

11.3. A trapezoidal earthen channel with bottom width 5 ft and $2H$ on $1V$ side slopes is carrying a discharge of 200 cfs at normal depth. If the channel is running on a slope of 0.0001 ft/ft and has a Manning n value of 0.025, find the normal depth.

11.4. An earthen canal in good condition having a bottom width of 4 m and side slopes of $2H$ on $1V$ is designed to carry a discharge of 6 m^3/s. If the slope of the canal is 0.39 m/km, what is the normal depth?

11.5. Determine the normal depth in a trapezoidal channel with side slopes of 1.5:1, bottom width of 25 ft, and channel slope of 0.00088, if the discharge is 1510 ft^3/s and $n = 0.017$.

11.6. A rectangular concrete lined channel 1.5 m wide is carrying 15 m^3/s at normal depth. If the channel is laid on a slope of 0.01 m/m, what would be the normal depth?

11.7. A trapezoidal channel is to be designed to carry a discharge of 250 cfs at a normal depth of 3.5 ft. The channel is running on a slope of 0.005 ft/ft and the side slopes can be no steeper than $2.5H$ on $1V$. The channel is earthen and will have an estimated Manning n value of 0.03. What is the minimum bottom width to effect the required normal depth?

11.8. A trapezoidal channel has a bed width of 10 ft and side slopes 2:1. The channel is paved with smooth cement surface of $n = 0.011$. If the channel is laid on a slope of 0.0001 and carries a uniform flow of depth 2 ft, determine the discharge.

11.9. Water flows uniformly in a 2-m-wide rectangular channel at a depth of 45 cm. Find the flow rate in m^3/s; discuss the results for the following cases.

a. The channel slopes 25 cm/km and $n = 0.025$
b. The channel slopes 5 cm/km and $n = 0.025$
c. The channel slopes 25 cm/km and $n = 0.05$

11.10. A partially filled pipe 1.0 m in diameter is laid on a uniform slope 1:2000. Calculate the maximum flow that can run through this pipe ($n = 0.016$).

11.11. A rectangular channel 15 ft wide flows at a normal depth of 6 ft. The discharge is 530 cfs and the Manning n value is 0.025. What is the rate of energy dissipation in ft per ft of channel length in this channel?

11.12. A rectangular canal with a bed slope 8 cm/km and a bed width of 100 m. If at a depth of 6 m, the canal carries a discharge of 860 m^3/s, find the Manning's roughness, n, Chezy coefficient, C, and the coefficient of friction, f. Also find the average shear stress on the bed.

11.13. Show that for a circular culvert of diameter D the velocity of flow will be maximum when the depth of flow y at the center is $0.81D$. Use the Chezy formula.

11.14. A sewer of diameter $D = 0.6$ m has a bed slope of 1:200. What is the possible maximum velocity of flow in this pipe? What is the discharge at this velocity? Take $C = 55$.

11.15. It is desired to design a trapezoidal channel with a bottom width of 10 ft and $2H$ on $1V$ side slopes. Sieve analyses revealed a grain size distribution which results in an allowable bed shear stress of 0.5 lb/ft^2 and a Manning n value of 0.03. If the channel is to be designed to run at a normal depth of 5 ft, what will be the resulting discharge?

11.16. It is desired to design a trapezoidal channel with a bottom width of 15 ft and $2H$ on $1V$ side slopes. Sieve analyses have revealed a grain size distribution that results in a maximum allowable bed shear stress of 0.75 lb/ft^2. The channel must be designed to carry a discharge of 1000 cfs at a normal depth of 7 ft.

a. At what slope should the channel be laid?
b. What is the Manning n value for this channel?

11.17. A 20-ft-wide rectangular channel carries a discharge of 400 cfs at a normal depth of 10 ft. If the roughness coefficient is 0.03, what shear stress in lb/ft^2 is imparted to the channel boundary by this flow?

11.18. Design a channel (assuming a rigid boundaries) to irrigate 54,430 hectares at a rate of 60 $m^3/ha/day$. The soil allows side slope of 1:1, $n = 0.025$, and the bed slope is 10 cm/km. Find the channel dimensions for each of the following cases:

a. the maximum allowable velocity being 0.7 m/s

b. the maximum allowable boundary shear stress = 0.05 lb/ft^2

c. $V = 0.36y^{0.64}$

11.19. Water is flowing in a rectangular channel with a bed slope of 0.0005 ft/ft. The channel width is 10 ft and the discharge is 300 cfs. Now, suppose that the depth at a given place in the channel is observed to be 6 ft, and the depth observed at a spot 1000 ft downstream is observed to be 4.5 ft. Estimate the Manning n coefficient of this channel.

11.20. A trapezoidal channel is designed to carry 25 m^3/s on a slope of 0.0015 m/m. The channel is unlined, and in order to prevent erosion, the maximum allowable velocity is 1.5 m/s. The side slope must be no steeper than $2H$ on $1V$ and the Manning n value is 0.03. In order to meet these requirements, what flow depth and bottom width should be used?

11.21. Show that for the most efficient trapezoidal channel, the side slopes must be equal to $1/\sqrt{3}$.

11.22. Show that the most efficient trapezoidal channel section will have a hydraulic radius equal to one-half the depth.

11.23. A rectangular flume is to be built to carry a discharge of 120 cfs on a slope of 0.001 ft/ft. If the flume is constructed of corrugated metal, what should be its dimensions to minimize the cost of the material?

11.24. Find the best hydraulic section for a rectangular section with cross-sectional area of 12 m^2.

11.25. A concrete-lined canal having one side vertical and other side is sloping at 2:1 carries a discharge of 10 m^3/s with velocity of 0.715 m/s. Determine the dimensions of the canal and the bed slope for minimum cost of construction if the cost of excavation is 2.0 \$/m^3 and the cost of lining is 6.0 \$/m^2 ($n = 0.014$).

11.26. A trapezoidal channel is designed to carry a discharge of 1800 cfs at normal depth on a slope of 0.0005 ft/ft. The side slopes of the channel must be $1.5\ H/1\ V$ and the Manning n value is 0.035. If the channel is to be a regular earthen canal in good condition, what should be its dimensions for maximum hydraulic efficiency?

11.27. Due to the topography of the ground surface, a long circular water tunnel is proposed to replace a 20-km-long escape channel of trapezoidal section, side slopes 1:1. The tunnel section is best discharging section ($\theta = 308°$) and the trapezoidal section is best hydraulic section ($R = y/2$) and both have the same roughness and bed slope. Determine whether the construction of the tunnel instead of the channel will reduce or increase the total cost of the project. The cost of construction of the trapezoidal section is \$4 per m^3, while the construction cost of the tunnel is \$20/m^3.

11.28. A trapezoidal channel is to be designed to carry a discharge of 75 m^3/s at maximum hydraulic efficiency. The side slopes of the channel are $2H/1\ V$ and the Manning n value is .030.

a. If the maximum allowable velocity in the channel is 1.75 m/s, what should be the dimensions of the channel?

b. What should be the slope of the channel?

11.29. A trapezoidal channel is designed to convey a discharge of 30 m^3/s and runs on a slope of 10 cm/km. Determine the dimensions of the most efficient section. Take Chezy's C as 40 (metric).

11.30. A rough timber flume ($n = 0.016$) with a cross section of a most efficient triangular section conveys water at a depth of 1.5 ft under uniform conditions. The channel has a bed slope of 0.001. Calculate the discharge.

References

Chow, Ven Te., 1959. *Open-Channel Hydraulics.* McGraw-Hill Book Company, New York.

Coyle, J. J., 1975. "Grassed waterways and outlets." *Engineering Field Manual,* United States Soil Conservation Service, Washington, D.C.

Hager, W. H., 1999. *Wastewater Hydraulics.* Springer-Verlag Publishing Company, Berlin.

Henderson, F. M., 1966. *Open Channel Flow.* Macmillan Publishing Company, New York.

Horton, Robert E., 1916. Some better Kutter's formula coefficients. *Engineering News* 75(8): 373–374.

Keulegan, G. H., 1938. Laws of turbulent flow in open channels. *J. of Research of National Bureau of Standards,* 21: 707–741.

Morris, H. M. and Wiggert, J. M., 1972. *Applied Hydraulics in Engineering,* Wiley, New York.

Palmer, V. J., 1945. A method for designing vegetated waterways. *Agricultural Engineering.* 26(12): 516–520.

Strickler, A., 1923. Some contributions to the problem of velocity formula and roughness coefficients for rivers, canals, and closed conduits. *Mitteilungen des eidgenossischen Amtes fur Wasserwirtschaft,* 16, Bern, Switzerland.

Sturm, T. W., 2001. *Open Channel Hydraulics.* McGraw-Hill Book Company, New York.

United States Army Corps of Engineers, 1993. HEC-2 Water Surface Profiles, Users Manual, U.S. Army Corps of Engineers, Hydrologic Engineering Center, Davis, Calif.

United States Geological Survey, 1967. Roughness Characteristics of Natural Channels. *USGS Water Supply Paper 1849,* Washington, D.C.

United States Geological Survey, 1989. Guide for Selecting Manning's Roughness Coefficients for Natural Channels and Flood Plains, United States Geological Survey Water-Supply Paper 2339, Washington, D.C.

Yen, B. C., 1992. Hydraulic resistance in open channels. *Channel Flow Resistance: Centennial of Manning's Formula.* B. C. Yen (Ed.). Water Resources Publications, Littleton, Colo.

CHAPTER 12

ENERGY PRINCIPLES IN OPEN-CHANNELS

Energy generated at an overfall (Niagra Falls). (Courtesy of Dr. Mohammad Al-Hamdan.)

The energy at any point in an open channel can be classified into position or elevation energy, pressure energy, and kinetic energy. The sum of the elevation energy and pressure energy is called the potential energy, also known as the stage. *However, in open-channel flow the surface of the water is open to the atmosphere and thus the pressure on the surface is relatively constant (atmospheric). For this reason, it is convenient to take the free-water surface as the pressure reference. In hydraulics, energy is usually expressed in terms of head (i.e., equivalent depth of water). Then, following normal practice, the pressure head at any point in the water body is just the vertical depth to that particular point (i.e., $p = \gamma h$, so $p/\gamma = h$). Thus, the total pressure head at a given cross section of the flow is just the water depth at that section. Of course, this representation is based on the application of the hydrostatic principle and would not be true in situations were the hydrostatic principle does not apply (curvilinear flow, for example). However, in most cases of open-channel flow the principle does apply. Thus, in terms of the total energy at a cross section, the static energy is expressed as the elevation of that point above a specified datum, pressure energy as the depth of flow, and kinetic energy is expressed as velocity head at that point. As discussed in Chapter 1, energy cannot be created or destroyed but it may transform from one form to the other.*

In open channels, energy is actually introduced to the system in static form. Water levels in reservoirs feeding open channels must be higher than the maximum water level anywhere in the entire channel. As flow proceeds from one reach to another having different geometric elements and different bed slopes, the energy may transform from one form to the other. Due to the frictional resistance between the flowing water and channel boundaries, some energy is transformed into heat and then lost to the atmosphere and dissipated through the boundaries. The available energy must be sufficient to overcome the resistance and deliver the water to its final destination.

12.1 Total Energy and Specific Energy

At a certain section in an open channel, the specific energy, E, is defined as the energy per unit weight of water measured from the channel bottom as a datum. It has the dimension of length. On the other hand, the total energy, E_T, is measured from a set horizontal datum (say, mean sea level). Therefore, the specific energy, E, and the total energy, E_T, are not generally equal.

The specific energy at section 1, as shown in Figure 12.1, is equal to the sum of the depth of flow and the velocity head, that is,

$$E_1 = y_1 + \alpha_1 \frac{V_1^2}{2g} \tag{12.1}$$

where α_1 is the kinetic energy correction factor or velocity coefficient. The total energy, E_{T_1}, at the same section is

$$E_{T_1} = z_1 + y_1 + \alpha_1 \frac{V_1^2}{2g} \tag{12.2}$$

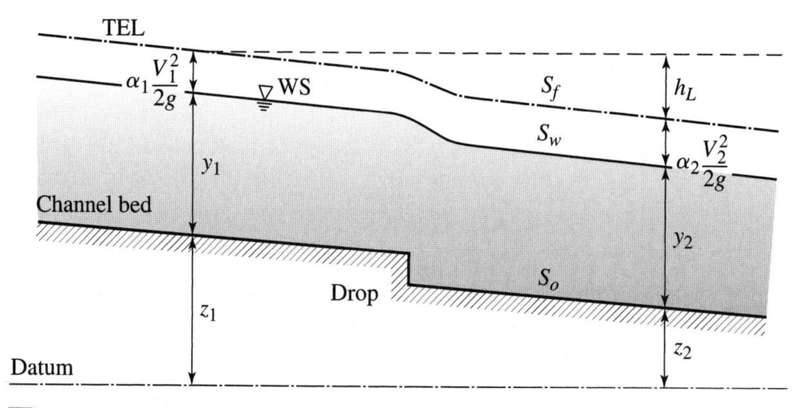

Figure 12.1 Specific energy and total energy in open channels.

As the datum for measurements of the specific energy is the bed line itself, it can vary considerably from one section to the other. It can also vary abruptly due to sudden drops or humps in the channel bed. On the other hand, the total energy should only vary slightly due to limited energy loss. However, in some cases, eddies, turbulence fields, or standing waves may cause a relatively abrupt drop in the total energy line. For section 2, the above equations are written as

$$E_2 = y_2 + \alpha_2 \frac{V_2^2}{2g} \tag{12.3}$$

and

$$E_{T_2} = z_2 + y_2 + \alpha_2 \frac{V_2^2}{2g} \tag{12.4}$$

As discussed in Chapter 5, the velocity coefficient (α) is used to account for the nonuniformity of the velocity distribution when using the average velocity to compute the kinetic energy term. Thus, it is a function of the channel geometry, roughness, and slope. The factor α can vary from about 1.05 for fairly uniform cross sections up to a value of 1.2 for very nonuniform sections (Hwang and Houghtalen, 1996). A common method for estimating α in natural channels is to break the channel into three sections where the velocity distributions are thought to be different as in Figure 12.2. The velocity coefficient is then estimated as (U.S. COE, 1982)

$$\alpha = \frac{V_1^3 A_1 + V_2^3 A_2 + V_3^3 A_3}{V_m^3 (A_1 + A_2 + A_3)} \tag{12.5}$$

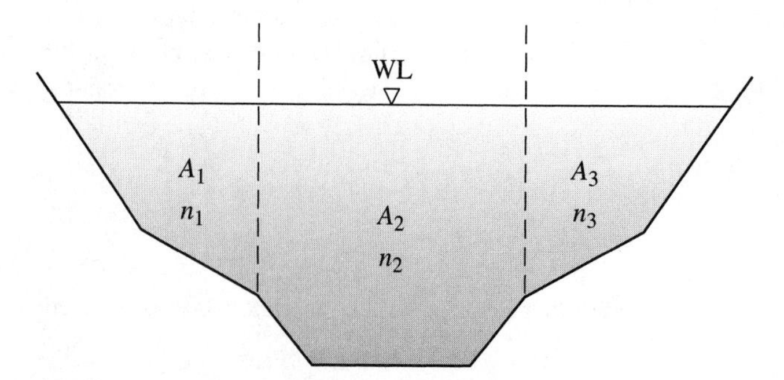

Figure 12.2 A channel section divided into three sections.

where V_m is the weighted mean velocity given by

$$V_m = \frac{V_1A_1 + V_2A_2 + V_3A_3}{A_1 + A_2 + A_3} \tag{12.6}$$

While the velocity coefficient is nearly always employed in practical situations, in this text uniform channel sections (e.g., trapezoids, rectangles, circles) will be employed in virtually all cases. In these situations, α is close enough to unity that it can be assumed to be so. Thus, in future equations involving the velocity head, the α factor will be assumed equal to one.

Substituting for the velocity as discharge divided by area, the specific energy, E, can also be expressed as

$$E = y + \frac{Q^2}{2gA^2} \tag{12.7}$$

It is often convenient in steady flow situations to write the energy by means of equation (12.7) since the discharge Q will remain constant while the average velocity, V may vary from section to section. Thus, since Q is known and the area A is a function of y, if the channel dimensions are known the only unknown in equation (12.7) is the flow depth.

Example 12.1

A trapezoidal channel, with a 1.0-m width, 3:2 side slope, and 2.0 m depth of flow, carries a discharge of 12.0 m³/s. Calculate the specific energy.

Solution: $A = (b + ty)y = (1.0 + 1.5 \times 2.0) \times 2.0 = 8.0 \text{ m}^2$

$$E = y + \frac{Q^2}{2gA^2} = 2.0 + \frac{(12.0)^2}{2 \times 9.81 \times (8.0)^2} = 2.11 \text{ m}$$

Example 12.2

A rectangular channel 2.0 m wide and 1.0 m deep, carries a discharge of 4.0 m³/s. A horizontal datum is assumed 2.0 m below the channel bed at a certain point, as shown in Figure 12.3. Find the total energy and the specific energy.

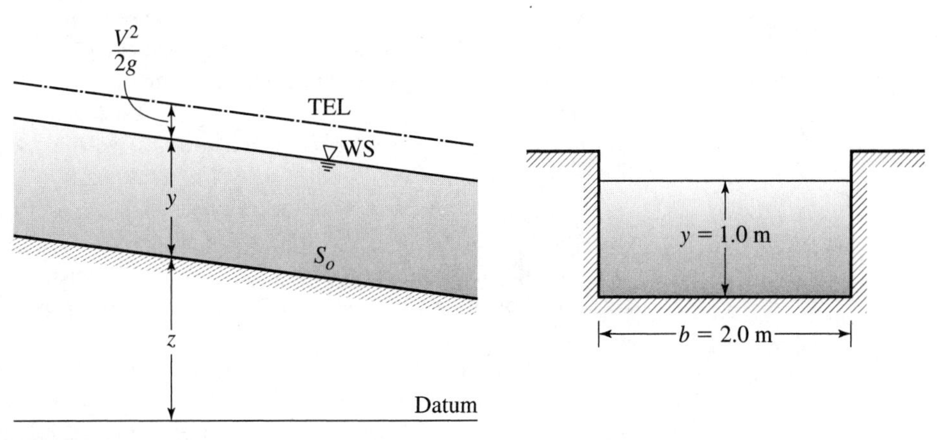

Figure 12.3 Rectangular section (Example 12.2).

Solution: $A = by = 2.0 \times 1.0 = 2.0 \text{ m}^2$

$$E_T = z + y + \frac{Q^2}{2gA^2} = 2.0 + 1.0 + \frac{(4.0)^2}{2 \times 9.81 \times 4} = 3 + 0.204 = 3.20 \text{ m}$$

$$E = y + \frac{Q^2}{2gA^2} = 1.0 + \frac{(4.0)^2}{2 \times 9.81 \times 4} = 1.0 + 0.204 = 1.20 \text{ m}$$

The total energy E_T will be equal to the specific energy, E, only when the arbitrary datum coincides with the channel bed. It should be noted that because channel beds are generally not horizontal, the difference between E_T and E varies longitudinally along the channel.

12.2 Specific Energy Diagram

The specific energy diagram (SED) is a graphical representation for the variation of the specific energy with the depth of flow in a given section. Equation (12.7) can be written as

$$E = E_s + E_k \tag{12.8}$$

where E_s and E_k are the static energy and the kinetic energy, respectively, where

$$E_s = y \tag{12.8a}$$

and

$$E_k = \frac{Q^2}{2gA^2} \tag{12.8b}$$

The static energy, E_s, varies linearly with the depth of flow, y. A straight line, OD, that passes through the origin and makes an angle of 45° with the horizontal is drawn, as shown in Figure 12.4, to represent equation (12.8a). As the depth of flow increases, the static energy increases as well. On the other hand, the kinetic energy, E_k, varies nonlinearly with the depth of flow, y. As y increases, E_k decreases and vice versa. Equation (12.8b) is represented by the curve kk', as shown in Figure 12.4. The specific energy is defined as the sum of the static energy and the kinetic energy. The horizontal sum of the line OD and the curve kk' produces the specific energy diagram, ACB. At any point on the specific energy diagram, ACB, the ordinate represents the depth of flow, while the abscissa represents the magnitude of the specific energy measured from the channel bed as a datum.

For a given cross section with a specified discharge, as the depth of flow increases, the static energy increases, and the kinetic energy decreases. Higher static heads are associated with lower kinetic energies and vice versa. The specific energy curve ACB approaches the line OD as the flow depth increases. Small depths of flow are associated with high kinetic energies. ACB approaches kk' for small values of y.

For a given specific energy, E_1, two different depths, y_1 and y_2 (*alternate* depths) are encountered, as shown in Figure 12.4. As the specific energy diagram is drawn for a specified (constant) discharge, Q, these two depths have the same discharge. Thus, alternate depths are

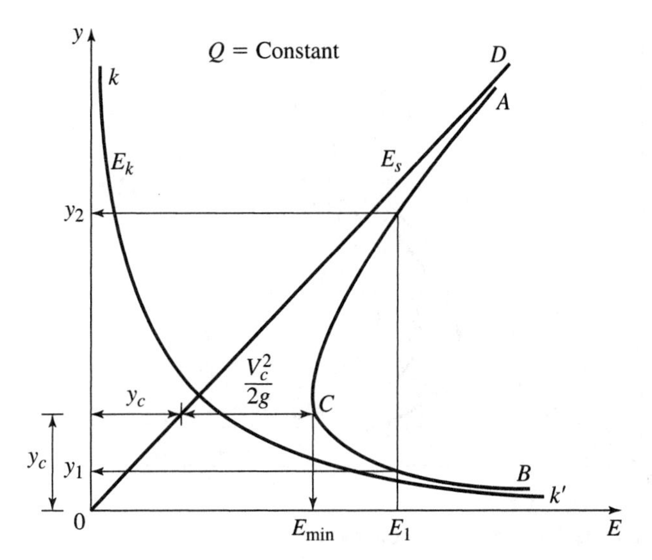

Figure 12.4 The specific energy diagram.

the two depths with the same specific energy and conveying the same discharge. The low stage depth, y_1, is the alternate depth of the high stage depth, y_2, and vice versa. The same flow can be transmitted, in a given channel section, with the same specific energy either by y_1 or by y_2.

At a certain point, *C*, the specific energy is minimum, E_{min}, as shown in Figure 12.4. E_{min} is the minimum required energy to transport the given discharge in the given cross section. One might surmise that this point plays a very important role in open-channel hydraulics. In fact, the depth associated with this minimum energy condition is called the *critical depth*, y_c. Critical depth plays a crucial role in the analysis of open-channel problems for many reasons. One such reason is evident at this point of the discussion, namely, that the point of critical condition (point *C* in Figure 12.4) is the only energy value on the SED for which there is only one associated depth (y_c). Other important aspects of critical depth and minimum energy will become evident as the discussion proceeds.

The specific energy diagram, as shown in Figure 12.4, is drawn for a specified discharge, *Q*. As the value of *Q* is increased to Q_1, Q_2, and so on, the diagram *ACB* is shifted toward the upper right corner, as shown in Figure 12.5, indicating an increase in the minimum specific required energy to yield bigger discharges on the same section. A family of similar curves can be plotted for various values of *Q*. A curve drawn from the origin *O* and passing through the points of minimum specific energy, *C*, C_1, C_2, . . . , for different discharges will divide the specific energy diagram into two flow fields. The upper flow field represents the *subcritical* flow conditions, where the depth of flow is higher than the critical depth ($y > y_c$) and thus the velocity of flow is lower than the critical velocity ($V < V_c$). On the other hand, the lower flow field represents the *supercritical flow* conditions, where $y < y_c$ and $V > V_c$. Starting in the supercritical zone the specific energy, *E*, decreases with the increase in the depth of flow, *y*, until the latter reaches the critical depth, y_c, after which the subcritical zone is entered and *E* starts to increase again.

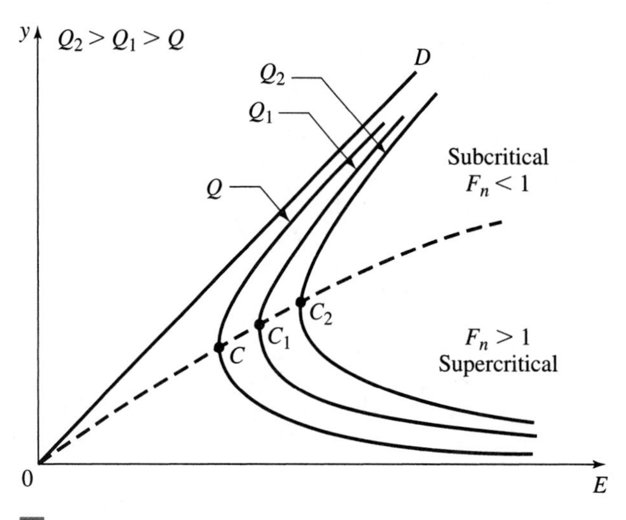

Figure 12.5 Specific energy diagram for various discharges.

Example 12.3

Draw the specific energy diagram for the problem given in Example 12.1. Find the alternate depth to the given depth of 2 m, the critical depth, and the minimum specific required energy to transmit the same discharge.

Solution: Assuming different values for the depth of flow and substituting into equations (12.8a,b), Table 12.1 is computed. The specific energy diagram (the plot of E versus y) is given in Figure 12.6. From this figure, the alternate depth, y_2 to y_1, of 2 m is 1.02 m; the critical depth, y_c, is 1.38 m; and the minimum specific energy, E_{min}, is 1.789 m. It should be noted that Figure 12.6 is drawn to a distorted scale to enhance the accuracy of measurements. From the figure it is clear that the given depth of 2 m is subcritical and its alternate depth of 1.02 m is supercritical.

TABLE 12.1 Calculations for the Specific Energy Diagram (Example 12.3)

y (m)	A (m^2)	$Q^2/2gA^2$ (m)	E (m)
0.2	0.26	108.57	108.77
0.6	1.14	5.65	6.25
0.8	1.76	2.37	3.17
1.0	2.5	1.174	2.174
1.2	3.36	0.65	1.85
1.3	3.835	0.499	1.8
1.4	4.34	0.390	1.79
1.5	4.875	0.309	1.809
2.0	8.0	0.115	2.115
2.5	11.875	0.052	2.552
3.0	16.5	0.027	3.027

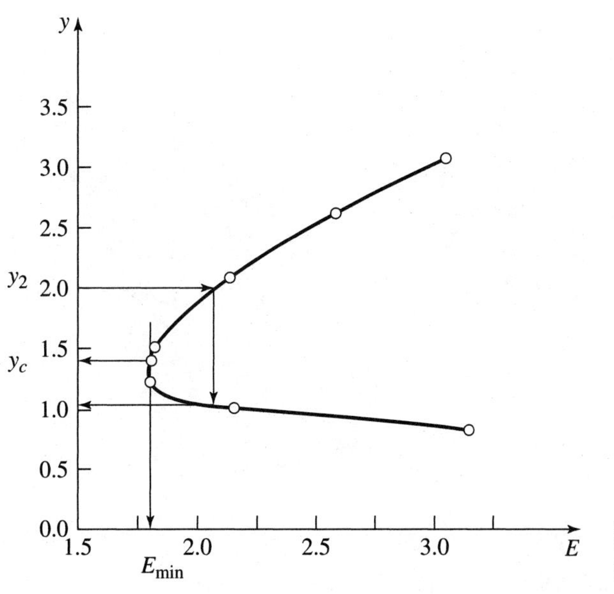

Figure 12.6 Specific energy diagram (Example 12.3).

Example 12.4

A 4.0-m-wide rectangular channel conveys a discharge of 12.0 m^3/s. The depth of flow is 2.5 m. Draw the specific energy diagram and find the critical depth and the alternate depth. If the channel width is reduced to 3.0 m while the discharge is maintained constant, draw the specific energy diagram and find the critical depth.

Solution: $E = y + \dfrac{Q^2}{2gA^2}$

For the rectangular section, define the specific discharge per unit width q equal to Q/B. Therefore,

$$E = y + \frac{q^2}{2gy^2}, \text{ where } q = \frac{12}{4} = 3 \text{ m}^2\text{/s}$$

Assuming different values for y, Table 12.2 is produced. From the specific energy diagram given in Figure 12.7, the critical depth, y_c, is 0.97 and the required alternative depth is

TABLE 12.2 Calculations for the Specific Energy Diagram (Example 12.4)

y (m)	$q^2/2gy^2$ (m)	E (m)	$q'^2/2gy^2$ (m)	E' (m)
0.2	11.468	11.668	20.39	20.59
0.4	2.867	3.27	5.10	5.50
0.5	1.835	2.335	3.26	3.76
0.75	0.815	1.565	1.45	2.2
1.0	0.459	1.459	0.815	1.815
1.5	0.204	1.704	0.362	1.862
2.0	0.115	2.115	0.204	2.204
2.5	0.073	2.573	0.130	2.63
3.0	0.051	3.051	0.091	3.091

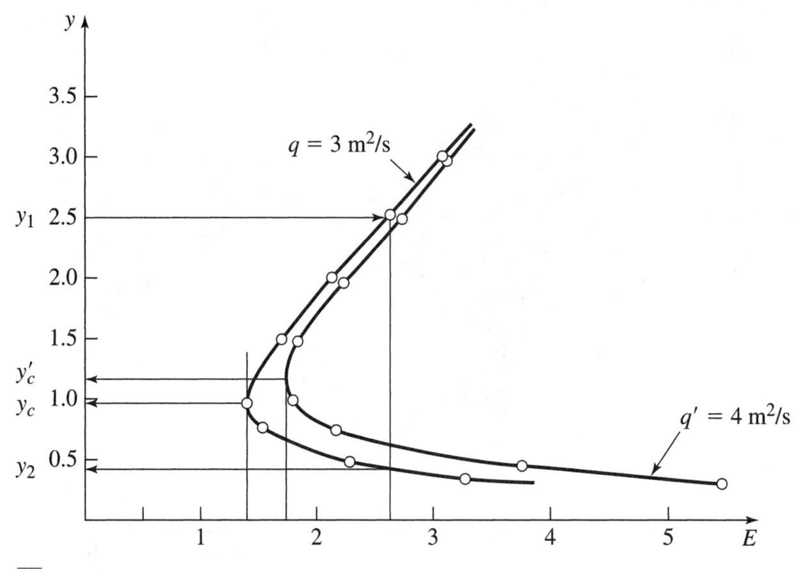

Figure 12.7 Specific energy diagram (Example 12.4).

0.47 m. For the second case, where the width is reduced to 3.0 m, the specific discharge, q', is 4.0 m^2/s. As shown in Figure 12.7 the critical depth, y_c, is 1.18 m.

12.3 Mathematical Solution of the Energy Equation

The use of the simplified expression for the specific energy in the case of rectangular channels makes a direct solution for the alternate depths easier. For instance, in the example given above, $y = 2.5$ m. Then $E = 2.5 + \frac{(3)^2}{2g(2.5)^2} = 2.57$ m. Then the alternate depth would be found from

$$2.57 = y_2 + \frac{(3)^2}{2gy_2^2} = y_2 + \frac{0.459}{y_2^2}$$

The above nonlinear expression can be solved for y_2 by some iterative solution technique, such as Newton's iterative method, false position, or even simple bisection. For illustrative purposes, we will employ Newton's iterative method. We convert the energy expression above into the equivalent cubic equation

$$f(y) = y_2^3 - 2.57y_2^2 + 0.459 = 0, \text{ and then}$$

$$f'(y) = 3y_2^2 - 5.14y_2$$

The algorithm is formulated as

$$y_{n+1} = y_n - \frac{f(y_n)}{f'(y_n)}$$

Since we know that $y_c = 0.97$ m, then we also know that the given depth of 2.5 m is the subcritical depth and thus we need to make sure that our algorithm converges to the smaller supercritical value. Therefore, we begin with an assumed depth $y = 0.5$ m (smaller than y_c). Then we calculate our next trial y as $y = 0.5 - \left(\frac{-0.0595}{-1.82}\right) = 0.467$ m. This value represents only a small change from our initial guess and if we substitute it back into the original function we obtain a residual of only -0.00064 m. Thus, we conclude that our estimate is essentially correct. It should be noted that the solution technique outlined above is applicable to any cross sectional shape as long as the area can be expressed as a unique function of the depth. For instance, in the case of trapezoidal channels, the expression would be

$$E = y + \frac{Q^2}{2g(by + ty^2)^2} \tag{12.9}$$

where t is the horizontal component of the channel side slope and b is the bottom width of the trapezoidal section. Equation (12.9) will reduce to the following fifth-order equation:

$$t^2y^5 + (2tb - Et^2)y^4 - (2tEb - b^2)y^3 - Eb^2y^2 + \frac{Q^2}{2g} = 0 \tag{12.10}$$

Note that equation (12.10) will still have only two real positive roots (i.e., the subcritical depth and its alternate supercritical depth). The equation can also be easily solved by spreadsheet as discussed in Chapter 11 with reference to the normal depth solution; however, care must always be taken to ensure that the solution converges to the proper root (i.e., either the subcritical or supercritical root as the case may be).

12.4 Critical Flow Conditions

It has been stated that for a given discharge in a specified cross section, the critical flow conditions are satisfied at the minimum specific energy, $E_{\min}$ as shown in Figure 12.4. A more general mathematical formulation for the critical flow conditions is given below.

For channels with small bed slopes, it has been shown that the specific energy is given as

$$E = y + \frac{Q^2}{2gA^2} \tag{12.11}$$

or in the case of rectangular sections,

$$E = y + \frac{q^2}{2gy^2} \tag{12.12}$$

The cross-sectional area, A, is a function of y (the only variable). Differentiating equation (12.11) with respect to y, one gets

$$\frac{dE}{dy} = 1 - \frac{Q^2}{gA^3}\frac{dA}{dy} \tag{12.13}$$

For the general cross section, as shown in Figure 12.8, the differential element dA/dy can be assumed equal to the top width of the channel, B. Substituting B for dA/dy into equation (12.13), one obtains

$$\frac{dE}{dy} = 1 - \frac{Q^2B}{gA^3} \tag{12.14}$$

At the critical flow conditions the specific energy is minimum (i.e., dE/dy is equal to zero), therefore,

$$\frac{Q^2B}{gA^3} = 1 \tag{12.15}$$

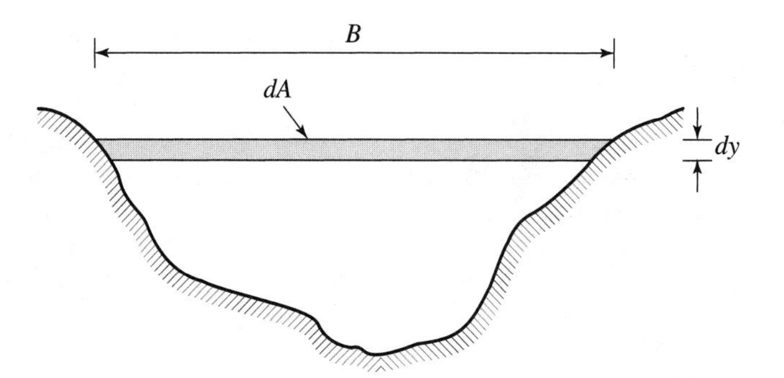

Figure 12.8 General cross section in open channels.

Equation (12.15) gives the mathematical expression for the critical flow conditions. Previously in Chapters 3 and 6 the dimensionless Froude number (F_n) has been introduced. There are a number of ways to approach a discussion of F_n, one of which is that it represents the ratio of the flow velocity to the celerity of a shallow gravity wave [i.e., $F_n = V/\sqrt{gy}$ (Henderson, 1966)]. As such, it is an indicator of the importance of gravity in open-channel flow. In fact, in dimensional analysis terms the Froude number essentially represents the ratio of the inertial forces to the gravity force (Morris and Wiggert, 1972). Noting that $V = Q/A$, equation (12.15) can also be expressed in the form

$$\frac{V^2 B}{gA} = 1 \tag{12.16}$$

The ratio A/B is often referred to in hydraulics as the *hydraulic depth*, D. Note that this depth is only equal to the actual flow depth in the case of rectangular channels and should not be confused with the depth of flow otherwise. Nevertheless, it is sometimes a convenient aid in illustrating certain hydraulic principles. If we employ D in this case, equation (12.16) becomes

$$\frac{V^2}{gD} = 1 = F_n^2 \tag{12.17}$$

where the Froude number squared is written in terms of the hydraulic depth. Thus, it is clear that the critical conditions are satisfied when the Froude number (or F_n^2) is equal to unity. Note that F_n^2 is also given by equation (12.15). Of course, in wide or rectangular sections the hydraulic depth, D, is equal to the depth of flow, y. Therefore, $F_n^2 = \dfrac{V^2}{gy_c} = 1$, and since $V = q/y$, then $\dfrac{q^2}{gy_c^3} = 1$ at critical depth. The general expressions for F_n^2 [i.e., $\dfrac{Q^2B}{gA^3}$ or $\dfrac{V^2}{gD}$ or $\dfrac{V^2}{gy}$ (rectangular)] can be used to determine the state of the flow relative to critical conditions. Elementary analysis will show that F_n^2 will be less than 1 for subcritical flow and F_n^2 will be greater than 1 for supercritical flow conditions.

12.4.1 Critical Velocity

The general expression for critical conditions states that $F_n^2 = \dfrac{V^2}{gD} = 1$. Therefore, the critical velocity must be given by $V_c = \sqrt{gD}$ for the general cross section. The velocity head at critical conditions would then be equal to $\frac{V_c^2}{2g} = \frac{D}{2}$. Again, since for wide or rectangular channels the hydraulic depth $D = y$, then $V_c = \sqrt{gy_c}$ and $\frac{V_c^2}{2g} = \frac{y_c}{2}$. Thus we see that the critical velocity is equal to the celerity of a shallow gravity wave on wide or rectangular surfaces.

12.4.2 Critical Depth

It has already been shown that for a certain section with a given discharge the critical depth, y_c, is defined as the depth of flow that requires the minimum specific energy, E_{min}, to convey the given discharge. To determine the critical depth for the general, or nonrectangular section, one could solve equation (12.15). For the trapezoidal section it might be convenient to rewrite equation (12.15) in a form such that all known quantities are on one side and the unknowns are on the other,

$$\frac{Q}{\sqrt{g}} = (b + ty_c)y_c\sqrt{\frac{(b + ty_c)y_c}{b + 2ty_c}} \tag{12.18}$$

where t is the horizontal component of the channel side slope per unit of vertical rise. This equation can be solved by trial and error in an iterative manner (say using a spreadsheet) as discussed in Section 12.4. However, over the past several years, a number of charts, or nomographs, have been developed to aid in the solution of critical depth, particularly for trapezoidal or circular cross sections. These charts are based upon a dimensionless representation of the SED that will be discussed later in this chapter. One such useful chart is given in Figure 12.9 (Sturm, 2001).

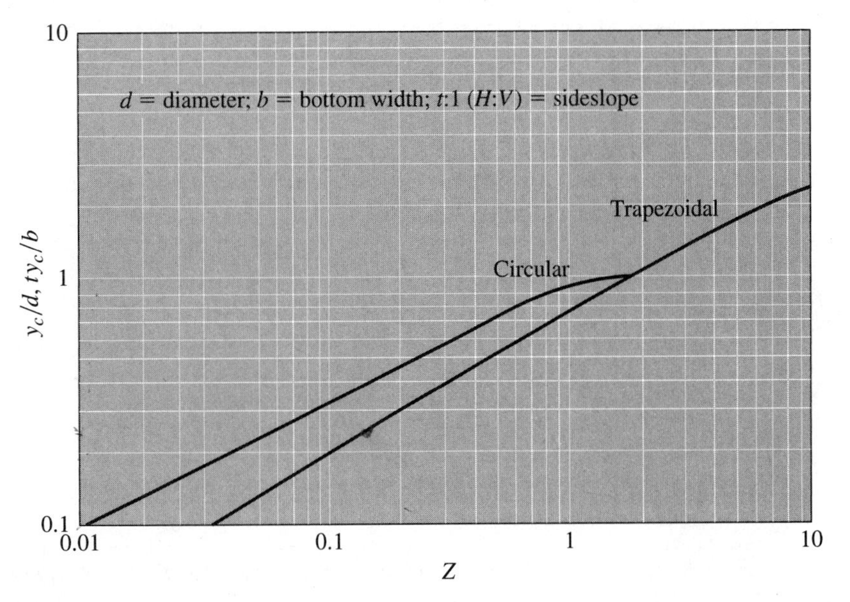

Figure 12.9 Critical depth for trapezoidal and circular sections; $Z_{circ} = Q/[g^{1/2}d^{5/2}]$; $Z_{trap} = Qt^{3/2}/[g^{1/2}b^{5/2}]$ (Henderson, 1966).

(*Source:* OPEN CHANNEL FLOW by Henderson, © 1966. Reprinted by permission of Prentice-Hall, Inc., Upper Saddle River, NJ.)

For rectangular or wide sections, a direct solution of the Froude number expression is possible to obtain y_c. Since we know that $q^2/gy_c^3 = 1$, then we can simply obtain the expression for critical depth as

$$y_c = \sqrt[3]{\frac{q^2}{g}} \tag{12.19}$$

It should be noted that equation (12.19) is valid for wide or rectangular sections only and should not be used for any other section. By "wide" it is meant that the width, B is very big compared to y, and thus $R = y$.

12.4.3 Critical Energy

Critical energy is the energy when the flow is under critical conditions.

$$E_{\min} = y_c + \frac{V_c^2}{2g} \tag{12.20}$$

It is the minimum required energy to maintain a specified flux in a given section.

In Section 12.4.1 it was shown that for any cross section, $V_c^2/2g = D/2$. Therefore,

$$E_{\min} = y_c + \frac{D}{2} \tag{12.21}$$

Again, it is important not to confuse the hydraulic depth D with the actual depth y in nonrectangular channels. Equation (12.21) is given for illustrative purposes and may not actually make computation of minimum energy easier in nonrectangular channels since the area and top width need to be computed. It may be easier to compute $V_c = Q/A_c$ and merely employ equation (12.20) directly.

Of course, for wide or rectangular sections, $D = y$. Therefore,

$$E_{\min} = y_c + \frac{y_c}{2} = \frac{3}{2}y_c \tag{12.22}$$

Although it is only valid for the rectangular representation cases, equation (12.22) is very useful in many situations. It can also be noted that if $E_{\min}$ is known (as is often the case), then y_c can be computed as $\frac{2}{3}E_{\min}$.

Example 12.5

A rectangular channel 5 ft wide carries a discharge of 50 ft^3/s. Find the critical depth and the minimum specific energy. If the depth of flow is 0.5 ft, determine the specific energy and the alternate depth. If the discharge is increased to 75.0 ft^3/s, find the corresponding critical depth.

Solution: For a rectangular section, $q = Q/B = 10$ ft^2/s.

$$y_c = \sqrt[3]{\frac{q^2}{g}} = \sqrt[3]{\frac{(10)^2}{32.2}} = 1.46 \text{ ft}$$

and $E_{\min} = 3/2y_c = 2.19$ ft. At the depth of 0.5 ft, $E = 0.5 + \dfrac{(10)^2}{2g(0.5)^2} = 6.71$ ft. Then, solving for the subcritical alternate depth, $6.71 = y_2 + \dfrac{(10)^2}{2g(y_2)^2} = y_2 + \dfrac{1.55}{(y_2)^2}$. Solving by

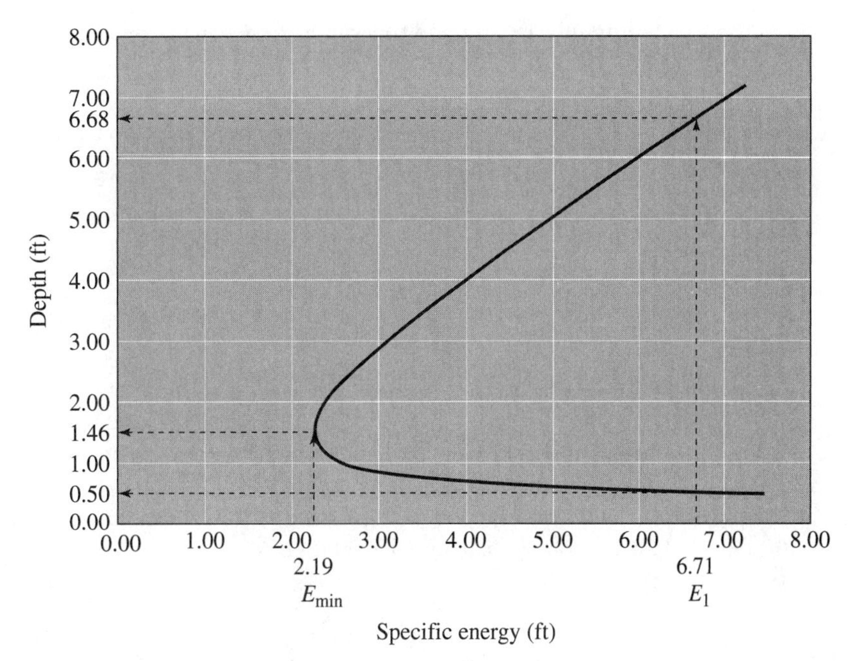

Figure 12.10 SED solution for Example 12.5.

trial and error, we obtain $y_2 = 6.68$ ft. If the discharge is increased to 75 ft^3/s, then $q = 15$ ft^2/s and

$$y_c = \sqrt[3]{\frac{q^2}{g}} = \sqrt[3]{\frac{(15)^2}{32.2}} = 1.91 \text{ ft}$$

A sketch of the SED with the relevant points labeled is shown in Figure 12.10.

Example 12.6

A 0.9-m concrete circular pipe carries a discharge of 0.566 m^3/s (Figure 12.11). Find the critical depth. If the discharge is increased to 1.0 m^3/s, what depth would give minimum energy?

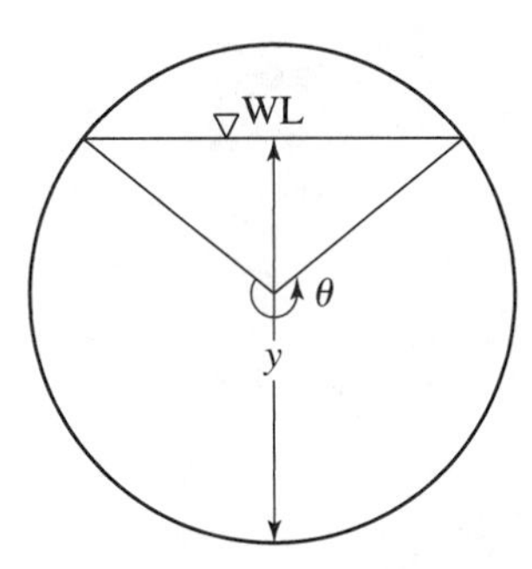

Figure 12.11 Concrete pipe for Example 12.6.

Solution: The area and top width of the circular section shown in the figure are given in Table 10.1.

$$A = d^2/8(\theta - \sin\theta) = 0.10125(\theta - \sin\theta)$$

$$B = d\sin\theta/2 = 0.9\sin\theta/2$$

The table also gives that $\theta = 2\cos^{-1}\left(1 - 2\frac{y}{d}\right)$, from which we can find that the depth

$$y = \frac{d}{2}(1 - \cos\theta/2)$$

The critical depth can be found from the equation (12.15):

$$\frac{Q^2B}{gA^3} = 1 \quad \text{or} \quad \frac{(0.566)^2(0.9\sin\theta/2)}{g(0.101(\theta - \sin\theta))^3} = 1$$

Solving this expression for θ by trial and error leads to an estimate $\theta = 3.09$ rad(177°). Substitution into the expression for y leads to $y_c = 0.44$ m.

Alternatively (and perhaps more easily) we can employ the chart given in Figure 12.9. Calculation of the required Z value gives

$$Z = \frac{Q}{g^{0.5}d^{2.5}} = \frac{0.566}{(9.81)^{0.5}(0.9)^{2.5}} = 0.235$$

With this value of Z, we find from Figure 12.9 (using the circular graph) that the factor $y_c/d = 0.5$. Then $y_c = (0.5)d = 0.45$ m. For a $Q = 1$ m³/s, $Z = 0.415$; and from Figure 12.9, $y_c/d = 0.68$, which leads to $y_c = 0.68(0.9) = 0.61$ m.

Example 12.7

A trapezoidal channel, with 3.0 m bed width and 3:2 side slope, carries a discharge of 10.0 m³/s. Find the critical depth and the minimum required energy to yield this flow. If the channel is replaced by a 3.0-m-wide rectangular channel, what is the minimum required energy to transport the same flow?

Solution: To determine the critical depth for the trapezoidal section, one can solve equation (12.18) by trial and error:

$$\frac{Q}{\sqrt{g}} = (b + ty_c)y_c\sqrt{\frac{(b + ty_c)y_c}{b + 2ty_c}}$$

Therefore,

$$\frac{10}{\sqrt{9.81}} = (3 + 1.5y_c)y_c\sqrt{\frac{(3 + 1.5y_c)y_c}{3 + 3y_c}}$$

or

$$\frac{[(3 + 1.5y_c)y_c]^{3/2}}{(3 + 3y_c)^{1/2}} = 3.19278$$

Solving by spreadsheet, we obtain the solution given below:

y	$(3 + 1.5y)y$	$[(3 + 1.5y)y]^{3/2}$	$3 + 3y$	$[(3 + 3y)]^{1/2}$	Ratio	Error
1	4.5	9.545942	6	2.44949	3.897114	0.704414
0.89	3.85815	7.578245	5.67	2.381176	3.182564	−0.01014

As in the case of Example 11.1, the solution is arrived at by setting up the calculations in the first line of the sheet using an assumed value of y, then the line is merely copied down and the y value toggled until an acceptable error is obtained.

Alternatively, we can employ Figure 12.9. Computing the required Z value, we obtain

$$Z = \frac{Qt^{1.5}}{g^{0.5}b^{2.5}} = \frac{10(1.5)^{1.5}}{g^{0.5}(3)^{2.5}} = 0.38$$

Then, using the trapezoidal graph given in the figure, we obtain the factor $ty_c/b = 0.45$, which leads to an estimate of y_c of 0.9 m. Allowing for the approximation in reading the graph, this value is in acceptable agreement with the analytically derived value of 0.89 m.

The area of this flow would be given by $A = by + ty^2 = 3(0.89) + 1.5(0.89)^2 = 3.858\ \text{m}^2$. The velocity would then be $V_c = Q/A_c = 10/3.858 = 2.59$ m/s. Then $E_{\min} = y_c + V_c^2/2g = 1.23$ m.

For the rectangular section, $q = 10.0/3.0 = 3.33\ \text{m}^2/\text{s}$. Then

$$y_c = \sqrt[3]{\frac{q^2}{g}} = 1.04\ \text{m}$$

$E_{\min} = 1.5y_c = 1.56$ m. Note that the minimum energy to transmit the given discharge in the rectangular section is greater than that for the trapezoidal section. This is an important point to which we will return later in the chapter.

12.4.4 Critical Slope

The critical slope (S_c) is the bed slope of the channel that produces critical flow conditions. It depends on the discharge, the channel geometry, and resistance or roughness. Recalling the Chezy resistance equation from Chapter 11, the velocity is given by

$$V = C\sqrt{RS_o} \tag{12.23}$$

The Chezy equation is applicable to the case of uniform flow. Assuming that the normal depth is equal to the critical depth, the slope S_o in equation (12.23) can be replaced with the critical slope S_c. Then

$$V_c = C\sqrt{RS_c} \tag{12.24}$$

Then, solving for the slope,

$$S_c = \frac{V^2}{C^2R} \tag{12.25}$$

Employing the Manning relation for C [i.e., $C = (1/n)R^{1/6}$ (SI units)], one obtains

$$S_c = \frac{V^2n^2}{R^{4/3}} \tag{12.26a}$$

or, in English units,

$$S_c = \frac{V^2n^2}{2.22R^{4/3}} \tag{12.26b}$$

It is sometimes convenient to employ the discharge directly in the computation of the critical slope. Since $V = Q/A$, substitution into equation (12.26a) leads to

$$S_c = \frac{Q^2 n^2}{A^2 R^{4/3}} \tag{12.27}$$

The concept of the critical slope is of extreme importance in open-channel hydraulics. Equations (12.25) and (12.26a,b) illustrate the fact that the critical slope is not merely a function of channel geometry, but also of discharge and roughness characteristics. If a given channel slope is less than the critical slope computed by equation (12.25) for a given discharge, then that slope cannot support critical depth and is said to be *mild*. Conversely, if the channel slope is greater than the computed critical slope, then the channel slope is called *steep*. The terms *steep* and *mild* always have these specific meanings in hydraulic engineering. It is clear from the above discussion that if a given channel slope is mild, then it can only support flow conditions that are subcritical (i.e., the normal depth in the channel must be greater than the critical depth). Likewise, if the channel slope is steep, then the channel will support velocities that are greater than the critical velocities; thus the normal depth is less than the critical depth and the flow will be supercritical. This discussion can be summed up as follows:

$S_o < S_c$	$y_o > y_c$	$V < V_c$	mild slope	subcritical flow
$S_o = S_c$	$y_o = y_c$	$V = V_c$	critical slope	critical flow
$S_o > S_c$	$y_o < y_c$	$V > V_c$	steep slope	supercritical flow

The synopsis given above encompasses much of the important concepts of both the energy and resistance principles as applied to open channels. It is important for the student to bear this information in mind in future work on open-channel flow.

Example 12.8

Water is flowing in a trapezoidal channel with a bottom width of 6 m and side slopes of $1V/1H$. The observed depth is 2.5 m for a discharge of 100 m^3/s and the channel slope is 0.0011 m/m. The estimated Chezy C value is 100 (metric).

a. Are the described flow conditions subcritical or supercritical?
b. Compute the critical slope for these conditions.
c. If the discharge is reduced to 25 m^3/s, what would be the critical slope?

Solution:

a. Employ equation (12.18) to solve for critical depth: Since the side slope $t = 1$, $Q = 100$ m^3/s, $g = 9.81$ m/s^2,

$$\frac{100}{\sqrt{9.81}} = [6y_c + y_c^2]\sqrt{\frac{6y_c + y_c^2}{6 + 2y_c}}$$

Solving by trial, $y_c = 2.61$ m. Thus, the given depth of 2.5 m is supercritical.

b. Critical slope is expressed by equation (12.25) as

$$S_c = \frac{V_c^2}{C^2 R_c}$$

$A_c = by_c + ty_c^2 = 6(2.61) + (2.61)^2 = 22.47$ m^2; $P_c = b + 2y_c\sqrt{1 + t^2} = 6 + 2(2.61)\sqrt{2} = 13.38$ m; $R_c = 22.47/13.38 = 1.68$ m; $V_c = 100/22.47 = 4.45$ m/s; and $S_c = \dfrac{(4.45)^2}{(100)^2(1.68)} = 0.001179$ m/m. The channel bed slope $S_o = 0.0011$, which is less than the critical slope. Therefore, the channel has a mild slope.

c. Now, for $Q = 25$ m^3/s, using Figure 12.9 to find y_c this time, we compute the required z value to be

$$Z = \frac{Qt^{1.5}}{\sqrt{g}b^{2.5}} = \frac{25(1)^{1.5}}{\sqrt{g}(6)^{2.5}} = 0.09$$

From this value we estimate the factor ty_c/b to be about 0.19. Thus, $y_c = 1.14$ m. Then $A_c = 8.14$ m^2, $P_c = 9.22$ m, $R_c = 0.88$ m, and $V_c = 3.07$ m/s. So $S_c = \dfrac{(3.07)^2}{(100)^2(0.88)}$ m/m $=$ 0.00109. Therefore, since the bed slope of 0.00109 is slightly greater than the critical slope, the bed slope is just barely steep.

12.5 Discharge-Depth Relation for Constant Specific Energy

In the preceding discussion, the discharge is assumed constant, while the specific energy is varied. Now, assuming that the specific energy is constant, E_o, the discharge, Q, is given from equation (12.7) as

$$Q = A\sqrt{2g}(E_o - y)^{1/2} \tag{12.28}$$

For the condition of the maximum discharge, equation (12.28) is differentiated with respect to y and then equated to zero:

$$\frac{dQ}{dy} = \sqrt{2g}\left[\frac{A}{2}(E_o - y)^{-1/2}(-1) + \frac{dA}{dy}(E_o - y)^{1/2}\right] = 0 \tag{12.29}$$

Substituting for $dA/dy = B$, the above equation can be written as

$$E_o - y = \frac{A}{2B} \tag{12.30}$$

Substituting equation (12.30) into equation (12.28), one can write

$$\frac{Q^2 B}{gA^3} = 1 \tag{12.31}$$

Equation (12.31) implies that for a certain specific energy the maximum discharge is encountered at the critical flow condition. Figure 12.12 presents the Q-y relation under constant specific energy.

For wide or rectangular sections, equation (12.28) is written as

$$q^2 = 2gy^2(E_o - y) \tag{12.32}$$

Differentiating equation (12.32) with respect to y and equating to zero, we get

$$2q\frac{dq}{dy} = 4gyE_o - 6gy^2 = 0 \tag{12.33}$$

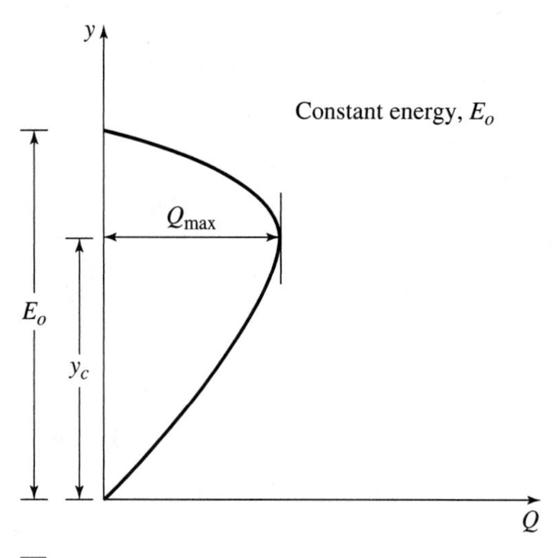

Figure 12.12 *Q-y* relation for constant specific energy.

which can be written as

$$y = \frac{2}{3}E_o \tag{12.34}$$

which is the same result obtained in Section 12.4.3 representing the critical flow condition in wide or rectangular sections. It is, therefore, concluded that the critical flow condition is satisfied either at the minimum specific energy for a given discharge or at the maximum discharge for a given specific energy.

12.6 Applications of Energy Principle

Many problems related to flow patterns in open channels can be solved using the specific energy formulation and illustrated with the specific energy diagram (SED). Some problems reduce to a simple application of the Bernoulli equation or analysis of the SED by ignoring small resistance losses associated with flow over short distances. The use of the SED in these cases consolidates the understanding of the relationship between flow and energy in open channels. Also, the SED reveals various important flow characteristics, such as the type of flow, the critical depth, the energy loss, and the alternative depths.

12.6.1 Transitions in Channel Beds

In natural situations, drops or rises in channel beds and expansions or contractions are frequently encountered. The depth and energy of flow after such a variation in channel conditions depend on the initial characteristics of the flow. Consider, for example, an open channel with a small drop, Δz, in its bed as shown in Figure 12.13a. The initial depth of flow (upstream of the drop) is y_1, which is greater than y_c. The flow is thus subcritical and the Froude number will be less than unity.

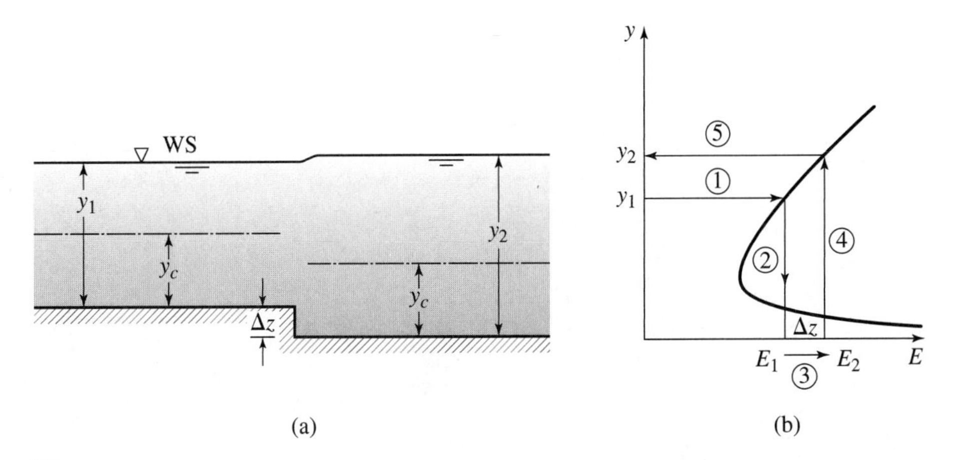

Figure 12.13 A small drop in the channel bed (subcritical flow): (a) change in water levels, and (b) steps for solution.

The specific energy, E_1, just upstream of the drop is

$$E_1 = y_1 + \frac{V_1^2}{2g}$$

Now, assume that the drop occurs over a sufficiently short channel distance such that the friction losses would be negligible. Also, assume that any minor losses due to the abrupt change in the channel can be ignored as well. These assumptions then imply that the total energy after the drop will be equal to the energy before the drop. Although these assumptions might seem to restrict the method to a small class of problems, this is in fact not the case. Most transitions in channel boundaries, such as bridge restrictions, rises and drops due to excavation requirements or to avoid utility crossings, and so on, do occur over relatively short distances. Although not totally precise, the methods discussed here can be used to obtain a good first approximation of the effects of the transition.

Under these conditions, one can use the SED as shown in Figure 12.13b to illustrate the steps to predict the new depth after the drop. The first step is always to compare the given conditions (depth, velocity, or Froude number) to critical conditions in order to determine if the initial state is subcritical or supercritical. In this example, we have already performed this step and determined that the initial state is subcritical. Now, using the known depth y_1, get the initial specific energy E_1. Since it is measured from the bed as a datum, the specific energy must be increased by Δz after the drop. Locate the new specific energy E_2 by adding Δz to E_1; then find the corresponding depth y_2. The new depth, y_2, after the drop is greater than the original depth y_1. One can see from this analysis that the water level may actually rise over a drop in the channel bed; however, this is not always the case. Obviously, the new water level (measured from an immutable datum) will depend on the magnitude of the drop.

Now, suppose an abrupt step-up or rise in the channel bed is encountered instead of a drop. This situation, depicted in Figure 12.14a, is more complicated than the drop problem. Assuming again that the upstream conditions are subcritical, the initial specific energy E_1 represents the available energy to push the flow across the step. However, this time in order to determine the energy after the step, E_2, the step dimension Δz is subtracted from E_1. Thus,

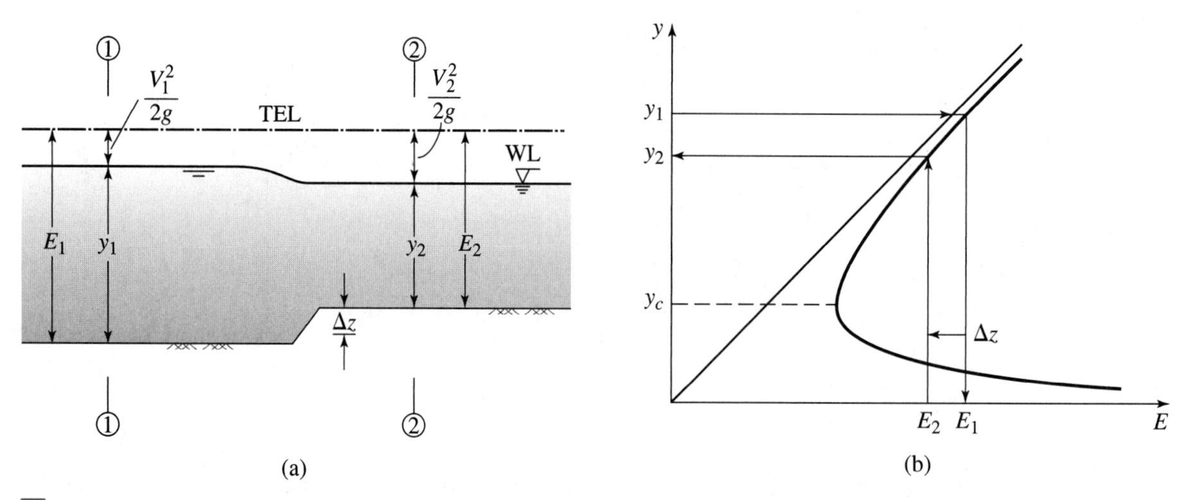

Figure 12.14 A small step in the channel bed (subcritical flow): (a) change in water level, and (b) specific energy diagram.

while the total energy available remains unchanged, the specific energy available for flow is reduced as shown in Figure 12.14b. Then, as the figure shows, the final depth will be less than the original depth.

Of course, the transition, or step, problem is actually solved analytically by employing the Bernoulli equation across the step, letting the step represent a change in datum. Using the original channel bed as the datum and writing the Bernoulli equation across the step, one obtains

$E_{1T} = E_{2T}$, where the E_T represents the total energy available.

Then

$$y_1 + \frac{V_1^2}{2g} = \Delta z + y_2 + \frac{V_2^2}{2g} \tag{12.35}$$

or

$$E_{1T} = \Delta z + y_2 + \frac{V_2^2}{2g} = \Delta z + E_2 \tag{12.36}$$

where the term $y_2 + (V_2^2/2g) = E_2$ or the specific energy after the step. Also note that the specific energy before the step is equal to the total energy there due to our use of the channel bed as the datum. Then, as illustrated on Figure 12.14b, the new specific energy is just the initial specific energy minus the step dimension, $E_2 = E_1 - \Delta z$.

Then, in terms of the total energy given by the Bernoulli equation, the depth after the step is given by

$$y_2 = E_{1T} - \left(\Delta z + \frac{V_2^2}{2g}\right) = E_{1T} - \left(\Delta z + \frac{Q^2}{2gA^2}\right) \tag{12.37}$$

For a rectangular channel section, equation (12.37) becomes

$$y_2 = E_{1T} - \left(\Delta z + \frac{V_2^2}{2g}\right) = E_{1T} - \left(\Delta z + \frac{q^2}{2gy_2^2}\right) \tag{12.38}$$

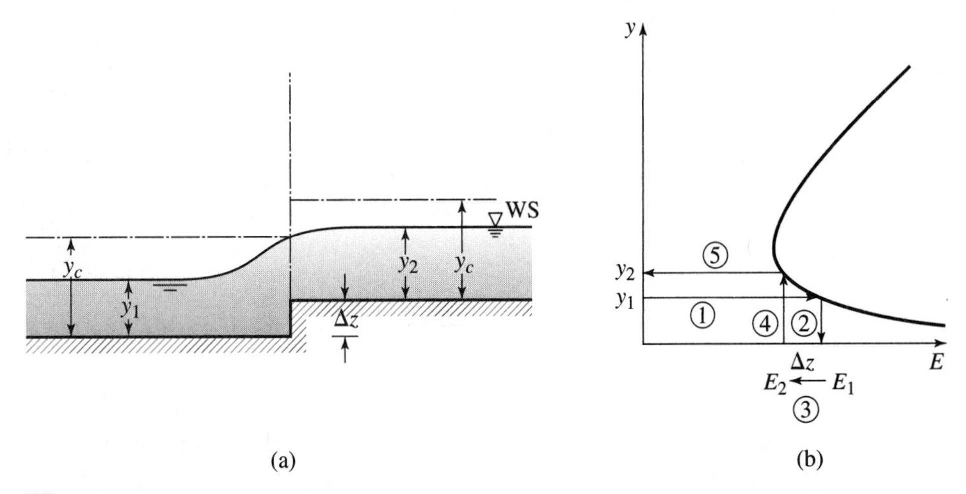

Figure 12.15 A small step in the channel bed (supercritical flow): (a) change in water levels, and (b) steps for solution.

It is observed that the total energy across the step remains unchanged since all losses are ignored. Thus, the energy available to force the flow across the step is limited by the available energy level upstream. This is an issue to which we will return presently.

However, first consider the supercritical situation where the depth of flow y_1 is less than the critical depth y_c and the Froude number is greater than one, as shown in Figure 12.15a. Initial specific energy is E_1. As shown in Figure 12.15a, an abrupt rise with a height of Δz is encountered in the channel bed. Figure 12.15b illustrates the steps for calculating the new depth y_2 after the rise. Again, the available specific energy is reduced by the step height Δz to predict E_2, which in turn can be used to compute the new depth y_2. The new depth is greater than the original depth, thus indicating that the water level must rise after the step as shown in Figure 12.15b.

12.6.2 Chokes

The discussion of the step-up or rise problem illustrates that the specific energy after the rise is actually less than that available before the step. Now, in Section 12.4.3, the concept of a minimum specific energy for flow (corresponding to critical depth) was discussed. Therefore, it is certainly possible to imagine a situation where the channel is constricted, or stepped up to such an extent that the available energy would not be sufficient to push the flow across the rise. Consider the situation depicted in Figure 12.16a. This condition is equivalent to that shown previously in Figure 12.14a. The upstream flow is subcritical and the rise in the channel bed forces the depth to decrease across the step. Now, suppose the step size is progressively increased as shown in Figures 12.16b and 12.16c. As the step increases, more and more energy is subtracted from the initial state as depicted in Figure 12.17, which has the effect of further decreasing the depth over the step until the critical depth is reached (Figure 12.16c). At this stage, the specific energy at the step has been reduced to the point that it equals the minimum energy associated with y_c as shown in Figure 12.17. Now, if the step is further increased, sufficient energy will not be present upstream to cause the flow to overcome the step. This is shown in Figure 12.17 as the point where the subtraction of Δz_d forces the energy state to be so far to the left that it is off

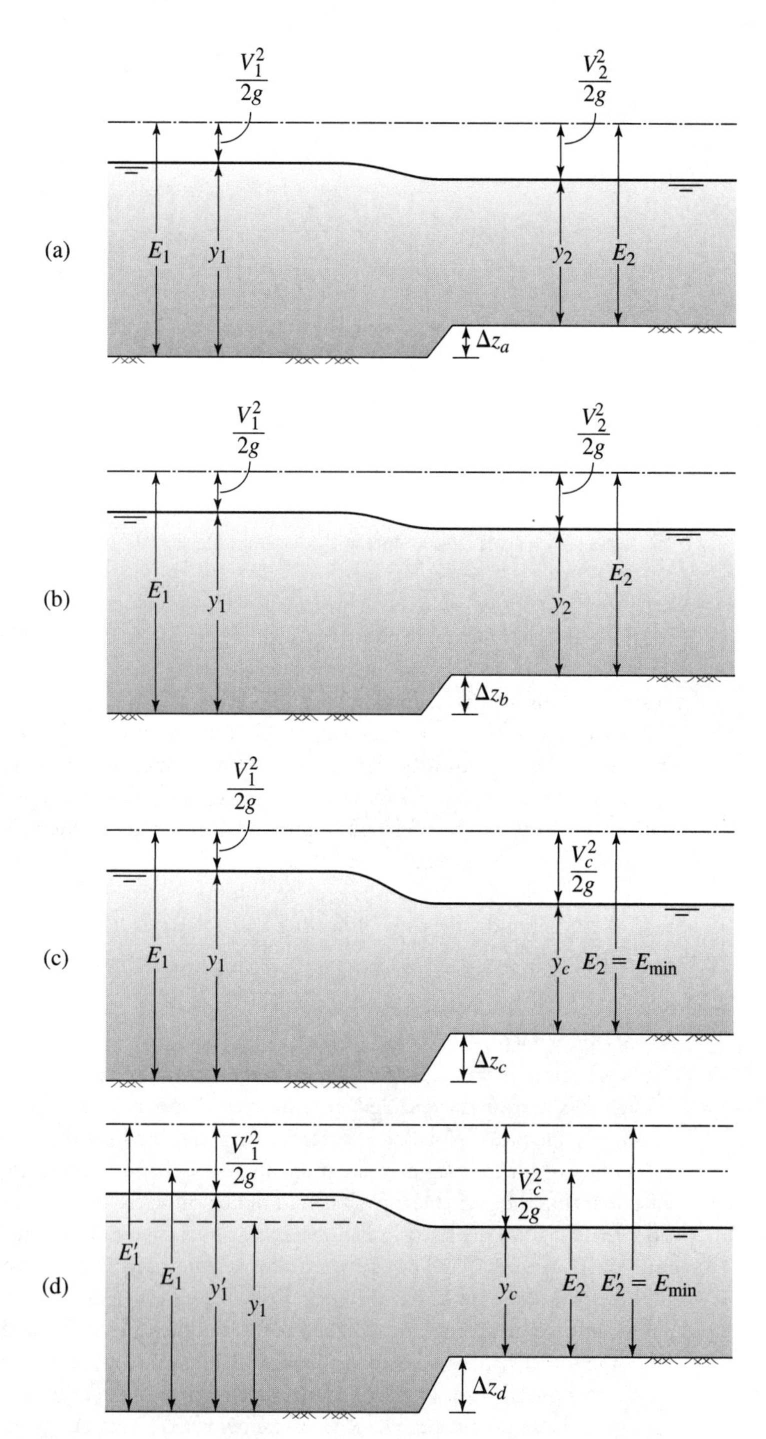

Figure 12.16 Rise in channel bed: (a) a small step-up, (b) a bigger step-up, (c) a still bigger step-up, and (d) changes in the specific energy.

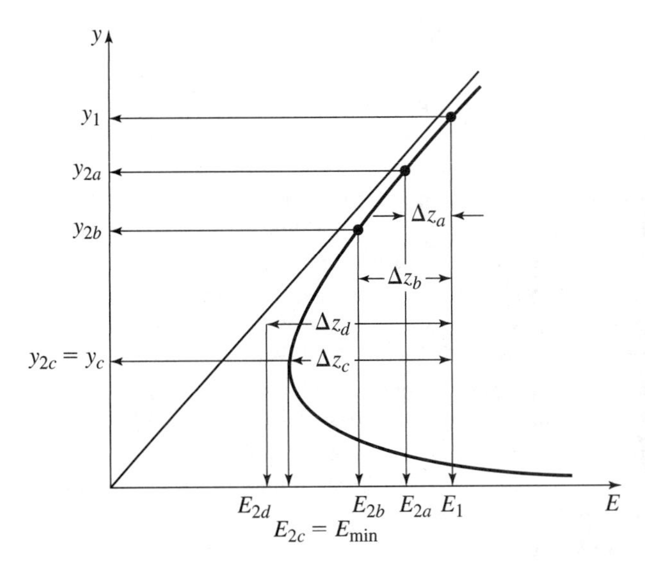

Figure 12.17 Specific energy diagram.

Toledo Bend Reservoir, LA. (Courtesy of Frank Dutton, www.toledo-bend.com.)

of the initial specific energy graph. In this case the depth of flow upstream of the step will increase to y_1', causing what is called the rising water profile (see Chapter 14). This is done to affect an increase in energy from E_1 to E_1' (Figure 12.16d). The water profile rises to the extent that the necessary minimum specific energy (critical flow condition) at the step is attained (i.e., $E_2' = E_{min}$) and the depth across the step will be critical.

The condition can be complicated in the case where the rising water profile, or zone of increased depth, extends back to the channel entrance or the source of the flow such as a reservoir. In such cases, either the reservoir level must rise or the discharge must decrease and the specific energy diagram will move to the left. The flow depth before the step and the critical depth over the step will adjust according to the new reduced discharge. Therefore, if the height of the step is bigger than that which would produce a critical flow condition above the step for the given discharge, two situations might be encountered. First, a rising water profile will be formed upstream of the step. Second, if the rising profile reaches the control point upstream of the step, the discharge must decrease or the control point will be drowned. In these two situations, the critical flow condition will be encountered above the step.

The situation discussed above corresponds to a condition known in open-channel hydraulics as a *choke*. Analysis of chokes, or potential chokes, is an extremely important topic in hydraulics. From the discussion given above, it should be clear that chokes can occur in any situation where the channel cross section is constricted, but will not occur where the flow area is expanded such as drops or expansions in channel widths. Obviously, in designing a channel transition that would tend to restrict the flow, the engineer wants to avoid forcing a choke to occur if at all possible. The situation is further complicated in the case of supercritical flow, in that if a choke does occur, the upstream flow may change from supercritical to subcritical if the depth rises sufficiently. This is a situation which will be discussed in the next chapter.

Based on the discussion so far, a procedure for handling the constriction problem can be summarized in a number of steps. First, check the upstream flow to determine if it is subcritical or supercritical. Then compute the initial energy upstream (E in Figure 12.16a) that will be available to push the flow through the constriction. Next, it is necessary to check the constricted area to determine if a choke will occur. To accomplish this, the critical depth is computed at the constriction and its corresponding minimum specific energy is calculated. Then the total minimum energy at the constriction is determined by adding in any datum shift that might occur, such as a step-up in the channel bed. This is the minimum required energy to force the flow through the constriction, and this value can then be compared to the total energy available upstream. If the initial energy is greater than the minimum energy, then the flow will pass through the constriction without choking and the depth at the constriction can be computed as outlined above. If the upstream energy is less than the minimum required, then a choke will occur. Under this condition, if the flow was initially subcritical, it will reduce to critical depth at the downstream constriction. If it was initially supercritical, then a hydraulic jump will most likely occur upstream. The transition examples illustrated thus far have primarily dealt with changes to the channel bed that tend to restrict or expand the cross-sectional flow area. Obviously, changes in the channel width or geometry (transitions from trapezoidal to rectangular, for example) can have the same effects. In fact, any constriction which is sufficient to cause a rise in the energy grade line above that of the initial flow will constitute a choke. For instance, hydraulic structures, such as weirs or underflow (sluice) gates that function as flow or elevation control devices, will function as chokes, as illustrated below.

Example 12.9

A uniform flow of 100 ft^3/s occurs in a rectangular channel of 5 ft width and 5 ft depth. A smooth rise of 0.5 ft is formed in the channel bed as shown in Figure 12.18a. Determine the depth of flow above the step. What is the height of the step that will produce a critical flow?

Solution: The discharge per unit width, $q = 100/5.0 = 20.0$ ft^2/s.

$$y_c = \sqrt[3]{\frac{q^2}{g}} = 2.32 \text{ ft}, \quad \text{and} \quad E_{\min} = 1.5(2.32) = 3.48 \text{ ft}$$

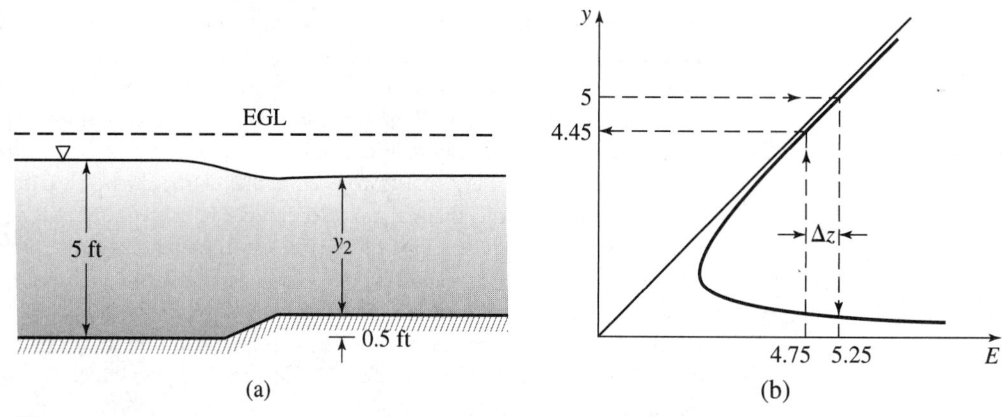

Figure 12.18 (a) Rising channel bed for Example 12.9. (b) SED for Example 12.9.

Therefore, the minimum energy necessary to force the water to flow across the step, with reference to the original (upstream) channel bed as the datum, would be $E_{min} + \Delta z = 3.48 + 0.5 = 3.98$ ft. We also note that since $y_c < y_1$, the initial depth of 5 ft is subcritical. Now we compute the initial energy available for the flow,

$$E_1 = y + \frac{q^2}{2gy^2} = 5 + \frac{(20)^2}{2g(5)^2} = 5.25 \text{ ft}$$

Since $E_1(5.25 \text{ ft}) > E_{min}(3.98 \text{ ft})$, there will be no choke and the solution can proceed. Applying the Bernoulli equation across the step and ignoring losses,

$$y_1 + \frac{q_1^2}{2gy_1^2} = \Delta z + y_2 + \frac{q_2^2}{2gy_2^2}$$

Due to the 0.5-ft rise in the bed level, the specific energy is reduced by 0.5 ft as it is measured from the bed as a datum. Therefore,

$$5 + \frac{400}{2g(5)^2} = y_2 + \frac{400}{2g(y_2)^2} + 0.5$$

$$4.75 = y_2 + \frac{6.21}{(y_2)^2}$$

Solving by trial (making sure to converge to the subcritical root), we obtain $y_2 = 4.43$ ft. Thus, the stage, or water level, after the step is $4.43 + 0.5 = 4.93$. Therefore, the water level relative to the original upstream datum has dropped by 0.07 ft.

Required Δz to cause critical depth at the step: Since we have already computed y_c and E_{min} to be 2.32 ft and 3.48 ft, respectively, we can use Figure 12.18b to easily find the solution to this problem. Critical depth will occur at the step when the available energy E_1 is equal to E_{min}. Therefore, simply making the total energy across the step equal for the given values of E_1, E_{min}, and the unknown Δz, we write

$$E_1 = E_{min} + \Delta z$$

$$5.25 = 3.48 + \Delta z, \quad \text{or } \Delta z = 1.77 \text{ ft}$$

Example 12.10

A 10-ft-wide rectangular channel carries a discharge of 500 cfs at a normal depth of 12 ft. A rectangular sill 10 ft high is placed across the channel as shown in Figure 12.19. Ignoring energy losses, what will be the depth of water behind the sill?

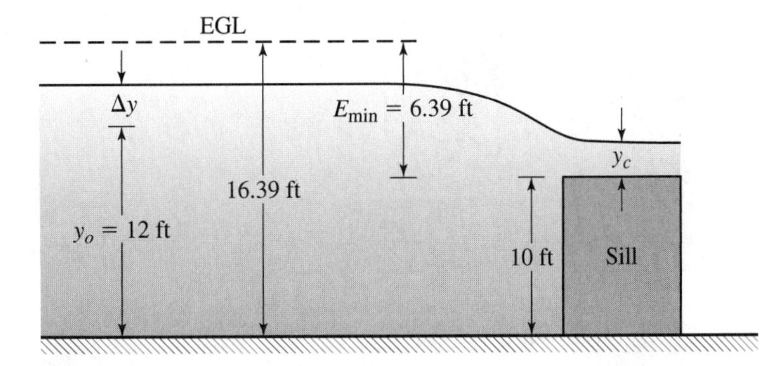

Figure 12.19 Sill for Example 12.10.

Solution: $Q = 500$ cfs so that $q = Q/b = 500/10 = 50$ cfs/ft. Therefore, $y_c = \sqrt[3]{q^2/g} =$ 4.26 ft, so that we know that y_1 is subcritical. Then $E_{min} = 3/2y_c = 1.5(4.26) = 6.39$ ft. The total required energy is $E_{min} + \Delta z = 6.39 + 10 = 16.39$ ft. Now the available energy is $E_1 = y_1 + \dfrac{q^2}{2g(y_1)^2} = 12 + \dfrac{50^2}{2g(12)^2}$; $E_1 = 12.27$ ft. Since $E_1 < E_{min}$, then the sill is functioning as a choke. We note that the minimum required energy to force the flow across the sill is 16.39 ft relative to the upstream bed as the datum. With this value now set as E_1, we solve for the required y_1. $E_1 = y_1 + \dfrac{q^2}{2g(y_1)^2} = 16.39$ ft. Therefore, $E_1 = y_1 + \dfrac{50^2}{2g(y_1)^2} = y_1 + \dfrac{38.82}{(y_1)^2} =$ 16.39 ft. From this we find $y_1 = 16.2$ ft.

Example 12.11

Referring to the situation in Example 12.10, a vertical underflow (sluice) gate is placed across the channel in lieu of the sill as shown in Figure 12.20. If the gate opening is 1.5 ft as shown, what will be the depth of water behind the gate? Ignore any energy losses associated with the gate.

Solution: We recall that $q = 50$ cfs/ft and $y_c = 4.26$ ft and $y_1 = 12$ ft (subcritical). From Figure 12.20 we can see that the depth behind the gate will merely be the alternate depth to the depth under the gate if we ignore losses. Therefore, finding the energy at the gate,

$$E_2 = y_2 + \frac{q^2}{2g(y_2)^2} = 1.5 + \frac{50^2}{2g(1.5)^2} = 18.75 \text{ ft}$$

Then, since $E_1 = E_2$, $18.75 = y_1 + \dfrac{50^2}{2g(y_1)^2} = y_1 + \dfrac{38.82}{(y_1)^2}$, from whence we obtain $y_1 = 18.6$ ft.

It can be noted that even though the gate is functioning as a choke (E_2 required = 18.75 ft > E_1 available = 12.27 ft), critical depth does not occur at the gate due to the fact that the physical gate setting of 1.5 ft is less than y_c of 4.26 ft. In fact, this is a necessary feature of the gate in order to force the flow from under the gate to be supercritical. Otherwise, since the energy remains fixed, the depth after the gate would be equal to the upstream depth and the gate would not be effective.

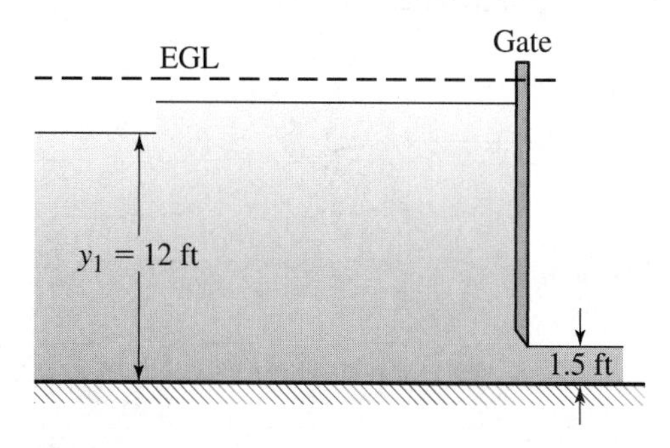

Figure 12.20 Sluice gate for Example 12.11.

12.6.3 Enlargements and Contractions in Channel Widths

Another type of transition is a change in the channel geometry itself. Consider the channel given in Figure 12.21a. For simplicity, let the section be rectangular and the channel width be reduced from b to b_1. The initial depth of flow is y_1 and is subcritical. The discharge per unit width, q, will increase after the contraction to q_1 and the SED should shift as shown in Figure 12.21c to represent the new flow conditions. Because the elevation of the channel bed has not been changed through the segment under consideration, the specific energy, E_o, will not vary, as shown Figure 12.21b. However, the critical depth will increase after the contraction (to y_c') due to the increase in q. Steps one, two, and three, shown in Figure 12.21c, are followed to determine the new depth (y_2). The water level will decrease after the channel contraction, as shown in Figure 12.21b.

Of course, it is obvious that this case is analogous to the step-up problem discussed in the previous sections. As the channel width is decreased and the specific discharge is increased, then the velocity is also increased. In addition, as q increases, then y_c increases and thus, since $E_{\min} = 3/2y_c$, $E_{\min}$ increases as well. Therefore, if the channel constriction is sufficient to force a situation where $E_{\min} > E_1$, then a choke will certainly result. Conversely, for the case of channel enlargement, the discharge per unit width will decrease and the SED will be shifted to the left. This situation is analogous to the step-down problem previously discussed and thus there can be no choke.

If the flow is supercritical (i.e., y_1 is less than y_c as shown in Figure 12.21d), the new depth, y_2, after the contraction will be bigger than the initial one and the water level will

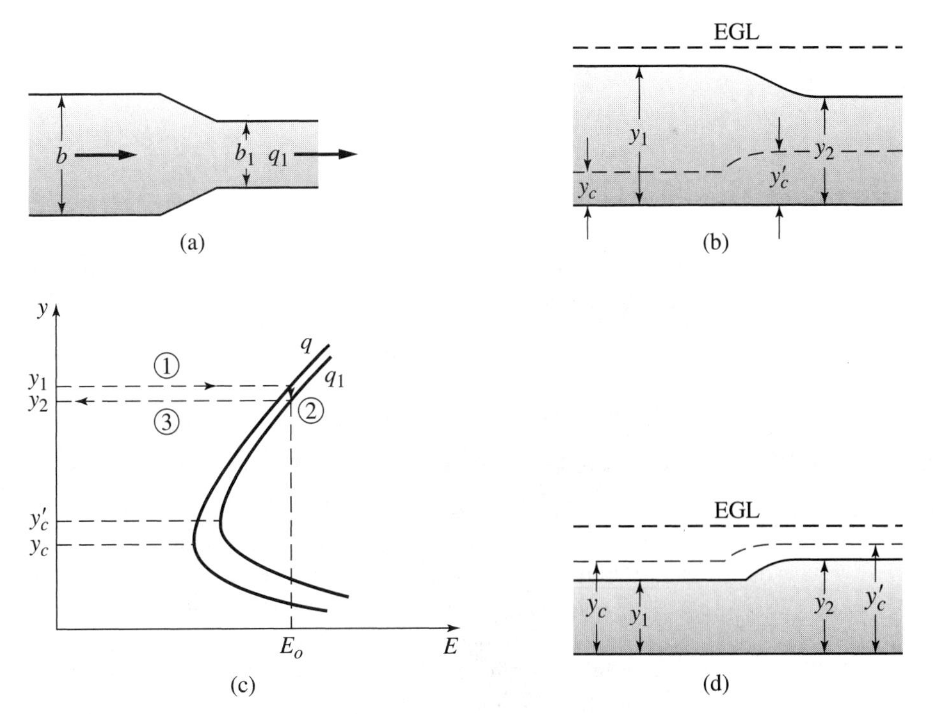

Figure 12.21 (a) A contracted channel. (b) Water levels in a contracted channel. (c) SED for a contracted channel. (d) Water level in a contracted channel–supercritical flow.

elevate after the contraction. As stated previously, this rise of the water depth within a constriction can be the source of further complications due to a sudden change from supercritical to subcritical flow (a hydraulic jump), a case that is discussed in Chapter 13.

Of course, in channel transitions such as contractions and expansions some minor losses will ensue just as in the case of pipes. These losses can be handled the same way as discussed in Chapter 8 (i.e., by multiplying the velocity head difference by a loss coefficient). Typical contraction coefficients for channels range from 0.1 for gradual contractions to 0.6 for abrupt contractions, while expansion coefficients vary from 0.3 for gradual expansions to 0.8 for abrupt expansions (U.S. Army Corps of Engineers, 1982). For instance, at transitions associated with bridge crossings it is common to employ values of 0.3 and 0.5 for contraction and expansion coefficients, respectively.

Example 12.12

A 6.0-m rectangular channel carries a discharge of 30 m^3/s at a depth of 2.5 m. Determine the channel width that produces critical depth.

Solution: As in the previous discussion, we want to make $E_1 = E_{min}$ in the constricted section. Therefore,

$$E_1 = y_1 + \frac{q_1^2}{2gy_1^2} = 2.5 + \frac{(5.0)^2}{2 \times 9.81 \times (2.5)^2} = 2.70 \text{ m}$$

referring to Figures 12.22a and 12.22b.

Then $E_1 = E_2 = 3/2y_{c2}$. Therefore, $y_{c2} = 1.80$ m. Now

$$\frac{q_2^2}{g} = y_{c2}^3 \quad \therefore \quad q_2 = \sqrt{gy_{c2}^3} = 7.56 \text{ m}^2/\text{s}$$

Then $b_2 = Q/q_2 = 30.0/7.56 = 3.97$ m.

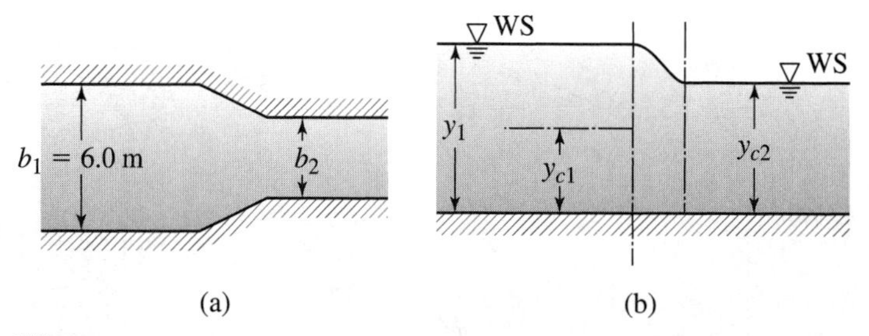

Figure 12.22 Solution for Example 12.12: (a) A plane for the rectangular channel. (b) Water surface and energy line.

It should be noted that a flow at or near the critical state will not be stable. Minor changes in the specific energy will cause considerable changes in the depth of flow. This is attributed to the vertical pattern of the specific energy diagram near the critical condition. The water surface appears wavy when the depth of flow is close to the critical depth as any variation in channel roughness, cross section, sediments, and so on will disturb the flow significantly. In other words, large changes in water surface levels may be introduced by small changes in bed levels or other geometric elements.

Consider the case where a step, with limited length, is of critical height, in the sense that it produces critical flow, as shown in Figure 12.23. In such a case, the downstream flow may

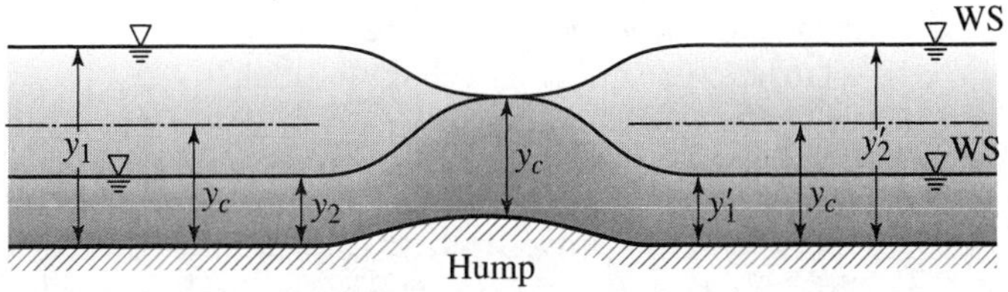

Figure 12.23 Water surface over a critical hump with limited length.

be either supercritical or subcritical depending on the flow conditions downstream. If there is no downstream control, the flow tends to continue the ongoing decrease or increase in the depth of flow. Supercritical flow tends to be subcritical and vice versa. However, it should be kept in mind that either of the upstream flow states can pass to either of the downstream flow states depending on downstream conditions. A similar analysis is applicable to channel contractions and enlargements.

In many practical situations, it may be feasible to introduce local constrictions into open channels before intersecting highways to reduce bridging costs. It is important to know the minimum section which sustains the given discharge with the given specific energy without affecting the flow conditions upstream. The smallest section is that which operates under critical flow conditions.

Example 12.13

Water flows in trapezoidal channel with a bottom width of 10 ft and side slopes of 2:1. The channel is running on a slope of 0.0005 with a discharge of 500 cfs and a Manning's n of 0.03. It is desired to transition to a 10-ft-wide rectangular channel. What will be the depth in the rectangle? Use a contraction coefficient of 0.3.

Solution: Applying Manning's equation to the trapezoidal channel,

$$(10y_o + 2y_o^2)\left[\frac{10y_o + 2y_o^2}{10 + 2y_o\sqrt{1+4}}\right]^{2/3} = \frac{0.03(500)}{1.49(0.0005)^{1/2}} = 450.2$$

This yields $y_o = 7.2$ ft. Then $A = 175.68$ ft^2, $P = 42.20$ ft, $R = 4.16$ ft, $V = 2.84$ ft/s, $B = 38.8$ ft, $E = 7.32$ ft, and $F_n = 0.236$, so the flow is initially subcritical.

Now, for the rectangular section, $q = Q/b = 500/10 = 50$ cfs/ft. Then $y_c = \sqrt[3]{q^2/g} = 4.27$ ft. Then $E_{min} = 1.5(y_c) = 6.40$ ft $< E_1$, so there is no choke. Then writing the energy equation across the transition,

$$E_1 = y_2 + \frac{q^2}{2gy_2^2} + h_L$$

$$E_1 = y_2 + \frac{q^2}{2gy_2^2} + 0.3\left(\frac{q^2}{2gy_2^2} - \frac{V_1^2}{2g}\right)$$

$$7.32 = y_2 + \frac{(50)^2}{2gy_2^2} + 0.3\left(\frac{(50)^2}{2gy_2^2} - \frac{(2.84)^2}{2g}\right) = y_2 + \frac{1.3(38.82)}{y_2^2} - 0.3(0.125)$$

$$7.32 = y_2 + \frac{1.3(38.82)}{y_2^2} - 0.0375 = 7.35 = y_2 + \frac{50.466}{y_2^2}$$

by trial and error, $y_2 = 5.9$ ft.

Example 12.14

If the rectangular section in Example 12.13 is set at a width of 8, how much must the channel bed need to be lowered to avoid a choke situation?

Solution: From the previous example, we know that $E_1 = 7.32$ ft. Now for a rectangular section 8 ft wide, $q = 62.5$ cfs/ft, $y_c = 4.95$ ft, and $E_{min} = 7.42$ ft. Thus, a choke would occur. To remove the choke, we want to set the E_{min} in the rectangle so that it is exactly equal to E_1. Thus, $7.32 + \Delta z = 7.42$ and $\Delta z = -0.1$ ft. The channel bed must be lowered by 0.1 ft.

Example 12.15

Water is flowing in a 10-ft-wide rectangular channel. The depth in the channel is 5 ft and the velocity is measured to be 3 ft/s. If a transition to a small rectangular channel is desired, what is the minimum bottom width of the rectangle so as not to affect the upstream conditions (i.e., not to force a choke)? Ignore minor losses.

Solution: $A = 50$ ft^2, $Q = 150$ cfs, and $q = 15$ cfs/ft. The critical depth is $y_c = 1.91$ ft and the minimum specific energy is $E_{min} = 2.865$ ft. The specific energy before the transition is 5.14 ft. The critical depth is $2/3 \times 5.14$ ft $= 3.42$ ft. Then the specific discharge is 35.88 cfs/ft $= Q/b$. Therefore, $b = 150/35.88 = 4.18$ ft.

12.6.4 Weirs and Spillways

Weirs are structures designed to control the elevation of the water upstream of the control. As such, a weir functions as a downstream choke control on the flow in the channel. Weirs are classified as sharp crested or broad crested depending on whether or not critical depth occurs on the crest. If critical depth does occur, then, of course, the critical section controls the discharge. In the case of the sharp-crested weir, y_c occurs off the crest due to hydrostatic pressure considerations and thus cannot be the control. The concept of the sharp-crested weir is the basis for the design of overflow spillways. Several configurations of sharp-crested weirs are possible: suppressed, end contractions, V-notched, rectangular, and trapezoidal.

12.6.4.1 Rectangular Weirs A suppressed weir is simply a rectangular sharp crested weir that runs the entire width of the channel. It can be constructed of sheet pile, timber or concrete. The suppressed weir is shown in profile in Figure 12.24. Now, we wish to derive an expression for the discharge over the weir. Referring to Figure 12.24, writing the energy equation from point 1 upstream prior to the advent of the drawdown curve to point 2 at the level of the weir crest and using the crest as the datum

$$y_1 + \frac{V_1^2}{2g} = y_2 + \frac{V_2^2}{2g} \tag{12.39}$$

Since $y_2 = 0$, we can write

$$V_2 = \sqrt{2gy_1 + V_1^2} \tag{12.40}$$

Letting $y_1 + (V_1^2/2g) = H$ (the head on the weir crest), equation (12.40) is reduced to the form of the orifice equation:

$$V_2 = \sqrt{2gH} \tag{12.41}$$

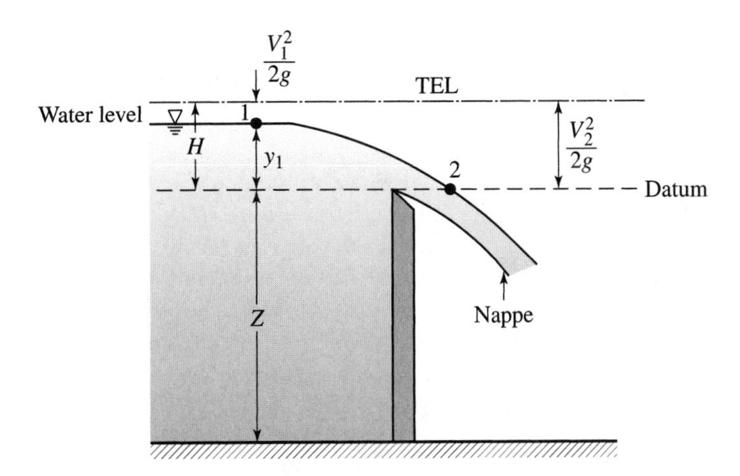

Figure 12.24 Principles of the suppressed sharp-crested weir.

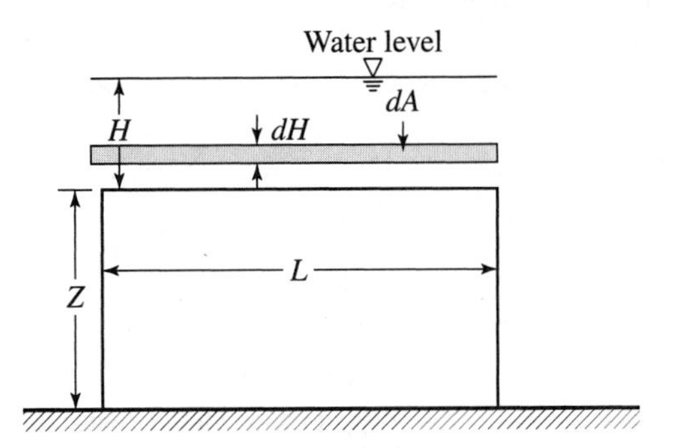

Figure 12.25 Flow cross section of a sharp-crested weir.

It is common practice to assume that V_1 is negligible when the head is measured a distance back from the wier (i.e, at the beginning of the rising water surface profile). In all of the following discussion this assumption is implicit.

Figure 12.25 depicts the immediate region of the weir crest, where L is the length of the weir across the channel (i.e., channel width) and dH is a differential vertical element of flow across the weir. Then the discharge dQ through the element would be given by

$$dQ = V\,dA = L\,dH\sqrt{2gH} \tag{12.42}$$

Then, integrating vertically across the head, H, we obtain the total discharge across the weir

$$Q = \sqrt{2g}L\int_0^H H^{1/2}dH = \frac{2}{3}\sqrt{2g}LH^{3/2} \tag{12.43}$$

Equation (12.43) represents the ideal flow across a suppressed sharp crested weir, neglecting losses. However, certain losses will occur due to the advent of the drawdown of the flow immediately upstream of the weir, as well as any other friction or contraction losses. For these reasons, the actual flow over the weir will not equal the ideal flow given by equation (12.43). To account for these losses, a coefficient of discharge, C_d, is introduced,

$$Q = C_d\tfrac{2}{3}\sqrt{2g}LH^{3/2} \tag{12.44}$$

Experimental work with C_d has shown that it can be approximated by the expression (Henderson, 1966)

$$C_d = 0.611 + 0.08\frac{H}{Z}, \text{ where } Z \text{ is the height of the weir.} \tag{12.45}$$

Although equation (12.45) is said to be applicable for H/Z ratios up to 2.0, care should be taken when using the equation for cases were $H/Z > 1$. In these cases, the tailwater is likely to be high and may interfere with the discharge nappe of the weir and thus interfere with its free operation. This is the so-called *drowned weir* case. The weir equation is more applicable in the *high head* weir case, where $H/Z < 0.4$ such as the overflow spillway.

Combining the constants with the discharge coefficient results in the well-known Francis weir equation due to experiments performed by J. B. Francis (1815–1892) in the late nineteenth century at Lowell, Massachusetts.

$$Q = CLH^{3/2} \tag{12.46}$$

The factor C is known simply as the Francis weir coefficient and may be further reduced below its computed value due to other losses associated with a particular weir environment or configuration. For instance, flow over a roadway section may behave as a sharp-crested weir with additional losses due to the guardrail, pavement roughness, and vegetation on the roadside. For example, if the H/Z ratio is on the order of 0.1, then equation (12.45) would result in a C_d value of 0.62, which, when combined with the other constants in equation (12.44), would yield a C value of 3.32 in English units (1.84 SI units). However, it is common practice to further reduce this value depending on the nature of other losses at the weir.

If the rectangular weir does not run the entire width of the channel, then it will have end contractions as shown in Figure 12.26. In this configuration, the flow lines are contracted from the full channel width to the width of the rectangular slot in the weir. Of course, this contraction adds to the head losses associated with the weir as a portion of the weir length becomes ineffective. Francis found that these effects could be handled in the manner of a minor loss by decreasing the effective weir length by an amount of $0.1H$ for each contraction. Thus, for two contractions the effective weir length is given by

$$L' = (L - 0.2H) \tag{12.47}$$

and

$$Q = CL'H^{3/2} \tag{12.48}$$

where L = the width of the notch as shown in Figure 12.26.

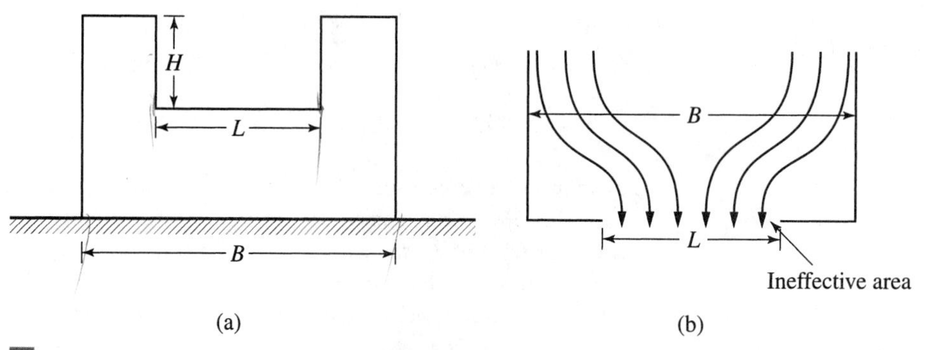

Figure 12.26 (a) Rectangular sharp-crested weir with end contractions. (b) Plan view of stream line convergence through a notched rectangular sharp-crested weir.

Example 12.16

A suppressed rectangular, sharp-crested weir is placed across a 20-ft-wide channel. The invert of the channel is at elevation 100 ft National Geodetic Vertical Datum (NGVD) and the weir is 15 ft high. Develop a headwater rating curve for this situation.

Solution: A range of solutions will be calculated for H values up to 5 ft, thus resulting in a maximum H/Z value of 0.333. Therefore, the discharge coefficient would range between

$$C_d = 0.61 + 0.08(0.066) = 0.615 \text{ (minimum)}$$

$$C_d = 0.61 + 0.08(0.333) = 0.636 \text{ (maximum)}$$

The Francis C value would then vary from $C = 0.667(8.02)(0.615) = 3.29$ to $C = 0.667(8.02)(0.636) = 3.40$. It is customary to use an average value of the Francis coefficient, so that a value of 3.34 might suffice. Using this value, the following headwater table is developed from application of equation (12.46) ($L = 20$ ft):

Headwater Elevation (ft)	H (ft)	Q (cfs)
116	1	66.8
117	2	188.9
118	3	347.1
119	4	534.4
120	5	746.8

It is important to note that the head on the weir is measured upstream of the crest at a sufficient distance so as not to be affected by the drawdown curve toward the critical depth just off the weir crest. This distance is sometimes estimated to be $4H$.

12.6.4.2 Triangular Weirs Rectangular weirs such as the suppressed weir are normally employed as elevation control devices in situations where relatively high discharge values are involved. In cases where small discharges are encountered, triangular or V-notch weirs are more often used. In particular, triangular weirs are excellent devices for measuring discharge in low flow cases. The small flow area available in these weirs requires the development of a greater upstream head that is easier to measure. For example, in measuring seepage through earth embankments or small flow in springs, artesian wells, and so on, triangular weirs are frequently employed due to their higher accuracy as compared to rectangular weirs.

The triangular weir is shown in Figure 12.27. From the geometry shown in the figure, it can be seen that the following relation holds:

$$\frac{X}{L} = \frac{H - y}{H}$$

or

$$X = \frac{L(H - y)}{H} \tag{12.49}$$

In terms of the notch angle θ, it can be seen that

$$\tan \theta/2 = \frac{L}{2H} \tag{12.50}$$

substituting equation (12.50) into equation (12.49), then

$$X = 2\tan(\theta/2)(H - y) \tag{12.51}$$

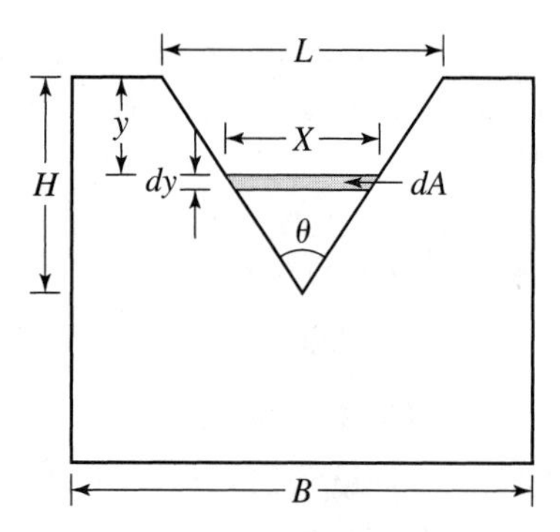

Figure 12.27 Principles of the triangular or V-notch weir.

Then the discharge through the differential element, dA is given by

$$dQ = V\,dA = \sqrt{2gy}\,(X\,dy) \tag{12.52}$$

$$= \sqrt{2g}\,Xy^{1/2}dy = \sqrt{2g}\,2\tan(\theta/2)(H - y)y^{1/2}dy \tag{12.53}$$

Hence, the total discharge through the notch becomes

$$Q = \int_0^H dQ = \int_0^H \sqrt{2g}\,2\tan(\theta/2)(Hy^{1/2} - y^{3/2})\,dy$$

which after integration yields

$$Q = \frac{8}{15}\sqrt{2g}\tan(\theta/2)H^{5/2} \tag{12.54}$$

The most common notch angles encountered in V-notch weirs are 90°, 60°, and 45°. For the 90° weir ($\tan 45° = 1$), the flow is just given by

$$Q = \frac{8}{15}\sqrt{2g}\,H^{5/2} = 0.533\sqrt{2g}\,H^{5/2} \tag{12.55}$$

while for the 60° weir ($\tan 30° = 0.577$), the discharge becomes

$$Q = 0.308\sqrt{2g}\,H^{5/2} \tag{12.56}$$

However, as in the rectangular case, the equations give the discharge under ideal conditions, so that a discharge coefficient must be employed to account for contraction and friction losses associated with the drawdown of the water surface in the vicinity of the weir. Discharge coefficients for triangular weirs appear to be on the same order as the rectangular weirs (i.e., around 0.60). For example, the C_d for 90° and 60° weirs vary from about 0.57 to 0.60 (Roberson et al., 1998).

12.6.4.3 Trapezoidal Weirs The trapezoidal weir, sometimes called the Cipolletti weir, is shown in Figure 12.28. It consists of a combination of the rectangular weir and the triangular weir. The standard Cipolletti weir has notch side slopes of $1H:4V$. It is used in cases where the rectangular contracted weir is not of sufficient capacity to carry the discharge to be handled. It is

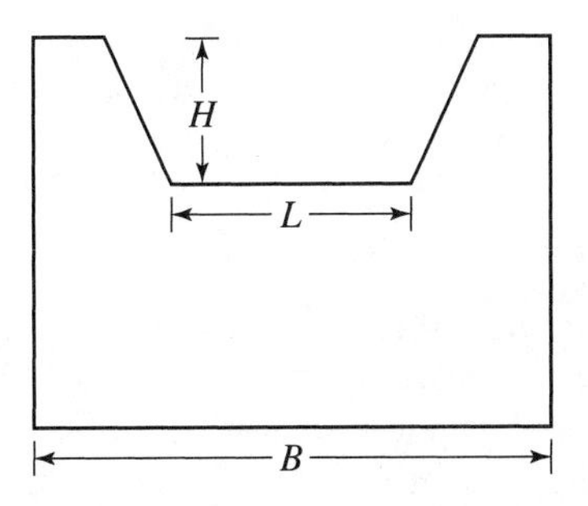

Figure 12.28 Trapezoidal or Cipolletti weir.

Flow over a weir. (Courtesy of Royalty-Free/ CORBIS.)

evident that flow through the trapezoidal weir could be approximated as the sum of the flows through the rectangular and triangular sections, that is,

$$Q = CLH^{3/2} + C_d\frac{8}{15}\sqrt{2g}\tan(\theta/2)H^{5/2} \tag{12.57}$$

However, the U.S. Bureau of Reclamation (USBR) recommends that the discharge capacity of the Cipolletti weir can also be obtained as (Hwang and Houghtallen, 1996)

$$Q = C_w LH^{3/2} \tag{12.58}$$

Experimentation by the USBR led to the recommendation of a weir coefficient (C_w) of 3.367 in standard BG units for the ideal case. However, as in the previous cases, this value must sometimes be reduced to account for other resistance losses.

12.6.4.4 Overflow Spillways The rectangular, sharp-crested weir provides the basis for the analysis and design of the overflow spillway as shown in Figure 12.29. The design head, H_d, is the actual head on the spillway crest, while the nominal head, H, is the head with respect to the weir crest. Assuming the high weir case (i.e., $H \ll Z$), the flow over the spillway would be given by

$$Q = CLH^{3/2} = 3.32LH^{3/2} \tag{12.59}$$

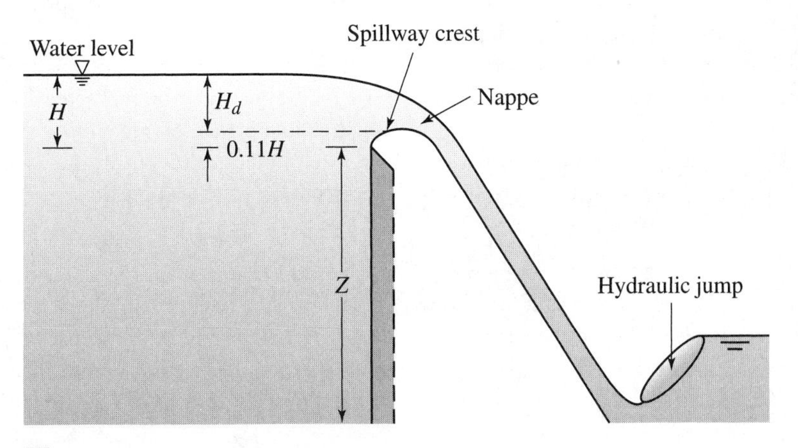

Figure 12.29 Principles of the overflow spillway.

Grand Coulee Dam. (Courtesy of VSBR and Charles Hubbard.)

while in terms of the actual head on the spillway crest, using the experimental value for the rise from the weir crest elevation to the top of the spillway crest ($0.11H$), the discharge can be expressed as (Henderson, 1966)

$$Q = 3.97LH_d^{3/2} \tag{12.60}$$

The face of the spillway is designed in the shape of the nappe of the jet over the weir such that the nappe will adhere to the concrete face as closely as possible. This is done in order to reduce the possibility of cavitation of the concrete due to negative pressures that would develop between the face and the high-velocity nappe. An experimentally derived curve is used to connect the crest segment to the face. The principal design of overflow spillways has been accomplished through experimentation at the USBR Hydraulics Laboratory in Denver Colorado and the U.S. Army Corps of Engineers Hydraulics Laboratory at Vicksburg, Mississippi (Morris and Wiggert, 1972).

The depth over the spillway reduces from the pseudocritical section near the crest toward the normal depth for the steep slope on the face. However, the depth may never actually reach y_o before reaching the toe, or bottom, of the spillway. An additional problem arises in the case of very high spillways in that the water attains a high velocity on the face and becomes entrained with air so that the flow is essentially a high speed jet of air/water mixture by the time it reaches the toe of the spillway. The USBR gave the "theoretical velocity" of this jet as (Bradley and Peterka, 1957)

$$V_t = \sqrt{2g(Z - H/2)} \text{ (ft/sec)} \tag{12.61}$$

where Z is the total fall from the upstream water surface to the downstream surface and H is the actual head on the spillway crest. However, the actual velocity must be obtained from experiments. Such experiments have been perfomed by the USBR and curves of actual versus theoretical velocities are available (e.g., Bradley and Peterka, 1957; Henderson, 1966).

In the case of very small dams, $Z < 6$ m (20 ft), experience has shown that losses on the face can be ignored altogether without appreciable error. In these cases, the depth at the toe can be approximated by treating the spillway as a step-up in the channel bed in the same manner as done in Section 12.6.1. If the depth behind the spillway is known, these problems are solved by merely writing the energy equation across the spillway and solving for the unknown depth at the toe by ignoring friction losses.

One other factor is the supercritical nature of the flow downstream of the spillway. In most cases, the channel in the immediate vicinity of the toe must be protected from scour and some measures must be taken to facilitate a hydraulic jump in this area in order to reduce excess kinetic energy and transition back to the normal subcritical depth in the channel that prevails in most practical cases. This issue will be discussed in the next chapter.

12.6.4.5 Broad-Crested Weirs If the breadth and friction characteristics of the weir are such that critical depth actually occurs on the weir crest, then the weir is considered to be broad crested. Of course, in this case, the critical section is actually the control on the discharge for the available energy above the crest elevation. Assuming a rectangular section (Figure 12.30), then we recall that at critical depth $V = \sqrt{gy_c}$, so that the discharge through the area of critical flow would be given by

$$Q = AV = Ly_c\sqrt{gy_c} = L\sqrt{g}\,y_c^{3/2} \tag{12.62}$$

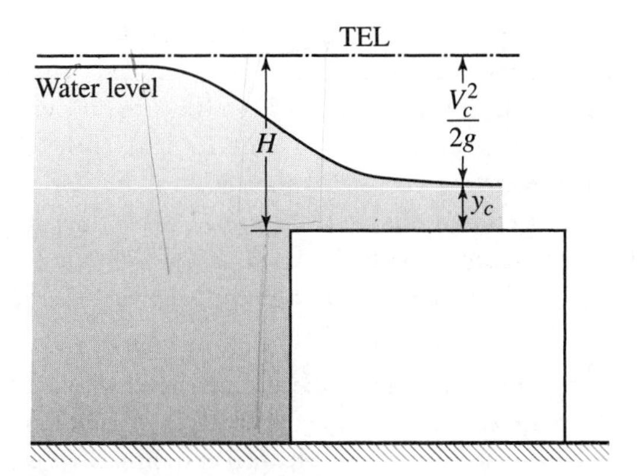

Figure 12.30 Principles of the broad-crested weir.

where L is the width of the channel section, or the length of the weir crest. We can write equation (12.62) in the more familiar form of the Francis equation by noting that for rectangular sections

$$E_c = H = 3/2y_c$$

Then simple substitution yields that

$$Q = \sqrt{g}\,L(2/3H)^{3/2} \tag{12.63}$$

In BG units, equation (12.63) reduces to

$$Q = 3.09LH^{3/2} \tag{12.64}$$

which is equivalent to the Francis weir formula with a C value of 3.09. However, since in the case of broad-crested weirs, contraction and friction losses are quite often very high, the C value is frequently reduced to values in the range of 2.5.

12.7 The Discharge Problem

Another aspect of hydraulic controls can be illustrated through the examination of an uncontrolled channel leading from a storage area, or lake, as shown in Figure 12.31. The discharge in the channel will be determined through a combination of the energy level in the lake and the resistance capacity of the channel. Since this situation represents a case of an extreme restriction in cross-sectional area as the flow goes from the wide lake to the more narrow

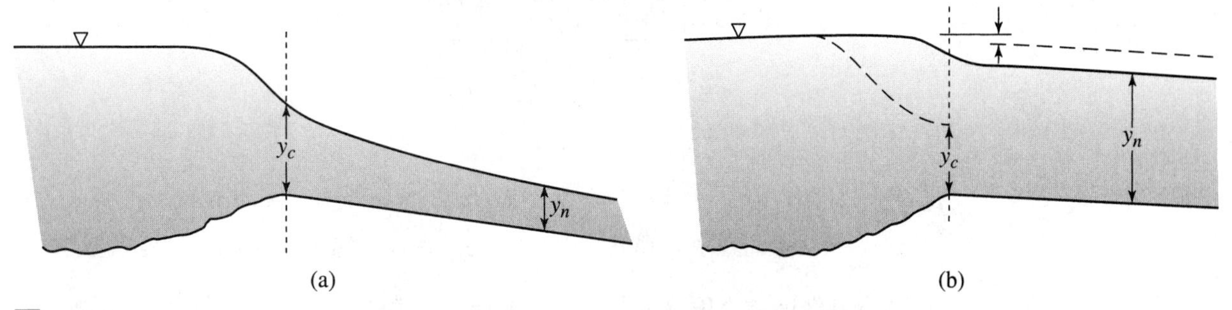

Figure 12.31 (a) Uncontrolled reservoir outlet to a steep slope. (b) Uncontrolled reservoir outlet on a mild slope.

channel, one would expect critical depth to occur at the lake outlet (channel entrance) and thus to control the flow in the channel. However, resistance considerations will show that this condition can only occur if the channel is steep (i.e., that it can accept all of the flow that the critical section would convey to it as in Figure 12.31a). If the channel is mild, then the subcritical normal depth in the channel would control and would thus tend to submerge or "drown" the critical section as shown in Figure 12.31b. Thus, in the case of the mild channel slope, the discharge would be reduced from that which would occur under the steep case. The situation can best be illustrated through the following example.

Example 12.17

A lake is drained by an uncontrolled trapezoidal channel with a bottom width of 10 ft and side slope of 2:1. The channel runs on a slope of 0.0015 ft/ft and has a Manning n of 0.035. If the lake level is 10 ft above the invert of the channel entrance, what is the discharge in the channel?

Solution: First, we must determine if the channel slope is steep or mild. This problem involves a trial-and-error solution. Let the critical depth be $y_c = (2/3)E = 6.67$ ft. Corresponding to this value of the flow depth, $A = 155.67\ \text{ft}^2$, and $B = 36.68$ ft. Therefore,

$$\frac{Q^2 B}{gA^3} = 1 = \frac{Q^2 \times 36.68}{32.2 \times (155.67)^3}$$

This yields $Q = 1819.78$ cfs. Then $V = 11.69$ ft/s and $E = 8.79$ ft, which is less than 10 ft. So another trial is needed.

Let $y_c = 7.5$ ft. Then $A = 187.5\ \text{ft}^2$, $B = 40$ ft, $Q = 2303.56$ cfs, $V = 12.28$ ft/s, and $E = 9.84$ ft, which is still less than 10 ft. Now, try $y_c = 7.6$ ft. Then $A = 191.5\ \text{ft}^2$, $B = 40.4$ ft, $Q = 2365$ cfs, $V = 12.35$ ft/s, and $E = 9.97$ ft, which is close to 10 ft. For this depth, $P = 43.97$, $R = 4.35$, and using Manning's equation, the critical slope is

$$S_c = \left[\frac{0.035 \times 12.35}{1.49 \times (4.35)^{2/3}}\right]^2 = 0.0118$$

which is greater than the channel slope of 0.0015, so our slope is mild. Then the flow in the channel will be controlled by the normal depth rather than the critical depth. However, the combination of the normal depth and its velocity head plus any losses at the channel entrance still cannot exceed the available energy in the lake. Thus, if we ignore any entrance losses, we can write

$$10 = y_2 + \frac{V_2^2}{2g} = y_2 + \frac{\left(\frac{1.49}{n} R^{2/3} S^{1/2}\right)^2}{2g}$$

$$= y_2 + \frac{\left(\frac{1.49}{0.035}\left(\frac{10y_2 + 2y_2^2}{10 + 4.47y_2}\right)^{2/3} (0.0015)^{1/2}\right)^2}{2g}$$

$$= y_2 + 0.0422\left(\frac{10y_2 + 2y_2^2}{10 + 4.47y_2}\right)^{4/3}$$

Solving by trial and error, we obtain $y_2 = 9.6$ ft. From this value we obtain that the area is $A = 280.32\ \text{ft}^2$. And since $E = 10 = y_2 + \frac{V_2^2}{2g}$, then $\frac{V_2^2}{2g} = 0.4$ ft (ignoring entrance losses). Thus, $V = \sqrt{0.4(2g)} = 5.07$ ft/s. Then $Q = VA = 1421$ cfs.

We can note that the actual discharge in the channel is only about 60% of the maximum that would occur if the channel had been steep and the critical depth had controlled. We can also note that the normal depth of 9.6 ft is 2 ft greater than the critical depth of 7.6 ft and thus we can easily see that the normal depth will simply drown the critical section in this case. Lastly, we note that a slightly more accurate answer could have been obtained if we had included some contraction, or entrance losses at the channel. A loss coefficient of 0.1 to 0.3 can be employed by merely multiplying the velocity head term in the energy equation above by the appropriate coefficient.

12.8 Dimensionless Representation of Specific Energy Diagram

In the general case, the discharge per unit width, q, varies considerably along the channel route. Therefore, we may need to construct an unlimited number of specific energy curves, each corresponding to a certain discharge, to cover the range of its variation. Crausse (1952) gave a dimensionless analysis for the specific energy equation, providing a single curve incorporating the various values of q. For wide and rectangular sections, the specific energy equation is written as

$$E = y + \frac{q^2}{2gy^2} \tag{12.65}$$

Dividing by y_c,

$$\frac{E}{y_c} = \frac{y}{y_c} + \frac{q^2}{2gy^2y_c} \tag{12.66}$$

Letting $E/y_c = E'$ and $y/y_c = y'$, and substituting $q = \sqrt{gy_c^3}$, equation (12.66) is written as

$$E' = y' + \frac{1}{2y'^2} \tag{12.67}$$

Equation (12.67) constitutes the dimensionless form of the specific energy equation. Plotting E' versus y', a variable scale E-y curve is obtained, as shown in Figure 12.32. This dimensionless curve is applicable to any problem once the critical depth is known. As stated previously, it is the basis for the chart given in Figure 12.9.

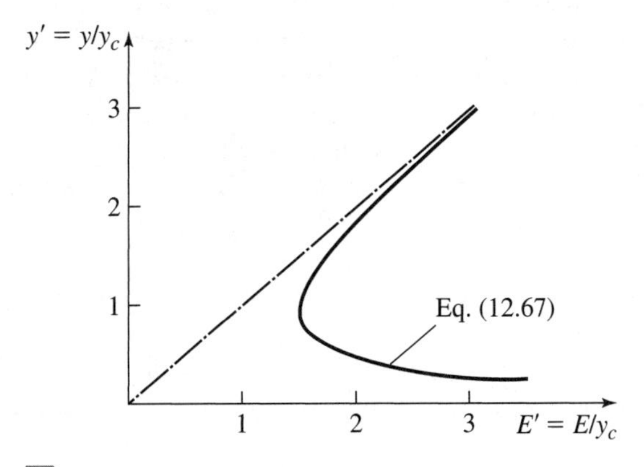

Figure 12.32 Dimensionless representation for the specific energy diagram.

READING AID

12.1. State the different types of energy which are generally encountered in open channels.

12.2. Define the total energy, E_T, in open channels and write the general equation for energy conservation. If the flow in an open channel is uniform, what is the relationship among the bed slope, the water surface slope, and the total energy line slope?

12.3. What is the difference between the total energy, E_T, and the specific energy, E, in open channels? Write down the specific energy equation in terms of y and Q.

12.4. What is meant by the specific energy diagram? How do the static energy and the kinetic energy vary with the depth of flow?

12.5. As the depth of flow increases, do you expect the kinetic energy to increase or decrease? Why or why not?

12.6. Define the alternative depths. What makes the flow travel with high stage or low stage depth?

12.7. At the minimum specific energy, what is the value of the Froude number?

12.8. If the depth of flow is greater than the critical depth, do you expect the Froude number to be greater or less than unity? Also, do you expect the velocity to be greater or less than the critical velocity?

12.9. Draw a family of curves of specific energy diagram for various values of Q. Explain the effect of increasing Q on the specific energy diagram.

12.10. Derive a general formula for the critical conditions in open-channel flow. Then develop an expression for the critical depth in rectangular channels.

12.11. Explain different approaches to determine the critical depth in open channels.

12.12. For a wide rectangular channel, express the minimum energy in terms of the critical depth.

12.13. Express the critical velocity in a trapezoidal section in terms of hydraulic depth and in a rectangular section in terms of flow depth.

12.14. Prove that the critical flow conditions are satisfied either when the energy is minimum for a given discharge or when the discharge is maximum for a given energy.

12.15. When a sudden drop is encountered in a channel bed, do you expect the specific energy to decrease or increase? Why?

12.16. If a small step is introduced to a channel bed where the flow is subcritical, do you expect the depth of flow to increase or decrease? Why or why not?

12.17. If a small drop is introduced to a channel bed where the flow is supercritical, do you expect the depth of flow to increase or decrease? Why or why not?

12.18. When the width of a rectangular channel increases while the discharge remains constant, how are the specific energy and total energy affected?

12.19. If the width of a rectangular section is increased, do you expect the depth of flow to increase or decrease in the following cases: (a) the flow is subcritical, and (b) the flow is super critical?

12.20. Explain why the flow at the critical condition is unstable and why it tends to either be subcritical or supercritical?

12.21. Derive the dimensionless form of the specific energy equation. Give some examples for the usefulness of the dimensionless representation of the specific energy diagram.

Problems

12.1. A trapezoidal channel having a bed width of 6 m and a side slope of 1:1 is discharging water at a rate of 8.0 m^3/s. Calculate the specific energy of water if the depth of flow is 2.0 m. What is the alternate depth to the observed depth?

12.2. For a trapezoidal channel with a bed width, $b = 6.0$ ft and a side slope, $t = 2$, find the critical depth if

a. $Q = 150$ ft^3/s
b. $Q = 75$ ft^3/s
c. $Q = 300$ ft^3/s

12.3. A cement lined rectangular channel 6.0 m wide carries water at the rate of 11.0 m^3/s. Determine the critical depth and the critical velocity.

12.4. A broad-crested weir of a height h is placed in a channel of width b. If the upstream depth is y_1 and the upstream velocity head and frictional losses are neglected, develop a theoretical equation for the discharge in terms of the upstream depth of flow. Assume the flow over the broad-crested weir to be critical.

12.5. A trapezoidal channel having a bed width of 7.0 ft and a side slope of 3:2 is discharging 300 ft^3/s of water. If the specific energy is equal 5.7 ft, calculate the alternate depths and their corresponding slopes to satisfy these conditions. Take $n = 0.025$.

12.6. A rectangular channel 5 m wide carries a discharge of 25 m^3/s. The observed velocity at a particular point in the channel is 2.5 m/s:

a. What is the specific energy of the flow at this point?
b. What are the alternate depths at this point?
c. What is the critical depth in the channel?

12.7. A trapezoidal channel with a bottom width of 5 m and side slopes $t = 2$ runs on a slope of 0.0005 and carries a discharge of 50 m^3/s at normal depth. Take Manning's n as 0.021 (metric).

a. What is the specific energy of the flow in the channel?
b. Is the flow subcritical or supercritical?
c. What is the alternate depth to the normal depth?
d. What is the critical depth in the channel?

12.8. A rectangular channel carries a discharge of 5.66 m^3/s. Find the critical depth y_c and critical velocity V_c for (a) a width of 3.66 m, and (b) a width of 2.74 m. What slope will produce critical velocity in case (a) if $n = 0.02$?

12.9. A channel of rectangular section, 7.5 ft wide, is discharging water at a rate of 200 ft^3/s with an average velocity of 3.0 ft/s. Find (a) the specific energy of the flowing water, (b) depth of water when the specific energy is minimum, (c) velocity of water when the specific energy is minimum, (d) minimum specific energy of the flowing water, and (e) type of flow.

12.10. Water flows with critical depth in a rectangular channel with a critical velocity of 2.0 m/s. What is the depth of flow?

12.11. Express the critical depth and critical velocity in terms of the specific energy for a triangular channel which has equal side slopes of n horizontal to 1 vertical.

12.12. A very wide section has a flow per unit width of 32 ft^2/s. Calculate the critical depth and the corresponding specific energy. If this specific energy is the minimum energy for another channel with a 90° triangular section, calculate the critical depth and the critical discharge for the triangular section.

12.13. Water flows with a critical depth in a circular section of radius 0.6 m under a velocity of 3.0 m/s. What is the depth of flow?

12.14. For a parabolic section, find the relation between y_c and E_{min}.

12.15. A 1.2-m-diameter concrete pipe culvert carries a water discharge of 0.8 m^3/s. Determine the critical depth and its specific energy.

12.16. A channel with a compound cross-section consists of a semicircular bottom and two vertical walls. The distance between the walls is 1.2 m. If the discharge is 0.8 m^3/s, calculate the alternate depths corresponding to a specific energy of 1.0 m. Find the corresponding slopes required to develop a uniform flow. Take $n = 0.02$.

12.17. A parabolic channel, with a perimeter equation of $y = 2x^2$, carries a discharge of 8 m^3/s. Calculate the critical depth and the critical velocity.

12.18. Referring to Problem 12.15, knowing that $n = 0.025$, determine the critical slope for $Q = 0.8$ m^3/s.

12.19. Referring to Problem 12.17, determine the critical slope. Take $n = 0.025$.

12.20. A channel has a triangular cross section with an apex angle of 90°. Draw the Q-y curve for a specified energy of 3.0 m. Determine the critical depth and the corresponding discharge.

12.21. Water flows in a rectangular channel at a rate of 250 ft^3/s. The depth of flow is 5 ft and the bed width is 10 ft. If the channel bed elevation is increased by 0.5 ft using a smooth step, what is the depth and type of flow at the raised section?

12.22. Water flows with a velocity of 3.0 m/s and a depth of 3.0 m in a rectangular channel. (a) What is the change in water surface elevation produced by a gradual upward change in bed elevation (hump) of 30 cm? (b) What would be the change in the water surface elevation if there were a gradual drop of 30 cm?

12.23. Referring to Problem 12.22, what is the maximum height of the step-up that would not affect the upstream flow condition? What is the minimum width that could exist with a 0.3-m step-up without changing the upstream condition, knowing that the upstream width is 6.0 m?

12.24. A trapezoidal channel has a bed width of 6 ft and side slope of 3:2 and conveys water 4.5 ft deep with a mean velocity of 3.5 ft/s. If the channel floor is raised 0.75 ft downstream at the particular section,

a. What would be the depth at the raised section?
b. What is the maximum permissible change in the floor elevation without changing the upstream condition?

12.25. A rectangular channel expands smoothly from a width of 5 ft to 10 ft. Upstream of the expansion, the depth of flow is 4 ft and the velocity of flow is 4 ft/s. Estimate the depth of flow after the expansion. Use an expansion coefficient of 0.50.

12.26. A trapezoidal channel with a bed width of 6.0 m and a side slope of 3:2 conveys a discharge of 20 m^3/s at a depth of 3.0 m. A smooth transition to a rectangular section 8.0 m wide, accompanied by a gradual lowering of the channel bed of 0.5 m, is introduced. (a) Find the water depth within the rectangular section neglecting all losses. (b) What is the type of flow in the rectangular section?

12.27. Water flows at a depth of 5 ft and a velocity of 3 ft/s in an 8-ft-wide rectangular channel. Find (a) the height of the hump required to produce critical flow without affecting the upstream depth, and (b) the depth of water over the hump when its height is one half of that in (a).

12.28. Water flows in a rectangular channel 8 m wide at a depth of 3.5 m. The discharge is 100 m^3/s. If the channel width is decreased gradually to a width of 7 m,

a. What is the depth in the constricted section? Use a contraction coefficient of 0.3.
b. What is the maximum height the channel bed can be raised without disturbing the upstream conditions?

12.29. A highway was subjected to a flood. The water surface elevation, at point *A* as shown in Figure P12.29, upstream of the highway was measured to be 30 m. The elevation, at point *B*, on the top of the crown of the pavement is 29.5 m. Estimate the discharge over a 50-m-long stretch of highway with this elevation. What was the depth of flow at the crown of the highway, knowing that critical flow conditions existed through the pavement's crown? Neglect the velocity head at point *A*.

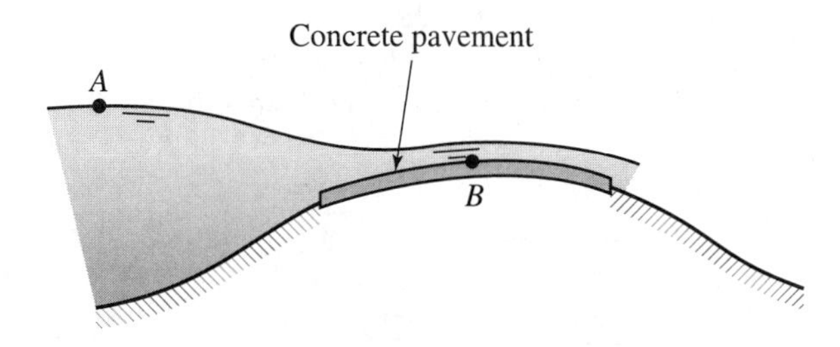

Figure P12.29 The highway given in Problem 12.29.

12.30. Water is flowing in a 10-ft-wide rectangular channel at a depth of 6 ft. A 1.5-ft step-up in the channel bottom is encountered and the water level over the step (with reference to the upstream channel invert) drops by 0.5 ft as shown in Figure P12.30. Ignoring losses, what is the discharge in the channel?

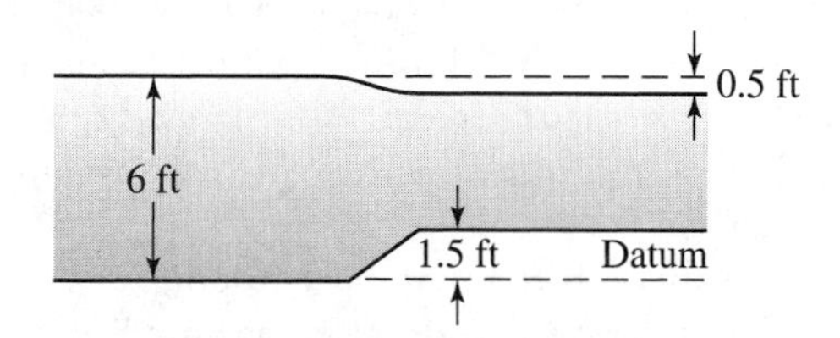

Figure P12.30 The channel rise for Problem 12.30.

12.31. A sharp-crested suppressed rectangular weir crosses a channel 10 m wide. The weir height is 7 m. If the head on the weir is measured to be 1.5 m, what is the discharge in the channel?

12.32. If the weir in Problem 12.31 was a rectangular notched weir with a notch opening of 3 m, what would be the discharge in the channel?

12.33. If the weir in Problem 12.31 functioned as a broad-crested weir, what would be the estimated discharge in the channel?

12.34. If the structure in Problem 12.31 was a rectangular notched weir (two end contractions) and the notch width was 6 m, what would be the estimated discharge in the channel?

12.35. If the structure in Problem 12.31 was a Cipolletti weir, and the observations were in BG units rather than SI, what would be the discharge in the channel? Take the base width of the weir as 5 ft.

12.36. A spillway is to be designed for a flood control reservoir. The design must pass the 200-year event of 50,000 cfs while not allowing the upstream water elevation to exceed 800 ft. The channel can be approximated by a rectangular section 500 ft wide.

a. What should be the elevation of the spillway crest in order to meet the design specifications?
b. What would be the elevation of the theoretical weir crest associated with the spillway?

12.37. A 7-m-high spillway is placed across a rectangular channel as a means of measuring discharge. If the actual head on the spillway crest is 0.92 m and the Francis *C* value is 3.32,

a. What is the specific discharge in the channel?
b. Ignoring losses across the spillway, what is the flow depth at the toe?

12.38. Water is flowing in a rectangular channel 20 ft wide at a discharge of 300 cfs. The channel is running on a slope of 0.0025 with a Manning n value of 0.030. A small spillway 15 ft high is placed across the channel. If the Francis *C* value for the spillway is 3.1 (with reference to the actual head on the crest),

a. What is the normal depth in the channel?
b. To what depth will the water rise behind the spillway?
c. Ignoring losses, what will be the depth at the toe of the spillway?
d. What is critical depth in this situation?

12.39. Water is flowing in a channel that can be approximated as a rectangular section 250 ft wide with a discharge of 10,000 cfs. A 100-ft-high spillway is erected across the channel with a Francis *C* value that is calibrated to be 3.25 (with reference to the actual head on the crest).

a. To what depth will the water rise behind the spillway?
b. What will be the theoretical velocity at the toe of the spillway?

12.40. The discharge from a small spring is to be measured with a triangular V-notch weir. If the head observed on the weir is 5 cm, what is the flow rate if

a. The notch angle is 90 degrees?
b. The notch angle is 60 degrees? Use a discharge coefficient of 0.58 in both cases.

12.41. Water is fed from a lake into an uncontrolled channel that runs for a long distance downstream as shown in Figure P12.41. The channel is an 8-ft rectangular section, running on a slope of 0.005 ft/ft, with a roughness coefficient of 0.030. If the water level in the lake is 12 ft above the channel invert at the lake outlet, what is the discharge in the channel?

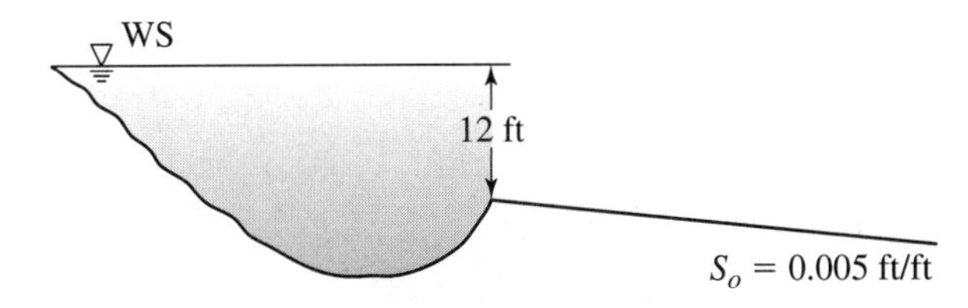

Figure P12.41 A lake feeding a channel for Problem 12.41.

References

Bradley, J. N. and Peterka, A. J., 1957. The hydraulic design of stilling basins, *Proceedings of the American Society of Civil Engineers*, 83(HY5): 1401–1406.

Crausse, E., 1952. Sur une Propriété des veines liquides horizontal en canal uniforme. *Compt. rend. de l'Acad. Française*, vol. 234.

Gerhart, P. M., Gross, R. J., and Hochstein, J. I., 1992. *Fundamentals of Fluid Mechanics*, 2nd ed. Addison-Wesley Publishing Company, Inc., New York, NY.

Henderson, F. M., 1966. *Open Channel Flow*, Macmillan, New York.

Hwang, N. H. C. and Houghtalen, R. J., 1996. *Fundamentals of Hydraulic Engineering Systems*, 3rd ed., Prentice Hall, Upper Saddle River, N. J.

Morris, H. M. and Wiggert, J. M., 1972. *Applied Hydraulics in Engineering*, Wiley, New York.

Roberson, J. A., Cassidy, J. J., and Chaudhry, M. H., 1998. *Hydraulic Engineering*, Second Edition, Wiley, New York, NY.

Sturm, T.W., 2001. *Open Channel Hydraulics*, McGraw-Hill, New York.

U.S. COE, 1982. HEC-2 Water Surface Profile Computations, User's Manual, U.S. Army Corps of Engineers, Hydrologic Engineering Center, Davis, Calif.

CHAPTER 13

MOMENTUM PRINCIPLES IN OPEN-CHANNELS

Flow over Toledo Bend spillway and downstream hydraulic jump. (Courtesy of Toledo-Bend.com/Frank Dutton.)

The momentum equation is derived from Newton's second law of motion. Its mathematical form shares some similarities with the energy equation when applied to certain flow problems. The two equations differ, however, in two respects: (1) The velocity distribution coefficients involved in these equations are different, and (2) the momentum equation represents an evaluation of the second law from a temporal perspective while the energy equation evaluates the second law spatially. Thus, the momentum equation can be applied in multiple dimensions while the Bernoulli equation is one dimensional.

It is worth recalling that energy is a scalar quantity whereas momentum is a vector quantity. In general, the energy principle is simpler and easier to apply. However, it is advantageous to apply the momentum principle to those problems that involve significant energy changes, as in the case of a hydraulic jump.

13.1 Momentum Function

For a body of water enclosed between two sections 1 and 2, the momentum equation, as shown in Chapter 5 is derived from an application of Newton's second law:

$$\mathbf{F_x} = m\mathbf{a_x} \tag{13.1}$$

where $\mathbf{F_x}$ = total force developed, m = mass of water, and $\mathbf{a_x}$ = acceleration of the water. In Chapter 5 it was shown that, for one dimensional flow in open channels, equation (13.1) is reduced to

$$F = \rho Q(V_2 - V_1) \tag{13.2}$$

where ρ = density of water, Q = discharge, and V = velocity. In Chapter 5, the concept that the velocity distribution in most channels and pipes is not uniform was discussed, and the momentum correction factor β was introduced to account for this fact. However, it was also mentioned that in many real-world situations, β is sufficiently close to unity that it can be considered so without appreciable error. This assumption is implied in the development of equation (13.2). It was further shown that in the case of free surface flow, considering the applicable forces to be hydrostatic, equation (13.2) can be written as

$$\frac{Q^2}{gA_1} + A_1\bar{h}_1 = \frac{Q^2}{gA_2} + A_2\bar{h}_2 \tag{13.3}$$

where A is the cross-sectional area of flow and $\bar{h}$ is the depth of the centroid of the flow area below the water surface, and g is the acceleration of gravity. Subscripts 1 and 2 refer to sections 1 and 2, respectively. The term $\frac{Q^2}{gA} + A\bar{h}$ is known as the momentum function (M) with dimensions of L^3.

The two sides of equation (13.3) are identical, but each refers to a different section. If both sides of the equation are multiplied by γ, then it can be written as

$$F_1 = F_2 \tag{13.4}$$

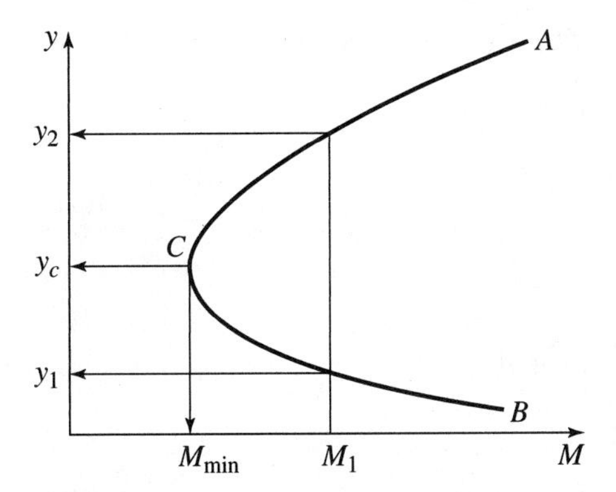

Figure 13.1 The momentum function.

in which

$$F = \gamma\left(\frac{Q^2}{gA} + A\bar{h}\right) \tag{13.5}$$

The first term in equation (13.5) (i.e., $\gamma Q^2/gA$) represents the dynamic force of the flow passing through the channel section per unit time, while the second term, $\gamma A\bar{h}$, represents the hydrostatic force. The quantity, F, is called the *pressure-momentum force,* and for a given channel section with specified discharge, it is a function of the depth of flow only.

The momentum-depth (M-y) diagram is a graphical representation of the momentum function, M, against the depth of flow, y. A typical M-y diagram is given in Figure 13.1. Like the specific energy diagram, this curve has two limbs, AC and BC. For any given value of M, say M_1, two depths are encountered, y_1 and y_2. These two depths of the same value of the momentum function and having the same discharge are called the conjugate or sequent depths. At point C, the momentum flux (and momentum-impulse force) is minimum and the depth of flow is critical, as shown in Figure 13.1. As will be demonstrated later, the two conjugate depths constitute the initial and final depths of the hydraulic jump. Similar to the alternate depths, one of the conjugate depths, y_1, is lower than the critical depth, y_c, while the other depth, y_2, is higher.

Differentiating equation (13.5) with respect to y and equating the result to zero to determine the condition of flow when the momentum flux is minimum, one obtains

$$\frac{dM}{dy} = -\frac{Q^2}{gA^2}\frac{dA}{dy} + \frac{d}{dy}(A\bar{h}) = 0 \tag{13.6}$$

The term $d(A\bar{h})$ is the change in the moment of area around the free surface due to a change of dy in the depth of flow. Hence,

$$\frac{d(A\bar{h})}{dy} = \frac{\left[A(\bar{h} + dy) + B\frac{(dy)^2}{2}\right] - A\bar{h}}{dy} \tag{13.7}$$

Neglecting the term $B(dy)^2/2$, equation (13.7) reduces to

$$\frac{d}{dy}(A\bar{h}) = A \tag{13.8}$$

Noting also that

$$\frac{dA}{dy} = B \tag{13.9}$$

and substituting equations (13.8) and (13.9) into equation (13.6), one gets

$$\frac{dM}{dy} = -\frac{Q^2B}{gA^2} + A = 0 \tag{13.10}$$

Equation (13.10) can be written as

$$\frac{Q^2B}{gA^3} = 1 \tag{13.11}$$

Equation (13.11) shows that the critical flow condition is satisfied at the minimum value of the momentum-impulse force. Thus, for a given section of constant discharge, the specific energy and pressure-momentum force are minimum at the critical flow condition. On the other hand, for a given specific energy or momentum force the discharge is maximum at the critical condition.

The M-y diagram is derived from the momentum function: $M = \frac{Q^2}{gA} + A\bar{h}$. For a given cross section and a known discharge, different values for the depth of flow, y, are assumed and then the corresponding values for the momentum function are calculated recalling that for small bed slopes, $\bar{y} \approx \bar{h}$.

Example 13.1

A 2.0-m-wide rectangular channel carries a discharge of 4.0 m^3/s with a depth of flow of 1.0 m. Determine the momentum-impulse force, the critical depth, and the conjugate depth.

Solution:

$$M = \frac{Q^2}{gA} + A\bar{h} = \frac{16.0}{9.81 \times 2} + 2 \times 0.5 = 1.815 \text{ m}^3$$

Then $F = \gamma M = 9810 \text{ N/m}^3(1.815 \text{ m}^3) = 17805 \text{ N}$.

To determine the critical depth and the conjugate depth, the momentum function diagram is constructed. Assuming different values for y, the momentum function is computed as shown in Table 13.1. Figure 13.2 presents the M-y diagram, from which the critical depth is 0.74 m and the conjugate depth is 0.54 m. The critical depth may also be calculated as

$$y_c = \sqrt[3]{\frac{q^2}{g}} = \sqrt[3]{\frac{4.0}{9.81}} = 0.74 \text{ m}$$

TABLE 13.1 Momentum Function for Different Values of y (Example 13.1)

y (m)	0.2	0.4	0.6	0.8	1.0	1.2	1.4	1.6
M (m^3)	4.12	2.2	1.72	1.66	1.815	2.12	2.54	3.07

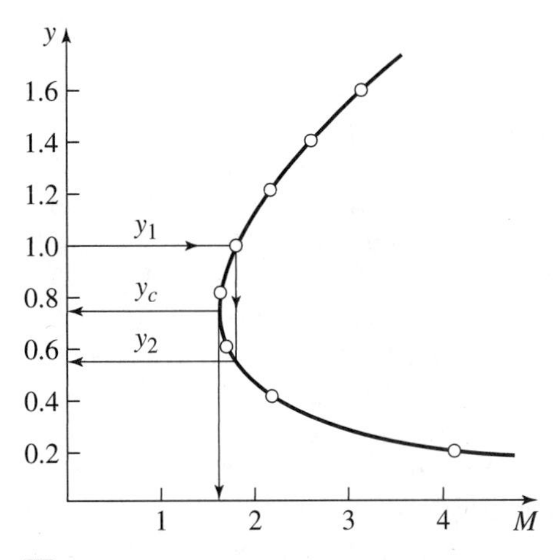

Figure 13.2 The *M*-*y* diagram (Example 13.1).

Example 13.2

Construct the *M-y* diagram for a triangular channel with an apex angle of 90° and a discharge of 10.0 m³/s.

Solution: The channel section is given in Figure 13.3a. Substituting into equation (13.5), excepting the γ, where $\bar{h}$ is equal to $y/3$, the momentum function is given as

$$M = \frac{10.194}{y^2} + \frac{y^3}{3}$$

Assuming different values for y and substituting in the above equation, the momentum function is computed as shown in Table 13.2. Figure 13.3b presents the *M-y* diagram.

TABLE 13.2 Momentum Function for Different Values of *y* (Example 13.2)

y	0.5	0.8	1.0	1.5	1.75	2.0	2.2	2.5	3.0	4.0	5.0
M	40.8	16.1	10.53	5.66	5.12	5.22	5.66	6.84	10.13	21.97	42.07

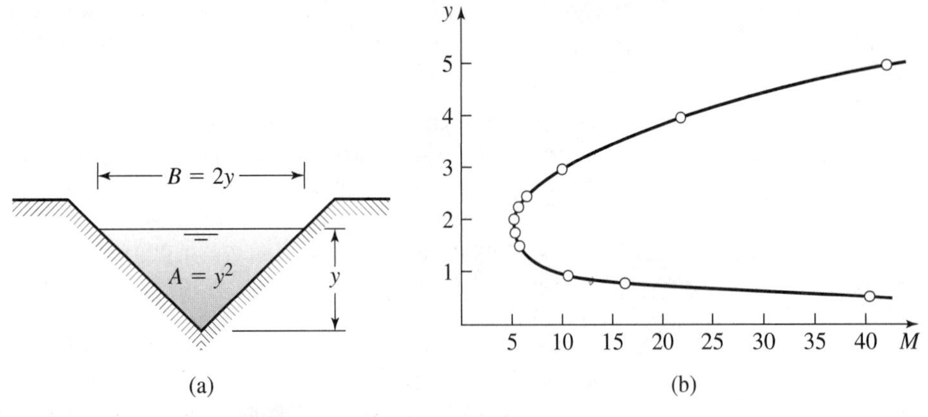

Figure 13.3 Triangular section and *M*-*y* diagram (Example 13.2).

13.2 Hydraulic Jump (Standing Wave)

The phenomenon of a sudden water rise, called hydraulic jump, or sometimes standing wave, has been investigated for more than two and a half centuries. Biden (1819) did an early study in which he reported some observations on the height of the hydraulic jump. Many pioneers have since contributed to the subject; outstanding among them are Bresse (1860), Darcy and Bazin (1865), Gibson (1913), Lindguist (1927), Citrini (1939), Blaisdell (1948), Forster and Skrinde (1950), to name but a few. The subject of the hydraulic jump is still under intensive investigation. Most of the analyses have been developed for horizontal and slightly sloping channels, where the weight of the hydraulic jump has a small effect on the overall driving force. Fortunately, many practical problems can be analyzed by neglecting the effect of bed slope. However, in some cases where the bed slope is relatively high, the component of the weight of the jump may affect the hydraulic balance of the system.

13.2.1 Condition for Formation of Hydraulic Jumps

A hydraulic jump is formed only if the depth of flow is forced to change from a depth y_1, which is lower than the critical depth, to another depth y_2, which is higher than the critical depth. In other words, the hydraulic jump takes place if the status of flow is changed from supercritical to subcritical flow (i.e., if the Froude number is changed from $F_n > 1$ to $F_n < 1$). Of course, the flow never changes its regime unless it is forced to do so. A flow may change its regime if the channel width is enlarged or reduced, the bed slope is altered, or some hydraulic structures are placed to regulate the flow.

Consider the case in which the bed slope, S_o, varies from $S_{o1} > S_c$ to $S_{o2} < S_c$, where S_c is the critical slope, as shown in Figure 13.4a. The initial flow is supercritical with y_{o1} less than y_c. After the point where the bed slope is reduced to S_{o2}, the flow is subcritical, with y_{o2} greater than y_c. The flow is thus forced to rise from $y_{o1} < y_c$ to $y_{o2} > y_c$ and a hydraulic jump will be formed. Other examples of the formation of the hydraulic jump are given in Figures 13.4b,c, and d.

13.2.2 Practical Applications of Hydraulic Jumps

A considerable amount of energy is lost through the hydraulic jump due to the rotational character of flow in the dead zone. Energy in open channels should be maintained unless the dissipation of energy is essential to meet certain considerations. Practical applications of the hydraulic jump include the following:

1. Dissipation of energy of water flowing over dams and weirs and below gates or through any other hydraulic structure to prevent possible destruction due to high velocities and hence eliminate scouring process.
2. Raising water levels in canals to enhance irrigation practices and reduce pumping heads.
3. Reducing uplift pressure under the foundations of hydraulic structures. This also involves reducing the thickness of concrete aprons by increasing the weight of the water above aprons.
4. Creating special flow conditions to meet certain needs at control sections; for example, gaging stations, flow measurement, and flow regulation.
5. Increasing discharges under gates by creating a bigger difference between water levels just upstream and downstream of the gates, as shown in Figure 13.4b.

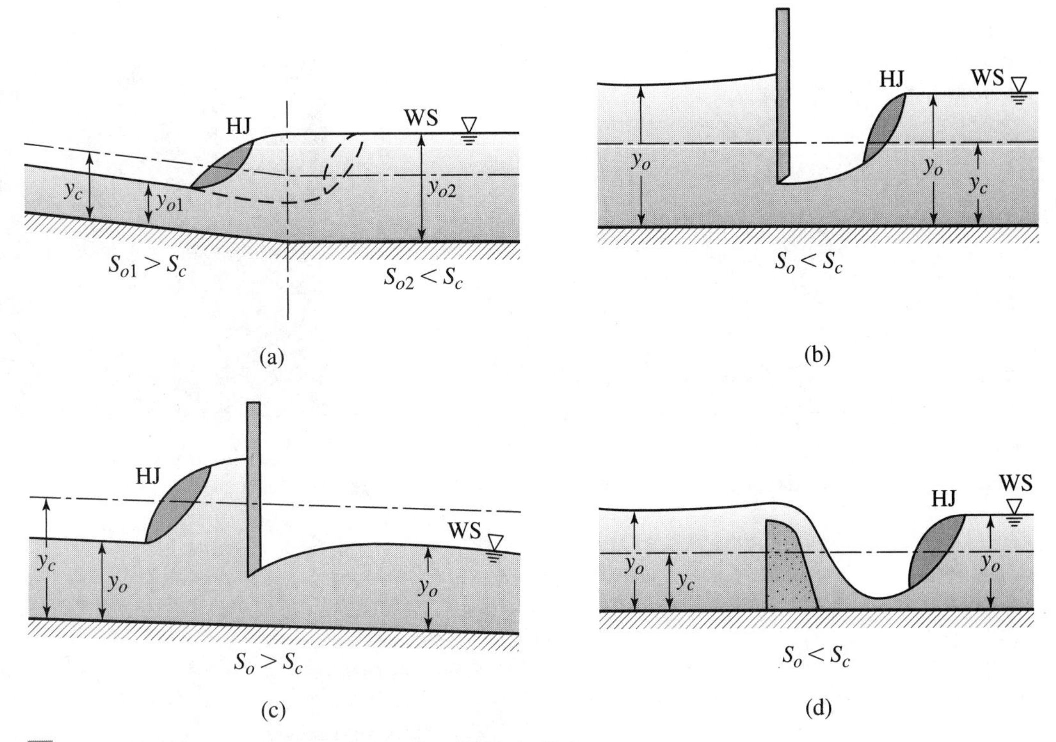

Figure 13.4 Different examples for the formation of the hydraulic jump.

6. Mixing of chemicals used for water purification and removing of air pockets in water supply systems to prevent air locking. Hydraulic jumps are used extensively in many industries which require mixing of different liquids, such as the dye and chemical industries.

13.3 Conjugate or Sequent Depths

Initial and final depths of a hydraulic jump are called conjugate or sequent depths in the sense that they occur simultaneously. A jump would only be stable if the sum of the momentum and pressure forces is exactly the same at the two conjugate depths, just upstream and downstream of the jump.

The relationship between the conjugate depths can be illustrated using the *M-y* diagram as shown in Figure 13.5. Let the initial supercritical depth be y_1 and its subcritical conjugate depth be y_1' associated with a momentum function value of M_1 as shown. Now, suppose that the actual subcritical depth in the channel is y_2, a value less than y_1'. As shown in the figure, the supercritical conjugate depth to y_2 is y_2', which is greater than the given depth y_1. Then, in order for the jump to occur, the supercritical depth must increase from y_1 to y_2' with a corresponding decrease in the momentum function from M_1 to M_2 as shown. Since it is possible for the flow to run downstream while slowing down and building up depth until y_2' is achieved, this situation is termed a "running jump." On the other hand, suppose the initial depth is still y_1 but a subcritical depth of y_3 is encountered. As shown in the figure, the supercritcal conjugate to this depth is y_3', which is smaller than y_1'. This corresponds to a case where the total forces in the subcritical region are greater than those in the supercritical zone. In this case, the jump would tend to be driven upstream until the

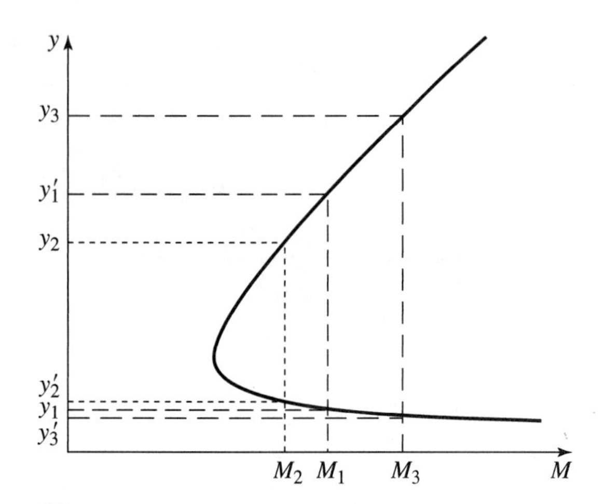

Figure 13.5 Momentum and conjugate depth relationships for the hydraulic jump.

Grand Coulee Dam, Washington State. (Courtesy of David Pearson/Visuals Unlimited.)

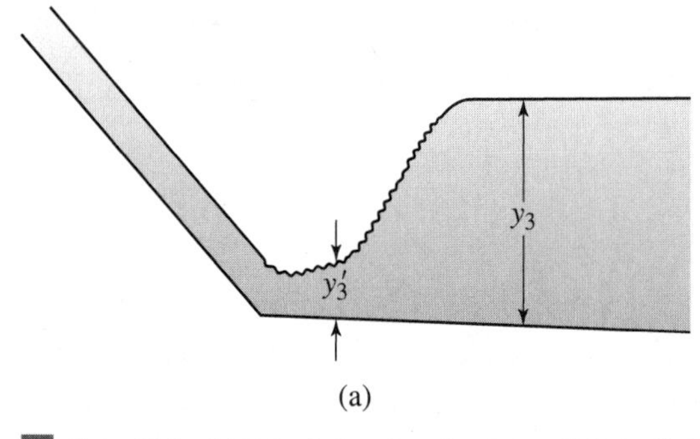

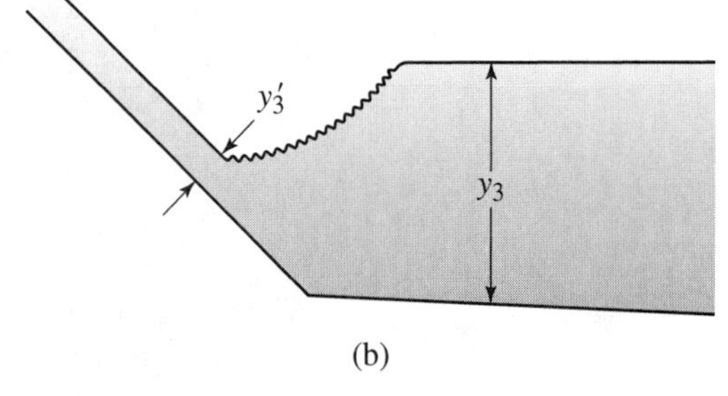

Figure 13.6 (a) Hydraulic jump forced upstream. (b) Hydraulic jump occurring on a steep slope.

required depth y_3' is encountered (Figure 13.6). As shown in Figure 13.6b, the jump may even be forced back to the extent that it occurs on the steep slope. However, it is quite possible that the jump would encounter the source of the supercritical flow (a sluice gate for instance) before the required depth is attained. In this case, the jump would become submerged, or "drowned."

For example, jump formation after a sluice gate (see Section 13.7) is considered in Figure 13.7. Let the bed slope of the channel under consideration be mild (i.e., the tailwater depth, y_2, is higher than the critical depth, y_c). Since y_1, the depth of flow at the vena contracted section just after the gate, is lower than y_c a hydraulic jump is possible.

There are different possibilities for the formation of the jump. The ideal case is encountered when the sequent depth of the jump, y_1', coincides with the tail water depth, y_2, as shown in Figure 13.7a. If y_2 is less than y_1', the jump moves downstream to a point where y_1' is equal to y_2. A rising curve will be formed after the gate to allow for such a rise in the water surface as shown in Figure 13.7b. If y_2 is bigger than y_1', the jump moves to the upstream and forms a submerged or drowned jump, as shown in Figure 13.7c.

One can draw the channel tailwater rating curve and the jump rating curve on the same graph as shown in Figure 13.8. If the two curves coincide for all values of y_2' and y_2, as shown in Figure 13.8a, then an ideal situation exists, which is rare in nature. The jump will always form at the same location without any need to stabilize it.

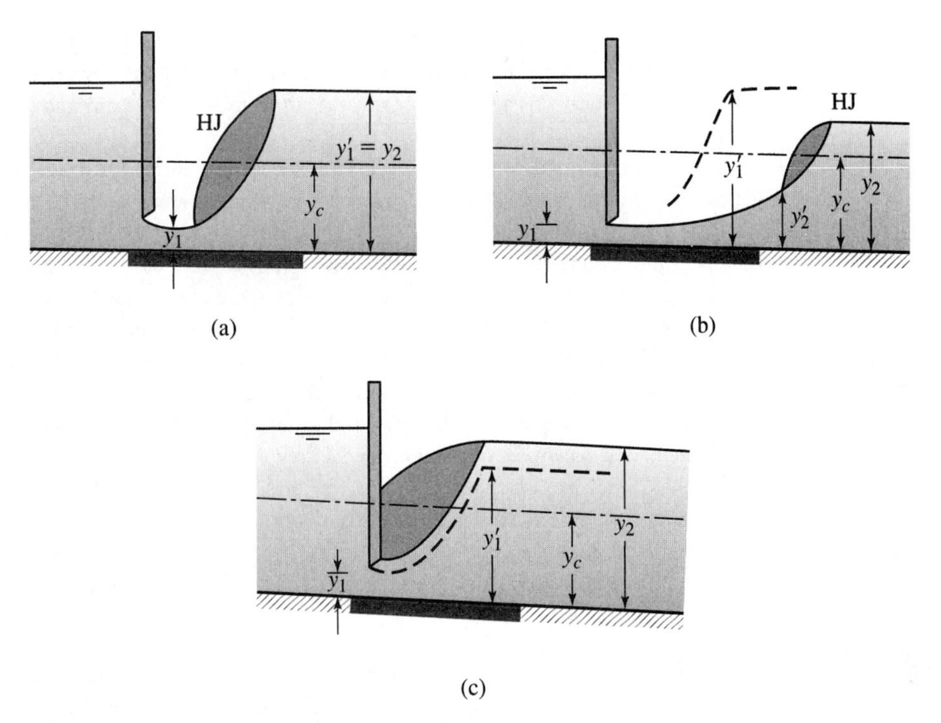

Figure 13.7 Different possibilities for jump location after a sluice gate.

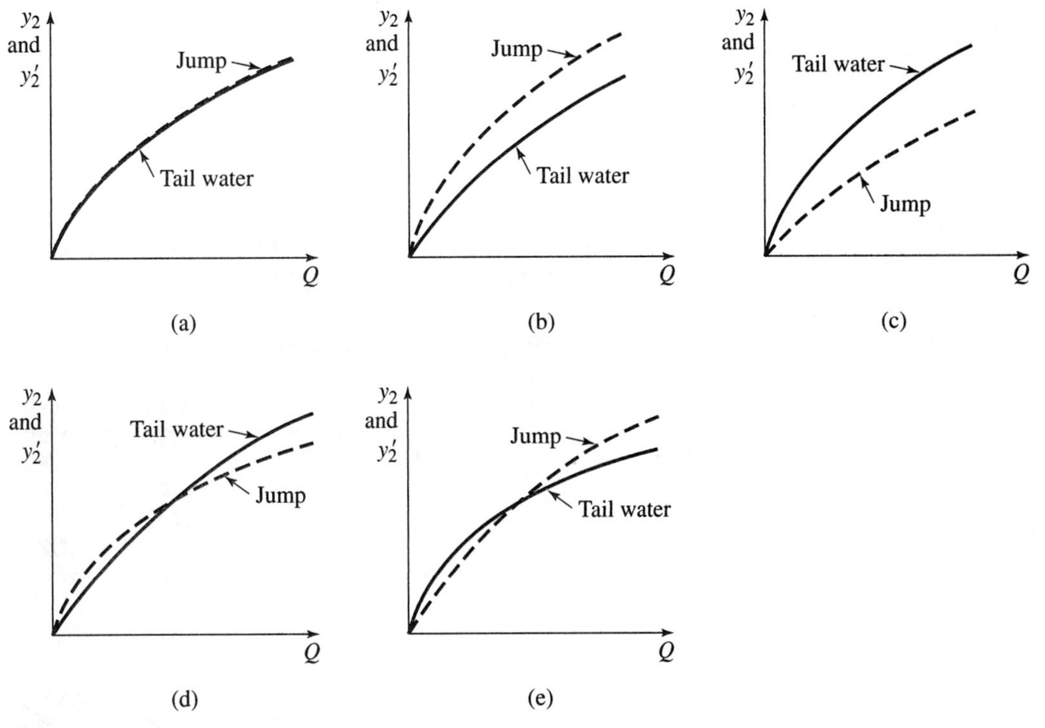

Figure 13.8 Different possibilities for tail-water and jump rating curves.

In Figure 13.8b, the jump rating curve (graph of discharge versus depth or elevation) is always higher than the tail-water rating curve and thus the jump runs downstream. In Figure 13.8c, conditions are adverse, and the jump moves upstream and is drowned as a submerged jump.

Sometimes the tail-water rating curve is below the jump rating curve at low values of discharge, whereas for higher discharges the tail-water rating curve is higher, as shown in Figure 13.8d. For low-discharge events the jump will run, while for high-discharge events the jump moves upstream or drowns. In the opposite case, as shown in Figure 13.8e, where the tail-water rating curve is higher at lower discharge events, and lower at higher discharge events, the jump would be submerged at low discharges and run for higher discharges.

For any cross section, and knowing either the initial or the final depth of a jump, equation (13.3) can be used to calculate the other depth. The M-y diagram can also be used to determine the unknown sequent depth. Graphical solutions of the jump equation for trapezoidal and circular channels have been developed by Massey (1961) and Thiruvengadam (1961) and are available in various open-channel hydraulics texts (see Sturm, 2001, for example). The two conjugate depths vary inversely. If the initial depth increases, the final depth decreases, and vice versa. In supercritical flow, as the depth of flow decreases, the specific force increases; while in subcritical flow, as the depth of flow decreases, the specific force decreases as well.

Example 13.3

A trapezoidal channel with 3.0 m bottom width and 1:1 side slope carries a discharge of 10.0 m^3/s at a slope of 12 cm/km. A hydraulic jump is formed in the channel with an initial depth of 0.6 m. Calculate the momentum function, the critical depth, and the conjugate depth.

Solution:

$$M_1 = \frac{Q^2}{gA_1} + A_1\bar{h}_1$$

$$= \frac{100}{9.81 \times 2.16} + (1.8 \times 0.3 + 2 \times 0.5 \times 0.36 \times 0.2)$$

$$= 5.33 \text{ m}^3$$

The critical depth is at the minimum value of the momentum function. Assuming different values for the depth of flow, the momentum function is calculated as shown in Table 13.3.

TABLE 13.3 Calculations for the M-y Diagram (Example 13.3)

y	A	$\bar{h}$	Q^2/gA	$A\bar{h}$	M
0.2	0.64	0.098	15.927	0.063	15.99
0.4	1.36	0.192	7.495	0.261	7.756
0.6	2.16	0.283	4.72	0.612	5.332
0.8	3.04	0.372	3.35	1.13	4.48
1.0	4.0	0.458	2.55	1.832	4.382
1.2	5.04	0.543	2.02	2.736	4.756
1.5	6.75	0.667	1.51	4.50	6.01
2.0	10.0	0.867	1.02	8.67	9.69
2.5	13.75	1.061	0.74	14.583	15.32

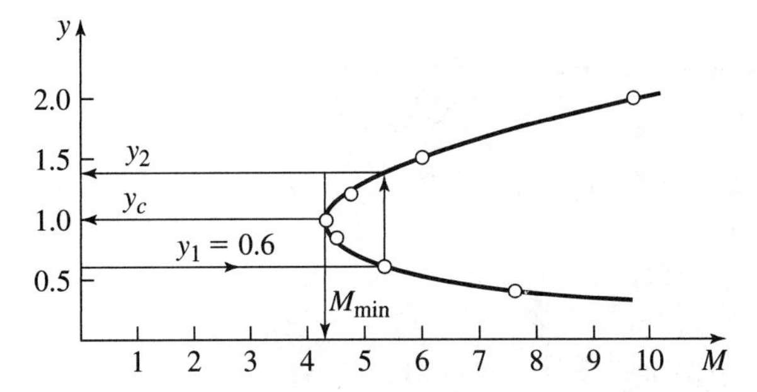

Figure 13.9 The M-y diagram (Example 13.3).

Figure 13.9 presents the momentum diagram, where $M_{\min} = 4.35$ m^3, $y_c = 0.934$ m, and the conjugate depth $= 1.36$ m.

13.3.1 Conjugate Depths in Rectangular or Wide Channels

Equation (13.3) is applicable to any cross section and can be used to evaluate the conjugate depths. However, for wide and/or rectangular sections, a direct relation between the two conjugate depths can be established that leads to a direct solution of the conjugate depths of a jump.

Consider a horizontal rectangular channel in which a hydraulic jump takes place as shown in Figure 13.10. Neglecting the friction forces, the momentum equation has been written as

$$\frac{Q^2}{gA_1} + A_1\bar{h}_1 = \frac{Q^2}{gA_2} + A_2\bar{h}_2 \tag{13.12}$$

Substituting the rectangular relationships by_1, by_2, bq, and $y/2$ for A_1, A_2, Q, and $\bar{h}$, respectively, and dividing by b, equation (13.12) is written as

$$\frac{y_1^2}{2} - \frac{y_2^2}{2} - \frac{q^2}{g}\left(\frac{1}{y_2} - \frac{1}{y_1}\right) = 0 \tag{13.13}$$

Equation (13.13) may be expressed as

$$\frac{y_1^2 - y_2^2}{2} - \frac{q^2}{g}\left(\frac{y_1 - y_2}{y_2 y_1}\right) = 0 \tag{13.14}$$

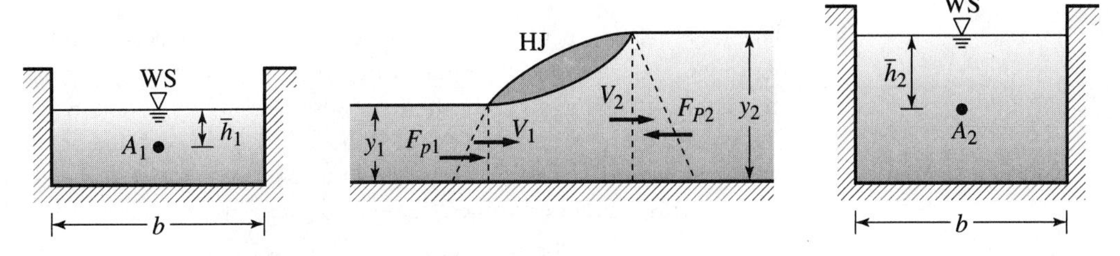

Figure 13.10 Conjugate depths in rectangular and wide sections.

or

$$\frac{(y_1 + y_2)(y_1 - y_2)}{2} - \frac{q^2}{g}\left(\frac{y_1 - y_2}{y_2 y_1}\right) = 0 \tag{13.15}$$

which can be written as

$$\frac{y_1 + y_2}{2} - \frac{q^2}{g}\left(\frac{1}{y_2 y_1}\right) = 0 \tag{13.16}$$

Multiplying equation (13.16) by $2y_1y_2$, one can write

$$y_2 y_1^2 + y_2^2 y_1 - \frac{2q^2}{g} = 0 \tag{13.17}$$

Therefore, by the quadratic formula

$$y_1 = \frac{y_2}{2}\left[-1 + \sqrt{1 + \frac{8q^2}{gy_2^3}}\right] \tag{13.18a}$$

or

$$y_1 = \frac{y_2}{2}[-1 + \sqrt{1 + 8F_{n2}^2}] \tag{13.18b}$$

or

$$y_1 = \frac{y_2}{2}\left[-1 + \sqrt{1 + 8\left(\frac{y_c}{y_2}\right)^3}\right] \tag{13.18c}$$

Equations (13.18b) and (13.18c) are just other forms of equation (13.18a). Equation (13.18b) is written in terms of the Froude number, F_n, while equation (13.18c) is written in terms of the critical depth, y_c. Of course, the two conjugate depths y_1 and y_2 in the set of equations (13.18) may replace each other. In other words, if the initial depth, y_1, is known and the final depth, y_2, is unknown, one can write

$$y_2 = \frac{y_1}{2}\left[-1 + \sqrt{1 + \frac{8q^2}{gy_1^3}}\right] \tag{13.19a}$$

$$y_2 = \frac{y_1}{2}[-1 + \sqrt{1 + 8F_{n1}^2}] \tag{13.19b}$$

$$y_2 = \frac{y_1}{2}\left[-1 + \sqrt{1 + 8\left(\frac{y_c}{y_1}\right)^3}\right] \tag{13.19c}$$

The above relation between y_1 and y_2 is known as the Belanger momentum equation. In its derivation, the following assumptions are made: (1) Uniform velocity distribution is assumed at the conjugate depths; (2) hydrostatic pressure distribution is assumed at the conjugate depths; (3) horizontal bed is assumed, or the slope is very small such that it can be neglected without affecting the overall balance; and (4) the boundary shear stress is negligible.

In terms of Froude numbers before and after a jump, equation (13.19b) can be written as

$$F_{n1}^2 = \frac{8F_{n2}^2}{(\sqrt{1 + 8F_{n2}^3} - 1)^3} \tag{13.20}$$

Example 13.4

A wide channel carries a specific discharge of 4.0 ft²/s. A hydraulic jump is formed with an initial depth of 0.5 ft. Find the final depth of the jump.

Solution: Applying equation (13.19a),

$$y_2 = \frac{0.5}{2}\left[-1 + \sqrt{1 + \frac{8(4.0)^2}{32.2(0.5)^3}}\right] = 1.18 \text{ ft}$$

The final depth of the jump is 1.18 ft.

Example 13.5

Water is stored behind a sluice gate to a depth of 10 feet. The downstream channel is a 10-ft-wide rectangular section running on a slope of 0.001 with a Manning n of 0.03. The depth in the vena contracted section after the gate is 1 ft. (a) Under these conditions is the gate clear or submerged? In other words, will a hydraulic jump occur, or be drowned? (b) Sketch the situation on the *M-y* diagram. (c) What is the minimum slope the channel can have for a jump to occur? (d) What is the maximum slope the channel can have for a jump to occur?

Solution: In the first case, applying the orifice equation to the gate,

$$V = \sqrt{2gH}$$

where H is the head on the gate $= y_1 - y_2 = 10 - 1 = 9$ ft. Then

$$V = \sqrt{2g(9)} = 24.07 \text{ ft/s}$$

Thus, $q = 24.07$ cfs/ft, $y_c = 2.62$ ft, and $Q = 240.7$ cfs. Using Manning's formula, the normal depth is computed in the channel as

$$10y_o\left[\frac{10y_o}{2y_o + 10}\right]^{2/3} = \frac{0.03 \times 240.7}{1.49(0.001)^{1/2}} = 153.25$$

This gives a value of the normal depth as 7.41 ft, which is greater than the critical depth. Now, to determine the conjugate depth for y_2,

$$y_2' = \frac{1}{2}\left[-1 + \sqrt{1 + \frac{8 \times (24.07)^2}{g(1)^3}}\right] = 5.52 \text{ ft} < y_o; \text{ so the jump is submerged.}$$

Likewise, we could find that $y_o' = 0.61 < y_2$, which leads to the same conclusion. Thus, the gate is not clear.

The *M-y* diagram is shown in Figure 13.11.

To compute the minimum slope such that the jump will not drown, it is required that $y_o = y_2' = 5.52$ ft; therefore, $A = 55.1$ ft², $P = 21.02$ ft, $R = 2.62$ ft, and $V = 4.37$ ft/s. Using Manning's formula, the slope is computed as

$$S_o = \left[\frac{0.03 \times 4.37}{1.49(2.62)^{2/3}}\right]^2 = 0.00214 \text{ ft/ft}$$

To compute the maximum slope, the required depth would be y_c since for slopes greater, then S_c the flow would remain supercritical. Thus, $S_{max} = S_c$, and $y_c = 2.62$ ft, $A = 26.2$ ft², $P = 15.24$ ft, $R = 1.72$ ft, and $V_c = 9.19$ ft/s. Using Manning's formula, the critical slope is computed as

$$S_c = \left[\frac{0.03 \times 9.19}{1.49(1.72)^{2/3}}\right]^2 = 0.0166 \text{ ft/ft}$$

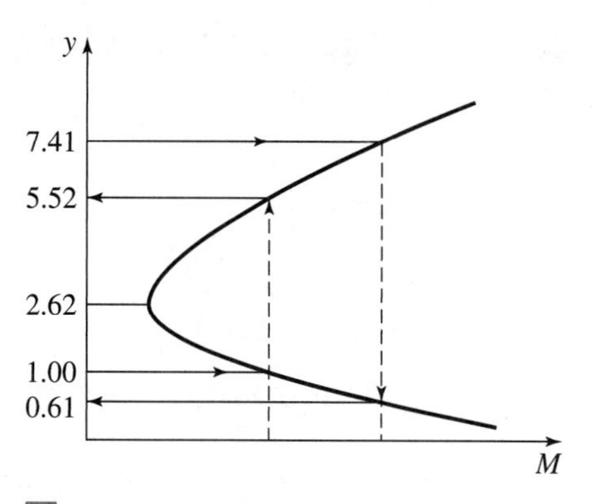

Figure 13.11 *M-y* diagram for Example 13.5.

13.3.2 Conjugate Depths versus Alternate Depths

Conjugate depths have the same pressure-momentum force, while alternate depths have the same specific energy. As the flow passes from a low-stage depth or supercritical condition to a high-stage depth or subcritical condition, considerable energy is dissipated or lost through the jump. Therefore, the two conjugate depths, just before and after the hydraulic jump, can never be alternate depths and vice versa.

Consider the momentum function and specific energy diagrams given in Figures 13.12a and b, respectively. Both curves are drawn for the same discharge. The two depths y_1 and y_2 are conjugate depths as shown in Figure 13.12a. However, they have different values for specific energy, E_1 and E_2, respectively. The low-stage depth, y_1, has bigger energy than the high-stage depth, y_2. The energy lost through the jump, ΔE, is thus equal to $E_1 - E_2$.

It may be useful at this point to study the opposite case where the depth of flow changes from subcritical to supercritical at a gate as shown in Figure 13.13a. In this case the specific

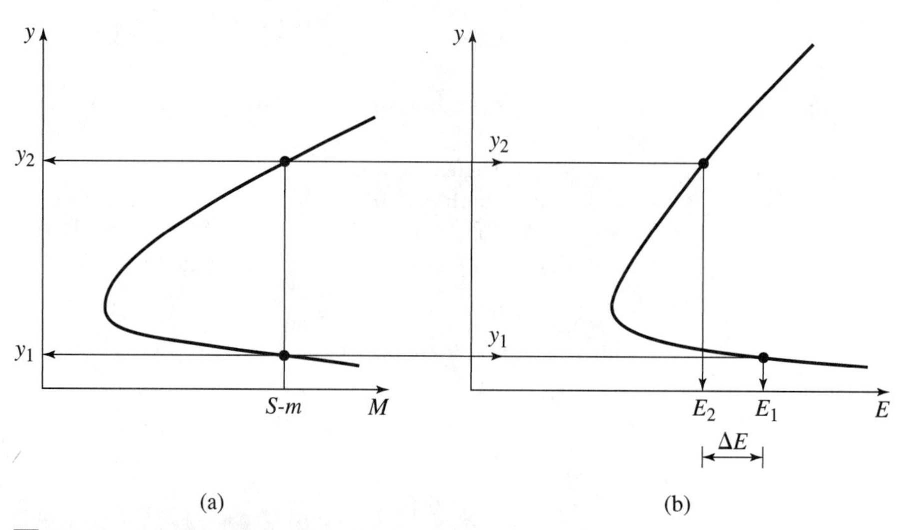

Figure 13.12 Relation between conjugate and alternative depths.

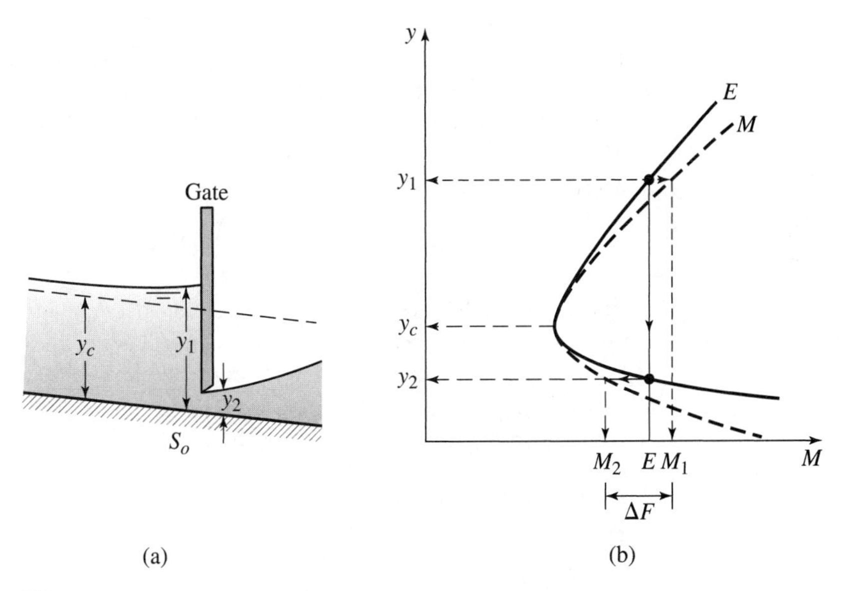

Figure 13.13 Specific energy and pressure-momentum force at gates.

energy is almost constant while the momentum function is reduced. The difference between the two pressure-momentum forces, before and after the gate, is, thus, acting on the gate itself.

Figure 13.13b presents two diagrams plotted on the same vertical but different horizontal scales. The two depths y_1 and y_2 just before and after the gate have an equal specific energy, E. However, they intersect with the M-y diagram at two different values, M_1 and M_2. Thus, the dynamic force acting on the gate is equal to $\gamma M_1 - \gamma M_2$ (i.e., $F_1 - F_2$). One should also recognize that the force on the gate resulting from the hydrostatic pressure is still present in this situation and must be included along with the dynamic force.

Example 13.6

Water is stored behind a sluice gate which is located in a rectangular channel 10 ft wide with a Manning n of 0.025 (B.G.). The gate depth after the gate is 0.75 ft and a hydraulic jump occurs a short distance downstream of the gate where the upstream and downstream depths are 1 foot and 5 feet, respectively. (a) What is the depth of water behind the sluice gate? (b) What is the minimum slope in the channel for a jump to occur? (c) What is the maximum allowable slope for the jump to still occur?

Solution: The specific discharge is computed as

$$q = \left[\frac{y_1 + y_2}{2} g y_1 y_2\right]^{1/2} = \left[\frac{1+5}{2}(32.2)(1 \times 5)\right]^{1/2} = 21.98 \text{ cfs/ft}$$

The normal velocity $V = 21.98/5 = 4.4$ ft/s, $A = 50$ ft^2, $P = 20$ ft, $R = 2.5$ ft. Using Manning's equation, the actual channel slope is

$$S_o = \left[\frac{0.025(4.4)}{1.49(2.5)^{2/3}}\right]^2 = 0.001606$$

Now, the specific energy of the flow coming from under the gate is

$$E_1 = 0.75 + \frac{(21.98)^2}{2g(0.75)^2} = 14.09 \text{ ft}$$

This is the alternate depth to the supercritical depth at the gate and thus is equal to the depth behind the gate since it is the ultimate energy source. This assumes that the velocity immediately upstream of the gate would be 0.

Now, the minimum channel slope that would not submerge the gate (drown the jump) would be such that the normal depth would be equal to the conjugate of the depth from under the gate:

$$y_2' = \frac{0.75}{2}\left(-1 + \sqrt{1 + \frac{8(21.98)^2}{g(0.75)^3}}\right) = 5.96 \text{ ft}$$

Note that this depth is greater than the actual jump depth of 5 ft. Thus it is possible to push the jump back to the gate opening. With the required depth of 5.96 ft, $A = 59.6$ ft^2, $P = 21.92$ ft, $R = 2.72$ ft, and $V = 3.69$ ft/s. Then the minimum bed slope for the jump to occur is

$$S_o = \left(\frac{0.025(3.69)}{1.49(2.72)^{2/3}}\right)^2 = 0.00101 \text{ ft/ft}$$

The critical depth is calculated as

$$y_c = \sqrt[3]{\frac{q^2}{g}} = 2.47 \text{ ft}$$

Then the critical velocity = 8.9 ft/s. Using Manning's equation, the critical slope is

$$S_c = \left[\frac{0.025(8.9)}{1.49(1.65)^{2/3}}\right]^2 = 0.0114$$

which is the maximum allowable slope for the jump to occur.

13.4 Energy Loss in Hydraulic Jump

The hydraulic jump involves considerable reduction in the velocity head associated with an increase in the static head. In the dead zone of the jump, rolling of surface, roiling of water, and turbulent eddies are recognized. These violent flow patterns are accompanied by a significant loss of energy, ΔE. The difference in the specific energy before and after the hydraulic jump can be evaluated graphically from the specific energy diagram, as explained in the former section. For horizontal channels, the loss of energy can be evaluated as

$$\Delta E = \left(y_1 + \frac{V_1^2}{2g}\right) - \left(y_2 + \frac{V_2^2}{2g}\right) \tag{13.21}$$

where ΔE is the energy loss per unit weight of water.

13.4.1 Energy Loss in Rectangular or Wide Channels

Equation (13.21) is the general equation for the energy loss not only for any type of section but also for any flow pattern. It can be applied between any two sections through an open channel. However, the bed elevation with respect to a specified horizontal datum should also be introduced as well for the more general case where the two sections are not close to each other. For rectangular and/or wide sections, a more simplified form for the energy loss can be deduced from the general form. Starting from equation (13.21), for wide rectangular sections the velocity can be replaced by q/y. Hence,

$$\Delta E = (y_1 - y_2) + \frac{q^2}{2g}\left(\frac{1}{y_1^2} - \frac{1}{y_2^2}\right) \tag{13.22}$$

From equation (13.16), one can write

$$\frac{2q^2}{g} = y_1 y_2 (y_1 + y_2) \tag{13.23}$$

Substituting into equation (13.22),

$$\Delta E = (y_1 - y_2) + \frac{y_1 y_2 (y_1 + y_2)}{4}\left(\frac{y_2^2 - y_1^2}{y_1^2 y_2^2}\right) \tag{13.24}$$

$$= (y_1 - y_2) + \left(\frac{y_1 + y_2}{4}\right)\left(\frac{y_2^2 - y_1^2}{y_1 y_2}\right) \tag{13.25}$$

$$= (y_2 - y_1)\left[-1 + \frac{(y_1 + y_2)^2}{4 y_1 y_2}\right] \tag{13.26}$$

$$= (y_2 - y_1)\left(\frac{y_2^2 - 2y_1 y_2 + y_1^2}{4 y_1 y_2}\right) \tag{13.27}$$

or

$$\Delta E = \frac{(y_2 - y_1)^3}{4 y_1 y_2} \tag{13.28}$$

Equation (13.28) presents the theoretical formula for the energy loss in the hydraulic jump. The ratio of the energy loss to the upstream energy can be expressed as (Ranga Raju, 1981)

$$\frac{\Delta E}{E_1} = \frac{8F_{n1}^4 + 20F_{n1}^2 - (8F_{n1}^2 + 1)^{3/2} - 1}{8F_{n1}^2 (2 + F_{n1}^2)} \tag{13.29}$$

Peterka (1963) conducted several experiments in which the energy loss, ΔE, normalized by the upstream energy, E_1, and the initial depth, y_1, was presented for various values of the approach Froude number, as shown in Figure 13.14.

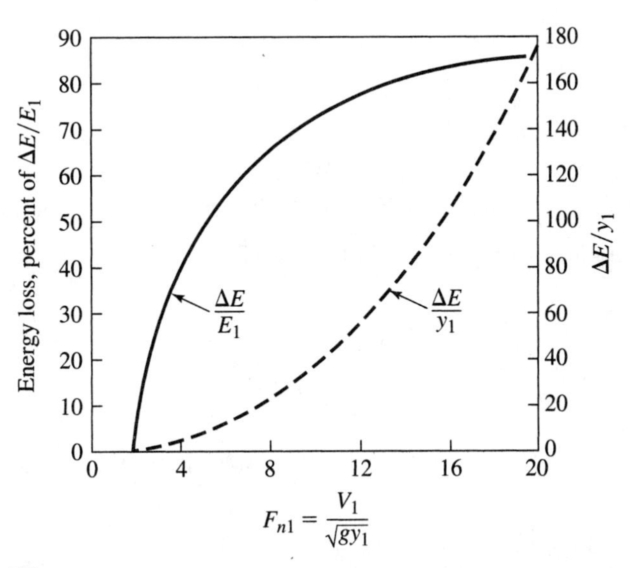

Figure 13.14 Energy loss versus Froude number (after Peterka, 1963).

Example 13.7

A wide channel carries a specific discharge of 2.2 m^2/s. A hydraulic jump is formed in the channel, after which the depth of flow is found to be 2.1 m. Find the critical depth, the initial depth of the jump, the energy lost, and the horsepower lost through the jump per unit width of the channel.

Solution:

$$y_c = \sqrt[3]{\frac{q^2}{g}} = \sqrt[3]{\frac{(2.2)^2}{9.81}} = 0.79 \text{ m}$$

$$y_1 = \frac{y_2}{2}\left[-1 + \sqrt{1 + 8\left(\frac{y_c}{y_2}\right)^3}\right]$$

$$y_1 = \frac{2.1}{2}\left[-1 + \sqrt{1 + 8\left(\frac{0.79}{2.1}\right)^3}\right] = 0.204 \text{ m}$$

$$\Delta E = \frac{(2.1 - 0.204)^3}{4 \times 2.1 \times 0.204} = 3.98 \text{ m}$$

$$\text{Horsepower lost} = \frac{\gamma q \Delta E}{745} = \frac{1000 \times 9.81 \times 2.2 \times 3.98}{745} = 115.3 \text{ hp}$$

The above horsepower is evaluated per unit width of the channel.

13.5 Geometry of Hydraulic Jumps

13.5.1 Height and Efficiency

The height of the hydraulic jump (standing wave), h_J, is defined as the difference between the water levels just after and before the jump, $y_2 - y_1$. Generally speaking, as y_1 decreases, y_2 increases and the height of the jump increases.

The efficiency of the hydraulic jump, η_J, can be defined as the energy after the jump E_2, divided by the energy before the jump, E_1,

$$\eta_J = \frac{E_2}{E_1} \tag{13.30}$$

Example 13.8

For the problem given in Example 13.7, what is the height and efficiency of the jump?

Solution: $h_J = y_2 - y_1 = 2.1 - 0.204 = 1.896$ m

$$\eta_J = \frac{E_2}{E_1} = \frac{2.1 + \dfrac{(1.048)^2}{2 \times 9.81}}{0.204 + \dfrac{(10.78)^2}{2 \times 9.81}} = 35.2\%$$

The efficiency of the hydraulic jump may also be expressed as

$$\eta_J = 1 - \frac{\Delta E}{E_1} \tag{13.31}$$

Substituting equation (13.29) into equation (13.31), η_J can be expressed in terms of the Froude number at the upstream side, F_{n1} as

$$\eta_J = \frac{1 + (8F_{n1}^2 + 1)^{3/2} - 4F_{n1}^2}{8F_{n1}^2(2 + F_{n1}^2)} \tag{13.32}$$

13.5.2 Length and Profile

Hydraulic jumps cause intensive scour at their locations. They should be contained in a specified location, known as stilling basins, which are secured against anticipated destruction. Apron length and height of side walls of a stilling basin are designed according to the expected length and profile of the hydraulic jump. As lengths and profiles may vary with discharge, stilling basins are required to contain the entire jump under all possible discharges.

In nature, it is difficult to mark precisely the beginning and the end of a hydraulic jump. Turbulent flow, rollers and eddies, random nature of surface disturbances, and air entrainment complicate the exact determination of the two ends of the hydraulic jump.

Many experimental investigations have been conducted to relate the jump length, L_J, to the upstream Froude number, F_{n1}. The jump length is normalized by dividing either by y_1 or y_2. However, because of the greater difficulty in measuring y_2 than in measuring y_1, satisfactory correlation has been reported for relating L_J/y_1 to the approach Froude number, F_{n1}. Figure 13.15 presents the variation of L_J/y_2 with F_{n1} as recommended by the U.S. Bureau of Reclamation. Some error in the jump length may be expected when using this figure as y_2 may not be accurately defined.

Hager (1992) developed the following equation using the criterion that the turbulence is diminished at the end of the jump,

$$\frac{L_J}{y_1} = 220 \tanh \frac{F_{n1}}{22} \tag{13.33}$$

Another relation reported in the related literature is written as

$$L_J = a(y_2 - y_1) \tag{13.34}$$

in which a varies from 5.0 to 6.9 according to different investigators.

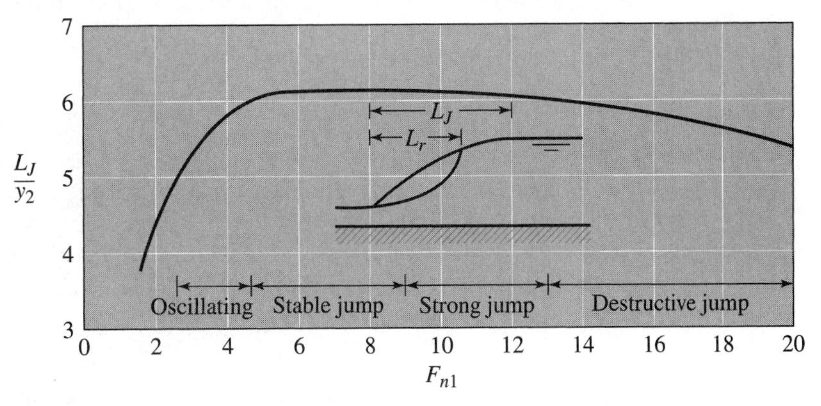

Figure 13.15 Length of the hydraulic jump (USBR).

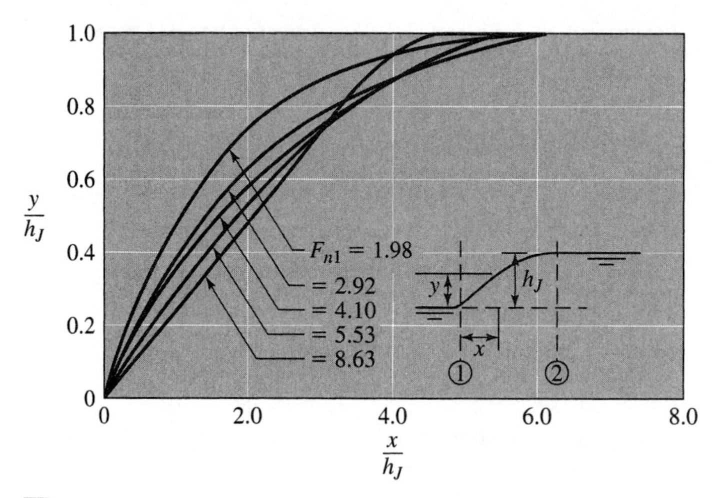

Figure 13.16 Profile of jumps (after Bakhmetef and Matzke, 1936).

The length of the roller, L_r, is generally smaller than the length of the jump, L_j, as shown in Figure 13.15. L_r varies between $0.4L_J$ for $F_{n1} = 3.0$ and $0.7L_J$ for $F_{n1} = 9.0$.

Provided that the channel width is more than ten times the initial depth of the jump, Hager (1992) gave the following equation for determination of L_r.

$$\frac{L_r}{y_1} = -1.2 + 160 \tanh \frac{F_{n1}}{20} \tag{13.35}$$

The jump profile through its length is important to determine the weight of water over the floor of a stilling basin which counteracts the uplift force if the basin is laid on a permeable foundation. Also, the height of the side walls may also vary in accordance with the jump profile to reduce construction costs.

Jump profiles for different approach Froude numbers are given in Figure 13.16 (Bakhmeteff and Matzke, 1936). Several investigators have shown that the vertical pressure on the floor may be reasonably obtained if hydrostatic pressures are assumed. The depth of flow, y, is empirically related to the distance, x, measured from the beginning of the jump as (Hager, 1992):

$$Y = \tanh(1.5X) \tag{13.36}$$

in which $X = x/L_r$, $Y = (y - y_1)/(y_2 - y_1)$, and L_r = length of roller.

An extensive study of the profile of the hydraulic jump was conducted by Rajaratnam and Subramanya (1968). Using data from different sources and with different values of the approach Froude number, they proved that there is a unique relation between $y/(0.75(y_2 - y_1))$ and $x/\bar{X}$ as shown in Figure 13.17, where $\bar{X}$ is the distance from the beginning of the jump to the section where the depth of flow, y, is $0.75(y_2 - y_1)$. The length $\bar{X}$ was empirically related to y_1 and F_{n1} as

$$\frac{\bar{X}}{y_1} = 5.08F_{n1} - 7.82 \tag{13.37}$$

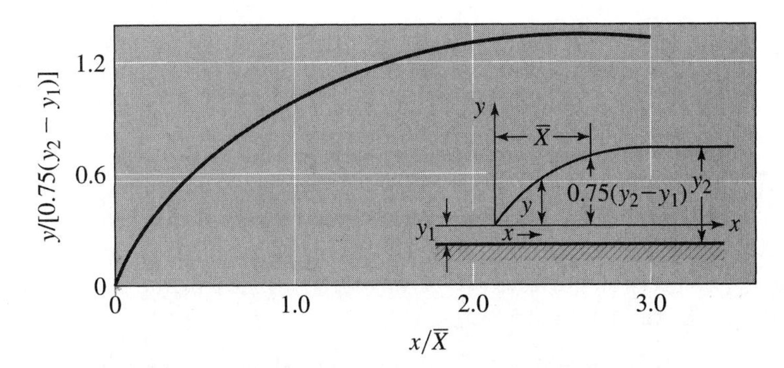

Figure 13.17 Profile of jumps in rectangular channels (after Bakhmetef and Matzke, 1936).

13.6 Classification of Hydraulic Jumps

Many investigators have classified hydraulic jumps according to the approach Froude number, which must be greater than 1.0. The range of F_{n1} in the different categories is not sharp as some overlap may be found between any two successive types. Generally speaking, the hydraulic jump occurs in five distinct forms, as shown in Figure 13.18. Each form is characterized by a certain flow pattern:

Undular jump ($1 < F_{n1} < 1.7$): The water surface exhibits slight undulation. The two conjugate depths are close and the transition from the supercritical state to the subcritical one is not abrupt and is encountered through the slightly ruffled water surface (Figure 13.18a).

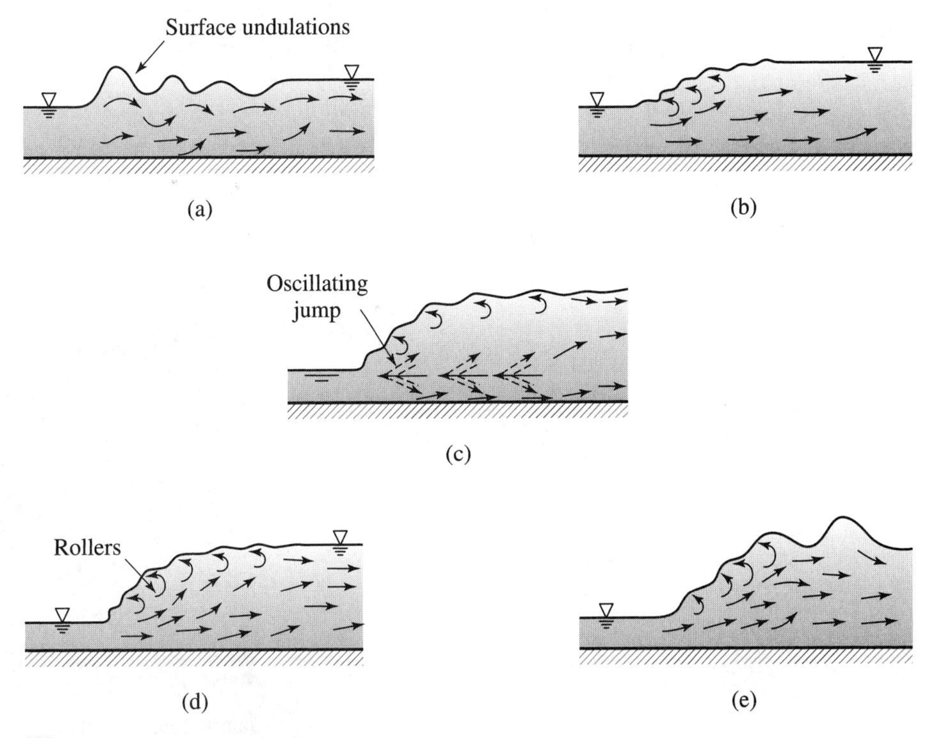

Figure 13.18 Different types of jumps.

Weak jump ($1.7 < F_{n1} < 2.5$): As F_{n1} exceeds 1.7, a number of small eddies and rollers are formed on the water surface (Figure 13.18b). The length of rolling surface, L_r, remains small as compared to the total length of the jump, L_j, and the energy loss is generally small. The ratio between the final depth to the initial depth of the jump, y_2/y_1, varies between 2.0 and 3.1.

Oscillating jump ($2.5 < F_{n1} < 4.5$): The incoming jet oscillates from the bottom to the top. This oscillating pattern causes surface waves which persist for a considerable distance beyond the end of the jump and may cause erosion to banks (Figure 13.18c). To prevent destructive effects of this category, it should be avoided in the design stage. The ratio y_2/y_1 is varied between 3.1 and 5.9.

Stable jump ($4.5 < F_{n1} < 9$): Many advantages are found in this category. Most importantly is its fixed position regardless of the downstream flow conditions (well-balanced jump). The jump location is least sensitive to any variation in y_2. It also results in a good dissipation of energy and in a considerable rise of the downstream water level (Figure 13.18d). The ration of y_2/y_1 varies between 5.9 and 12.0.

Strong or rough jump ($F_{n1} > 9$): The jump is effective but becomes increasingly rough as F_{n1} increases. F_{n1} should not be allowed to exceed 12.0 as the required stilling basins would be very massive and expensive. The jump ability to dissipate energy is high and y_2/y_1, is over 12 and may exceed 20 (Figure 13.18e).

13.7 Underflow (Sluice) Gates

Sluice, or underflow, gates have been used in previous examples to illustrate energy principles and to examine characteristics of hydraulic jumps. We can now discuss these structures in more detail. The sluice gate is a control device frequently employed with storage impoundments. In contrast to the weir or spillway discussed in Chapter 12, the sluice gate is a discharge control device rather than an elevation control. The gate is used to exert a tight control on the discharge that is allowed to flow over the spillway, or through any outlet associated with a control structure. In a typical situation, the gate controls the flow rate during a certain range of low and medium discharges, and then the weir, or overflow device, takes, control for large discharges so that the elevation of the lake will not rise precipitously. As in the case of weirs, the gate acts as a downstream choke control.

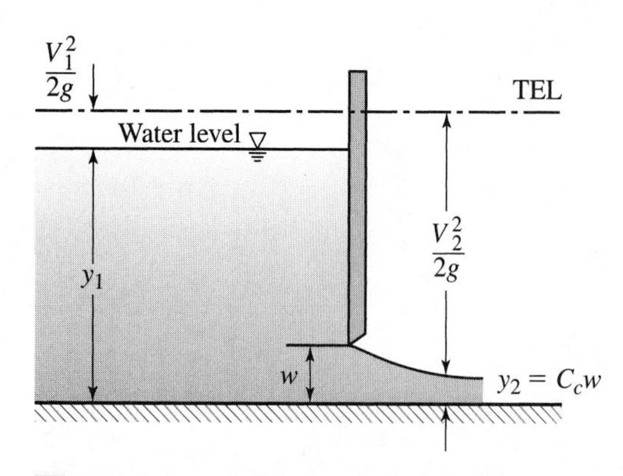

Figure 13.19 Hydraulic principles of the sluice gate.

Tainter Gate at Chief Joseph Dam and Spillway, Columbia River. (Courtesy of Robert Ashworth, Bellingham, WA.)

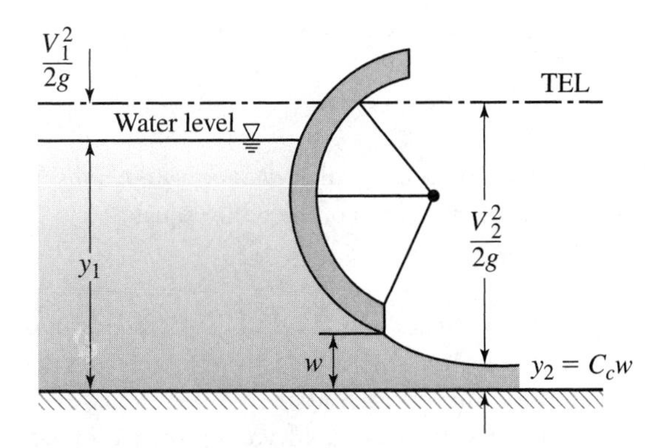

Figure 13.20 The Tainter gate.

Underflow gates are typically of two designs: the vertical lift gate (Figure 13.19) and the radial or tainter gate (Figure 13.20). The basic hydraulic principles are the same for the two gates designs; the difference being that the circular gate is easier to manipulate than is the vertical gate in that it rotates to raise rather than lifting vertically. For this reason, tainter gates are normally employed in situations where the gate needs to be very large and heavy.

The principles of the sluice gate are demonstrated in Figure 13.19. The gate functions essentially as an orifice and thus lends itself to the same type of analysis. The discharge capacity can be determined by writing the energy equation across the gate (neglecting losses)

$$y_1 + \frac{Q^2}{2gA_1^2} = y_2 + \frac{Q^2}{2gA_2^2} \tag{13.38}$$

Then

$$\frac{Q^2}{2g} = \frac{(y_2 - y_1)(A_1^2 A_2^2)}{A_2^2 - A_1^2} \tag{13.39}$$

In most cases the gate is rectangular in section, thus reducing equation (13.39) to

$$\frac{q^2}{2g} = \frac{y_1^2 y_2^2}{y_2 + y_1} \tag{13.40}$$

whence

$$q = y_1 y_2 \sqrt{\frac{2g}{y_1 + y_2}} \tag{13.41}$$

As in previous cases, equation (13.41) represents the ideal flow through the gate neglecting local energy losses due to the contraction of the flow lines. As usual, these losses are incorporated through the introduction of a contraction coefficient C_c. Contraction coefficients vary from about 0.59 to 0.62 in most cases (Henderson, 1966). In this case, as shown in Figure 13.19, the contraction will be evident in the reduction of the depth out from under the gate to a level less than the gate opening, w. Thus, the downstream depth y_2 can be written as a function of the gate opening and the contraction coefficient

$$y_2 = C_c w$$

Then equation (13.41) becomes

$$q = C_c w y_1 \sqrt{\frac{2g}{y_1 + y_2}} \tag{13.42}$$

Equation (13.42) can be further reduced to

$$q = C_c w \sqrt{\frac{y_1}{y_1 + y_2}} \sqrt{2gy_1} \tag{13.43}$$

It is common practice to combine the first two terms into a discharge coefficient, C_d, and write equation (13.43) in the familiar form of the orifice equation,

$$q = C_d w \sqrt{2gy_1} \tag{13.44}$$

Written in the form of equation (13.44), the discharge under the gate can be computed from knowing only the depth behind the gate and the gate opening. The contraction coefficient used to compute C_d hovers around a value close to 0.60 (Henderson, 1966).

In previous examples, we have used an approximation of the discharge computed by equation (13.44) through the use of the orifice equation directly,

$$q = Vy_2 = y_2 \sqrt{2gH} \tag{13.45}$$

where $H = y_1 - y_2$. Using the relation $y_2 = C_c w$, the contraction of the flow from under the gate will be accounted for. Minor losses associated with this approach can be included through the traditional manner [i.e., $\Delta H = k_e(V^2/2g)$, where k_e represents the entrance loss coefficient equal to 0.1 or 0.2 velocity heads].

In the above discussions it is assumed that the outlet of the gate will remain unsubmerged during its operation. Certainly, when the underflow gate is used as a discharge control on spillway crests, this assumption will always be justified. In most other cases where the gate is employed, the downstream conditions are designed in such a way that the outlet will remain clear if at all possible. Otherwise, the efficiency of the gate will be significantly impaired. From perusal of the energy relationship across the gate shown in equation (13.38), it is clear that the two depths y_1 and y_2 simply represent the alternate depths associated with the given energy level of the water behind the gate if losses are ingnored. Since the upstream depth y_1 will be subcritical, then the depth from under the gate must be supercritical. Thus, if the downstream channel slope is mild, a hydraulic jump must occur downstream of the gate. If the downstream slope is such that the jump can form, either immediately at the gate, or some distance downstream, the gate outlet is clear and the discharge is given by equations (13.43) or (13.45). However, if the downstream conditions will not allow the jump to form (i.e., the jump is drowned as shown in Figure 13.21), then the gate opening will be submerged and the discharge significantly reduced.

An approximation of the discharge in the drowned gate case can be obtained by the following method given by Henderson (1966). The method involves simultaneous solution of the energy equation written across the gate (prior to the jump) and the momentum equation

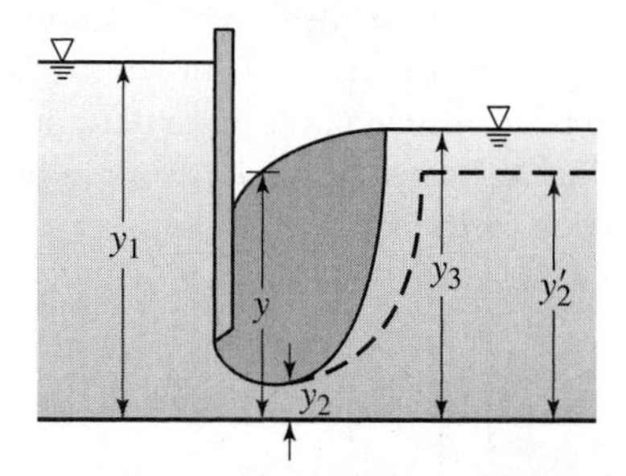

Figure 13.21 The drowned (submerged) sluice gate.

written across the (drowned) jump. Referring to Figure 13.21, for the rectangular case the equations become

$$y_1 + \frac{q^2}{2gy_1^2} = y + \frac{q^2}{2gy_2^2} \tag{13.46}$$

$$\frac{y^2}{2} + \frac{q^2}{gy_2} = \frac{y_3^2}{2} + \frac{q^2}{gy_3} \tag{13.47}$$

In the above equations, the depth y is the hypothetical depth of the drowned jump and represents the hydrostatic energy and force term. The common term, q, is eliminated from the equations, and with the other terms known, the solution is for the hypothetical y. Then this value can be substituted back into either equation to determine q.

Example 13.9

A vertical sluice gate is placed across a 10-m-wide rectangular channel running on a slope of 0.0025 m/m with a Manning n value of 0.025 (metric) as shown in Figure 13.22. The discharge in the channel is 55 m^3/s. Now, under these conditions, what is the maximum gate opening for which the gate will be clear (i.e., a hydraulic jump will occur downstream)? What would be the depth behind the gate under these conditions? Use a contraction coefficient of 0.6 for the gate.

Solution: First, we must find the normal depth in the channel. Writing the Mannning equation,

$$AR^{2/3} = \frac{0.025(55)}{(0.0025)^{1/2}} = 27.5 = 10y_o\left(\frac{10y_o}{2y_o + 10}\right)^{2/3}$$

Solving by trial and error, we find that $y_o = 2.1$ m. Next, we note that $y_c = \sqrt[3]{q^2/g} = 1.46$ m (noting that $q = 55/10 = 5.5$ m^3/m) and thus the channel slope is mild and a hydraulic jump is possible. Now, the maximum depth that can be produced by the gate and not drown the jump will simply be the conjugate of the normal depth. Thus, $y_o' = \frac{2.1}{2}\left(-1 + \sqrt{1 + \frac{8(5.5)^2}{g(2.1)^3}}\right)$ and we find that $y_o' = 0.96$ m. Then the maximum gate opening would be $w = 0.96/0.6 = 1.6$ m. Note that for smaller gate openings, the depth will be less and thus the jump will run downstream, while for greater openings the jump will drown. Next, we must find the depth behind the gate. Employing equation (13.42), we write

$$q = C_c w y_1 \sqrt{\frac{2g}{y_1 + y_2}} = 5.5 = 0.6(1.6)y_1\sqrt{\frac{19.62}{y_1 + 0.96}}$$

or $32.82 = \frac{19.62(y_1)^2}{y_1 + 0.96}$. Solving by the quadratic formula, we find that $y_1 = 2.35$ m.

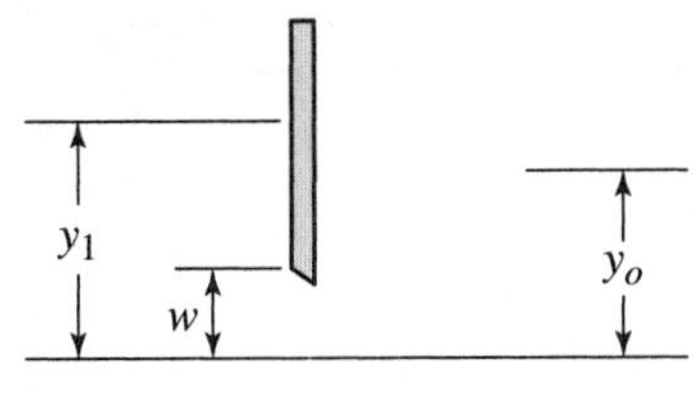

Figure 13.22 The sluice gate in Example 13.9.

13.8 Forced Jumps

Water levels in channels may vary according to various water events and applications. Rainfall events may cause considerable rise in the water level at certain reaches along a channel. Rain water in various catchments may drain into channels causing continuous fluctuations in water levels.

Because the total force (momentum and pressure) is dependent on the water depth before and after the jump, a jump would lose its stability when any of the two conjugate depths is changed. From the *M-y* diagram, as shown in Figure 13.1, if the downstream depth, y_2, increases, the momentum function will increase and the jump will move to the upstream side to a point where the two forces exerted on both sides of the jump are equal. If y_2 decreases, the jump will move to the downstream side. A similar analysis may be done if the supercritical depth of flow, y_1, is changed.

In practical applications, we need to stabilize jumps at specified locations (stilling basins). Otherwise, if jumps are allowed to move in channels, many destructive effects to channel beds and sides would be inevitable. Therefore, it is generally required to take certain measures to ensure that the jump will not move away from its designated location. Such jumps are known as forced jumps. Various methods are commonly employed to control jump locations, the major of which are discussed below.

13.8.1 Abrupt Drop and Abrupt Rise in Channel Bottom

The jump location can be controlled at a specified distance by creating an abrupt rise or an abrupt drop in the channel bottom. If such a variation is introduced, the jump will be formed over or around this sudden change in the bed level.

Consider a sudden drop as shown in Figure 13.23a, with three cases for the water level at the downstream side. Higher tail-water levels (Case 1) force the jump to the upstream side of

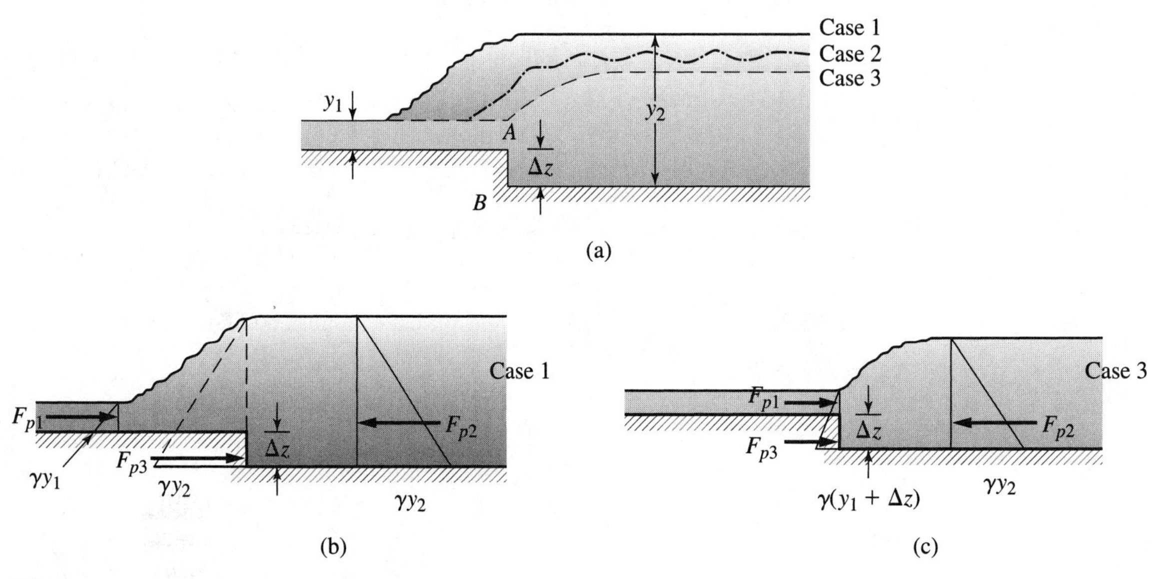

Figure 13.23 Hydraulic jump at an abrupt drop.

the drop, while lower tail-water levels (Case 3) allow the jump to form after the drop. Case 2, where the tail-water level is located between Case 1 and Case 3, represents the wavy jump which is generally formed over the drop. From the continuity and momentum equations and referring to Figures 13.23a,b and c, one can write for a rectangular channel

$$q = y_1V_1 = y_2V_2 \tag{13.48}$$

where

$$F_{p1} + F_{p3} - F_{p2} = \frac{\gamma}{g}q(V_2 - V_1) \tag{13.49}$$

$$F_{p1} = \gamma\frac{y_1^2}{2} \tag{13.50a}$$

$$F_{p2} = \gamma\frac{y_2^2}{2} \tag{13.50b}$$

$$F_{p3} = \gamma\left(y_2 - \frac{\Delta z}{2}\right)\Delta z \quad \text{(for Case 1)} \tag{13.50c}$$

and

$$F_{p3} = \gamma\left(y_1 + \frac{\Delta z}{2}\right)\Delta z \quad \text{(for Case 3)} \tag{13.50d}$$

The above equations can be solved to evaluate Δz. The Froude number can be related to y_1, y_2 and Δz as (Ranga Raju, 1981)

$$F_{n1}^2 = \frac{[(y_2/y_1) - (\Delta z/y_1)]^2 - 1}{2(1 - y_1/y_2)} \quad \text{(for Case 1)} \tag{13.51a}$$

and

$$F_{n2}^2 = \frac{(y_2/y_1)^2 - (1 + \Delta z/y_1)^2}{2(1 - y_1/y_2)} \quad \text{(for Case 3)} \tag{13.51b}$$

In practical applications, the specific discharge, q, and the downstream depth, y_2, are generally known and one is required to obtain the drop, Δz, or to determine the floor elevation before the drop in order to hold the jump between the two extreme cases. The jump position should be checked for all possible discharges. Introducing drops in channel bottoms should only be considered when tail-water levels are relatively high.

An abrupt rise in a channel bed may be considered where there is a deficiency in the tail water level. The hump would hold the jump from proceeding to the downstream end. An analysis similar to that done for the case of a sudden drop can be carried out for the case of an abrupt rise. Empirical relations have been developed through experimental work with an abrupt rise in which the beginning of the jump was held fixed at a distance of $5(\Delta z + y_2)$. Forster and Skrinde (1950) developed a diagram to predict the jump performance on an abrupt rise knowing Δz, y_1, and y_2, as shown in Figure 13.24. A point being plotted on the corresponding $\Delta z/y_1$ line means that the jump is exactly located at a distance of $5(\Delta z + y_2)$ upstream of the hump. If the point falls on the left of the corresponding line, the jump is forced to move upstream and vice versa.

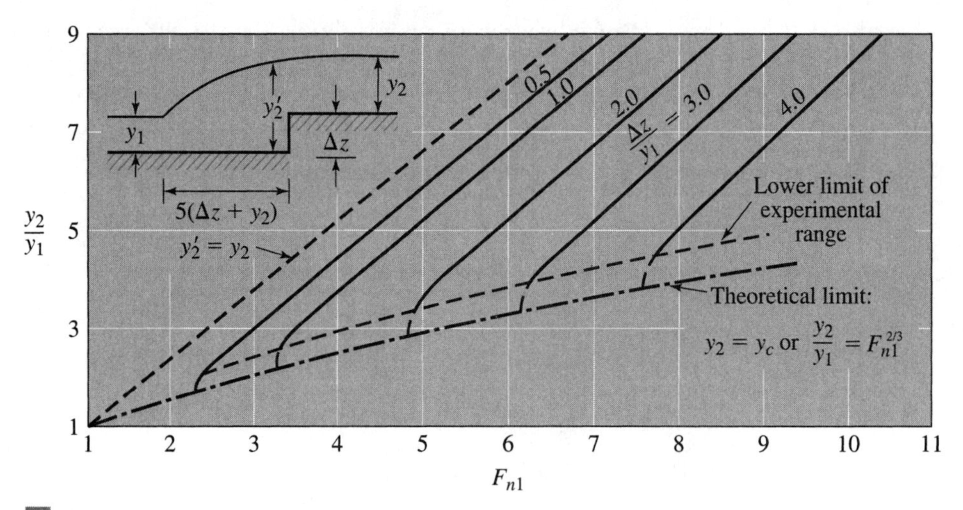

Figure 13.24 Hydraulic jump at an abrupt rise (after Forster and Skrinde, 1950).

13.8.2 Baffle Blocks

Hydraulic jumps are encountered between two sections having equal specific forces. Baffle blocks are frequently placed in stilling basins to counterbalance any difference in the forces at the two sides of the hydraulic jump. They are mostly trapezoidal in shape and may be placed in a single row or staggered in two rows.

Applying the momentum equation to the single row of baffle blocks in a rectangular channel as shown in Figure 13.25 and considering a unit width of the channel, one can write

$$F_{p1} - F_{p2} - F_B = \frac{\gamma}{g}q(V_2 - V_1) \tag{13.52}$$

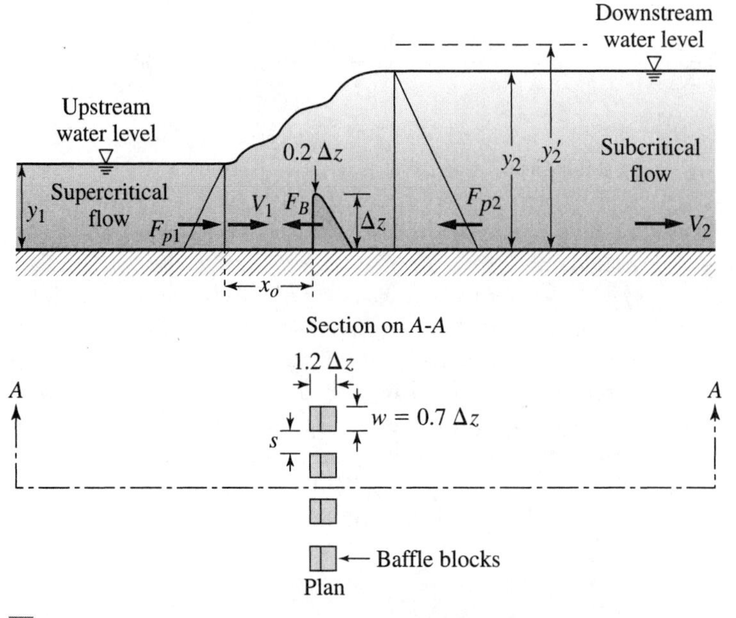

Figure 13.25 Control of hydraulic jump by baffle blocks.

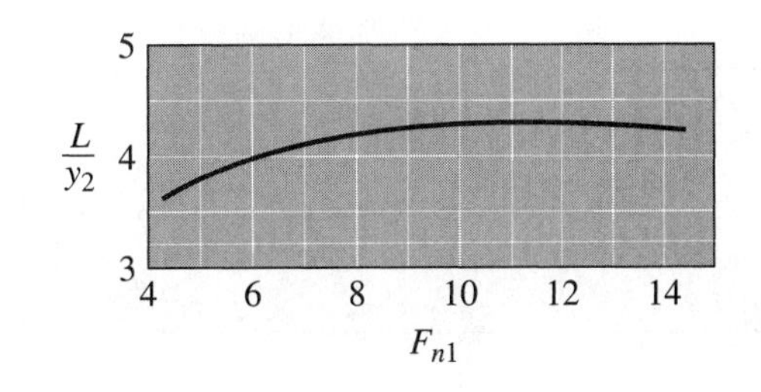

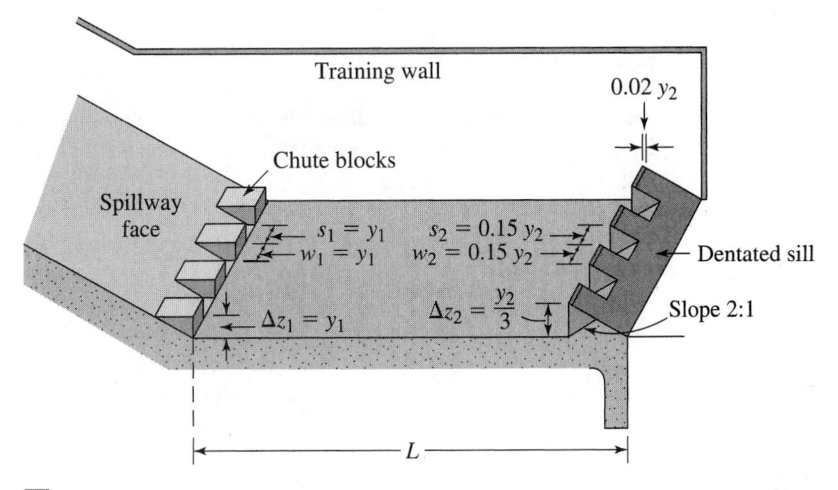

Figure 13.26 Design of stilling basin for $F_{n1} > 4.0$; spillway height ≤ 60.0 m and $q \leq 45.0\ \mathrm{m^3/s}$; Peterka, 1963.

The force on the baffle blocks, F_B, per unit width of the channel is thus

$$F_B = \frac{\gamma}{g} q(V_1 - V_2) - \gamma \frac{y_2^2 - y_1^2}{2} \tag{13.53}$$

Many empirical and experimental relations between the height of the baffle blocks, Δz, the upstream depth of flow, y_1, the distance at which a jump is formed upstream of the blocks, x_o, and the blockage ratio, defined as $w/(w + s)$ (see Figure 13.25) have been developed. Basco and Adams (1971) and Ranga Raju et al. (1980), among others, conducted numerous investigations in this respect. Ranga Raju et al. (1980) outlined a procedure for the design of stilling basins provided with baffle blocks and an end sill on lines. Commonly used nowadays are the designs after the U.S. Bureau of Reclamation which were proposed for different ranges of the upstream Froude number. A typical design is given in Figure 13.26.

13.9 Hydraulic Jumps in Expanding Sections

In the previous chapter (Section 12.5) the changes in the water surface due to drops, steps, enlargements, and contractions were demonstrated. However, the type of flow was kept unchanged, in the sense that the flow maintained its original state, either supercritical or subcritical, throughout the studied region. In this section we consider a supercritical flow in a section subject to a considerable expansion, either abruptly or gradually, in such a way that the flow is altered to a subcritical one. A hydraulic jump will, thus, be formed. In some cases,

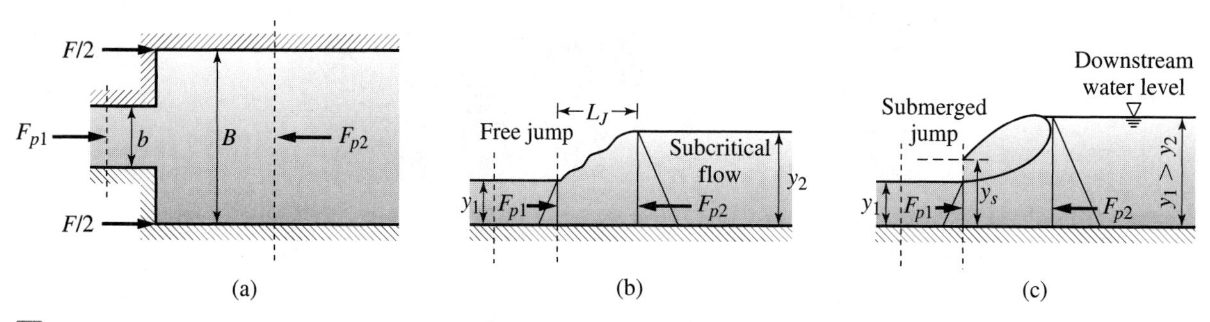

Figure 13.27 Hydraulic jump in expanding sections.

the tail-water depth in the stilling basin does not support the formation of the jump unless a lateral expansion is formed. Hydraulic jumps created under such circumstances are often called spatial hydraulic jumps.

13.9.1 Spatial Hydraulic Jumps at Abrupt Expansions of Rectangular Channels

Consider the case shown in Figure 13.27a. Due to a sudden enlargement in the channel width, the discharge per unit width is reduced and the flow is changed to a subcritical flow. Let the initial depth, before the enlargement, be y_1. Then the tail-water depth, y_2, can be evaluated from the continuity and momentum equations as

$$F_{p1} + F - F_{p2} = \frac{\gamma}{g}Q(V_2 - V_1) \tag{13.54}$$

or

$$\gamma b\frac{y_1^2}{2} + \gamma(B - b)\frac{y_1^2}{2} - \gamma B\frac{y_2^2}{2} = \frac{\gamma}{g}Q^2\left(\frac{1}{By_2} - \frac{1}{By_1}\right) \tag{13.55}$$

which can be written as

$$y_1^2 - y_2^2 = \frac{2Q^2}{gB}\left(\frac{A_1 - \dfrac{A_2 b}{B}}{A_1 A_2}\right) \tag{13.56}$$

Let ψ_y equal y_2/y_1 and ψ_b equal b/B. Then equation (13.56) may be expressed as (Ranga Raju, 1981)

$$\psi_y^3 - \psi_y(1 + 2F_{n1}^2\psi_b) + 2F_{n1}^2\psi_b^2 = 0 \tag{13.57}$$

Figure 13.28 presents the solution of equation (13.57). It is obvious that ψ_y decreases as ψ_b increases until it reaches the case of a prismatic channel. Herbrand (1973) gave the following empirical relation for the ratio between the conjugate depth in an expanding channel, y_2, and the corresponding conjugate depth if the channel is left prismatic, y_{2b}:

$$\frac{y_2}{y_{2b}} = \psi_b^{3/8} \tag{13.58}$$

The above discussion is applicable for the free jump where the toe is located at the section of channel enlargement, as shown in Figure 13.27b. If the tail-water depth is larger than the sequent (or conjugate) depth obtained from Figure 13.28, the toe of the jump will move

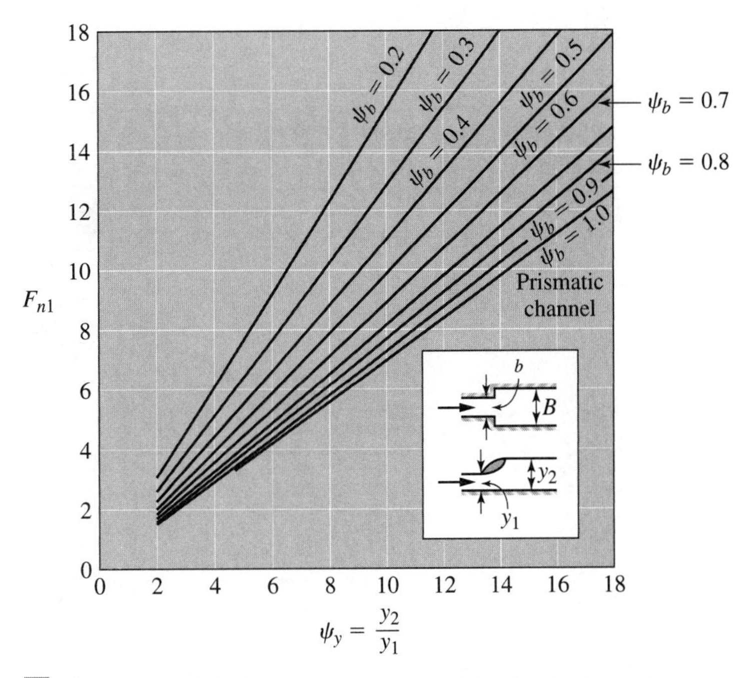

Figure 13.28 Ratio between conjugate depths in sections of an abrupt expansion.

inside the reduced section. To prevent such conditions a sluice gate may be placed just before the sudden expansion in the channel section. This will result in a submerged jump as shown in Figure 13.27c. The ratio between y_s and y_t is given as (Ranga Raju, 1981)

$$\frac{y_s}{y_t} = \sqrt{1 - 2F_{nt}^2\left(\frac{y_t}{\psi_b y_1} - 1\right)} \tag{13.59}$$

where F_{nt} is the tail (or downstream) Froude number.

13.9.2 Spatial Hydraulic Jumps in Gradually Expanding Rectangular Channels

Consider the case of a gradual expansion in the channel width with straight walls, shown in Figure 13.29. The stream lines are assumed radial and the water surface through the jump forms a quarter ellipse. Arbhabhirama and Abella (1971) showed that

$$r_o y_o = \tfrac{1}{2}(\sqrt{1 + 8F_{nc}^2} - 1) \tag{13.60}$$

where, $r_o = r_2/r_1$, $y_o = y_2/y_1$, and

$$F_{nc}^2 = (F_{n1}^2 r_o + C_p) \tag{13.61}$$

in which C_p is the side pressure correction factor, given as

$$C_p = \frac{r_o y_o (r_o - 1)\left\{r_o\left(\frac{y_o^2}{3} + 0.118 y_o + 0.0480\right) + 0.5\right\}}{(r_o y_o - 1)} \tag{13.62}$$

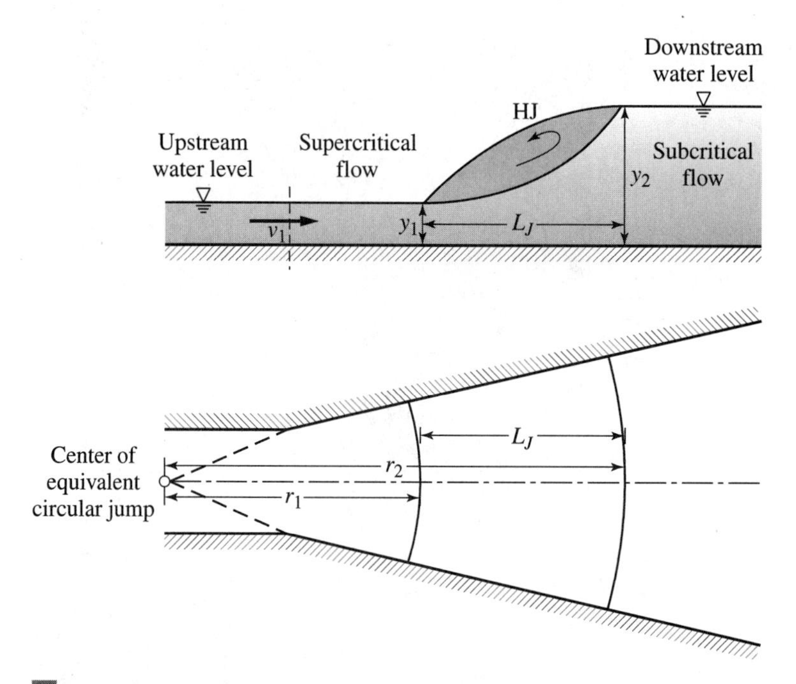

Figure 13.29 Hydraulic jump in a gradually expanding channel.

The length of the jump, L_J, is given as $(r_2 - r_1)$. The relation between L_J and y_1 is given empirically as

$$\frac{L_J}{y_1} = 3.70F_{n1}^{1.35} \tag{13.63}$$

Good results were obtained from equation (13.63) in the range $3 \leq F_{n1} \leq 10$. Knowing r_1 and y_1, equations (13.60) and (13.63) can be solved by trial and error to determine y_2.

13.10 Hydraulic Jumps in Rectangular Channels with Sloping Beds

Hydraulic jumps may occur in channels with very steep bed slopes. Other examples include overflow weirs with sloping faces and spillways. In such cases, the weight component of the liquid in the flow direction cannot be ignored as in the previous discussions. Neglecting the boundary friction and assuming the hydrostatic pressure distribution, the momentum equation for the element under consideration shown in Figure 13.30, is written as

$$\gamma\frac{y_1^2}{2}\cos^3\theta - \gamma\frac{y_2^2}{2}\cos^3\theta + W\sin\theta = \frac{\gamma}{g}q(V_2 - V_1) \tag{13.64}$$

where

$$V_1 = \frac{q}{y_1 \cos\theta}, \quad \text{and} \quad V_2 = \frac{q}{y_2 \cos\theta}$$

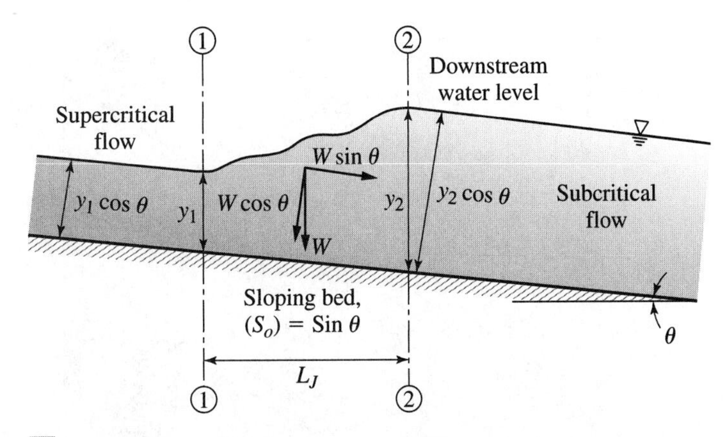

Figure 13.30 Hydraulic jump in a sloping channel.

The weight of the jump, W (element under consideration), is equal to

$$W = K\gamma L_J\left(\frac{y_1 + y_2}{2}\right) \tag{13.65}$$

where K is a factor introduced to account for the nonuniformity of the jump surface.

Substituting equation (13.65) into (13.64) and knowing that $F_{n1} = V_1/\sqrt{gy_1 \cos\theta}$, equation (13.64), after simplification, can be written as

$$\left(\frac{y_2}{y_1}\right)^3 - (2G^2 + 1)\frac{y_2}{y_1} + 2G^2 = 0 \tag{13.66}$$

where

$$G = \frac{F_{n1}}{\sqrt{\cos\theta - \dfrac{KL_J \sin\theta}{y_2 - y_1}}} \tag{13.67}$$

The solution of equation (13.66) is given as

$$\frac{y_2}{y_1} = \frac{1}{2}(\sqrt{1 + 8G^2} - 1) \tag{13.68}$$

It is generally believed (Stevens, 1944) that K and $L_J/(y_2 - y_1)$ are dependent on F_{n1} and the bed slope, S_o. One can also conclude from equation (13.66) that $G = \varphi(F_{n1}, \theta)$. Hence, from equation (13.68), y_2/y_1 is also a function of F_{n1} and θ. Experimental data on sloping channels have provided empirical relations between y_2/y_1, L_J, and F_{n1} under different bed slopes as shown in Figures 13.31 and 13.32, respectively. The dashed lines in Figure 13.32 are obtained by interpolation as data were limited. The length of the jump, L_J, shown in Figure 13.30 was taken as the roller surface length, L_r, which is obviously smaller than the actual length of the jump. The water surface has, thus, continued to rise after the end of the jump.

Rajaratnam (1967) proposed the following empirical relations which may be used instead of Figure 13.31 to determine G and hence get the conjugate depth ratio using equation (13.66):

$$G^2 = K_1^2 F_{n1}^2 \tag{13.69}$$

$$K_1 = 10.0^{0.027}\theta \tag{13.70}$$

where θ is the angle of the channel bed with the horizontal in degrees.

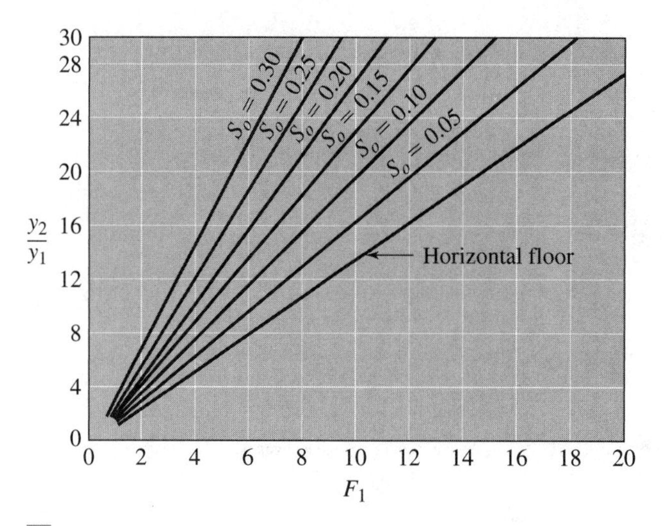

Figure 13.31 Relation between conjugate depths on sloping floors.

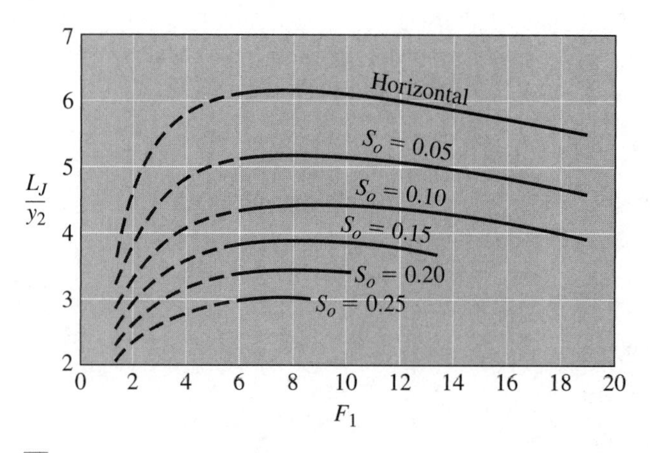

Figure 13.32 Length of jumps on sloping floors.

13.11 Oblique Jumps

If the vertical boundary of a channel is deflected inward to the course of a supercritical flow, as shown in Figure 13.33, an oblique or shock wave jump will be produced. The initial depth of flow, y_1, will increase to y_2, through a wave front AB, where A is the point at which the boundary is deflected. The wave angle β depends on the angle of the boundary deflection θ. If θ is reduced to zero, the wave becomes a standard hydraulic jump in which the wave front is normal to the flow direction. Referring to Figure 13.33, one can write

$$V_1 \cos\beta = V_2 \cos(\beta - \theta) \tag{13.71}$$

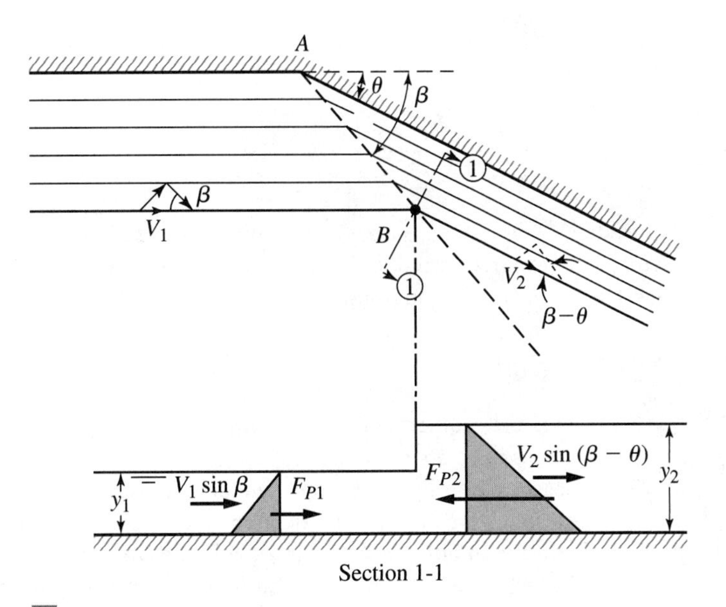

Figure 13.33 The oblique jump.

The left and right terms in the above equation represent the tangential velocity components on either side of the front. The velocity components normal to the wave front are $V_1 \sin \beta$ and $V_2 \sin(\beta - \theta)$. Applying the continuity equation, one can write

$$y_1 V_1 \sin \beta = y_2 V_2 \sin(\beta - \theta) \tag{13.72}$$

Applying the momentum equation, one can prove (Chaudhry, 1993):

$$\sin \beta = \frac{1}{F_{n1}} \sqrt{\frac{1}{2} \frac{y_2}{y_1} \left(\frac{y_2}{y_1} + 1 \right)} \tag{13.73}$$

Also,

$$\frac{y_2}{y_1} = \frac{\tan \beta}{\tan (\beta - \theta)} \tag{13.74}$$

The Froude number before the jump, F_{n1}, is

$$F_{n1} = \frac{V_1}{\sqrt{g y_1}} \tag{13.75}$$

The Froude number normal to the wave front before jump is given as

$$(F_{n1})_N = \frac{V_1 \sin \beta}{\sqrt{g y_1}} = F_{n1} \sin \beta \tag{13.76}$$

Substituting equation (13.76) into equation (13.19b), the ratio between the conjugate depth and the initial depth is

$$\frac{y_2}{y_1} = \frac{1}{2}\left(-1 + \sqrt{1 + 8F_{n1}^2 \sin^2 \beta}\right) \tag{13.77}$$

Eliminating y_2/y_1 from equations (13.74) and (13.77), a relationship among F_{n1}, θ, and β is developed:

$$\tan\theta = \frac{\tan\beta(\sqrt{1 + 8F_{n1}^2\sin^2\beta} - 3)}{2\tan^2\beta + \sqrt{1 + 8F_{n1}^2\sin^2\beta} - 1)} \tag{13.78}$$

Ippen (1951) developed several graphs showing the relationship expressed by equation (13.77). Comprehensive studies and investigations on oblique jumps are found in Ippen (1951), Ippen and Harleman (1956), and Rouse (1938).

READING AID

13.1. Write the momentum equation for a water body in balance between two specified sections in an open channel.

13.2. What is meant by momentum function in open-channel flow? Explain the two components of the momentum function and define its dimensions.

13.3. From the Impulse-momentum equation, prove that the critical flow condition is satisfied when the pressure-momentum force is at a minimum.

13.4. What does the momentum function diagram represent? Can you determine the critical depth from it?

13.5. Define conjugate depths and how are they related to each other. If the first conjugate depth decreases, do you expect the second one to decrease or increase?

13.6. Discuss how the change in discharge affects the momentum function diagram, assuming that other variables are constant.

13.7. What is meant by a standing wave? How does the balance of a standing wave in a horizontal channel differ from that in a steep channel?

13.8. State the condition of formation of the hydraulic jump. Give some examples with illustrations.

13.9. In a supercritical flow, can a hydraulic jump be expected to be formed if the bed slope of the channel is sharply increased or reduced. Why or why not?

13.10. Discuss the velocity distribution in the hydraulic jump. Draw the total energy and hydraulic gradient lines through a hydraulic jump.

13.11. Give examples illustrating the practical usefulness of creating a hydraulic jump. Does the hydraulic jump have any application in the industry?

13.12. Prove that the relation between initial and final water depths of a hydraulic jump in a horizontal wide or rectangular channel can be expressed as a function of the Froude number. What assumptions are employed to develop this relation?

13.13. Cite the differences between conjugate and alternate depths. Can two conjugate depths be alternate depths as well? Why or why not?

13.14. Write the general equation for energy loss in a hydraulic jump. Then deduce an equation for energy loss in rectangular and wide channels.

13.15. Define the height and efficiency of a hydraulic jump. When the height of a jump increases, do you expect its efficiency to increase or decrease? Why or why not?

13.16. What is the rationale for determining the length and profile of hydraulic jumps? Give a practical value for the length of the hydraulic jump, L_j, and the length of roller, L_r, in terms of its height.

13.17. Classify hydraulic jumps according to the upstream Froude number. If the approach Froude number of two different jumps is 3.5 and 6, which of them is more likely to be contained within the basin and why?

13.18. Discuss different possibilities for the jump location after a sluice gate. How does a jump move if its rating curve is always higher than that of the tail water?

13.19. Using illustrations, explain the practical use of creating a hump or a sudden drop in the bed of a channel.

13.20. Explain the main function of the baffle blocks. Write down the equation used to determine the force acting on baffle blocks per unit width of wide channels.

13.21. Explain why the forces acting on a jump over a sloping bed differ from those acting on a jump over a horizontal bed. Give two examples for hydraulic jumps formed over a steep bed.

13.22. How are oblique jumps formed? Derive the relationship between conjugate depths of an oblique jump.

Problems

13.1. A hydraulic jump is formed in a trapezoidal channel of 2.0-m bed width, 1 : 1 side slope, and carrying a discharge of 6.0 m^3/s. The initial and final depths of the jump are 0.5 m and 2.0 m, respectively. Concrete blocks are placed to stabilize the jump. Calculate the minimum weight of these blocks if the coefficient of friction between the blocks and channel bed, C_f, is 0.6.

13.2. Construct the *M-y* diagram for the channel section given in Problem 13.1 under a given discharge. Find the critical depth.

13.3. Construct the *M-y* diagram for a parabolic channel having a peripheral equation of $y = 2x^2$. The channel carries a discharge of 6 m^3/s. Find the conjugate depth for an initial depth of 1.25 m.

13.4. For the channel section given in Problems 13.1 and 13.2, construct the *M-y* diagram on the same graph for $Q = 4$, 6, and 8 m^3/s. Find the critical depth for each case.

13.5. For Problem 13.4, determine the two conjugate depths for the three different discharges at a momentum of 5.0 m^3.

13.6. A wide channel carries a specific discharge of 30 ft^2/s at a bed slope of 0.0003 ft/ft. A gate with an opening of 0.2 ft is placed in the channel. Determine whether a hydraulic jump will occur after the gate or not. Take $C = 40$.

13.7. A trapezoidal channel has a side slope of 1:1 and a bed width of 6.0 m. A hydraulic jump with a height of 0.5 m is encountered in the channel. Determine the initial and ultimate water depths of the jump and the critical depth if the channel has a discharge of 10.0 m^3/s.

13.8. A trapezoidal channel has a bed width of 3.0 m and a side slope of 1:1 and carries a discharge of 8.0 m^3/s. Construct the specific force diagram. A hydraulic jump is to be formed in this channel such that the downstream depth is not to exceed 1.75 m. Find the limiting value of the upstream depth. Calculate the head lost and the horsepower dissipated in the jump.

13.9. A triangular channel with an apex angle of 90° carries a discharge of 10.0 m^3/s. What is the minimum upstream depth for a jump such that the downstream depth does not exceed 3.0 m? Calculate the power dissipated under such a condition.

13.10. A hydraulic jump is formed in a triangular channel with a side slope of 2 : 1. The height of the jump is 1.0 m, and the Froude number is equal to 2.5 at the upstream side of the jump. Determine the Froude number after the jump and the power lost in the jump.

13.11. A rectangular channel 6.0 ft wide carries a discharge of 35 ft^3/s. Calculate the critical depth. If a standing wave is formed at a point, where the depth of flow is 0.5 ft, what would be the rise in the water level?

13.12. A sluice gate is located in a horizontal rectangular channel of constant width. When the gate is partially opened, water issues with a velocity of 20.0 ft/s and forms a standing wave with an initial depth of 0.60 ft. Determine the critical depth, the height of wave, and the energy lost in the jump.

13.13. A hydraulic jump is formed in a rectangular channel 6.5 ft wide. The conjugate depths are 0.5 ft and 4.0 ft, respectively. Calculate the specific discharge, energy lost in the jump, critical depth, and the minimum specific energy.

13.14. A 15-ft-high spillway is placed across a 15-ft-wide rectangular channel which is carrying a discharge of 600 cfs as shown in Figure P13.14. The water is flowing at an initial normal depth of 6 ft. The Francis C for the spillway is 3.33 relative to the actual head on the spillway.

a. To what depth will the water rise behind the spillway?

b. Ignoring losses on the face, what will be the water depth at the toe?

c. Will a hydraulic jump form downstream of the spillway?

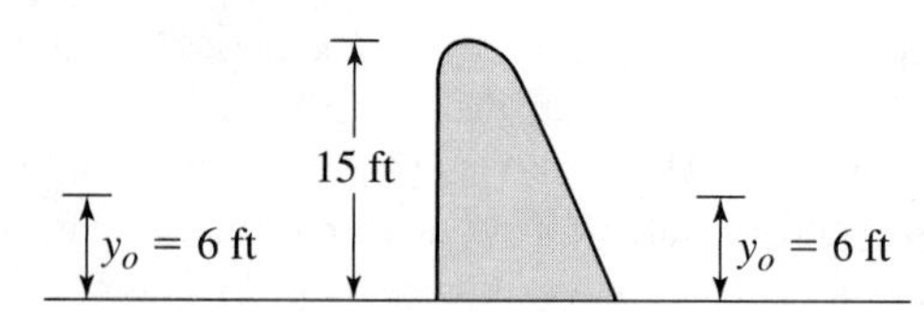

Figure P13.14 The spillway for Problem 13.14.

13.15. A 25-ft-high spillway is placed across a 20-ft-wide rectangular channel. The slope of the spillway face is $1V/4H$. The downstream channel is running on a slope of 0.000845 with an estimated Manning n value of 0.03. If the head on the spillway is measured to be 3 ft and the C value is 3.1,

a. Assuming losses on the spillway face will be $0.1\left(\frac{V_2^2 - V_1^2}{2g}\right)$, what is the water depth at the spillway toe?

b. Will a hydraulic jump form downstream of the spillway, and, if so, will it occur right at the toe or run?

c. If there is a jump, what would be its conjugate depths?

13.16. Water is stored behind a sluice gate to a depth of 3 m as shown in Figure P13.16. The gate opening is 0.55 m and the channel is wide rectangular with a Manning n value of 0.035. Use a contraction coefficient of 0.6 for the gate.

a. If the gate is clear (i.e., unsubmerged), what is its discharge?

b. What is the minimum channel slope downstream for this case to prevail?

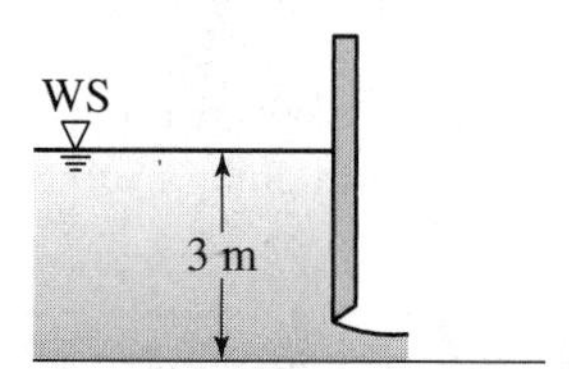

Figure P13.16 The sluice gate for Problem 13.16.

13.17. Let the actual channel slope in Problem 13.16 be 0.0015 m/m and its width be 50 m. All other data are as specified in the problem.

a. Will the discharge remain the same as computed in Problem 13.16?

b. If a hydraulic jump forms downstream of the gate, will it run? What are its conjugate depths?

13.18. Water is stored behind a sluice gate to a depth of 10 ft. The channel is rectangular with a width of 15 ft. The contraction coefficient for the gate is 0.60.

a. If the specific discharge (q) through the gate is 20 cfs/ft, what two different gate openings would allow for this?

b. Which of these two possibilities is more reasonable?

13.19. Water is stored behind a sluice gate to a depth of 14 ft. The channel is a rectangular section 10 ft wide running on a slope of 0.0175 with a Manning n value of 0.025. The specific discharge (q) through the gate is 28.93 cfs/ft and the contraction coefficient is 0.60. Will a hydraulic jump form downstream of the gate, and, if so, what will be its conjugate depths?

13.20. A spillway has a specific discharge of 70.0 ft^2/s. A hydraulic jump is formed over the apron of the spillway as shown in Figure P13.20. If the water depth just upstream of the spillway is 15.0 ft, determine the sequent depths of the jump. Neglect friction losses over the spillway.

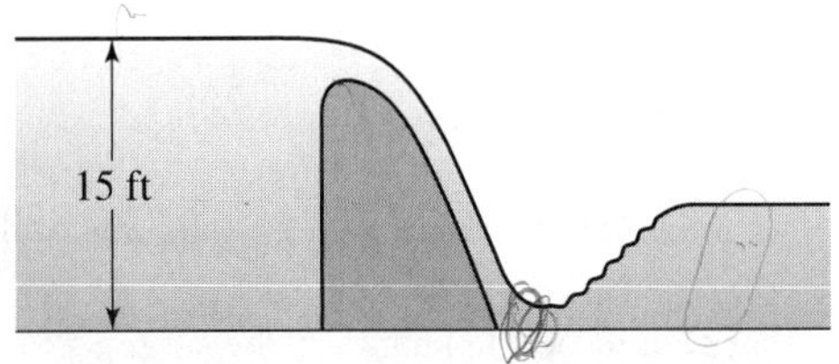

Figure P13.20 The spillway and jump for Problem 13.20.

13.21. Calculate the value and direction of the horizontal force acting on the spillway given in the previous problem, and knowing that the width of the spillway is 20 ft.

13.22. Cite the conditions which when any of them are found, the flow would be critical.

13.23. Using the momentum principle, derive a relation for conjugate depths and the flow rate in a triangular open channel having a side slope of 1:1.

13.24. If y_1 and y_2 are the two conjugate depths of a hydraulic jump formed in a channel of triangular section with an apex angle of 90°, prove that the Froude number at the upstream side, F_{n1}, can be expressed as

$$F_{n1}^2 = \frac{2}{3}\left(\frac{y_2}{y_1}\right)^2 \left\{ \frac{1 - \left(\frac{y_2}{y_1}\right)^3}{1 - \left(\frac{y_2}{y}\right)^3} \right\}$$

13.25. For the triangular section given in Problem 13.9, calculate F_{n1} for the following values of (y_2/y_1): 2, 3, 4, 5, 6; and then plot the relation between (y_2/y_1) and F_{n1}. Determine the head loss in a hydraulic jump if the initial depth and upstream Froude number are 0.5 m and 5.0, respectively.

13.26. A rectangular channel 5.0 m wide carries a specific discharge of 4.0 m²/s. A hydraulic jump is fixed in a certain position by placing concrete blocks as shown in Figure P13.26. The initial depth of the jump is 0.5 m. Determine the force acting on the blocks if the depth of water after the jump is 2.0 m. In case that the concrete blocks are removed, what is the new water depth?

13.27. Knowing the two conjugate depths of a jump and the flow rate in a triangular channel with an apex angle of 90°, derive a formula for the energy dissipation in the jump.

13.28. The discharge through a 6.0-ft-wide rectangular channel is 200.0 ft³/s. Find the conjugate depths of the jump formed in such a channel if 3.5 ft of energy head is lost through it.

13.29. Water flows at a rate of 45 ft³/s along a horizontal rectangular channel 4.5 ft in width. A hydraulic jump is formed at a point where the water depth is 0.25 ft. Determine the rise in the water level and the horse power lost in the jump.

13.30. Water flows at a depth $y_1 = 0.4$ ft in a concrete channel which is 10 ft wide. The discharge per unit width is 10.0 ft²/s. Estimate the height of the hydraulic jump that will be formed if the bed slope is equal to 0.0 and 0.15, respectively.

13.31. Applying the momentum principle and the continuity equation to the analysis of a submerged hydraulic jump which occurs after the gate opening in rectangular channels, prove that

$$\frac{y_s}{y_2} = \sqrt{1 + 2F_{n2}^2\left(1 - \frac{y_2}{y_1}\right)}$$

where y_s is the submerged depth, y_1 is the height of sluice-gate opening, y_2 is the tail-water depth, and $F_{n2}^2 = q^2/gy_2^3$. Neglect the bed friction.

13.32. A sluice gate is located in a 6.0-ft-wide rectangular channel as shown in Figure P13.32. The difference in the water level just before and after the gate is 4.8 ft. Concrete blocks are placed on the bed to

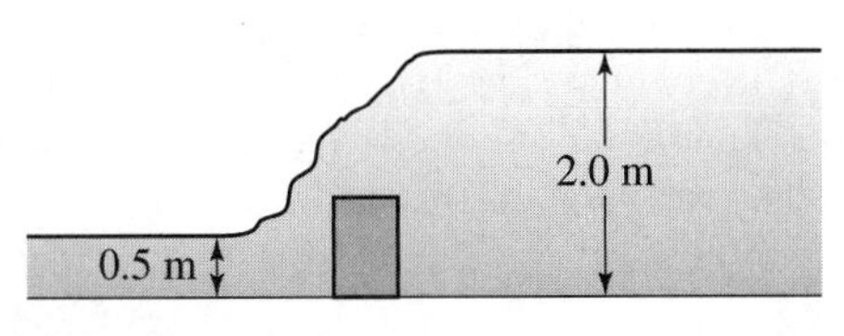

Figure P13.26 The Hydraulic jump for Problem 13.26.

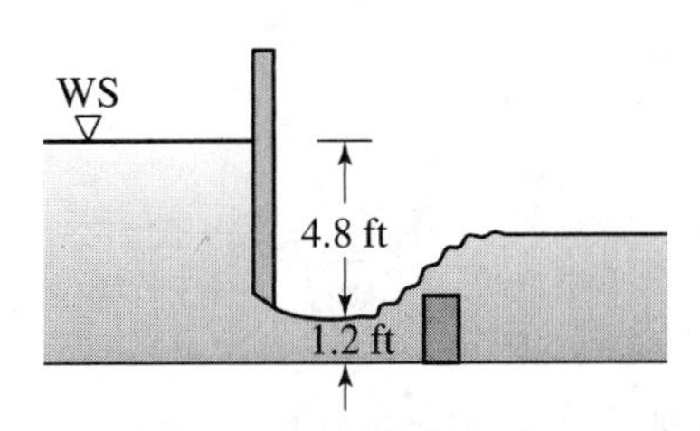

Figure P13.32 The sluice gate in Problem 13.32.

form a hydraulic jump. Determine the force acting on the blocks if the upstream depth of the jump is 1.2 ft. Use a contraction coefficient for the gate of 0.60.

13.33. Water is stored behind a sluice gate to a depth of 5 m. The gate opening is 0.5 m and the contraction coefficient is 0.60. The channel is rectangular with a normal depth for the flow through the gate of 1.5 m.

a. Will a hydraulic jump form downstream of the gate, and, if so, what will be its conjugate depths?
b. If an end sill, functioning as a broad-crested weir, is constructed downstream of the gate as shown in Figure P13.33, what would be its maximum height so as to not *affect* the jump?
c. To what height can be sill be constructed so that it will not *drown* the jump?

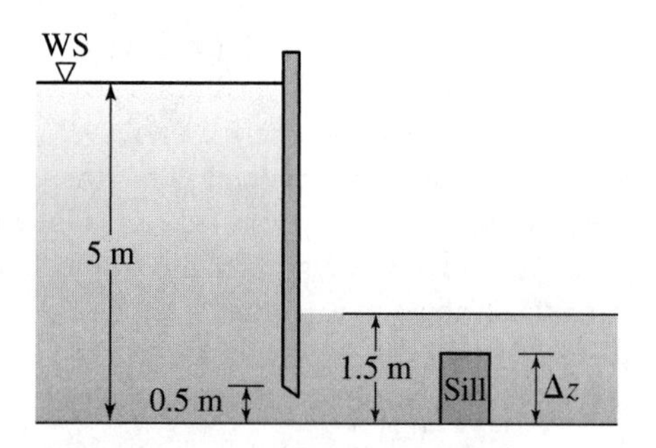

Figure P13.33 The gate and sill for Problem 13.33.

13.34. A small 12-ft-high spillway is placed across a 8-ft-wide rectangular channel carrying a discharge of 200 cfs as shown in Figure P13.34. The Francis C value for the spillway is 3.30. A hydraulic jump will be formed downstream of the spillway using an end sill that will function as a sharp-crested weir with a Francis C value of 3.33 relative to the actual head on the weir. Ignore energy losses over the spillway face.

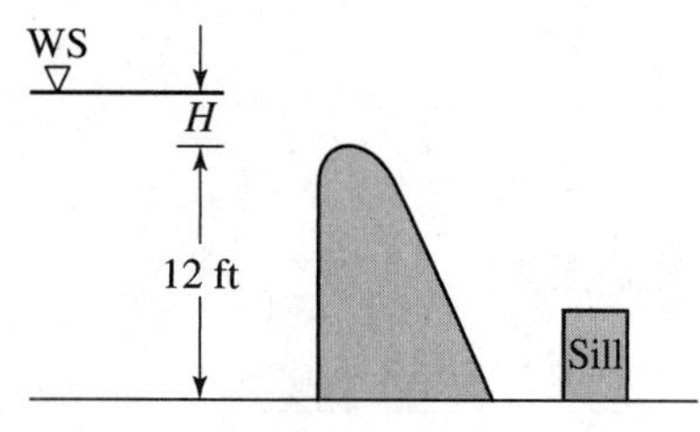

Figure P13.34 Spillway and sill for Problem 13.34.

a. What should be the height of the sill so as to force the jump to occur right at the toe of the spillway?
b. What will be the force acting on the sill?

13.35. Referring to Problem 13.34, if we want the upstream depth of the jump to be 1.1 ft, what would be the required height of the sill? What would be the force acting on the sill in this case?

13.36. A hydraulic jump is formed in a 7.5-ft-wide rectangular channel which carries a specific discharge of 25.0 ft^2/s. Determine the jump height and efficiency if the downstream depth is 6.5 ft.

13.37. A trapezoidal channel of 2.0 m bed width and 3:2 side slope carries a discharge of 10.0 m^3/s with a depth of flow of 0.5 m. Determine whether the flow is supercritical or subcritical. If a hydraulic jump were to form, what would be its height and efficiency?

13.38. A 3.0-m rectangular channel carries a discharge of 8.0 m^3/s with a water depth of 4.0 m. A free jump is forced to form at a section where the channel width increases to 4.0 m. The relation between y_2 and y_{2b} is given as

$$\frac{y_2}{y_{2b}} = \psi_b^{3/8}$$

where y_2 is the conjugate depth in the expanding section of the channel, y_{2b} is the conjugate depth if the channel width has not been changed, and ψ_b is equal to the initial width, b, divided by the expanded width B. Determine the conjugate depth at the enlarged section.

13.39. Suppose in Example 13.6 we need to place a sill just downstream of the jump. If the height of the sill is 1.5 ft, what will be the effect, if any, on the jump? Assume the sill is acting as a broad-crested weir.

13.40. A gate is located in a rectangular open channel 2.0 m wide. The depth of flow just before and after the gate is 3.0 m and 0.40 m, respectively. The channel carries a specific discharge of 2.85 m^3/s. Find the force acting on the gate assuming no energy loss through the gate.

References

Arbhabhirama, A. and Abella, A. U., 1971. Hydraulic jump within gradually expanding channel. *Jour of Hyd. Div., Proc. ASCE*, Vol. 97, No. NY1, pp. 31–41.

Bakhmeteff, B. A. and Matzke, A. E., 1936. The hydraulic jump in terms of dynamic similarity. *Transactions, American Society of Civil Engineers*, Vol. 101, pp. 630–647.

Basco, D. R. and Adams, J. R., 1971. Drag forces on baffle blocks in hydraulic jumps, *Jour of Hyd. Div., Proc. ASCE*, Vol. 97, No. 12, pp. 2033–2035.

Blaisdell, F. W., 1948. Development and hydraulic design, Saint Anthony Fall-stilling basin. *Transactions, American Society of Civil Engineers*, Vol. 123, pp. 483–520.

Bresse, J. A. Ch., 1860. Cours de mécaníque appliquée, 2e partie, Hydraulique (Course in Applied Mechanics, pt. 2, Hydraulics), Mallet-Bachelier, Paris.

Biden, G., 1819. Observations sure hauteur du raced hydraulic en 1818 (Observations on the height of the hydraulic jump in 1818); a report presented at the Dec. 12, 1819, meeting of the Royal Academy of Science of Turin and later incorporated as a part of (2), pp. 21–80.

Chaudhry, M. H., 1993. *Open Channel Flow*, Prentice Hall, Englewood Cliffs, N.J.

Citrini, D., 1939. R1 Salto di Biden (The hydraulic jump). *L'Energia electrica*, Milano, Vol. 16, no. 6, pp. 441–465, June and no. 7, pp. 517–527, July. Contains a resume of work done up to 1939.

Darcy, H. and Basin, H., 1865. Recherche experimentalism relatives aux Remus et a propagation des odes (Experimental research on backwater and wave propagation). Vol. 12, *Recherche Hydraulic es* (Hydraulic Researches), Academie des Sciences, Paris.

Forster, J. R. and Skrinde, R. A., 1950. Control of hydraulic jump by sills. *Trans., Amer. Soc. Civil Engr.*, 125: 973–1022.

Gibson, A. H., 1913. The formation of standing waves in on open stream. paper 4081, Minutes of Proceedings of the Institution of Civil Engineers, London, Vol. 3, pp. 233–242.

Hager, W. H., 1992. *Energy Dissipaters and Hydraulic Jump*. Kluwer Academic, Dordrecht, The Netherlands.

Henderson, F. M., 1966. *Open Channel Flow*, Macmillan, New York.

Herbrand, K., 1973. The special hydraulic jump. *Jour. of Hyd. Research*, Vol. 11, No. 3, pp. 205–210.

Ippen, A. T., 1951. Mechanics of supercritical flow, 1st paper of high-velocity flow in open channels: A symposium. *Transactions, American Society of Civil Engineers*, Vol. 126, pp. 268–295.

Ippen, A. T. and Harleman, D. R. F., 1956. Verification of theory for oblique standing waves. *Transactions, American Society of Civil Engineers*, Vol. 121, pp. 678–694.

Lindquist, E. G. W., 1927. Anordningar for effektive energiomvandling vid foten av over falls dammar (Arrangements for effective energy dissipation at the toes of dams). Anniversary Volume, Royal Technical University, Stockholm, Sweden.

Massey, B. S., 1961. Hydraulic jump in trapezoidal channels, and improved method. *Water Power* (June 1961): 232–237.

Peterka, A. J., 1963. Hydraulic Design of Stilling Basins and Energy Dissipators, Engineering Monograph No. 25, U.S. Bureau of Reclamation, Denver, Co.

Rajaratanam, N., 1967. Hydraulic jumps. Chapter in *Advances in Hydroscience*, Vol. 4, Chow V. T. (ed.). Academic Press, New York and London.

Rajaratnam, N. and Subramanya, K., 1968. Profile of the hydraulic jump. *Jour. of Hyd. Div., Proc. ASCE,* Vol. 94, pp. 55–112.

Ranga Raju, K. G., 1981, *Flow through Open Channels.* Tata McGraw-Hill Publishing Company Limited, New Delhi.

Ranga Raju, K. G., Mittal, M. K., Verma, M. S., and Ganeshan, V., 1980. Analysis of flow over baffle blocks and end sills. *Jour. of Hyd. Research,* Vol. 18, No. 2, pp. 000.

Rouse, H., 1938. *Fluid Mechanics for Hydraulic Engineers.* McGraw-Hill Book Company, Inc., New York.

Stevens, J. C., 1944. Discussion on the hydraulic jump in sloping channels by Carl E. Kinsvaster. *Transactions, American Society of Civil Engineers,* Vol. 109, pp. 1225–1246.

Sturm, T. W., 2001. *Open Channel Hydraulics,* McGraw-Hill Book Company, New York.

Thiruvengadam, A., 1961. Hydraulic jump in circular channels. *Water Power* (December, 1961): 496–497.

CHAPTER 14

GRADUALLY VARIED FLOW

Non-uniform flow at Bonne Carre Spillway. (Courtesy of Colleen Perilloux Landry.)

Flow in an open channel tends to be uniform under natural conditions where the channel cross section, roughness, and bed slope are constant along its length. Under such conditions, the driving force or the component of the fluid weight in the flow direction is balanced by the resisting force or the friction force along the bottom and sides of the channel. This results into a constant depth of flow in space and hence a uniform flow is developed. If any changes are introduced to the channel, the flow would proceed through a transition zone until it reaches a new uniform condition in which the forces are balanced again. The new uniform depth may differ significantly or slightly from the initial one. Even the state of flow may also change. Varied flow is generally encountered between two uniform depths or in the vicinity of hydraulic structures, such as gates, weirs, dams, and so on.

In the previous chapter, rapidly varied flow, in which the depth of flow varies considerably along short distances, was discussed. If the variation in the depth of flow is smooth, the flow is gradually varied and the curvature of streamlines can be ignored. As the depth of flow increases or decreases, the average velocity decreases or increases as well. Due to this smooth variation, the pressure distribution in the vertical direction is assumed hydrostatic. Unlike rapidly varied flow, the loss of energy in gradually varied flow is mainly attributed to the boundary friction.

In this chapter, the discussion will be limited to prismatic channels with constant discharge along the sections under consideration. Spatially varied flow, in which lateral inflow and outflow exist, is not considered.

14.1 Gradually Varied Flow Equation

In uniform flow the slope of the water surface is equal to that of the bed and the total energy. At any point along the channel the water levels can, thus, be easily evaluated. However, the same is not true in the case of nonuniform flow. Prediction of the depth of flow or water level at any point in open channels is of prime importance. Knowledge of the depth of flow is essential for the calculation of other geometric and hydraulic parameters of a channel section.

A typical case of gradually varied flow is found upstream of gates, as shown in Figure 14.1a. As the flow approaches the gate, the depth of flow increases gradually and the average velocity decreases. Figure 14.1 provides various examples for common cases of gradually varied flow.

14.1.1 Assumptions

In the analysis of gradually varied flow, the following assumptions are generally made: (1) The channel is prismatic and the flow is steady (i.e, the depth of flow is constant with time). (2) The bed slope, S_o, is relatively small. Therefore, the vertical depth of flow is almost equal to the depth measured normal to the channel bed. (3) The velocity distribution in the vertical section is uniform and the kinetic energy correction factor is close to unity. (4) Streamlines are

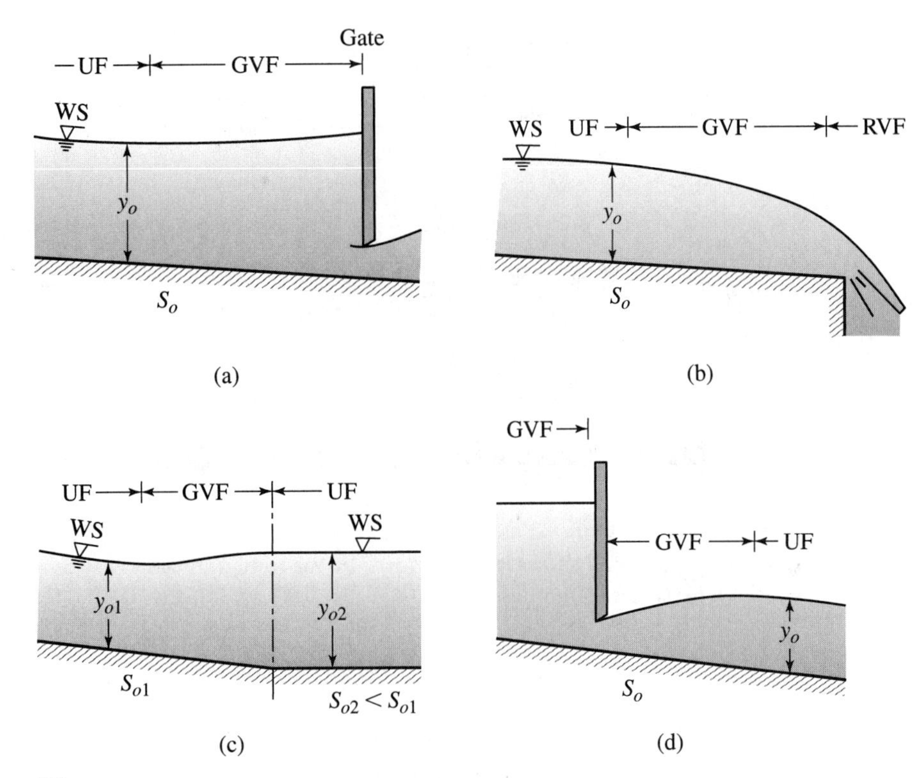

Figure 14.1 Examples for gradually varied flow in open channels.

parallel and the pressure distribution is hydrostatic. (5) The channel roughness is constant along its length and does not depend on the depth of flow.

14.1.2 Derivation of Gradually Varied Flow Equation

Consider a channel segment of length dl as shown in Figure 14.2(a). The depth of flow, y, varies by dy while the average velocity, V, varies by dV along the given segment. Under these conditions of steady, gradually varied flow, the one-dimensional longitudinal water surface profile can be described by a simple ordinary differential equation. From energy considerations, we have

$$E = y + \frac{Q^2}{2gA^2} \tag{14.1}$$

and from Section 12.4 we know that for a general cross section

$$\frac{dE}{dy} = \frac{d}{dy}\left(y + \frac{Q^2}{2gA^2}\right) = 1 - \frac{Q^2 B}{gA^3} = 1 - F_n^2 \tag{14.2}$$

Referring to Figure 14.2b, the total energy per unit weight of water measured from a horizontal datum, or the total head, is given as

$$H = z + y + \frac{V^2}{2g} = z + E \tag{14.3}$$

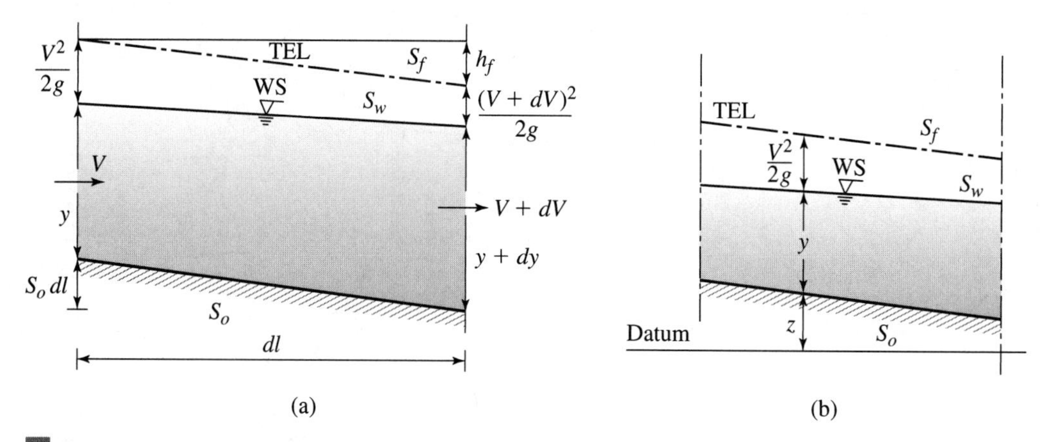

Figure 14.2 Notation for derivation of the gradually varied flow equation.

where E is the specific energy. The variation in total head with respect to longitudinal distance is written as

$$\frac{dH}{dl} = \frac{dz}{dl} + \frac{dE}{dl} = \frac{dz}{dl} + \frac{dE}{dy}\frac{dy}{dl} \tag{14.4}$$

Substituting $-S_f$ and $-S_o$ for dH/dl and dz/dl, respectively, as both H and z decrease in the flow direction, one gets

$$-S_f = -S_o + \frac{dE}{dy}\frac{dy}{dl} \tag{14.5}$$

Then, substituting $1 - F_n^2$ for $\frac{dE}{dy}$, one can write

$$S_o - S_f = (1 - F_n^2)\frac{dy}{dl} \tag{14.6}$$

Hence, equation (14.6) can be written as

$$\frac{dy}{dl} = \frac{S_o - S_f}{1 - F_n^2} \tag{14.7}$$

Equation (14.7) is the general governing equation for gradually varied flow in open channels and forms the core of the discussion in this chapter. The left-hand side of this equation, dy/dl, denotes the variation in flow depth with respect to distance, along the flow direction.

Substituting for S_f in equation (14.7) from the Chezy or Manning equation, one can write

$$\frac{dy}{dl} = \frac{S_o - \dfrac{V^2}{C^2R}}{1 - F_n^2} \tag{14.8a}$$

and in SI units,

$$\frac{dy}{dl} = \frac{S_o - \dfrac{V^2 n^2}{R^{4/3}}}{1 - F_n^2} \tag{14.8b}$$

or in BG units,

$$\frac{dy}{dl} = \frac{S_o - \dfrac{V^2 n^2}{2.22 R^{4/3}}}{1 - F_n^2} \tag{14.8c}$$

where C and n are the Chezy and Manning coefficients, respectively. At this time, the discussion in Section 11.1 concerning the reliance of the Chezy C coefficient and the Manning n factor on the system of units being used in a particular problem should be recalled. Tables of C and n are normally given in BG units (including those in Chapter 11).

For rectangular cross sections the Froude number can be expressed as $F_n = \dfrac{V}{\sqrt{gy}}$. Therefore, equation (14.8) is written as

$$\frac{dy}{dl} = \frac{S_o - S_f}{1 - \dfrac{V^2}{gy}} = S_o \left[\frac{1 - \dfrac{S_f}{S_o}}{1 - \dfrac{V^2}{gy}} \right] \tag{14.9}$$

For wide sections, the hydraulic radius, R, is assumed equal to y and the wetted perimeter, P, is assumed equal to the top width, B. Using the Chezy equation, one can write

$$S_f = \frac{V^2}{C^2 R} = \frac{q^2}{C^2 y^3} \tag{14.10a}$$

and similarly

$$S_o = \frac{q^2}{C^2 y_o^3} \tag{14.10b}$$

Therefore,

$$\frac{S_f}{S_o} = \left(\frac{y_o}{y}\right)^3 \tag{14.11}$$

Also,

$$\frac{V^2}{gy} = \frac{q^2}{gy^3} = \left(\frac{y_c}{y}\right)^3 \tag{14.12}$$

Substituting equations (14.11) and (14.12) in equation (14.9), one obtains

$$\frac{dy}{dl} = S_o \left\{ \frac{1 - \left(\dfrac{y_o}{y}\right)^3}{1 - \left(\dfrac{y_c}{y}\right)^3} \right\} \tag{14.13}$$

Following the same procedure but using Manning's formula, equation (14.11) is written as

$$\left(\frac{S_f}{S_o}\right) = \left(\frac{y_o}{y}\right)^{10/3} \tag{14.14}$$

Substitution of equations (14.12) and (14.14) in equation (14.9) yields

$$\frac{dy}{dl} = S_o\left\{\frac{1 - \left(\frac{y_o}{y}\right)^{10/3}}{1 - \left(\frac{y_c}{y}\right)^{3}}\right\} \tag{14.15}$$

Equations (14.13) and (14.15) are valid only for wide sections; however, the general principle embodied in equations (14.7)–(14.8) is true for any cross section. In these equations three depths are found: the normal depth y_o, the critical depth y_c, and the depth of varied flow y. Equations (14.7), (14.8), and (14.15) are known in hydraulics as the process equation and can be used to examine the shape of one-dimensional, steady, gradually varied water surface profiles qualitatively and quantitatively. In a qualitative sense, we need only be concerned with the sign of the derivative dy/dl in order to determine if we have a rising or falling profile in the longitudinal direction.

Example 14.1

A rectangular channel 10.0 ft wide and 3.0 ft water depth has a bed slope of 5×10^{-5} ft/ft. The channel carries a discharge of 100 ft^3/s. Find the rate of change in the depth of flow, assuming $C = 50$.

Solution: The rate dy/dl is given by equation (14.7), where

$$V = \frac{Q}{by} = \frac{100}{10 \times 3} = 3.33 \text{ ft/s}$$

$$R = \frac{A}{P} = \frac{10 \times 3}{10 + 6} = 1.875 \text{ ft}$$

$$S_f = \frac{V^2}{C^2 R} = 2.36 \times 10^{-3}$$

$$F_n^2 = \frac{V^2}{gy} = 0.114$$

Hence,

$$\frac{dy}{dl} = \frac{5(10^{-5}) - 2.36(10^{-3})}{1 - .114} = \frac{-2.31 \times 10^{-3}}{0.886} = -2.61 \times 10^{-3}$$

The negative sign means that the depth is decreasing in the flow direction.

Example 14.2

A wide channel carries a specific discharge of 1.0 m^2/s. The depth of flow is 2.0 m and the bed slope is 15 cm/km. Determine the rate at which the flow depth is varied and whether it is increasing or decreasing. Assume $n = 0.021$ (metric).

Solution: Using the equations of a wide section, one can write

$$y_c = \sqrt[3]{\frac{q^2}{g}} = \sqrt[3]{\frac{(1.0)^2}{9.81}} = 0.467 \text{ m}$$

$$y_o = \left(\frac{qn}{(S_o)^{1/2}}\right)^{3/5} = \left(\frac{1(.021)}{(0.00015)^{1/2}}\right)^{3/5} = 1.38 \text{ m}$$

Substituting in equation (14.15), one gets

$$\frac{dy}{dl} = (0.00015)\left(\frac{1 - \left(\frac{1.38}{2}\right)^{10/3}}{1 - \left(\frac{0.467}{2}\right)^3}\right) = 0.000108 \text{ m/m}$$

The depth of flow is increasing at a rate of 10.8×10^{-5} m/m.

14.1.3 Water Surface Slope, S_w, Related to a Horizontal Line

In the former analysis the term dy/dl denotes the rate of variation of the depth of flow with respect to distance. In cases where dy/dl is found equal to zero, the line of water surface will be parallel to the bed line and the flow will be uniform (no variation in depth with respect to distance). In this case S_w will be equal to S_o. In other words, the term dy/dl defines the water surface slope with respect to the bed line, which is not horizontal. The following relation can thus be developed to evaluate S_w, with respect to a horizontal line:

$$S_w = S_o + \frac{dy}{dl} \tag{14.16}$$

14.2 Water Surface Profiles

Classification of flow profiles allows for the analysis of the behavior of steady, nonuniform flow in open channels and facilitates the association of observed patterns of behavior with specific hydraulic controls. In this context, the term *control* assumes a more general meaning (i.e., the cause and originator of a specific longitudinal water surface profile associated with a given discharge). Of course, the specific hydraulic definition (i.e., the location of a known relationship between the depth and discharge) still applies in this context. Thus, through classification, one is able to identify the nature and location of the control of any observed water surface profile if the value of the discharge can be estimated. In the above section, the governing equation which defines the flow transition from one level to the other under gradually varied flow conditions was developed. The equation is valid for any type of flow and any bed slope.

The process equation as illustrated by equation (14.15) implies a classification scheme using three pieces of information relevant to the flow (i.e., how does normal depth compare to critical depth for the given discharge; how do the given flow characteristics compare to critical conditions; and how do the given flow conditions compare to normal conditions?). With these pieces of information known, the equation can be used in a qualitative manner to determine the behavior of the water surface in the longitudinal direction. In evaluating the flow profiles it is useful to recall some basic slope-depth relationships:

1. If $y > y_o$, then $S_f < S_o$
2. If $y < y_o$, then $S_f > S_o$
3. If $y_o > y > y_c$, then $S_o < S_f < S_c$
4. If $y < y_c$, then $S_f > S_c$ and $F_n > 1$
5. If $y > y_c$, then $S_f < S_c$ and $F_n < 1$

where S_f is the slope of the total energy line and S_c is the critical slope.

It is now recognized that there are five possible cases of nonuniform depths and slopes (and thus profiles) which are of interest; that is, $y_o > y_c$ (mild), $y_o = y_c$ (critical), $y_o < y_c$ (steep), $S_o = 0$ (horizontal), and S_o of opposite sign to the prevalent bed slope (adverse).

14.2.1 Classification of Channels According to Bed Slope

According to the bed slope, five different categories are defined:

1. The mild slope *M*: The bed slope, S_o, is less than the critical slope, S_c, as shown in Figure 14.3a. In this category, the normal depth, y_o, is located above the critical depth, y_c, and the Froude number, F_n, will be less than unity under normal flow conditions.
2. The steep slope *S*: The bed slope is steeper than the critical slope, as shown in Figure 14.3b. The normal depth is less than the critical depth and the Froude number will be greater than unity under normal flow conditions.
3. The critical slope *C*: In this case, the bed slope is exactly equal to the critical slope, as shown in Figure 14.3c, and the Froude number will be exactly equal to unity under normal flow conditions.
4. Horizontal slope *H*: The bed slope is equal to zero (i.e., the channel has a horizontal bed, as shown in Figure 14.3d). The critical depth can be evaluated while the normal depth will be equal to ∞. Substituting S_o as zero in the Chezy or Manning equation will give infinite value for the normal depth.
5. Adverse slope *A*: The bed slope S_o is less than zero, as shown in Figure 14.3e. Like the horizontal slope, the normal depth is undefined.

On should note that the last two categories are just special cases of the first category. They are mild by definition ($S_o < S_c$), but the normal depth is not defined exactly. However,

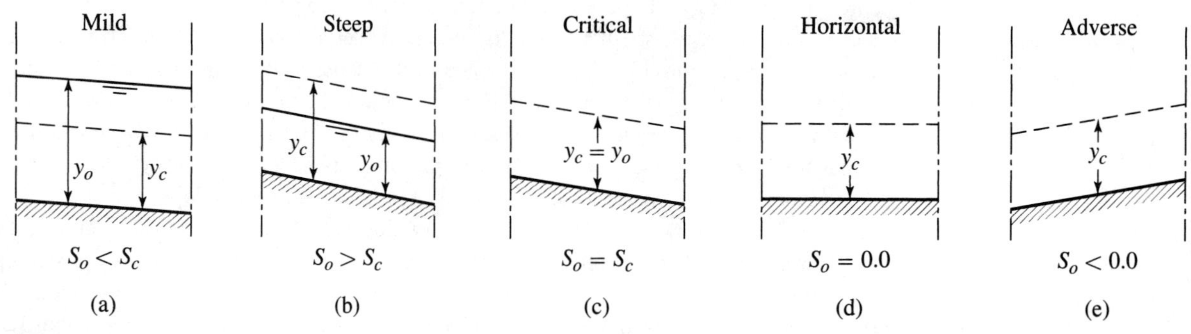

Figure 14.3 Classification of channels according to bed slope.

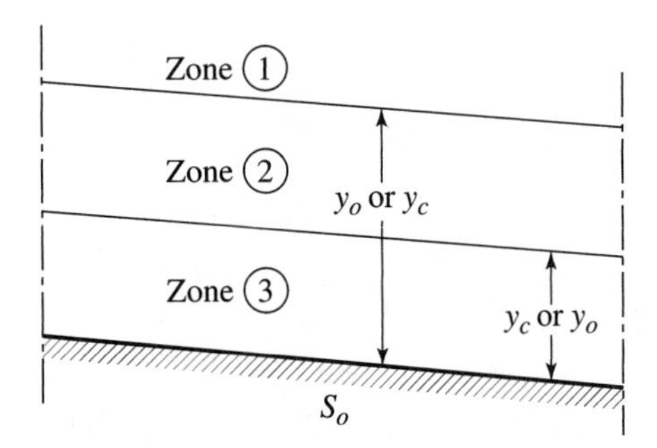

Figure 14.4 Classification of zones according to y_o and y_c.

the water level in the H and A categories should be located above the critical depth. Reducing the bed slope should result in higher depths of flow. Therefore, the water level in a horizontal channel reach should be higher than that in the proceeding and following reaches. The same concept is valid for channel reaches with adverse bottom slope.

14.2.2 Stage Zone According to Normal and Critical Depths

Three zones are defined according to the stage of the water surface with respect to normal and critical depths. Zone 1 is defined above both the normal depth and critical depths, as shown in Figure 14.4. Zone 2 is defined between the normal depth and the critical depth, while Zone 3 is located below both the critical and normal depths, as shown in Figure 14.4. This classification holds regardless of the positions of the normal and critical depths relative to each other (i.e., it is valid for any type of bed slope).

One should note that for the horizontal and adverse bed slopes the normal depth is not defined. However, the water surface is anticipated to be located at a relatively high stage. In this case, Zone 1 could not be defined, while Zone 2 will be located above the critical depth. Also for the critical slope, the normal depth is exactly equal to the critical depth; thus Zone 2 vanishes. In conclusion, three zones are defined as shown in Figure 14.4:

Zone 1: y is located above both y_o and y_c, i.e., $y > y_o > y_c$, or $y > y_c > y_o$
Zone 2: y is located between y_o and y_c: $y_o > y > y_c$ or $y_c > y > y_o$
Zone 3: y is located below both y_o and y_c, i.e., $y_o > y_c > y$ or $y_c > y_o > y$

14.2.3 Classification of Profiles According to (*dy*/*dl*)

According to the rate of change of the flow depth with distance, dy/dl, the trend of the water surface in gradually varied flow is divided into six classes or regimes:

1. dy/dl is equal to a positive value. This means that the depth of flow is increasing with distance. The water surface forms a rising curve, Figure 14.5a.
2. dy/dl is equal to a negative value. This indicates that the depth of flow decreases with distance. The water surface profile forms a falling curve, Figure 14.5b.
3. dy/dl is equal to a zero. This means that the flow is uniform and the water surface is parallel to the bottom of the channel, Figure 14.5c.
4. dy/dl is equal to negative infinity. This means that the water surface forms a right angle with the channel bed, Figure 14.5d. This condition can never be exactly encountered in nature.

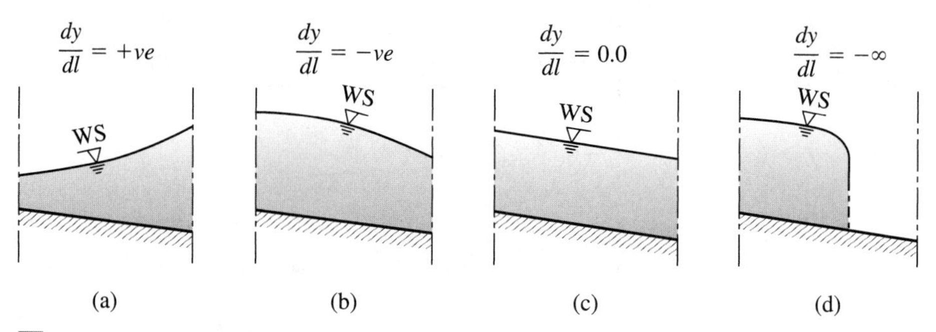

Figure 14.5 Classification of profiles according to dy/dl.

However, this type of profile may occur at the free overfall of a mild channel or at the transition between a mild reach to a steep or critical reach.

5. dy/dl is not defined [i.e., is equal to (∞/∞)]. The depth of flow approaches zero to satisfy this condition. Such a condition is of limited application since a zero depth of flow seldom occurs in open channels.
6. dy/dl is equal to S_o. This means that the water surface profile forms a horizontal line. This is a special case of the rising water profile.

14.2.4 Graphical Representation of the Gradually Varied Flow Equation

A graphical representation of equations (14.7), (14.8), and (14.15) is presented in Figure 14.6. The water surface should follow exactly one of these profiles as it moves from one stage to the other. Consider a channel having a mild bed slope as shown in Figure 14.6a. The three zones in this case are defined as

Zone 1: $y > y_o > y_c$
Zone 2: $y_o > y > y_c$
Zone 3: $y_o > y_c > y$

In Zone 1, the gradually varied flow equation yields the $M1$ curve. For Zones 2 and 3 it yields $M2$ and $M3$ curves, respectively. For simplicity of illustration, consider equation (14.13) for wide or rectangular channels recognizing that the principle applies for all sections.

$$\frac{dy}{dl} = S_o \left\{ \frac{1 - \left(\frac{y_o}{y}\right)^3}{1 - \left(\frac{y_c}{y}\right)^3} \right\} \tag{14.17}$$

As the depth of flow, y, approaches the normal depth, the numerator in the right-hand side of equation (14.17) will be equal to zero. Thus, dy/dl will be zero and the flow tends to be uniform. The water surface will thus always approach the normal depth asymptotically, as shown in Figures 14.6a, b, and c.

The general method for classification will be illustrated in detail using the mild slope situation ($S_o < S_c$; $y_o > y_c$). In the first case (Zone 1), let $y > y_o$. Then $S_f < S_o$ and $F_n < 1$, so

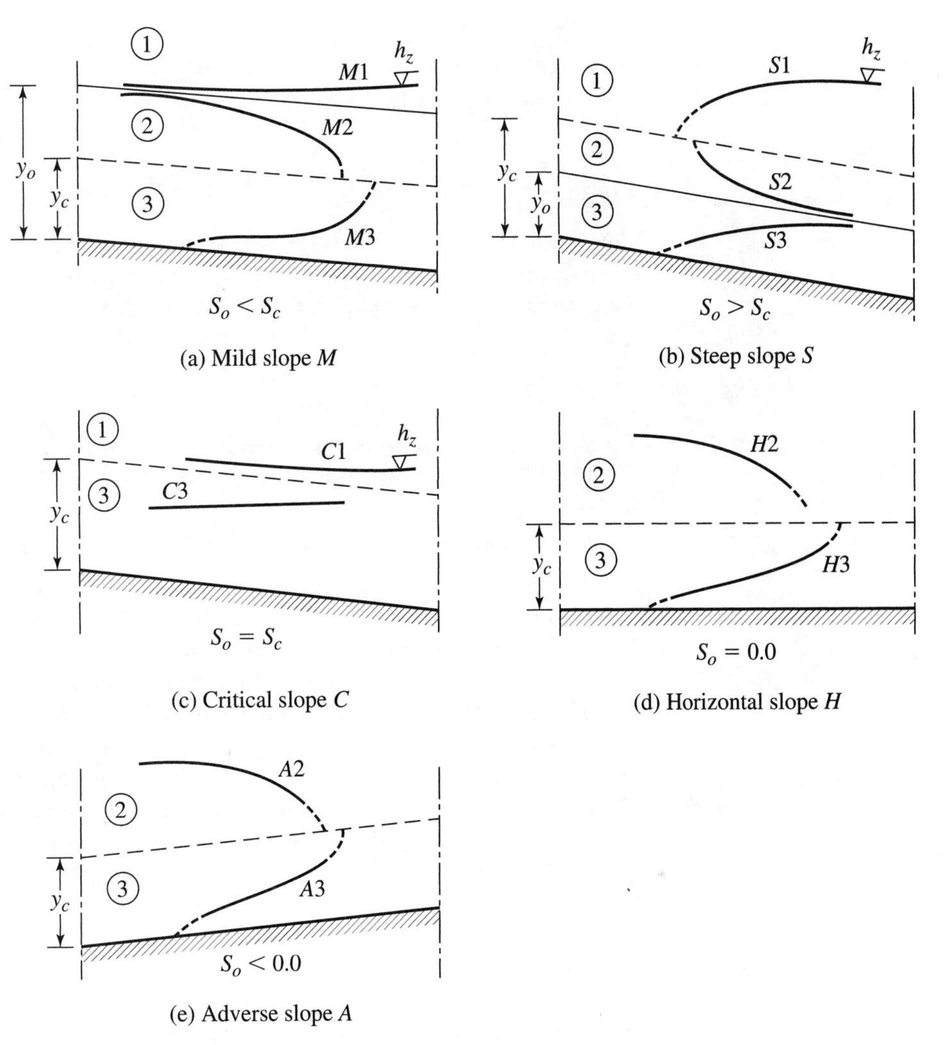

Figure 14.6 Water surface profiles in open channels.

for the general cross section using equation (14.7) in a qualitative sense to determine the behavior of the profile,

$$\frac{dy}{dl} = \frac{S_o - S_f}{1 - F_n^2} = \frac{+}{+} = +$$

Thus, the depth is increasing in the positive (downstream) direction. Indicating that this is case 1 of the mild slope situation, we denote this case as the $M1$ profile as mentioned above. It is called the "backwater proper" profile in that it arises from a downstream control which has the effect of a choke (constriction), thus forcing a depth greater than normal depth upstream. This is the classic case of a backwater situation due to a downstream choke such as a dam or sluice gate.

This means that the $M1$ is a rising curve. At a length tending to infinity the depth of flow, y, will also be infinity (the depth increases with distance). For example, substituting for $y = \infty$ in equation (14.17).

$$\frac{dy}{dl} = S_o \left\{ \frac{1 - 0}{1 - 0} \right\} = S_o \tag{14.18}$$

Thus, the $M1$ curve tends to be horizontal at infinity.

In Zone 2, like the $M1$, the $M2$ curve is tangent to the normal depth. The depth of flow at any point on $M2$ is less than y_o and greater than y_c. On using equation (14.7).

$$\frac{dy}{dl} = \left\{ \frac{-ve}{+ve} \right\} = -ve \tag{14.19}$$

where the ($-ve$ or $+ve$) represents negative and positive signs, respectively.

The $M2$ curve is thus a falling curve in which the depth decreases in the flow direction, Figure 14.6a.

At the critical depth, F_n will be equal to 1. Therefore, as the $M2$ approaches y_c, $\frac{dy}{dl}$ tends to $-\infty$. Thus, the $M2$ profile tends to be normal to the critical depth. Using the same notation as previously defined, we denote this case as the $M2$ profile, called the drawdown curve. Clearly, this profile arises as the result of a downstream control which is tending to draw the profile down from normal depth upstream toward the downstream depth, which would be critical in the limiting case. This is the situation which prevails upstream of an overfall, for example.

A similar analysis can be followed for the $M3$. The curve is a rising one. It approaches the critical depth with a right angle (not fully satisfied in nature) and approaches the channel bed with an acute angle. In Zone 3, the flow depth at any point on the $M3$ profile is less than both y_c and y_o ($y_o > y_c > y$). Therefore, supercritical flow is encountered on a mild slope ($S_f > S_o$, $F_n > 1$). In this case, it is deduced from equation (14.7) that the depth must be increasing downstream, but in supercritical flow. Denoting this case as the $M3$ profile, it is called the slowdown curve because it results from the necessity of the supercritical flow to slow down and thus build up depth as it encounters a mild slope. As a rule, the control of all supercritical flow profiles is placed upstream at the critical section. However, as a practical matter it will be seen that the downstream subcritical depth can also play a vital role in the evolution (or existence) of the $M3$ profile.

It is also useful to note what equation (14.17) implies about the boundaries of steady, nonuniform flow profiles (i.e., as the depth gets very large or approaches either normal or critical depth). In the case of normal depth, as $y \rightarrow y_o$, then $S_f \rightarrow S_o$ and thus $dy/dl \rightarrow 0$, implying a constant depth and thus a fixed asymptotic boundary at y_o. Likewise, as $y \rightarrow \infty$, then S_f and $F_n \rightarrow 0$ and $dy/dl \rightarrow S_o$; thus the profile approaches a horizontal boundary. In the case of critical depth, as $y \rightarrow y_c$ then $F_n^2 \rightarrow 1$ and $dy/dl \rightarrow \infty$, thus implying a vertical water surface as a theoretical boundary at a critical section. Since this is impossible in practice, the water surface will merely approach the critical depth at a large finite angle (Henderson, 1966). With this in mind, one can then sketch the complete water surface profiles for the mild case as shown in Figure 14.6a.

Of particular note is the $M3$ profile (supercritical flow), which cannot end in either the normal depth or critical depth downstream. The profile is rising toward y_c downstream but must encounter a hydraulic jump before it reaches y_c since there must be a transition between the supercritical flow of the $M3$ profile and the subcritical flow downstream. The discussion in the previous paragraph shows that as the depth approaches y_c, it must lead to a discontinuity (hydraulic jump) in the profile. This profile is usually found on a mild slope downstream of a structure physically causing the supercritical flow conditions such as a spillway or sluice gate.

The same procedure can be followed with the rest of water surface profiles. For the steep slope, the *S* profiles are sketched in Figure 14.6b. For the critical slope, the *C* profiles are sketched in Figure 14.6c. Similarly, Figures 14.6d and e present the profiles for horizontal and adverse slopes, respectively. One should always note that these figures are just the graphical representation of the gradually varied flow equation. Water surfaces should closely follow the appropriate profile to satisfy the governing equation.

It is of particular interest to examine the steep profiles in greater detail. The *S*1 profile ($y > y_c > y_o$) is called the backwater proper profile as in the mild slope case. As such, its downstream control must be a choke of some kind but its upstream boundary cannot simply be the normal depth since y_o in this case is supercritical. Clearly, a transition between the subcritical flow downstream and the supercritical normal depth upstream must be present. Thus the *S*1 profile must lead into a hydraulic jump upstream unless is it interrupted by some other physical boundary (such as a dam, sluice gate, etc.) beforehand.

Also of interest is the *S*2 profile ($y_c > y > y_o$). It is called the dropdown curve because it provides the transition between a critical section upstream and the supercritical normal depth associated with the steep downstream slope.

In the cases of the horizontal and adverse slopes, the normal depth does not exist and thus there can be no type 1 profiles ($y > y_o$). In these cases the type 2 profile must approach the horizontal boundary for $y \rightarrow \infty$ upstream. Similarly, in the critical slope case, $y_o = y_c$ and thus there can be no type 2 profile ($y_o > y > y_c$).

14.2.5 Summary

A close examination of the gradually varied flow equation reveals that regardless of the bed slope any water surface profile located in either Zone 1 or Zone 3 should be a rising curve $\{[dy/dl] = +ve\}$. Also, all profiles in Zone 2 are falling curves $\{[dy/dl] = -ve\}$. On the other hand, all water surface profiles approach the normal depth asymptotically and approach the critical depth perpendicularly. They also approach the channel bottom mostly with acute angles. All curves located in Zone 1 tend to be horizontal at their infinity. *H*1, *A*1, and *C*2 profiles do not exist.

These rules constitute the basis for constructing water surface profiles presented in Figure 14.6. For example, the *S*2 curve in Figure 14.6b is located in Zone 2; then it must be a falling curve. It approaches the normal depth asymptotically and the critical depth perpendicularly. Satisfying these three constraints, one will be able to draw it correctly, as shown in Figure 14.6b. This analysis holds for the rest of the profiles.

Example 14.3

A rectangular channel 10 ft wide is carrying 150 ft³/s at a velocity of 1.85 ft/s. The channel slope is 0.0009 ft/ft and the Manning roughness coefficient is 0.025. Classify the profile which prevails in this vicinity and determine the nature and location of the control.

Solution:

$$Q = 150 \text{ ft}^3\text{/s}; \quad \text{then } q = 150/10 = 15 \text{ cfs/ft}$$

$$150 = 1.85(A)$$

$$A = 81.08 \text{ ft}^2$$

$$y = (81.08/10) = 8.1 \text{ ft}$$

$$y_c = \sqrt[3]{q^2/g} = 1.91 \text{ ft} < y, \text{ so the given depth is subcritical}$$

$$A_c = 19.1 \text{ ft}^2$$
$$R_c = 1.38 \text{ ft}$$
$$V_c = (gy_c)^{0.5} = 7.85 \text{ ft/s}$$
$$S_c = \frac{(0.025)^2 \times (7.85)^2}{(1.49)^2 \times (1.38)^{4/3}} = 0.01129 \text{ ft/ft}$$

Since $S_c > S_o$, the slope is mild. Thus $M1$, $M2$, and $M3$ profiles are possible. To distinguish among these profiles we compare the given depth to the normal depth. To do this we will use the friction (energy) slopes as surrogates for the depths.

$$y = 8.1 \text{ ft}$$
$$A = 81.08 \text{ ft}$$
$$R = 3.09 \text{ ft}$$
$$V = 1.85 \text{ ft/s}$$
$$S_f = \frac{(0.025)^2 \times (1.85)^2}{(1.49)^2 \times (3.09)^{4/3}} = 0.000214 \text{ ft/ft}$$

Since $S_f < S_o$, then $y > y_o$ and thus we have an $M1$ profile. The normal depth, y_o, may also be calculated directly using the Manning equation:

$$q = y_o \frac{1.49}{n} R^{2/3} S_o^{1/2}$$
$$15 = \frac{1.49}{0.025} y_o \left(\frac{10y_o}{10 + 2y_o} \right)^{2/3} (0.0009)^{1/2}$$
$$8.39 = y_o \left(\frac{10y_o}{10 + 2y_o} \right)^{2/3}$$

Solving by trial and error, $y_o = 4.65 \text{ ft} < y\,(8.1 \text{ ft})$.

The control of this profile is located downstream and is a choke, such as a dam, sluice gate, or confluence with a larger stream. Upstream, the profile would approach the normal depth in the channel or, alternatively, it could transition into another profile of some type.

Example 14.4

A 5-m-wide rectangular channel carries a discharge of 20 m^3/s at a normal depth of 2.5 m. The channel roughness coefficient (n) is 0.025 and the bottom slope is 0.0005 m/m. If an 8-m-high spillway is placed across the channel, will an $M3$ profile develop downstream of the spillway?

Solution: From the given data, $A = 2.5(5) = 12.5 \text{ m}^2$.

$$V = 20/12.5 = 1.6 \text{ m/s}$$
$$q = Q/B = 4 \text{ m}^2\text{/s}$$
$$y_c = \sqrt[3]{q^2/g} = 1.18 \text{ m}$$
$$A_c = 1.18(5) = 5.9 \text{ m}^2$$
$$V_c = q/y_c = 3.39 \text{ m/s}$$
$$P_c = B + 2(y_c) = 7.36 \text{ m}$$
$$R_c = A_c/P_c = 0.80 \text{ m}$$
$$S_c = \frac{(0.025 \times 3.39)^2}{(0.8)^{4/3}} = 0.00967$$

Since $S_c > S_o$, the channel slope is mild and an $M3$ profile is possible. One may also calculate the normal depth, y_o, and compare with y_c to decipher the possible profiles. At the spillway,

$$E_1 = H + E_c = 8 + 3/2(y_c) = 9.77 \text{ m}$$

So

$$9.77 = y_2 + \{q^2/[2gy_2^2]\} = y_2 + (0.815/y_2^2)$$

$$y_2 = 0.294 \text{ m}$$

Now the conjugate depth to the normal depth is

$$y_0' = \frac{2.5}{2}\left[-1 + \sqrt{1 + \frac{8(4)^2}{g(2.5)^3}}\right]$$

$$y_0' = 0.44 \text{ m}$$

Since the conjugate depth of 0.44 m is greater than the depth at the spillway toe of 0.29 m, and the slope is mild, then an $M3$ profile will form and run from the spillway toe to the depth which is the conjugate of the normal depth in the channel.

14.3 Outlining Water Surface Profiles

A large number of combinations for water surface profiles can be found in nature. The following are the main points to be considered while outlining such profiles.

1. Determine the type of bed slope (mild, steep, critical, horizontal, or adverse) in each reach of the channel according to the bed slope as compared to the critical slope. One can also compare the normal depth with the critical depth if the bed slope is not given.
2. Plot the critical depth which is constant along the entire channel. It does not depend on the bed slope. Also plot the normal depths in the different reaches away from any hydraulic structures and/or points of variation in bed slope, as shown in Figure 14.7a. Water surface profiles should bridge these normal depths.
3. According to the type of bed, select the appropriate curves from the corresponding profiles shown in Figure 14.6. If the water depth needs to increase above the critical depth, then consider the corresponding curve from Zone 1. If this increase is encountered below the critical depth, then pick the corresponding curve from Zone 3. If the depth needs to decrease, one should always select the corresponding curve from Zone 2, as shown in Figure 14.7b.
4. Various types of curves should not be allowed to extend beyond their own category. For example, M curves should not extend beyond mild reaches in the channels. C curves should only exist within reaches of critical slopes.
5. There is no way to cross the critical depth from $y < y_c$ to $y > y_c$ (i.e., from supercritical to subcritical conditions) without a hydraulic jump, as shown in Figure 14.7c. However, crossing the critical depth from subcritical to supercritical conditions is only encountered through falling profiles, as shown in Figure 14.7b.
6. Before any gate or dam, the elevation of the water surface must be above the critical and normal depths. A backwater curve must be formed.
7. Before any hydraulic jump in mild, horizontal or adverse slopes, $M3$, $H3$, or $A3$, respectively, should exist (Figure 14.7c). After any jump in a steep slope, the $S1$, profile is found (Figure 14.7d). These curves are essential to adjust the two conjugate depths of the jump.

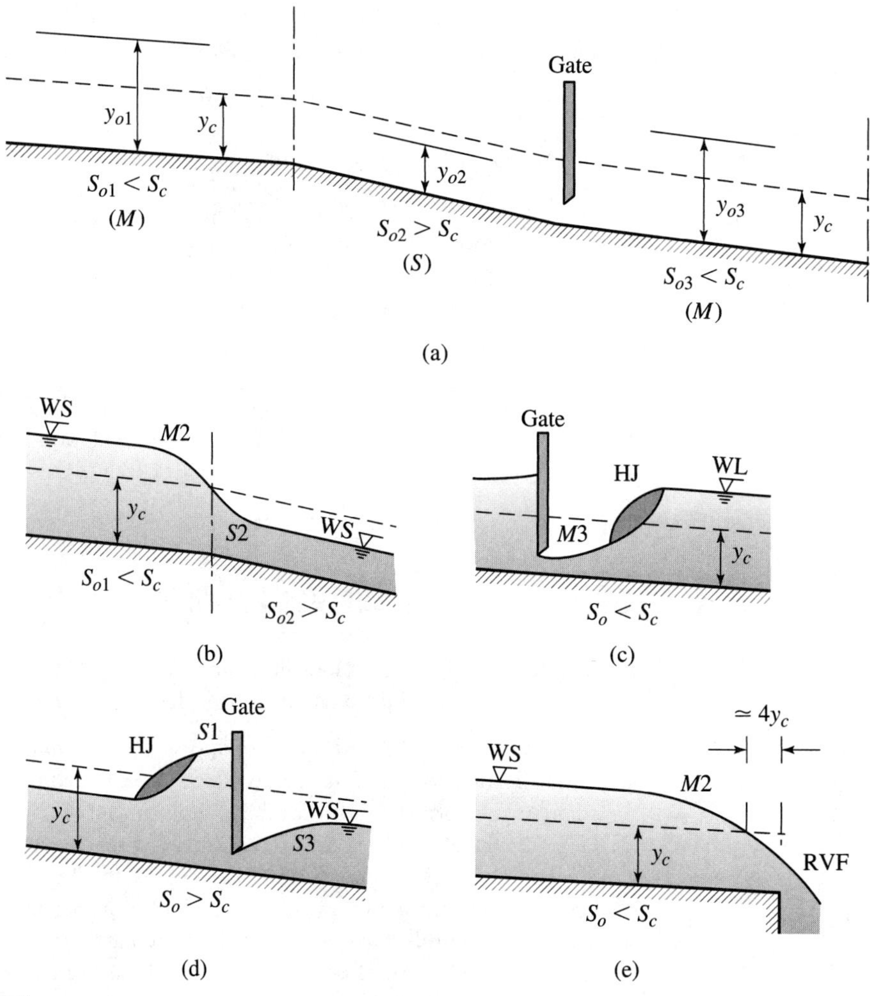

Figure 14.7 Outlining water surface profiles.

A part of the curve is formed to ensure that the two depths of the jump have the same pressure-momentum force. In very rare cases there can be a jump without any water curve at its two ends, although this is theoretically possible if the two depths just after and before the jump have the same force.

8. Subcritical flows are governed by the downstream control while supercritical flows are dominated by the upstream control. An intermediate gate may act as a control for both the upstream and downstream sides.
9. At the free overfall in mild channels, the water surface crosses the critical depth at a distance of about four times the critical depth upstream of the fall, as shown in Figure 14.7e.
10. Water surface profiles do not cross between any two successive reaches even though these reaches may be of the same category but with slightly different bed slopes. They attain normal depths asymptotically.

Example 14.5

Draw the water surface profile for a wide open channel with three different reaches as given in Figure 14.8a.

Solution: The bed slope of the first reach, S_{o1}, is less than critical slope S_c. It is a mild slope and hence y_{o1} will be bigger than y_c. The critical depth, y_c, is constant throughout the channel, regardless of the bed slope in the different sections. The slope in the second reach is less than that in the first but not horizontal. The second reach is thus milder then the first and y_{o2} will be bigger than y_{o1}, as shown as in Figure 14.8b. The third reach has a steep bed slope and its normal depth y_{o3} will be less than the critical depth. Figure 14.8b presents the first step to the solution. The normal depths and the critical depth are located relative to each other. A vertical line is placed between successive reaches to ensure that no curve has crossed from one reach to the other. Regardless of the reach category (mild, steep, critical, horizontal, or adverse), no profiles can continue across these vertical lines. This holds true even for any two successive reaches of the same category but having slightly different bed slopes.

A rising profile is needed to evaluate the water surface from y_{o1} to y_{o2}. Both of the first and second reaches have a mild bed slope. Therefore, a rising profile from the M category and defined above the critical depth would fit. The only possible selection is the $M1$ profile. The question now is whether to place the profile in the first reach to end at the vertical line between the first two reaches or in the second reach to start at this vertical line. The second

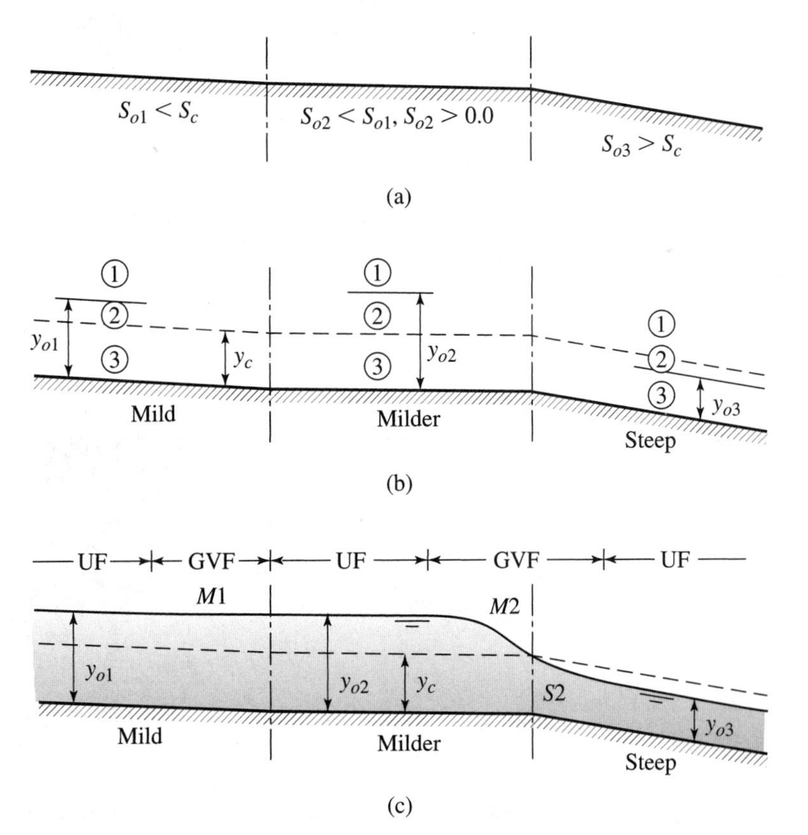

Figure 14.8 Water surface profile for the wide channel given in Example 14.5. (a) An open channel with three reaches. (b) Critical and normal depths and category of bed slope. (c) Water surface profile.

option is not correct because $M1$ should always be in Zone 1 above the normal depth. The second option would place an $M1$ curve in Zone 2. Therefore, the $M1$ will fit in the first reach as shown in Figure 14.8c.

Falling profiles are needed to bridge the water surface between the second and third reaches. Falling profiles are only found in Zone 2. Hence $M2$ followed by $S2$ profiles will fit for this case, as shown in Figure 14.8c. The two profiles meet at the intersection between the line of the critical depth and the vertical line between the two reaches.

Example 14.6

Draw water surface profiles for the two reaches of the open channel given in Figure 14.9a. A gate is located between the two reaches and the second reach ends with a sudden fall.

Solution: The first reach has a mild bed slope while the bed slope of the second reach is steep. The critical depth, y_c, is constant along the channel while y_{o1} is located above y_c and y_{o2} is located below it, as show in Figure 14.9b. As the flow approaches the gate, the water velocity reduces and the depth of flow increases. A rising profile from the M family (defined above both the critical and normal depths) is thus needed. The $M1$ curve fits as shown in Figure 14.9c. Water then flows from the gate opening, which is less than y_c, into the steep reach. A rising curve is needed from the S family (defined below both the critical and normal depths). The $S3$ curve fits, after which the flow continues with y_{o2} until the sudden fall where a rapidly varied flow is encountered, as shown in Figure 14.9c.

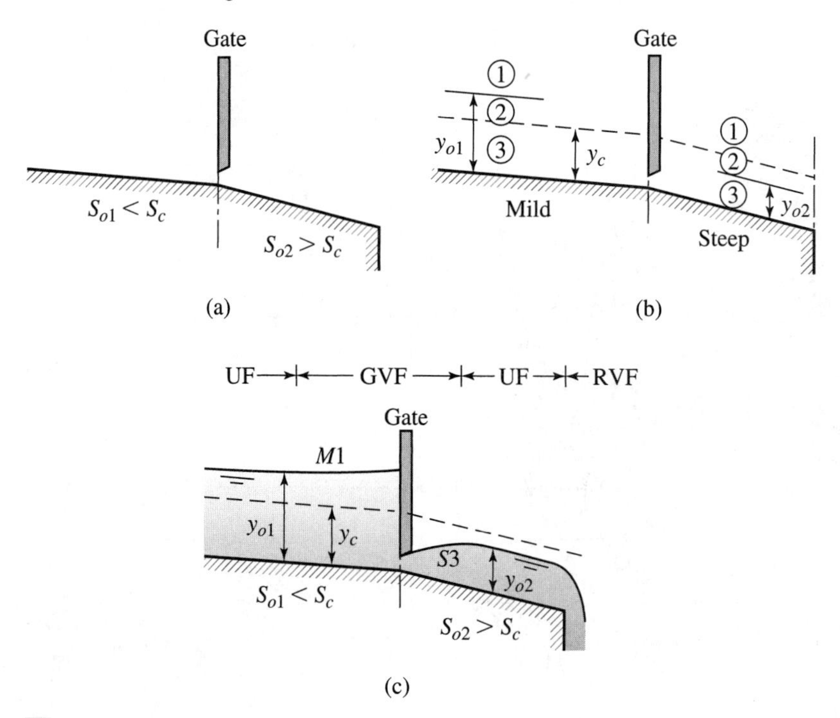

Figure 14.9 Water surface profile in Example 14.6. (a) The open channel and gate location. (b) Critical and normal depths. (c) Water surface profile.

Example 14.7

For the previous example, draw the water surface profile considering the first and second reaches have steep and mild bed slopes, respectively, as shown in Figure 14.10a.

Solution: The normal depth in the first reach, y_{o1}, will be less than y_c, while y_{o2} will be greater than it. Before the gate the water surface must be located above the critical depth. A hydraulic jump will be formed and the $S1$ curve will follow after the high stage conjugate depth. The $S1$ curve would provide an increasing flow depth as the flow approaches the gate to accommodate the velocity reduction before the gate, as shown in Figure 14.10b.

Water issues from the gate opening (which is less than y_c) into the mild reach. To cross the critical depth, a hydraulic jump must be formed before which an $M3$ curve should exist. Approaching the sudden fall at the end of the channel a falling curve, an $M2$ profile is formed. The $M2$ curve ends at its intersection with the line of critical depth at a distance of approximately $4y_c$ upstream of the fall. A rapidly varied flow will take place after this point. The water profile is given in Figure 14.10b.

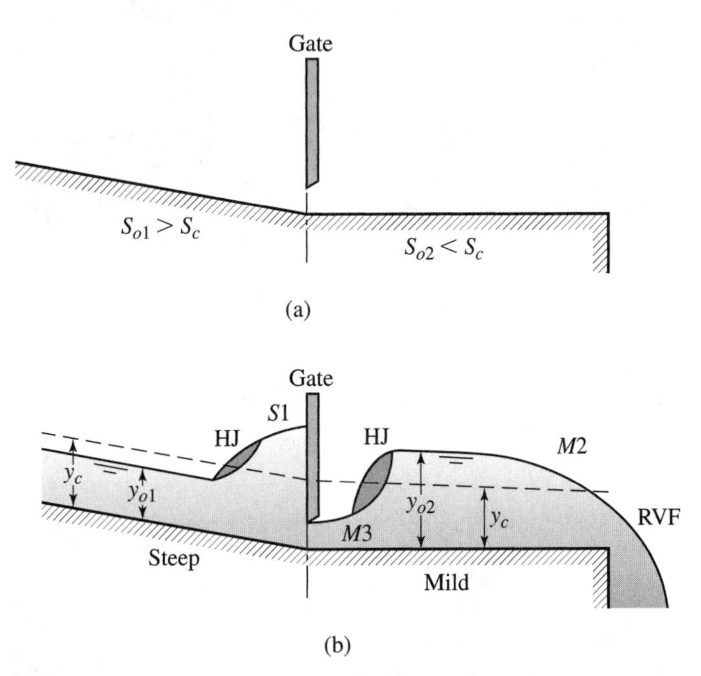

Figure 14.10 Water surface profile in Example 14.7. (a) The open channel and gate location. (b) Water surface profile.

Example 14.8

A wide channel consists of three long reaches with a large reservoir at its beginning, as shown in Figure 14.11a. Two gates are located midway of the first and last reaches. Bed slopes for the three reaches are $S_{o1} = 0.008$, $S_{o2} = 0.00309$, and $S_{o3} = 0.0005$, respectively. The channel has a specific discharge (discharge per unit width) of 0.675 m^2/s. Take the Manning roughness coefficient n as 0.015 (metric). (i) Find the critical and normal depths in each reach. (ii) Sketch the water surface profile along the channel.

Solution: To draw the water surface profiles, one must evaluate the critical and normal depths first even if not required.

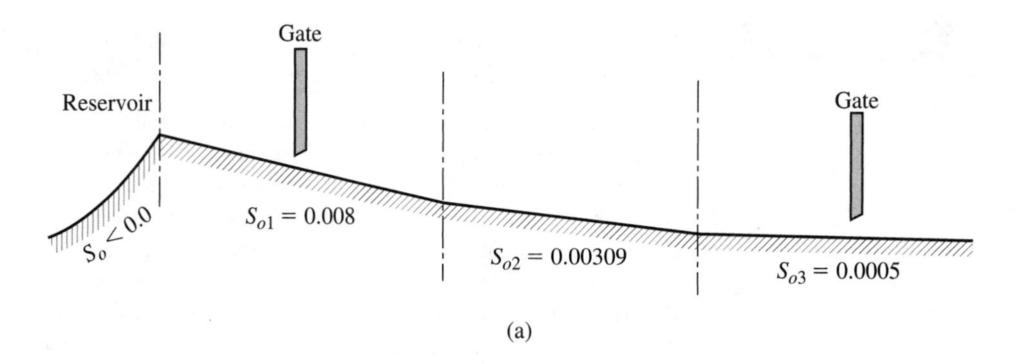

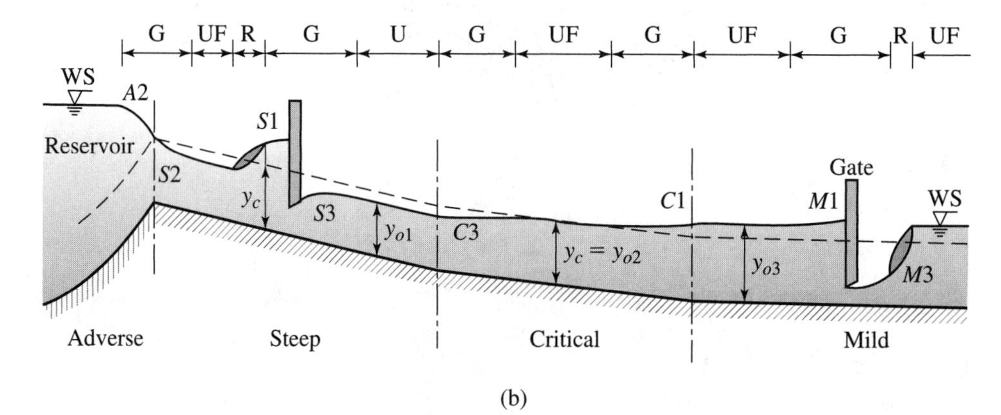

Figure 14.11 Water surface profile in Example 14.8. (a) Layout for the channel. (b) critical and normal depths and water surface profile.

(i) Therefore, using the equations of wide sections,

$$y_c = \sqrt[3]{\frac{q^2}{g}} = \sqrt[3]{\frac{(0.675)^2}{9.81}} = 0.36 \text{ m}$$

Also, using Manning equation for wide sections, the normal depth is given as

$$y_o = \left(\frac{q^2 n^2}{S_o}\right)^{0.3}$$

Hence,

$$y_{o1} = \left(\frac{(0.675)^2(0.015)^2}{0.008}\right)^{0.3} = 0.27 \text{ m}$$

$$y_{o2} = \left(\frac{(0.675)^2(0.015)^2}{0.00309}\right)^{0.3} = 0.36 \text{ m}$$

$$y_{o3} = \left(\frac{(0.675)^2(0.015)^2}{0.0005}\right)^{0.3} = 0.62 \text{ m}$$

Comparing the normal depths in the three reaches with the critical depth, it is obvious that the first reach has a steep bed slope, and the second has a critical bed slope, while the last one has a mild bed slope.

(ii) The two gates are placed in the middle of the first and last reaches. Critical and normal depths are first plotted in the middle of the reaches and away from the gates. The water level in the reservoir is generally higher than any other level through the channel. Also, before the gates one must plot the water level above the critical depth as shown in Figure 14.11b.

As the water issues from the reservoir to the steep reach, falling profiles must be formed. The *A*2 will form (above y_c) just before the exit of the reservoir and the *S*2 will form (below y_c) at the channel entrance. Therefore the *A*2-*S*2 will form in the transition zone between the reservoir and the channel and the flow will then be uniform with a depth $y_{o1} < y_c$. Before the gate a hydraulic jump will be formed and will be followed by the *S*1 profile as shown in Figure 14.11b.

While approaching the second reach, a rising curve is needed to elevate the water depth from y_{o1} to y_c. The *C*3 curve is the only profile (below y_c) which can elevate the water surface from y_{o1} to y_{o2}, which is equal to the critical depth. It should take place in the second reach and if the length of the second reach is long enough the flow would continue uniformly with a critical depth, as shown in Figure 14.11b. Another rising curve is needed (above y_c) to elevate the depth from y_{o2} to y_{o3}. The *C*1 profile fits and should end at the beginning of the last reach where water flows with y_{o3}. The *M*1 curve will be formed before the second gate and water will issue from the gate with an *M*3 curve followed by a hydraulic jump to cross the critical depth. Figure 14.11b shows the required profile throughout the channel.

14.4 Jump Location and Water Surface Profiles

In many open channels, the bed slope is reduced from steep to mild to elevate water levels and enhance irrigation practices by reducing pumping heads. In such cases, a hydraulic jump must be formed, as shown in Figure 14.12. Two different locations are expected for the jump according to the normal depths y_{o1} and y_{o2}. The following procedure is made to determine whether case I or case II is going to happen, as shown in Figure 14.12.

1. Knowing the low stage depth y_{o1}, the high stage conjugate depth of the jump (y') is calculated.
2. If y' is less than y_{o2}, then case I should occur and an *S*1 curve will follow the jump to bridge y' with y_{o2} at section *A-A*.
3. If y' is greater than y_{o2}, then case II should occur. The normal depth y_{o1} will continue until section *A-A*, after which an *M*3 curve will be formed. One can use y_{o2} to determine the low-stage conjugate depth of the jump in case II. The *M*3 curve will bridge y_{o1} at section *A-A* with the low-stage depth of the jump, as shown in Figure 14.12.

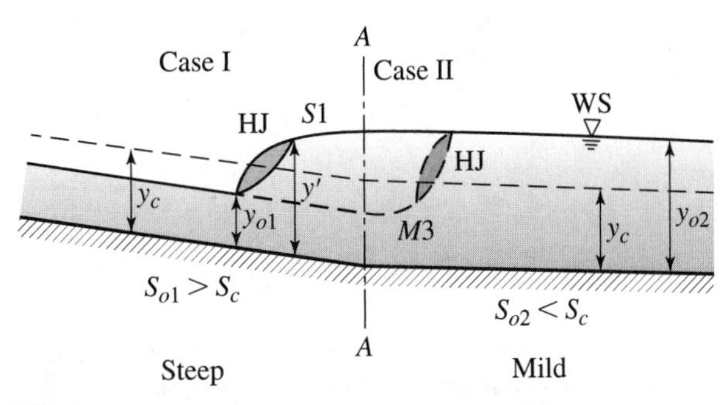

Figure 14.12 Jump location between steep and mild reaches.

Example 14.9

A trapezoidal channel of 5.0 m bed width and 1:1 side slope issues a discharge of 13.0 m³/s. The channel consists of three long reaches having the following bed slopes, $S_{o1} = 0.0001$, $S_{o2} = 0.007$, and $S_{o3} = 0.0009$, respectively. A gate is located just between the first and second reaches and the last reach is ended with a sluice gate to control the flow. Taking the Chezy coefficient as 50 (metric), sketch the water surface profile along the channel.

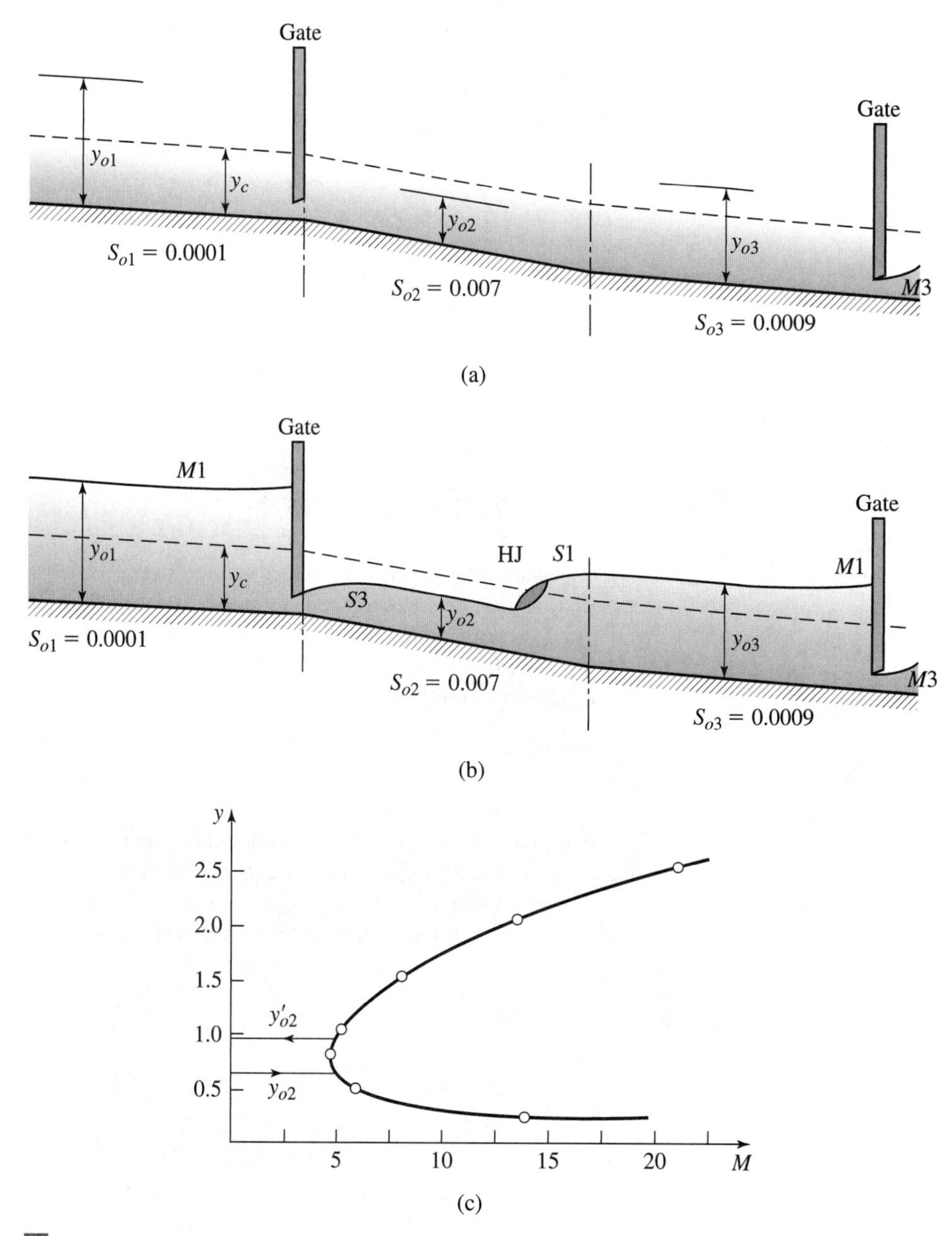

Figure 14.13 Water surface profile in Example 14.9. (a) Layout and normal and critical depths. (b) Water surface profile. (c) Momentum function diagram to determine the jump location.

Solution: The first step is to calculate the critical depth and the normal depth for the three reaches. Because the channel is of trapezoidal section, the simplified rectangular or wide forms are not applicable. First, using Figure 12.9 to find the critical depth, $Z = \frac{Qt^{3/2}}{g^{0.5}b^{2.5}} = \frac{13(1)^{3/2}}{(9.81)^{0.5}(5)^{2.5}} =$ 0.074. From Figure 12.9 we estimate the factor $\frac{ty_c}{b} = 0.17$; thus $y_c = 0.85$ m. The Chezy equation can be written as $Q = AC\sqrt{RS_o}$. Then

$$Q = (by_o + ty_o^2)C\sqrt{\left(\frac{by_o + ty_o^2}{b + 2y\sqrt{1 + t^2}}\right) \times S_o}$$

$$0.26 = (5y_o + y_o^2)\sqrt{\left(\frac{5y_o + y_o^2}{5 + 2\sqrt{2}y_o}\right) \times S_o}$$

Solving by trial and error for y_o in the three reaches using S_{o1}, S_{o2}, and S_{o3}, one gets $y_{o1} = 2.55$ m, $y_{o2} = 0.76$ m, and $y_{o3} = 1.38$ m. Comparing the respective normal depths with the critical depth, it is found that the first and third reaches have mild slopes while the second reach has a steep slope. The critical and normal depths are plotted in Figure 14.13a. A hydraulic jump must be formed either in the second reach or in the third reach to cross the critical depth. Knowing that $y_{o2} = 0.76$ m and $y_{o3} = 2.55$ m, one can calculate the high-stage depth (y'_{o2}) of the jump if it starts from y_{o2}.

From the momentum diagram, as shown in Figure 14.13c, the high-stage depth (y'_{o2}) is approximately equal to 0.95 m (i.e., less than y_{o3}). The jump will take place in the steep slope and an $S1$ curve will flow the jump, as shown in Figure 14.13b. One should always make sure that each profile is located in its appropriate zone.

Example 14.10

A wide rectangular channel carries a specific discharge of 4.0 m^2/s. The channel consists of three long reaches with bed slopes of 0.008, 0.0004, and S_c, respectively. A gate is located at the end of the last reach. Draw the water surface profile along the channel using the Manning roughness coefficient $n = 0.016$ (metric).

Solution: For wide channels,

$$y_c = \sqrt[3]{\frac{q^2}{g}} = \sqrt[3]{\frac{(4.0)^2}{9.81}} = 1.177 \text{ m}$$

$$y_{o1} = \left(\frac{q^2n^2}{S_{o1}}\right)^{0.3} = \left(\frac{(4.0)^2(0.016)^2}{0.008}\right)^{0.3} = 0.818 \text{ m}$$

$$y_{o2} = \left(\frac{q^2n^2}{S_{o2}}\right)^{0.3} = \left(\frac{(4.0)^2(0.016)^2}{0.0004}\right)^{0.3} = 2.01 \text{ m}$$

$$y_{o3} = y_c = 1.177 \text{ m}$$

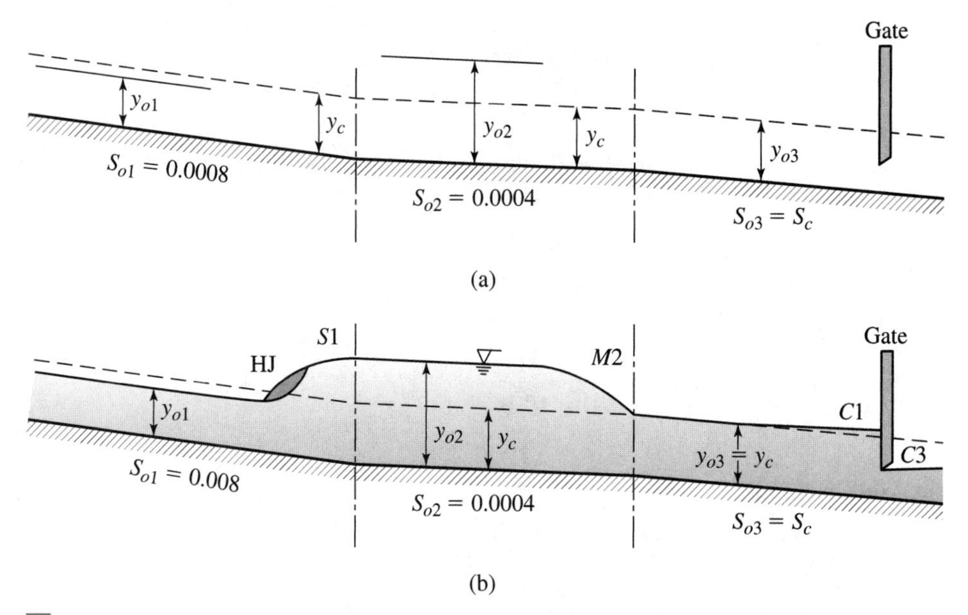

Figure 14.14 Water surface profile in Example 14.10. (a) Layout and normal and critical depths. (b) Water surface profile.

Figure 14.14a presents the critical and normal depths along the channel. To know whether the hydraulic jump will occur in the first or the second reach, the high-stage depth (y'_{o1}) of the jump is calculated using y_{o1}. Therefore,

$$y'_{o1} = \frac{y_{o1}}{2}\left\{-1 + \sqrt{1 + \frac{8q^2}{gy_{o1}^3}}\right\}$$

$$= \frac{0.818}{2}\left\{-1 + \sqrt{1 + \frac{8(4.0)^2}{9.81(0.818)^3}}\right\} = 1.629 \text{ m}$$

y'_{o1} is less than y_{o2}. Hence, the jump will take place in the first reach and will be followed by an $S1$ curve to elevate the water surface to y_{o2} at the beginning of the second reach. The flow will continue with y_{o2} until it approaches the last reach of the critical bed slope. A falling curve is needed to reduce the water level. A $C2$ profile does not exist. An $M2$ profile will take place at the end of the second reach. Before the gate, a $C1$ profile "horizontal" will take place, while after the gate $C3$ profile "also horizontal" will be formed, as show in Figure 14.14b.

Example 14.11

Give some examples for various cases of water surface profile which may be encountered in nature, and illustrate them graphically.

Solution: See Figure 14.15a through 14.15i.

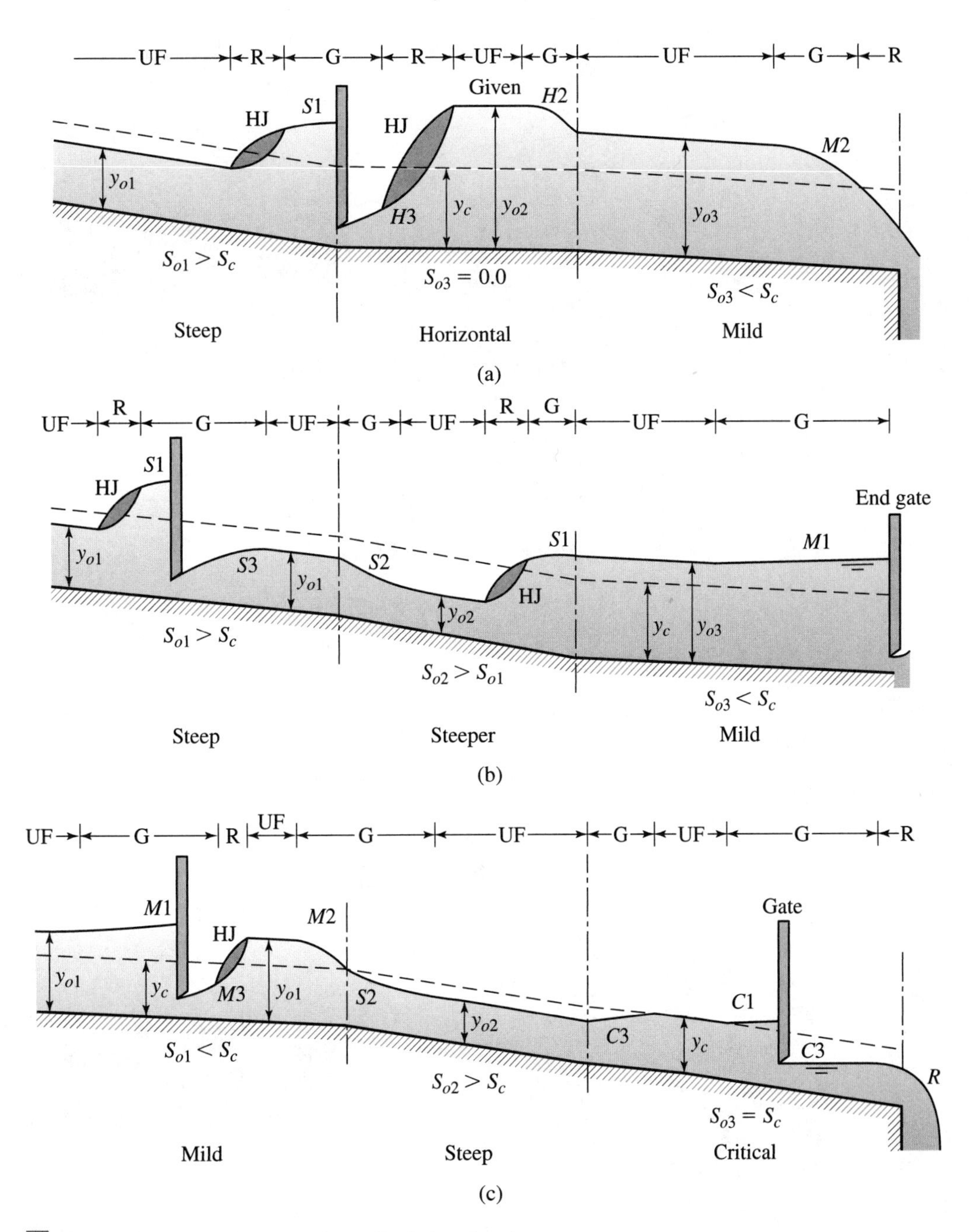

Figure 14.15 Examples for water surface profiles in open channels.

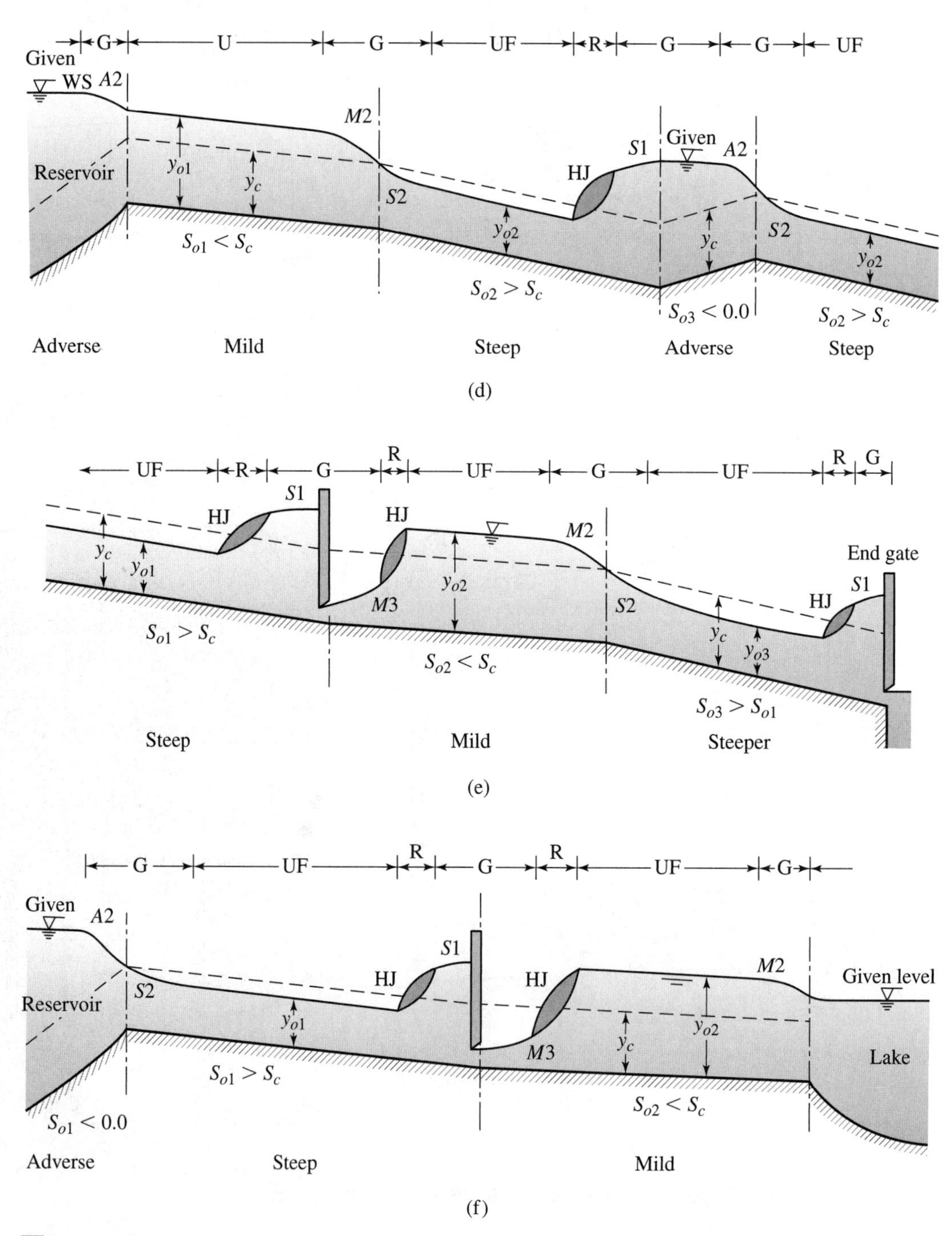

Figure 14.15 *(Continued)*

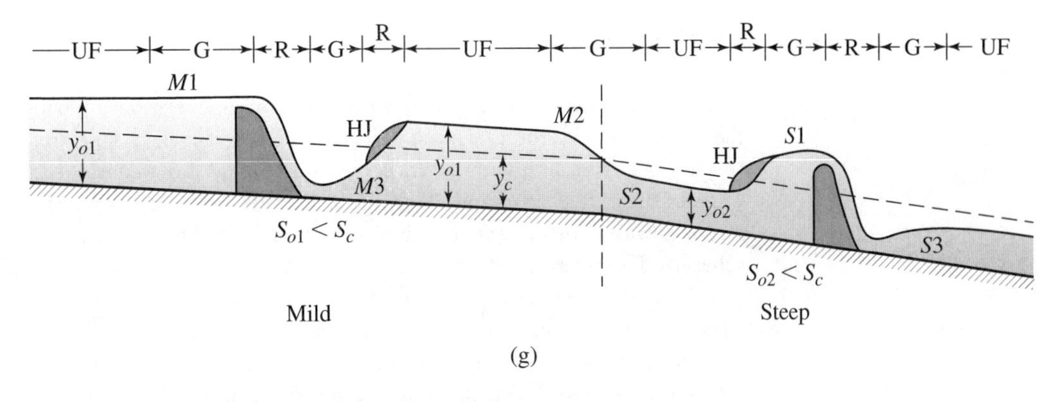

(g)

(h)

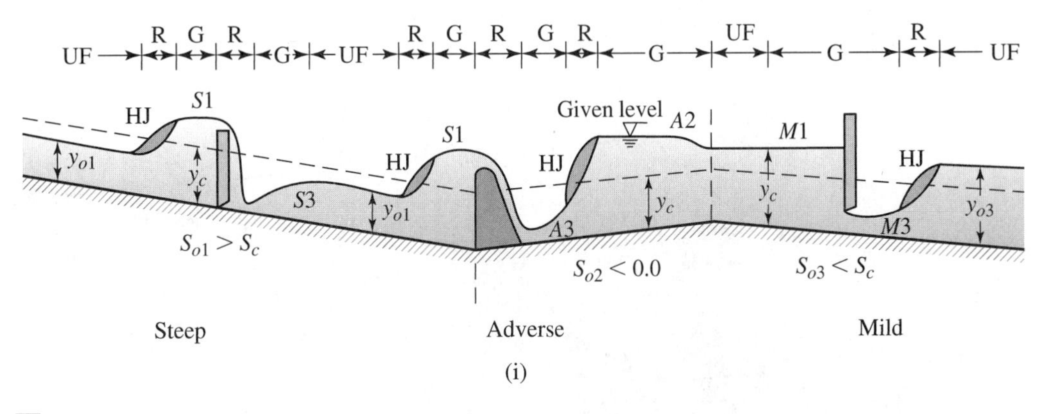

(i)

Figure 14.15 *(Continued)*

14.5 Control Sections

A control section is a section where a definite or unique relationship between the discharge and the depth of flow is found. Gates, weirs, and sudden falls in channel beds are examples of control sections. A section with a critical depth of flow is also a control section, since the relation between the critical depth and discharge is defined. Subcritical flows have their control sections at the downstream side, as they are dominated by the flow conditions downstream. On the other hand, supercritical flows are generally dominated by the flow conditions upstream.

A gate located between two reaches may control flow in both the upstream and downstream reaches. A section where the bed slope changes from mild to steep is also a control section. Control sections may not necessarily be critical flow sections. However, critical flow sections are certainly control sections. Several examples of control sections are provided in Figure 14.16 where the control sections are shown as bold squares. The figure demonstrates the fact that water surface profiles essentially connect controls along the channel. Thus, the essence of drawing the profiles is to determine the control sections at each end of the profile and sketching the profile between them as done in the figure.

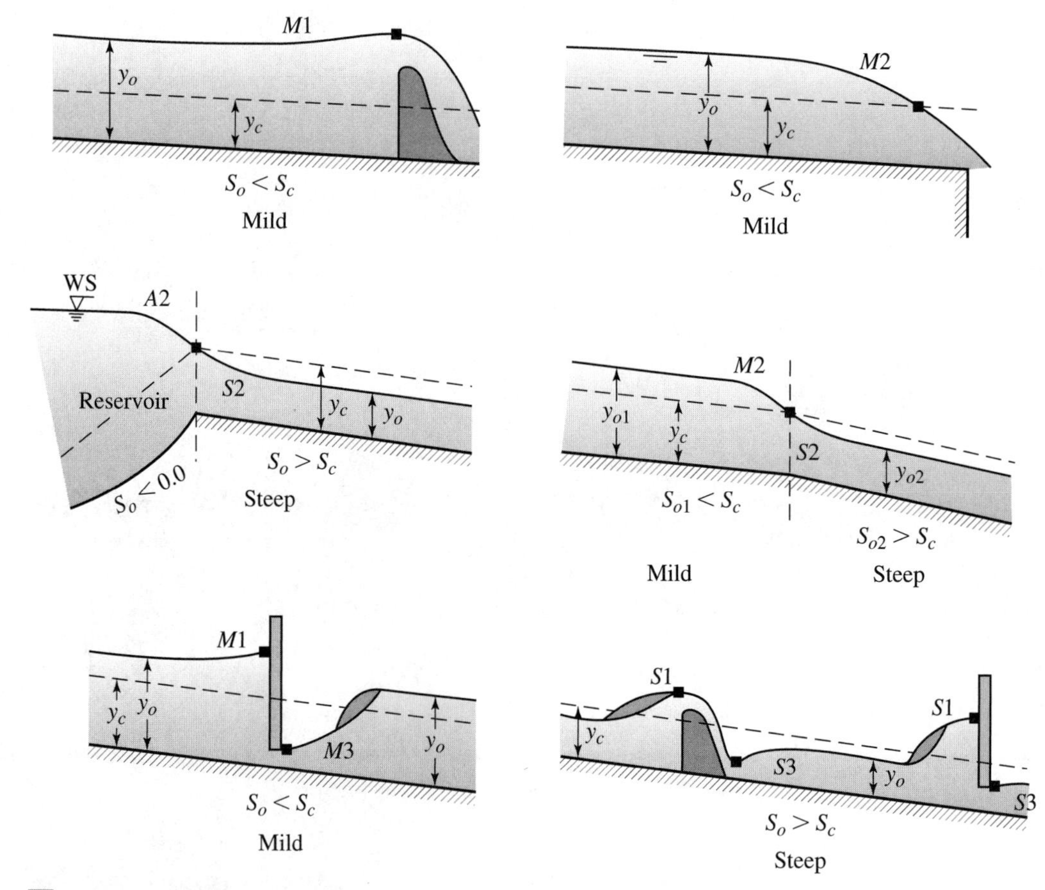

Figure 14.16 Examples for control sections.

READING AID

14.1. What is meant by gradually varied flow (GVF) in an open channel? Give some examples, using sketches, for the occurrence of GVF.

14.2. Discuss the major differences between gradually varied flow and rapidly varied flow in open channels.

14.3. State some conditions that force the flow to vary gradually. Is the gradually varied flow considered a steady or unsteady flow?

14.4. How can you identify the gradually varied flow in a prismatic channel having a constant bed slope in the field?

14.5. Does gradually varied flow occur in frictionless open channels with constant bed slope? Why or why not?

14.6. Write down the main assumptions used in the derivation of the gradually varied flow equation. Develop the GVF equation from the energy principle in the general case. Then derive formulas for rectangular and wide sections.

14.7. Prove that the variation in depth with respect to distance in an open channel with a specified bed slope can be expressed in terms of the gradient of the total energy line and Froude number.

14.8. Derive a relation to determine the water surface slope, S_w, with respect to a horizontal datum. Would you expect a water profile with a negative value of S_w to be a rising or a falling curve?

14.9. Classify water surface profiles according to (i) the channel bed slope and (ii) dy/dl.

14.10. Define different zones in an open channel as classified according to the critical and normal depths. State the missing zones in the various categories of bed slopes and discuss why they are missing.

14.11. What is meant by rising and falling curves? To what category does a horizontal curve belong?

14.12. Give some practical examples for the following conditions:

$$\text{(i)}\ \frac{dy}{dl} = 0, \qquad \text{(ii)}\ \frac{dy}{dl} = +ve, \qquad \text{(iii)}\ \frac{dy}{dl} = -ve, \qquad \text{(iv)}\ \frac{dy}{dl} = \infty$$

14.13. Illustrate the classification of all water profiles in open channels under the various types of bed slopes. Discuss the characteristics of these profiles using the gradually varied flow equation.

14.14. Sketch the following types of water profiles and indicate whether they are rising or falling profiles and give practical examples for three of them: (i) $M1$ curve (ii) $S1$ curve (iii) $S3$ curve (iv) $C1$ curve (v) $A2$ curve (iv) $A3$ curve (vii) $H3$ curve.

14.15. Explain why an $M3$ curve should exist before a hydraulic jump in the mild slope and $S1$ curve should exist after a jump in the steep slope. Is the control section located upstream or downstream of a subcritical flow?

14.16. What is meant by control sections in open channels? Give some examples of control sections encountered in nature.

14.17. Explain how a gate can control the flow on both the upstream and downstream sides. Give examples where the control sections are located at the channel entrance and end.

14.18. Why do the control sections have special importance in the study of open-channel hydraulics? Give an example in which a control section is encountered between two channel reaches.

14.19. Explain how the channel delivery from reservoirs can be affected by the channel bed slope. If two identical channels are taking their water from one reservoir, do you expect the critical depth to differ or not (and explain) in the following cases? (i) The two channels have different slopes but both are steep. (ii) The two channels have different slopes but both are mild.

Problems

14.1. A trapezoidal channel of 2.0 m bed width, 2.0 m water depth, and 1 : 1 side slope carries a discharge of 20.0 m^3/s. The channel has a bed slope of 8 cm/km and a Chezy roughness coefficient of 60 (metric). To what rate does the depth of flow change with distance?

14.2. A wide rectangular channel carries a specific discharge of 6.0 m^2/s with a bed slope of 12 cm/km. The depth of flow is 1.2 m and the Manning roughness coefficient is 0.017 (metric). Find the rate at which the depth changes with distance.

14.3. For Problem 14.2, determine the water surface slope with respect to the horizontal.

14.4. Water is flowing in a 12-ft-wide rectangular channel running on a slope of 0.002 ft/ft with a Manning n value of 0.025. The discharge is known to be 750 cfs. At a particular point the velocity is measured to be 8 ft/s.

a. Classify the profile that prevails in this vicinity.
b. What is the control of this profile and where is it located (upstream or downstream)?
c. What would you expect to find upstream and downstream of this profile?

14.5. Water is flowing in a 5-m-wide rectangular channel running on a slope of 0.0065 m/m with a Manning n value of 0.035 (metric). The discharge is estimated to be 21 m^3/s. At a certain point the depth is observed to be 3 m.

a. Classify the profile that prevails in this vicinity.
b. What is the control of this profile and where is it located (upstream or downstream)?
c. What would you expect to find upstream and downstream of this profile?

14.6. Water is flowing in a trapezoidal channel with a 10-ft bottom width and 2 on 1 side slopes. The Manning n value is 0.025, the discharge is estimated at 1000 cfs, and the channel bed slope is 0.009 ft/ft. At a certain point the depth is observed to be 7 ft.

a. Classify the profile that prevails in this vicinity.
b. What is the control of this profile and is it located upstream or downstream?
c. What would you expect to find upstream and downstream of this profile?

14.7. Water is flowing in a wide rectangular channel with a specific discharge of 10 m^2/s that is running on a slope of 0.025 m/m. The estimated Manning n value is 0.015 (metric). If the observed velocity at a point is 3 m/s,

a. Classify the profile that predominates in this vicinity.
b. What is the control of this profile and where is it located (upstream or downstream)?
c. What do you think is located upstream and downstream of this profile?

14.8. Water is flowing in a trapezoidal channel with a bottom width of 4 m and 2/1 side slopes. The channel is running on a slope of 0.0025 m/m with an estimated Manning n value of 0.025 (metric). The discharge is estimated to be 10 m^3/s. The depth at a particular point is observed to be 0.5 m,

a. Classify the profile that prevails in this vicinity.
b. What is the control of this profile and where is it located (upstream or downstream)?
c. What do you think is located upstream and downstream of this profile?

14.9. Water is flowing in a concrete circular aqueduct with a diameter of 10 ft at a discharge of 100 cfs. The pipe is running on a slope of 0.0025 ft/ft with a Manning n value of 0.013. At a particular point the depth is observed to be 3 ft.

a. Classify the profile that is predominating in the aqueduct.
b. What is the control of this profile and is it located upstream or downstream?
c. What would you expect to find upstream and downstream of this profile?

14.10. An open channel of mild slope takes its water from a reservoir and terminates into a lake. Draw the water profile in the channel if the water level in the lake is (i) above the normal depth, (ii) at the normal depth, or (iii) below the channel bed.

14.11. Solve Problem 14.10 considering that the channel has a steep bed slope.

14.12. An open channel consists of two long reaches of mild bed slope as shown in Figure P14.12. The second reach is milder than the first and terminates into a free overfall. Draw the water surface profile for the following cases: (i) no gates exist, (ii) a gate is placed between the two reaches, and (iii) two gates are placed midway in the first and second reach, respectively.

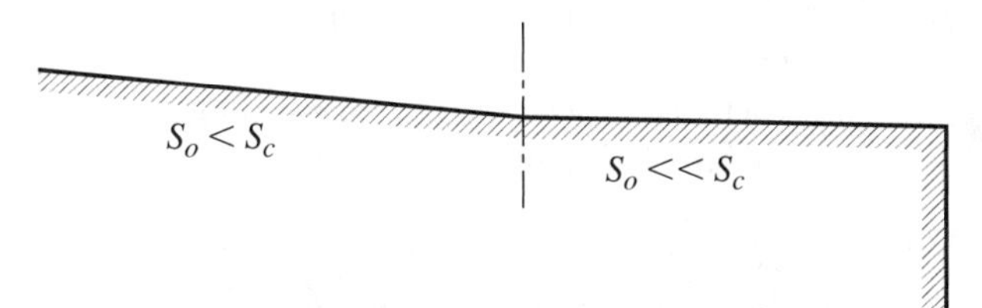

Figure P14.12 Two channels for Problem 14.12.

14.13. Solve Problem 14.12 while the mild slope is replaced with the steep slope. The second reach is steeper than the first one.

14.14. A long mild reach in an open channel is followed by another reach of steep slope as shown in Figure P14.14. Draw the water surface profile in the following cases: (i) no gates exist, (ii) a gate is located between the two reaches, and (iii) the gate is moved to the end of the second reach.

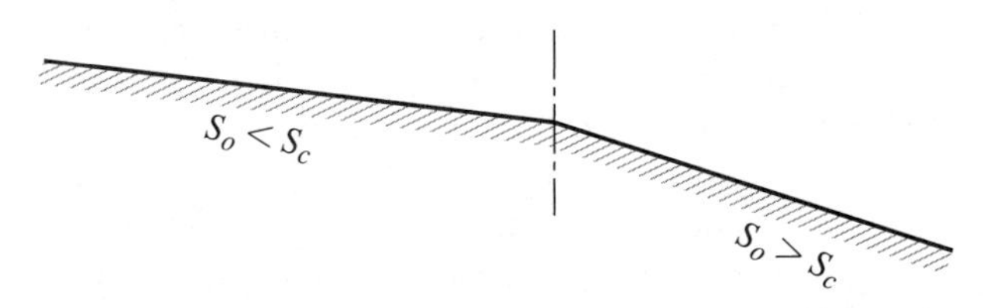

Figure P14.14 Two channels for Problem 14.14.

14.15. Solve Problem 14.14 by interchanging the bed slopes of the two reaches.

14.16. An open channel consists of two long reaches separated by a short horizontal reach. The first reach is steep while the last reach is mild and two gates are located at the end of the first and last reaches. Assuming that the flow depth in the horizontal reach is higher than the normal depth in the mild reach by 1.5 m, draw the water surface profile.

14.17. A huge reservoir of constant water level feeds a wide channel of three reaches as shown in Figure P14.17. The first reach has a steep bed slope, the second has a mild bed slope, and the bed slope in the last one is adverse. The channel terminates into a free overfall. Two gates are located at the end of the first and second reaches, respectively. Assuming that the water level in the adverse reach is above the water level just before the second gate by 0.8 m, sketch the water surface profile.

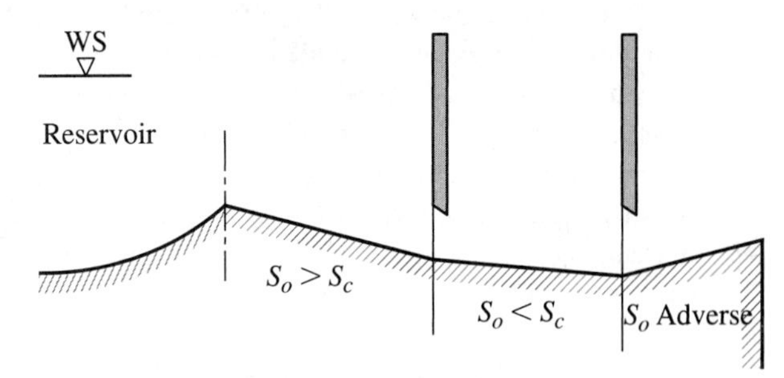

Figure P14.17 A lake feeding a channel for Problem 14.17.

14.18. An open-channel reach of the critical bed slope is located after a steep reach and before a mild reach. Two gates are placed at the points where the bed slope is varied. Draw the water surface profile along the channel.

14.19. Three reaches of steep, mild, and critical bed slopes, respectively, constitute an open channel which is connected to a large reservoir as shown in Figure P14.19. Two gates are located in the middle of the steep and critical reaches. Sketch the water surface profile along the channel.

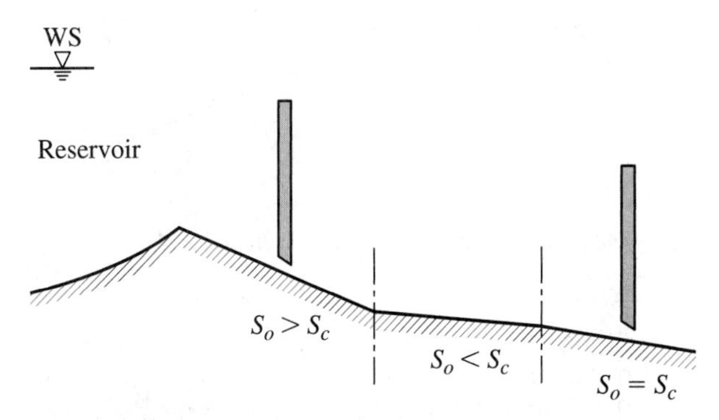

Figure P14.19 A lake feeding a channel with gates for Problem 14.19.

14.20. A weir is located in the midway of a long steep channel to measure the flow. A gate is placed at the end of the channel to elevate the water level just before the gate. Draw the water surface profile.

14.21. Water flows underneath a sluice gate to a mild reach in an open channel. At the end of this reach a weir is built to develop the required navigation depth upstream. After the weir, the channel has a steep slope. Sketch the water surface profile along the channel.

14.22. A wide rectangular channel consists of two reaches. The first reach has a slope of 0.005 ft/ft while the second one has a slope of 0.0002 ft/ft. The discharge per unit width is 25 ft^2/s. Determine whether a hydraulic jump will be formed or not and if so, where. Sketch the water surface profile in the channel. The Manning coefficient $n = 0.015$.

14.23. A triangular channel is divided by a gate into two reaches having a bed slope of 0.01 m/m and 0.0004 m/m, respectively. The channel issues a discharge of 10 m^3/s, and the side slope is 3:2 Find the critical and normal depths in the channel and draw the water surface profile. Let $C = 50$ (metric).

14.24. A trapezoidal channel of 8 ft bed width and 1:1 side slope carries a discharge of 300.0 ft^3/s. The channel consists of two reaches with a bed slope of 0.0004 ft/ft and 0.03 ft/ft, respectively as shown in Figure P14.24. A gate is located in the middle of the second reach to elevate the water level at its upstream. Determine the critical and normal depths and sketch the water surface profile. Let $n = 0.03$.

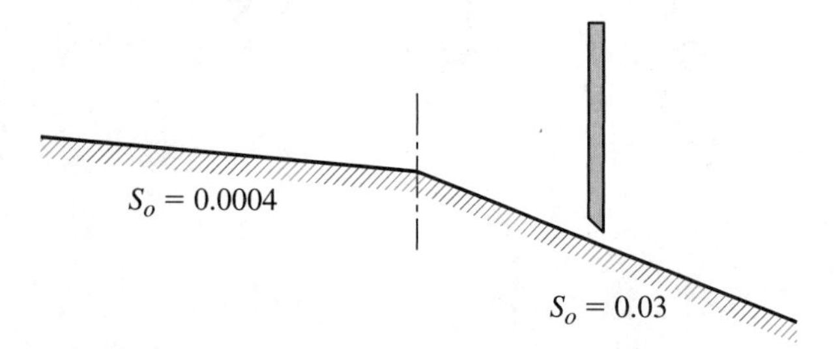

Figure P14.24 Two channels with a gate for Problem 14.24.

14.25. Water is stored behind a vertical sluice gate to a depth of 4 m. The gate opening is 0.5 m and the downstream channel is running on a slope of 0.005 m/m. The channel is rectangular in shape with a width of 3 m and a Manning n value of 0.035.

a. Will a hydraulic jump form downstream of the gate in this case and, if so, what will be the conjugate depths?
b. Sketch the water surface profiles that will develop on either side of the gate.

14.26. A 25-ft-high spillway is placed across a 20-ft-wide rectangular channel. The downstream channel is running on a slope of 0.000845 with an estimated Manning n value of 0.03. If the spillway is passing a discharge of 300 cfs,

a. What is the total depth of water behind the spillway? Assume a C value of 3.3.
b. Ignoring losses, what is the depth at the toe?
c. Will a hydraulic jump form downstream of the spillway, and, if so, what will be its conjugate depths?
d. Sketch the situation of this problem, both upstream and downstream, being careful to label all of the profiles.

14.27. A trapezoidal channel of 5.0 m bed width and 3:2 side slope carries a discharge of 16.0 m^3/s. Find the critical depth. The channel exhibits changes in the bed slope in three reaches. Normal depths of flow are 1.2 m, 0.7 m, and 0.94 m, respectively. A gate is located in the midway of the second reach. Draw the water surface profile.

14.28. A wide channel is formed of two reaches. The first reach has a mild bed slope, while the second reach has a steep bed. The depth of flow at the point of change in the bed slope is 1.5 ft. Depths of flow in the first and second reaches are 1.8 ft and 0.6 ft, respectively. Find the discharge per unit width and the bed slope in the first and second reaches. The Manning roughness coefficient $n = 0.016$.

14.29. A triangular channel with a side slope of 3 : 2 takes its discharge from a large reservoir of a constant water level. Assuming that the channel has a steep bed slope and the Manning roughness coefficient, n, 0.014, find the discharge and the bed slope if the depths of flow are 1.2 m and 0.9 m at the channel entrance and at the uniform flow condition, respectively.

14.30. A trapezoidal channel of 2.5 m bed width and 1:1 side slope consists of three long reaches with a gate located between the first and second reaches as shown in Figure P14.30. The bed slopes for the three reaches are 0.0045, 0.00008, and 0.009, respectively. The bed slope of the first reach is equal to the critical slope and the Chezy roughness coefficient, C, is 45 (metric). Find the discharge and draw the water surface profile knowing that the depth of flow at the point of change in the bed slope between the second and third reaches is 0.9 m.

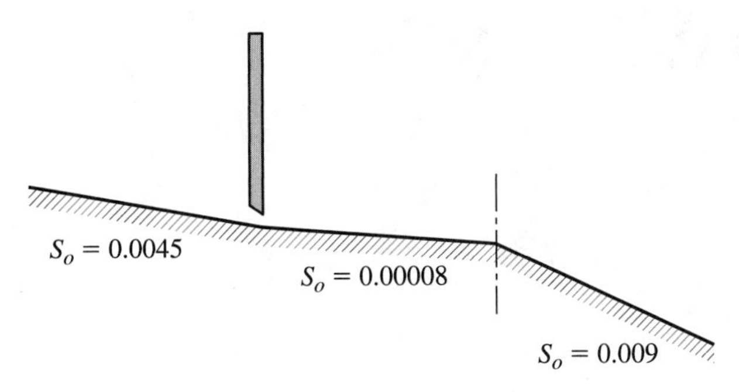

Figure P14.30 Three reaches with a gate for Problem 14.30.

14.31 A large reservoir issues water to a long wide channel with a bed slope of 0.01 ft/ft. The water level in the reservoir is 10 ft above the bed level at the channel entrance. The Chezy coefficient is 60. Evaluate the channel delivery from the reservoir.

14.32. In Problem 14.31, evaluate the channel delivery if the bed slope is 0.004 and find the critical depth. Let the entrance loss to the channel be $0.1\frac{V^2}{2g}$.

14.33. A trapezoidal channel of 2.0 m bed width and 3:2 side slope takes water from a large reservoir. The bed level at the channel entrance is 2.8 m below the water level in the reservoir and the Manning roughness coefficient, n, 0.018. Find channel delivery if its bed slope is 0.008.

Reference

Henderson, F. M., 1966. *Open Channel Flow.* MacMillan Publishing Company, New York, NY.

CHAPTER 15

Computation of Water Surface Profiles

Tennessee River and Highway Bridge. (Courtesy of Dr. Mohammad Al-Hamdan.)

In the previous chapter various water surface profiles that allow for gradual increase or decrease in water levels under different categories of bed slopes, were discussed. In general, twelve profiles (M1, M2, M3, S1, S2, S3, C1, C3, H2, H3, A2, and A3) can form to allow for gradual variation in the depth of flow. Each profile corresponds to a specific case. Any profile designated as either 1 or 3 is a rising profile (i.e., the depth of flow increases in the flow direction), while any profile designated as 2 is a falling one. It has also been demonstrated that these profiles follow the general differential equation of gradually varied flow [equation (14.7)] and all profiles are just graphical representations of this governing equation under different flow conditions and different bed slopes.

This chapter is devoted to the computation of the lengths and depths of water surface profiles, that is, (1) determination of the distance between any two given depths on a water surface profile, and (2) determination of the depth at a certain distance from a given depth within the water surface profile. The determination of either the distance or the depth along a water surface profile involves integration of the differential equation of gradually varied flow. As both the depth of flow and the velocity of flow vary with distance under gradually varied flow conditions, the exact solution of the differential equation in conjunction with the continuity equation is not tractable for most realistic conditions. A numerical solution is, therefore, required. Other simplified methods are also adapted which yield, under certain conditions, reliable solutions and involve less computation. It should be noted that although there are several procedures for computation of these profiles, for a particular case, one procedure might be more practical or precise than others. It is important to consider the problem under investigation carefully before selecting the computational scheme. Generally, all computations should begin with the depth at a control section (given depth or where the depth-discharge relation is defined) and proceed in the direction in which the control operates. Many methods are used for computation of water surface profiles. Three of them are explained in what follows, and the use of HEC-RAS, a computer software package developed by the U.S. Army Corps of Engineers for the computation of profiles is discussed.

15.1 Numerical Integration Method

The Manning formula is assumed to be sufficiently accurate to evaluate the slope of the total energy line, S_f, as shown in Figure 15.1. Therefore, one can write (in SI units):

$$S_f = \frac{Q^2 n^2}{A^2 R^{4/3}} \tag{15.1}$$

where Q is the discharge, A is the cross-sectional area, R is the hydraulic radius, and n is Manning's roughness coefficient. Combining equation (15.1) with equation (14.7), one gets

$$\frac{dy}{dl} = \frac{S_o - \dfrac{Q^2 n^2}{A^2 R^{4/3}}}{1 - \dfrac{Q^2 B}{g A^3}} \tag{15.2}$$

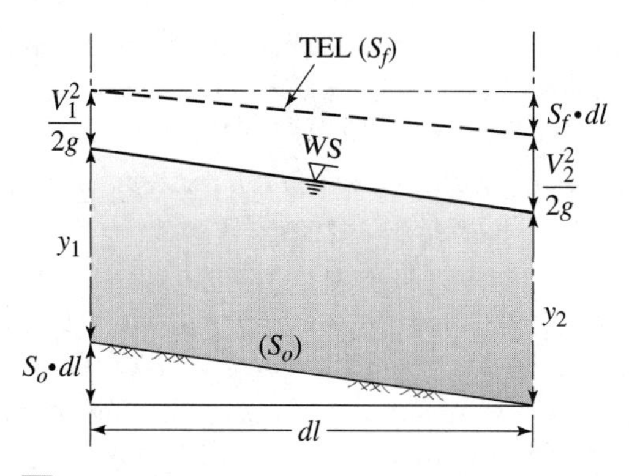

Figure 15.1 A nonuniform water surface profile.

Equation (15.2) is a nonlinear equation as *A*, *B*, and *R* on the right-hand side of the equation are dependent on *y* for a given section. For wide rectangular sections equation (15.2) is written as

$$\frac{dy}{dl} = \frac{S_o - \dfrac{q^2 n^2}{y^{10/3}}}{1 - \dfrac{q^2}{gy^3}} \tag{15.3}$$

Equation (15.3) can be solved numerically by assuming that the depth of flow, *y*, is a function of the distance, *l*, that is, $y = \varphi(l)$. Hence,

$$y_{i+1} = y_i + dy \tag{15.4a}$$

or

$$y_{i+1} = y_i + \frac{dy}{dl}dl \tag{15.4b}$$

Subscript *i* refers to a specified section along the channel. If the distance *dl* is chosen small enough, then *dy*/*dl* can be satisfactorily assumed to vary linearly over the length increment *dl*. Therefore,

$$y_{i+1} = y_i + \frac{y'_{i+1} + y'_i}{2}dl \tag{15.5}$$

where $y'_i = dy_i/dl$ and $y'_{i+1} = dy_{i+1}/dl$. Equations (15.3) and (15.5) can be solved numerically to outline the water surface profiles as follows:

1. Start from the given depth y_i, and calculate y'_i using equation (15.3).
2. As a first trial and to initiate the solution, let $y'_{i+1} = y'_i$.
3. Using equation (15.5), calculate the value of y_{i+1}; this is an approximate value.
4. Using equation (15.3), calculate y'_{i+1}. Substitute the new value of y'_{i+1} into equation (15.5) to get y_{i+1}.
5. If the values of y_{i+1} as obtained from steps 3 and 4 are not close to each other, repeat steps 3 and 4.
6. The correct solution for y_{i+1} is obtained when its values as calculated from steps 3 and 4 are the same. Proceed to the next section located after a distance *dl*. The last computed y_{i+1} now becomes y_i. The above steps are repeated until the profile is completed.

Prasad (1970) recommended that dl should be assumed between 30 and 150 m with much smaller intervals when the flow approaches the critical depth. Various computer programs are available for the calculation of water surface profiles using the above scheme, even though hand calculations are not complicated. As in previous trial and error procedures, hand calculations can be greatly facilitated through the use of a spreadsheet program.

Example 15.1

A wide rectangular channel with a slope of 0.0006 conveys a specific discharge of 3.5 m^2/s. A gate is located midway in the channel length, as shown in Figure 15.2. The depth of water at the vena contracted section, just after the gate, is 0.25 m. Draw the water surface profile and estimate the length of the rising curve formed after the gate. Take the Manning roughness coefficient as 0.02 (metric).

Solution: To draw the water surface profile, the critical and normal depths are calculated first:

$$y_c = \sqrt[3]{\frac{q^2}{g}} = \sqrt[3]{\frac{(3.5)^2}{9.81}} = 1.08 \text{ m}$$

For wide rectangular sections, the normal depth can be calculated as

$$y_o = \left(\frac{q^2 n^2}{S_o}\right)^{0.3} = \left(\frac{(3.5)^2(0.02)^2}{0.0006}\right)^{0.3} = 1.88 \text{ m}$$

The water surface profile is drawn as shown in Figure 15.2. The normal depth constitutes the high stage depth of the jump. The low stage depth can be calculated as

$$y_1 = \frac{y_o}{2}\left(-1 + \sqrt{1 + 8\left(\frac{y_c}{y_o}\right)^3}\right) = \frac{1.88}{2}\left(-1 + \sqrt{1 + 8\left(\frac{1.08}{1.88}\right)^3}\right) = 0.55 \text{ m}$$

Then an $M3$ curve will form through which the depth of flow increases from 0.25 m to 0.55 m as shown in Figure 15.2. To estimate the length of $M3$ curve, the above steps are followed. Let $y_i = 0.25$ m and take $dl_1 = 40$ m, for example. Then

$$y_i' = \frac{0.0006 - \dfrac{(3.5)^2 \times (0.02)^2}{(0.25)^{10/3}}}{1 - \dfrac{(3.5)^2}{9.81 \times (0.25)^3}} = \frac{-0.497}{-78.92} = 0.0063$$

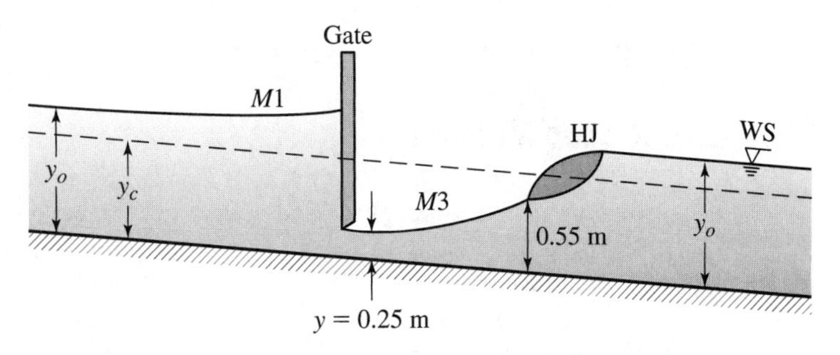

Figure 15.2 Water surface profile (Example 15.1)

Assuming $y'_{i+1} = y'_i$,

$$y_{i+1} = 0.25 + 0.0063 \times 40 = 0.502 \text{ m}$$

$$y'_{i+1} = \frac{0.0006 - \dfrac{(3.5)^2(0.02)^2}{(0.502)^{10/3}}}{1 - \dfrac{(3.5)^2}{9.81 \times (0.502)^3}} = \frac{-0.0481}{-8.87} = 0.0054$$

Then

$$y_{i+1} = 0.25 + \left(\frac{0.0054 + 0.0063}{2}\right) \times 40 = 0.484 \text{ m}$$

Using the new value of y_{i+1}, y'_{i+1} is evaluated again,

$$y'_{i+1} = \frac{0.0006 - \dfrac{(3.5)^2(0.02)^2}{(0.484)^{10/3}}}{1 - \dfrac{(3.5)^2}{9.81 \times (0.484)^3}} = \frac{-0.0544}{-10.014} = 0.0054$$

Then the last value of y_{i+1} is correct. Now let $y_i = y_{i+1} = 0.484$ m and take dl_2 as 15 m only as we need to reach the depth of 0.55 m. Then

$$y_{i+1} = 0.484 + 0.0054 \times 15 = 0.565 \text{ m}$$

$$y'_{i+1} = \frac{0.0006 - \dfrac{(3.5)^2(0.02)^2}{(0.565)^{10/3}}}{1 - \dfrac{(3.5)^2}{9.81 \times (0.565)^3}} = \frac{-0.032}{-5.923} = 0.0054$$

Then the calculated value for $y_{i+1} = 0.565$ m is correct.

One may get the exact length to the depth of 0.55 by interpolation. The last 15 m on the *M*3 curve gave an increase in the flow depth by $0.0054 \times 15 = 0.081$ m. However, one needs to get the curve length which gives an increase in the depth of $0.55 - 0.484 = 0.066$ m. Then

$$dl_2 = 15 \times \frac{0.066}{0.081} = 12.2 \text{ m}$$

Then the total length of *M*3 curve before the hydraulic jump $= dl_1 + dl_2 = 40 + 12.2 = 52.2$ m.

15.2 Direct Step Method

The direct step (or step-by-step) method is applicable to any uniform prismatic channel and can be employed using either the Manning or Chezy formula. Using the notations given on Figure 15.1, one can write

$$S_o\, dl + y_1 + \frac{V_1^2}{2g} = y_2 + \frac{V_2^2}{2g} + S_f\, dl \tag{15.6}$$

Hence,

$$(S_o - S_f)\,dl = \left(y_2 + \frac{V_2^2}{2g}\right) - \left(y_1 + \frac{V_1^2}{2g}\right) = E_2 - E_1 \tag{15.7}$$

Unlike the bed slope, S_o, the slope of the total energy line, S_f, may vary from one section to the other. Because we are calculating the length of a water surface profile between two sections, S_f in equation (15.7) should be replaced with the average value $\overline{S_f}$, where

$$\overline{S_f} = \frac{S_{f1} + S_{f2}}{2} \tag{15.8}$$

where the individual point estimates of S_f are computed from equation (15.1). Then equation (15.7) can be written as

$$dl = l_{1-2} = \frac{E_2 - E_1}{S_o - \overline{S_f}} = \frac{\Delta E}{S_o - \overline{S_f}} \tag{15.9}$$

According to the location of the control section, the computations should start and proceed in the direction in which the control operates. For the subcritical flow the computation should proceed in the upstream direction due to the fact that the control section (where the depth is known) is located downstream. For supercritical flow, the computation should proceed in the downstream direction for the same reason.

To calculate the distance between any two given depths, one may divide the reach under consideration into several subreaches. The lengths of profiles are calculated for subreaches and then summed to give the required length. The change in the depth of flow (*dy*) between which the length of the profile is calculated in any subreach should, in general, not exceed 50 cm (1.5 ft). However, if the depth and hence the velocity of flow changes rapidly, smaller values of *dy* [about 10 to 20 cm (0.5 ft) or even less] should be considered.

Since the direct step method requires many repetitive calculations, it is ideal for solution on a computer spreadsheet such as Microsoft Excel or Corel Quattro Pro. One merely needs to set up the computations once in the spreadsheet and then use the pull-down function to perform the computations for all other assumed depths automatically. The following examples illustrate spreadsheet solutions of the computation of water surface profiles using the direct step method.

Example 15.2

A rectangular channel has a width of 30.0 ft and a bed slope of 1:12100. The normal depth is 6 ft. A dam across the river elevates the water surface and produces a depth of 9.8 ft just upstream of the dam. Find the length of the backwater curve; take $n = 0.025$.

Solution: The discharge in the river can be calculated from the uniform flow condition. $A = 30 \times 6 = 180\ \text{ft}^2$, $P = 30 + 2 \times 6 = 42$ ft, $R = A/P = 4.28$ ft.

$$V = \frac{1.49}{n} R^{2/3} S_o^{1/2} = \frac{1.49}{0.025} \times (4.28)^{2/3} \times \left(\frac{1}{12100}\right)^{1/2} = 1.43\ \text{ft/s}$$

$$Q = AV = 180 \times 1.43 = 257.4\ \text{ft}^3/\text{s}$$

One needs to calculate the critical depth to determine the type of flow:

$$y_c = \sqrt[3]{\frac{(8.58)^2}{32.2}} = 1.31\ \text{ft}$$

TABLE 15.1 Spreadsheet Calculations for the Length of $M1$ Profile (Example 15.2)

y (ft)	A (ft^2)	V (ft)	E (ft)	ΔE (ft)	R (ft)	$S_f \times 10^{-5}$	$\bar{S}_f \times 10^{-5}$	$(S_o - \bar{S}_f) \times 10^{-5}$	dl (ft)
9.8	294	0.87	9.81		5.92	1.98	2.26		
				−0.8			1.905	6.00	−13333
9.0	270	0.95	9.01		5.62	2.54			
				−0.99			3.05	5.21	−19002
8.0	240	1.07	8.02		5.21	3.56			
				−1.0			4.39	3.07	−25840
7.0	210	1.22	7.02		4.77	5.22			
				−0.99			6.78	1.520	−65131
6.0	180	1.43	6.03		4.28	8.26			
Total									−123306 ft

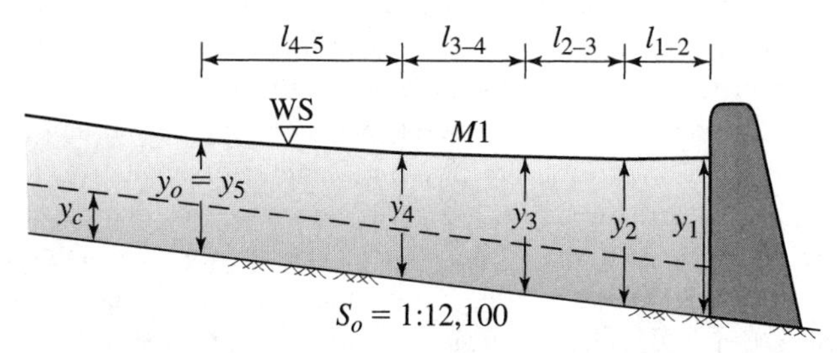

Figure 15.3 Water surface profile and intermediate depths (Example 15.2).

Since $y_o > y_c$, the river has a mild bed slope and the $M1$ profile will be formed to elevate the water surface. The depth of flow just upstream of the dam (control section) is 9.8 ft. It is required to evaluate the length of the $M1$ profile between $y_1 = 9.8$ ft and $y_5 = y_o = 6.0$ ft, as shown in Figure 15.3. Consider three other depths y_2, y_3, and y_4 between the two given depths y_1 and y_5. Hence, calculations are done for the $M1$ profile, as shown in the spreadsheet (Table 15.1), to calculate the required length.

The total length of the $M1$ profile is, therefore, 23.4 miles.

One may note that equal depth intervals (dy) do not develop equal lengths (dl) as the slope of the total energy line varies from one point to the other in gradually varied flow. As the flow depth approaches the normal depth, the water surface profile tends to be asymptotic to the uniform flow. The slope of the total energy line S_f approaches the bed slope S_o. An equal dy gives longer dl as the flow approaches the normal depth. In the former example, the same dy (1 ft) gave more distance as one moves toward the uniform flow. Although the final interval ($y_4 - y_5$) was 1 ft, it gave a length of more than twice the former dy interval.

Example 15.3

A 1.83-m reinforced concrete pipe culvert, 72.5 m long, is laid on a slope of 0.02 with a free outlet. Compute the flow profile if the culvert discharges 7.14 m^3/s, $n = 0.012$ (metric). Assume the flow to be critical at the culvert inlet.

Solution: The normal depth can be obtained from the Manning equation:

$$Q = A\frac{1}{n}R^{2/3}S_o^{1/2}$$

Now, using the relationships for circular sections from Table 10.1,

$$7.14 = \frac{r^2}{2}(\theta - \sin\theta)\frac{1}{0.012}\left[\frac{(r^2/2)(\theta - \sin\theta)}{r\theta}\right]^{2/3} S_o^{1/2}$$

where $r = 0.915$ m. Solving by trial and error, $\theta = 2.87$ rad, $A = 1.0884$ m^2.

$$y_o = r(1 - \cos(\theta/2)) = 0.793 \text{ m}$$

The critical depth can be obtained by solving the critical flow equation:

$$\frac{Q}{\sqrt{g}} = A\sqrt{\frac{A}{B}} = \frac{A^{3/2}}{B^{1/2}}$$

$$\frac{7.14}{\sqrt{9.81}} = \left(\frac{r^2}{2}(\theta - \sin\theta)\right)\sqrt{\frac{(r^2/2)(\theta - \sin\theta)}{2r\sin\theta/2}}$$

$$2.28 = \frac{((r^2/2)(\theta - \sin\theta))^{3/2}}{(2r\sin\theta/2)^{1/2}}$$

or

$$11.388 = \frac{(\theta - \sin\theta)^{3/2}}{(\sin\theta/2)^{1/2}}$$

Solving by trial, $\theta = 4.075$ rad. Hence, $y_c = 1.327$ m. Of course, the critical depth could have been obtained for Figure 12.9 as well. Since $y_c > y_o$, the channel has a steep bed slope.

Water enters the culvert with a critical depth and thereafter the flow tends to approach the normal depth. $S2$ (a falling profile) is formed to bridge y_c with y_o. It is now required to determine the depth of flow at the culvert outlet. The computation is initiated from the control section (critical flow) and proceeds downstream. Different depths are assumed and lengths are computed until the outlet section is reached, as shown in Table 15.2.

Referring to Table 15.2, one may note that a distance of 90.5 m is required to reach a depth of 0.84 m. However, the total length of the culvert is 72.5 m. Therefore, this

TABLE 15.2 Spreadsheet Computations for the Length of $S2$ Profile (Example 15.3)

y (m)	A (m^2)	V (m/s)	E (m)	ΔE (m)	R (m)	$S_f \times 10^{-5}$	$\bar{S}_f \times 10^{-5}$	$(S_o - \bar{S}_f) \times 10^{-5}$	dl (m)	Σl (m)
1.32	2.04	3.50	1.95		0.55	392				
				0.11			529	1471	7.47	7.47
1.10	1.65	4.34	2.06		0.51	666				
				0.36			933	1067	33.74	41.21
0.91	1.31	5.44	2.42		0.46	1200				
				0.29			1412	588	49.32	90.53
0.84	1.18	6.05	2.71		0.43	1624				
				0.20			1343	657	30.44	71.65
0.86	1.21	5.88	2.62		0.44	1487				

step ($y = 0.84$ m) is ignored and another bigger depth ($y = 0.86$ m) is assumed. Another alternative (but approximate) method is to evaluate the depth by the proportionality of the distance obtained from the last step of calculations. In the above example a dy of 7.0 cm (from 91 cm to 84 cm) gave a distance of 49.32 m. But one needs the value of dy that gives a distance of 31.29 m (72.5 − 41.21). Then $dy = (7 \times 31.29)/(42.32) = 4.44$ cm. The depth at the outlet will, therefore, be (91.0 − 4.44) = 86.56 cm.

Example 15.4

Solve the problem given in Example 15.1 using the direct step method.

Solution: We now consider the two depths corresponding to the beginning and end of the $M3$ profile as 0.25 m and 0.55 m, respectively. We consider two other intermediate depths. Then $y_1 = 0.25$ m, $y_2 = 0.35$ m, $y_3 = 0.45$ m, and $y_4 = 0.55$ m. Calculations are then organized as shown in the following spreadsheet (Table 15.3).

The resulting length of the $M3$ profile is 49.59 m, which is quite close to the calculated length (52.2 m) in Example 15.1 via the numerical integration method. According to the method of solution, slightly (sometimes notably) different lengths may be obtained.

TABLE 15.3 Spreadsheet Calculation for the Length of *M3* Profile Given in Example 15.4

y (m)	$V = q/y$ (m/s)	E (m)	ΔE (m)	$S_f \times 10^{-3}$	$\bar{S}_f \times 10^{-3}$	$(S_o - \bar{S}_f) \times 10^{-3}$	dl (m)
0.25	14	10.24		498			
			−4.79		330	−329.4	14.54
0.35	10	5.45		162			
			−1.92		116	−115.4	16.63
0.45	7.771	3.531		70			
			−0.920		53	−52.4	17.56
0.55	6.361	2.611		36			
							$\Sigma l = 48.73$ m

Example 15.5

A trapezoidal channel with a bed width of 5.0 ft and a side slope of 2:1 carries a discharge of 150 ft³/s. The channel has a bed slope of 0.0008 ft/ft and terminates in a free overfall. Determine the type and length of the water surface profile formed before the fall. Let $n = 0.025$.

Solution: One calculates the normal and critical depths first to identify the type of the water surface profile to be formed.

$$Q = A\frac{1.49}{n}R^{2/3}S_o^{1/2} = (by_o + ty_o^2)\frac{1.49}{n}\left(\frac{by_o + ty_o^2}{b + 2y_o\sqrt{1 + t^2}}\right)^{2/3} S_o^{1/2}$$

$$150 = (5y_o + 2y_o^2)\frac{1.49}{0.025}\left(\frac{5y_o + 2y_o^2}{5 + 2y_o\sqrt{5}}\right)^{2/3}(0.0008)^{1/2}$$

$$88.98 = \frac{(5y_o + 2y_o^2)^{5/3}}{(5 + 4.472y_o)^{2/3}}$$

Solving by trial, or spreadsheet, $y_o = 4.0$ ft.

TABLE 15.4 Spreadsheet Calculations of $M2$ Curve (Example 15.5)

y (ft)	A (ft^2)	P (ft)	R (ft)	$S_f \times 10^{-3}$	$\bar{S}_f \times 10^{-3}$	V(ft/s)	E (ft)	ΔE (ft)	$(S_o - \bar{S}_f) \times 10^{-3}$	dl	l (ft)
2.25	21.37	15.05	1.42	8.68		7.01	3.01				
					6.255			0.16	5.75	−29.4	−27.82
2.75	28.87	17.29	1.67	3.83		5.19	3.17				
					2.870			0.33	2.070	−159.4	−187.2
3.25	37.37	19.53	1.91	1.91		4.01	3.50				
					1.475			0.41	0.675	−607.4	−794.6
3.75	46.87	21.76	2.15	1.04		3.20	3.91				
					0.934			0.17	0.134	−1269	−2064
3.95	50.95	22.65	2.25	0.828		2.94	4.08				
					0.814			0.05	0.014	−3571	
4.0	52	22.88	2.27	0.800		2.88	4.13				

For the critical flow condition, solving using Figure 12.9,

$$Z = \frac{Qt^{3/2}}{g^{0.5}b^{2.5}} = \frac{150(2)^{3/2}}{g^{0.5}(5)^{2.5}} = 1.34$$

Using Figure 12.9, we estimate the factor $\frac{ty_c}{b} = 0.90$, thus leading to an estimate $y_c = 2.25$ ft. Therefore, the channel has a mild bed slope and the $M2$ curve will be formed between y_o and y_c. Since the $M2$ curve approaches y_o asymptotically, the length of the $M2$ is evaluated between $y_c = 2.25$ ft and $y = 3.95$ ft. The calculations are organized in Table 15.4.

To demonstrate the sensitivity of the obtained length to the final selected depth as the flow approaches the normal condition, the computations in this example were continued up to the normal depth. This was accomplished easily using the Excel spreadsheet. As can be seen from the table, the first 1.5 ft difference in the water level (3.75 − 2.25) gave a distance of about 800 ft, while the last 0.05 ft (4.0 − 3.95) gave a distance of about 3571 ft. It is always recommended to avoid the calculations as the flow tends to be uniform, because the accuracy will be reduced significantly. The channel design will not be affected by just assuming a uniform flow at this stage.

Example 15.6

A dam is built near the downstream end of a rectangular channel of 3.0 m width and 0.001 bed slope as shown in Figure 15.4. The channel carries a discharge of 8.5 m^3/s with a normal depth of 1.5 m. If the depth of flow just before the dam is 2.5 m, how far upstream will the backwater curve cause a velocity reduction of 20% as compared to the velocity of the uniform flow? Take the Chezy coefficient as 69 (metric).

Solution: The normal depth, y_o, is equal to 1.5 m. One may evaluate the critical depth, y_c, to know the category of the bed slope. The specific discharge is given as $q = Q/b = 2.83$ m^2/s.

$$y_c = \sqrt[3]{\frac{q^2}{g}} = 0.935 \text{ m}$$

Since $y_o > y_c$, then the bed has a mild slope and an $M1$ curve will be formed before the dam. Applying the continuity equation between a section of uniform flow and the section where

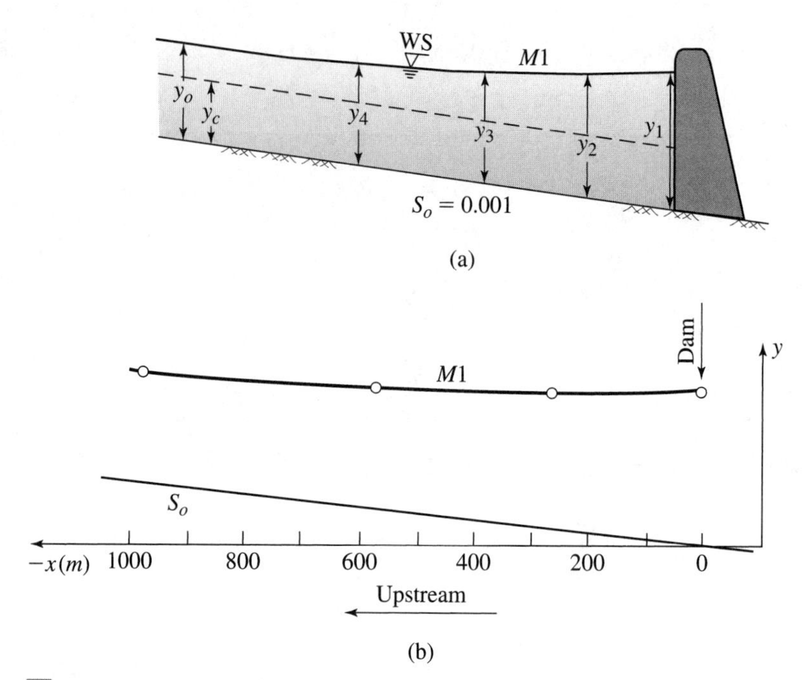

Figure 15.4 Computation and plot of *M*1 profile (Example 15.6): (a) water surface profile, and (b) plot of the resulting profile.

the velocity is reduced by 20%,

$$y_o \times V_o = y \times 0.8\,V_o, \quad \text{or} \quad y = y_o/0.8 = 1.875 \text{ m}$$

It is now required to calculate the length of the *M*1 profile enclosed between $y_1 = 2.5$ m and $y_4 = 1.875$ m, as shown in Figure 15.4a. Two intermediate depths are considered: $y_2 = 2.3$ m and $y_3 = 2.1$ m. The computations start from the section at the dam and proceed upstream to compute the required length, as shown in Table 15.5.

The required length of the *M*1 profile is 987 m. The profile is plotted in Figure 15.4b. The reader may practice by selecting other intervals of *dy* and comparing the final results.

TABLE 15.5 Spreadsheet Computations for *M*1 Profile (Example 15.6)

y (m)	A	V	E	ΔE	P	R	$S_f \times 10^{-5}$	$\bar{S}_f \times 10^{-5}$	$S_o - \bar{S}_f \times 10^{-5}$	dl
2.5	7.5	1.133	2.565		8.0	0.937	28.76			
				−0.188				31.93	68.1	−276
2.3	6.9	1.232	2.377		7.6	0.907	35.11			
				−0.184				39.39	60.6	−303
2.1	6.3	1.349	2.193		7.2	0.875	43.68			
				−0.202				50.62	49.4	−408
1.875	5.6	1.511	1.991		6.7	0.833	57.57			
										$\Sigma l = -987$ m

15.3 Standard Step Method

The direct step method for computation of nonuniform water surface profiles is applicable only to uniform and regular channel cross sections. In other words, it is applicable for computing *nonuniform* flow profiles in *uniform* channels. While there are a great number of instances where this situation arises, a more general numerical method for computing water surface profiles is necessary for the majority of cases. The technique most commonly employed in practice is the standard step method. The standard step method is completely general and can be employed in any steady flow situation regardless of the cross-sectional shape or channel conditions.

The execution of the standard step technique requires that surveyed channel cross sections be available at known locations along the stream. The one-dimensional energy equation (Bernoulli) is then written from section to section starting at the control section where the flow depth and velocity are known. Of course, in subcritical flow this section will be in the downstream location, while for supercritical flow it will be upstream. The computations then proceed to the next upstream section in subcritical flow and vice versa in supercritical flow.

Referring to Figure 15.5, it is assumed that the flow is subcritical and all conditions are known at section 1. Then the conditions (depth and velocity) at the upstream section 2 are to be determined. In the execution of the standard step method, the stage (WS) is usually employed for the static energy term. Thus, the Bernoulli equation is written as

$$WS_1 + \alpha \frac{V_1^2}{2g} + h_L = WS_2 + \alpha \frac{V_2^2}{2g} \tag{15.10}$$

where WS = water surface elevation (stage) and h_L = total head loss between sections 1 and 2.

The head loss includes the friction loss and any minor losses (expansion or contraction) that develop between the two sections so that the roughness condition of the channel reach must also be estimated. Note that since the elevation of the inverts of the sections are known, the depth at each section can be computed from the stage, or vice versa. Next, a value of stage is assumed at section 2 (WS_2) and all hydraulic variables are computed for this section from the assumed stage. Thus, every term in equation (15.10) will be known except the head loss, which can be computed from the equation. This value of the head loss will be termed the observed head loss, although it is based on an assumed stage at section 2. The reader will note that the head loss computed in this

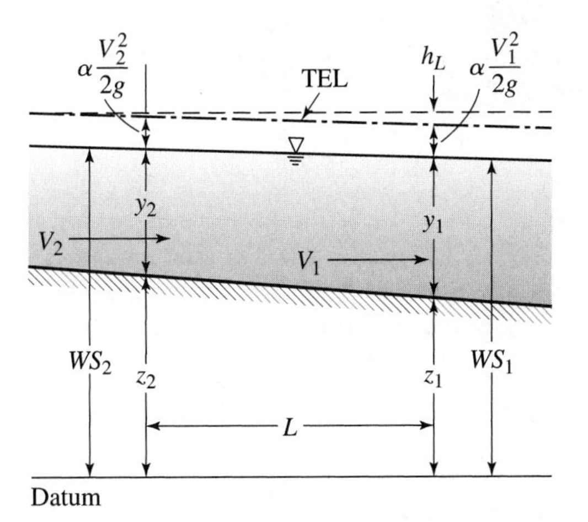

Figure 15.5 Notation diagram for the standard step method.

manner is not based on any evaluation of the actual resistance in the channel between the two sections. This head loss can be computed from a resistance equation (usually Manning) and any minor losses can be added in to result in the calculated head loss for the channel reach, that is,

$$h_L = \bar{S}_f L + C\left|\alpha_2 \frac{V_2^2}{2g} - \alpha_1 \frac{V_1^2}{2g}\right| \tag{15.11}$$

where $\bar{S}_f$ = average friction slope between the two sections, L = length of the channel between the two sections, and C = expansion or contraction coefficient, whichever is applicable. The average friction slope is usually computed using the Manning equation and can be calculated either from an average of the point estimates at the two sections or by averaging the relevant hydraulic variables and inserting these values into the equation.

The two estimates of the head loss are then compared and if they do not match within a specified tolerance, a new stage value is assumed at section 2 and the computations are repeated. The next estimate of the stage at section 2 can be obtained from the following:

$$WS_1 + \alpha \frac{V_1^2}{2g} + h_{L\,\text{cal}} = WS_2 + \alpha \frac{V_2^2}{2g} \tag{15.12}$$

where $h_{L\,\text{cal}}$ is the head loss calculated from the friction slope and minor losses in the previous step and $V_2^2/2g$ is also the velocity head at section 2 obtained from the previous step. Then, with all conditions at section 1 known, equation (15.12) is solved for WS_2. This would be the necessary stage at section 2 if the head loss was really equal to the calculated value as opposed to the observed value. The new assumed stage is then estimated as the average of the previous value and the value computed from equation (15.12).

In execution of the standard step method, as in the direct step method, it is important to remember that the procedure is based upon a linearization of the energy grade line between the two cross sections. Thus, the sections should not be taken too far apart and should not represent significantly different channel conditions. Sections should be taken as closely spaced as possible, but in no case should they be more than 1 mile apart, and sections should be taken wherever the channel geometry, roughness, or slope changes significantly.

Also as in the direct step method, because the procedure consists of many repetitive trials and computations, the solution is greatly facilitated by the use of spreadsheets. However, in the example given below, a spreadsheet solution will not be used in order to better illustrate the method. Once the student understands the technique thoroughly, then problems can be worked with the aid of spreadsheets, or using the computer software package discussed in the next section.

Example 15.7

Consider the two cross sections shown in Figure 15.6. All dimensions and elevations are as shown on the figure. Section 1 is in the downstream position and section 2 is located 1000 ft upstream in a straight channel. Let the Manning roughness coefficient for the channel be 0.05 and the discharge be 1000 cfs. If the water level at the downstream section is 10 ft as shown, what will be the depth and stage at the upstream section?

Solution: Since the cross sections are of regular shape, we will set the velocity coefficient, α equal to one. First, all relevant parameters at section 1 are computed.

$$y_1 = 5\text{ ft}, A_1 = 175\text{ ft}^2, V_1 = 1000/175 = 5.71\text{ ft/s}, P_{w1} = 47.4\text{ ft},$$

$$R_{h1} = 175/47.4 = 3.69\text{ ft}$$

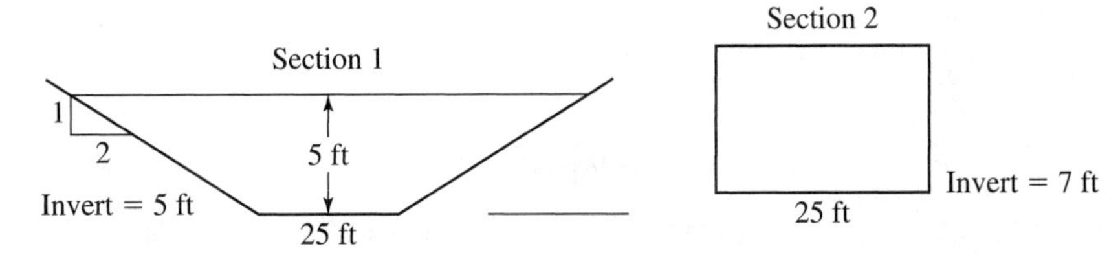

Figure 15.6 Two channel cross sections for the standard step method.

$$E_1 = Z_1 + y_1 + \frac{V_1^2}{2g} = 5 + 5 + \frac{(5.71)^2}{2g} = 10.506 \text{ ft}$$

$$S_{f1} = \left(\frac{nV_1}{1.49(R_{h1})^{2/3}}\right)^2 = \left(\frac{0.05(5.71)}{1.49(3.69)^{2/3}}\right)^2 = 0.00644 \text{ ft/ft}$$

Now, as a first assumption, let the depth at section 2 be 5 ft as well. Then, using this value, $y_2 = 5$ ft, $A_2 = 125$ ft^2, $V_2 = 8$ ft/s, $P_{w2} = 35$ ft, $R_{h2} = 3.57$ ft.

$$E_2 = 7 + 5 + \frac{(8)^2}{2g} = 12.994 \text{ ft}$$

$$S_{f2} = \left(\frac{0.05(8)}{1.49(3.57)^{2/3}}\right)^2 = 0.0132 \text{ ft/ft}$$

Now, comparing the head loss estimates, the observed head loss, $h_{L\,\text{obs}} = E_2 - E_1 = 12.994 - 10.506 = 2.488$ ft.

And the calculated head loss, $h_{L\,\text{cal}} = \bar{S}_f L + C\,\text{abs}\left(\frac{V_2^2}{2g} - \frac{V_1^2}{2g}\right)$.

Using an expansion coefficient of 0.3, one gets

$$h_{L\,\text{cal}} = \frac{0.00644 + 0.0132}{2}(1000) + 0.3(0.994 - 0.506) = 9.966 \text{ ft}$$

Since the two estimates of head loss do not compare favorably, we must assume a new stage at section 2 and try again. Employing the algorithm discussed above,

$$WS_1 + \frac{V_1^2}{2g} + h_{L\,\text{cal}} = WS_2 + \frac{V_2^2}{2g}$$

$$10 + 0.506 + 9.966 = WS_2 + 0.9937$$

$$WS_2 = 19.48 \text{ ft}$$

Then the next assumed stage at section 2 is computed from $\frac{19.48 + 12.0}{2} = 15.74$ ft.

Using this estimate for WS_2, the relevant parameters at section 2 are recomputed.

$$y_2 = 8.74 \text{ ft},\ A_2 = 218.5 \text{ ft}^2,\ R_{h2} = 5.14 \text{ ft},\ V_2 = 4.58 \text{ ft/s}$$

$$E_2 = 7 + 8.74 + \frac{(4.58)^2}{2g} = 16.065 \text{ ft}$$

$$S_{f2} = \left[\frac{0.05(4.58)}{1.49(5.14)^{2/3}}\right]^2 = 0.00266 \text{ ft/ft}$$

Then

$$\overline{S_f} = \frac{0.00644 + 0.00266}{2} = 0.00455 \text{ ft/ft}$$

Comparing head loss etimates, $h_{L\text{obs}} = 16.065 - 10.506 = 5.56$ ft.

$$h_{L\text{cal}} = 0.00455(1000) + 0.3 \text{ abs}(0.325 - 0.506) = 4.61 \text{ ft}$$

Since the two estimates still do not quite match, another value for WS_2 is estimated:

$$10 + 0.506 + 4.61 = WS_2 + 0.325$$

$$WS_2 = 14.79 \text{ ft}$$

$$\text{Then } \frac{14.79 + 15.74}{2} = 14.26 \text{ ft} = WS_2.$$

Then, at section 2, $y_2 = 7.26$ ft, $A_2 = 206.5$ ft^2, $R_{h2} = 4.97$ ft, $V_2 = 4.84$ ft/s,

$$E_2 = 7 + 7.26 + \frac{(4.84)^2}{2g} = 14.62 \text{ ft}$$

and

$$S_{f2} = \left[\frac{0.05(4.84)}{1.49(4.97)^{2/3}}\right]^2 = 0.00311 \text{ ft/ft}$$

Then

$$\overline{S_f} = \frac{0.00644 + 0.00311}{2} = 0.00478 \text{ ft/ft}$$

Comparing head losses, $h_{L\text{obs}} = 14.62 - 10.506 = 5.12$ ft.

$$h_{L\text{cal}} = 0.00478(1000) + 0.3 \text{ abs}(0.364 - 0.506) = 4.82 \text{ ft}$$

The two estimates now agree to within 0.3 ft, which is about 5%. Another trial would result in an estimate of water surface of 14.11 ft at section 2. This value would result in estimates of head loss as $h_{L\text{obs}} = 4.98$ ft and $h_{L\text{cal}} = 4.90$ ft. However, the difference in water surface elevations, WS_2, between these two trials is only $14.26 - 14.11 = 0.15$ ft, which is well within the error of the method.

15.4 HEC-RAS

HEC-RAS (Hydrologic Engineering Center River Analysis System) is a software package developed by the U.S. Army Corps of Engineers primarily for computation of steady-state water surface profiles (HEC, 2001). It is a modification of an earlier package known as HEC-2. These programs have become the industry standard for water surface profile computations in hydraulics work throughout the United States and many other countries around the world. The basic computation method employed is the standard step procedure given above. Because the method is one dimensional, HEC-RAS employs a computational technique that makes the method quasi two dimensional. The procedure works as follows.

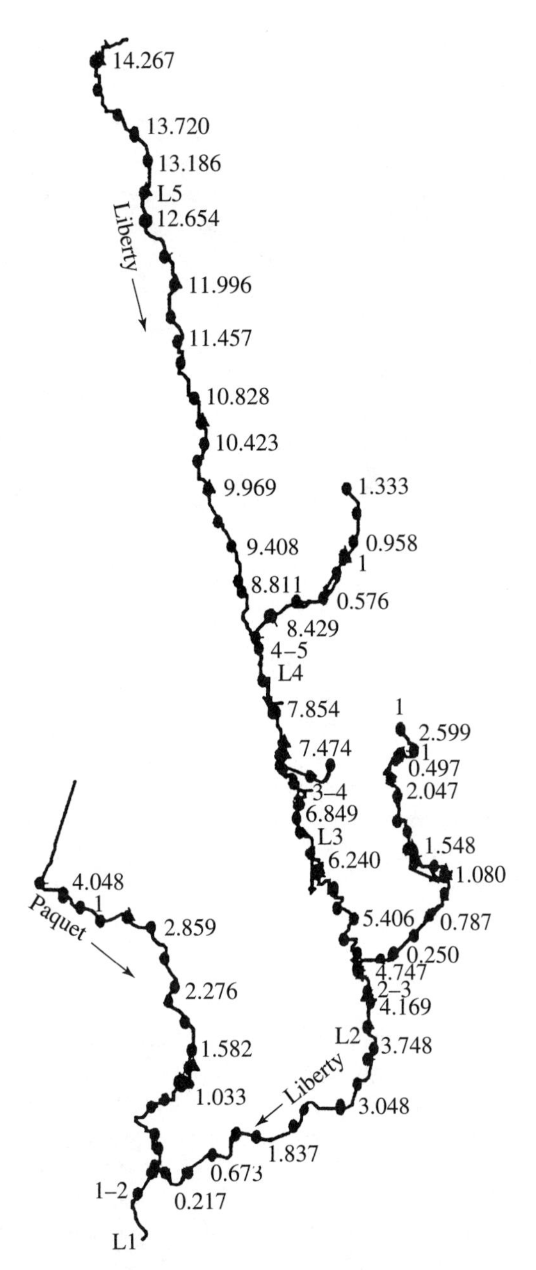

Figure 15.7 HEC-RAS schematic of Bayou Liberty, Louisiana.

A river system is idealized in the program as shown in the schematic in Figure 15.7. This schematic shows the layout of all of the surveyed cross sections of the main channel and tributaries to be modeled. In subcritical flow (the normal case) the procedure starts at the most downstream cross section (where starting conditions must be specified) and proceeds upstream via the standard step method. The program can be started using known water surface elevations (such as a lake), normal depth, or critical depth. The elevations in the main

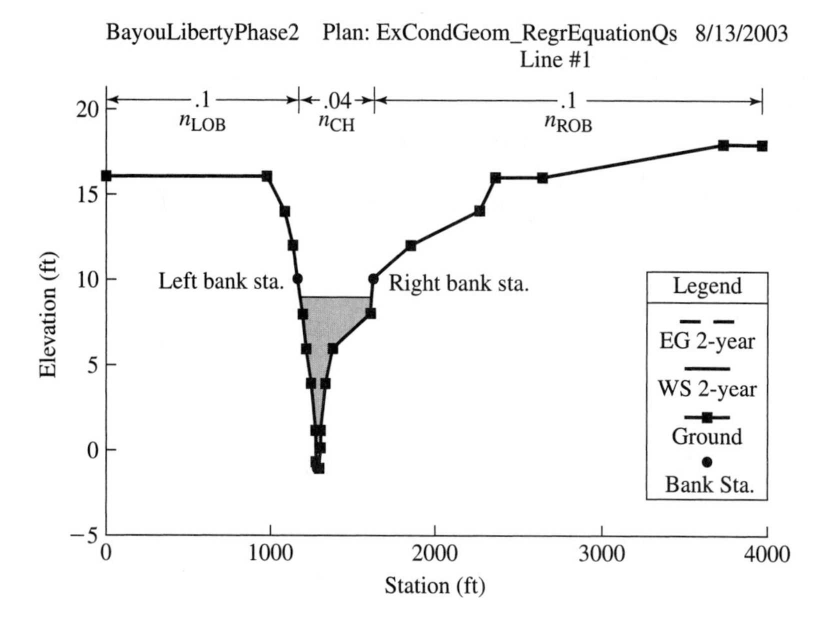

Figure 15.8 Typical cross section from HEC-RAS.

channel at the confluence with tributaries then provide the starting elevations for the tributaries, and so on.

The surveyed cross sections must be input into the program starting with the most downstream section. Now, consider the cross section shown in Figure 15.8. The section can be divided into three parts as shown depending on roughness and channel geometry. The total discharge is divided among the three areas based on the conveyance (see Chapter 11). The conveyance is computed from the Manning equation as

$$K = \frac{1}{n} A R_n^{2/3} \text{ (SI units)} \tag{15.13}$$

The discharge is then given by $Q = KS_f^{1/2}$. Use of the conveyance facilitates the operation due to the fact that the energy slope (S_f) must be the same for all three parts of the section. The discharge is balanced among the three portions of the section such that the energy and water surface elevations (but not depth) are equal across the section in the transverse direction. Thus, although the one-dimensional nature of the Bernoulli equation is maintained, some adjustment can be made for variation in channel characteristics across the section. It is also possible for the user to specify variations in Manning n values laterally across the section, even within the three designated sectional areas.

In setting up an input file for HEC-RAS, the user specifies the stations at which the section is to be subdivided. These points are designated the overbank stations. They are normally placed in a manner to separate the main part of the channel from the more heavily vegetated overbank areas as shown in the figure. Some judgment is necessary in selecting these points. In Figure 15.8 Manning n values of 0.1 are used for the overbank roughness, while a value of 0.04 is used for the channel.

The program has the capability to handle constrictions and obstructions in the channel such as bridge openings, piers, and culverts. A bridge cross section from the program is shown

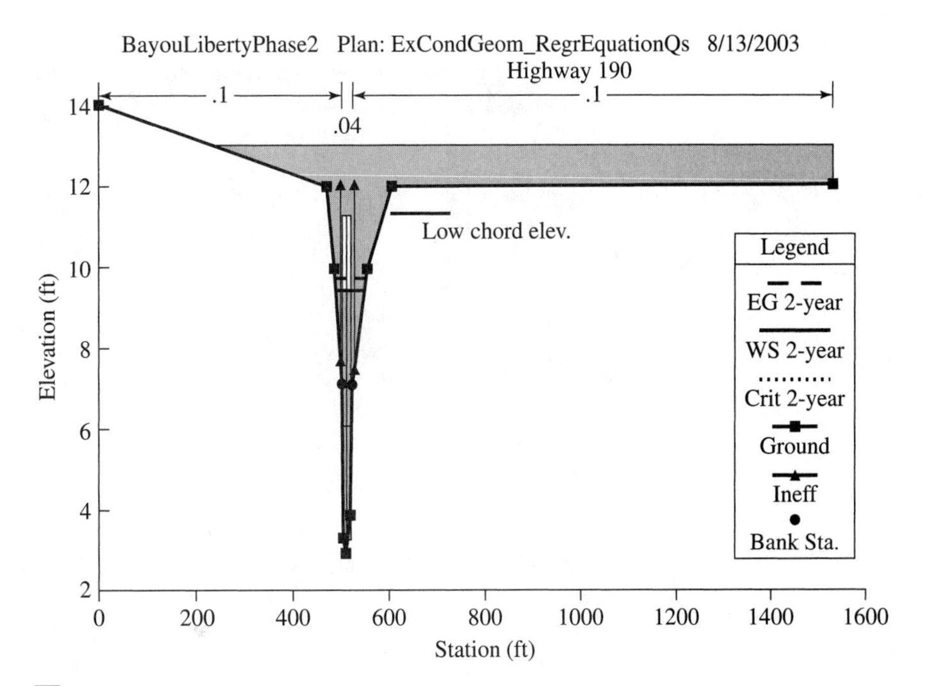

Figure 15.9 Bridge cross section in HEC-RAS.

in Figure 15.9, which actually depicts two culvert openings in a road embankment. The user can select from a number of options for handling these cases. If the flow will pass through the bridge opening without the water level reaching the underside of the superstructure or bridge deck, or the top of the culverts (known as the low chord, Figure 15.9), one can merely employ the standard step method as in any channel situation, except that the obstructions to the flow such as piers are subtracted from the available flow area. On the other hand, if the water surface does reach the low chord, or goes over the bridge, then another approach is usually selected. In this case, pressure and weir flow are computed as discussed in Chapters 12 and 16. If the low flow case prevails in this routine (i.e., the water surface does not reach the low chord), then the energy equation can be employed or a quasi-empirical energy equation known as the Yarnell equation (HEC, 2001) can be employed

$$\Delta H = 2K(K + 10\omega - 0.6)(\alpha + 15\alpha^4)\frac{V_2^2}{2g} \tag{15.14}$$

where $\Delta H =$ the drop in water surface elevation from the upstream side to the downstream side of the bridge, $K =$ the pier shape coefficient (0.9 for semicircular noses; 1.25 for square piers); $\omega =$ the ratio of the velocity head to the downstream depth; $\alpha =$ the ratio of the obstructed cross-sectional area to the total unobstructed area; $V_2 =$ the velocity downstream of the bridge. If significant obstructions to the flow exist, in the form of piers, for example, then a full momentum balance approach is also available in HEC-RAS.

The subcritical low flow case is known as class A low flow. If the constriction is sufficient to force critical depth to occur at the site, or the tail water is sufficiently low, then the class B case prevails (Figure 15.10). If the flow within the bridge or culvert does go supercritical,

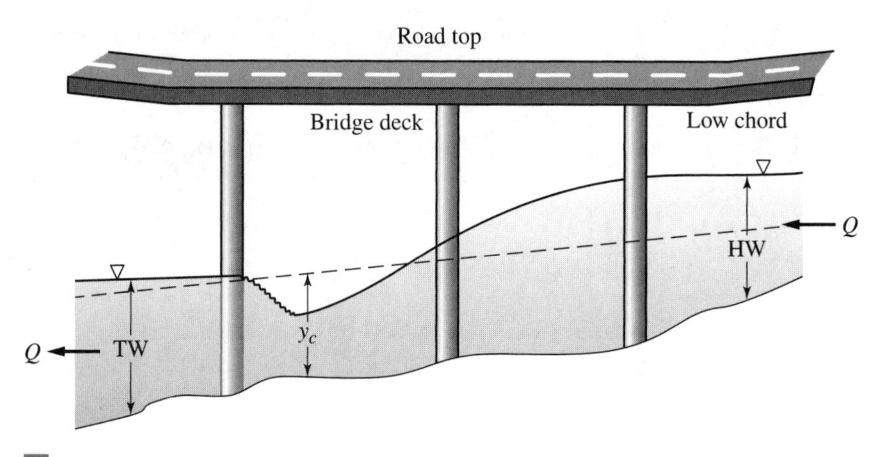

Figure 15.10 Class B low flow in a bridge.

then a hydraulic jump will most likely occur either somewhere within the crossing, or immediately downstream as shown in the figure. Usually, the engineer wishes to avoid the case of class B low flow for obvious reasons. As stated above, HEC-RAS has improved on the methodology for handling troublesome cases by the addition of a momentum balance operation if the energy losses cannot be calculated accurately. An example of this situation would be when the steady flow assumption is no longer valid at the bridge section. The user has the option to select whichever routine he or she desires to handle the bridge constriction case. A convenient menu allows the user to merely choose which options to be employed for both the high-flow and low-flow cases.

It is also possible to select a one-dimensional unsteady flow option for the overall profile analysis if conditions warrant. This routine was modified from an earlier package known as UNET (HEC, 1993). In this routine, the unsteady shallow flow equations are solved in their one-dimensional approximation.

The HEC-RAS package includes a graphical Windows-based user interface not available in the earlier HEC-2 package. The interface allows for cross sections to be plotted in real time as the user inputs the data so that input errors can be immediately discerned (as in Figure 15.8). The water surface profiles can also be plotted within the interface as shown in Figure 15.11. The program will run up to 15 trials of the standard step procedure at each section. The critical depth, critical velocity, flow area, actual velocity and velocity head are calculated at each step. Errors or undesirable profiles can be immediately identified and mistakes corrected. The package also allows the user to modify the cross section for channel improvements and floodplain encroachments. The impact of proposed hydraulic structures can also be assessed. Dredging quantities are automatically computed for channel modification.

HEC-RAS is public domain software and easily available from the Hydrologic Engineering Center Web site, along with complete documentation. It is used by most professionals engaged in hydraulic work and is required for most projects funded by the federal government and for the federal flood insurance program. It is completely general for steady-state work, its major weaknesses being the one-dimensional nature of the Bernoulli equation and the empirical aspects of the Manning resistance formula. It is thought to be accurate to $+/-1.0$ ft of stage.

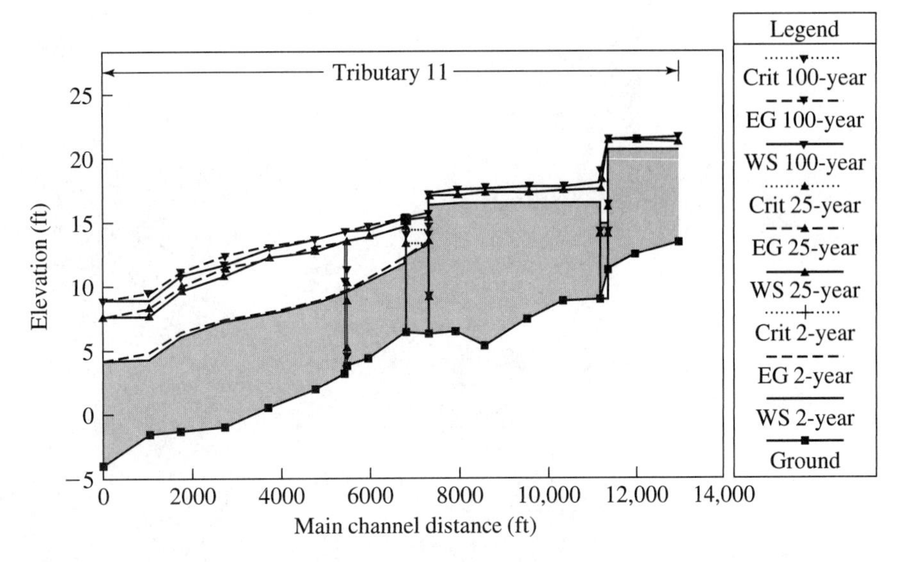

Figure 15.11 Plotted water surface profiles from HEC-RAS.

15.5 Geographical Information Systems Applications

Hydraulic modeling is greatly facilitated by the use of Geographical Information System (GIS) technology. GIS allows for all of the data necessary for modeling to be displayed and spatially analyzed concurrently. Figure 15.12 shows a GIS data display of land cover, topographic coverage, and cross section cut lines for a watershed in southeastern Louisiana. With modern instruments, it is now possible to obtain spatial data of sufficient resolution and accuracy to perform hydraulic modeling. High resolution aerial photographs (1–5 m resolution), land cover images derived from satellite data (such as Landsat Thematic Mapper), and detailed topographic data can be ingested, analyzed, and spatially manipulated to facilitate hydraulic modeling. Powerful image processing and compression software, such as LizardTech's Multi-Resolution Seamless Image Database (known as MrSid), is combined with technologically advanced data storage and spatial analysis algorithms in popular GIS packages such as Arcview and Geomedia. An airborne laser system, known as Light Detection and Ranging (LiDaR), has recently become a very popular technique for acquiring accurate and high-resolution topographic data. Processed topographic coverage from digital terrain model applications at submeter spatial resolutions and 10–30 mm accuracy have become standard products from LiDaR sources. It would not be possible to utilize these data without the advent of GIS.

The HEC-RAS hydraulic modeling package was modified some time ago in order to facilitate interaction with GIS. HEC-GeoRAS consists of the HEC-RAS program linked to the ArcView (or ArcInfo) GIS package through a graphical user interface (GUI). The interface facilitates the preparation of the geometric input data for HEC-RAS and processes the simulation results from the program. To use HEC-GeoRAS, the user creates a series of line themes that are necessary for running HEC-RAS in the ArcView environment. The required vectors are the stream centerline and cross-sectional cut lines. The flow path centerlines and

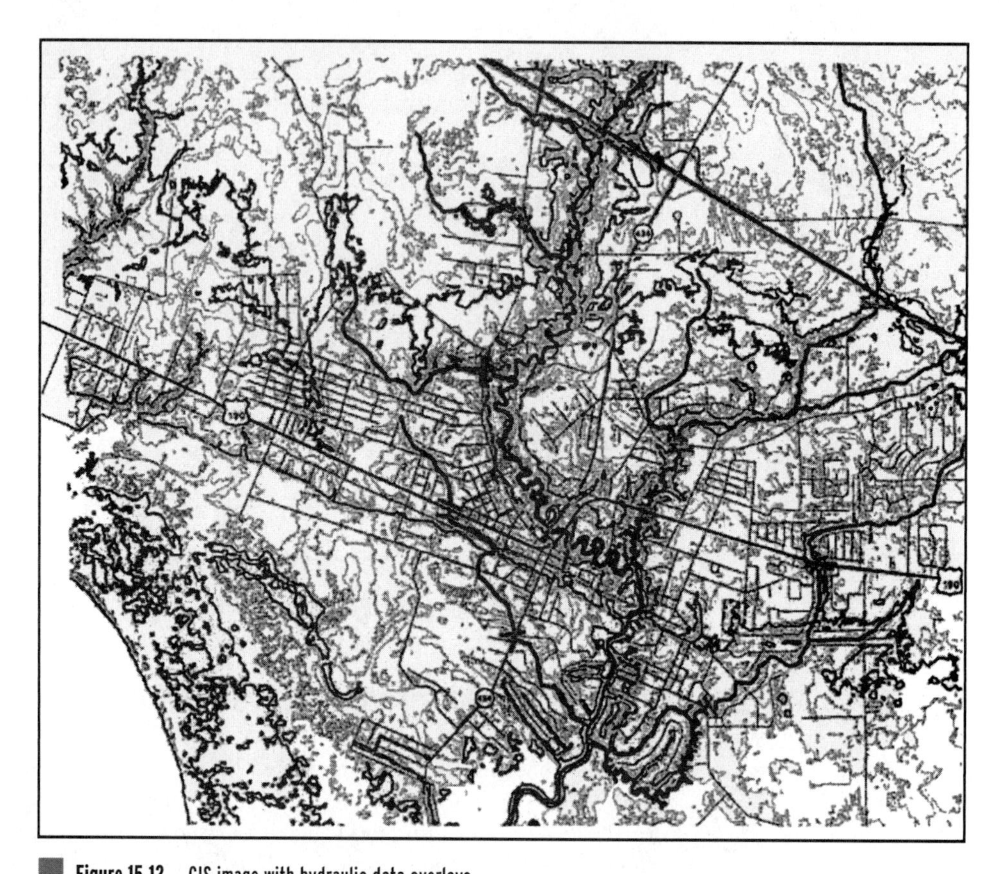

Figure 15.12 GIS image with hydraulic data overlays.

Source: Camp, Dresse, and McKee, 2004. St. Tammany Parish Bayou Lacombe Watershed Management Plan, Final Report, Figure 2-2.

main channel banks are optional geometric themes. Additional vector and raster themes that are useful in running and analyzing HEC-RAS may be created if desired, including land use and topography. Of particular interest are the cross-sectional cut lines. These consist of vectors that are spatially registered with the stream centerline and the topographic data. Once this is accomplished, the cross sections can be automatically extended using the topographic data, new cross sections can be interpolated from the data, and a three-dimensional image of the flow fields can be created (Figure 15.13). In fact, it is becoming quite common for modelers to survey only the actual stream channel itself, then create the cross-sectional cut lines in GEO-RAS and have the program create the overbank portion of the sections.

After running the program, simulation results including water surface profiles and velocity data are exported back to the GIS environment and may be processed for floodplain mapping, flood damage computations, and flood warning response and preparedness. The program is particularly effective in facilitating model calibration, in that observed water surface elevations associated with known storm events can be placed in the topographic coverage. Simulations can then be run iteratively with different values of the Manning *n* coefficient (for example), and comparisons made in real time to the observed elevations until the proper roughness coefficient is obtained.

Figure 15.13 Three-dimensional image of a river and floodplain.
The source of this figure in the HEC website.

READING AID

15.1. Name different water surface profiles of gradually varied flow in open channels.
15.2. What is the main usefulness of computations of water surface profiles?
15.3. For subcritical flow, should the computations start from the upstream or downstream side? Why?
15.4. Write down the steps of calculations of the lengths of water surface profile using the numerical integration method.
15.5. For a given discharge, what is the relation between the channel conveyance and the bed slope?
15.6. To which type of sections and to what water surface profiles can the direct step (step-by-step) method be applied?
15.7. Derive an equation for the length of a water surface profile using the direct step method.
15.8. Explain why it is recommended to consider equal depth increments (dy) in the calculations of the profile lengths.
15.9. Do you expect the same depth increment (dy) to lead to longer or shorter profile lengths near the normal depth? Why or why not?
15.10. In general, which one of $M1$ and $M2$ profiles is of shorter length? Why?
15.11. Discuss the advantages and limitations of the different methods of computations of water surface profiles.
15.12. What is the maximum distance allowed between cross sections in the standard step method and what is the reason for this?
15.13. Explain the difference between the observed head loss and the calculated head loss in the standard step method.
15.14. Explain the similarities and differences between the direct step method and the standard step method of computing water surface profiles.
15.15. Based on your own search through the internet and other resources, write a summary on the capabilities, advantages and limitations of HEC-RAS.

15.16. Discuss the data and information needed to run HEC-RAS. How can such information be obtained?

15.17. In which case, the momentum balance is considered in the solution of water surface profiles by HEC-RAS?

15.18. Perform an internet search to elaborate the usefulness of Geographical Information Systems in various hydraulic applications.

Problems

15.1. A wide rectangular river with a bed slope of 14×10^{-4} carries a specific discharge of 3.6 $m^3/s/m$. A dam is built across the river at which the afflux is found to be 1.0 m. Using the numerical integration method, estimate the water depth at a distance of 500 m upstream of the dam. Take the Manning roughness coefficient as 0.025.

15.2. A wide channel, having a bed slope of 0.01 as shown in Figure P15.2, carries a specific discharge of 5.0 m^2/s. A gate is located in the midway of the channel where the depth of water at the vena-contracted section is 0.4 m. Draw the water surface profile and estimate the length of the rising curve formed after the gate using the numerical integration method. Let $n = 0.025$.

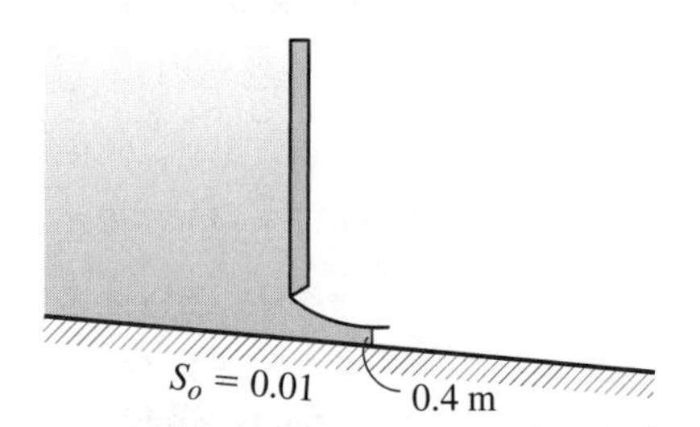

Figure P15.2 A channel with a gate for Problem 15.2.

15.3. A wide channel carrying a specific discharge of 32 $ft^3/s/ft$ is laid at a slope of 12×10^{-4}. The channel has a Manning roughness coefficient of 0.018. The channel terminates into a free overfall at which the depth of flow is equal to $1.1y_c$. Determine the depth of flow 400 ft upstream of the fall using the numerical integration method.

15.4. A wide rectangular channel under a bed slope of 14×10^{-4} carries a specific discharge of 3.6 m^2/s. An afflux of 1.0 m is measured at a dam located downstream of the river. Determine the distance upstream of the dam where the water depth is 2.2 m. Use the direct step method.

15.5. A wide river having a bed slope of 20 cm/km conveys a specific discharge of 4.1 m^2/s. A dam is built across the river where the afflux at the dam site is 1.55 m. The Chezy roughness coefficient is 45 (metric). Find the length of the backwater curve.

15.6. A weir is constructed across a wide channel to elevate the water level as shown in Figure P15.6. The channel conveys a specific discharge of 3 m^2/s and has a Chezy roughness coefficient of 50 (metric). Assuming that the water depth at the weir remains constant at 4.5 m, estimate the length of the backwater curve using the direct step method for the following two cases:

(a) The bed slope = 5 cm/km
(b) The bed slope = 10 cm/km

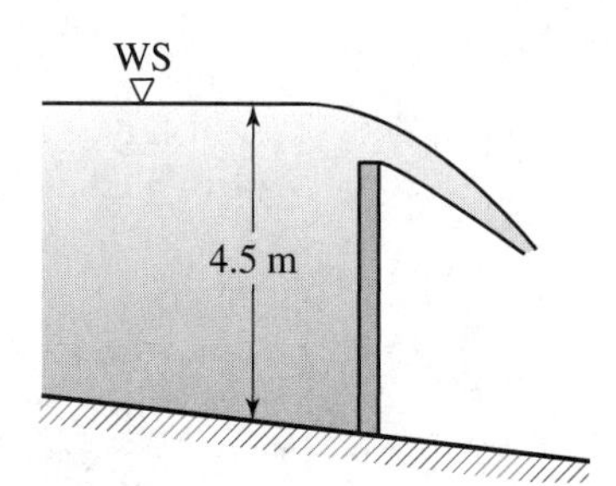

Figure P15.6 A weir across a channel for Problem 15.6.

15.7. A wide rectangular channel has a bed slope of 8 cm/km and carries a specific discharge of 1.5 $m^3/s/m$. The channel terminates into a sudden fall. Determine the length of the water surface profile, assuming a critical depth at the sudden fall. Take the Chezy coefficient as 40 (metric).

15.8. A triangular channel with a bed slope of 20 cm/km and a side slope of 1:1 carries a discharge of 0.6 m^3/s.

The Chezy coefficient is 50 (metric) and the channel ends with a sudden fall. Determine the normal and the critical depths. Hence, find the length of the water profile if the flow has a critical depth at the fall.

15.9. A wide channel having a bed slope of 20 cm/km carries a specific discharge of 2.2 m^2/s. A gate is located along the channel where the depth of flow at the vena-contracted section is 0.1 m as shown in Figure P15.9. Determine whether a hydraulic jump will be formed after the gate or not. At what distance will the jump be formed downstream of the gate? Take the Manning roughness coefficient as 0.02 (metric). Use the direct step method.

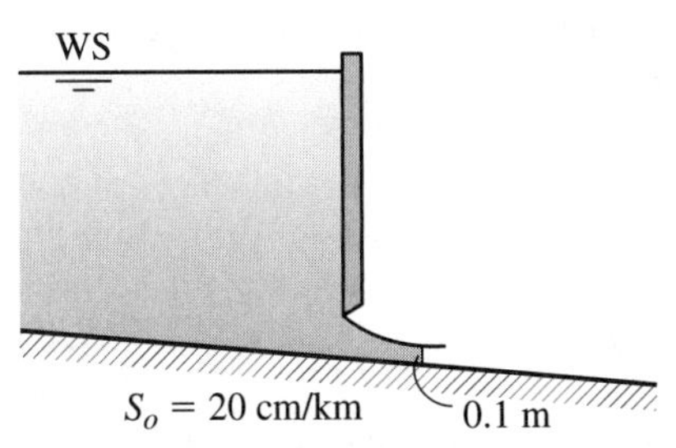

Figure P15.9 A gate across a wide channel for Problem 15.9.

15.10. A trapezoidal channel having a bed slope of 15 cm/km conveys water at a rate of 12.0 m^3/s. The channel has a bed width of 6.0 m and a side slope of 2:1. The Chezy coefficient is 40 (metric). A dam is built across the channel at which the water depth is 3.2 m as shown in Figure P15.10. Sketch the water surface profile and estimate its length.

15.11. In Problem 15.10, calculate the depth of flow at a distance of 8.0 km upstream of the dam.

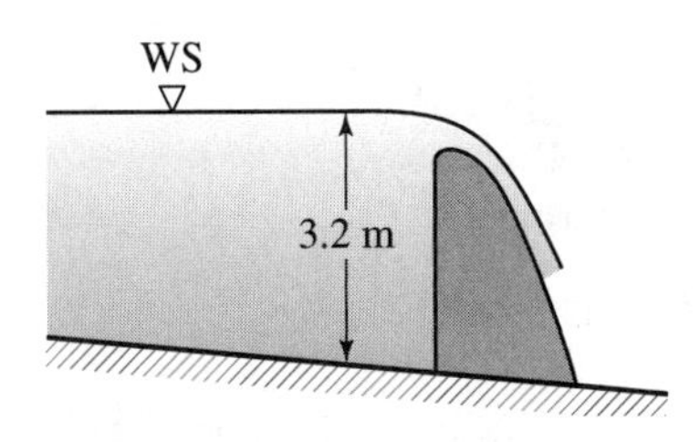

Figure P15.10 A dam across a channel for Problem 15.10.

15.12. A wide channel having a bed slope of 0.045 and the Manning roughness coefficient of 0.025 carries a discharge of 2.0 m^3/s. A dam is built across the channel at which the depth of water is 1.6 m. Draw the water surface profile and estimate the distance between the dam and the high stage depth of the formed hydraulic jump.

15.13. A wide rectangular channel, with a bed slope of 0.008 and the Manning roughness coefficient of 0.025, carries a specific discharge of 2.5 $m^3/s/m$. At a certain point the bed slope changes to become 0.05. Sketch the water surface profile and estimate the length of the water profile.

15.14. A trapezoidal channel with a bed width of 2.0 m and a side slope of 3:2 carries a discharge of 300,000 m^3/d. The channel has a bed slope of 12×10^{-5} and a Manning roughness coefficient of 0.018. At a certain point the bed slope increases to 20×10^{-3}. Draw the water surface profile and determine the length of the water profiles in the two reaches.

15.15. Water is issued into a wide rectangular channel from a large reservoir where the water level is 6.0 m above the channel bed as shown in Figure P15.15. The channel has a bed slope of 0.01 and a Manning roughness coefficient of 0.021 (metric). Draw the water surface profile and estimate the length of the water profile formed after the channel inlet.

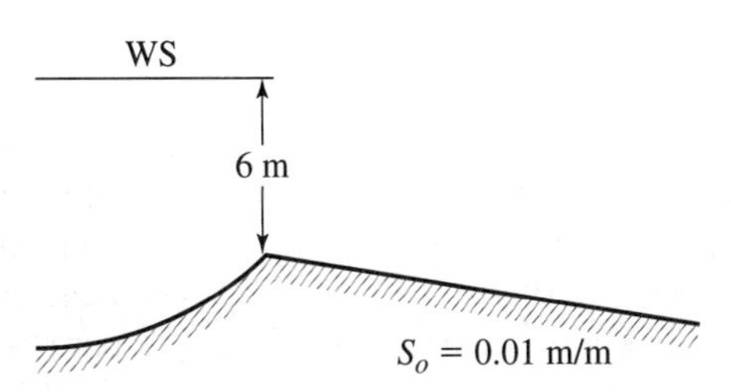

Figure P15.15 A reservoir feeding a channel for Problem 15.15.

15.16. A specific discharge of 2.1 m^2/s is issued into a wide channel through a sluice gate where the depth of flow at the vena contracted section is 0.2 m. The channel has a bed slope of 8×10^{-4} and a Chezy coefficient of 40 (metric). Determine whether a hydraulic jump will be formed or not. What is the minimum length of the concrete apron required after the gate to contain the jump assuming that the length of the jump is seven times its height?

15.17. A wide rectangular channel having a bed slope of 4.85×10^{-3} carries a specific discharge of 2.5 m^2/s. The Chezy roughness coefficient is 45 (metric). A dam is built across the channel at which the depth of water is measured as 1.8 m. Draw the water surface profile and estimate the length of the backwater curve.

15.18. A rectangular channel with a bed width of 6.0 ft and a normal depth of 4.0 ft conveys a discharge of 30 ft^3/s. Determine the bed slope assuming the Manning roughness coefficient $n = 0.025$. At a certain point along the channel, the bed slope is reduced such that the normal depth is increased to 6.0 ft. Draw the water surface profile and estimate the distance through which the depth of flow varies.

15.19. A trapezoidal channel has a bed width of 4.0 ft, a side slope of 2:1, and a bed slope of 12×10^{-4}. The channel carries a discharge of 90 ft^3/s and its Manning roughness coefficient is 0.021. A dam is built across the channel where the heading up (afflux) is 2.22 ft. Determine the depth of flow 1500 ft upstream of the dam. What is the total length of the profile?

15.20. A 12.0-m rectangular channel, having a bed slope of 50 cm/km and a Manning roughness coefficient of 0.015, conveys water at a normal depth of 1.0 m. A dam is built across the channel at which the water depth is 2.0 m. Calculate the depth of flow 2.0 km upstream of the dam using the step-by-step method.

15.21. A trapezoidal channel with a bed width of 7.0 m and a side slope of 3:2 is laid on a slope of 0.001 and carries a discharge of 30.0 m^3/s. The channel terminates to a freefall at which the depth is reduced to the critical depth. If the Manning roughness coefficient is 0.02, compute and plot the flow profile upstream from the overfall to a section where the water depth is 0.9 times the normal depth. Use the step-by-step method.

15.22. The flow in a very wide channel tends to a freefall at which the depth is reduced to 0.67 times the normal depth. The channel conveys a specific discharge of 4 m^3/s and has a bed slope of 10 cm/km. Estimate the distance upstream of the fall where the water depth is 0.9 times the normal depth. Take the Manning coefficient as 0.025 and use the step-by-step method.

15.23. A broad-crested weir is located in a 40-m-wide rectangular channel, as shown in Figure P15.23. The channel conveys a discharge of 500 m^3/s under a bed slope of 1.8×10^{-3}. The Manning roughness coefficient, n, is 0.032. The sill height of the weir is 2.0 m. The weir discharge is given as

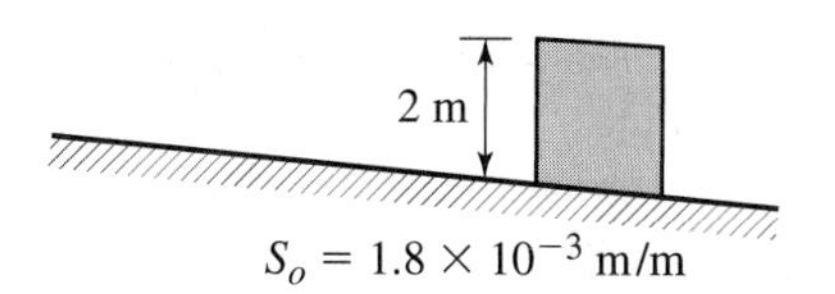

Figure P15.23 A broad-crested weir for Problem 15.23.

$$Q = 1.705 C_d B h^{3/2}$$

where B is the weir width, C_d is the discharge coefficient = 0.86, and h is the water head above the weir. Determine the length of the backwater curve using the step-by-step method.

15.24. Considering the data given in Problem 15.3, estimate the length of the $M2$ profile using the step-by-step method.

15.25. A trapezoidal channel with a bed slope of 18×10^{-5} and a side slope of 2:1 carries a discharge of 600 ft^3/s. The channel has a best hydraulic section and terminates into a free overfall. Determine the water depth 2750 ft upstream of the fall using the step by step method. Take the Manning roughness coefficient as 0.025 and assume a critical depth at the fall.

15.26. A trapezoidal channel of 2.5 m bed width and 1:1 side slope has a bed slope of 0.0045. Under the normal conditions the depth of flow is equal to the critical depth, which is 0.9 m. A gate is located in the midway of the channel length as shown in Figure P15.26. The depth of water just upstream of the gate is 1.4 m. Determine the length of the backwater curve using the step by step method. Take the Chezy roughness coefficient, C, as 45 (metric).

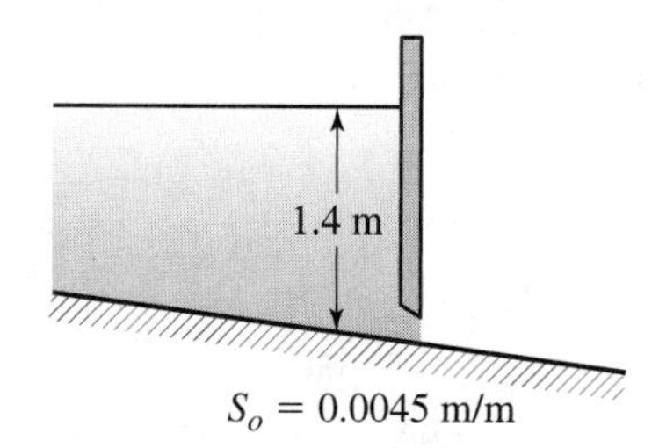

Figure P15.26 A gate across a channel for Problem 15.26.

15.27. A trapezoidal channel with a side slope of 1:1 and a bed width of 3.0 m is comprised of two reaches with bed slopes of 0.0004 and 0.03, respectively. The channel has a Chezy roughness coefficient of 55 (metric) and coveys a discharge of 12.0 m^3/s. Determine the length of the gradually varied flow reach in this channel.

15.28. A triangular channel with a side slope of 1:1 and a bed slope of 0.02 carries a discharge of 100 ft^3/s. A gate is located at the end of the channel at which the water depth is 7.0 ft. Determine the length of the rising curve before the gate. Take the Manning roughness coefficient as 0.02 and use the step-by-step method.

15.29. A wide channel terminates into a sudden fall. The channel conveys a specific discharge of 4.2 $m^3/s/m$ and has a bed slope of 16 cm/km. Assuming that the depth of flow at the fall is equal to the critical depth and the Manning roughness coefficient is 0.022, find the length of the water surface profile ahead of the fall.

15.30. A trapezoidal channel with a bed width of 7.0 m and a side slope of 3:2 has a bed slope of 0.001. The channel conveys a discharge of 30 m^3/s and has the Manning roughness coefficient of 0.02. A dam is constructed to elevate the water level. The water depth at the dam is 2.5 m. Find the distance upstream from the dam where the water depth is 2.0 m.

15.31. A wide channel with a bed slope of 15×10^{-4} conveys a specific discharge of 25 ft^2/s. A gate is located at the channel inlet where the depth of water at the vena-contracted section is 1 ft. Determine the length of the concrete apron to be constructed after the gate to contain the jump. Take the Manning roughness coefficient as 0.018 and assume the length of the jump to be 7 times its height.

15.32. A wide rectangular channel having a critical bed slope carries a specific discharge of 40 ft^2/s. Determine the bed slope knowing that the Manning roughness coefficient is 0.022. A gate is placed midway in the channel. Draw the water surface profile and determine the length of the gradually varied flow reach after the gate. The depth of water at the vena-contracted section after the gate is 1.5 ft.

15.33. A barrage is constructed across a wide river with a bed slope of 16×10^{-4}. The specific discharge of the river is 150 ft^2/s. The afflux at the barrage is 5.0 ft. Find the distance upstream of the dam where the depth of flow is 14.0 ft. Take the Manning roughness coefficient as 0.026.

15.34. A sharp-crested weir is constructed across a wide rectangular channel that carries a specific discharge of 6.5 m^2/s as shown in Figure P15.34. The channel has a bed slope of 20 cm/km and a Manning roughness coefficient of 0.026. The weir has a sill height of 3.0 m and the discharge per unit width over the weir is given as

$$q = \tfrac{2}{3} C_d \sqrt{2g} h^{3/2}$$

where $C_d = 0.7$ is the discharge coefficient and h is the water head over the weir. Determine the normal and critical depths. Estimate the distance upstream of the weir where the depth of flow is 4.75 m.

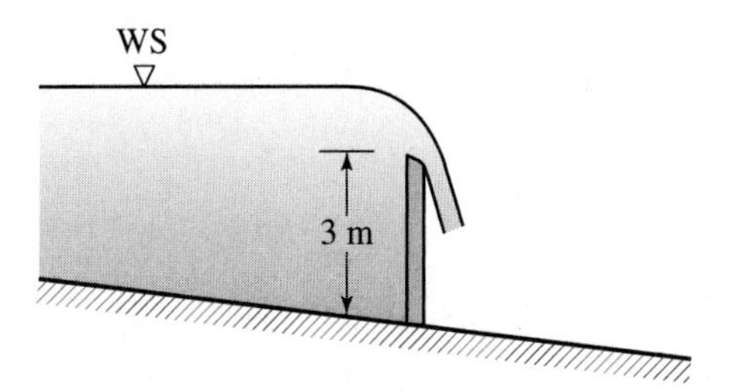

Figure P15.34 A sharp-crested weir for Problem 15.34.

15.35. A dam is built across a wide rectangular river with a bed slope of 10^{-4}. The river conveys a specific discharge of 2.833 m^2/s. If the depth of flow just upstream of the dam is 3.5 m, estimate the length of the backwater curve. How far upstream of the dam will the backwater curve cause a velocity reduction of 18% as compared to the velocity under normal flow conditions? Take Chezy's roughness coefficient as 70 (metric).

15.36. A very wide river, with a bed slope of 10 cm/km and a Manning roughness coefficient of 0.032, conveys a specific discharge of 10 m^2/s. A dam is constructed across the river so that the water depth behind it becomes 14.0 m with a water surface elevation of 55.0 m measured from an arbitrary datum. Two villages A and B are to be built 20 km and 45 km upstream of the dam, respectively. Find the respective land elevation (relative to the datum) of the two villages so that they are not to be drowned. Use the standard step method.

References

Chow, V. T., 1959. *Open Channel Hydraulics.* McGraw-Hill Book Company, New York.

French, R. H., 1986. *Open Channel Hydraulics.* McGraw-Hill Book Company, New York.

Hydrologic Engineering Center, 1993 UNET, One-Dimensional Unsteady Flow through a Full Network of Open Channels, Users Manual, U.S. Army Corps of Engineers, Davis, Calif.

Hydrologic Engineering Center, 2001, HEC-RAS River Analysis System, Users Manual, U.S. Army Corps of Engineers, Davis, Calif.

Prasad, R., 1970. Numerical method of computing flow profiles. *Jour. of Hyd. Div., Proc. ASCE.* Vol. 96, No. Ay1, pp. 75–86.

CHAPTER 16

DESIGN OF HYDRAULIC CONTROLS AND STRUCTURES

Highway culverts and bridge crossing.

Design of hydraulic structures and systems is a complex task which incorporates many elements of engineering practice. The primary elements involved in any design are safety, efficiency, reliability, cost effectiveness, and environmental concerns. Of course, the safety issue is paramount among these concerns. Any engineering project must be designed with the safety of the user and the public in mind. Associated with this is the issue of reliability (i.e., the ability of the project to efficiently perform the function for which it was designed for the specified length of time). The project should be designed and constructed in the most cost-efficient manner, consistent with its safety and functionality requirements. Lastly, the project should reflect the environmental and social values of the community and profession.

16.1 Basic Principles

Since the primary factor involved in the design of a water resources project (discharge) is usually a random variable, the design must be done in a risk-based manner. This means that, necessarily, there is a finite probability that the project will fail sometime during its lifetime. Also, in light of the dual issues of safety and functionality (efficiency and reliability), there can be two design standards and associated failure probabilities. Normally, the design standard for project reliability is lower than that for safety. That is, we are willing to accept a higher probability that a particular project will fail to perform its function sometime during its lifetime than that it will be subject to a catastrophic failure that may result in harm to the public. For instance, a water supply reservoir may be designed with storage requirements to meet a certain required user demand subject to design inflows. The design inflows are normally low flows associated with a design drought. However, additional storage and an emergency spillway may be added for safety purposes so that the dam will not be overtopped and fail during a design flood event. Thus, in this example there are two design standards and associated failure probabilities (drought and flood).

Another technique to account for uncertainties in the design of water resources projects is to allow additional height for the structure beyond that computed strictly by the design equations. This additional margin of safety is termed *freeboard.* Freeboard is defined as the difference between the top elevation of a hydraulic structure (including channel banks) and the maximum water surface elevation obtained from the design procedure. Estimates of appropriate freeboard allowances vary depending on the type of project. For example, the U.S. Army Corps of Engineers (1970) recommends 2.5 ft of freeboard for trapezoidal flood control channels (lined or rip-rapped) and 3 ft for earth levees. The American Society of Civil Engineers (1992) recommends a minimum freeboard for channel design as approximately 1 ft. Freeboard is designed to account for errors in the design process (or the data that went into the design, such as discharge measurements) and other unforeseen factors such as wave run-up during high winds.

In the design of a water resources project, the risk is the probability that the project will fail to perform its function, or fail physically sometime during its economic lifetime. However, we formulate the risk in terms of the probability of failure in any given year. The *exceedance probability* is the probability that a given discharge value will be equaled or exceeded in any year. The goal is to determine that probability with a known degree of accuracy. Sophisticated statistical techniques have been developed to estimate annual failure probabilities and the level of confidence in those estimates. The reciprocal of the annual exceedance probability is known as the *return period* and is defined as the average interval between recurrences of a discharge above a specified value. Thus, the discharge value whose annual probability of exceedance is 10% would be the 10-year flood; that with an exceedance probability of 2% is the 50-year flood; and the particular discharge whose annual exceedance probability is 1% is the 100-year flood.

Mathematically, it can be shown that the probability that a particular design discharge will be exceeded in a given span of years (and thus the project will fail, either functionally or physically):

$$R = 1 - (1 - p)^N \tag{16.1}$$

where R = the risk that a discharge of annual exceedance probability p will be exceeded in N years. By employing equation (16.1), it is a simple matter to estimate the risk associated with any design discharge.

Design discharges for water resources projects are selected by various statistical techniques which are beyond the scope of this text. In this chapter we will be concerned with the principles of design for a number of hydraulic projects and structures which are commonly encountered in open-channel situations.

Example 16.1

A levee project is to be designed so as not to be overtopped at a frequency of more than once every 200 years. What is the probability that the levee will overtop at least once in the first 50 years after its construction?

Solution: Exceedance probability $p = 1/200 = 0.005$.

Risk: $R = 1 - (1 - 0.005)^{50} = 0.222$ or 22.2%

16.2 Design of Hydraulic Drainage and Control Structures

16.2.1 Storm Sewer Design

An important aspect of many engineers' job is in the area of urban development activities. This work can encompass construction of streets, parking lots, buildings, and utilities in existing urban environments, or in the development of new commercial or residential areas. An important part of any such construction is site development. This work involves the preparation of the physical site where a construction project is to be built. Projects such as housing subdivisions, shopping centers, office complexes, and so on frequently require extensive site preparation, including topographic grading of the site (cut and fill); clearing of trees and underbrush; installation of utilities including sewer, power, and water; construction of parking areas; and, of course, installation of drainage facilities. Provision must be made to collect the runoff from the site and transmit it either off-site or to a storage facility.

The issue is complicated due to the requirements of most municipalities that peak runoff after development of an area cannot exceed that which occurred prior to the development. This requirement normally means that some water must be stored on the site to reduce the postdevelopment runoff peak. Thus, in many cases the runoff is collected and carried to a detention pond located at some convenient point within the site. Water is then released from the detention pond at a rate no greater (or possibly lower) than that which would have occurred prior to development. In this section, the design of the storm water collection system via an underground pipe network will be discussed.

16.2.1.1 Calculation of Flow Rates The discharges to be carried by storm sewers are usually computed by a formula known as the rational method:

$$Q = CIA \tag{16.2}$$

where Q = peak flow rate (ft^3/s), I = rainfall intensity (in/hr), A = area (acres), and C = runoff coefficient. If SI units are employed (i.e., I = cm/hr, $A = m^2$), then the peak flow can be expressed in m^3/s by multiplying equation (16.2) by the conversion factor 1/360000 (Wurbs and James, 2002). Equation (16.2) represents a direct translation of rainfall intensity to flow rate and thus does not include the effects of delay or attenuation or storage of water on the land surface. For this reason it is only considered to be applicable to small areas. Opinions vary on the limits of applicability of the method with estimates ranging from 10 acres to 3000 acres (Wurbs and James, 2002). Current opinion appears to favor the lower side, with areas < 50 acres recommended in many instances. If a larger area is involved, then it is advisable to decompose the area into smaller subareas and compute peak discharge for each subarea and route the peak flows to the outlet of the larger area.

The rational method is based on the premise that the peak flow from an area subjected to a constant rainfall intensity will occur at a time when the entire area is contributing to the outfall point. This time is simply the time required for water to flow from the farthest point in the area to the outfall point. In hydrology, that time is known as the *time of concentration.* Of course, this time is merely given by L/V, where L is the longest distance the water must travel and V is the average velocity of the flow. The duration for which rainfall occurs is at least equal to the time of concentration. It is obvious then the time of concentration is strongly related to the topography of the area (i.e., length and slope) as well as the state of the ground cover that the water is flowing over. Over time, a number of empirical equations have been proposed for estimating the time of concentration, many of which have been summarized by Mishra and Singh (2003). One of the oldest and most widely used is the Kirpich formula:

$$T_c = \frac{0.06628 l^{0.77}}{Y^{0.385}} \tag{16.3}$$

where T_c = time of concentration (hr), l = longest flow length (km), and Y = slope (m/m). Another commonly employed formula is that recommended by the National Resources Conservation Service for sheet overland flow:

$$T_c = \frac{0.007(nl)^{0.8}}{(P_2)^{0.5} S^{0.4}} \tag{16.4}$$

where n = Manning roughness coefficient for overland flow, l = longest flow length (ft), P_2 = 2-year frequency, 24-hr duration rainfall for the area (in), and S = land slope (ft/ft). Note that either equation (16.3) or (16.4) can be converted to the other set of units by the use of proper conversion coefficients. The above formulas have been found to be effective in cases where the overland flow distances are fairly short ($\approx$300 ft).

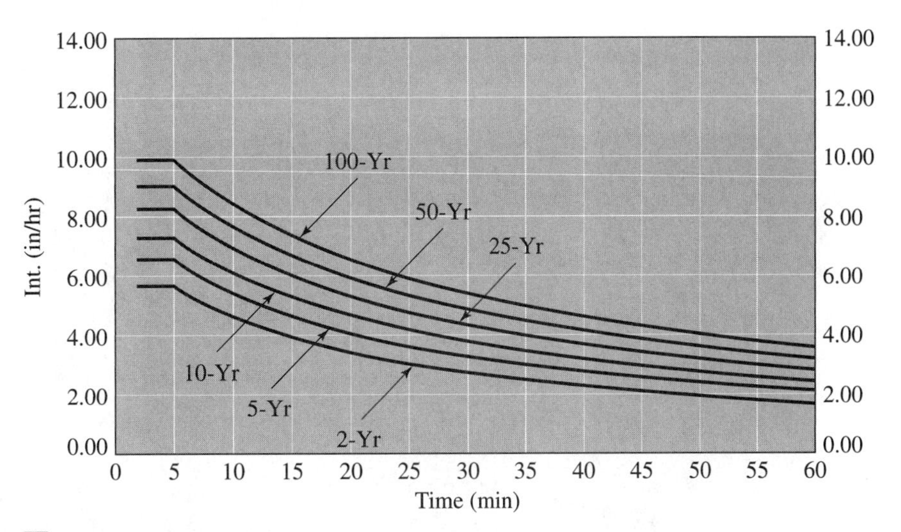

Figure 16.1 Typical IDF curves.
Source: http://bama.ua.edu/~rain

The time of concentration enters into the analysis through the selection of the rainfall intensity. The design intensity is selected from a rainfall intensity-duration-frequency (IDF) relationship that has been developed from observed rainfall data for the area of interest. IDF curves have been developed and published in various National Weather Service publications, and IDF curves or tables are also often available in drainage manuals or publications of regulations for municipalities or counties. The IDF relationship is developed through statistical analysis of rainfall data recorded for the various durations of interest, usually from around 5 minutes to 24 hours. Probability density functions (pdf) are fitted to the observed data and then the desired quantiles are determined from the pdf. These quantiles are plotted, or tabled, against the duration to obtain the desired relationship. A typical IDF curve is shown in Figure 16.1. The figure demonstrates the basic inverse relationship between rainfall intensity and duration that common sense would dictate.

The runoff coefficient (C) accounts for the losses from the raw rainfall to produce excess runoff over the area. Of course, it is related to the land cover of the area, so it is also frequently employed in the computation of the time of concentration as well. Standard tables of C values versus land development classes are available in most hydrology texts as well as drainage manuals and regulations. Typical values range from 0.15 to 0.20 for undeveloped areas such as forests or parks, and from 0.7 to 0.95 for fully developed land such as parking lots or business districts.

The concept behind the rational formula requires that the duration selected for the design rainfall be equal to the time of concentration of the area. Under this concept, a constant rainfall of the selected intensity is applied to the area, so then the peak of the outflow would occur exactly at the time of concentration of the area. Guidance usually exists in the drainage regulations as to which frequency to select for the design. A 10-year design (i.e., a 10% chance of exceedance in any given year) is commonly employed in storm drain design, although lower frequencies (lower probabilities) are sometimes mandated.

16.2.1.2 Inlet Design The water that runs off the surface of the area as sheet flow must be collected at the surface and transmitted to the subsurface pipe. This is accomplished through the storm drain inlet. Several types of inlet designs and configurations exist. The most common

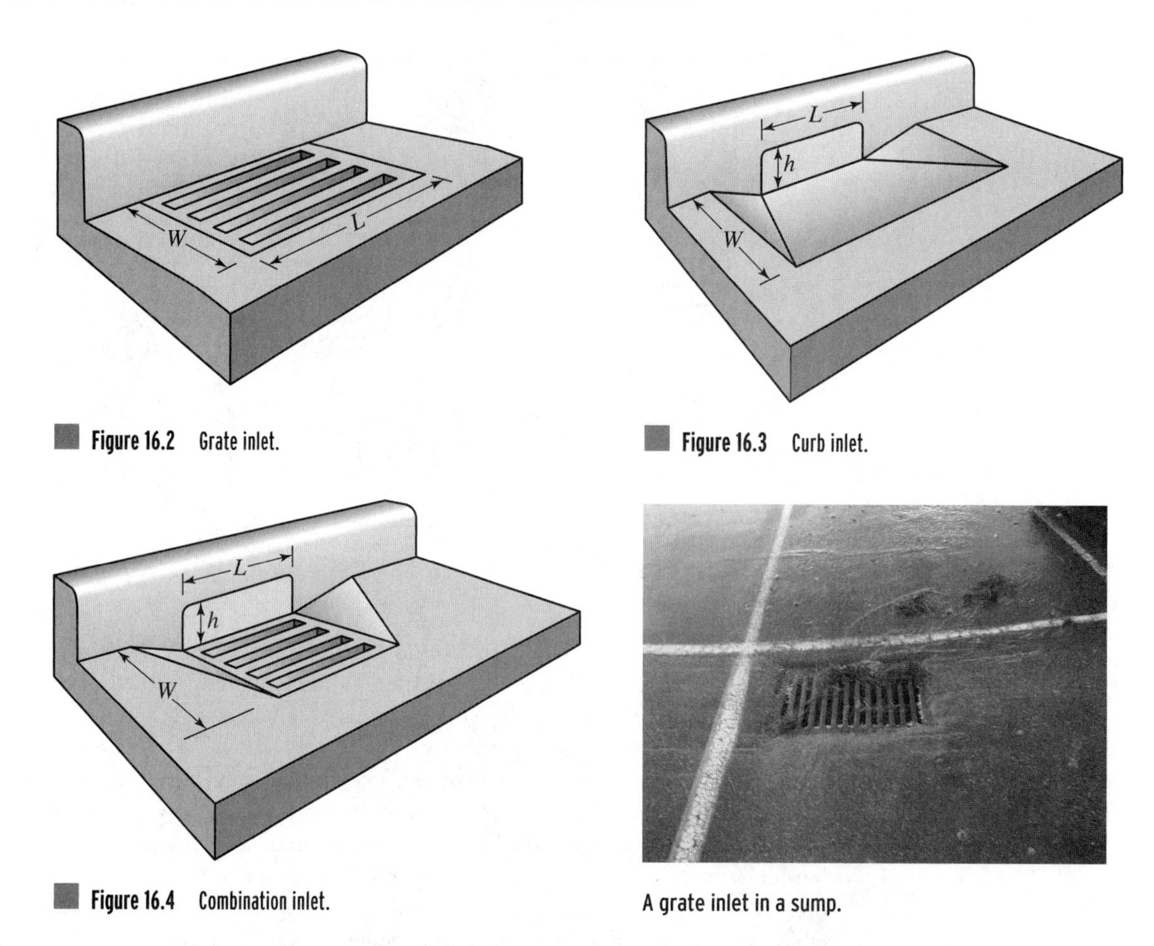

Figure 16.2 Grate inlet.

Figure 16.3 Curb inlet.

Figure 16.4 Combination inlet.

A grate inlet in a sump.

inlets are grate, curb, and combination type. Grate inlets (Figure 16.2) are situated in a lot, parking area, or roadside such that they can receive flow from the lot or road surface. A grate is placed over the inlet for protection from debris. The inlet may receive flow from three sides or four sides depending on whether or not it is set in a road curbside where one side is blocked by the curb. If the inlet is set in a parking lot, for example, it is usually set below grade in an area between the parking zones. Such inlets are sometimes known as "sump" or "sag" inlets and are designed to accept the entire amount of flow coming to them. Curb inlets (Figure 16.3) are simply openings in the side of curbs along roadways and therefore receive flow only on one side. The height of the opening should not exceed 0.5 ft for safety purposes (Wurbs and James, 2002). A combination inlet (Figure 16.4) is set in a curb and thus has an unobstructed opening in the curb but also has a grate opening for debris protection. Inlets are often placed on the grade rather than in a sag and therefore will only intercept a portion of the flow. The design criteria for storm inlets is given in the Federal Highway Administration (FHWA) Hydraulic Engineeing Circular 22 (2001).

Sump Inlets In the case of an inlet located in a low place in the ground, the flow over the lip of the inlet is governed by the mechanics of a broad crested weir (see Chapter 12):

$$Q = CLH^{3/2} \tag{16.5}$$

where L = total combined weir length (all sides that receive flow), C = weir coefficient ($C < 3.09$), and H is the head on the weir. The principle of the weir with end contractions is usually applied here such that the weir length is taken as $L + 1.8W$ where L and W are shown in Figures 16.2–16.4 (Wurbs and James, 2002). The hydraulic head on the weir in this case is the allowable ponding depth on the parking lot or street that is being drained by the inlet. The inlet is usually designed for a ponding depth of no more than six inches (16.24 cm). A clogging factor is also sometimes applied, especially with grate inlets, to account for the fact that some of the weir length may be blocked by debris. The weir length may be decreased by 10% (Wurbs and James, 2002) or more if experience dictates that debris flow is a problem at the particular location.

In extreme cases of very high flow, the weir flow over the lip may exceed the capacity of the opening and the inlet may become submerged and thus will function as an orifice. In these cases, the flow is governed by the orifice equation as discussed in the section on culvert hydraulics. The ponding depth is considered to be the head on the orifice in that case (see Section 16.2.3.3).

Inlets on Grade If the inlet is located on the grade, then the key computation involves the amount of the total incoming flow that will be intercepted by the inlet. This ratio is known as the inlet efficiency (FHWA, 2001). If the inlet is of the grate or combination type, then flow can be intercepted from both the front and side directions. The ratio of the frontal flow that is intercepted is given by (FHWA, 2001)

$$Q_f = 1 - C_f(V - V_o) \tag{16.6}$$

where

Q_f = ratio of the the available frontal flow intercepted by the inlet;
C_f = contraction coefficient ≈ 0.3 (SI) or ≈ 0.1 (Imperial)
V = average velocity of the approach flow (L/T)
V_o = bypass (or splash-over) velocity (i.e., the minimum velocity that will cause flow to pass over the inlet).

A curb inlet.

V_o is a function of the inlet type and geometry. Figure 16.5, taken from FHWA Circular 22 (2001), gives values of V_o for various inlet configurations, grate lengths, and approach velocities and can also be used to graphically solve equation (16.6). The side flow interception ratio is given by

$$Q_s = \frac{1}{\left(1 + \frac{k_s V^{1.8}}{S_o L^{2.3}}\right)} \tag{16.7}$$

Street flooding when an inlet overflows.

where

Q_s = ratio of the available flow approaching the side of the inlet that will be intercepted;
k_s = side coefficient ≈ 0.08 (SI) or ≈ 0.15 (Imperial);
S_o = cross slope of the pavement (i.e., slope toward the inlet);
L = grate length of the inlet.

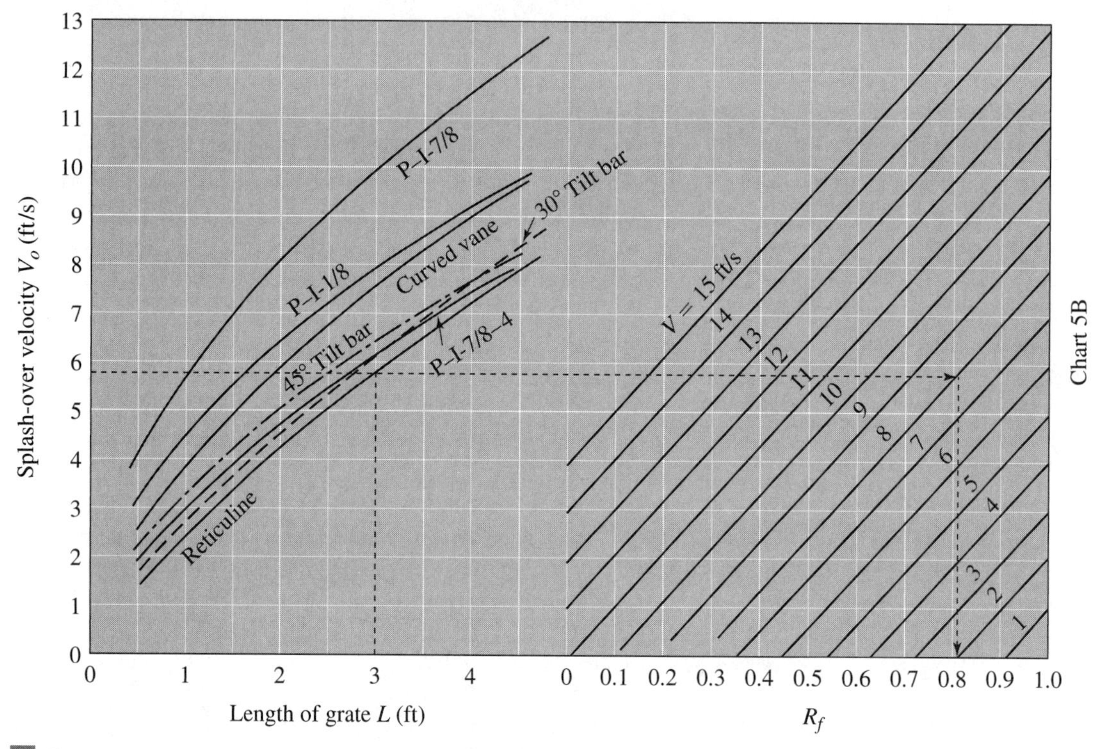

Figure 16.5 Grate inlet frontal flow interception efficiency. (FHWA, 2001.)

Example 16.2

A 10-ha undeveloped area is to be developed into a shopping center. Suppose the following characteristics were measured from a topographic map for this area: length = 3500 m and average slope = 5%. Using Figure 16.1, estimate the 10-year discharge from this area under both existing and future conditions and design a sump inlet for the future conditions flow.

Solution: Using the Kirpich formula [equation (16.3)] to find the time of concentration for the area from the given conditions, $l = 3500$ m = 3.5 km, slope (Y) = 5% = 0.05 m/m:

$$T_c = \frac{0.06628 l^{0.77}}{Y^{0.385}} = \frac{0.06628(3.5)^{0.77}}{(0.05)^{0.385}} = 0.55 \text{ hr} = 33 \text{ minutes}$$

Then, from Figure 16.1, for the 10-year rainfall intensity for this duration, we find $I = 3.5$ in/hr = 8.89 cm/hr. Then, for the undeveloped case, using a typical C value of 0.25 and $A = 10$ ha = 100,000 m^2,

$$Q = CIA = 0.25(8.89)(100000)(1/360000) = 0.617 \text{ m}^3/\text{s}$$

For the developed condition, using $C = 0.6$, we find that

$$Q = 0.6(8.89)(100000)(1/360000) = 1.48 \text{ m}^3/\text{s}$$

A weakness of the Kirpich formula for time of concentration is that it does not relate the runoff time to the land use or cover and thus does not allow for the concentration time to be changed to reflect the new land use.

Now, to size the inlet, we will apply the broad crested weir formula:

$$Q = CLH^{3/2}$$

Letting the ponding depth (H) be 6 inches (15.24 cm), and using a $C = 3.1$, $1.48 = 3.1L(0.15)^{3/2}$. Solving for the total weir length L, we find $L = 8.22$ m. From this, the inlet dimensions would be roughly 2 m × 2 m if we ignore the end contractions and assume the inlet remains unclogged. If we take these factors into account, then the inlet dimensions would need to be made even larger. Since this is larger than one would ordinarily see, it would probably be more appropriate to provide sag locations at more than one point in the parking lot, intercept smaller amounts of the flow, and thus use smaller inlet dimensions.

Example 16.3

Design an inlet for the situation described in Example 16.2 to be on the grade.

Solution: We know that the design discharge is 1.48 m^3/s, but since our design curve (Figure 16.5) is in FPS units, we will convert this to obtain a design Q of 52.3 cfs. To determine the amount of this flow that could be intercepted by a single inlet, we will need to estimate the velocity of the approaching flow. Since the design rainfall rate is 3.5 in/hr for a period of 33 minutes, and with a C value of 0.6, then we will estimate the flow depth, y to be $0.6(3.5) = 2.1$ inches. This seems reasonable for a paved parking lot. Next, to estimate the velocity, since we are dealing with sheet flow over the parking lot, we will make the wide rectangular assumption ($y = R$). Thus,

$$V = \frac{1.49}{n} y^{2/3} S^{1/2}$$

With a typical Manning n value for rough paved surfaces of 0.015, and the given slope of 0.05, we determine that the approach velocity $V = 6.94$ ft/s. Now, we must select a trial inlet type from the ones shown in Figure 16.5. If we select the reticuline configuration with a grate length of 3 ft, then Figure 16.5 indicates that our spash-by velocity $V_o = 5.8$ ft/s. Then, using equation (16.6),

$$Q_f = 1 - C_f(V - V_o) = 1 - 0.1(6.94 - 5.8) = 88.6\%$$

Now, the percent of the side flow intercepted is given by equation (16.7) as

$$Q_s = \frac{1}{\left(1 + \dfrac{k_s V^{1.8}}{S_o L^{2.3}}\right)} = 11.3\%$$

Now, since we are assuming sheet flow approaching the inlet, it is reasonable to assume that most of the flow will be frontal, so the small percentage of the side interception is not of concern in this case. So it appears that a single 3-ft reticuline grate inlet will handle around 90% of the flow. If desired, the grate length could be lengthened so that one grate would handle all of the flow, or, conversely, it could be shortened and another grate could be added.

16.2.1.3 Pipe Design As a general rule, it is not considered advisable to design the pipe to flow full (i.e., in pressure flow). There are several reasons for this practice: The geometry of circular sections results in undesirable characteristics as the flow approaches the crown of the section, and the possibility that if pressure flow occurs the energy grade line could rise to the point that water might actually back out of the sewer inlet under certain circumstances. In many municipalities the requirements are that the flow be no more than 50% full, although Sturm (2001) suggests that 80% full flow makes maximum use of the flow capacity of the pipe.

Another requirement to be considered in the design is that the low flows generate sufficient velocity such that sediment and other solids will be flushed out, and thus clogging of the pipe will not occur. Historically, velocities in excess of 2 ft/s (0.61 m/s) have been considered sufficient for this purpose, although velocities of 3 ft/s (0.91 m/s) have been suggested by others (ASCE, 1982; Sturm, 2001). Of course, these minimum velocities refer to minimum expected flows rather than to maximum discharges. Thus, the problem is to select a pipe size and material that meets the requirement that the maximum flow not exceed 50% (or 80%) of the pipe cross section and that the minimum flow generate velocities of at least 2 (or 3) ft/s.

The design begins with the initial selection of a pipe size to meet the maximum design flow requirements. This selection can be done by simply solving Manning's equation for diameter:

$$Q = \frac{1}{n} A R^{2/3} S^{1/2} \quad \text{(SI units)} \tag{16.8}$$

For circular sections, $A = \dfrac{\pi D^2}{4}$ and $R = \dfrac{D}{4}$, where D is the pipe diameter. Then

$$\begin{aligned} Q &= \frac{1}{n}\frac{\pi D^2}{4}\left(\frac{D^{2/3}}{4^{2/3}}\right)S^{1/2} \\ &= \frac{1}{n}(0.312)D^{8/3}S^{1/2} \end{aligned}$$

or, solving for D, we get

$$D = 1.55\left(\frac{Qn}{S^{1/2}}\right)^{3/8} \quad \text{(SI)} \quad \text{or} \quad D = 1.55\left(\frac{Qn}{(1.49)S^{1/2}}\right)^{3/8} \quad \text{(BG units)} \tag{16.9}$$

Concrete pipes are normally used for storm sewer work, but steel or PVC pipes can also be used. The selection of the pipe material depends on the requirement of longevity, maintenance, and cost. However, concrete is usually the desired choice if at all possible.

The next step is the computation of the normal depth for the pipe that has been selected. This is done in a manner similar to that used in the cases of rectangular or trapezoidal sections. First, the Manning equation is written as

$$AR^{2/3} = \frac{nQ}{S^{1/2}} \quad \text{(SI units)}, \quad \text{or} \quad AR^{2/3} = \frac{nQ}{1.49S^{1/2}} \quad \text{(BG units)} \tag{16.10}$$

In the following discussion it will be helpful to refer to Figure 16.6. First we compute the angle θ from the expression given in Table 10.1:

$$\theta = 2\cos^{-1}\left(1 - 2\frac{y_o}{D}\right) \text{ radians, where } y_o \text{ is the normal depth}$$

Then, also referring to Table 10.1, we can compute the area and wetted perimeter as

$$A = (\theta - \sin\theta)\frac{D^2}{8} \quad \text{and} \quad P = \theta\frac{D}{2}$$

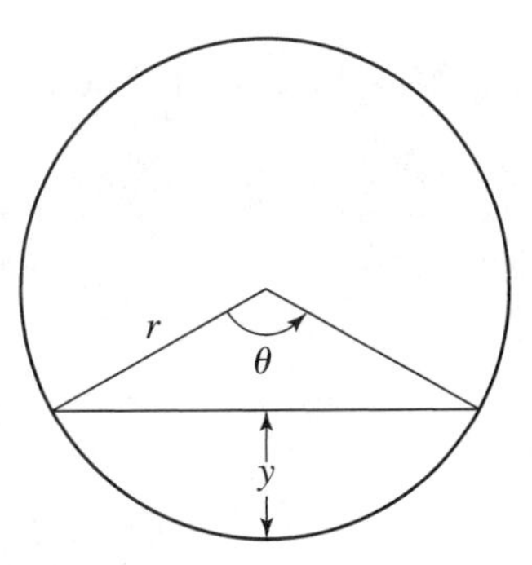

Figure 16.6 Cross section of a storm sewer.

The solution then proceeds in an iterative manner by assuming y/D ratios and computing the resulting left side of equation (16.10) and comparing with the right side, which, of course, remains fixed. As mentioned in similar situations, the procedure is ideal for solution by spreadsheet. Once the correct normal depth is obtained, it can be determined if the y/D ratio exceeds the required value. Other desired hydraulic properties, including velocity and critical depth, can also be computed for comparison to determine if the design is suitable. The normal depth is then also found for the minimum expected flow, and this leads to the computation of the area and resulting velocity for this condition. If the velocity is not suitable (i.e., too small), then a smaller pipe size would have to be selected, and, of course, that would also impact the requirement at the maximum discharge. The process continues until the pipe size is selected that meets both sets of requirements.

If noncircular pipe is to be used, then the solution for the pipe size is not as direct as that given above. Typical noncircular pipe employed for storm drains or culverts are elliptical or arch pipe. Since the geometry of these sections is not as simple as for circular sections, the geometric characteristics and flow capacities for various pipe configurations have been worked out in advance and published. For example, the American Concrete Pipe Association (ACPA) publishes a concrete pipe manual with all of the relevant data for concrete ellipital and arch pipes (APCA, 2000). Noncircular pipes are normally employed in situations where the minimum ground cover, or maximum loads on the pipe, will not allow for circular pipe. Figure 16.7 shows typical configurations of elliptical pipe installation. The most common installation is the horizontal configuration due to cover considerations. The geometric relationships of

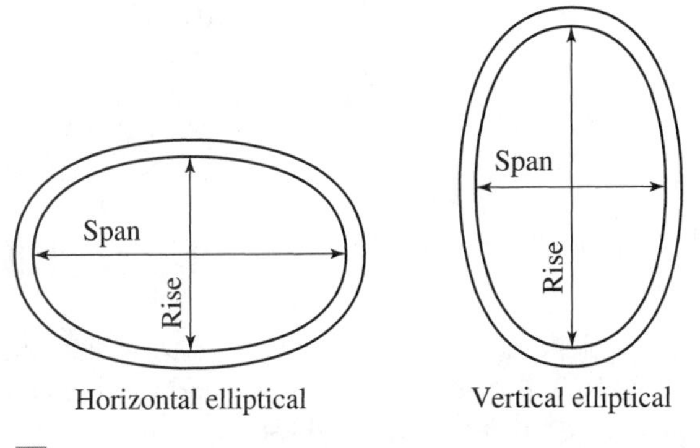

Figure 16.7 Elliptical pipe configurations.

TABLE 16.1 Dimensions and Approximate Weights of Elliptical Concrete Pipe

ASTM C 507—Reinforced Concrete Elliptical Culvert, Storm Drain and Sewer Pipe

Equivalent Round Size, inches	Minor Axis, inches	Major Axis, inches	Minimum Wall Thickness, inches	Water-Way Area, square feet	Approximate Weight, pounds per foot
18	14	23	2¾	1.8	195
24	19	30	3¼	3.3	300
27	22	34	3½	4.1	365
30	24	38	3¾	5.1	430
33	27	42	3¾	6.3	475
36	29	45	4½	7.4	625
39	32	49	4¾	8.8	720
42	34	53	5	10.2	815
48	38	60	5½	12.9	1000
54	43	68	6	16.6	1235
60	48	76	6½	20.5	1475
66	53	83	7	24.8	1745
72	58	91	7½	29.5	2040
78	63	98	8	34.6	2350
84	68	106	8½	40.1	2680
90	72	113	9	46.1	3050
96	77	121	9½	52.4	3420
102	82	128	9¾	59.2	3725
108	87	136	10	66.4	4050
114	92	143	10½	74.0	4470
120	97	151	11	82.0	4930
132	106	166	12	99.2	5900
144	116	180	13	118.6	7000

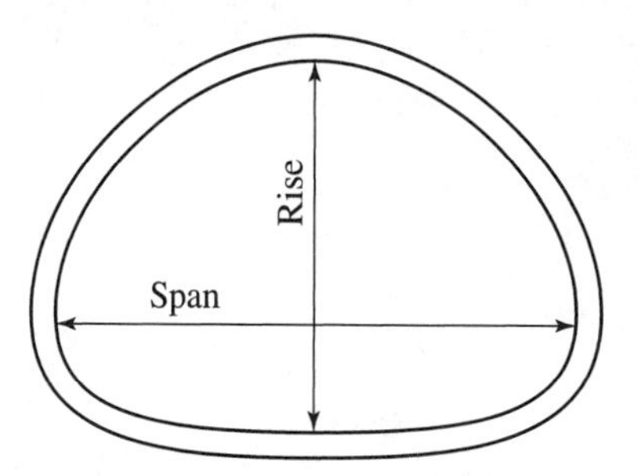

Figure 16.8 Typical arch pipe cross section.

various sizes of elliptical pipe are given in Table 16.1 while the hydraulic characteristics are given in Table 16.2. Arch-shaped pipe is used in similar circumstances to those that call for elliptical pipe, except that the cover and load characteristics of arch pipe are even more favorable than for elliptical. Figure 16.8 shows a typical section of arch pipe and the geometric and hydraulic characteristics are given in Tables 16.3 and 16.4, respectively.

TABLE 16.2 Full Flow Coefficient Values Elliptical Concrete Pipe

Pipe Size $R \times S$ (HE) $S \times R$ (VE) (Inches)	Approximate Equivalent Circular Diameter (Inches)	A Area (Square Feet)	R Hydraulic Radius (Feet)	Value of $C_1 = \frac{1.486}{n} \times A \times R^{3/4}$			
				$n = 0.010$	$n = 0.011$	$n = 0.012$	$n = 0.013$
14 × 23	18	1.8	0.367	138	125	116	108
19 × 30	24	3.3	0.490	301	274	252	232
22 × 34	27	4.1	0.546	405	368	339	313
24 × 38	30	5.1	0.613	547	497	456	421
27 × 42	33	6.3	0.686	728	662	607	560
29 × 45	36	7.4	0.736	891	810	746	686
32 × 49	39	8.8	0.812	1140	1036	948	875
34 × 53	42	10.2	0.875	1386	1260	1156	1067
38 × 60	48	12.9	0.969	1878	1707	1565	1445
43 × 68	54	16.6	1.106	2635	2395	2196	2027
48 × 76	60	20.5	1.229	3491	3174	2910	2686
53 × 83	66	24.8	1.352	4503	4094	3753	3464
58 × 91	72	29.5	1.475	5680	5164	4734	4370
63 × 98	78	34.6	1.598	7027	6388	5856	5406
68 × 106	84	40.1	1.721	8560	7790	7140	6590
72 × 113	90	46.1	1.845	10300	9365	8584	7925
77 × 121	96	52.4	1.967	12220	11110	10190	9403
82 × 128	102	59.2	2.091	14380	13070	11980	11060
87 × 136	108	66.4	2.215	16770	15240	13970	12900
92 × 143	114	74.0	2.340	19380	17620	16150	14910
97 × 151	120	82.0	2.461	22190	20180	18490	17070
106 × 166	132	99.2	2.707	28630	26020	23860	22020
116 × 180	144	118.6	2.968	36400	33100	30340	28000

TABLE 16.3 Dimensions and Approximate Weights of Concrete Arch Pipe

ASTM C 506—Reinforced Concrete Arch Culvert, Storm Drain, and Sewer Pipe

Equivalent Round Size, inches	Minimum Rise, inches	Minimum Span, inches	Minimum Wall Thickness, inches	Water-Way Area, square feet	Approximate Weight, pounds per foot
15	11	18	2¼	1.1	—
18	13½	22	2½	1.65	170
21	15½	26	2¾	2.2	225
24	18	28½	3	2.8	320
30	22½	36¼	3½	4.4	450
36	26⅝	43¾	4	6.4	595
42	31 5/16	51⅛	4½	8.8	740
48	36	58½	5	11.4	880
54	40	65	5½	14.3	1090
60	45	73	6	17.7	1320
72	54	88	7	25.6	1840
84	62	102	8	34.6	2520
90	72	115	8½	44.5	2750
96	77¼	122	9	51.7	3110
108	87⅛	138	10	66.0	3850
120	96⅞	154	11	81.8	5040
132	106½	168¾	10	99.1	5220

TABLE 16.4 Full Flow Coefficient Values Concrete Arch Pipe

Pipe Size $R \times S$ (Inches)	Approximate Equivalent Circular Diameter (Inches)	A Area (Square Feet)	R Hydraulic Radius (Feet)	Value of $C_1 = \frac{1.486}{n} \times A \times R^{2/3}$			
				$n = 0.010$	$n = 0.011$	$n = 0.012$	$n = 0.013$
11 × 18	15	1.1	0.25	65	59	54	50
13½ × 22	18	1.6	0.30	110	100	91	84
15½ × 26	21	2.2	0.36	165	150	137	127
18 × 28½	24	2.8	0.45	243	221	203	187
22½ × 36¼	30	4.4	0.56	441	401	368	339
26⅝ × 43¾	36	6.4	0.68	736	669	613	566
31 5/16 × 51⅛	42	8.8	0.80	1125	1023	938	866
36 × 58½	48	11.4	0.90	1579	1435	1315	1214
40 × 65	54	14.3	1.01	2140	1945	1783	1646
45 × 73	60	17.7	1.13	2851	2592	2376	2193
54 × 88	72	25.6	1.35	4641	4219	3867	3569
62 × 102	84	34.6	1.57	6941	6310	5784	5339
72 × 115	90	44.5	1.77	9668	8789	8056	7436
77¼ × 122	96	51.7	1.92	11850	10770	9872	9112
87⅛ × 138	108	66.0	2.17	16430	14940	13690	12640
96⅞ × 154	120	81.8	2.42	21975	19977	18312	16904
106½ × 168¾	132	99.1	2.65	28292	25720	23577	21763

Noncircular sections can also have a significant impact on the hydraulic characteristics of the pipe. Of course, friction losses can still be computed from the Manning equation as above using the actual hydraulic radii of the pipes as given in Tables 16.2 and 16.4.

Example 16.4

A 5-acre area on a 1% slope is to be developed into a shopping center. Assuming a postdevelopment runoff coefficient (C) of 0.85 and a 10-year design rainfall intensity of 3.7 in/hr, determine the optimum size concrete storm sewer to carry this runoff. Develop the design for a minimum flow of 2 cfs.

Solution: For the maximum flow case: $Q = CIA = 0.85(3.7)(5) = 15.7$ cfs. Therefore, $Q_{max} =$ 15.7 cfs, $Q_{min} = 2$ cfs, $S = 1\%$, concrete pipe ($n = 0.012$).

First, compute the initial estimate of diameter:

$$D = 1.55\left(\frac{Qn}{(1.49)S^{1/2}}\right)^{3/8} = 1.55\left(\frac{15.7(0.012)}{(1.49)(0.01)^{1/2}}\right)^{3/8} = 1.69 \text{ ft, or } 20.28 \text{ inches}$$

So we round up to the closest available pipe which might be 21 inches, or 1.75 ft. Now, we compute the normal depth for this pipe.

$$AR^{2/3} = \frac{nQ}{1.49S^{1/2}} = \frac{0.012(15.7)}{1.49(0.01)^{1/2}} = 1.26$$

Assume $y_o/D = 0.8$ (the maximum allowable). Then

$$\theta = 2\cos^{-1}\left(1 - 2\frac{y_o}{D}\right) = 4.429 \text{ rad}$$

Therefore $A = (\theta - \sin\theta)\frac{D^2}{8} = 2.06\ \text{ft}^2$ and $P = \theta\frac{D}{2} = 3.875$ ft. Then $R = A/P = 0.53$ ft. So $AR^{2/3} = 2.06(0.53)^{2/3} = 1.35$. Now, comparing the two estimates, we see that the value computed from the assumed depth ratio is greater than the constant computed from the problem specifications of flow, slope, and roughness. Thus, the assumed depth is greater than the actual depth and we are assured that the normal depth will be less than 80% of the pipe diameter. Continuing to find normal depth, we assume $y_o/D = 0.7$. Then $\theta = 3.96$ rad, $A = 1.8\ \text{ft}^2$, and $P = 3.465$ ft. Thus $AR^{2/3} = 1.16$, which is too small. So we assume $y/D = 0.75$, from which we compute $\theta = 4.19$ rad, $A = 1.935\ \text{ft}^2$, and $P = 3.666$ ft. Then $AR^{2/3} = 1.26$, which checks. Thus, $y_o = (0.75)(1.75) = 1.31$ ft.

Now we wish to check if this depth will be supercritical. Using Figure 12.9, we compute the factor:

$$\frac{Q}{\sqrt{g}D^{5/2}} = \frac{15.7}{\sqrt{g}(1.75)^{5/2}} = 0.68$$

From Figure 12.9 we find that $y_c/D = 0.82$, or $y_c = 1.43$ ft. Then the normal depth of 1.31 ft is indeed supercritical. The fact that our pipe is going to flow supercritical may or may not be a cause of concern. If we are certain that the outfall of the sewer is going to remain free so that a hydraulic jump will not occur in the pipe, and the outfall region is designed to handle the supercritical velocities that occur, then we may accept the supercritical conditions. Otherwise, we may want to redesign the pipe to operate in subcritical conditions.

Now, check the minimum flow conditions. We need to find the normal depth for the minimum flow of 2 cfs. So

$$AR^{2/3} = \frac{nQ}{1.49S^{1/2}} = \frac{0.012(2)}{1.49(0.01)^{1/2}} = 0.16$$

Now, assume $y_o/D = 0.2$; then $\theta = 2\cos^{-1}\left(1 - 2\frac{y_o}{D}\right) = 1.85$ rad. Then

$$A = (\theta - \sin\theta)\frac{D^2}{8} = 0.342\ \text{ft}^2, \quad P = \theta\frac{D}{2} = 1.618\ \text{ft}, \quad \text{and} \quad R = A/P = 0.21\ \text{ft}$$

Then $AR^{2/3} = 0.12 < 0.16$. Next assume $y_o/D = 0.25$ from whence we compute $\theta = 2.09$ rad, $A = 0.47\ \text{ft}^2$, $P = 1.828$ ft, and $R = 0.257$ ft. Then $AR^{2/3} = 0.20 > 0.16$. Then assume $y_o/D = 0.225$ and accordingly compute $\theta = 1.977$ rad, $A = 0.405\ \text{ft}^2$, $P = 1.729$ ft, and $R = 0.234$ ft. Then we find $AR^{2/3} = 0.153 \approx 0.16$, so we accept this assumption. Then $V = Q/A = 4.94$ ft/s, which is acceptable for self-cleansing purposes.

Example 16.5

What would be the equivalent arch pipe to be used for the situation in Example 16.4?

Solution: The computations for Example 16.4 have already determined that a 21-in circular concrete pipe will suffice for the conditions stated in the problem. Now from Table 16.4 we see that the equivalent concrete arch pipe would have a span of 26 inches and rise of 15.5 inches. The area of this pipe is 2.2 ft^2 and the hydraulic radius is 0.36 ft. Then the capacity of this pipe is found from the Manning equation:

$$Q = \frac{1.49}{n} A R^{2/3} S_o^{1/2} = Q = \frac{1.49}{0.012} 2.2(0.36)^{2/3}(0.01)^{1/2} = 13.8 \text{ cfs}$$

Since this capacity is less than the design flow of 15.7 cfs, a larger pipe will have to be selected. If we move up to the next largest pipe (span = 28.5 in, rise = 18 in), then the capacity is again calculated by the Manning equation as above to be 20.4 cfs. Now, since the pipe capacity at full flow is greater than the design discharge of 15.7 cfs, the pipe will obviously not flow full during the design storm. In fact, since the design flow is only about 76% of the pipe capacity, this appears to be a good selection under the criterion that the pipe should not flow more than 80% full.

16.2.2 Culverts

The design and anlaysis of culverts are among the most important issues faced by the practicing civil engineer. Culvert design is central, not only in highway construction, but also in site development and dam design. Culverts are frequently employed as outlet structures for empoundments, both for flood control and detention storage. In the following discussions, culvert hydraulics are developed for design cases, where the culvert is expected to carry a significant discharge. The simple low-flow case, where the flow in the culvert is merely the normal depth for the culvert, is not of interest here.

In high-discharge cases, a culvert functions essentially as a closed conduit with the energy supplied by the water level on the upstream side of the embankment as shown in Figure 16.9. Generally, the control on the culvert hydraulics can be in any of three states: the outlet (tailwater), the culvert structure itself (barrel), or the inlet. If the control is the tail water or the culvert barrel, then the culvert will flow full and be operating in pressure flow conditions. Because of the similarities of these cases, many references combine the tail water and barrel control cases into one case, usually referred to as outlet control. In most practical cases, particulary in road design or dam construction, the culvert will be designed to carry the design discharge while flowing full. Of course, when operating in pressure flow, the velocity through the culvert is given by the orifice equation:

$$V = \sqrt{\frac{2gH}{k}} \tag{16.11}$$

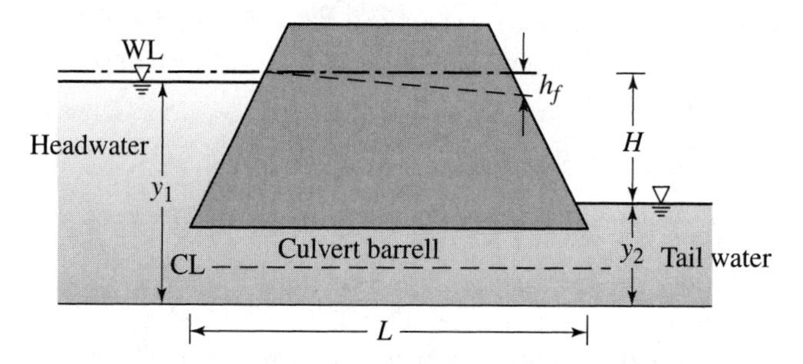

Figure 16.9 A culvert operating as a closed conduit.

and the discharge for circular pipes is given by

$$Q = A\sqrt{\frac{2gH}{k}} = \frac{\pi D^2}{4}\sqrt{\frac{2gH}{k}} \tag{16.12}$$

where D = culvert diameter, g = acceleration of gravity, H = head on the culvert, and k is the total energy loss coefficient in units of velocity head. The loss coefficient is the sum of the entrance loss, the exit loss and the friction loss through the culvert barrel. An alternate form of the equation is sometimes written

$$Q = CA\sqrt{2gH} \tag{16.12a}$$

where the loss coefficient $C = \sqrt{1/k}$. Values of C for fully submerged conditions have been found to range from 0.7 to 0.9 (Corps of Engineers, 1982). The head on the culvert is computed as the difference between the upstream energy level and the downstream control. Thus, in this case, the problem reduces to one of determining the proper control.

16.2.2.1 Culverts Flowing under Tail-Water Control If the culvert is forced into pressure flow due to the elevation of the water surface downstream of the outlet, then the culvert is said to be under tail-water control. This situation is shown in Figure 16.10. Note that the downstream water level does not have to actually submerge the outlet for this case to control. A common assumption is that if the tail-water elevation is above the center line of the outlet, then the culvert will function in tailwater control. On the other hand, the Federal Highway Administration (FHWA, 1985) recommends the use of the more conservative factor $\frac{y_c + D}{2}$ (where y_c = critical depth in the culvert and D = culvert diameter) as the control depth for tail-water control. It is also common practice to compute the head on the culvert as the difference between the headwater and tail-water (or control) elevations, rather than as the difference at energy grade line. This approximation assumes that the upstream velocity head is negligible when compared to the tailwater velocity head.

It should be clear that in the analysis of culverts, the first step must be to compute the depth (or elevation) of the water in the channel downstream of the culvert outlet. This requires knowledge of the channel conditions as well as any potential controls (such as chokes) located downstream that may affect the water level at the culvert outlet. Thus, the engineer must examine the channel for a sufficient distance downstream to determine the control on the flow at the outlet location. If no potential controls are located in the vicinity of

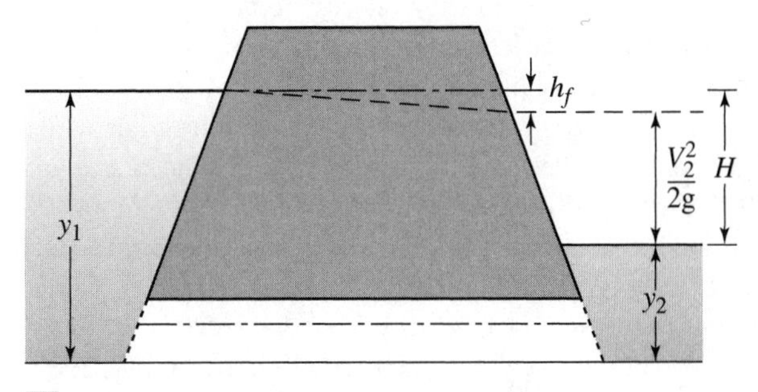

Figure 16.10 Culvert under tail-water control.

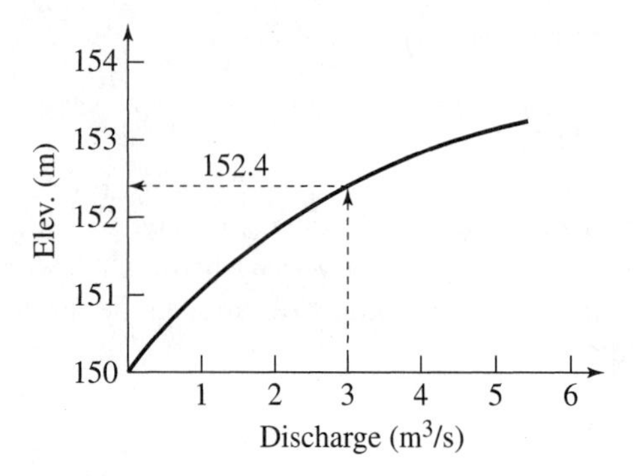

Figure 16.11 Tail-water rating curve.

the culvert, then the flow conditions there will be in channel control and normal depth can be used to compute the water surface level. However, if some potential control is located downstream, then the water surface profile must be computed back from that control to the culvert location by one of the methods discussed in Chapter 15. The usual technique is to develop a tail-water rating curve for the channel at the culvert outlet, either from normal depth or water surface profiles. A tail-water (TW) rating curve (discharge vs. elevation, or depth) is shown in Figure 16.11. This curve can then be used to determine when the downstream water depth will exceed the control depth at the culvert outlet and thus send the culvert into tailwater control, as well as to determine TW depths associated with various discharges through the culvert.

The usual situation is to compute the headwater elevation that will result from a given design discharge through the culvert. This computation may be carried out either with an existing culvert, or as part of the design of a new culvert. The TW rating curve is used to determine the elevation of the water in the channel downstream, and assuming that it does exceed the control depth at the culvert outlet, the following procedure would be employed. Letting Q be the design discharge, then

$$Q = \frac{\pi D^2}{4}\sqrt{\frac{2gH}{k}} \tag{16.13}$$

Now, assuming that the culvert length and material are known, the loss coefficient, k is computed as

$$k = k_{\text{ent}} + k_{\text{exit}} + k_f \tag{16.14}$$

where k_{ent} is the entrance loss (in velocity heads), k_{exit} = exit loss, and k_f= friction loss coefficient. The minor energy loss at the entrance is a function of the entrance design. If the entrance is flush with the embankment or head wall, then $k_{\text{ent}} = 0.1$ (COE, 1982), or 0.2 (Roberson et al., 1998). The loss coefficient will increase for protruding entrances (as discussed in Chapter 8; see Figure 8.5) to reflect the decrease in effective opening due to the interference of the flow lines as they enter the culvert. Loss coefficients from 0.3 to 0.9 might be appropriate depending on the amount of protrusion or whether the entrance is square or rounded (COE, 1982). Rounded entrances generally have lower loss coefficients. The loss associated with the culvert exit is normally taken as 1 velocity head for culverts in outlet control. Thus, for the flush, rounded entrance, the total loss coefficient becomes (COE, 1982)

$$k = 1.1 + k_f$$

The friction loss coefficient can be determined using the Manning equation for fully turbulent flow through rough pipes as described in Chapter 8. Under this formulation, we can set the friction loss, $k_f \frac{V^2}{2g}$ equal to that determined from application of Manning's equation,

$$k_f \frac{V^2}{2g} = S_f L \tag{16.15}$$

where L is the culvert length and S_f is the friction slope obtained from the Manning relation. Thus, using SI units,

$$S_f = \frac{n^2 V^2}{R^{4/3}} \tag{16.16}$$

Then

$$k_f = \frac{S_f L 2g}{V^2} = \frac{\frac{n^2 V^2}{R^{4/3}} L 2g}{V^2}$$

$$= \frac{n^2 L 2g}{R^{4/3}} \tag{16.17}$$

For SI units, equation (16.16) becomes

$$k_f = \frac{19.62 n^2 L}{R^{4/3}} \tag{16.18}$$

while in English units, it becomes (recalling the 1.49 conversion factor, which would be squared)

$$k_f = \frac{29.1 n^2 L}{R^{4/3}} \tag{16.19}$$

Then, using equation (16.18) for k_f, the total loss coefficient becomes

$$k = 1.1 + \frac{19.62 n^2 L}{R^{4/3}} \tag{16.20}$$

Hence the discharge through the culvert can be computed from

$$Q = A \sqrt{\frac{2gH}{1.1 + \frac{19.62 n^2 L}{R^{4/3}}}} \tag{16.21}$$

The most common culverts are reinforced concrete pipes (RCP) and corregated metal pipes (CMP), although other materials such as steel are sometimes used.

As discussed in the previous section, noncircular pipes (such as elliptical and arch) are used where minimum cover requirements and/or structural loads are such that circular culverts will not suit. The relevant geometric and hydraulic properties of these pipes are given in Tables 16.1–16.4. Manning roughness coefficients for some culvert materials are given in Table 16.5. Inlet loss coefficients for elliptical and arch pipe are similar for those of circular pipe.

TABLE 16.5 Manning Roughness Coefficients for Various Culvert Materials

Culvert Material	*n* Value
Reinforced concrete pipe	
new, smooth finish	0.011–0.012
old, unfinished, rough	0.013–0.017
Concrete box	0.014–0.018
Corrugated metal pipes	
2.66 × 0.5 in corrugations	0.022–0.027
6 × 1 in corrugations	0.022–0.025
5 × 1 in corrugations	0.025–0.026
3 × 1 in corrugations	0.027–0.028
Cast iron	
good condition	0.011–0.012
poor condition	0.013–0.015
Riveted steel	0.017–0.020

Example 16.6

A 10-m-long RCP (circular) culvert traverses an embankent as shown in the Figure 16.12. The culvert diameter is 1 m and the invert of the outlet is at elevation 150 m. For a design discharge of 3 m^3/s, and using the tailwater rating curve shown in Figure 16.11, determine if the culvert will be in tailwater control, and compute the resulting headwater elevation.

Solution: Using the tail-water rating curve, it is seen that for a discharge of 3 m^3/s, the tail-water elevation is about 152.4 m. Since the centerline of the culvert will be at elevation 150 m + $D/2$, for $D = 1$ m the centerline will be at elevation 150.5 m. Alternatively, using the nomograph given in Figure 12.9, the critical depth for the culvert is determined to be about 0.8 m. Then, the FHWA criterion, $\frac{y_c + D}{2}$, equals 0.9 m, or 150.9 m elevation. Thus, the tail water is well above the culvert centerline and the FHWA control elevation, and in fact will submerge the outlet altogether. So the culvert is clearly operating in tail-water control. Now, the culvert area is given by

$$A = \pi D^2/4 = 0.785 \text{ m}^2$$

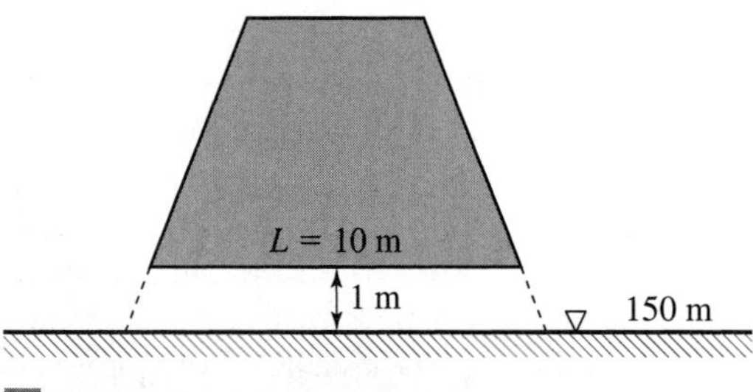

Figure 16.12 Culvert for Example 16.6.

The culvert length is 10 m and the material is smooth concrete, so we will take the Manning n value to be 0.012 (Table 16.5). Then the friction loss coefficient will be

$$k_f = \frac{19.62(0.012)^2(10)}{(\frac{1}{4})^{4/3}} = 0.18$$

Then the total loss coefficient, for a flush inlet, would be $k_t = 0.1 + 0.18 + 1 = 1.28$. Then the discharge through the culvert is given by equation (16.21), which, when substituting the known quantities and solving for H, results in a design head of 0.95 m. Thus, the headwater elevation for a discharge of 3 m^3/s would equal $152.4 + 0.95 = 153.35$ m.

16.2.2.2 Culverts under Barrel Control Even if the tail-water elevation is below the downstream centerline of the outlet, the culvert may still flow full due to the head loss that accumulates within the culvert itself. This is the situation of barrel control depicted in Figure 16.13. This situation usually prevails in very rough culvert materials such as corregated metal. Of course, the head loss in the culvert, as given in the analysis in the previous section, is a function of the culvert roughness, diameter (or hydraulic radius), and length. The head loss can be computed from either of the relations discussed, that is, $h_f = k_f(V^2/2g)$ or $h_f = S_f L$, where the friction slope S_f is computed from Manning. It is a simple matter to start from the downstream water level and compute the head loss in the culvert barrel. The energy grade line can be used to compute an estimated water surface profile through the culvert starting with the tailwater level, and if the water level inside the culvert reaches the soffit (i.e., top), then pressure flow will result. Culverts of sufficient length to cause pressure flow for this reason are known as "hydraulically long" culverts, while those of insufficient length for this result are termed "hydraulically short." Corregated metal culverts of any practical length (5 m or more) are normally hydraulically long. Smoother culverts may not function in this manner unless the culvert length is significantly longer.

If the culvert is known to be functioning in barrel control, the flow is computed in the same manner as in the previous section, except that the head, H, is taken as the difference of the upstream water level and the centerline elevation of the outlet as shown in Figure 16.13. The fact that the control elevation is still taken in the downstream position is the reason that many references combine this case with tailwater control. The only difference is in the selection of the control elevation as either the tail water itself or the alternate control elevation (i.e., the culvert centerline or $\frac{y_c + D}{2}$).

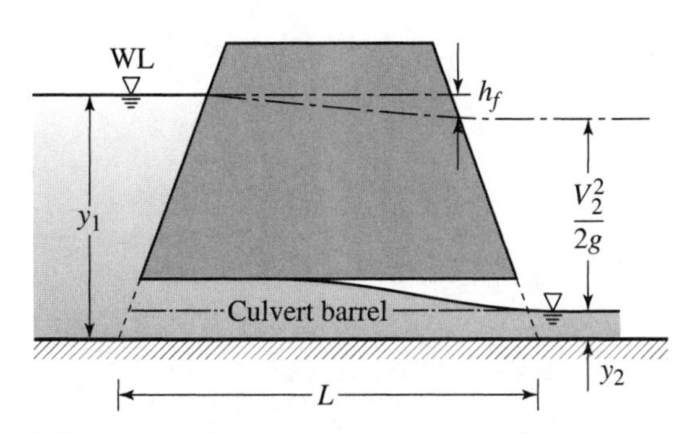

Figure 16.13 A culvert operating under barrel control.

Two elliptical concrete pipes.

Example 16.7

A circular culvert is to be designed to traverse a 2-m-high embankment with side slopes of 1.5H to 1V as shown in Figure 16.14. The top width of the embankment is 1 m. The 10-year design discharge for this area is 4 m^3/s and the water surface elevation in the channel downstream is 25.5 m for this flow. The channel invert at the outlet is 25 m above datum. Determine the minimum size CMP such that the embankment will not overtop under design conditions.

Solution: From the geometry of the embankment, we determine that the culvert length will be 7 m, assuming flush inlet and outlet. From Table 16.5 we estimate the Manning n for CMP to be 0.0225. Then the total loss coefficient for the culvert will be

$$k_t = 0.1 + 1 + \frac{19.62(0.0225)^2(7)}{(D/4)^{1.333}}$$

$$= 1.1 + \frac{0.0695}{(D/4)^{1.333}} = 1.1 + \frac{0.441}{D^{1.333}}$$

The culvert discharge is given by equation (16.21). Substituting known quantities,

$$4 = \frac{\pi D^2}{4}\sqrt{\frac{2gH}{1.1 + \dfrac{0.441}{D^{1.333}}}}$$

or

$$5.09 = D^2\sqrt{\frac{2gH}{1.1 + \dfrac{0.441}{D^{1.333}}}}$$

In a first approximation, assume tail-water control, so that the maximum head would be given by $27 - 25.5 = 1.5$ m. Then solving the for D by trial, we get $D = 1.1$ m approximately. Under this design, the centerline of the culvert would be at 25.55 m, which is above the tail-water elevation so the culvert would be in barrel control. Using this control, we get the maximum head to be $2 - 0.55$ m $= 1.45$ m. Using this head, we determine that the culvert will carry 4.15 m^3/s, which meets specifications. As a final check, we determine the critical depth in the culvert to be 0.99 m from Figure 12.9. Using this value, we will approximate the available head loss in the culvert before it would start to flow full as $1.1 - 0.99 = 0.11$ m, which would work out to a friction slope of $0.11/7 = 0.015$ m/m. Note that in this calculation depths were used instead of energies to estimate the head loss, so this is an approximation. We will now check using energy loss coefficients. From the culvert loss data, we see that the head loss in the culvert will be $k_f(V^2/2g)$. Using information given in Table 10.1, we determine that the velocity at critical depth would be 4.68 m/s. Then the head loss would be at least

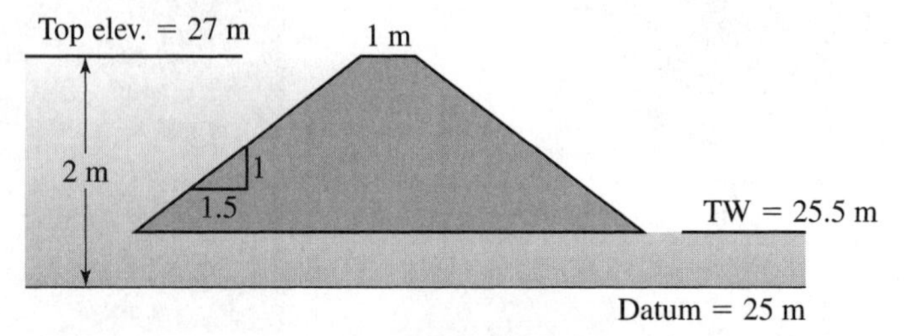

Figure 16.14 The culvert and embankment for Example 16.7.

$\left(\frac{(0.441)}{(1.1)^{1.333}}\right) \times \left(\frac{(4.68)^2}{2g}\right) = 0.44\text{ m} > 0.11\text{ m}$. Thus, the culvert cannot flow under either inlet control or tail-water control (as discussed above). Hence the assumption of barrel control and our computations are correct.

One final note: We have not accounted for any freeboard in the design. It is common for up to 2 ft (0.66 m) of freeboard to be used in the design of embankments that contain water of varying flow rates in order to correct for any errors in the calculation of the design discharge, or to account for wave run-up on the structure.

A circular concrete culvert. (Courtesy of Dr. Mohammad Al-Hamdan.)

Two concrete box culverts under a highway.

16.2.2.3 Culverts Flowing Partially Full in Inlet Control Culverts may flow partially full under either low-flow conditions or due to the presence of inlet control. Inlet control occurs when the constriction due to the culvert opening forces critical depth to occur at the inlet and the culvert slope is steep, such that flow inside the barrel will be supercritical. This situation is shown in Figure 16.15. A number of factors determine whether or not a culvert can function in inlet control:

1. The constriction must be sufficient to force y_c at the inlet.
2. The culvert must be laid on a steep slope.
3. The culvert must be hydraulically short.
4. The culvert outlet must be unsubmerged, that is, either the downstream channel slope remains steep, or a hydraulic jump occurs clear of the outlet.

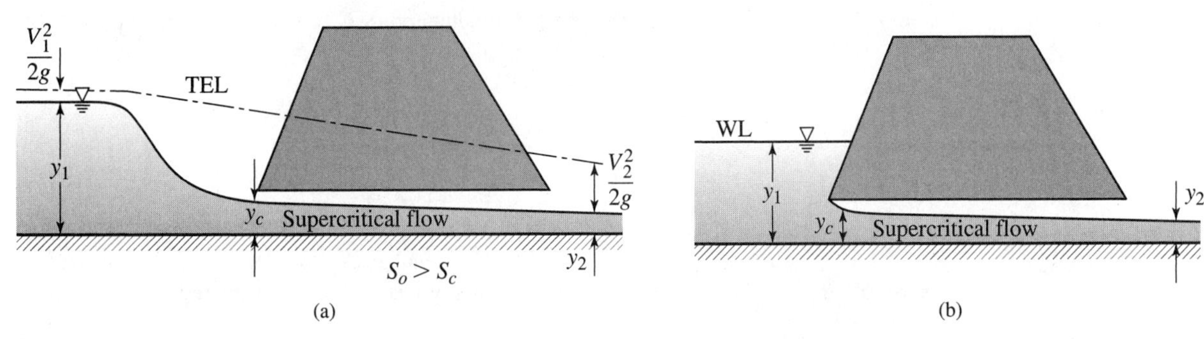

Figure 16.15 (a) Culvert operating in inlet control–unsubmerged inlet. (b) Culvert operating in inlet control–submerged inlet.

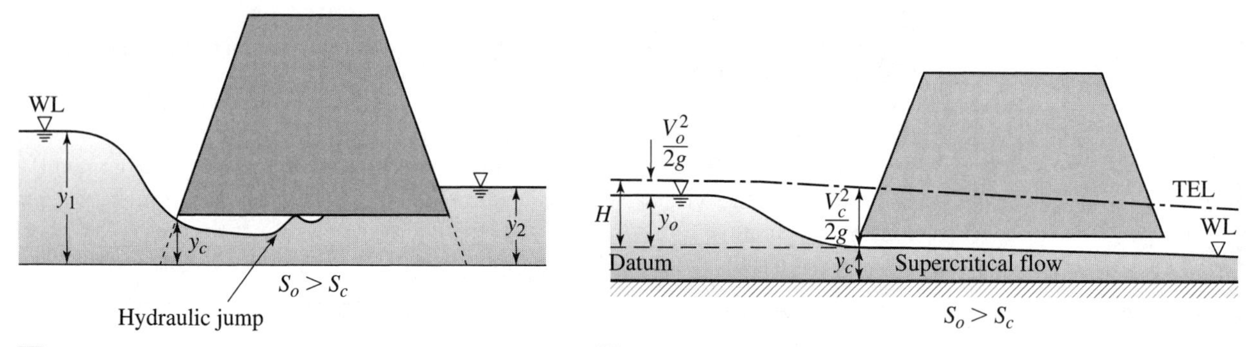

Figure 16.16 Hydraulic jump within a culvert barrel under inlet control.

Figure 16.17 Principles of inlet control at culverts.

As depicted in Figure 16.15b, the culvert inlet need not be unsubmerged for inlet control to prevail. In fact, it has been stated (Roberson et al., 1998) that only if the headwater is less than $1.2D$, will the inlet be unsubmerged.

An interesting case results when the first two conditions above prevail, but the outlet is submerged (or above the control elevation), or the culvert is hydraulically long. In that case, the culvert will be thrown into pressure flow and a hydraulic jump will form within the barrel as shown in Figure 16.16. If the downstream depth is great enough, the jump will be forced back to the inlet, thus drowning the critical depth. In this case the control is shifted from the inlet to either the tail-water or the barrel. When this happens, the culvert will commonly surge (i.e., the flow will back up raising the headwater level until inlet control is reestablished, the hydraulic jump forms again, and the process starts over). Needless to say, this is a situation that should be avoided if at all possible.

Of course, in the inlet control case, the critical depth at the inlet controls the discharge capacity of the culvert. Following the custom of measuring head with respect to the control point, the head on the culvert is measured with respect to the upstream energy level. Referring to Figure 16.17, the energy equation can be written with respect to a datum running through the inlet as

$$H = y_o + \frac{V_o^2}{2g} = \frac{V_c^2}{2g} = \frac{Q^2}{2gA_c^2} \tag{16.22}$$

As in previous cases of constrictions, a contraction loss coefficeint must be employed to account for the loss of effective area as the flow lines enter the culvert. For flush, rounded entrances, the same coefficient as in the sluice gate case is normally used (i.e., $C_c = 0.62$). So equation (16.22) becomes (Morris and Wiggert, 1972)

$$H = y_o + \frac{V_o^2}{2g} = \frac{Q^2}{2g(0.62A_c)^2} \tag{16.23}$$

Thus,

$$Q = 0.62A_c\sqrt{2gH} \tag{16.24}$$

where, of course, H is merely the upstream energy level measured with respect to the inlet datum.

The situation is simplified considerably for the box culvert case. Of course, in that situation the control at the culvert inlet is given by relation for critical depth in rectangular sections,

$$\frac{V_c^2}{gy_c} = 1, \quad \text{or} \quad V_c = \sqrt{gy_c} \tag{16.25}$$

Recalling that for the rectangle, $y_c = 2/3H$, then $V_c = \sqrt{\frac{2gH}{3}} = 0.577\sqrt{2gH}$. Since the culvert opening may represent a significant constriction to the channel cross-sectional area, it may be necessary to include contraction coefficients in the analysis. The values usually employed in practice for this coefficient is 1.0 for flush rounded entrances and 0.90 for the square (box) flush inlet.

The situation is different still for the cases of elliptical or arch pipe. Although the actual computations at the inlet are not as straightforward as for the box culvert case, the situation differs from that of circular entrances to the extent that inlet control can be significantly affected if a change is made from circular to elliptical or arch pipe. In all nonrectangular cases, the critical depth can be determined from the general Froude Number relation:

$$\frac{Q^2 B}{gA^3} = 1 \tag{16.26}$$

For elliptical or arch pipe, the relevant geometrical properties can be determined from the data given in Tables 16.1–16.4 and equation (16.26) solved iteratively for y_c. The critical depth at elliptical or arch pipe inlets will normally be less than that for circular pipe, so that a round culvert might be operating in inlet control under flow conditions where an elliptical or arch pipe would not.

Example 16.8

A 3-m-long RCP culvert with a diameter of 1 m is laid on a slope of 0.02 m/m. The channel upstream and downstream of the culvert is a rectangular section 4 m wide running on a slope of 0.005 m/m with a Manning n value of 0.025. The channel runs for a considerable distance downstream of the culvert maintaining channel control. For a design discharge of 2.5 m^3/s, determine if the culvert will operate in inlet control and, if so, determine the upstream water level for this case.

Solution: First determine the tailwater depth at the culvert outlet by computing the normal depth in the channel.

$$4y\left(\frac{4y}{2y+4}\right)^{2/3} = \frac{nQ}{S^{1/2}} = \frac{0.025(2.5)}{(0.005)^{1/2}} = 0.88$$

By trial and error, we compute $y_o = 0.45$ m, which places the tail water below the centerline of the culvert. Next, determine the critical depth in the culvert from Figure 12.9 to be about 0.81 m, so that the outlet is also clear from the FHWA criterion $\left(\frac{y_c + D}{2} = 0.9 \text{ m}\right)$. So inlet control is possible. Also, if the depth in the culvert at its outlet is near critical, or less, then we can approximate that there will be more than 1 m − 0.81 m or more than 0.19 m of head loss available in the barrel of the culvert before it would switch to barrel control. This would require an approximate friction slope of $0.19/3 = 0.063$ m/m, which is three times steeper than the culvert slope. It is also known from the discussions of water surface profiles that the friction slope must always be less than the bed slope for the S2 profile. Therefore, barrel control is not possible.

Now, to eliminate the possibility of the culvert operating merely under low-flow conditions, we must check the occurrence of the critical depth control at the culvert inlet. For this to be the case, the culvert slope must be steep (i.e., y_c must be greater than y_o). Then, using Manning's equation with $n = 0.012$ for smooth concrete,

$$AR_h^{2/3} = \frac{0.012(3)}{(0.02)^{1/2}} = 0.250$$

Solving for y_o by trial using the relationships for circular sections given in Table 10.1, the normal depth is estimated to be about 0.52 m. Since $y_o < y_c$, then the culvert slope is indeed steep and critical depth will control at the inlet.

The upstream water depth above the inlet datum is computed from equation (16.24):

$$Q = 0.62A_c\sqrt{2gH}$$

$$3 = 0.62(0.636)\sqrt{2gH}$$

$H = 2.95$ m above the inlet datum

Since $H > 1.2D$, the inlet is expected to be submerged in this case.

16.2.3 Drop Inlets

The drop inlet is shown schematically in Figure 16.18. This structure is used in detention situations where water is to be stored for aesthetic or recreation purposes but the outflow must still be controlled during flooding conditions. The drop inlet was developed for this purpose by the National Resources Conservation Service (formerly Soil Conservation Service) and so is usually known as the SCS drop inlet. It combines the concepts of the overflow weir and the culvert and is based on the premise that the water surface will be maintained at a minimum level by the weir during low-in flow conditions, while high inflows will be controlled by the culvert. As shown in Figure 16.18, the system consists of a riser that controls the water surface

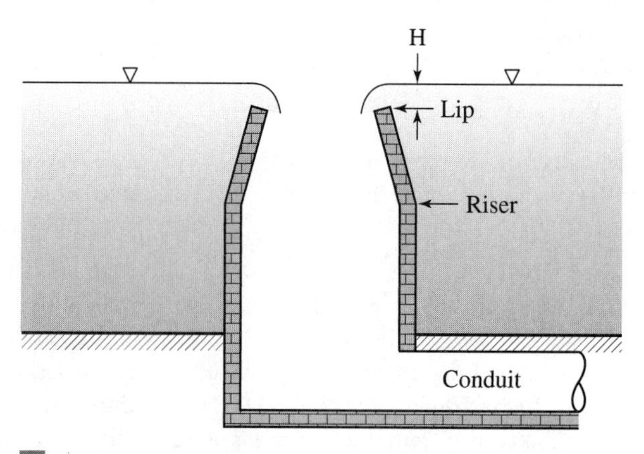

Figure 16.18 Drop inlet structure.

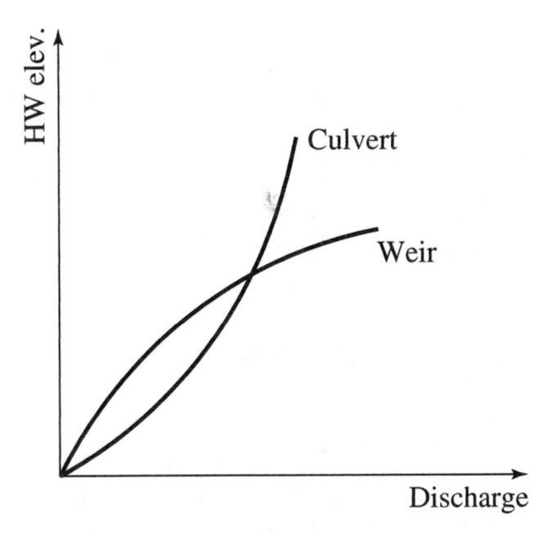

Figure 16.19 Headwater rating curves for weirs and culverts.

elevations during low flow situations and the conduit which controls the high flows. The lip or rim of the riser functions as the weir with flow over three or four of its sides. In some cases, one side may be obstructed by a device that controls the gate on the culvert. Headwater rating curves are shown for weirs and culverts in Figure 16.19. It can be observed that at a certain point the two curves intersect (i.e., the weir will pass more water to the culvert than the culvert can handle). At this point, the control switches from the weir to the culvert. Then the discharges are computed by the methods discussed for culverts above. Of course, the culvert will be in one of the outlet control cases and the head will be computed from the outlet control level (tail water or centerline) to the upstream level. During early stages of culvert control, the effective headwater on the culvert may not be the actual lake level in that the headwater may not exceed the height of the box.

When determining the flow capacity of a drop inlet, a practical solution proceeds as follows: The flow rate through the culvert is determined for a minimal head, say 0.25 ft (0.08 m). Then it can be assumed that for inflows substantially less than this value, the control will be at the weir or lip of the riser. When inflows reach values near this discharge, then the control will shift to the culvert. Then, using the proper outlet control elevation, headwater elevations can be computed until the headwater exceeds the top of the box. Once the riser fills up, the lake elevations become the headwater values.

The design of a drop inlet involves the determination of the dimensions and elevation of the riser structure for the inlet control of the low flows, as well as the required size of the pipe for the high flow control situation. The final design is determined by circumstances, cost, and safety.

Example 16.9

A drop inlet structure is to be designed to handle outflows from a small dam that is 30 ft high with a base width of 50 ft. The invert elevation of the downstream channel is 150 ft with a channel rating curve as given in Figure 16.20. The estimated low flow into the reservoir is 10 cfs while the maximum inflow is estimated at 1000 cfs. It is desired to control the elevation of the lake to a level that is no lower than 7 ft below the top of the embankment and no higher than 5 ft below the embankment crest.

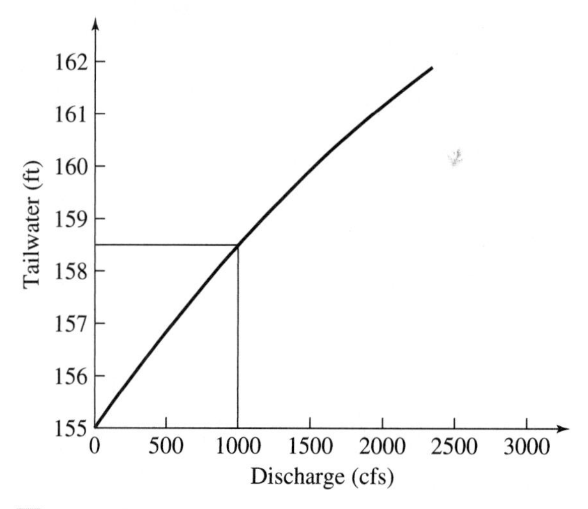

Figure 16.20 Tail-water rating curve for Example 16.9.

Solution: Handling the low-flow case first, we must determine the dimensions and elevation of the riser. Using the Francis equation with an assumed C value of 3.1 for the sharp-crested wier, $Q = CLH^{3/2}$. Now we select the dimensions of the riser in order to get the weir length L. If we select reasonable dimensions of the lip of the riser to be 10 ft $\times$ 10 ft and let one side be used for appurtenances and control devices so that flow will only occur on three sides, then $L = 30$ ft. Then, solving for H for the low-flow case, $10 = 3.1(30)\ H^{3/2}$, thus leading to a value of $H =$ 0.226 ft, or 2.71 inches. This seems reasonable for a project of this scale. Thus, since the top of the embankment is at $150 + 30 = 180$ ft, then the lip of the riser is set at $180 - 7 - 0.226 =$ 172.77 ft. Now, for the high-flow case, we don't want the water to rise to a level higher than 175 ft. Using the tail-water rating curve given in Figure 16.20, we determine that for a flow of 1000 cfs, the tailwater level would be 158.5 ft. Then, for a maximum headwater level of 175 ft, the available head on the pipe will be $H = 175 - 158.5 = 16.5$ ft, assuming tail-water control. Then, our problem is to size a culvert to carry 1000 cfs with a head of 16.5 ft. Now, we know that the base width of the embankment is 50 ft; however, the box that connects to the pipe will not be exactly against the side of the embankment. For our purposes, we might assume that the pipe would need to extend an extra 15 ft through the embankment making a total length of 65 ft. Then, using the culvert equation,

$$Q = A\sqrt{\frac{2gH}{k_t}}$$

Now, to determine the total loss coefficient, an exit loss of 1 velocity head is used and the entrance loss is increased to 0.25 to account for the fact that the flow will not proceed directly into the culvert entrance but must make a bend from the bottom of the riser into the culvert. Then the friction loss is given by $k_f = \dfrac{29.1n^2L}{R^{4/3}}$. If it is assumed that the pipe will be concrete, then $k_f = \dfrac{29.1(0.012)^2(65)}{(D/4)^{4/3}} = \dfrac{0.272}{(D/4)^{4/3}}$. Then $k_t = 1.25 + \dfrac{0.272}{(D/4)^{4/3}}$. Thus,

$$1000 = \frac{\pi D^2}{4}\sqrt{\frac{2g(16.5)}{1.25 + \dfrac{0.272}{(D/4)^{4/3}}}}, \quad \text{or} \quad 1273.24 = D^2\sqrt{\frac{1062.6}{1.25 + \dfrac{0.272}{(D/4)^{4/3}}}}$$

Solving for D, we find that for $D = 6.5$ ft the left side of the equation is 1167 and for $D = 7$ ft, the left side is 1360. So, in reality, D is between 6.5 and 7 ft, but to select a culvert size that is readily available we would probably take $D = 7$ ft (84 in). Then, checking our control, we find that the outlet of the pipe will be submerged so that tail water will control.

16.2.4 Design Considerations for Impoundment Facilities

Extreme care must be taken in the design of facilities that will impound water and thus originate a type I (backwater) profile upstream of the structure. All of the hydraulic structures that have been discussed thus far in this chapter fall into this category. In addition, in most of these cases (excluding some culvert and weir cases), supercritical flow results at the outlet of the structure and thus some measures must be taken to deal with this situation. Energy dissipators such as stilling basins are often designed to set up a controlled hydraulic jump and thus facilitate the transition from the high velocity supercritical conditions to the normal subcritical conditions in the downstream channel. A stilling basin (Figure 16.21) is merely a concrete-lined section placed immediately downstream of the structure to contain the supercritical flow. The stilling basin is often rectangular in shape. The goal of the design is to set up the supercritical conditions such that a stable jump will occur within the basin and will safely meet the subcritical depth in the downstream channel. Thus, the design is to produce the supercritical conjugate depth to the downstream subcritical depth at a specified distance from the structure for the design discharge. Concrete blocks, known as *baffle blocks* or *piers* (Figure 16.21b), may be placed on the bottom of the stilling basin in order to absorb the momentum of the supercritical jet and allow the depth to build up more rapidly, thus shortening the length of the *M3* profile. An *end sill* may also be employed as a safety measure in order to insure that the jump will never run off of the end of the stilling basin. The design of stilling basins involves applications of the hydraulic jump principles discussed in Chapter 13, as well as water surface profile information given in Chapter 14. The specific design of baffle blocks and end sills is a matter of experimentation for different outlet scenarios and has principally been accomplished by Peterka (1964). Further information about stilling basins is given in Chapter 13.

Design of stilling basins is primarily a field for the specialist in hydraulic engineering. In this section, we are more interested in the design of common hydraulic structures (such as culverts) such that the backwater that is created will not exceed some specified limit. The reader will appreciate that engineers would not want to design a culvert under an access road to a subdivision in such a manner that it would cause water to back up and flood the houses in the subdivision. Thus the computation of the backwater depths and distances associated with the design of hydraulic empoundment structures is a crucial element of the design process.

The design usually incorporates the basic principles of the structure as described in the previous sections as well as the computation of the appropriate water surface profile. For

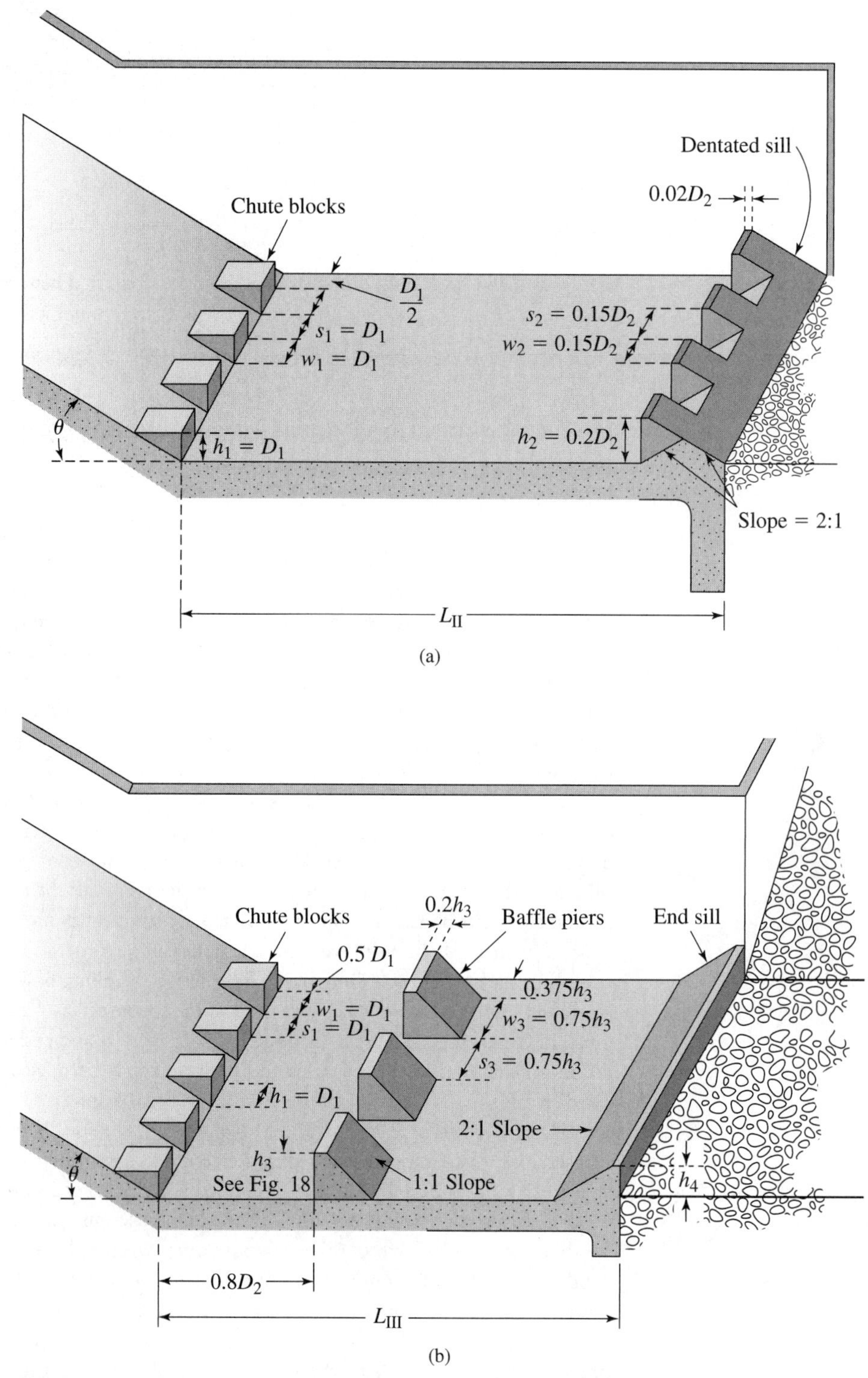

Figure 16.21 (a) Stilling basin without baffle blocks. (b) Stilling basin with baffle blocks.

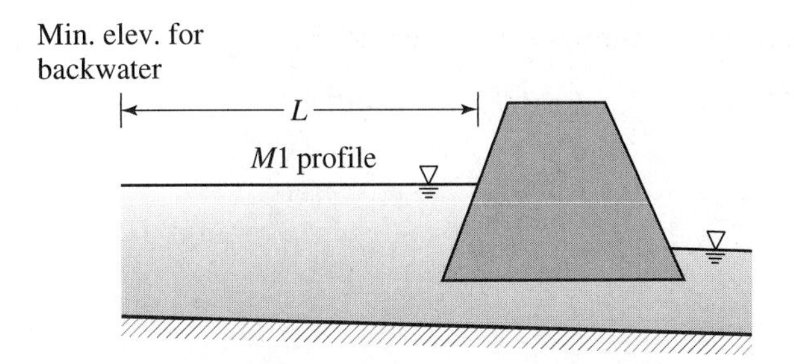

Figure 16.22 Backwater profile upstream of a culvert.

example, suppose we wish to design a culvert to carry a given design discharge under the roadway such that the water surface not only will not overtop the road, but also will not exceed some specified level at a location a given distance upstream of the road (Figure 16.22). In fact, this is the normal situation encountered in the design of culverts. The first step is to determine which of the elevations (top of road or upstream specification) will control the culvert design. To accomplish this, one merely computes the profile upstream of the road to the specified location using the top of the road as the downstream control elevation. If the water surface elevation exceeds the specified level at the given location, then the control location will be the upstream spot. Otherwise, the control is the roadway elevation. The next step is to determine the design head on the culvert such that the required water level will result at the upstream location. If the upstream location controls, this is an iterative procedure, beginning with trial elevations at the culvert and profiling upstream until the desired elevation is attained at the specified location. Usually, some freeboard is allowed below the elevation of the houses. For instance, if the first floor of the houses are at 100 ft, then the control elevation may be 97 or 98 ft. Once the design head is computed at the culvert location, then the design proceeds as given in the previous examples.

Example 16.10

A 30-ft-long RCP is to be designed to traverse an embankment whose base is at elevation 100 ft and whose crest is at elevation 115 ft as shown in Figure 16.23. Additionally, a shopping center is located 1500 ft upstream of the embankment and the maximum allowable flood elevation at that location is 115.5 ft (including freeboard). The channel is a 10-ft-wide rectangular section running on a slope of 0.0075 ft/ft with a Manning n value of 0.035. Determine the minimum diameter RCP to carry the 10-year design discharge of 150 cfs.

Solution: The situation is sketched in Figure 16.23. Since the channel is a uniform rectangular section we will use the direct step method (distance from depth) to perform the water surface computations, and for simplicity we will use only one step in the calculations. Beginning with

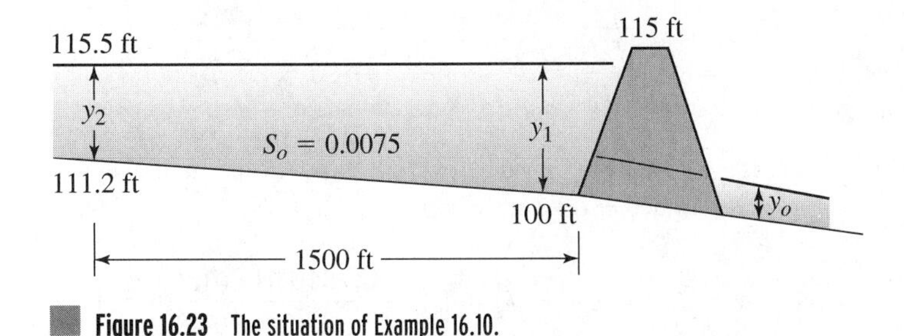

Figure 16.23 The situation of Example 16.10.

an assumed downstream elevation of 115 ft ($y = 15$ ft), we compute the following:

$$Q = 150 \text{ cfs}; \quad q = Q/B = 15 \text{ cfs/ft}$$

$$y_c = \sqrt[3]{\frac{q^2}{g}} = 1.91 \text{ ft}$$

Let $y = 15$ ft, $V = q/y = 1.0$ ft/s, $E = y + \dfrac{V^2}{2g} = 15.01$ ft

$$A = 150 \text{ ft}^2; \quad P = 2y + B = 40 \text{ ft}; \quad R_h = A/P = 3.75 \text{ ft}$$

$$S_f = \left(\frac{nV}{1.49R_h^{2/3}}\right)^2 = \left(\frac{0.035(1.0)}{1.49(3.75)^{2/3}}\right)^2 = 0.0000947 \text{ ft/ft}$$

1500 ft upstream, $y = 4.3$ ft, $V = 3.49$ ft/s, $E = y + \dfrac{V^2}{2g} = 4.49$ ft

$$A = 43 \text{ ft}^2, \quad P = 18.6 \text{ ft}, \quad R_h = 2.31 \text{ ft}, \quad S_f = 0.00219 \text{ ft/ft}$$

Then $\Delta E = 15.01 - 4.49 = 10.52$ ft and $\overline{S_f} = (0.0000947 + 0.00219)/2 = 0.00114$ ft/ft. Then $\Delta L = \Delta E/(S_o - \overline{S_f}) = 10.52/(0.0075 - 0.00114) = 1655$ ft.

Thus, the backwater profile has not played out to the extent that the depth has reduced to 4.3 ft a distance of 1500 ft upstream as assumed. Therefore, the control depth at the embankment must be reduced to a value which will result in the required depth of 4.3 ft at the shopping mall. Therefore a depth of 14 ft is assumed at the embankment. Hence,

$$y = 14 \text{ ft}; \quad V = 1.07 \text{ ft/s}; \quad E = 14.02 \text{ ft}$$
$$A = 140 \text{ ft}^2; \quad P = 38 \text{ ft}; \quad R_h = 3.68 \text{ ft}; \quad S_f = 0.000111$$
$$\Delta E = 9.53 \text{ ft}; \quad \overline{S_f} = 0.00115; \quad \Delta L = 1500 \text{ ft}$$

Thus, by this rudimentary, one-step computation, we estimate that the depth behind the embankment cannot exceed 14 ft in order to remain at or below the required depth at the mall. The design of the culvert to accomplish this task proceeds as in Example 16.6. First, we determine the tailwater depth in the channel downstream of the embankment. Assuming no downstream constrictions, normal depth will control:

$$10y_o\left(\frac{10y_o}{2y_o + 10}\right)^{2/3} = \frac{nQ}{1.49S_o^{1/2}} = \frac{0.035(150)}{1.49(0.0075)^{1/2}} = 40.68$$

By trial, we determine that $y_o = 2.7$ ft. Then, by first assuming tail-water control,

$$Q = A\sqrt{\frac{2gH}{k_t}} = \frac{\pi D^2}{4}\sqrt{\frac{2g(11.3)}{k_t}}$$

Letting the Manning n value for RCP be 0.012, taking the culvert length of 30 ft, and assuming a flush rounded entrance, we compute

$$k_t = k_{\text{ent}} + k_{\text{exit}} + k_f = 1.1 + \frac{29.1(0.012)^2(30)}{(D/4)^{4/3}} = 1.1 + \frac{0.798}{D^{4/3}}$$

Then

$$150 = \frac{\pi D^2}{4}\sqrt{\frac{727.72}{1.1 + \frac{0.798}{D^{4/3}}}}$$

whence, by trial, we determine $D = 2.85$ ft, approximately. With this diameter, the centerline of the downstream end of the culvert will be at elevation 101.2 ft, (accounting for the culvert slope of 1:134), while the tail water (depth = 2.7 ft) will be at elevation 102.475, which is well above the centerline. In fact, the tail water will nearly submerge the end of the culvert (2.7 ft vs. 2.85 ft) and thus our assumption of tail-water control is confirmed.

The discussion and example given above illustrate the ways in which hydraulic control can shift locations based on conditions in the flow or channel. This issue almost always arises in cases of design of open channel structures and controls. For example, a common situation in site develoment is to retain excess storm runoff on-site in small detention ponds that might also serve aesthetic and recreational purposes. However, these facilities must be designed in a manner such that any backwater produced does not interfere with the free operation of the storm drainage systems or with flow through culverts or other drainage structures. Thus, the computation of the water surface profile lengths associated with different design alternatives is a key element of the process. In fact, in most instances the entire design is done in the context of the profile calculation using some standardized computer package such as HEC-RAS (see Chapter 15).

16.2.5 Channel Design

In today's world, works such as flood control channels or water distribution canals must be designed not only for hydraulic efficiency but also to meet the environmental standards of the local community and the nation. This means that the channel should be as aesthetically pleasing as possible and fit in well with the surrounding environment. It should not cause any undue harm to its contiguous ecosystem, and should even enhance the ecosystem if possible. This often necessitates the channel to be constructed for ease of maintenance, which might result in side slopes that are very shallow. It is also now quite common for buffer zones consisting of water-resistant vegetation to be planted along the channel to interrupt the flow of nutrients and other pollutants (such as hydrocarbons and metals) from the surrounding area into the channel. These environmental issues often receive equal attention with the purely hydraulic concerns and may sometimes even take precedence over them.

16.2.5.1 Erodible Channels–Minimum Tractive Force Often in channel design the goal is to determine the optimal channel dimensions such that erosion of the channel boundary will be minimized so that the cross section will remain stable. The purpose is to minimize the cost of construction and maintenance through the design of a stable unlined channel. Channels designed on these principles may meet both the engineering and environmental standards discussed above. This problem reduces to one of design such that the shear stress applied to the boundary will be less than the force that would be required to initiate movement of the particles on the boundary. If this maximum allowable shear stress can be determined, then the basic resistance relationships derived previously can be utilized to determine the design velocity for the channel. Then the methods used in Chapter 10 can be employed to determine the required dimensions. These issues will be explored further in this section and some other issues relative to channel design will be introduced.

The basic resistance relationships for channels were discussed in Chapter 11. It was demonstrated that the basic shear stress formulation developed in Chapter 7 is valid for channels, that is,

$$\tau_o = \gamma R_h S \tag{16.27}$$

where τ_o = shear stress, R_h = hydraulic radius (L), and S = slope of the energy grade line = bottom slope for steady, uniform flow.

On the basis of the validity of equation (16.27) it was determined that the Chezy equation would also be valid, and this led to the development of the Manning form of the equation for the design of channels operating under steady uniform flow conditions:

$$Q = \frac{1}{n} A R_h^{2/3} S^{1/2} \tag{16.28}$$

where n is the Manning roughness coefficient with typical values given in Table 11.1. Equation (16.28) was employed to determine channel dimensions in simple design scenarios in Chapter 11. It was also demonstrated how minimum dimensions could be determined to result in maximum hydraulic efficiency and in design parameters such as maximum permissible velocities.

The basic formulation for application of shear stress principles to erodible channel design was presented by Lane (1955). Application of these simple concepts can lead to some important and interesting results regarding stability of channel cross sections. The first deals with the concept of incipient motion of particles on the channel boundary. As in the resistance discussions in Chapter 7, the basic problem is to relate the velocity of water in the channel to the shear stress applied to the boundary. It might then be possible to relate this shear stress to the size of particles that would be dislodged by the flow.

The most common cross-sectional shape used in channel design is the trapezoid. Other shapes are sometimes employed where circumstances allow including rectangular and triangular. Rectangular sections are often employed in cases were the channel is to be lined with concrete so that the vertical side slopes can be supported. Triangular shapes are employed in cases, such as storm gutters, where relatively small flow rates are to be conveyed. Assuming trapezoidal cross sections (Figure 16.24), the shear stress applied to the channel bed would be given by equation (16.27); however, the full stress vector would not apply to the sides. Lane

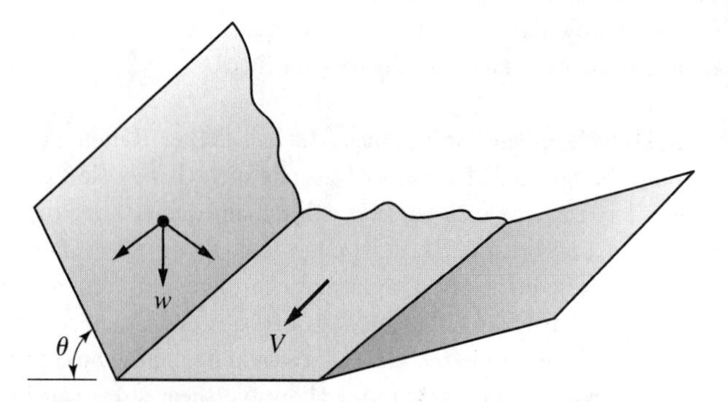

Figure 16.24 Forces acting on particles on a channel bed and banks.

found that the maximum shear applied to the banks is about three quarters of that applied to the bed (Chin, 2000):

$$\tau_o = 0.76\gamma R_h S \tag{16.29}$$

Experience has shown that in these situations the wide rectangular representation can often be employed to a first approximation (Chin, 2000; Henderson, 1966) (i.e., $R_h = y$). The average shear force applied to a given area of channel bed or side slope is known as the *tractive force* and is computed by equations (16.27) or (16.29) (usually with the substitution of the depth y for hydraulic radius).

Particles submerged on the channel bed or sides resist these forces through the friction that is generated with the surrounding granular material that composes the boundary. This resistance force can be expressed as (Chin, 2000)

$$F_f = cw_s \tag{16.30}$$

where F_f is the friction force, c is the coefficient of friction, and w_s is the submerged specific weight of the particle (i.e., $\gamma_s - \gamma$). Then, just at the point before the particle is dislodged, the shear force applied would be equaled by the resistance force, that is,

$$cw_s = \tau_o A_s \tag{16.31}$$

where A_s is the effective surface area of the particle under consideration. Then the critical shear stress, τ_c, would be given by

$$\tau_c = \frac{cw_s}{A_s} \tag{16.32}$$

The coefficient of friction is often expressed as simply the tangent of the angle of repose of the soil material (i.e., $c = \tan\phi$). Then equation (16.32) becomes (Chin, 2000)

$$\tau_c = \frac{w_s}{A_s}\tan\phi \tag{16.33}$$

The effective area to be used in equation (16.33) is usually defined as the "area of influence" of the particle under consideration. This area of influnce can itself be defined in terms of the sediment diameter and a "packing factor," r, which is a function of the number of grains per unit area of exposed surface. The area of influence would be equal to d^2/r; $r < 1$. Then, remembering that the submerged specific weight of the particle is $\gamma_s - \gamma$, equation (16.33) becomes (Henderson, 1966)

$$\tau_c = \frac{\pi}{6} rd(\gamma_s - \gamma)\tan\phi \tag{16.34}$$

From the work of Shields (1936), discussed in the next section, it can be deduced that the limiting value of r for coarse granular material would be approximately 0.15.

Equation (16.34) can be used directly to estimate the critical shear stress for the channel bed; however, for particles on the banks, the additional force associated with gravity tends to make the particle slide down the bank. If the bank angle is given by θ (Figure 16.24), then the component of this force acting down the bank would be given by $w_s \sin\theta$ as shown in the figure. Then the total force acting to move the particle would be (Chin, 2000).

$$F = ((\tau_o A_s)^2 + (w_s \sin\theta)^2)^{1/2} \tag{16.35}$$

As in the previous case, the frictional forces resisting this force would be given by the weight of the particle multiplied by the friction coefficient, and the component of this force in the direction of movement would be $w_s \tan\phi \cos\theta$. Then, at the moment of particle movement, we have (Chin, 2000)

$$[(\tau_o A_s)^2 + (w_s \sin\theta)^2]^{1/2} = w_s \tan\phi \cos\theta \tag{16.36}$$

From this expression, the critical shear stress on the channel banks can be expressed as (Chin, 2000)

$$\tau_{bc} = \frac{w_s}{A_s} \tan\phi \cos\theta \left(1 - \frac{\tan^2\theta}{\tan^2\phi}\right)^{1/2} \tag{16.37}$$

Dividing equation (16.37) through by $(w_s/A_s)\tan\phi$, the ratio of the critical bank shear stress to the critical bed stress can be expressed by the tractive force ratio, K, given by (Henderson, 1966)

$$K = \cos\theta \left(1 - \frac{\tan^2\theta}{\tan^2\phi}\right)^{1/2} \tag{16.38}$$

Also, employing some trigonometric relationships between sin θ and tan θ, K is often given as (Chin, 2000; Henderson, 1966; COE, 1970)

$$K = \frac{\tau_{bc}}{\tau_c} = \left(1 - \frac{\sin^2\theta}{\sin^2\phi}\right)^{1/2} \tag{16.39}$$

Lane (1955) presented data to support relationships between particle size and angularity and angle of repose and thus related these sediment characteristics to maximum permissible tractive force. These relationships are given in graphical form in many texts, including Chow (1959) and Chin (2000).

Observation has shown that the critical shear relationships are somewhat affected by the degree of sinuosity of the channel as well (Lane, 1955). Lane recommended correction factors for the above expressions, ranging from 1.0 for straight channels to 0.9 for only slightly sinuous channels to a value of 0.6 for very meandering channels.

Example 16.11

A trapezoidal channel is to be laid on a straight line through an alluvial terrain with a d_{75} sediment size of 1.5 in and specific weight 2.65. The natural slope of the terrain is 0.002 ft/ft and the angle of repose of the soil is estimated to be 35°. A preliminary design calls for a bottom width of 10 ft and $2H$: $1V$ side slopes. The design discharge is 500 cfs. Will this design result in a stable channel?

Solution: First, determine the normal depth in the channel. Using the Strickler equation to estimate the Manning n, that is, $n = 0.034d^{1/6}$ (d in ft), we determine the n value for a sediment size of 1.75 in to be 0.024. Then, from Manning,

$$(10y_o + 2y_o^2)\left(\frac{10y + 2y_o^2}{10 + 4.47y_o}\right)^{2/3} = \frac{0.024(500)}{1.49(0.002)^{1/2}} = 180.09$$

By trial, we find y_o to be about 4.6 ft. The hdraulic radius R_h is 2.89 ft. Then the shear applied to the channel bed by this flow would be

$$\tau_o = \gamma R_h S = \gamma(2.89)(0.002) = 0.36 \text{ lb/ft}^2$$

where γ is the specific weight of water $= 62.4$ lb/ft^3. The shear applied to the banks would be 76% of that or 0.27 lb/ft^2.

Now, the critical allowable shear on the channel bed would be given by equation (16.34), with $r = 0.15$,

$$\tau_c = 0.0785 d_{75}(\gamma_s - \gamma)\tan\phi = 0.707 \text{ lb/ft}^2$$

And the critical value for the channel banks is given by equation (16.39):

$$\tau_{bc} = \tau_c \left\{1 - \frac{\sin^2\theta}{\sin^2\phi}\right\}^{1/2}$$

$$= 0.707\left\{1 - \frac{\sin^2 26.5}{\sin^2 35}\right\}^{1/2} = 0.707\left\{1 - \frac{0.1991}{0.329}\right\}^{1/2} = 0.44 \text{ lb/ft}^2$$

Now, comparing the critical tractive force on the banks of 0.44 lb/ft² to the applied shear on the banks of 0.27 lb/ft², the factor of safety for this design would be 1.63.

16.2.5.2 Allowable Shear and Scour Relationships—Shields's Analysis Equations (16.33) and (16.37) give the means to compute the critical or allowable shear stress, or tractive force, for a channel bed or banks if the soil characteristics and flow depth are known. However, the usual case is that the soil characteristics are known and it is desirable to compute the maximum velocity or depth that the channel can be allowed to develop before scour begins. Thus, the boundary shear must be related to the flow parameters and the characteristics of the soil particles. Of course, as discussed in Chapter 7, one of the key factors in this analyis is the shear velocity:

$$v^* = (\tau_o/\rho)^{1/2} \tag{16.40}$$

The reader will recall from Chapter 7 that v^* has units of velocity and that this fact, along with its inclusion of the boundary shear τ_o, as well as the fluid density ρ, makes it ideal for relating physical forces to velocity distributions. Further application of this concept in Chapter 7 led to the development of a dimensionless number called the *wall Reynolds number*, R_n^*, used to relate the shear stress at the wall to the velocity at distances from the wall:

$$R_n^* = \frac{yv^*}{\nu} \tag{16.41}$$

where y = distance from the boundary and ν is the fluid kinematic viscosity. However, the reader is aware by now that in dealing with energy concepts in free surface flow, the state of the flow conditions (subcritical or supercritical) must also be considered, so that an analogous Froude number must also be employed Henderson (1966):

$$F^* = \frac{v_*^2}{gy} \tag{16.42}$$

Since our concern is with the shear stress applied to soil particles on the channel boundary with some diameter, *d*, then it follows that the length factor in equations (16.41) and (16.42) can be replaced with this sediment diameter. Obviously, one would expect a relationship to exist between the characteristics of the particles that comprise the channel boundary (diameter and weight or specific gravity) and their ability to withstand the shear forces applied by the water without being dislodged. Thus, it should be possible to gain insight into this relationship by plotting R_n^* versus F^* for different sediments. The experimental anaylysis to confirm this relationship was first performed by Shields (1936). However, Shields found that experimental results could be reduced into an even more simplified form when the sediment specific gravity,

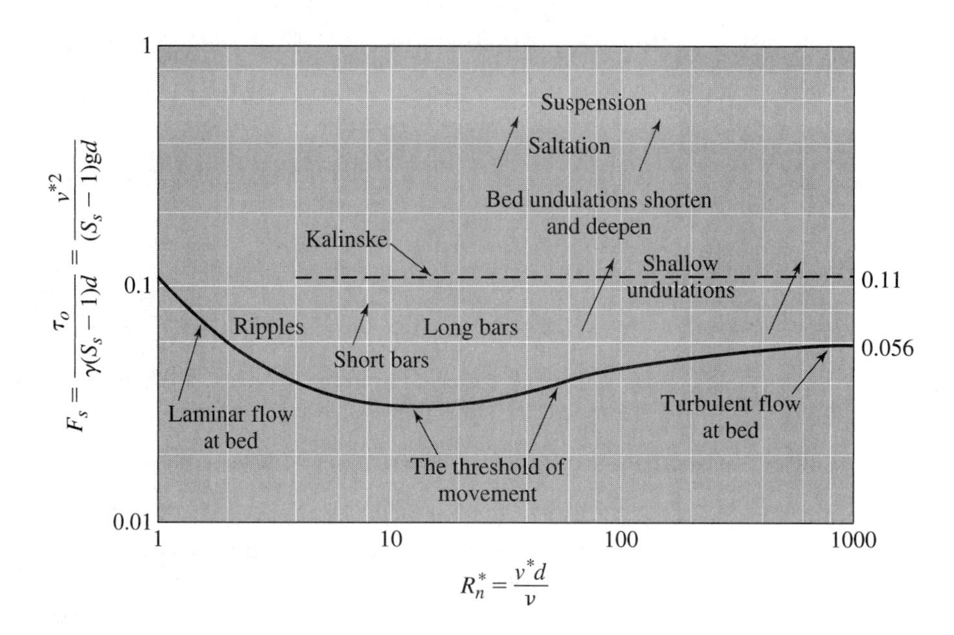

Figure 16.25 The Shields entrainment function.

S_s, was combined with the particle Froude number in the form:

$$F_s = \frac{v_*^2}{gd(S_s - 1)} \tag{16.43}$$

He found that in this form, experimental results for all sediments could be represented with a single line on the $R_n^* - F_s$ diagram. This famous plot, known as the *Shields diagram*, is shown in Figure 16.25. The modifed particle Froude number, F_s, is known as the *Shields entrainment function* due to the fact that it gives the particle diameter (d) of a sediment with specific gravity S_s that would be dislodged, or entrained at given shear stresses.

The Shields diagram forms a useful tool in the design of earthen channels for stable cross sections. Examination of Figure 16.25 reveals that, not surprisingly, it has some characteristics similar to those of the Moody diagram of Chapter 8 in that the curve becomes asymptotic to a horizontal line on the right side of the graph (i.e., for larger sediment sizes). In this case the asymptotic value is approximatly 0.056 and can be used for R_n^* values larger than about 400. It can be shown (Henderson, 1966) that the particle diameter at this point is about 0.25 in (0.089 cm). Then

$$F_s = \frac{v_*^2}{gd(S_s - 1)} = \frac{\tau_o}{\rho gd(S_s - 1)} = \frac{\tau_o}{\gamma d(S_s - 1)} = 0.056 \tag{16.44}$$

Now, using $d = 0.25$ in and $S_s = 2.65$, assuming quartz as the base sediment material, and recalling equation (16.27), then equation (16.44) reduces to the simple expression as given by Henderson (1966):

$$d = 11RS \tag{16.45}$$

Equation (16.45) gives the limiting sediment diameter for a stable channel with given hydraulic radius and slope. Employing the wide rectangular approximation, the expression can often be written as

$$d = 11yS \tag{16.46}$$

Example 16.12

A straight canal is to be designed to carry a discharge of 300 cfs through a material with a d_{75} value of 1 inch and an angle of repose of 32°. The natural slope of the terrain is 0.0045 ft/ft. Determine minimum dimensions of a trapezoidal channel to carry the design discharge without scour.

Solution: Using equation (16.45), $d = 0.0833$ ft; $S = 0.0045$ ft/ft, the required R is found to be 1.68 ft. If we wish to employ the wide rectangular assumption as a first approximation, then the normal depth will also be 1.68 ft. Using Strickler to estimate the Manning coefficient,

$$n = 0.034(0.0833)^{1/6} = 0.022$$

Then the velocity would be given by $V = \frac{1.49}{0.022}(1.68)^{2/3}(0.0045)^{1/2} = 6.42$ ft/s. Then the required area would be $300/6.42 = 46.73$ ft^2, from which we derive the wetted perimeter as $46.73/1.68 = 27.81$ ft.

Now, forgoing the wide rectangular assumption, we will find the true dimensions by the methods discussed in Chapter 11. Using the geometry for a trapezoidal section,

$$A = by + ty^2, \quad \text{or} \quad b = \frac{A - ty^2}{y} = \frac{46.73 - ty^2}{y}$$

where t is the horizontal side slope component. Also,

$$P = b + 2y\sqrt{1 + t^2} = \frac{46.73 - ty^2}{y} + 2y\sqrt{1 + t^2}$$

Now, selecting a side slope such that the angle will be less than the angle of repose of the soil, say let $t = 2$ (i.e., $\theta = 26.6°$). Then we have

$$P = 27.81 = \frac{46.73 - 2y^2}{y} + 4.47y$$

Using the quadratic formula, we find that $y = 2.05$ ft or $y = 9.2$ ft. Using $y = 2.05$, we get $b =$ 18.7 ft and using $y = 9.2$ ft, we get $b =$ negative, which cannot be used.

Making one last check, we find that for this situation, $y_c = 2.05$ ft. Since our design calls for a normal depth of 2.05 ft as well, this would not be an ideal situation. To relieve this, we might modify the channel dimensions in an appropriate manner such that the normal depth would remain subcritical.

16.2.5.3 U.S. Corps of Engineers Approach Of course, alternative approaches using the same basic principles are possible. The U.S. Army Corps of Engineers (1970) employs an expression for the average local shear stress applied to channel boundaries as a function of velocity by substituting the Chezy equation into equation (16.27), that is, $S_f = \frac{V^2}{C^2R}$. Then

$$\tau_o = \gamma\frac{V^2}{C^2} \tag{16.47}$$

Then an empirical expression is substituted for the Chezy C equivalent to those discussed in Chapter 7, that is,

$$C = 32.6 \log\frac{12.2R}{k} \tag{16.48}$$

An equation for local boundry shear is thus derived as a function of the flow velocity,

$$\tau_o = \gamma \frac{V^2}{\left(32.6 \log \dfrac{12.2R}{k}\right)^2} \tag{16.49}$$

In equation (16.49), the factor k represents an equivalent boundary roughness element as usual. In practical work, the Corps recommends that the d_{50} sediment diameter be used for this factor. As before, the expression is often cast in terms of the flow depth, y by use of the wide rectangular approximation,

$$\tau_o = \gamma \frac{V^2}{\left(32.6 \log \dfrac{12.2y}{d_{50}}\right)^2} \tag{16.50}$$

Equation (16.50) neatly gives the relationship between flow parameters (depth and velocity) and the shear that will develop on the channel boundary. The velocity employed in this equation is supposedly the local average value in the vicinity of the particular part of the boundary under investigation, including the channel banks. The Corps manual (1970) recommends that this value be determined simply as the ratio of the discharge and the cross sectional area at the particular point in the flow field. Table 16.6 gives maximum velocities recommended by the Corps of Engineers for unlined channels of various soil characteristics.

16.2.5.4 Lined Channels In many situations it is not possible to design a channel such that permissible velocities will not be exceeded. In these cases, the earthen channel boundary must be protected by some type of lining in order to prevent scour. The lining may be simply natural grass, or made of some resistant material such as concrete, stone rip-rap, or geosynthetic polymers. Each of these options presents the engineer with a different set of design standards. For example, in the case of grass lining, the problem is not simply the determination of permissible velocities for various types of grass, but also of knowing the roughness characteristics and

TABLE 16.6 Suggested Maximum Permissible Mean Channel Velocities (USCOE, 1970)

Channel Material	Mean Velocity (ft/s)
Fine sand	2.0
Coarse sand	4.0
Earth	
Sandy silt	2.0
Silt clay	3.5
Clay	6.0
Poor rock	
Soft sandstone	8.0
Soft shale	3.5
Good rock (igneous or hard metamorphic)	20.0

TABLE 16.7 Permissible Velocities for Grass-Lined Channels (Coyle, 1975)

		Soil Resistance Level	
Species	**Slope Range (%)**	**Erosion Resistant (m/s)**	**Erodible (m/s)**
Bermuda grass	0–5	2.4	1.8
	5–10	2.1	1.5
	>10	1.8	1.2
Bahia, Buffalo grass	0–5	2.1	1.5
Kentucky blue grass	5–10	1.8	1.2
Smooth brome, blue	>10	1.5	0.9
Grama, tall fescue			
Grass mixtures, reed	0–5	1.5	1.2
Canary grass	5–10	1.2	0.9
Lespedeza sericea,	0–5	1.1	0.8
Weeping lovegrass			
Yellow bluestem,			
Redtop, Alfalfa			
Red fescue			
Common lespedeza	0–5	1.1	0.8

hydraulic efficiency of each species as well. Design factors for grass-lined channels have been derived primarily through experimentation.

The most common and popular technique for determination of roughness factors for grass linings was developed by Palmer (1945) and is based on the relationship between the Manning n and the product of velocity and hydraulic radius as discussed in Chapter 11. Experiments have shown (Singh, 1996) that this relationship is relatively invariant to the channel slope that is used for a single grass species. Figure 16.26 represents a plot of the n-VR_h relationship for various grass species as determined by the NRCS (Coyle, 1975). In this graph, each individual line represents a grouping of species according to "retardance" or resistance to flow due to thickness and height (Chin, 2000). Typical species associated with each retardance group are given below the graph. Table 16.7 gives the NRCS recommended values for permissible velocities for various grass linings (Coyle, 1975).

The design of a grass-lined channel necessitates a trial and error procedure. First one selects the species of grass to be employed based on the soils, slopes, and expected hydraulic conditions of the project. After this selection the retardance class can be determined from Figure 16.26 and the permissible velocity for the selected species can be obtained from Table 16.7. The procedure then begins with the assumption of a roughness coefficient (Manning n) that can then be employed in preliminary computations. Channel dimensions are estimated as described in Chapter 11. Once the hydraulic radius is determined, then the VR ratio is computed and Figure 16.26 is used to determine the Manning n value associated with the computed ratio. If this value does not match the assumed value, then the n value obtained from the figure is employed to compute new channel dimensions and VR ratio from which the n value is checked again. The procedure continues until the two roughness values match.

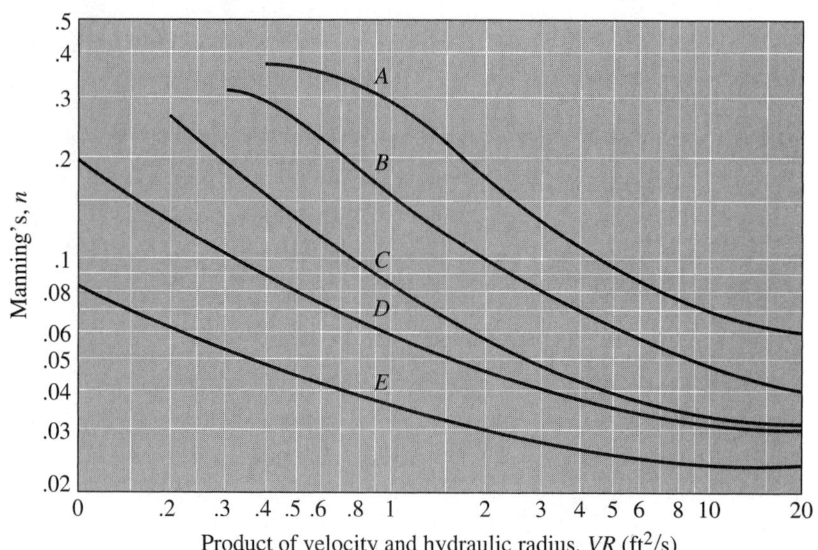

Retardance	Species	Condition
A	red canary grass	90 cm avg.
	yellow bluestem	90 cm avg.
B	smooth bromegrass	30–40 cm avg.
	Bermuda grass	30 cm avg.
	native grass mixture	
	bluestem, blue grama	unmowed
	tall fescue	45 cm avg.
	Lespedeza sericea	50 cm avg.
C	bahia	15–18 cm avg.
	Bermuda grass	15 cm avg.
	redtop	40–60 cm avg.
	centipede grass	15 cm avg.
	Kentucky bluegrass	15–30 cm avg.
D	Bermuda grass	6 cm avg.
	red fescue	30–45 cm avg.
E	Bermuda grass	4 cm avg.
	Bermuda grass	Burned stubble

Figure 16.26 Plot of VR versus Manning n for grass species. (Coyle, 1975)

Example 16.13

A trapezoidal channel is to be designed to carry a maximum discharge of 1000 cfs. The channel is to be grass lined and will run on a slope of 0.005 ft/ft through material with an angle of repose ϕ of 37°. Conditions dictate that Bermuda grass (30 cm height) will be the optimum lining. Determine the channel dimensions for this case.

Solution: With Bermuda grass (30 cm height) the retardance class is determined to be B from Figure 16.26. From Table 16.7 we find that the permissible velocity for Bermuda grass on a slope less than 5% is 1.8 m/s, or 5.9 ft/s for erodible soil. Then the allowable cross-sectional area of the channel is $A = Q/V = 1000/5.9 = 169.5$ ft^2. Now, we begin by assuming that the Manning n value will be 0.05. Then

$$V = \frac{1.49}{n} R^{2/3} S^{1/2} = 5.9 = \frac{1.49}{n} R^{2/3} (0.005)^{1/2}$$

from which we find $R = 4.68$ ft. Then $VR = 5.9(4.68) = 27.6$ ft^2/s. Using Figure 16.26, we find that this value is slightly off the scale, but that for a class *B* retardance the limiting *n* value from the graph is about 0.04 for a *VR* ratio of 20 ft^2/s. Using this value for *n*, we solve the Manning equation again as above and find that $R = 3.35$ ft for this case. Checking, we find that $VR = 5.9(3.35) = 19.77 \sim 20$, so our Manning value is good. Then we employ the geometry of trapezoidal sections to determine the dimensions.

$$A = 169.5 = by + ty^2; \; P = b + 2y\sqrt{1 + t^2}$$

Additionally, since $R = A/P = 3.35$ ft, then $3.35 = 169.5/P$, or $P = 50.6$ ft. Next, noting that the angle of repose of the material is 37°, we select a bank angle of 30°, which corresponds to a side slope component (t) of 1.73/1. Then

$$169.5 = by + 1.73y^2; \; 50.6 = b + 2y\sqrt{1 + t^2} = b + 3.996y; \; b = 50.6 - 3.996y$$

Then $169.5 = (50.6 - 3.996y)y + 1.73y^2$, from which $50.6y - 2.266y^2 - 169.5 = 0$

From the quadratic formula, we find that $y = 18.22$ ft or $y = 4.1$ ft. Plugging back into the wetted perimeter formula, we find that the value of 18.22 ft is impossible, so that $y = 4.1$ ft. Then $b = 50.6 - 3.996y = 34.2$ ft.

The same concepts of resistance and retardance can be used to design grass buffers between the channel and the surrounding landscape. These buffers serve multiple purposes, including channel protection, aesthetics, storage of excess runoff, and interception of pollutants from the watershed. Besides the hydraulic issues discussed in the previous paragraph relative to permissible velocities and flow resistance, hydrologic issues of storage and attenuation or delay of flood hydrographs also play important roles in the design of buffer strips or zones. An additional consideration is the ability of the plants to handle and safely dispose of the pollutants carried in the runoff.

16.2.5.5 Rip-Rap Lining Erodible channel banks may also be protected with blocks of stone or cement known as rip-rap. According to the Corps of Engineers manual (1970), factors affecting rip-rap design include material characteristics (shape, size, gradation, layer thickness) and channel parameters (side slopes, roughness, shape, alignment, and bed slope) as well as the flow velocity. The stone or cement rip-rap blocks should be nearly cubicle in shape with clean sharp edges in order for them to adhere together and resist movement. In gradation, the Corps recommends that not more than 25% of the stones should have a length more than 2.5 times their breadth or

A rip-rap lined channel.

thickness. Of course, the weight of the blocks is a key parameter in resistance to erosion and the following relationship is recommeded for calculation of block weight from size (COE, 1970):

$$D_s = \left(\frac{6W}{\pi\gamma_s}\right)^{1/3} \tag{16.51}$$

where D_s = equivalent stone diameter (ft), W = stone weight (lb), and γ_s = specific weight of the stone (or cement).

The side slopes of the channel are also key factors in rip-rap design and placement. As previously shown in the discussion of the erosion of soil particles, the forces tending to erode the banks (and thus move the rip-rap blocks) are a combination of the shear stress applied by the flowing water and the gravity force tending to slide the particle (or block) down the bank. Of course, the force tending to slide the block down the bank increases with slope steepness, while conversely, the component of gravity tending to hold the block in place decreases with increasing steepness. For this reason, the Corps recommends that machine dumped rip-rap not be placed on side slopes steeper than $1H$ to $2V$ and that hand placed rip-rap not be used in cases of side slopes steeper than $1H$ to $1.5V$.

The same basic analysis techniques as used in the erosion case applies in the design of rip-rap blocks. The average, or local shear stress applied to the blocks is computed by use of equations (16.49) or (6.50). The rip-rap design shear is defined as the amount of the local boundary shear that the rip-rap can safely withstand and is given by the following expression for level channel beds:

$$\tau_c = a(\gamma_s - \gamma)d_{50} \tag{16.52}$$

where the factor a is given as 0.04 (COE, 1970) and the other parameters are as defined previously. The design shear for channel banks is given by equation (16.39):

$$\frac{\tau_{bc}}{\tau_c} = \left(1 - \frac{\sin^2\theta}{\sin^2\phi}\right)^{1/2} \tag{16.53}$$

According to the Corps design, the stone or cement blocks should be graded so that the lower limit of the W_{50} weight class is not less than that required to withstand the design shear forces as calculated above. The lower limit of the W_{100} blocks should not be less than 2 times the lower limit of the W_{50} blocks. The rip-rap layer thickness should normally be at least 12 inches and should never be less than the spherical diameter of the largest blocks, or 1.5 times the diameter of the W_{50} blocks, whichever is larger. The layer thickness should be increased by 50% if the rip-rap is placed underwater to counter uncertainties associated with underwater work.

Example 16.14

A trapezoidal channel has been designed to carry a discharge of 650 cfs. The channel has a bottom width of 10 ft, side slopes of 1.5/1, and running on a slope of 0.00262 through material with a median equivalent diameter of 1 inch and an angle of repose of 38°. Determine if rip-rap lining is required, and, if so, determine the size of the stone on both the bed and sides of the channel.

Solution: First, we must determine the depth and velocity at which the water will flow. Writing the Manning equation,

$$AR^{2/3} = \frac{nQ}{1.49\sqrt{S_o}} = (10y_o + 1.5y_o^2)\left(\frac{10y_o + 1.5y_o^2}{10 + 3.6y_o}\right)^{2/3} = \frac{n(650)}{1.49\sqrt{0.00262}}$$

Using the Strickler equation to find n with a grain size of 1 in (0.0833 ft), $n = 0.034d_m^{1/6} = 0.022$. Thus the right side of the above expression becomes 187.5. Solving by trial, we find the $y_o = 5$ ft. Thus the cross-sectional area, $A = 87.5$ ft² leading to a velocity, $V = 7.42$ ft/s. From the values of allowable velocity given in Table 16.6, we see that this value exceeds that given for coarse earth, so rip-rap lining will be required. Next, we find the shear stress that will be applied to the channel bed using equation (16.49):

$$\tau_o = \gamma \frac{V^2}{\left(32.6 \log \frac{12.2R}{d_{50}}\right)^2} = \gamma \frac{(7.42)^2}{\left(32.6 \log \frac{12.2(3.125)}{0.0833}\right)^2} = 0.456 \text{ lb/ft}^2$$

Now, if we place a factor of safety of 1.5 on this value, that would lead to a desired $\tau_c = 0.684$ lb/ft^2. Then, finding the median stone diameter by equation (16.52); $\tau_c = a(\gamma_s - \gamma)d_{50}$, or $0.684 = 0.04(2.65\gamma - \gamma)d_{50}$, assuming a specific gravity of the stone of 2.65. Solving for d_{50}, we find that $d_{50} = 0.166$ ft = 2 inches. Finding the critical shear stress on the banks, using the angle of repose $\phi = 38°$ and with *SS* of 1.5, the bank angle $\theta = 33.69°$, using equation (16.38),

$$\frac{\tau_{bc}}{\tau_c} = \left(1 - \frac{\sin^2\theta}{\sin^2\phi}\right)^{1/2} = \frac{\tau_{bc}}{0.684} = \left(1 - \frac{\sin^2(33.69)}{\sin^2(38)}\right)^{1/2} = 0.434.$$

Thus, $\tau_{bc} = 0.296$ lb/ft^2

However, according the relationship given in equation (16.28), the applied shear to the banks would be about 76% of that applied to the bed; thus $\tau_{obc} = 0.76(0.456) = 0.346$ lb/ft^2, which exceeds the critical value of 0.296 lb/ft^2; thus rip-rap will be necessary on the banks as well. Again, applying a factor of safety of 1.5 on the applied value, we find that $\tau_{bc} = 1.5(0.346) = 0.519$ lb/ft^2. Thus, $0.519 = 0.04(2.65\gamma - \gamma)d_{50}$, from which, $d_{50} = 1.5$ in.

READING AID

16.1. Discuss the primary elements that should be considered in the design of any hydraulic structure.

16.2. Which comes first in designing hydraulic structures, the design standards and functionality of the structure or the safety consideration?

16.3. Why are most water structures designed according to the risk-based approach? Why are the uncertainties relatively high in the water resources field?

16.4. Give an example for a safety consideration while designing a dam with a certain storage capacity.

16.5. What is meant by freeboard in water resources projects? Give examples for recommended freeboard values.

16.6. From your point of view, would you expect the life time of hydraulic structures to be on the order of years, decades, or centuries? Discuss the expected life time of different types of hydraulic structures.

16.7. Would the size of the hydraulic structure affect the expectancy of its life time and safety consideration?

16.8. What is meant by *risk* in the design of water resources projects? What is the accedence probability? Define *return period* of a certain discharge.

16.9. Write an expression for the probability that a particular discharge will be exceeded in a given lifetime of a hydraulic structure.

16.10. Give examples for the types of preparations needed for housing or shopping center projects.

16.11. Using the rational method, write an equation to estimate the discharge to be carried by a storm sewer system. Define each parameter of the equation and check the dimensional homogeneity of the equation.

16.12. Discuss the applicability of the rational method for different areas. Elaborate the limitations of this method.

16.13. Define the term *time of concentration.* What are the main factors that affect the time of concentration?

16.14. Define the runoff coefficient and discuss the parameters that might affect its value. Give a practical range for the runoff coefficient in developed and undeveloped areas.

16.15. Write an equation for the design of sump inlets and define the different parameters, in your equation. What is the relevant value for the clogging factor?

16.16. For the design of storm pipes, why it is not recommended to design the pipe to flow under pressurized conditions? What is the suggested flow depth in such systems?

16.17. Why are storm pipes designed to maintain the flow velocity above a certain value? What is the recommendation for the minimum velocity in stormwater systems?

16.18. Which equation is generally used in the design of circular stormwater systems? How is the bed slope of the pipe selected? Whenever possible, would you recommend to increase or decrease the slope of the pipe?

16.19. Define the three different control states in designing culverts. Under which control state will be the culvert operating as a closed conduit?

16.20. Write the discharge equation for the culvert provided that the culvert is flowing full and define the different parameters in the equation.

16.21. For culverts flowing under tail-water control, what are the design considerations and steps? Classify the energy losses in the culvert.

16.22. Define the different terms of equation (16.21) and check its homogeneity.

16.23. State the factors affecting the head loss inside culverts. What is meant by "hydraulically long" and "hydraulically short" culverts?

16.24. What are the factors that would cause the culvert to function in inlet control?

16.25. For culverts functioning under inlet control conditions, how is the critical depth calculated for rectangular and nonrectangular sections?

16.26. Under which situations are drop inlets structures used? What is the advantage of this system?

16.27. Write the steps for determining the flow capacity of a drop inlet.

16.28. What is the main function of stilling basins? Why are baffle blocks usually placed on the bottom of stilling basins?

16.29. Discuss the main design considerations for impoundment facilities.

16.30. Discuss the environmental and ethical standards that should be considered while designing a hydraulic structure.

16.31. What is the main criterion that should be considered in designing erodible channel sections?

16.32. Define the term *tractive force* and write an equation to compute the tractive force on a give area of the channel bed and sides.

16.33. What is meant by packing factor, r? What is the limiting value of r for the coarse granular material?

16.34. Discuss the Shields diagram and state its uses.

16.35. Write an equation to calculate the modified practical Froude number, F_s and define each parameter in the equation. Check the dimensional homogeneity of the equation.
16.36. Given the channel depth and slope of a wide rectangular channel, what is the limiting sediment diameter for a stable channel?
16.37. Discuss the usefulness and assumptions of equation (16.50) and define the different parameters in the equation.
16.38. Why it is necessary to use lining for channels bottom and sides? Give examples for possible kinds of lining.
16.39. What are the main factors affecting rip-rap design? Would you recommend to use irregular shapes or cubicle shapes with clear sharp edges? Explain.
16.40. What is the minimum thickness of the rip-rap layer when placed under water?

Problems

16.1. A spillway on a flood control dam is designed to pass a flood with an exceedance probability of 0.015. About how often will a flood of this magnitude or larger occur on this watershed? What is the risk that the spillway will fail at least once in 100 years?

16.2. It is desired to design a hydraulic structure to a standard such that its risk of failure will be no greater than 5% in a time period of 50 years. What will be the exceedance probability and recurrence interval of the design discharge?

16.3. A sharp-crested suppressed rectangular weir is to be used to regulate the depth of water in a navigation project. The depth in the channel must be maintained at least 10 m in order for vessels to navigate the channel. The channel slope is essentially horizontal, its width is 6 m, and its invert is at elevation 10 m above sea level. The design low flow discharge is 1.5 m^3/s for a recurrence interval of 20 years.

a. What is the exceedance probability of the design?
b. What is the probability that vessels will not be able to navigate the channel at least once in the next 10 years?
c. What must be the elevation of the weir crest in order to meet the design specifications? Assume the discharge coefficeint is 0.61.

16.4. A 1-acre lot is to be developed for residential dwellings. The original undeveloped condition is such that the longest watercourse length is 300 ft, the avearage slope is 3%, and the Manning n value of the ground cover is 0.1. After development, the n value reduces to 0.05. Assuming that the 24-hr two year rainfall depth for the area is 3.5 inches; use Figure 16.1 to estimate the peak stormwater runoff for the 10-year event for this area under both present and future conditions. Use runoff coefficients of 0.25 and 0.6 for existing and improved conditions, respectively.

16.5. Design a sag inlet and a pipe to carry the runoff developed in Problem 16.4 under future conditions. Assume the pipe will be concrete and will run on the same slope as the ground slope.

16.6. Suppose we wanted inlets to function on grade for the situation in Problem 16.4. Design an inlet system (one or more inlets on grade) to collect the runoff from Problem 16.4 for future conditions.

16.7. A 5-ha area is to be converted into a subdivision. The initial conditions are: longest watercourse length 1500 m and average slope 3.5%. Use Figure 16.1 to estimate the existing and future peak storm runoff values for both the 10- and 25-year events. Use a C value of 0.3 for exisiting conditions and 0.7 for future conditions.

16.8. Assuming all of the flow from Problem 16.7 is diverted to a single access road, design a curb inlet system and concrete pipe to collect and carry the runoff for the 10-year event for future conditions. Use a concrete pipe and assume that the slope is the same as the ground slope.

16.9. A concrete sewer pipe is to be laid on a slope of 2.5% and must carry a maximum design discharge of 5 cfs. Also, it must carry a minimum discharge of 0.5 cfs with an acceptable cleaning velocity. What size pipe should be used?

16.10. It is desired to design a concrete storm sewer to carry a maximum discharge of 1.5 m^3/s on a slope of 3%. The pipe must also carry a minimum flow of 0.1 m^3/s with an acceptable cleaning velocity. Determine the minimum pipe diameter to meet these requirements.

16.11. Determine the equivalent arch pipe for the situation in Problem 16.10.

16.12. An embankment is traversed by a 6 ft × 6 ft concrete box culvert 100 ft long as shown in Figure P16.12. The upstream invert is at an elevation of 8 ft and the downstream invert is 4 ft. If the depth at the upstream end of the culvert is 8 ft, what is the flow through the culvert? Assume the culvert is not in tail-water control.

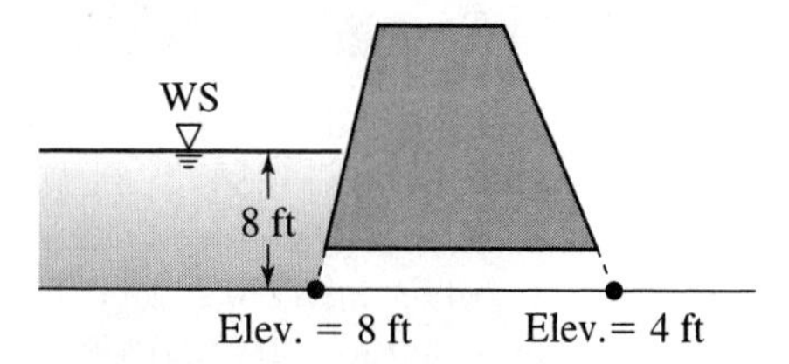

Figure P16.12 Culvert for Problem 16.12.

16.13. A small pond is controlled by a sluice gate. The depth of water behind the gate is 14 ft and the channel is a rectangular section 10 ft wide running on a slope of 0.0175 ft.ft with a Manning n value of 0.025. The specific discharge through the gate is 28.93 cfs/ft.

a. Will a hydraulic jump form downstream of the gate?

b. How far upstream will the effects of the gate extend?

16.14. A highway embankment must cross the stream downstream of the gate in Problem 16.13 as shown in Figure P16.14. The embankment is traversed by a 48-in RCP that is 75 ft long. The channel downstream is the same as in Problem 16.13 and runs on for a long distance. How far downstream of the gate must the embankment be placed so as not to effect the discharge through the gate? Assume the culvert inlet is flush with the embankment.

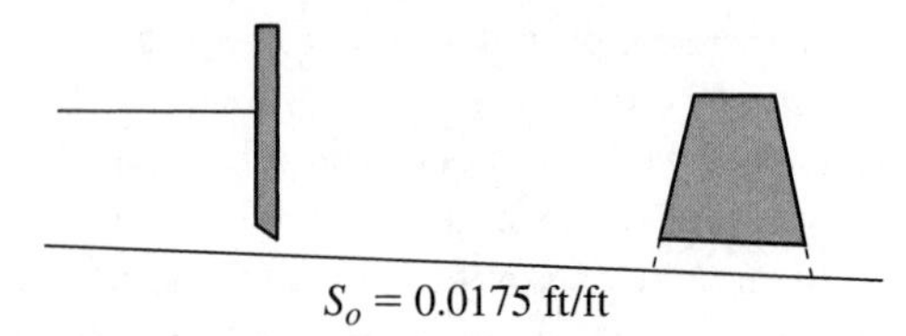

Figure P16.14 The sluice gate and embankment for Problem 16.14.

16.15. An 8-ft-wide rectangular channel is crossed by a highway embankment which is traversed by a 36 in corregated metal pipe (CMP) 75 ft long. The channel runs on a slope of 0.001 ft/ft with a roughness coefficient of 0.035 and runs on for a long distance downstream of the embankment with no other obstructions. For a design discharge of 100 cfs,

a. What will be the design head on the culvert, assuming a flush entrance?

b. How far upstream will the effects of the culvert extend?

16.16. A rectangular channel 12 feet wide is crossed by a highway embankment as shown in Figure P16.16. The design discharge is 200 cfs and the downstream channel runs a long distance on a slope of 0.001 ft/ft with a Manning n value of 0.030. The downstream invert of the culvert will be at elevation 0.00 ft and the culvert must be 100 ft long. The elevation of the top of the embankment is 22 ft. It is desired to design a culvert to carry the design discharge with a maximum upstream water level no higher than 2 feet below the top of the embankment. Determine the minimum diameter RCP to accomplish this task.

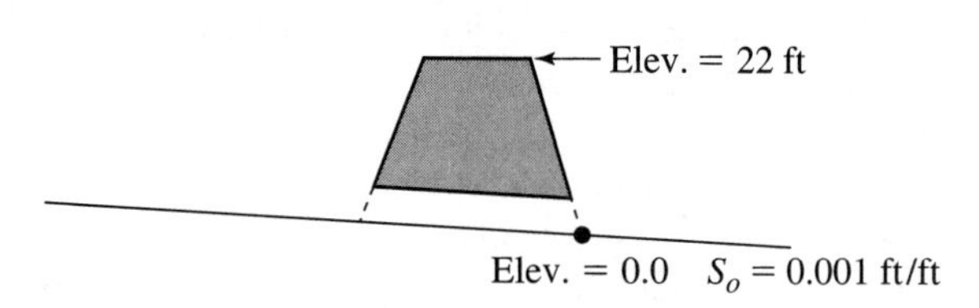

Figure P16.16 High way embankment for Problem 16.16.

16.17. What would be the equivalent arch pipe for the situation in Problem 16.16?

16.18. It is desired to design a culvert to traverse a highway embankment that crosses a trapezoidal channel with a bottom width of 12 feet and $1V/2H$ side slopes. The channel is running on a slope of 0.005 ft/ft with a Manning n value of 0.035. The culvert must be a 75-ft-long RCP with downstream invert of 0.0 ft. A shopping center is located 2000 ft upstream of the embankment with a parking lot at elevation 17 ft as shown in Figure P16.18. Design the culvert to carry a design discharge of 700 cfs without flooding the parking lot of the shopping center. Use the step method of profile calculation.

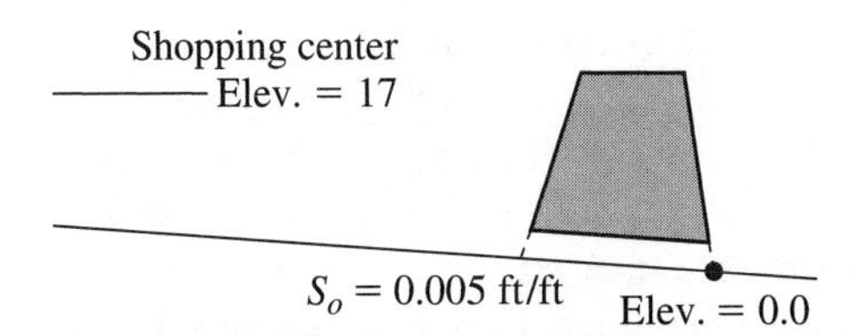

Figure P16.18 Embankment and shopping center for Problem 16.18.

16.19. A highway embankment is traversed by a 36-in-diameter RCP 90 feet long. Another highway is located 1000 ft downstream of the first and is traversed by a 75-ft-long 48 in CMP. The channel between the two embankments is rectangular running on a slope of 0.005 ft/ft with a roughness coefficeint of 0.04 and a bottom width of 15 feet. The invert of the downstream end of the first (most upstream) culvert is 0.0 ft. What is the discharge through the system when the headwater on the first culvert is at elevation 10 ft and the tail water on the second (downstream) culvert is 5 feet deep? Assume both culverts have flush entrances.

16.20. The outflow from a detention pond is controlled by the SCS drop inlet shown in the sketch below. The quantities and elevations are as shown in the sketch in English units. Derive a headwater rating curve up to elevation 45 ft for this structure assuming that when the pipe controls it is in barrel control (i.e., ignore the tail-water).

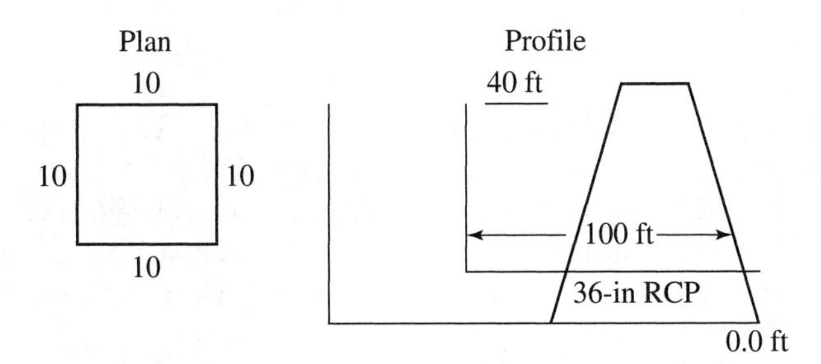

Figure P16.20 The Drop Inlet for Problem 16.20.

16.21. The outflow from a reservoir is to be controlled by the SCS drop inlet as shown in the sketch below. The design discharge is 450 cfs and the tail-water elevatioin is 10 ft as shown. If the top of the embankment is 42 ft and the water level is not to get higher than 2 ft below the top, design the riser and RCP to achieve the desired result.

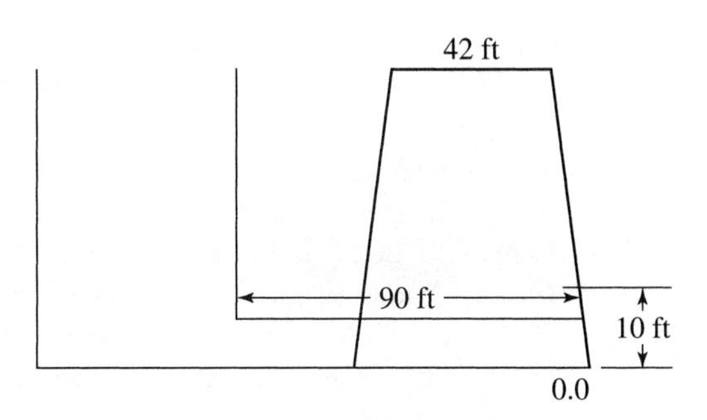

Figure P16.21 The Drop Inlet for Problem 16.21.

16.22. A drop inlet is to be designed to control the discharge from a small reservoir. The design discharge is 10 m^3/s and the target water level in the reservoir is 15 m above datum. If the tail water under design conditions is 5 m, design the riser and RCP to meet the desired conditions. Assume the pipe will be 25 m long.

16.23. It is desired to design a trapezoidal channel with a bottom width of 10 ft and $1V/2H$ side slopes. Sieve analyses have revealed a grain size distribution which results in an allowable shear stress of 0.5 lb/ft^2 on the bed and a Manning n value of 0.03. If the channel is designed to run at a normal depth of 5 ft, what will be the resulting discharge?

16.24. A canal is to be excavated through alluvial terrain with a d_{75} grain size of 2 cm. The natuaral slope of the terrain is 0.00025 m/m and the angle of repose of the material is estimated to be 32°. Using the maximum tractive force method, determine the minimum channel width and appropriate side slopes to carry a design discharge of 50 m^3/s with a design depth of no more than 2 m. The canal will be slightly sinuous.

16.25. A trapezoidal channel is to be cut through alluvial material with a d_{75} grain size of 0.33 in. The channel must carry 1000 cfs and run on a general grade of 0.001 ft/ft. The angle of repose of the material is estimated to be 30°. Using the limiting value from the Shields entrainment diagram (Figure 16.25), determine the required channel width to result in a stable channel.

16.26. A trapezoidal channel is to be designed to carry a maximum discharge of 800 cfs. The channel is to be grass lined and will run on a slope of 0.0035 ft/ft through material with an angle of repose ϕ of 35°.

Conditions dictate that Lespedeeza grass (50 cm height) will be the optimum lining. Determine the channel dimensions for this case.

16.27. It is desired to design a trapezoidal channel to carry a maximum discharge of 1770 cfs through an alluvium on a slope of 0.002 ft/ft and with an angle of repose of 30°. Conditions are such that the optimum lining will be Bermuda grass with an average height of 6 cm. Determine the channel dimensions for the canal to meet these requirements.

16.28. A trapezoidal channel is to be designed through a silty clay alluvium on a slope of 0.0035 ft/ft. The angle of repose of the material is 35° and the Manning coefficient is 0.025. If the channel is to be designed to carry a maximum discharge of 700 cfs at a depth of 4.9 ft, determine if rip-rap will be required, and, if so, compute the stone size for both the bed and sides.

16.29. A trapezoidal channel is to be designed to carry a maximum discharge of 500 cfs through an alluvium with a d_{50} of 0.25 in. The channel is to run on a slope of 0.0025 ft/ft and land constrictions are such that the bottom width cannot be more than 7.5 ft. Determine if the channel will have to be lined with rip-rap, and if so size the stone for both the bed and sides.

References

American Concrete Pipe Association, 2000. Concrete Pipe Design Manual, compact disc available from APCA.

American Society of Civil Engineers (ASCE), 1982. Gravity Sanitory Server Design and Construction, Manual No. 60, ASCE, New York, NY.

American Society of Civil Engineers, 1992. *Design and Construction of Urban Storm Water Management Systems.* ASCE, New York.

Chow, V. T., 1959. *Open Channel Hydraulics.* McGraw-Hill Book Company, New York.

Chin, D. A., 2000. *Water-Resources Engineering.* Prentice Hall, Upper Saddle River, N.J.

Coyle, J. J., 1975. Grassed waterways and outlets, Engineering Field Manual, United States Soil Conservation Service, Washington, D.C.

Department of the Army, Corps of Engineers, 1982. HEC-2 Water Surface Profiles, Users Manual, Corps of Engineers, Hydrologic Engineering Center, Davis, Calif.

Department of the Army, Corps of Engineers, 1970. Hydraulic Design of Flood Control Channels, Engineering Manual EM 1110-2-1601, Corps of Engineers, Office of the Chief of Engineers, Washington, D.C.

Federal Highway Administration, 1985. Hydraulic Design of Highway Culverts, Hydraulic Design Series No. 5, U.S. Department of Transportation, Washington, D.C.

Federal Highway Administration, 2001. Urban Drainage Design Manual, 2nd ed. Hydraulic Engineering Circular No. 22, U.S. Department of Transportation, Washington, D.C.

Haestad Methods, 1999. *Computer Applications in Hydraulic Engineering* Third Edition, Haestad Press, Waterbury, CT.

Henderson, F. M., 1966. *Open Channel Flow,* Macmillan Publishing Company, New York.

Hwang, N. H. C. and Houghtalen, R. J., 1996. *Fundamentals of Hydraulic Engineering Systems.* Prentice Hall, Upper Saddle River, N.J.

Lane, E. W., 1955. Design of stable channels. *Transactions of the American Society of Civil Engineers,* 120(1955): 1234–1279.

Mishra, S. K. and V. P. Singh, 2003. Soil Conservation Service Curve Number (SCN–CN) Methodology, Kluwer Academic Publishers, Dordrecht, The Netherlands.

Morris, H. M. and Wiggert, J. M., 1972. *Applied Hydraulics in Engineering.* Wiley, New York.

Palmer, V. J., 1945. A method for designing vegetated waterways. *Agricultural Engineering,* 26(12): 516–520.

Peterka, A. J., 1964. Hydraulic Design of Stilling Basins and Energy Dissipators, U.S. Bureau of Reclamation, Engineering Monograph 25, United States Department of Interior, Washington, D.C.

Roberson, J. A., Cassidy, J. J., and Chaudhry, M. H., 1998. *Hydraulic Engineering.* Wiley, New York.

Shields, A., 1936. Application of similarity principles and turbulence research to bed-load movement, Mitteilungen der Preuss, Versuchsanst fur Wasserbau und Schiffbau, No. 26, Berlin, Germany.

Singh, V. P., 1996. *Kinematic Wave Modeling in Water Resources.* Wiley, New York.

Sturm, T. W., 2001, Open Channel Hydraulics, McGraw-Hill, Boston, MA.

United States Army Corps of Engineers, 1970. Hydraulic Design of Flood Control Channels, Engineering Manual 1110-2-1601, Department of the Army, Office of the Chief of Engineers, Washington, D.C.

Wurbs, R. A., and James, W. P., 2002. *Water Resources Engineering.* Prentice Hall, Upper Saddle River, N.J.

Glossary

Air vessel—a closed chamber attached to both the suction and delivery pipes of a reciprocating pump to equalize the velocity and reduce friction losses.

Alternate depths—two depths (one subcritical and the other supercritical) that have the same specific energy.

Axial flow pump—a pump in which momentum is conserved in the direction of flow and force is provided to the fluid by the impeller blades alone.

Backwater profile—a region of a channel where the depth increases and is greater than normal depth due to a restriction downstream.

Baffle blocks—concrete blocks located within a zone of supercritical flow to absorb momentum and facilitate the formation of a hydraulic jump.

Boundary shear—shear stress developed in a fluid at the boundary with the conduit wall.

Boundary layer—a thin layer of laminar flow at the boundary between the fluid and the conduit wall.

Broad-crested weir—a weir whose crest has sufficient breadth such that critical depth occurs on the crest.

Bulk modulus—coefficient measuring the compressibility of a fluid or the ease with which its volume and density change with pressure.

Buoyancy—state in which a submerged body is subjected to an uplift force equal to the weight of the liquid displaced by the body.

Buoyancy force—the net upward force applied to a submerged body due to the pressure distribution on the body.

Celerity—the speed of a wave or disturbance traveling in a fluid.

Centroid—the center of mass of an object.

Center of buoyancy—the point of action of the buoyancy force on an object submerged in a fluid.

Center of pressure—centroid of the pressure distribution exerted on an object.

Centrifugal pump—a pump whose pressure head is generated primarily through conservation of angular momentum.

Channel efficiency—a resistance term concerned with the energy loss in a channel flowing with a specified discharge the ability of the channel to convey water under a unit loss of energy head.

Choke—a constriction in a channel such that the energy level upstream of the constriction is forced to rise above its initial level.

Cippolletti (trapezoidal) weir—a weir with a trapezoidal notch through which the flow passes.

Conjugate depths—two depths (one subcritical and the other supercritical) associated with the same pressure-momentum force.

Continuity (conservation of mass)—principle that all mass entering a conduit must be accounted for through outflow or storage.

Control—location in a channel where a precise and unique relationship between depth and discharge is known.

Conveyance—A resistance term measuring the capacity of a channel, as computed by Manning's equation, divided by the square root of the bed slope.

Compressibility—a fluid property that describes a change in volume or strain induced under an applied stress.

Corregated metal pipe (CMP)—a thin-walled steel pipe with corregations in the wall in order to increase the strength.

Critical depth—depth in a channel where the energy required to transmit a certain flow is at a minimum.

Critical discharge—discharge associated with the minimum specific energy.

Critical energy—minimum energy required to transmit a certain flow.

Critical slope—channel slope that results in the normal depth equal to the critical depth.

Culvert barrel—a large pipe usually placed through an embankment, such as a highway, to transmit water through the embankment.

Density—measure of total mass of a volume of fluid or the number of individual molecules per unit volume multiplied by their respective masses (mass per unit volume).

Depth—distance from the water surface to the channel bed measured in the vertical direction.

Discharge—volumetric time rate of flow past a given point in space.

Drawdown curve—water surface profile in subcritical flow connecting normal depth to critical depth.

Dropdown curve—a water surface profile in supercritical flow connecting critical depth to normal depth.

Drop inlet—hydraulic structure consisting of combination of a riser that functions as a weir control during low flows and a culvert that controls during high flows.

Drowned jump—situation where a potential hydraulic jump is prevented from occurring due to the downstream (subcritical) depth being greater than the conjugate of any possible upstream (supercritical) depth.

Dynamic viscosity—measure of a fluid's resistance to shear or angular deformation.

End sill—barrier placed at the end of a stilling basin to contain the hydraulic jump.

Energy grade line—longitudinal gradient of total energy head along a conduit.

Floodplain—land area adjacent to a channel that is inundated only during flows that exceed the channel capacity.

Forced jump—hydraulic jumps that are forced to occur within certain designated locations through the use of abrupt changes in channel geometry or baffle blocks.

Forewater—the computation of water surface profiles in supercritical flow beginning upstream and moving in the downstream direction.

Freeboard—height added to a hydraulic structure or to the depth of a channel to account for unforeseen circumstances and risk inherent in the design process.

Friction loss—energy loss due to friction or boundary shear with the conduit wall.

Friction slope—the rate of energy loss in flowing water due to boundary shear or friction.

Froude number—dimensionless number representing the ratio of inertia forces of the flow to gravity forces.

Geographic Information System—computer software package for display, analysis, and manipulation of spatial data.

Global Positioning System—a positioning system whereby a particular point on the surface of the earth can be located precisely through triangulation among three satellites.

Gradually varied flow—flow situation where, even though depth changes longitudinally, change is so gradual that hydrostatic forces on each side of a reach in the flow direction are still considered equal.

Head—energy measured in units of length (ft, m).

Hydraulics—the study of the mechanical behavior of water (or other fluids) at rest or in motion.

Hydraulic depth—the area of a flow section divided by its top width.

Hydraulic grade line—longitudinal gradient of the sum of the position and pressure heads along a conduit.

Hydraulic jump—abrupt transition of flow from a supercritical state to a subcritical state (also called standing wave).

Hydraulic radius—ratio of the cross-sectional area of a flow section to the wetted perimeter of the section.

Hydrostatic force—force applied to an object by the pressure of a stationary body of fluid.

Inlet—location at which water enters a pipe, culvert, or storm drain.

Kinematic viscosity—ratio of dynamic viscosity and density.

Kinetic energy—energy developed in flowing water due to the velocity of the flow.

Laminar flow—flow situation wherein the deformation of fluid is on a macroscopic level.

Loss coefficient—a constant that is applied to the velocity head in a pipe or channel situation to compute minor losses.

Low chord—the lowest point on the underside of a bridge deck or a culvert.

Manometer—instrument for measurement of pressure or pressure differences in a closed conduit.

Metacenter—the intersection of the axis passing through the center of mass of a submerged object and the line of action of the buoyancy force.

Mild slope—channel slope that is not sufficiently steep to support supercritical flow conditions.

Minor loss—energy losses in the flow through a pipe or channel due to causes other than friction.

Modulus of elasticity—measure of the elastic force of a substance expressed as the ratio of the stress applied to the substance to the accompanying distortion or strain.

Moment of inertia—the second moment of an area about an axis (i.e., the integral of the product of an element of an area times the square of its distance from the relevant axis).

Momentum coefficient—numerical coefficient applied to the dynamic force in the momentum equation to account for non-uniform velocity distributions in conduits.

Momentum function—the total momentum developed by water flowing through a cross-sectional area with the mass term is exempted.

Net positive suction head—the difference between the total energy developed at the suction side of a pump and the vapor pressure of the fluid at the specified temperature.

Normal depth—the constant longitudinal depth in a channel associated with the situation of uniform flow.

Piezometer—device for measuring the total static head in a closed conduit.

Piezometric head—total static head in a closed conduit equal to the sum of the position and pressure heads.

Position (station) head—energy attributed to a unit volume of water due to its elevation with reference to the fixed datum.

Potential energy—sum of the position energy and pressure energy.

Pressure head—energy developed in a fluid in a closed conduit due to the presence of greater than atmospheric pressure.

Pressure momentum force—the sum of the hydrostatic and dynamic forces resulting from the flow of water through a cross-sectional area per unit time.

Pump characteristic (performance) curve—a graph showing the relationship between various pump characteristics (head, horsepower, efficiency, etc.) versus pumping discharge.

Pump efficiency—ratio of the total out power generated by a pump to the total energy or power input to the pump by the motor.

Pumping head—total energy head that must be developed by a pump in order to meet the specifications required by a particular situation.

Reciprocating pump—a pump in which the flow is directed from the suction to delivery side by displacement. Also called plunger (piston) pump.

Reinforced concrete pipe (RCP)—a pipe (usually a culvert or storm drain) constructed of reinforced concrete material.

Relative roughness—ratio of the height of the roughness elements on a conduit wall to the depth of flow in the conduit.

Remote sensing—method of determining certain characteristics of a surface or object without physically contacting the object.

Resistance—energy lost in the flow due primarily to friction between the water particles and the conduit boundary.

Rip-rap—material (usually stone or concrete) used to line a channel to prevent erosion.

Risk—probability of failure of a design in some specified number of years.

Reynold's number—dimensionless number representing the ratio of the inertia forces acting on a fluid to the viscous forces.

Rough pipe—a pipe with fluid flowing under sufficient conditions that the height of the wall roughness elements are significantly greater than the thickness of the laminar boundary layer for the particular discharge and temperature.

Roughness element—the roughness of a conduit wall defined in terms of an equivalent sand grain diameter.

Running jump—situation where the pressure momentum force of the supercritical upstream flow is greater than that of the downstream subcritical flow, thus forcing the hydraulic jump to move in the downstream direction.

Sharp-crested weir—a weir whose crest is of insufficient breadth for critical depth to occur on the crest.

Shear stress—force per unit area caused by velocity differences between adjacent layers in a flow field or between the conduit boundary and the adjacent fluid layer.

Shear velocity—square root of the ratio of boundary shear to fluid density (units of velocity).

Side slope—horizontal component of incline slope of a channel bank.

Slowdown curve—a water surface profile in supercritical flow wherein the depth increases in the downstream direction.

Sluice gate—a water control structure that releases flow from underneath a gate, that either rises vertically or rolls on a central spindle, thus controlling the release rate of the water for a particular elevation behind the gate.

Smooth pipe—a pipe with fluid flowing under sufficient conditions such that the laminar boundary layer thickness is significantly greater than the height of the wall roughness elements.

Spillway—a water control structure wherein water is released over the top of the structure, thus controlling the elevation of the water for particular release rates.

Specific discharge—volumetric rate of flow per unit width of an open channel.

Specific energy—energy head available in a channel measured relative to the channel bed.

Specific energy diagram—graphical representation of specific energy against the flow depth.

Specific gravity—ratio of the density of a given fluid to the density of water at a specified temperature (4°C).

Specific weight—weight per unit volume of a fluid.

Stage—elevation of a water surface relative to a standard reference as a datum.

Steady flow—state of flow characterized by no change in flow characteristics with respect to time.

Steep slope—channel slope that is steep enough to maintain supercritical flow conditions.

Stilling basin—concrete structure located downstream of a spillway or sluice gate to contain and control the hydraulic jump.

Storm frequency—the average period between recurrences of a given rainfall amount occurring over a specified duration of time.

Subcritical flow—flow condition in which the depth of flow is greater than that at which the energy required to transmit the flow is a minimum (critical).

Supercritical flow—flow condition in which the depth of flow is less than that at which the energy required to transmit the flow is a minimum (critical).

Suppressed weir—a rectangular weir that runs across the entire width of a channel.

Surface tension—intensity of the fluid's molecular attraction per unit length.

Surge tank—tank placed in a water line upstream of a valve in order to relieve the water hammer pressure when the valve is closed.

Tailwater—water level (or depth) downstream of a hydraulic structure or feature.

Tainter gate—underflow (sluice) gate that is controlled by rolling on a central axel.

Thermal expansion—represents the ability of a fluid to expand or compress when the temperature is changed.

Time of concentration—time required for water to flow from the most distant point in a basin to the outlet point.

Torque—external force equal to that generated in flow through application of angular momentum.

Total energy head—sum of the potential energy and kinetic energy developed in a body of fluid relative to a fixed datum.

Tractive force—average shear stress applied to a given area of channel bed or sides.

Transition—an abrupt change in the geometry or datum of a conduit.

Turbulent flow—flow condition characterized by fluid deformation at the microscopic level.

Uniform flow—flow condition characterized by spatially constant properties in both longitudinal and transverse directions.

Unsteady flow—flow condition characterized by temporally changing properties.

V-notch weir—weir with a notch (usually triangular) cut through with the flow passes.

Velocity coefficient—numerical coefficient applied to the velocity head to account for non-uniform velocity distributions in conduits.

Velocity head—energy measured in length units due to the velocity of the fluid.

Viscosity—a property characterizing the fluidity of the fluid (see Dynamic viscosity).

Water hammer—process by which extraordinary forces are developed in a pipe due to pressure surges caused by an abrupt change in velocity such as a sudden valve closure.

Water surface profile—region in a channel where the depth of flow is non-uniform and gradually changing.

Wetted perimeter—the boundary of a conduit (pipe or channel) that is in contact with the fluid.

Weight—mass of a fluid multiplied by the acceleration of gravity.

Wide channel—a channel whose width is so much greater than its depth that the hydraulic radius can be assumed equal to the depth.

Credits

Index

R

S